Animal Parasites

AF772175

Heinz Mehlhorn

Animal Parasites

Diagnosis, Treatment, Prevention

Springer

Heinz Mehlhorn
Department of Parasitology
Heinrich Heine University
Düsseldorf, Germany

This is an updated translation of the 7th edition of the German book "Die Parasiten der Tiere" (2012) published by Springer Spektrum.

ISBN 978-3-319-83512-9 ISBN 978-3-319-46403-9 (eBook)
DOI 10.1007/978-3-319-46403-9

© Springer International Publishing Switzerland 2016
Softcover reprint of the hardcover 1st edition 2016
This work is subject to copyright. All rights are reserved by the Publisher, whether the whole or part of the material is concerned, specifically the rights of translation, reprinting, reuse of illustrations, recitation, broadcasting, reproduction on microfilms or in any other physical way, and transmission or information storage and retrieval, electronic adaptation, computer software, or by similar or dissimilar methodology now known or hereafter developed.
The use of general descriptive names, registered names, trademarks, service marks, etc. in this publication does not imply, even in the absence of a specific statement, that such names are exempt from the relevant protective laws and regulations and therefore free for general use.
The publisher, the authors and the editors are safe to assume that the advice and information in this book are believed to be true and accurate at the date of publication. Neither the publisher nor the authors or the editors give a warranty, express or implied, with respect to the material contained herein or for any errors or omissions that may have been made.

Cover illustration: Left: Scanning electron micrograph of the anterior end of the hookworm *Ancylostoma caninum* showing typical teeth. Right: Scanning electron micrograph of the anterior end of a tapeworm of the genus *Taenia* showing a species specific crown of hooks besides the suckers. Photos Heinz Mehlhorn

Printed on acid-free paper

This Springer imprint is published by Springer Nature
The registered company is Springer International Publishing AG
The registered company address is: Gewerbestrasse 11, 6330 Cham, Switzerland

The Structure of the Book and How to Use the Diagnostic Keys

Contents

Chaps. 1–3 explain what parasitism means, where such animals live and how they can be diagnosed most easily.

Chaps. 4–6 present the most important parasites of animals. Using the tables in Chap. 2, where the different parasites are listed according to their **site of occurrence**, it becomes easy to find the relevant chapter in the book where a detailed description is given in addition to **micrographs** and **summarizing tables**.

How to Use the Diagnostic Keys

In several chapters, a key is provided, which helps to diagnose some important parasites. The keys should be used as follows:

- The user of the book gets in general 2 (rarely 3) choices of answers on questions related to morphologic aspects of a given parasitic stage (explained in the text and/or shown on micrographs).
- If the user has decided, he/she finds at the right side of the question a number in bold appearance.
- Then he/she goes to the next question written behind this number in the next line and so on. It is, however, important to read always all possible answers before a decision is made!
- If at the right side the name of a parasite or a group of parasites occurs, the diagnosis is completed and the user follows to the given page number, where the searched parasite is described in detail.

Text

The parasites of all animals in the close surroundings of humans are described in detail in Chaps. 4–6, which are presented in a sequence with respect to their organization as **protozoans, worms (helminths) and ectoparasites**. Each parasite and its closely related groups are described in the following subheadings:

1. **Name**: The origin of the species names is explained in order to understand the often complicated names.
2. **Geographic distribution/epidemiology**: This section gives insights into where the parasite occurs (worldwide, locally, etc.), and it is shown whether they are of high or low importance.
3. **Biology/morphology**: This section covers aspects of the appearance and reproduction of the parasites.
4. **Symptoms of disease**: The most important symptoms of disease introduced by this parasite are listed.
5. **Diagnosis**: This topic contains information how to find most easily a supposed parasite in the body of a host or in its feces, etc.
6. **Pathway of infection**: This section describes how the host can be infected with a supposed parasite.
7. **Prophylaxis**: This section gives short information about how infections with the diagnosed parasite can be avoided.
8. **Incubation period**: This is the period ranging from the day of infection until first symptoms of disease occur.
9. **Prepatent period**: This is the period from the day of infection with the diagnosed parasite until first parasitic stages can be diagnosed, which can be transmitted. The prepatent period is often shorter than the **incubation period** so that other hosts can be infected without knowing it.
10. **Patency**: Period during which a parasite can be transmitted to another host.
11. **Therapy**: Presentation of methods/compounds that are in common use to treat infections. **Note**: Medication may differ in different countries due to local laws.
12. **Further reading**: This section gives some rather new and/or older but important papers, which help to approach relevant literature concerning a diagnosed parasite.

Addenda

Addendum A: Antiparasitic Drugs Used for Animals

This addendum contains tables where different medicaments are listed that can be used for the treatment of infections occurring in the different host groups. It gives more details than in the special Chaps. 4–6.

Addendum B: Diagnostic Stages

This addendum encloses comparative tables with additional micrographs of stages of several parasites of the different hosts.

Addendum C: Questions to Test Obtained Knowledge

This addendum lists 100 questions concerning important animal parasites offering always five potential answers, but only one of them is correct. This allows us to control our own knowledge and is recommended especially for students before examinations. Correct answers are listed at the end of this addendum.

Addendum D: Origin of Figures

The origins of the macro- or microscopical figures of the parasites used in this book are listed.

Preface

Parasites endanger not only the health of humans but also that of animals which live together with humans in homes and on farms. Many groups of parasites are in addition transferable from animals to humans and back. The pathways of transmissions of endoparasites and ectoparasites have obtained increasing importance in our days of **globalization** and **global warming**. While globalization offers easy transportation of agents of diseases from one end of the world to the other, global warming increases the chances of many parasites (especially ectoparasites) to enter new biotopes in formerly cold regions. Before this background, it seems necessary to stabilize and to increase the knowledge on parasites with respect to occurrence, transmission and control. Thus, the contents of this volume are addressed to veterinarians and students of veterinary medicine but also to all people who keep farm animals and especially to those who are owners of pet animals.

In order to make it easy to find the wanted information, the book presents the relevant knowledge on each parasite under the following 12 subheadings:

1. Name
2. Geographic distribution/epidemiology
3. Biology/morphology
4. Symptoms of disease
5. Diagnosis
6. Pathway of infection
7. Prophylaxis
8. Incubation period
9. Prepatent period
10. Patency
11. Therapy
12. Further reading

The correct **diagnosis** of the different parasitic stages of animals is not only possible by the description of the relevant methods but also supported by more than 600 micrographs, by many diagrammatic representations as well as by comparative tables. The listed control measurements represent the actual knowledge in the year of the appearance of this book and should always be checked.

Each chapter on a given parasite ends with the presentation of some recent papers, which allow a retrospective of important features of the parasite.

Hundred questions—each with five choices—allow personal control of our own knowledge. They had been tested in many student examinations to confirm a solid knowledge.

This book represents the first English edition after seven previous ones in German.

Düsseldorf, Germany
June 2016

Heinz Mehlhorn

Acknowledgements

The representation of a book such as this one, considering a broad spectrum of many features of the same topic, is not possible without the help of experienced colleagues and friends. Thus, this book is based on common books and articles published with renowned colleagues such as Dieter Düwel, Dieter Eichenlaub, Alfred Otto Heydorn, Thomas Löscher, Werner Peters (†), Gerhard Piekarski (†), Wolfgang Raether and Eberhard Schein (†). Furthermore, I am deeply indebted to my colleagues Johannes Eckert (Zürich), Axel Haberkorn (Bayer Leverkusen), Heinz Hänel (Hoechst AG, Frankfurt), Alfred Otto Heydorn (Berlin), Sven Klimpel (Frankfurt), Gerd Lehmann (Aalbaum), Brigitte Loos-Frank (Stuttgart), Gerhard Piekarski (†) (Bonn), Rainer Pospichel (Bayer AG), Regine Ribbeck (Leipzig) and Eberhard Schein (†) (Berlin) for their contribution of beautiful pictures, their advices and/or nice glass slides to prepare rare aspects of parasites. Dr. Volker Walldorf (Düsseldorf) and Mr. Fried Theissen (†) (Essen) contributed nice drawings especially prepared from drafts.

My son Tim, Mrs. Diehl and my wife Birgit helped in translating the text into English. The preparation of the text was done by Mrs. Inge Schaefers and Mrs. Susanne Walter. Presentation of the figures was accomplished by Mrs. Isabelle Mehlhorn, Mrs. Susanne Walter and Mr. Bernd Prümm.

The staff at Springer Heidelberg (Drs. Andrea Schlitzberger and Lars Körner) finished this book in its hopefully very useful and agreeable final version.

Düsseldorf, Germany
June 2016

Heinz Mehlhorn

Contents

About the Author

Prof. Dr. Heinz Mehlhorn has investigated parasites, their transmission pathways and significant control measures for over 40 years. He has published more than 20 books and 250 original publications and received 25 patents on antiparasitic drugs, some of which he uses at his university spin-off company Alpha-Biocare (founded in 2000). As a university instructor, he had the pleasure to introduce many students to the topics in parasitology. Many of them are now professors or in leading industrial positions. In television and radio broadcasts, he regularly informs the public about relevant parasitological problems in our days of globalization and global warming.

Phenomenon Parasitism 1

1.1 Who Are Parasites?

The name "parasite" comes from the Greek term *parasitos* (*para* = close by, besides; *sitheisthai* = eating), thus describing an individual, which participates at a meal. This term was used in Greek times to characterize a professional tester of food, which was cooked for noble persons. These tests should avoid that the VIPs of these times could be killed by any poisonous compounds added erroneously or intentionally to the food. Since other people noted that these "parasitos" got their food without work, the term obtained a negative meaning which stands even today.

All animals and also humans have to settle the common problem: to get food in sufficient amounts and quality which allows to overcome the daily "**struggle for life**". Apart from plant feeding animals, the smaller, weaker species serve as food for the stronger ones. Exceptions are made by so-called **commensal species**, which participate at the meal (or at the remnants of such meals) of carnivorous specimens. Another possibility to obtain their food is used by the so-called **ectoparasites** which suck blood or lymph at the surface of stronger animals (hosts). Since it is very dangerous for smaller animals to approach and to stay close to larger ones, many of them have developed strategies to enter either the skin or body cavities of larger animals and humans in order to become **endoparasites**. These species, however, are always endangered by the immune system of these hosts and thus must develop sophisticated defence and disguise systems. Thus, the parasites of our days are the survivors of this permanent struggle for survival.

1.2 Parasites and Their Hosts

According to the activities of the recent spectrum of parasites, it can be differentiated between ectoparasites and endoparasites depending on their preferred sites of feeding. **Ectoparasites** can live exclusively **stationary** (e.g. mites, lice) or **temporary** (e.g. mosquitoes) on their hosts. However, there exist also

© Springer International Publishing Switzerland 2016

H. Mehlhorn, *Animal Parasites*, DOI 10.1007/978-3-319-46403-9_1

intermediate groups such as ticks and some flea species (e.g. sand fleas), which stay for longer periods on the same hosts. The pathway to **endoparasitism** was probably started by parasitic stages like today's existing species such as scabies mites and fly larvae that penetrate into the skin like *Gasterophilus* species or *Hypoderma* species.

Intracellular parasites represent a peculiar form of the **endoparasitism**. However, they need to fulfil some conditions such as to have a rather small size and to develop the ability to enter a cell without destruction of the cell membrane.

Parasites have developed life cycles which may include one or several hosts in a fixed order. One-host parasites have **monoxenous life cycles**, while those with several hosts are described as **heteroxenous**. Both types of life cycle require rather special hosts, while some other groups of parasites are not very fastidious when selecting their hosts. Again other parasites are able to select whether they obtain their food from hosts (e.g. some female mosquitoes) or from plant fluids and thus are described as **facultative parasites**.

If a sexually defined endoparasite uses during its lifetime different hosts in a fixed sequence, those hosts are called **final hosts** or **definitive hosts**, within which the sexual stages of the parasite are located. For example, for the tapeworm *Taenia saginata* humans are final hosts, since the adult (bisexual) worm lives in their intestine.

Intermediate hosts are those hosts, within which only an asexual reproduction of a multihost parasite occurs (e.g. cattle, which harbour the "bladder stage" of *Taenia saginata*, belong to this host type).

Erroneous (or blunder) hosts may host a parasite, but transmission from humans to other hosts is practically excluded due to several reasons. For example, humans are at the same time erroneous, final and intermediate hosts for *Trichinella spiralis*. Transmission to other hosts can, however, only occur, if carnivorous large cats or human "maneaters" do their work.

Some ectoparasites (mosquitoes, flies, ticks, mites) are termed **vectors** in the case they transmit agents of disease to humans and/or animals. Originally, it was thought that they would transmit only on a mechanical pathway by entering their contaminated mouthparts into the skin of a host. However, nowadays thousands of examples are known, where parasites, viruses or bacteria reproduce themselves inside the vectors, which thus become either final or intermediate hosts.

During the developmental cycle of most multihost parasites, mostly only a single type of **final hosts** occurs (e.g. carnivores), while several types of **intermediate hosts** may follow each other. For example, in the case of the trematode *Dicrocoelium dendriticum* grass feeding ruminants act as final hosts, while snails act as first intermediate hosts and ants second intermediate hosts. However, in the case of the protozoan parasite *Caryospora bigenetica*, there exist two different final hosts, since there are sexual processes in the **primary final host** (snakes) as well as in a **secondary final host** (rodents). If these stages are ingested by dogs, oocysts can also become developed in their skin. Thus, this parasite is extremely flexible and cannot be submitted clearly into the normal final and intermediate host system.

With respect to the **spreading of parasites,** some host types have obtained a considerable geographic distribution and are thus important for the persisting of a parasite species in a given biotope.

Looking from the position of humans, some vertebrate animals act as **reservoir hosts** for a given parasite. For example, dogs and rodents may act as reservoir hosts for stages of the *Leishmania* species. From there, they are transmitted during blood sucking of sand flies to other main hosts. On the other side, human *Plasmodium* species do not have reservoir hosts apart from some monkey species.

Transport hosts (=**paratenic hosts**) are intermediate hosts, wherein which no reproduction of an included parasite occurs but only a transformation to reach infectivity. This mosquitoes of the genera *Aedes* or *Culex* belong to this host type since they ingest larvae 1 of filarial species and inject larvae 3 of these species into other vertebrates.

Erroneous or accidental hosts are hosts which have been entered by parasites, which have therein no chance to develop further on and are unable to leave again this host. There are two groups existing:

(a) Humans act as true host (e.g. as intermediate host for *Echinococcus* species or *Toxoplasma gondii*), but transmission does not occur, since humans are in general ingested neither by dogs nor cats.
(b) Humans are entered by fish tapeworm larvae (e.g. **spargana**), by larvae of the dog nematode *Toxocara canis* or by cercariae of bird schistosomes, but these stages cannot develop further on and thus die within human tissues inducing inflammations.

Host specificity is often developed by adaptations of a parasitic species to a host and has reached different grades during evolution. This specificity may be:

(a) **Very close**, so that only one single host species is accepted (e.g. *Isospora hominis*; pig tapeworm in humans);
(b) **Very loose**, so that many hosts are accepted (e.g. many blood sucking ectoparasites, many trematodes or *Cryptosporidium* species);
(c) **Loose** with respect to intermediate hosts, but **very close** with respect to final hosts (e.g. *Toxoplasma gondii* has only felids as final hosts, but humans and many mammals and birds may act as intermediate hosts);
(d) **Close** with respect to intermediate hosts and **loose** with respect to final hosts (e.g. *Plasmodium* species occur only in humans as intermediate hosts, but many *Anopheles* species may be final hosts).

The **individual development** of parasite species may occur on two pathways:

(a) **Direct** from egg via larval stages that look rather similar to the adult stages (**metamorphosis**, e.g., in insects, nematodes);
(b) **Indirect** including different reproduction processes (e.g. Coccidia, trematodes) with several generations. This follow-up of different generations

may occur **obligatorily** (e.g. *Sarcocystis* species) or **facultatively** (e.g. *Strongyloides stercoralis*).

During the life cycle of parasites, very often an **alternation of generations** occurs. In the case of many protozoan species, a so-called **primary follow-up** of generations can be observed, since cell divisions induce a repeated reproduction of single cells, while in the case of metazoans cell reproduction leads only to the increase of an individual. Only in cases, where such multicellular organisms divide, a new generation occurs **secondary follow-up** of generations.

(a) Typical **primary follow-ups** of generations occur in Coccidia covering a sexually reproducing generation and one up to several generations with an asexual reproduction.
(b) In the case of the **secondary follow-up** of generation two different types can be observed:
  - **Metagenesis**: Here occurs a follow-up of a bisexual generation and a single one or of even several generations which are reproduced asexually.
  - **Heterogony**: This term describes the follow-up of a single female = parthenogenetic generation and a bisexual generation (e.g. *Strongyloides stercoralis*).

Since in many species the knowledge is still poor on sexuality and/or on the occurrence and types of chromosomes, many species/groups of parasites can only be placed with difficulties into one of the above-listed groups (e.g. trematodes). Furthermore, it is possible that the larvae of some parasites may become mature (**neoteny**) without reaching the full appearance of adults (e.g. some Monogenea). **Polyembryogony** occurs as well when larval stages start asexual divisions.

The parasitic worms are **males**, **females** or **hermaphrodites** (harbouring both female and male sexual organs). Hermaphrodites, which harbour both male and female sexual organs, are named according to the Greek god Hermes and the goddess Aphrodite. It is the rule that in these animals the sperms are produced first. This phenomenon is named **protandry**: In general, specimens of **hermaphroditic species** try to mate with other specimens in order to avoid self-insemination. However, in cases where a sexual partner is absent, large tapeworms produce fertile eggs by self-insemination (=the sperms being produced by anterior proglottids are injected into the posterior proglottids, where the female gametes have reached maturity).

The time needed for the **individual larval development** of **ectoparasites** depends on the temperatures inside their biotopes, while larval **endoparasites** have to overcome the defence systems of their hosts. Thus, in both cases the needed time for the individual development may vary considerably. Higher temperatures shorten needed developmental phases, so that masses of parasites may occur in a very short time. The same phenomenon can be observed when internal parasites have infected a host with a weak immune system. The term **prepatency** (prepatent period) defines the period from the moment of infection and the first occurrence (excretion) of infectious stages (eggs larvae). The period of the first appearance of eggs or larvae until the last day of such an excretion is called **patency**, which can last a few days (e.g. Coccidia) up to years (tapeworms). The period between an infection until the first occurrence of clinical symptoms is described as **incubation period**.

Adaptations Ectoparasites and endoparasites have peculiar morphological structures that allow them to get their food from the host's surface and/or to enter the host. While recent species of **ectoparasites** show a broad spectrum of sophisticated mouthparts, **endoparasites** have developed systems to solve the following problems:

- To obtain effective mechanisms to enter a host
- To become anchored inside a host
- To obtain sufficient food inside the host
- To survive the attacks of the host's defence system
- To protect their progeny inside their hosts
- To place their progeny inside the host's body at places which allow an easy exit in order to reach other hosts

Thus successful parasites have established the **following mechanisms**:

(a) **Invasion**:
The infestation of a host by an endoparasite may proceed passively by oral uptake of **persisting stages** such as eggs, cysts or tissue cysts or by **active invasion** during bites of ectoparasites, which enter their highly specialized mouthparts and introduce saliva components which block coagulation of blood, thus keeping open the fine channels of the mouthparts. Penetration may be supported with the help of the excretion fluids of rather large glands dissolving the surface of their potential hosts (e.g. as it is done by miracidium larvae of trematodes and larvae 1 or 3 of nematodes).

(b) **Anchoring and food uptake**:
Many groups of parasites have developed a broad spectrum of sophisticated systems which allow firm attachment at outer or inner surfaces of their hosts (e.g. hooks, thorns, suckers, peculiar protrudible systems, cuticular foldings, etc.). Most parasites ingest their food via different types of mouths. However, several intestine-less groups (Acanthocephala, tapeworms) take up their food via their surface.

(c) **Protection from host reactions (immune evasion)**
In case that endoparasites parasitize inside vertebrates, which mostly have developed sophisticated immune defence reactions in order to protect

themselves, they will only survive and thus get the chance to reproduce themselves therein, if they succeed to develop measurements that offer protection from attacks of the host (e.g. from digestive enzymes, antibodies, etc.). A very large number of endoparasites protect themselves by a layer of **mucopolysaccharides**, which are described as **surface coat**. This layer was first observed in the case of trypanosomes, but is now documented for many parasites of vertebrates. It is characteristic for this surface coat that the parasite changes constantly its **antigenic properties**. For example, trypanosomes have about 1000 genes that are used to produce different surface coats in a non-predictable sequence. Due to the presence of such a surface coat parasites, which live in the blood of their hosts, remain undiscovered by this mode of "**eclipse**" from the **specific** (=antibodies = immunoglobulins, e.g. IgE, IgG, IgM) and **non-specific** (=phagocytic or lytic cells) **defence systems** of the host.

The so-called **molecular mimicry** was additionally developed by many parasites. This term describes the phenomenon that parasites have developed the ability to use host substances and to enter them into their surface coat so that hosts do not recognize this parasite in disguise. Similar systems are, e.g., known from schistosomes, *Fasciola hepatica* and several species of filarial worms. Other parasites mask themselves leading to the final effect that the parasitic antigenic material is fully covered by host antibodies (e.g. *Fasciola*). Again other parasites mask themselves by suppressing or reducing the formation of the MHC antigens (major histocompatibility complex), so that they cannot become recognized by the T-lymphocyte system. Finally, again other parasites settle in regions of their hosts where immune activities are low (e.g. the cysticerci of tapeworms prefer often the brain of their hosts). This phenomenon of searching regions with low immune activities is described as **sequestration**. However, all these above-described systems are often not fully sufficient to overcome the immune system of a host. Therefore, they have developed additional systems, which may lead to a nearly full or even full **immune suppression** of the host. This may occur by the production and presentation of overwhelming amounts of **antigens**, which bind practically all **antibodies** of the host leading to the fact that not enough of them may become attached at the surface of the parasite. Another strategy is used by parasites like the trypanosomes which stimulate the B-lymphocytes of the host as intensely so that they produce such enormous amounts of antibodies that this system is finally fully exhausted. Again other parasites excrete substances, which block antibodies and/or immune-competent defence cells. As soon as the immune system has become weakened by an agent of disease, further so-called **opportunistic agents** may **overcrowd** hosts, since then the balance between host and parasite obtained in a long-lasting co-evolution has been shifted in advantage to the parasites.

Nematodes and insects as well as ticks are protected by a very stiff and resistant cuticle, which in addition is moulted from time to time. Other parasites enter host cells and thus are protected therein from the host's defence system (e.g. coccidian merozoites, schizonts and gamonts as well as *Trichinella* larvae). If inside such cells a reproduction occurs (e.g. in *Toxoplasma gondii* or *Sarcocystis* species), large **tissue cysts** may be formed within which the parasites may survive even for years. Such a long survival enhances considerably the chances of a transmission to another host. In these cases, the parasite apparently steers the physiology of their host cells; otherwise the often complete transformation of the former undifferentiated host cell into a large completely different containment would not be possible.

Parasites are also able to stimulate their host cells to start division processes. For example, the schizonts of *Theileria* species, which develop, e.g., inside cattle lymphocytes, that do not divide anymore start again divisions, so that their number and the permanently dividing parasites (schizonts) occur in larger numbers in the blood. This leads to the fact that the vector ticks may become infected at higher rates.

The **host specificity** of parasites may be strong or rather loose depending on the species. This phenomenon is not fully understood; however, some physiological phenomena may enlighten the pathways. Some parasitic worms have lost their ability to produce de novo lipid complexes. Then they depend on the lipids of the host and thus depend strictly on this host species. If another host does not offer the needed lipids, they cannot survive therein. On the other side, a dependence on peculiar carbohydrates and proteins is much less common among parasites, since these compounds may be produced using rather non-specific molecules.

Brood Care Endoparasites have to protect their progeny as well from defence systems of their hosts but also as from influences outside of the body. This protection is achieved by the formation of thick covers and/or thick egg shells. Furthermore, it is needed to place the offspring at places inside a host's body, from where they have the chance to enter another host. Therefore, the eggs of schistosomes are placed inside blood vessels close to hollows such as intestine or bladder, from where they may reach the outside when being excreted within feces or urine. *Plasmodium* gamonts and microfilariae of filarial worms are thus found often in peripheral blood vessels at which blood suckers may ingest them. Adult *Paragonimus* specimens depone their eggs in the alveolus of the lung, from where they may become expectorated. Another possibility to increase the own progeny and thus to increase the chance to become transmitted to another host is the production of enormous amounts of progeny by flooding infested organs as it occurs during the asexual schizogony of Coccidia. Another example is the mass production of cercariae inside snails infected with schistosomal sporocysts. All these different methods help to guarantee that the chain of repeated infections and thus the survival of the parasite in a given biotope is not interrupted.

I. Pathogenicity Parasites live per definition and de facto on costs of their hosts. They are deleterious even if their effects are low graduated. In principle, they induce the following damages:

- **Destruction** of cells or organs (e.g. *Plasmodium* species, *Onchocerca volvulus, Ancylostoma*, hydatids of tapeworms);

- **Stimulation** of malign tumours (e.g. liver flukes, *Echinococcus* species);
- **Withdrawing of essential compounds** (blood flukes, *Diphyllobothrium* = withdrawal of vitamin B_{12});
- **Intoxication** by excretion of metabolically active substances (*Trypanosoma* species, *Plasmodium* species, ticks);
- **Transmission** of agents of diseases (viruses, bacteria, protozoans, worms).

The pathogenic effects of the same parasite may differ considerably in different hosts reaching different grades of **virulence**, which apparently depend on adaptations, that may differ from host to host and among strains of the same host. In principle, parasites "must be interested" not to harm their hosts significantly since their death would be also the end of their own life. Thus evolutionary old parasites mostly induce only low graded damages inside their hosts, so that both hosts and parasites may live together for many years (e.g. tapeworms, filarial worms). Diseases due to parasites start mostly with an acute phase, which is decreased due to adaption and activity of the host's immune system reaching often a symptomless cohabitation.

Zoonosis are diseases, which are induced by the same parasites in humans and animal hosts (e.g. toxoplasmosis, trichinosis). The direction of the infections is explained by the topics **anthropozoonosis** and **zooanthroponosis**. As true **anthroponosis** diseases are considered, which are induced by parasites, which are exclusively transmitted from humans to humans (e.g. *Enterobius vermicularis*). A peculiar form of zoonosis are diseases where arthropods transmit agents of diseases. Hereby it has been differentiated between the **direct mechanical transmission** (without reproduction) of a parasite (e.g. some trypanosomes) and **cyclic transmission** (metazoonosis) with reproduction inside the insect vector (e.g. malaria, filariae).

Diagnosis of Parasites The complicated host–parasite relationships, which in addition vary considerably from species to species, cannot be described in detail in this book. However, each chapter on a given parasite contains the needed information for a quick and safe diagnosis, which is mainly based on **morphological criteria**. However, in cases of low-level infections direct methods may not give reliable results. Then it is needed to use **serological methods**. These methods are based on the fact that each parasite represents an **antigen** (antigenic complex), which stimulates the host to react by the production of **antibodies**.

The results of such serological methods are obtained by the fact that the antibodies of the host become attached at the surface of the antigens (=parasites). These complexes must be made visible and their amounts must become measurable. To reach these goals, direct and indirect methods have been developed. A **direct method** would be considered when the antigen (e.g. cercariae of *Schistosoma* species) is brought into contact with a fresh immune serum obtained from a *Schistosoma*-infected mouse. The result would be a precipitation layer around the cercariae. This "cercarial covering method" is used in human medicine. The **indirect methods** need a mediator which makes the antigen–antibody reaction macroscopically or microscopically visible. One simple method is based on the agglutination of the antigen at very fine plastic particles, which agglutinate in the presence of the specific antibodies (=**indirect latex-agglutination, ILAT**). The results become visible by coagulation of the tiny plastic globules. If the reaction is negative, these globules remain dispersed in the solution.

Another classic indirect method is the so-called **complement binding reaction (KBR)**. Further indirect methods are the **indirect immunofluorescence test (IIFT)** and the **indirect haemagglutination test (IHAT)**. The **radioimmunoassay (RIA)** and the **enzyme-linked-immuno-sorbent assay (ELISA)** deliver in many cases results which are superior to those with other test systems, since they are able to indicate even smallest amounts of antibody/antigen complexes. The details and technical equipment should be taken from special literature.

Biological System Parasitic species of animals had to develop a broad spectrum of adaptations in order to survive in/on their hosts. Thus, the morphologic appearance of parasite species may be very similar although they belong to completely different groups. Thus, this book presents the parasites not according to their often doubtful phylogenetical relationships but in a sequence, which allows a quick and easy recognition of the single members of each group. The parasites of veterinary importance are presented in the following chapters:

1.	Protozoa	Chapter 4
2.	Trematodes	Section 5.1
3.	Cestodes	Section 5.2
4.	Nematodes	Section 5.3
5.	Pentastomida	Section 5.4
6.	Acanthocephala	Section 5.5
7.	Leeches	Section 5.6
8.	Chelicerata	Section 6.1
9.	Ticks	Section 6.2
10.	Mites	Section 6.3
11.	Insects	Section 6.4
12.	Crustaceans	Section 6.5

In order to make it easy to find desired information, the presentation of a given parasite includes the following subheadings:

1. Name of the parasites
2. Geographic distribution/epidemiology
3. Biology/morphology
4. Symptoms of the disease
5. Diagnosis
6. Pathway of infection
7. Prophylaxis
8. Incubation period

9. Prepatent period
10. Patency
11. Therapy
12. Further reading

Where Are Which Parasites? 2

The parasites attack their hosts from different positions. Some **endoparasites** enter their hosts from soil and migrate eventually for a while inside the skin before final settlement inside and interior organs, where they start reproduction or wait until they are transmitted to another host. Other parasites are ingested as stages such as eggs or as larvae inside food of predator hosts and remain there lifelong.

Again other parasites obtain their food while ingesting blood, lymph or wound fluids from the surface of their hosts. These parasites are described as **ectoparasites**. They endanger their hosts not only by blood sucking but also by their ability to transmit a broad spectrum of agents of diseases such as prions, viruses, fungi, bacteria or even other parasites belonging to the group of endoparasites.

Parasites may enter practically all organs of their human and animal hosts, wherein they are attacked by the defence system of the hosts. Thus, only those parasites have survived during evolution which had been able to develop methods to escape such host attacks. In order to ameliorate the first diagnosis of the parasites, Chap. 2 shows their occurrence inside or on the different host organs with the help of Tables 2.1, 2.2, 2.3, 2.4, 2.5, 2.6, 2.7, 2.8, 2.9, 2.10, 2.11, 2.12, 2.13, 2.14, 2.15, 2.16, 2.17 and 2.18. Chapter 3 presents the most common (easy) diagnosis methods, while the Chaps. 4–6 offer deeper insights into the most important parasites of dogs, cats, ruminants, horses, birds, rodents, rabbits, reptiles, amphibians and fish. However, in order not to get lost in too much details, the selected parasites were presented in the following groups:

© Springer International Publishing Switzerland 2016

H. Mehlhorn, *Animal Parasites*, DOI 10.1007/978-3-319-46403-9_2

1.	Protozoans	Chapter 4
2.	Trematodes	Chapter 5.1
3.	Tapeworms	Chapter 5.2
4.	Nematodes	Chapter 5.3
5.	Tongue worms	Chapter 5.4
6.	Acanthocephalans	Chapter 5.5
7.	Annelids	Chapter 5.6
8.	Spiders, scorpions	Chapter 6.1
9.	Ticks	Chapter 6.1.1
10.	Mites	Chapter 6.1.2
11.	Insects	Chapter 6.2
12.	Crustaceans	Chapter 6.3

Table 2.1 Parasites in feces and in the lumen of the intestine

Parasite groups	In feces	In the lumen of the intestine
Giardia sp.	Cysts	Trophozoites
Trichomonads	Evtl. cysts	Trophozoites
Amoebae sp.	Cysts	Amoebae
Eimeria sp.	Oocysts	All stages
Isospora, Cystoisospora sp.	Oocysts	All stages
Sarcocystis sp.	Oocysts, sporocysts	All stages
Toxoplasma gondii, Hammondia sp.	Oocysts	All stages
Cryptosporidium sp.	Oocysts	All stages
Ciliates	Cysts	Trophozoites
Microsporidia	Spores	All stages
Myxosporidia sp.	Spores	All stages
Blastocystis sp.	Cysts	Trophozoites
Trematodes (Digenea)	Worm eggs	Adult worms
Trematodes (Monogenea)	Worm eggs	Adult worms
Tapeworms = cestodes	Proglottids, eggs	Adult worms
Nematodes	Eggs, larvae	Adult worms
Acantocephala	Eggs	Adult worms, larvae
Pentastomida	Eggs	–
Insects	Fly larvae	Fly larvae

2.1 Parasites Within Feces or Intestine

The stages of many parasitic species of this group are transmitted **orally-fecally** without the involvement of an intermediate host from **final host** to final host. The transmitted stages represent so-called spores, cysts, oocysts or worm eggs. These stages live intracellularly in the intestinal cells or in the lumen of the intestine

Table 2.2 Parasites in blood

Parasite groups	Free in blood	Intracellular in E, L, M
Flagellates	*Trypanosoma* stages	*Leishmania* stages (M)
Toxoplasma gondii	Tachyzoites	In macrophages (M)
Plasmodium sp.	–	In red blood cells (E)
Babesia sp.	–	In red blood cells (E)
Theileria sp.	–	In red blood cells (E)
Hepatozoon sp.	–	In leucocytes (L)
Leucocytozoon sp.	–	In leucocytes (L)
Haemoproteus sp.	–	In red blood cells (E)
Microsporidia	Spores	–
Myxosporidia	Spores	–
Schistosoma sp.	Adults, larvae	–
Trichinella sp.	Larvae	–
Dirofilaria sp.	Larvae, adults	–
Angiostrongylus sp.	Larvae, adults	–
Ascarids	Larvae	–
Filaria	Larvae	–
Hookworms	Larvae	–

E Erythrocytes; *L* Leucocytes; *M* Macrophages

Table 2.3 Parasites in saliva

Parasites	Parasitic stage
Pneumocystis carinii	Cysts
Trematodes in lung	Eggs
Capillaria sp.	Larvae, eggs
Angiostrongylus sp.	Larvae
Lung nematodes	Larvae
Pentastomids	Eggs
Lung mites	Larvae, adults

Table 2.4 Parasites in lymph fluid

Parasites	Parasitic stage
Leishmania sp.	Amastigote stage
Hepatozoon sp.	Schizonts
Toxoplasma gondii	Tachyzoites
Echinococcus sp.	Protoscolices
Nematodes (Filariae)	Larvae, microfilariae

giving rise to sexual or asexual developmental stages, which after growth and maturation produce again transmittable stages being excreted within the feces. This type of propagation involves only similar types of hosts, wherein the whole life cycle occurs. Besides this type of transmission, many species have developed a life cycle involving final and intermediate hosts, which belong always to different biological groups. For example, *Sarcocystis* species develop their sexual stages in

Table 2.5 Parasites in urine

Parasites	Parasitic stage
Entamoeba invadens	Cysts
Cryptosporidium parvum	Oocysts
Klossiella sp.	Oocysts
Encephalitozoon cuniculi	Spores
Monogenea (trematodes)	Eggs
Schistosomes	Eggs
Trichosomoides sp.	Eggs
Capillaria sp.	Eggs
Dictyophyme renale	Eggs
Flies	Larvae

Table 2.6 Parasites in/on mucous layers

Parasites	Parasitic stage
Trichomonads	Trophozoites
Leishmania sp.	Micromastigotes (=amastigotes)
Amoebae	Amoebic stages
Coccidia	All stages
Cryptosporidium sp.	All stages
Myxozoa	All stages
Digenea (trematodes)	Sucking adults
Hookworms	Sucking adults
Acanthocephala	Larvae, adults
Leeches	Larvae, adults
Flies	Larvae

Table 2.7 Parasites in/on tissues

Parasites	Parasitic stage
Leishmania sp.	Amastigotes (=micromastigotes)
Toxoplasma gondii	Tachyzoites
Leucocytozoon sp.	Gamonts
Hepatozoon sp.	Gamonts

intestinal cells of predators (=final hosts), while their tissue cysts are found in muscle fibres of plant and/or general feeders (=intermediate hosts). The same follow-up of the life cycle occurs in *Trichinella* species. Examples of fecally transmitted species are listed in Table 2.1.

2.2 Parasites in Blood

Several parasites parasitize constantly inside the blood or even constantly inside blood cells. Others, however, use the blood or lymph fluids exclusively as a pathway to reach target organs and thus are only accidentally found inside blood or lymph. In Table 2.2, examples of such blood parasites are listed.

Table 2.8 Parasites in liver and spleen

Parasites	Parasitic stage
Leishmania sp.	Amastigotes
Entamoeba sp.	Abscesses
Hepatozoon sp.	Schizonts
Eimeria sp.	All stages
Toxoplasma gondii	Tachyzoites
Frenkelia sp.	Tissue cysts
Plasmodium sp.	Schizonts
Microsporidia	Spores
Myxozoa	Spores
Liver flukes	Adults
Schistosoma sp.	Granulomes
Echinococcus sp.	Tissue cysts, hydatids
Tapeworms	Cysticercus, plerocercoids
Ascaris sp.	Larvae
Capillaria sp.	Larvae, adults
Filariae	Microfilariae
Linguatula sp.	Larvae

Table 2.9 Parasites in muscles

Parasites	Parasitic stage
Amoebae	Amoebic stages
Sarcocystis sp.	Tissue cysts
Hammondia sp.	Tissue cysts
Besnoitia sp.	Tissue cysts
Microsporidia	Spores
Myxozoa	Spores
Trematodes	Metacercariae
Tapeworms	Cysticercus
Trichinella sp.	Larvae in cysts
Acanthocephala	Larvae
Pentastomids	Larvae

Table 2.10 Parasites in lung, trachea and gills

Parasites	Parasitic stage
Pneumocystis carinii	All stages
Trematodes	Adults
Tapeworms	Cysticercus
Nematodes (e.g. *Ascaris* sp.)	Wandering larvae
Capillaria aerophila	Adults, larvae
Angiostrongylus vasorum	Adults, larvae
Mites	Adults, larvae, eggs

Table 2.11 Parasites in sexual organs

Parasites	Parasitic stage
Trichomonads	Trophozoites
Trypanosoma equiperdum	Trophozoites
Histomonas meleagridis	Trophozoites
Cryptosporidium parvum	All stages
Prosthogonimus sp.	Adults
Enterobius sp.	Adults

Table 2.12 Parasites in bones

Parasites	Parasitic stage
Leishmania sp.	Amastigotes
Toxoplasma gondii	Tachyzoites
Microsporidia	Spores
Myxozoa	Spores

Table 2.13 Parasites in kidneys

Parasites	Parasitic stage
Eimeria sp.	All stages
Klossiella muris	All stages
Encephalitozoon cuniculi	Spores
Myxozoa sp., e.g. *Hoferellus, Sphaerospora*	Spores
Trichodina sp.	Trophozoites
Sanguinicola sp.	Eggs
Trematodes	Metacercariae
Nematodes	Wandering larvae
Trichosomoides crassicauda	Adults, eggs
Capillaria sp.	Adults, eggs
Dioctophyme renale	Adults, eggs
Flies	Larvae

Table 2.14 Parasites in swim bladder

Parasites	Parasitic stage
Amoebae	Amoebic stages
Trypanoplasma sp.	Trophozoites
Eimeria sp.	All stages
Microsporidia	Spores
Myxozoa	Spores
Cystidicola sp.	Adults, larvae
Anguillicola crassus	Adults

Table 2.15 Parasites in the brain

Parasites	Parasitic stage
Toxoplasma gondii	Tissue cysts
Hammondia sp.	Tissue cysts
Sarcocystis sp.	Tissue cysts
Neospora sp.	Tissue cysts
Babesia	Trophozoites in clumps of erythrocytes
Tapeworms	Cysticercus, tissue cysts, larvae
Nematodes	Wandering larvae
Angiostrongylus cantonensis	Larvae
Microsporidia	Spores
Myxozoa	Spores, cysts

Table 2.16 Parasites in eyes

Parasites	Parasitic stage
Toxoplasma gondii	Tissue cysts
Myxozoa	Spores
Trematodes (Monogenea)	Adults
Trematodes, e.g. *Diplostoma, Clinostomum* sp.	Metacercariae
Tapeworms	Cysticercus
Thelazia sp.	Adults, larvae
Onchocerca sp.	Microfilariae
Nematodes	Wandering larvae
Loa loa	Adults
Flies	Larvae

Table 2.17 Parasites in skin

Parasites	Parasitic stage
Ichthyophthirius and related species	Trophozoites
Ciliates	Trophozoites
Caryospora sp.	Tissue cysts
Besnoitia sp.	Tissue cysts
Microsporidia	Spores
Myxozoa	Spores
Schistosomes	Schistosomula
Collyriclum faba	Adults
Onchocerca sp.	Adults, larvae
Parafilaria sp.	Adults, larvae
Demodex sp.	Adultes, larvae
Sarcoptes sp.	Adultes, larvae
Nest mites	Hypopus
Nodule mites	Adults
Laminosioptes cysticola	All stages
Oestridae	Larvae

Table 2.18 Parasites on skin, fur or plumage

Parasites	Parasitic stage
Flagellates	Trophozoites
Ciliates	Trophozoites
Myxosporidia	Spores
Trematodes (Monogenea), e.g. *Dactylogyrus* sp.	Adults
Trematodes (Digenea), e.g. *Azygia*	Adults
Nematodes	Adults
Mites	Larvae, nymphs, adults
Ticks	Larvae, nymphs, adults
Fleas	Adults
Bugs	All stages
Mosquitoes	Adult females
Louse flies	Adults, larvae
Flies	Larvae, adults
Sucking lice (Anoplura)	All stages
Biting lice (Mallophaga)	All stages
Parasitic crustaceans, e.g. *Argulus*	All stages
Leeches	All stages

2.3 Parasites in Saliva

In cases of lung infections, several stages of transmittable parasites are expectorated or ingested together with saliva, so that they can be found in the saliva or within the feces (Table 2.3).

2.4 Parasites Inside Lymph Fluid

The lymph system is besides the blood vessels the most important transportation system used by parasites to reach their final site of parasitation. Table 2.4 contains some examples of important parasites which use lymph fluid as main transportation system.

2.5 Parasites in Urine

The urogenital system is well supplied with blood. Therefore, it is the final goal of a series of parasites which are set free either by active penetration of the bladder wall (e.g. worm larvae) or by inflammation process (e.g. *Schistosoma* eggs). Table 2.5 collects some important examples.

2.6 Parasites in/on Mucous Layers

Body cavities are often covered by mucous layers, which protect these surfaces from drying. These smooth cells are often attacked by parasites, which are listed in Table 2.6.

2.7 Parasites in/on Tissues

2.7.1 Parasites in Macrophages

Macrophages ingest remnants of cells or even whole (defective) host cells. However, some parasites developed means to survive inside such control cells (see Table 2.7).

2.7.2 Parasites in Liver and Spleen

Liver and spleen are important filter systems in bodies of humans and animals. Therefore, it is not astonishing that parasites drifted by body fluids may stick inside liver and/or spleen. Examples of such parasites are listed in Table 2.8.

2.7.3 Parasites Inside Muscles

The muscles are well supplied with blood. Therefore, many parasites have chosen there their usual site of living, since from there they are easily transmitted to predators (=carnivorous animals). Table 2.9 lists some important parasites transmitted on this pathway.

2.7.4 Parasites in Lung, Trachea and Gills

Since these organs have direct connections to the surroundings of a potential host, parasites may enter or leave the body on these pathways. Examples are presented in Table 2.10.

2.7.5 Parasites in Sexual Organs

Sexual organs offer various chances for parasites to enter or to leave hosts. Table 2.11 lists some important parasites, which are transmitted on these pathways.

2.7.6 Parasites in Bones

Zones wherein the bone cells are developed are well supplied with blood and thus offer good conditions for several parasites to become established there. Table 2.12 shows some important examples.

2.7.7 Parasites in Kidneys

The kidneys are filter systems for blood. Thus, many parasites are located there since they find there besides shelter and sufficient food also the chance to leave a host within urine. Table 2.13 lists some important examples of parasites living inside kidneys.

2.7.8 Parasites in the Swim Bladder of Fish

The swimbladder is densely surrounded by blood vessels, from where oxygen may enter into its interior. Thus, the swimbladder offers good conditions for a series of parasites which are found in its interior (Table 2.14).

2.7.9 Parasites Inside the Brain

The brain is a peculiar protected region of the body, where aggressive body cells are scarce if at all present. Therefore, the brain is very attractive for a series of parasites, which may induce fatal effects on parasitized specimens (see Table 2.15).

2.7.10 Parasites in Eyes

The eyes are very attractive for a series of parasites due to the fact that there are only few defensive cells. Table 2.16 lists some common examples.

2.7.11 Parasites in the Skin

The skin is the protection shield of each body against potential invaders. Nevertheless, many parasites are able to overcome this considerable barrier. Some stay therein, while others take their way from there to other "well-loved" organs. Table 2.17 summarizes some important parasites which live in the skin of their hosts.

2.8 Parasites on the Skin, in the Fur or in Plumage of Birds

The skin with its varying derivatives offers food for a broad spectrum of so-called ectoparasites, which may stay there permanently or just for a short feeding period. Table 2.18 lists some important parasites which attack animals.

3 Methods to Diagnose Parasites

3.1 Investigation of the Feces

The feces of animals contain several stages of parasites, which are visible either macroscopically or only with the help of a microscope. Especially the stages of the second group may be overlooked in the masses of excreted feces. Therefore, it is needed to prepare the probes according to a protocol based on significant and widely tested methods.

(a) **Obtaining of individual fecal probes**

In the cases of larger animals, where the appearance of the feces (slimy, bloody and/or fluid aspects) gives rise to suspect an ongoing parasitaemia, it is recommended to obtain fresh probes directly from the anus by using plastic gloves. In cases of small animals, it is recommended to enter a terminally rounded tiny glass or plastic rod. Fresh feces diluted in a warm saline solution allow to detect motile stages, while in older and cold feces mostly only cysts or worm eggs are stable enough to remain visible even after many hours. If a direct removal of feces is not possible, only portions of feces should be used which are obtained from the upper side, since feces in contact with the floor of the stable might be contaminated with parasites from other hosts.

(b) **Obtaining mixed fecal probes**

If it is not possible to get fecal probes from single animals, since several or even larger numbers of animals are kept together in cages, in a stable or on a common pasture, it is recommended to collect portions of many separate feces and to mix them thoroughly in order to find parasitic stages that are common in this group.

(c) **Storage of fecal probes**

If it is not possible to start the examination shortly after obtaining the fecal probes, they should be stored in closed plastic tubes in a refrigerator at about 8–10 °C in order to avoid reproduction of aggressive bacteria.

© Springer International Publishing Switzerland 2016

H. Mehlhorn, *Animal Parasites*, DOI 10.1007/978-3-319-46403-9_3

3.1.1 Macroscopical Procedures

In the cases of accidentally excreted single parasites or after anthelminthic therapy, the whole feces are diluted in 0.85 NaCl and squeezed through a sieve with a maximal mesh width of 1 mm. The macroscopically visible stages remain on the surface of the sieve and can be obtained from there in order to become examined in a stereomicroscope or even by a simple magnifying glass.

However, the species determination of trematodes and cestodes is based on the structure of inner organs. Therefore, it is needed to fix and to stain them with the help of different chemicals. Species of nematodes on the other hand can mostly be significantly diagnosed by determination of their outer morphological features.

The following procedures have been successfully established in order to stain **trematodes** and **cestodes** or to clear up the interior of nematodes.

3.1.1.1 Acetic Acid–Carmine Coloration According to Rausch

1. Trematodes and proglottids of tapeworms are placed into cold tap water.
2. Fixation and staining of the probes is done for 1–3 h in a 1:1 mixture of acetic acid–carmine solution and 70 % ethanol. The acetic acid solution contains 50 % glacial acetic acid and 50 % aqua dest. This solution becomes saturated with carmine, is cooked for 30 min, and is then cooled down and filtered.
3. Differentiation is done by entering the probes into 1 % hydrochloric acid until the probes appear slightly rosy.
4. Neutralization is proceeded for 2–3 h in tap water until the wanted contrast has been reached.

3.1.1.2 Lactic Acid and Carmine Coloration According to Rukhadze and Blajin

1. Trematodes and proglottids of tapeworms are placed into cold tap water.
2. Fixation and staining for 1–3 h inside the staining solution, which is produced as follows: 0.3 g carmine is cooked in 100 ml of lactic acid, cooled down and filtered. Then 1 ml ferrochloride is added (1 % in H_2O).
3. Neutralization in tap water until the red colour has been changed into blue violet.
4. Differentiation in 1 % hydrochloric acid until the wanted colour is reached.

The stained parasites (according to any of the above-listed methods) are then transferred (placed between two glass slides in 70 % ethanol). After this, dehydration is proceeded within different alcohol concentrations followed by an incubation in xylol in order to clear up the internal organs of the worms. Afterwards, the parasites are embedded in Eukitt and covered by a glass slide.

3.1.1.3 Clear-Up Method for Nematodes

1. Fixation within Becker solution (24 % methanol, 15 % formalin, 5 % glacial acetic acid, 10 % glycerine and 46 % aqua dest.).
2. Dehydration via increasing stages of ethanol followed by transfer into a solution of 70 % ethanol and 5 % glycerine.

3. Evaporation of the alcohol.
4. Drying of the probes at 50°–60 °C.
5. Investigation of the worms being cleared up in 100 % glycerine.
6. Eventual inclusion of the worms by covering them with a cover slide and surrounding it with wax of fluid plastic.

3.1.2 Microscopical Procedures

3.1.2.1 Anus Tests (Glueing Tape Method)

Eggs of *Oxyuris* species (e.g. from horses, donkeys, monkeys) or eggs of tapeworms of carnivores are often glueing in the surroundings of the anus of infected animals. If an infection is presumed, a transparent adhesive tape is used to dab around the anus. Then this piece of tape is placed with its contaminated side adjacent to the glass slide, becomes covered by a drop of xylol and is studied with the help of a light microscope. If the test remains negative, it should be repeated on the following days. The eggs and the adults of the *Oxyuris* species are shown in Sect. 5.3.3.10.

3.1.2.2 Fresh Preparations

A small portion of fresh (warm) feces is placed onto a glass slide and mixed with a drop of physiological salt solution (0.85 NaCl) before being covered with a thin cover slide and examined with the help of a light microscope. Motile stages will be easily diagnosed due to their movements. By addition of a drop of a so-called **Lugol's solution** and after a waiting period of 3–5 min, protozoans and their cysts can be easily diagnosed. Their cytoplasm appears slightly brown and citron like; contained glycogen appears red until dark brown; flagella and nuclei become well discernible. Using phase contrast microscopy, contents of protozoan cysts and of worm eggs become clearly differentiated. Lugol's solution consists of 7.5 g iodine–potassium diluted in 18 ml aqua dest added to 5 g iodine and finally filled up with aqua dest at 100 ml prior to use.

3.1.2.3 Enrichment Methods

Since many parasitic stages occur often only in rather low numbers inside masses of feces and since they are mostly rather tiny, they can be easily missed. Therefore, it is advisable to use at least one concentration method.

A. **Universal methods**
   1. **M.I.F.C. (Merthiolate-iodine-formaldehyde concentration method)**
      This method has the advantage that protozoan cysts as well as worm eggs can be diagnosed. The following basic solutions (**a**, **b**) are needed and should be stored in a bottle of brownish glass:
      (a) 250 ml Aqua dest,
      200 ml Thimerosal (1:1000 diluted in aqua dest),

25 ml Concentrated formaline (40 %),
5 ml Glycerine.

(b) 5% Lugol's solution (not older than 3 weeks; see above).

Immediately prior to the microscopical investigation 4 ml of (**a**) and 1 ml of (**b**) are mixed. Then a small amount of feces (size of a pea) is mixed with this **a** plus **b** solution and squeezed through a fine net in order to take away gross fecal particles. Then the remnant material is transferred into a centrifuge tube, filled up with cold ether and intensely shaken by hand. Then the tube stands open for 1–2 min. Then it is closed and centrifuged at 5 500–1600 g. Inside the closed tube, finally a clear separation of 4 zones can be seen (Fig. 3.1a):

- The ether layer,
- The detritus material,
- The M.I.F. layer,
- The sediment containing the parasitic stages in case they had been present in the feces.

After the detritus material has been removed from the wall of the tube, the fluid components are poured down carefully so that the sediment remains inside the tube and can be taken out in order to be examined with the help of a microscope. The original recipe of this method was published by Blagg et al. (1995) in the Am J Trop Med Hyg 4:23–28. It uses instead of thimerosal the compound "merthiolate" of Fa Eli Lilly. However, in our routine practice and in those of other labs, the above-described variation/modification of the Bonn Medical Parasitology Laboratory was very efficacious. Thimerosal

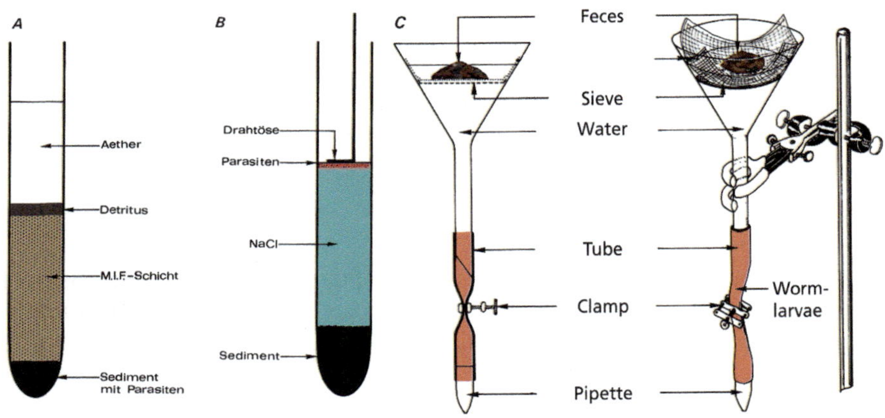

Fig. 3.1 (**a**) Diagrammatic representation of the layers after use of the M.I.F.C. or S.A.F.C. protocol. (**b**) Aspect of the layers after use of the flotation method. The parasites are obtained with the help of a wire loop. (**c**) Diagrammatic representation of the use of the so-called Baermann funnel technique. The larvae leave the feces and enter the water, from where they can be obtained. Translation of German terms: Drahtöse = wire ring, eyelet; Schicht = layer; Sediment = condensed material

($C_9H_9HgO_2SNa$) is distributed by company Sigma, Deisenhofen, order number T5125.

The S.A.F. concentration, which is the basis of the ParasiTrap® diagnostic system of the company Biosepar, Mühldorf, Germany, is chemically less dangerous than the components of the M.I.F.C. but is as effective.

B. **Special tests (showing different groups of parasites)**

These tests are useful in those cases where at least a rather large amount of the feces must be checked (e.g. the minimum amounts are as follows: horses, 30 g; small ruminants and pigs, 20 g; dogs and cats, 5–10 g; rodents, 5 g; etc.).

1. **Flotation** to detect coccidian oocysts, eggs of tapeworms (except for those of *Diphyllobothrium*) and eggs of nematodes.

   This test system is based on the light weight of the parasitic stages when entered in concentrated solution of salt and $ZnCl_2$.

   - 800 ml H_2O
   - 220 g $ZnCl_2$
   - 310 g NaCl

   About 5 g of the feces are entered and mixed with 100 ml of the above-described solution and then filtered through a sieve with a mesh size of about 1 mm. Then this solution becomes centrifuged for about 3–5 min at 300 g. Afterwards, probes were taken from the surface with the help of a wire loop with a diameter of about 5–7 mm and investigated microscopically (Fig. 3.1b).

2. **Egg counting method (according to McMaster)**

   The basic process of the flotation technique is also used for the determination of the amount of excreted eggs.

   - 3 g feces are mixed with 45 small glass pearls and 42 ml water in order to mince the fecal material.
   - The mixed material is filtered with the help of a sieve with a mesh diameter of about 0.15 mm.
   - 15 ml of the solution is filled into a centrifuge tube and centrifuged at 200 rpm for 2 min. After the fluids have been decanted, the sediment is filled up with about 15 ml saturated NaCl solution and slightly mixed.
   - With the help of a glass pipette, fluid is obtained and entered into the two chambers of a McMaster system.
   - Then both chambers are covered by the marked cover slides and investigated with the help of a microscope at magnifications of about 20×.
   - Counting of the eggs which are seen inside the marked squares.
   - The multiplication of the observed number of eggs by 50 gives an approximate idea on the number of eggs within one gram of feces.

3. **Sedimentation method**

   This method is used to detect eggs of trematodes and (rarely) larvae of worms. However, results are poor in the case of *Dicrocoelium dendriticum*, since these eggs sink too slowly. In contrast to this, the large oocysts of *Eimeria leuckarti* can easily be documented with the help of this method.

About 5–10 g feces are mixed with 100 ml physiological salt solution and sieved in order to eliminate gross particles. Then the suspension is kept untouched for about ½ h until a sediment is produced at the bottom of the vessel. By repeated discharging the supernatant fluid and filling up fresh water, the sediment is cleaned. Finally, the sediment is centrifuged and studied with the help of the light microscope.

4. **Larval concentration method (so-called Baermann funnel)**

   A glass funnel, which is equipped with a rubber tube and a clamp system, is provided with metal net system and filled by warm water (Fig. 3.1c). About 20 g of fresh feces (obtained from different sites of the feces) are entered into the funnel. Then the contained larvae wander into the warm water and finally sink into the closed rubber region. Within 1 h (up to 6–15 h), the larvae are concentrated in the rubber region. If portions of the bottom water are investigated, contained larvae can be diagnosed.

5. **Feces cultures**

   This method is used to determinate worm larvae. According to the method of Harada and Mori (see Mulisch and Welsch 2010), the following procedure is recommended:

   – A centrifugation tube becomes filled with 3 ml aqua dest.
   – A piece of filter paper (1 × 12 cm) is covered in its centre with a 1–2 mm thick layer of feces.
   – The lower margin of the contaminated paper is brought into contact with the water, while the upper margin is fixed.
   – Then this paper is incubated for 7–10 days at 24–28 °C under constant control whether the filter paper has still contact with the water.
   – Larvae hatching from worm eggs move into the direction of the water and can be removed with the help of a pipette and determined using a light microscope.

   In cases when larger amounts (e.g. 10 g) of feces are available, Eckert's (Zürich, Swiss) method is recommended: feces are mixed with sawdust within a glass vessel closed by a glass cover and placed for 10–12 days at 24°–28 °C inside an incubator. The vessel is opened daily for about 30 min and eventually moistened with water. After 10–12 days, the sawdust/feces mixture is intensively pressed, covered by water and a petri dish and finally placed top downwards on a table. After about 12 h, eventually present larvae can be obtained with the help of a pipette and determined in a microscope.

6. **Cultures to diagnose amoebae**

   Entamoeba histolytica and other amoebae can be cultured on different growth media. The method of Dobbel and Laidlaw (see Mulisch and Welsch 2010) is rather easy:

   (a) **Production of a solid component of the growth medium**:

   Sterile horse serum becomes filled into a glass tube, which is then brought into an oblique position and stored at 80 °C until the serum is stiffened.

(b) **Production of the fluid component of the growth medium**:
This fluid component is based on a so-called protein-Ringer solution. The "white" content of a chicken egg is obtained in a sterile status and mixed with 500 ml of a sterile Ringer solution at pH 7.4–7.5. This mixture receives additions of penicillin (5000–10,000 units/100 ml), streptomycin (5000–10,000 units/100 ml) and amphotericin (250 μg/ 100 ml).

(c) **Final production**
The fluid component is then filled into the vessel with the solid component in a manner that the fluid compound reaches one-third of the total weight. Then a little amount of sterile rice starch is added.

(d) **Incubation**
This ready-to-use compound in the vessel is then incubated with about 0.5 g fresh feces and closed with the help of absorbent cotton. Cultures containing material of homoeothermic animals (dogs, cats, monkeys, etc.) are incubated at 37 °C, while probes of poikilothermic animals (e.g. reptiles, frogs) are stored at 20–22 °C. After 2–3 days samples of the stored material are taken with the help of a sterile pipette and investigated with the help of a light microscope, which makes visible motile stages such as amoebae, flagellates, etc.

7. **Cultures of *Balantidium coli* and *Entamoeba invadens***
1–2 g feces and some mucous layer material are entered in a tube (10 ml), which is filled with a medium that consists of one portion horse or pig serum, 9 portions of Ringer solution and 0.2–0.5 rice starch. Incubation is done at 37 °C and 22 °C (in the case of *E. invadens*). The stages of *Balantidium* amoebae or related species can be found after 3–5 days at the bottom of the culture vessel.

8. **Cultures of *Giardia* and *Trichomonas* species**
The culture of the stages of these species is rather complicated. The details are described by Jensen (1983).

9. **Coloration/staining of protozoan stages**
To obtain significant results, it is needed to fulfil the following criteria:
- Use of fresh feces;
- Sufficient fixation;
- Avoidance of drying of the smear preparation;
- Complete dehydration of the preparation before embedding.

The following four staining techniques are common:

A. **Haematoxyline staining according to Heidenhain.**
B. **Trichrome staining according to Wheatley.**
C. **Staining of cryptosporidia in feces.**
D. **Methylene green staining.**

For details, see Mulisch M, Welsch U (2010) Romeis. Mikroskopische Technik, 18th edn. Springer Spektrum Ahad Verlag, Heidelberg.

3.1.2.4 Postmortem Investigations

Parasites inside the intestine or other organs can even be diagnosed after the death of the animal. The methods are described in detail by Herlich (1956), Williams et al. (1979) and Mulisch and Welsch (2012).

3.2 Investigation of Parasites Inside Blood

(a) The "**Thick Drop**": A drop of blood is spread out (not too thin) over an area of about the size of the thumb nail on a clean slide that has been freed as far as possible from grease and is dried in the air. Then it is placed for 5–10 min into ordinary water in a dish to eliminate the haemoglobin (Fig. 3.2 B).

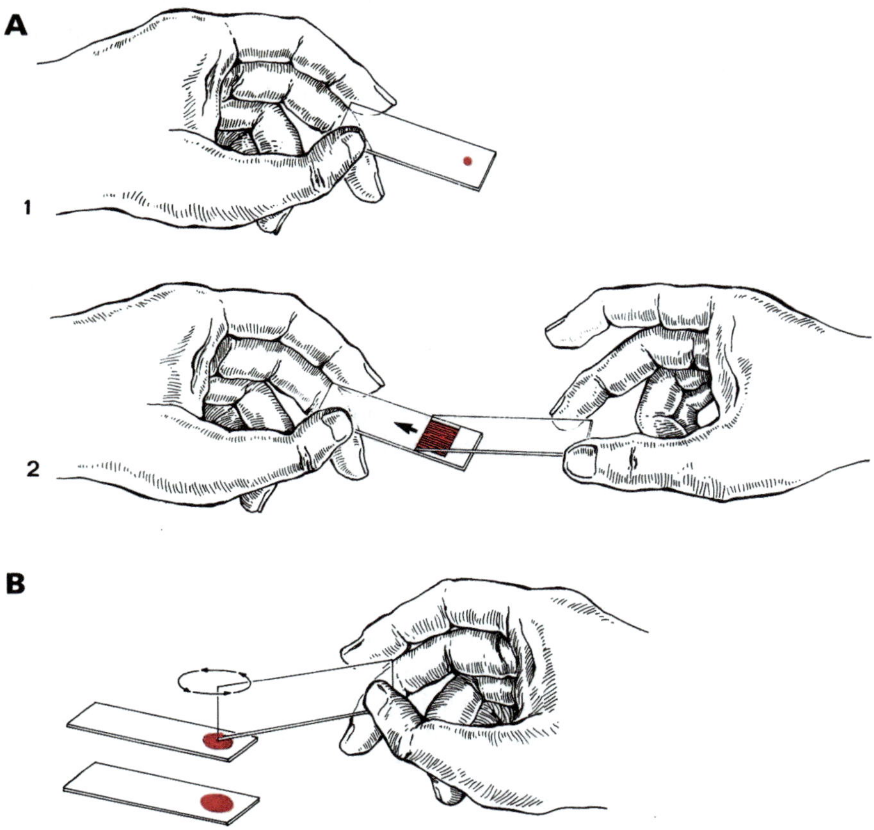

Fig. 3.2 Diagrammatic representation of the preparations of a blood smear (**a1**, **a2**) and of a so-called thick droplet (**b**). In case of the blood smear a very thin layer is produced, while in case of the "thick droplet" a larger amount of blood is at first defibrinated, dried and finally lysed by entering into water

Freshly drawn blood is haemolysed in only a few minutes; it is recommended that, if circumstances permit, a few drops of acetic acid should be added to preparations older than 3 weeks; when this is done the acid must be removed ("neutralized") by tap water, before the slide is stained. After removal of the haemoglobin, the slide is stained (without fixation!) according to Giemsa as follows:

The blood smear is covered with the Giemsa solution (stock solution 1:20 diluted with water) for 30 min. Then the stain is washed off with distilled or buffered water or in emergencies (if necessary) with rainwater. Afterwards, the slide is dried in the air (not between filter papers!).

Important rule: Use neutral water, pH 7.0. Use either twice-distilled water (aqua bidestillata) or Weise's original buffer solution, according to the instructions.

(b) The **Blood Smear** (Fig. 3.2A) (this staining method can also be used for smears taken from organs): A drop of blood, not too big, is smeared onto a clean glass slide with a cover glass. Then process as follows:
1. Fixation of the air-dried smear for 3 min in methyl alcohol;
2. Drying in the air (not between filter papers!);
3. Staining with Giemsa solution (for each smear, add 5 drops of the stock solution to 5 ml of neutral or buffered distilled water); stain for 30 min;
4. Wash off the stain solution with a brisk stream of water; then allow to dry.
5. Add a drop of immersion oil on the dry glass slide and examine it with the help of the oil immersion lens. If filariae are suspected, it is better to examine the smear with a dry lens after the smear has been covered with a layer of oil. For examination of microfilariae, it is better to use Delafield's haematoxylin staining rather than Giemsa.

3.3 Further Cytological and Serological Methods

There exist a broad spectrum of other methods to detect parasites in other organs of animals and in saliva, lymph, urine, mucous layers, etc., which are presented in detail within books of several authors:

Blagg et al. (1995), Dada (1979), Deplazes et al. (2013), Herlich (1956), Jensen (1983), MAFF (1986), Martini (1952), Mehlhorn (2016a, b), Mulisch and Welsch (2010), Reichenow et al. (1969), Williams et al. (1979), Woo (1969).

Further Reading

Aspöck H (ed) (2010) Sick due to arthropods. Denisia 10, Linz.
Blagg EN et al (1995) Diagnostic methods. Am J Trop Med Hyg 4:23–28.
Dada BJO (1979) A new technique for the recovery of *Toxocara* eggs from soil. J Helminthol 53:141–144.
Deplazes PM et al (2014) Parasitologie für Tiermediziner, 3rd edn. Enke, Stuttgart.

Herlich H (1956) A digestion method for post-mortem recovery of nematodes from ruminants. Proc Helminthol Soc Wash 23:102–103.

Jensen JB (1983) In vitro cultivation of protozoan parasites. CRC Press, Boca Raton.

Lane RP, Crosskey RW (1995) Medical insects, 2nd edn. McMillan Press, London.

MAFF (1986 Manual of veterinary parasitological laboratory techniques, 3rd edn. Ministery Agricult, Fishery, Food, London.

Martini E (1952) Lehrbuch der medizinischen Entomologie. Fischer, Jena.

Mehlhorn H (ed) (2016) Encyclopedia of parasitology, 4th edn. Springer, Berlin, New York.

Mehlhorn H (2016) The parasites of humans, 4th edn. Springer, Heidelberg.

Mulisch E, Welsch U (2010) Microscopical techniques, 18th edn. Spektrum Akad Verlag, Heidelberg.

Reichenow WE et al (1969) Leitfaden zur Untersuchung der tierischen Parasiten des Menschen und der Haustiere, 4th edn. Ambrosius Barth Verlag, Leipzig.

Sonenshine DE, Roe RM (2014) Biology of ticks, vol 2, 2nd edn. Oxford University Press, Oxford.

Williams J et al (1979) Activity of fenbendazole against inhibited early fourth-stage larvae of *Ostertagia ostertagi*. Am J Vet Res 40:1087–1090.

Woo PTK (1969) The haematocrit centrifuge for the detection of trypanosomes in blood. Can J Zool 47:921–923.

Protozoan Parasites

4

4.1 Trichomonads

This group of unicellular parasites, which live in the sexual organs and inside the intestine, is characterized by the possession of a species-specific number of anterior free flagella and a so-called recurrent flagellum, which is attached at the cell surface and acts like a rudder. The stability of the cell is supported by a so-called axostyl, which consists of a bunch of microtubule enclosing the nucleus and forming a terminal tip (Figs. 4.1, 4.2 and 4.5). The trichomonad species of cattle, horses, donkeys and birds are important and may lead to considerable economic losses (see Sects. 4.1.1–4.1.4).

4.1.1 *Tritrichomonas foetus*

1. **Name**: Greek: *tri* = three; *thrix* = hair; *monas* = living organism. Latin: *foetus* = unborn individual, foetus.
2. **Geographic distribution/epidemiology**: Worldwide in ruminants and cats; however, rather rare in Europe.
3. **Biology, morphology**: There exist two morphologically not distinguishable strains, one in cattle and one in cats. The 10–25 × 3–15 μm sized flagellates appear pear shaped, showing 3 free flagella at the apical pole and in addition a lateral recurrent flagellum (Table 4.1; Figs. 4.1, 4.2, 4.3 and 4.4). The axostyl, which consists of microtubules, surrounds the anteriorly situated nucleus and leads to a pointed protrusion at the terminal end of the cell (Fig. 4.1). **Reproduction** occurs by longitudinal **binary fission**. In **cows** they settle mainly in the vagina, in the oviduct and in the uterus, while in bulls they are mainly found inside penis and related channels and thus are transmitted within the sperm. These parasites ingest bacteria and/or cell remnants by phagocytosis, which occurs along the cell surface. **Note**: Trichomonads do not possess mitochondria, but obtain their energy with the help of so-called **hydrogenosomes**.

© Springer International Publishing Switzerland 2016

H. Mehlhorn, *Animal Parasites*, DOI 10.1007/978-3-319-46403-9_4

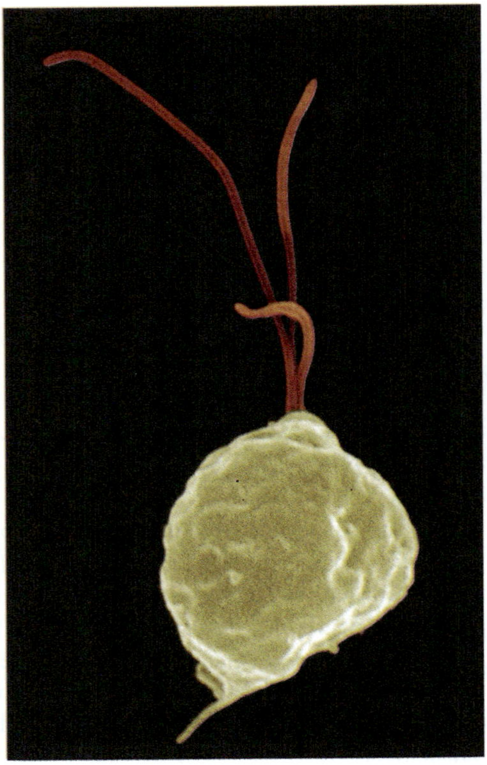

Fig. 4.1 Scanning electron micrograph of a trophozoite of *Trichomonas foetus* of cattle. Note the tree terminal flagella, the lateral recurrent flagellum and the terminal tip, which represents the axostyl covered by the single cell membrane

4. **Symptoms of disease (Trichomoniasis)**: In about 50 % of infected males, cattle symptoms remain low grade or are even absent. In cows, however, these parasites induce severe symptoms, which may endanger or even kill the foetus. Very common are vaginitis and endometritis (producing yellowish-green, sweet smelling, slimy fluids inside cervix and vulva). These symptoms may introduce abortion of the foetus after about 6–16 weeks. In general, the foetus is ejected; however, also a maceration process may occur in the uterus (pyometra). In the latter case, pregnancy may be pretended due to persistence of the corpus luteus. In addition, bacterial superinfections occur inside the sexual organs of cows.

   Colitis and chronic diarrhoeas occur in cats.
5. **Diagnosis**: In cows the parasitic stages can be diagnosed either with the help of PCR or due to their motility in fresh microscopical preparations made from probes obtained from materials of the mucous layers of the sexual organs or from pyometra within uterus and vagina. Fixed and Giemsa-stained smear preparations can be made, when fresh material is not available. In male hosts, fresh or Giemsa-stained preparations can be made from secretions obtained

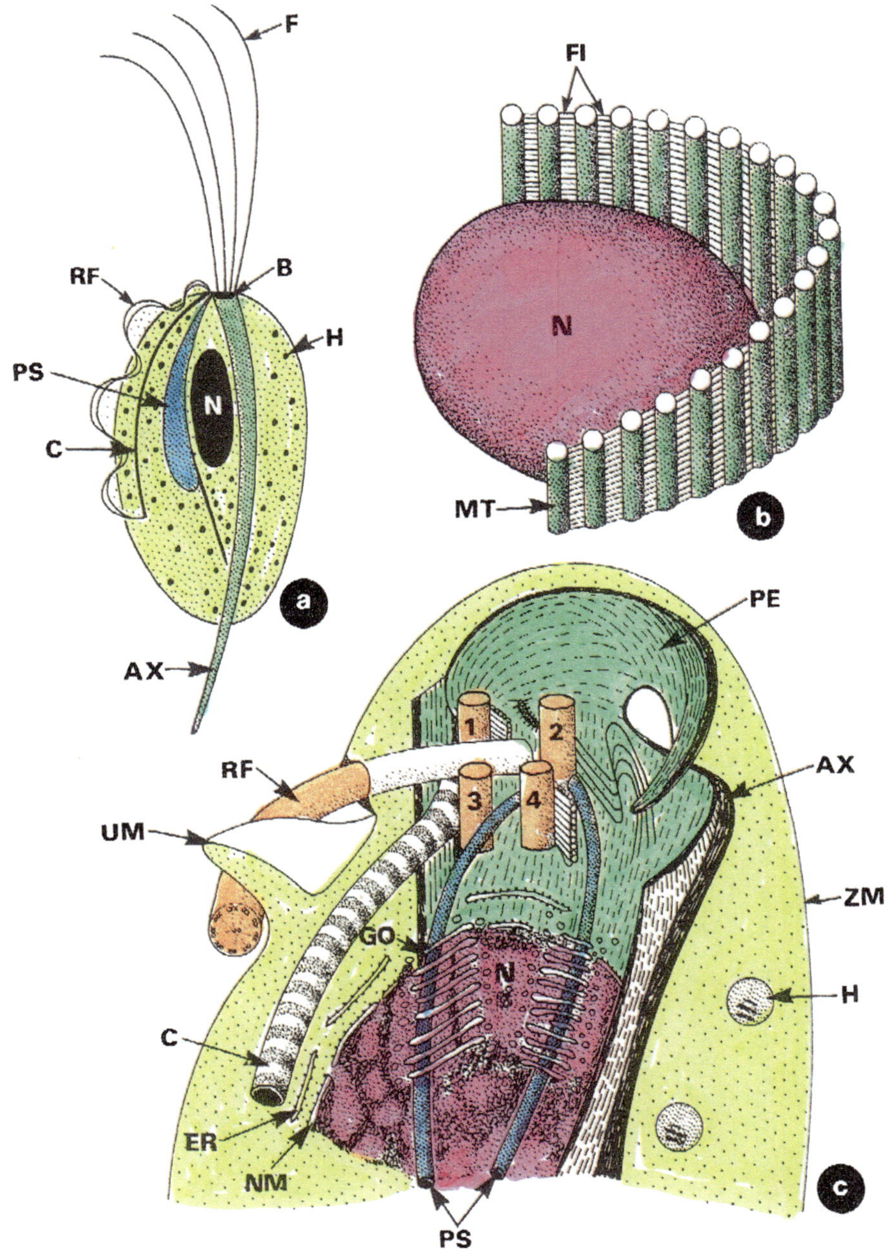

Fig. 4.2 Diagrammatic representation of the inner organization of typical trichomonad specimens (here *T. vaginalis* of humans). (**a**) Total aspect; (**b**) Aspect of the axostyle close to the nucleus; (**c**) Apical pole of the cell. *1–4* basal bodies of the flagella; *AX* axostyl; *B* basal bodies; *C* costa; *ER* endoplasmic reticulum; *F* flagellum; *FI* fibrils; *GO* golgi apparatus; *H* hydrogenosomes; *MT* microtubules; *N* nucleus; *NM* nuclear membrane; *PE* pelta; *PS* parabasalian strands; *RF* attached flagellum; *UM* undulating membrane; *ZM* cell membrane

Table 4.1 Important species of flagellates—Trichomonadida

Species	Size	Infected organs	Characteristics	Important symptoms of disease	Hosts
Tritrichomonas foetus	10–15 × 3–15 μm	Sexual organs	3 free flagella, 1 recurrent flagellum	Abortion	Cattle
Tritrichomonas equi	10 × 5 μm	Caecum, colon, small intestine	3 free flagella, 1 recurrent flagellum	Diarrhoea, colitis	Horses, donkeys
Trichomonas equibuccalis	9 × 6 μm	Caecum, colon	4 free flagella, 1 recurrent flagellum	Diarrhoea, colitis	Horses, donkeys
Tritrichomonas muris	26 × 16 μm	Caecum, colon	3 free flagella, 1 recurrent flagellum	None	Rodents
Trichomonas gallinae	7–15 × 5–10 μm	Craw, caecum, colon, internal organs	Axostyl clearly visible, 4 free flagella, 1 recurrent flagellum	Yellow button, generalized form	Pigeons, chickens, raptors, parrots, sparrows, etc.
Trichomonas gallinarum	8 × 5 μm	Caecum	Axostyl unclear	Pale yellow diarrhoeas	Chickens, turkeys
Histomonas meleagridis	5–30 μm in diameter	Intestine, cloaca, internal organs	1–4 flagella, cyst stages exist	Blackhead disease	Turkeys, chickens, peacocks, pheasants
Tritrichomonas, Trichomonas and Pentatrichomonas species	10 × 15 μm	Intestine, excretion system	3, 4 respectively 5 free flagella, 1 recurrent flagellum	Loss of weight, often inapparent	Amphibians, reptiles
Tetratrichomonas ovis	9 × 6 μm	Rumen, caecum	4 free flagella, 1 recurrent flagellum	Hardly, mostly inapparent	Sheep
Trichomonas suis	16 × 8 μm	Small intestine	4 free flagella, 1 recurrent flagellum	Hardly, mostly inapparent	Pigs
Trichomonas vaginalis	10–25 × 8 μm	Urogenital system	4 free flagella, 1 recurrent flagellum	Discharge, potential bacterial superinfection	**Humans**
Pentatrichomonas hominis	20 × 8 μm	Small intestine	5 free flagella, 1 recurrent flagellum	Apathogen	**Humans**

from preputium slime. If an infection is suspected, it is recommended to start a culture using material obtained from the sexual organs of the male or female cattle. If the test is positive, it is obligatory in many countries with a strict law for animal protection (e.g. in Germany) to inform the veterinary health authorities, which will steer the following obligatory measurements depending on the grade of infection. Similar **obligatory announcements** are needed (e.g. in Germany) in cases of infections of horses with *Trypanosoma equiperdum* (dourine), infections of bees with *Varroa* species (varroasis) and infections with *Toxoplasma gondii* in pigs, dogs and cats.

Diagnosis of trichomonads in **cats** may be done by fecal culture using the product InPouch®, wherein 1 ml fresh feces were incubated for 24 h at 37 °C.

6. **Pathway of infection**: The parasites are mainly transmitted during sexual intercourse of cows and bulls or (rarely) by artificial fecundation.
7. **Prophylaxis**: Regular examinations of the sperma of bulls and control of the vaginal fluids of cows—especially in animals which had been newly added to the herd.
8. **Incubation period**: Very variable.
9. **Prepatent period**: Very variable.
10. **Patency**: Up to several years.
11. **Therapy**: Treatment of the **cows** has to be steered by the local veterinary authorities to avoid an endemy. While infected **bulls** are in general slaughtered, worthful ones might be treated under the survey of the local veterinary authorities. **Cows**, however, are mostly treated. However, the process takes time and is expensive, since the whole sexual system has to be cleaned. Depending on the different countries, control and treatment are steered by the local veterinary health authorities. Treatment of cats can be done by oral application of ronidazole (30 mg/kg bodyweight) all 12 h for 2 weeks.

Further Reading

Benchimol M et al (2016) The fungal metabolite gliotoxin inhibits proteasome proteolytic activity and induces an irreversible pseudocystic transformation and cell death in *Tritrichomonas foetus*. Parasitol Res (in press).

Frey CF et al (2009) Intestinal *Trichomonas foetus* infection in cats. Parasitol Res 104:783–788.

Marming K (2010) Update on the diagnosis and management of *Tritrichomonas foetus* infections in cats. Top Comp Anim Med 25:145–148.

Michi A et al (2016) A review of sexually transmitted bovine trichomoniasis. Theriogenology 85:781–791.

Yao C (2013) Diagnosis of *Trichomonas foetus*. J Med Microbiol 62:1–9.

Yao C, Köster LS (2015) *Trichomonas foetus* infection. Vet Res 46:35.

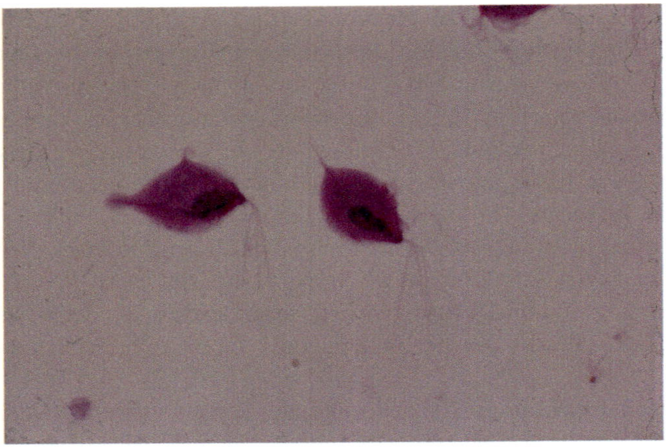

Fig. 4.3 Light micrograph of two trophozoites of *Tritrichomonas equi*

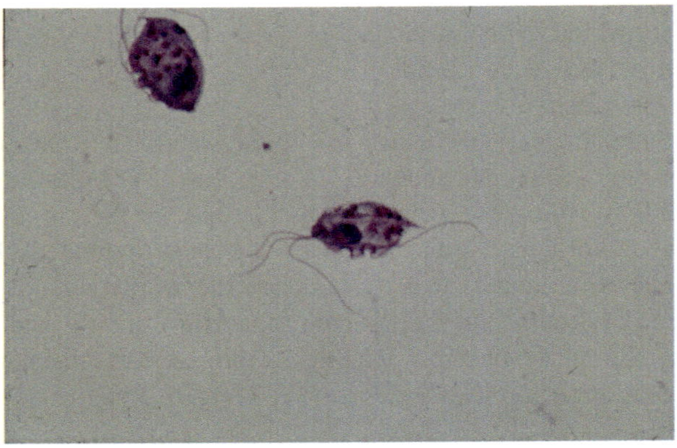

Fig. 4.4 Light micrograph of trophozoites of *Trichomonas equibuccalis*

4.1.2 Trichomonads of Donkeys and Horses

1. **Name**: Latin: *equus* = horse; *buccalis* = belonging to the cheek. Greek: *tri* = three; *trix* = hair; *monas* = living organism.
2. **Geographic distribution/epidemiology**: Worldwide.
3. **Biology, morphology**:
   (a) ***Tritrichomonas equi***: About 11 × 6 μm in size, 3 free flagella and one recurrent one (Fig. 4.3). Stages parasitize inside the caecum and colon, cysts are not formed. For general shape, see *Tritrichomonas foetus*.
   (b) *Trichomonas equibuccalis*: About 6–9 μm (mean 7.6 × 6.3 μm) in length and is found onto mucous layers inside the mouth; the stages possess 4 free anterior flagella and one recurrent flagellum (Fig. 4.4).

   The reproduction of both species occurs by repeated longitudinal binary fission.

4. **Symptoms of disease (Trichomoniasis)**: In the case of *Tritrichomonas equi*, diarrhoeas have been occasionally observed as well as colitis, while *T. equibuccalis* apparently does not introduce symptoms of disease.
5. **Diagnosis**: Microscopically detection of the living stages in fresh, warm feces or by fecal cultures.
6. **Pathway of infection**: Not sure in the case of *Tritrichomonas equi* (probably via unknown cyst-like stages). Mouth-to-mouth contact in the case of *T. equibuccalis*.
7. **Prophylaxis**: Unknown.
8. **Incubation period**: Unknown.
9. **Prepatent period**: Unknown.
10. **Patency**: Years.
11. **Therapy**: Mostly not needed. However, reports showed that dimetridazole and other 5-nitroimidazoles (e.g. ronidazole, metronidazole) are effective but several are no longer registered in some European countries.

Further Reading

Damron GW (1976) Gastrointestinal trichomonads in horses: occurrence and identification. Am J Vet Res 37:25–28.

Deplazes P et al (2013) Lehrbuch der Parasitologie für die Tiermedizin. Enke, Stuttgart.

Levine ND (1973) Protozoan parasites of domestic animals and man, 2nd edn. Burgess, Minneapolis.

4.1.3 *Trichomonas* Species and *Tetratrichomonas* Species of Birds

1. **Name**: Greek: *thrix* = tiny hair; *monas* = living organism, individual. Latin: *gallina* = chicken; *columba* = dove; *tetra* = four.
2. **Geographic distribution/epidemiology**: Worldwide.
3. **Biology, morphology**: In some European countries (e.g. Germany), 60–100 % of untreated chickens in stables are infected with different species of flagellates. The most important parasites belong to two species:
   (a) ***Trichomonas gallinae* (syn. *T. columbae*)** parasitizes at the mucous layer of the upper region of the intestinal tract of birds such as doves, chickens, raptor birds, parrots, sparrows, seagulls, etc. This parasite (Fig. 4.5a) reaches a size of 7–15 × 5–10 μm and appears pear like (Fig. 4.5b). Characteristics are the presence of 4 free anterior flagella and a fifth, which appears as recurrent flagellum and does never surpass the posterior terminal pole, where, however, the protruding axostyl remains always visible (Fig. 4.5a, b).
   (b) ***Tetratrichomonas gallinarum*.** This species is found inside the caeca of doves, chickens, turkey and related birds. The stages are rather small and their axostyl appears often very indistinct (Figs. 4.5a and 4.7). *Tetratrichomonas anseris* lives in the caeca of geese and *T. anatis* in the intestine of ducks.

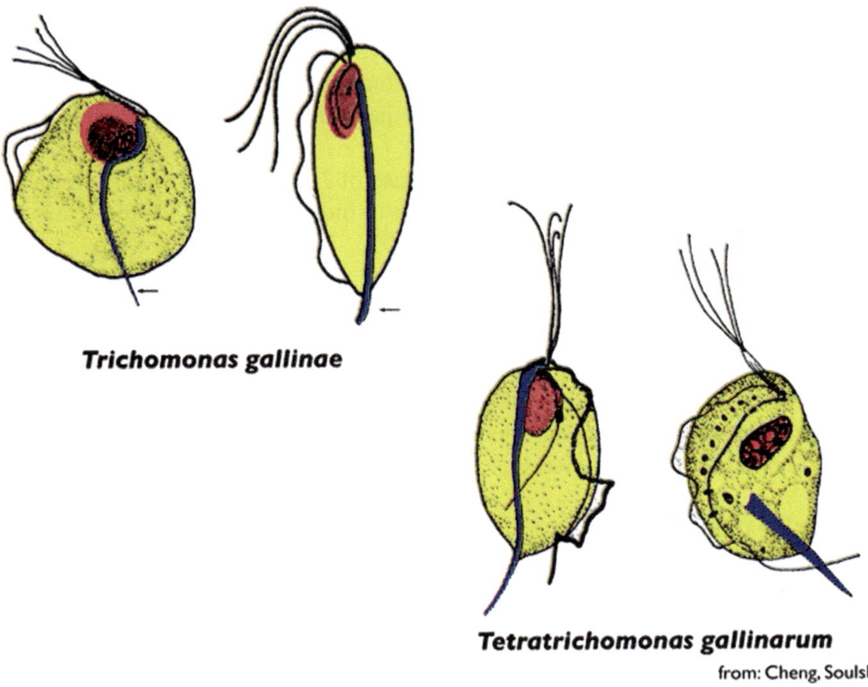

Fig. 4.5 Diagrammatic representations of stages of *T. gallinae* and *Tetratrichomonas gallinarum*

4. **Symptoms of disease (Trichomoniasis)**: Although adult birds may be permanently infected without showing distinct symptoms of disease, younger ones may suffer severely due to the existence of strains with different levels of virulence.
   (a) *Trichomonas gallinae* infections may show three different grades of virulence:
      – The **throat-crop infection** (so-called **yellow button** of doves) is characterized by yellowish cheesy layers (diphtheroid type) inside the beak, throat and crop (Fig. 4.6). Smear preparations of these slimy layers show masses of parasites and bacteria (Fig. 4.7). Due to these masses of slimy material, the uptake of water and food is strongly reduced as well as the feeding activity. Furthermore, also the breathing activity is considerably reduced, so that homing and racing pigeons are considerably impeded.
      – The **inner generalized infection** is characterized by the fact that spots of parasites occur in/on liver, lung, heart, trachea, etc. Heavy infections may lead to death.
      – **Infections of the navel region**, which then appears yellowish, reduce the general fitness of very young animals, lead to loss of bodyweight and may be followed by high mortality rates.
   (b) Infections with *Tetratrichomonas gallinarum* lead to diarrhoeas (light yellow, fluid caecal excretions) due to a fibrous enteritis. In rare cases,

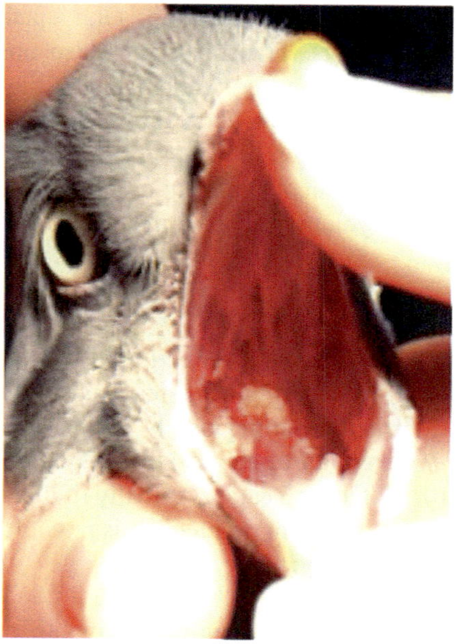

Fig. 4.6 Macrophoto of a bird suffering from a throat-crop infection

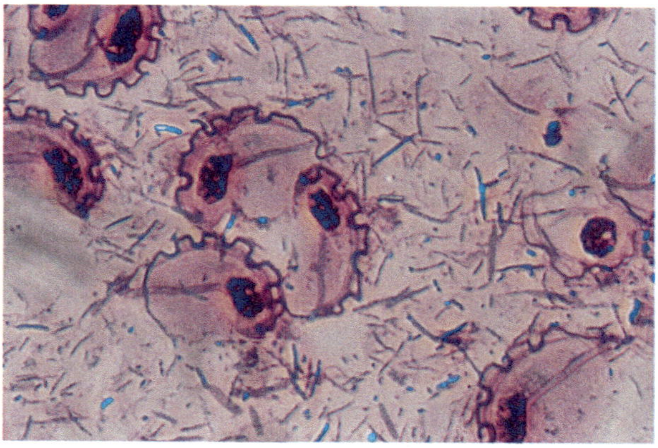

Fig. 4.7 Light micrograph of a Giemsa-stained smear preparation containing stages of *T. gallinae*

also the liver may be involved leading to necrotic alterations of the parenchyma. Further clinical symptoms are food refusal and heavy losses of weight. However, chickens suffer mostly only from slight symptoms, while young pheasants may show severe reactions and may die. On the other side, parrots show high infection rates, but very often only low-grade symptoms of disease. In young or very old animals, however, symptoms such as apathia, crop infections, breathing problems or slimy layers inside the beak have been described in relevant literature.

5. **Diagnosis**: Preparations of material obtained from slimy layers inside the mouth show the parasites in fresh or coloured status (Figs. 4.5a, b and 4.7). However, with respect to differential diagnosis, infections due to viruses, fungi, bacteria (salmonellosis, tuberculosis, etc.) have to become excluded.

6. **Pathway of infection**: Oral uptake of agents of disease during beak–beak contacts. Transmission is also possible during oral uptake of contaminated drinking water, while parasites die on dry food.
7. **Prophylaxis**: The following protection procedures should follow regularly according to a fixed scheme:
   - Medical treatment of the adult birds about 8–14 days before hatching of the young ones from the eggs. Treatment has eventually to be repeated.
   - Treatment of new birds has to be done before entering them into the stock.
   - Treatment of animals should be done when coming back from bird expositions.
   - Repeated disinfection of water vessels.
   - Cleaning procedures inside the stables at regular intervals.
8. **Incubation period**: 1–4 days.
9. **Prepatent period**: 1–4 days.
10. **Patency**: Years, eventually lifelong. If infection breaks down or disappears due to treatment, reinfection is possible since immunity lasts only for a few weeks.
11. **Therapy**:

    Doves:

    (a) Do not remove the diphtheroid layers in the throat (danger of severe bleedings).

    (b) Lesions along the navel should be opened smoothly and cleaned.

    (c) In the case of an **inner trichomoniasis**, parenteral antibiotics (e.g. chloramphenicol, streptomycin) and vitamins A, E, B complex should be given for 1–2 weeks.

    (d) Therapy of animals, which are not used for human consumption, may be done by application of 5-nitroimidazoles diluted in the drinking water. Dosage must consider the daily water need of homing doves: in summer 50–60 ml per day and in autumn/winter 20–30 ml per day. Older animals need less water than younger ones. During the treatment period, flight times should be limited to about 1–2 h per day.

    (e) The duration of the treatment should vary depending on the location of the parasites. In the case of **throat-crop infections**, treatment is recommended for 6–7 days up to 12 days in heavy infections. In cases of an **inner trichomoniasis**, the prognosis for a successful treatment via drinking water is rather bad, since the weak birds will not drink enough of the drug containing water. Therefore, valuable birds should get the medication with the help of a pipette for the first week and later within drinking water for up to two further weeks.

    Parrots:

    Literature reports recommend treatment with metronidazole in the drinking water in similar dosages as used for doves. Individual treatment of large parrots can be done using carnidazole (10 mg/large parrot). Additional application of vitamins is also recommended.

Further Reading

Amini A et al (2014) Trichomonads in birds—a review. Parasitology 141:733–747.

Duarte M et al (2009) Cytotoxicity of solubilisation vehicles for *Trichomonas gallinae* and *Tritrichomonas foetus*. Vet Parasitol 166:167–170.

Hess M et al (2006) Clonal cultures of *Histomonas meleagridis, Tetratrichomonas gallinarum* and *Blastocystis* sp. Parasitology 133:547–554.

Liebhardt D et al (2014) A single strain of *Tetratrichomonas* causes fatal typhlohepatitis in red legged pariges (*Alectoris rufa*). Avian Pathol 43:473–480.

Pham AND et al (2013) Establishing monoeukaryotic *Histomonas meleagridis* cultures from in vivo infection contaminated with *Tetratrichomonas gallinarum* and *Blastocystis* spp. Parasitology 140:1266–1274.

Richter B et al (2010) First report of typhlitis/typhlohepatitis cuased by *Tetratrichomonas gallinarum*. Avian Pathol 39:499–503.

Zimre-Grabensteiner E et al (2011) Genetically different clonal isolates of *Trichomonas gallinae*. Parasitol Int 60:213–215.

4.1.4 *Histomonas meleagridis* (Histomoniasis)

1. **Name**: Greek: *histos* = tissue; *monas* = living organism; *meleagridis* = pearl chicken. The name has its origin in a Greek fairy tale, wherein the sisters of a fighter named Meleagros, who was a follower of Odysseus in the Trojan War, were transformed into chickens while mourning for his death.
2. **Geographic distribution/epidemiology**: Epidemics are common and long-lasting. Worldwide, in extreme large animal houses.
3. **Biology, morphology**: The trophozoites of *H. meleagridis* live in the intestine of birds such as chickens, pearl chickens, turkeys, pheasants, peacocks, etc.). They reach sizes between 5 and 30 μm (Fig. 4.8) and show mostly 1 flagellum and only occasionally 2–4 flagella (apparently due to failed divisions). The total life cycle is still under discussion. However, it is now widely accepted that apparently changings of the intestinal fluid due to the presence of peculiar bacteria, coccidians or fungi lead to the fact that spherical *H. meleagridis* stages, which have reduced their flagellum and measure 8–15 μm in diameter, enter the mucous layer of the caeca and enter via the portal vein the liver. There these amoeboid stages start divisions and induce lesions and necrosis. Some of these stages appear encysted (Fig. 4.9) as was shown by common investigations of the groups of Hess (Wien) and Mehlhorn (Düsseldorf). Some authors claim that trophozoites of *H. meleagridis* inside the intestine become included into eggs of the nematode *Heterakis gallinarum* and are thereby transmitted. Such transmission may occur in rare cases, but a successful life cycle like that of *H. meleagridis* cannot be based on such accidental events although it was claimed that such stages might be stored for up to 4 years inside eggs.
4. **Symptoms of disease (Histomoniasis, black head disease, necrotic enteritis, enterohepatitis syndrome)**: During the acute phase of the disease, typical diarrhoeas occur showing a yellowish, slimy consistence and leading to

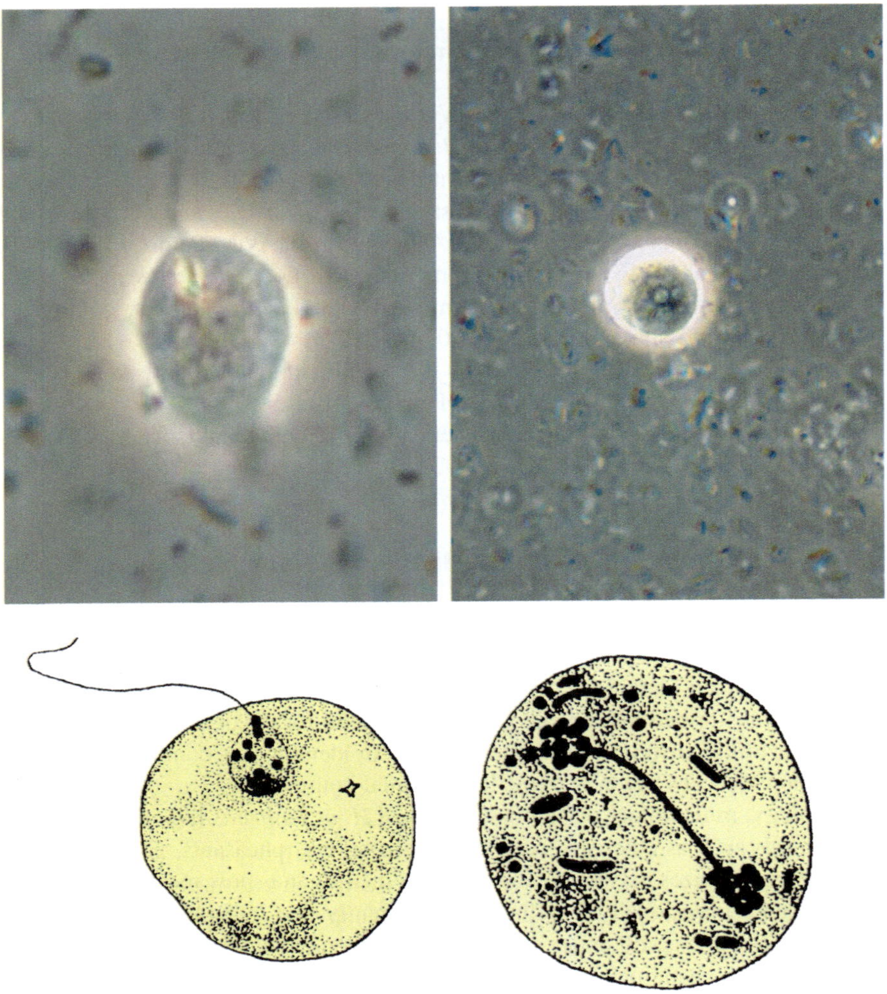

Fig. 4.8 *Histomonas meleagridis*. Light microscopic and diagrammatic representations of intestinal stages

weakness and intense sleeping, which are due to an increasing necrotizing hepatitis (Fig. 4.10). Further symptoms are breathing problems, shaggy plumage and the fact that the skin (especially that of the crest) appears blackish. In the case of chronic infections, anaemia may occur. Especially young animals die at high rates (15–100 % !). Turkeys may be infected during any stage of life, however, predominantly during the first 4 months after hatch. Pearl chickens show tissue damages inside the caeca. Increasing infection rates of chickens were reported during the last years since 2014.

5. **Diagnosis**: Infections with *H. meleagridis* can be verified at early stages by microscopical examination of fresh feces (Figs. 4.8 and 4.9). In cases of *post-*

Fig. 4.9 Light micrographs of cysts of *H. meleagridis* within the colon of birds

mortem examinations, the liver shows pea-sized, defined yellowish necrosis sites (Fig. 4.10). Squeezing preparation of the liver tissues shows amoeboid trophozoites (Fig. 4.8). Inside the feces, cysts occur in low numbers (Fig. 4.9). For differential diagnostic purposes, systemic fungal infections have to be considered (PAS coloration shows details), which are supported by several PCR tests.

6. **Pathway of infection**: Oral by uptake of cysts and vegetative stages within fresh faeces (e.g. in cases of intense stable rearing of poultry).
7. **Prophylaxis**: Keeping young turkey away from chickens, as well as separation of young turkeys from older ones. Keeping the animals on wire gratings helps to minimize contacts with potentially infected feces. Trials to immunize turkeys and chickens failed.
8. **Incubation period**: About 6–9 days.
9. **Prepatent period**: After about 2 days (during experimental infections), first trophozoites and cysts were seen in the feces. Five days after an experimental infection, cysts were already proven inside the mucous tissues of the caeca.
10. **Patency**: Months until lifelong (in inapparently infected adult animals). This fact indicates that there exist strains with a different pathogenicity.
11. **Therapy**: Actually there does not exist a registered medicinal product in Europe, so that prophylactic measurements are very important in order to avoid mass infections.

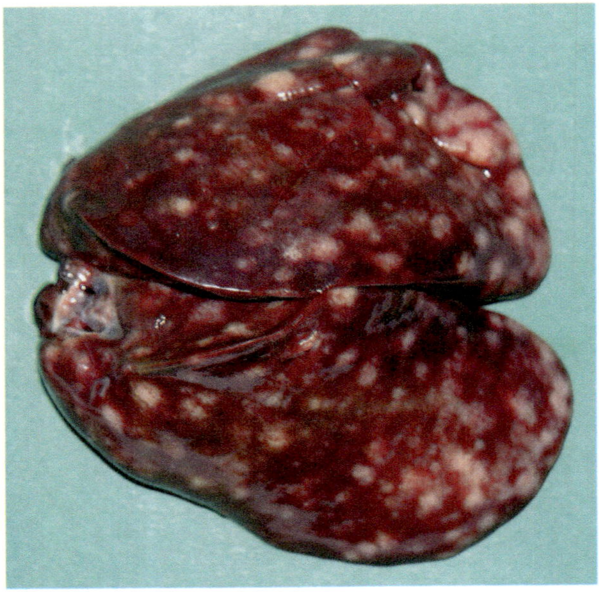

Fig. 4.10 Macrophoto of a turkey liver showing the typical symptoms of typhlohepatitis due to an infection with *H. meleagridis*

Further Reading

Grabensteiner E et al (2006) Broad dissemination of *Histomonas meleagridis*. Parasitol Int 55:317–322.

Hauck R, Hafez HM (2013) Experimental infections with the protozoan parasite *Histomonas meleagridis*: a review. Parasitol Res 112:19–34.

Hess M et al (2015) *Histomonas meleagridis*—new insights into an old pathogen. Vet Parasitol 208:67–76.

Liebhart D et al (2013) Vaccination against histomoniasis prevents a stop in egg production. Avian Pathol 42:79–84.

McDougald LR (2005) Blackhead disease (histomoniasis) in poultry: a critical review. Avian Dis 49:462–476.

Zaragatzki E et al (2010) Light and transmission electron microscopic studies on the encystation of *Histomonas meleagridis*. Parasitol Res 106:977–983.

4.2 *Giardia* Species

This genus was named honouring the French scientist Alfred Mathieu Giard (1846–1908). It contains protozoans belonging to the order Diplomonadida, which possess at least a double set of each organelle (e.g. two nuclei instead of one). The evolution of this group is explained by a lacking binary fission after the inner and outer organelles had been duplicated. This phenomenon can also be seen in the case of tissue cultures of flagellates of the genus *Enteromonas*. Today seven genotypes (assemblages; Table 4.2) are differentiated in the genus *Giardia*, so that species names pointing on hosts are superfluous.

Table 4.2 *Giardia* species

Groups	Size (μm)	Hosts
G. duodenalis (Assemblage A)	10 × 22	Many wild and domestic animals, **humans**
G. enterica (Assemblage B)	10 × 22	**Humans**, primates, dogs, some wild mammals
G. agilis (Assemblage C/D)	9 × 4	Amphibians
G. muris (Assemblage C/D)	13–19 × 8–11	Rodents
G. ardeae (Assemblage C/D)	8 × 15	Birds
G. psittaci (Assemblage C/D)	8 × 15	Birds, parrots
G. microti (Assemblage C/D)	12 × 8	Rodents
G. canis (Assemblage C/D)	10–17 × 7–10	Dogs, canids
G. cati (Assemblage F)	10–17 × 7–10	Cats
G.bovis (Assemblage E)	11 × 19	Cattle, other ungulates
G. simondi (Assemblage G)	14 × 10	Rats

The members of the diplomonadid species live inside the intestine of their hosts. In general, there occur flagellated vegetative stages, which are reproduced by a longitudinal binary fission (starting in general at the apical pole after reduplication of the species-specific numbers of flagella). Their ventral side is in general attached with the help of a holdfast system at the microvilli of their host cells, while they phagocytize intestinal fluids along their upper side. The ingested carbohydrates are stored as glycogen, which can be degraded to ethanol, acetate and CO_2. The species of the genus *Giardia* and related ones possess neither mitochondria nor peroxisomes. Since they may use oxygen (if it is present), they are called **aerotolerant anaerobians** as are the members of the trichomonads and amoebae. The transmission from host to host occurs with the help of cysts, which are set free within the feces (Figs. 4.11 and 4.12). The wall of the cysts is excreted from the surface by a process called **exocytosis** containing filamentous and chitinous elements, which build up a very stable and resistant **cyst wall**. Inside the cysts, each nucleus divides once, so that excreted cysts contain already four nuclei (Figs. 4.11 and 4.12).

Light microscopical investigations of the species of the Diplomonadina do not give enough significant features to differentiate between the species, since the stages are mostly too small. However, electron microscopy and molecular biological data are today the methods of choice to obtain species-specific features. This is especially needed in cases when animals have ingested larger amounts of cysts within feces of another host type. In these cases, the cysts would be excreted unchanged and could lead to a false-positive reaction.

The unique diagnostic feature of the members of the genus *Giardia* is the typical ventral sucker (ventral disc) (Figs. 4.12 and 4.13), which is used to attach themselves at the intestinal wall. In scanning electron micrographs, depression patterns

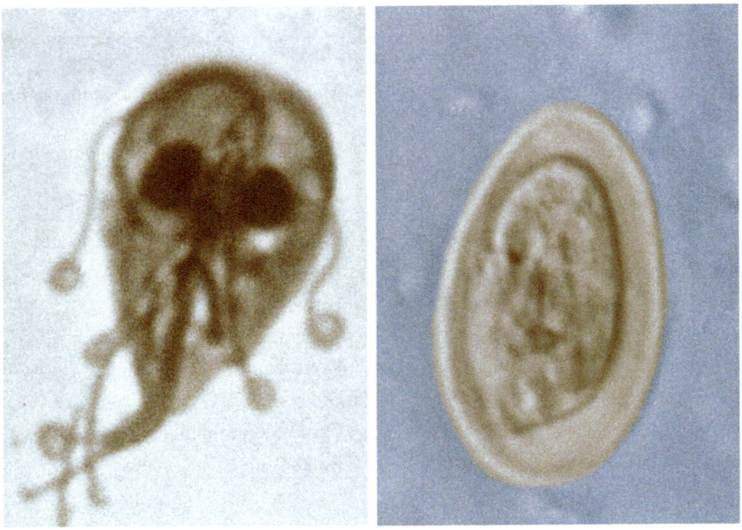

Fig. 4.11 Light micrographs of a trophozoite (*left*) and a cyst of *Giardia duodenalis*. *F* flagellum; *N* nucleus; *Z* cyst wall

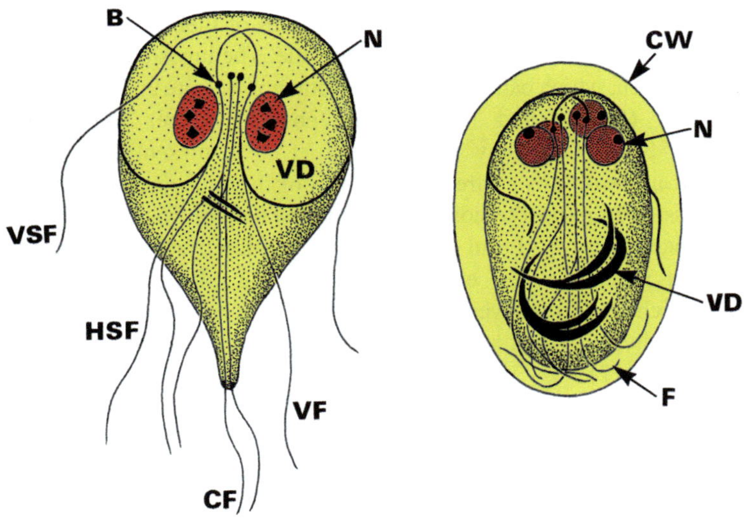

Fig. 4.12 Diagrammatic representation of a trophozoite (*left*) and a cyst of *Giardia* sp. *B* basal body; *CF* caudal pair of flagella; *CW* cyst wall; *F* flagellum; *HSF* posterior-lateral flagella; *N* nucleus; *VD* ventral disc = sucker and remnants of it inside the cyst; *VF* ventral flagella; *VSF* ventral-lateral flagella

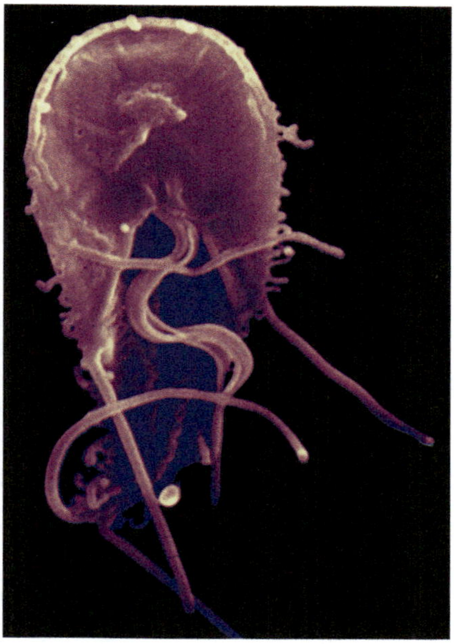

Fig. 4.13 Scanning micrograph of the ventral side of a trophozoite of *Giardia duodenalis*

can be seen after the parasites had been detached from there (Fig. 4.14). Further typical diagnostic features are eight flagella being arranged in a genus-specific pattern (Fig. 4.12, left side) besides the typical position of the two nuclei in the trophozoites (Figs. 4.11 and 4.12, left) and the four nuclei in excreted cysts. All species of the genus *Giardia* have a considerable potency to become pathogenic to their hosts. However, pathogenic effects depend also on the immune status of their hosts, so that the typical slimy, but non-bloody diarrhoeas occur mostly in very young animals (e.g. up to 17 % of pups may suffer from such infections). Since a common serotype of *Giardia* exists in humans, dogs and cats, children may be endangered if pet dogs and cats are invisibly infected in households. In some countries (e.g. Canada, Australia, some regions of the USA, but also in Europe, etc.), drinking water may be a source of infection, if this water is obtained from lakes and not sufficiently treated as was shown by documented outbreaks during the last years. Especially young children and immunosuppressed people (AIDS patients, drug immunosuppressed patients, etc.) are highly endangered.

Molecular-biological investigations showed that at least seven "assemblages" of *Giardia* species exist, which often have a broad spectrum of potential hosts (Table 4.2).

Besides *Giardia* species animals and humans may be infected by members of species of the genera *Enteromonas*, *Trepanomonas*, *Hexamita* (=*Spironucleus*) and *Octomitus* (Table 4.3).

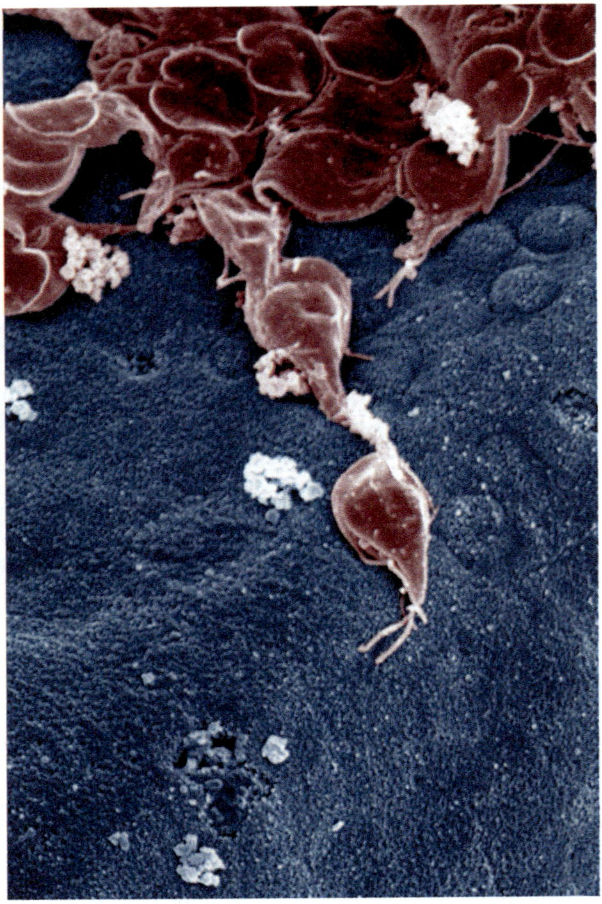

Fig. 4.14 Scanning electron micrograph of the intestinal surface with attached *Giardia* trophozoites and impressions of detached ones (showing the surroundings of the ventral sucker)

Table 4.3 Examples of the genera *Enteromonas, Trepomonas, Hexamita* (=*Spironucleus*) and *Octomitus*

Groups	Size (μm)	Hosts
Enteromonadina		
Enteromonas hominis *E. intestinalis*	10 × 4 8 × 6	**Humans** Rabbits
Diplomonadina		
Octomitus species *O. intestinalis*	12 × 6 9 × 6	Birds Mice
Hexamita (=*Spironucleus*) *muris*	10 × 6	Rodents
Spironucleus species	6–14 × 3–5	Birds
Trepomonas species	12 × 7	Fishes

4.2.1 *Giardia* Species of Dogs and Cats

1. **Name**: Alfred Mathieu Giard (1846–1908): French zoologist. Greek: *katte* = cat. Latin: *canis* = dog.
2. **Geographic distribution/epidemiology**: Worldwide; in Europe up to 90 % of the animals are infected and in the USA up to 50 % according to some reports.
3. **Biology, morphology**: *Giardia canis* (dog, fox), *G. cati* (cat); both species cannot be differentiated morphologically from each other and from *G. duodenalis* of humans. Reproduction occurs by longitudinal reproduction of the pear-shaped vegetative flagellated stages, which reach a size of 10–17 × 7–10 μm and possess a depressed ventral side (Figs. 4.11 and 4.13). The flagellated stages are attached with the help of their ventral discs at the microvilli of the small intestine and of the colon. Trophozoites possess eight flagella at their ventral side. The round cyst stages measure about 15 mm in length and contain four nuclei and are excreted within the feces (Fig. 4.12). This species belongs to the genotype A like humans but also to genotype C/D (dogs) and F (cats).
4. **Symptoms of disease (Giardiasis)**: Heavy infections are characterized by long intermittent periods of fluid (watery) excretions, which may contain only traces of blood. Occasionally, vomiting and abdominal cramps/spasms occur. However, even latent infections may be symptomless but may endanger other animals and humans.
5. **Diagnosis**: Demonstration of cysts inside feces by concentration methods (see Chap. 3 of the book). Antibodies occur only in cases of alterations of the mucous layers (e.g. during malabsorption). Demonstration of antigenic material inside the feces can be done with the help of PCR.
6. **Pathway of infection**: Oral by uptake of the infectious 4-nucleated cysts within fecally contaminated food or by licking at contaminated hair of other animals.
7. **Prophylaxis**: Desinfections of floors of animal rooms by hot steam or by the use of registered and tested chemicals such as Chevi 75, P3-anticoc, etc., for at least 2–5 h.
8. **Incubation period**: Highly variable (depending on the immune status or fitness of the animal: 5–14 days).
9. **Prepatent period**: 5–14 days.
10. **Patency**: Months up to years (due to constant reinfections).
11. **Therapy**: Fenbendazole 1× daily 50 mg/kg bodyweight orally on 3–5 days. However, two repetitions at intervals of 5 days are often needed due to reinfection. Fenbendazole or ronidazole 2×× daily 30—50 mg orally at 7 days are also effective.

Further Reading

Adriana G et al (2016) *Giardia duodenalis* genotypes in domestic and wild animals. Vet Parasitol 217:71–73.

Bouzid M et al (2015) The prevalence of *Giardia* infection in dogs and cats. Vet Parasitol 207:181–202.
Dado D et al (2012) Prevalence and genotype of *Giardia duodenalis* from dogs in Spain: possible zoonotic transmission and public health importance. Parasitol Res 111:2419–2422.
Flechter R et al (2012) Control of *Giardia* infections with rodinazole and intensive hygiene management in dog kennel. Vet Parasitol 187:93–98.
Leonhard S et al (2007) The molecular characterization of *Giardia* from dogs. Vet Parasitol 150:33–38.
Pallant L et al (2015) The epidemiology of infections with *Giardia* species and genotypes in well cared dogs and cats. Parasites Vectors 8:2.
Slapeta J et al (2015) Differences in the fecal microbiome of non-diarrhoeic clinically healthy dogs and cats associated with *Giardia duodenalis* infection. Int J Parasitol 45:585–594.
Tysnes KR et al (2014) Subclinical *Giardia* in dogs: a veterinary conundrum relevant to human infection. Trends Parasitol 30:520–528.

4.2.2 *Giardia* Species of Ruminants

1. **Name**: Latin: *bos, bovis* = cattle, cow; *capra* = goat; *ovis* = sheep.
2. **Geographic distribution/epidemiology**: Worldwide.
3. **Biology, morphology**:
   The species system has been replaced by so-called assemblages (see Table 4.2). However, literature still refers to species:
   (a) ***G. bovis***: The vegetative stages of cattle measure 19 × 11 μm, while the cysts reach a size of 16 × 10 μm. Their nuclei are ovoid; the parabasal bodies appear as two bended sticks. The trophozoites are mainly found along the surface of the duodenum and jejunum. This species belongs to the genotype E (Table 4.2).
   (b) ***G. caprae, G. ovis*** parasites inside goats and sheep. The stages of this species are somewhat smaller than those of *G. bovis*, measuring only 14 × 18 μm. However, it is not yet clear whether they are a separate species.
4. **Symptoms of disease (Giardiasis)**: Severe infections are scarce. Only rarely symptoms of a catarrhalic enteritis combined with diarrhoeas occur.
5. **Diagnosis**: Demonstration of cysts with the help of flotation technique or M.I. F.C./S.A.F.C. and microscopically observation of trophozoites in fluid feces.
6. **Pathway of infection**: Oral within contaminated food; **Attention**: cysts may survive for 3 months in humid food.
7. **Prophylaxis**: Regular removal of feces from stable floors and their cleaning with hot steam and hot water.
8. **Incubation period**: Very variable depending on the fitness of the infected animal.
9. **Prepatent period**: Single cysts are already excreted 4–5 days after an infection.
10. **Patency**: Months.

11. **Therapy**: Fenbendazole: oral application of 10–20 mg/kg bodyweight at 3–5 days, eventually repeated after 1 week.

Further Reading

Abeywardena H et al (2015) A perspective on *Cryptosporidium* and *Giardia*. Adv Parasitol 88:243–301.

Ehsan AM et al (2015) Assessment of zoonotic transmission of *Giardia* and *Cryptosporidium* between cattle and humans. PLoS One 10:e0118.

Geurden T, Olson M (2011) *Giardia* in pets and farm animals and their zoonotic potential. In: Luja H, Svard S (eds) *Giardia*: a model organism. Springer, Vienna, pp 71–85.

Kakandelwa C et al (2016) Prevalence of *Giardia* in dairy cattle. Vet Parasitol 215:114–116.

Thompson RC. Monis PT (2011) Taxonomy of *Giardia* species. In: Lujan HD, Svard S (eds) *Giardia*: a model organism. Springer, New York.

4.2.3 *Giardia* Species of Birds

1. **Name**: Latin: *ardea* = heron. Greek: *psittakos* = parrot. French: peroquet.
2. **Geographic distribution/epidemiology**: Worldwide.
3. **Biology, morphology**: Especially budgerigars and parrots are often infected by several not yet fully described species which parasitize inside the small intestine and which may introduce high mortality rates. Their cysts measure about 15×8 μm.
4. **Symptoms of disease (Giardiasis)**: Apathia, loss of weight, weakness, rough feather aspects, pale skin along throat skin, excretion of large amounts of whitish, fluid feces, strongly reduced uptake of water. Up to 50 % of the infected birds may die in stables.
5. **Diagnosis**: Microscopical demonstration of cysts with the help of the flotation technique.
6. **Pathway of infection**: Oral by uptake of cysts from floor or within fecally contaminated food in cages.
7. **Prophylaxis**: Hot steam cleaning of cages and cleaning cages from feces.
8. **Incubation period**: Variable, depending on the grade of infection: a few days until some weeks.
9. **Prepatent period**: 4–6 days
10. **Patency**: Months due to repeated self-infection during latent infections.
11. **Therapy**: Application of dimetridazole in cases of birds, which are not eaten by humans.

Further Reading

Deplazes P et al (2013) Lehrbuch der Parasitologie für die Tiermedizin. Enke, Stuttgart.

4.2.4 *Spironucleus* Species (syn. *Hexamita*)

1. **Name**: Greek: *speira* = spiral; *hexa* = six; *mitos* = small, grain; *meleagridis* = pearl chicken. Latin: *nucleus* = core, nucleus; *columba* = dove.
2. **Geographic distribution/epidemiology**: Worldwide.
3. **Biology, morphology**: *Spironucleus* species (formerly *Hexamita* species), e.g. *S. meleagridis* and *S. columbae*, occur in the small intestine as well as in the colon. The parasites appear bilaterally symmetric, show body shapes ranging between eggs and pears and reach a size of 6–14 × 3–5 μm (Fig. 4.15a, b). They are provided with two nuclei, six anterior flagella and two recurrent flagella. In contrast to *Giardia* species, they do not possess a ventral sucker (adhesion disc). The ovoid infectious cysts have a length of 5–7 μm and thus are rather small compared to *Giardia* species. These cysts are excreted with the feces (Table 4.3).
4. **Symptoms of disease (Spironucleosis, syn. hexamitosis)**: Doves and turkeys show only rarely symptoms of disease. However, acute infections appear as a catarrhalic enteritis being accompanied ulcerations along the surfaces of the ileum and rectum. Mortality rates of 10–90 % have been described in animals which were <8–10 weeks old. Older animals may act as reservoir hosts. Imported cranes are often severely infected and show high infection rates.

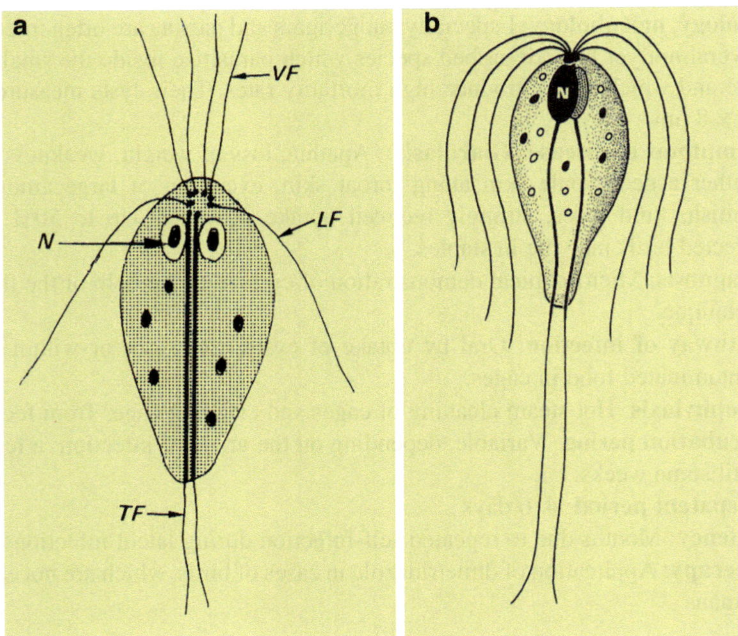

Fig. 4.15 *Hexamita* (syn. *Spironucleus*) *meleagridis*. (**a**, **b**) Diagrammatic aspects of a trophozoite in dorsal (**a**) and lateral (**b**) aspects. *LT* lateral flagella; *N* nucleus containing a nucleolus; *TF* terminal flagella; *VF* anterior, apical flagella

5. **Diagnosis**: Demonstration of cysts in the feces with the help of flotation methods and/or observation of motile stages in fresh, warm feces (e.g. in Zoological Gardens).
6. **Pathway of infection**: Oral uptake of cysts from soil and/or of contaminated food or drinking water.
7. **Prophylaxis**: Separation of young turkeys and other birds from adult ones; separate holding of new animals; regular cleaning using hot steam/water of the floors and drinking facilities.
8. **Incubation period**: 4–8 days.
9. **Prepatent period**: 2–4 days.
10. **Patency**: Months until lifelong due to unapparent repeated infections.
11. **Therapy**: Literature shows that 5-nitroimidazoles as well as furan derivates are effective; however, these compounds are not allowed for food birds in the countries of the European Union and in the Swiss.

Further Reading

Dezfoulian O et al (2010) *Hexamita* infection associated with diarrhoea and stunting in native turkey poults. Trop Biomed 27:504–508.

Keeling PJ, Brugerolle G (2006) Evidence from SSUrRNA phylogeny that *Octomitus* is a sister lineage to *Giardia*. Protist 157:2005–2012.

Kolisko M et al (2008) Molecular phylogeny of diplomonads and enteromonads. BMC Evol Biol.

Levi MG et al (2015) *Spironucleus meleagridis*. Vet Parasitol 208:169–173.

Lloyd D, Williams CF (2014) Comparative biochemistry of *Giardia*, *Hexamita* and *Spironucleus*. Mol Biochem Parasitol 197:43–44.

Lloyd D et al (2015) Motility of the diplomonadid fish parasite *Spironucleus vortens*. Microbiology 161:213–218.

Poyton SL et al (2004) Ultrastructure of *Spironucleus salmonis* and *Hexamita salmonis*. Dis Aquat Org 60:49–64.

4.3 Trypanosomes

The group of so-called trypanosomes belongs to the systematical order **Kinetoplastida**, which got its name due to the possession of a so-called **kinetoplast**. This is not a peculiar separate organelle but represents a peculiar region of the large mitochondrion where extremely large amounts of DNA are concentrated so that it appears as a dense red dot when stained according to the Feulgen stain method. Electron microscopical studies, however, show that this DNA concentration is a peculiar site inside the mitochondrion being situated closely below the basal body of the single flagellum (Figs. 4.16, 4.17 and 4.18). This spot, at which extranuclear DNA was first demonstrated, was in 1917 named **kinetoplast** by the Russian scientist **Alexeieff** keeping it for the "motor" of the flagellum. Other authors described this spot as micronucleus, parabasal body, centrosome, kinetosome or even as blepharoplast.

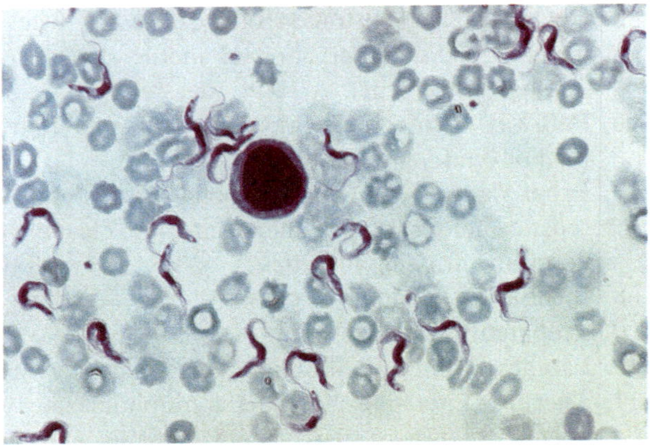

Fig. 4.16 Light micrograph of a Giemsa-stained blood smear showing trypomastigote stages, red blood cells and a large monocyte

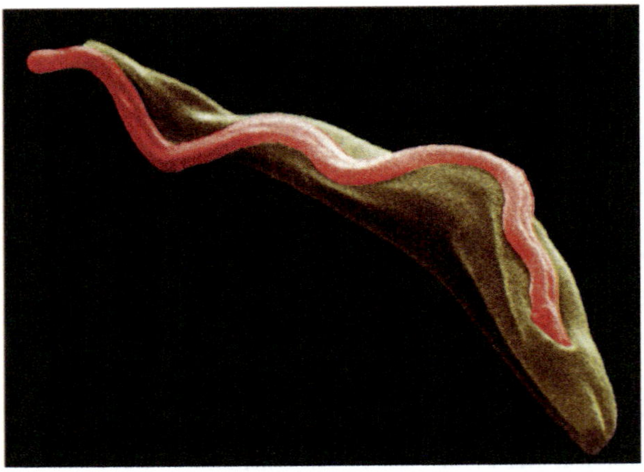

Fig. 4.17 Scanning electron micrograph of a trypomastigote stage, the flagellum of which starts from the interior of a flagellar pocket and is attached at some places at the surface of the parasite's cell membrane, thus leading to the aspect of an "undulating membrane" in light micrographs

The pathogenic species belong to the family Trypanosomatidae. Other related species belong to the family of Bodonidae which includes either non-parasitic free-living species or very important parasitic species of fish (e.g. genera *Cryptobia*, *Trypanoplasma*).

The species belonging to the Trypanosomatidae are characterized by the occurrence of differently shaped stages within their life cycle, which in the case of the heteroxenous species in general involves two types of hosts (e.g. a vertebrate and an insect as vector). Monoxenous species, however, have mostly only a single vertebrate host (an insect). In the case of heteroxenous species, the stages may appear with a flagellum which is attached at different sites of the cell surface (Figs. 4.17, 4.18 and 4.19).

In early investigations, the polymorphous stages had been considered each as a separate species and thus had been named differently. When studying the life cycles, however, it was found that many shapes may belong to a single species just presenting a different stage in an often complicated life cycle. For example, in species of the genus *Trypanosoma* formerly so-called *Leishmania* stages, *Leptomonas* stages,

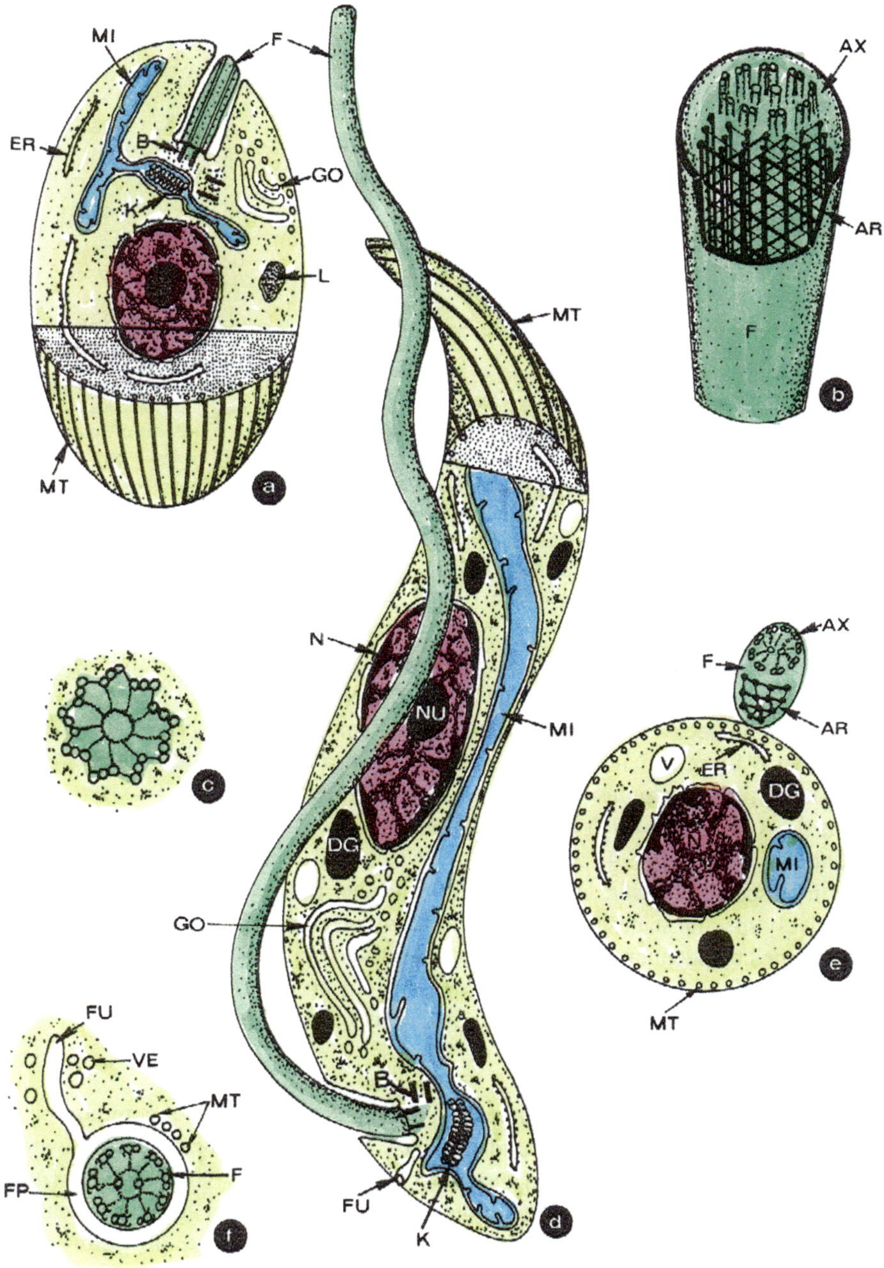

Fig. 4.18 Diagrammatic representation of the ultrastructures of *Trypanosoma* stages. (**a**) Amastigote and cryptomastigote stage. (**b**) Flagellum. (**c**) Basal body cross-sectioned. (**d**) Trypomastigote stage longitudinally sectioned. (**e**) Trypomastigote stage cross-sectioned. (**f**) Flagellar pocket in cross section. *AR* axial rod; *AX* axoneme; *B* basal body; *DG* dense inclusion (glycosome); *ER* endoplasmic reticulum; *F* flagellum; *FP* flagellar pocket; *FU* finger-like channel; *GO* Golgi apparatus; *K* kinetoplast; *L* lipid; *MI* mitochondrion; *MT* microtubule; *N* nucleus; *NU* nucleolus; *PM* peritrophic membrane; *SCO* surface coat; *UM* cell membrane; *V* vacuole; *VE* vesicle

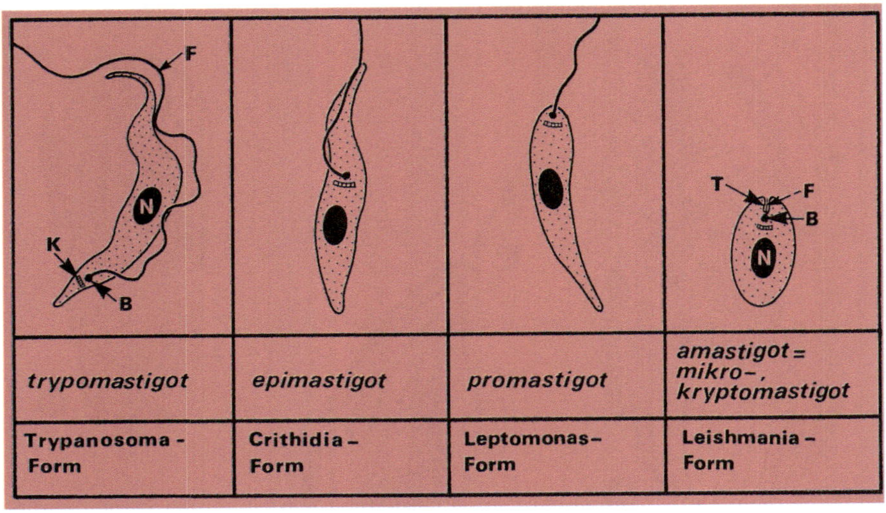

Fig. 4.19 Diagrammatic representation of important developmental stages of the Trypanosomatidae according to older and recent terminology. *B* basal body; *F* flagellum; *K* kinetoplast; *N* nucleus; *T* flagellar pocket (Early German publication; Form = stage)

Crithidia stages or even *Trypanosoma* stages may belong to a single life cycle and follow each other. Thus, since 1966 the different stages in a life cycle of a trypanosomal species are named according to a proposal of the English scientists Hoare and Wallace with respect to the situation of the basal apparatus of the flagellum as **a-(or crypto)mastigotes** (=*Leishmania* stage), **trypomastigotes** (=*Trypanosoma* stage) (Figs. 4.17 and 4.18), **promastigotes** (=*Leptosoma* stage) or **epimastigotes** (=*Crithidia* stage) (Fig. 4.19). In the life cycle, these different stages follow each other in a species-specific, unchangeable sequence. Bloodsucking vectors transmit exclusively specific stages which have been developed inside their bodies and which are equipped by peculiar surface coats that allow survival inside the new host. Today most stages of the life cycle of trypanosomes can be produced within tissue cultures, so that physiological process can be studied more easily.

Cytologic Features

The stages of the Trypanosomatidae are structured according to a common basic plan (Fig. 4.20). They are surrounded by a single cell membrane, below which spirally arranged microtubules occur and stabilize the body shape as a peculiar cytoskeleton. In the case of stages living in the blood of their hosts, a so-called, 10–15 nm thick **surface coat (glycocalyx)** occurs on their surface protecting them from attacks of antibodies and complement products of the host. In the case of African *Trypanosoma* species, they are covered by a layer of glycoproteins, which is always changed after each of the subsequent longitudinal divisions. Thus, this layer is described as **variant surface protein (VSG)**, where each molecule consists of about 500 amino acids. The genome of several *Trypanosoma* species (e.g. those of the *T. brucei* group, *T. congolense, T. evansi, T. equiperdum*, where **antigenic**

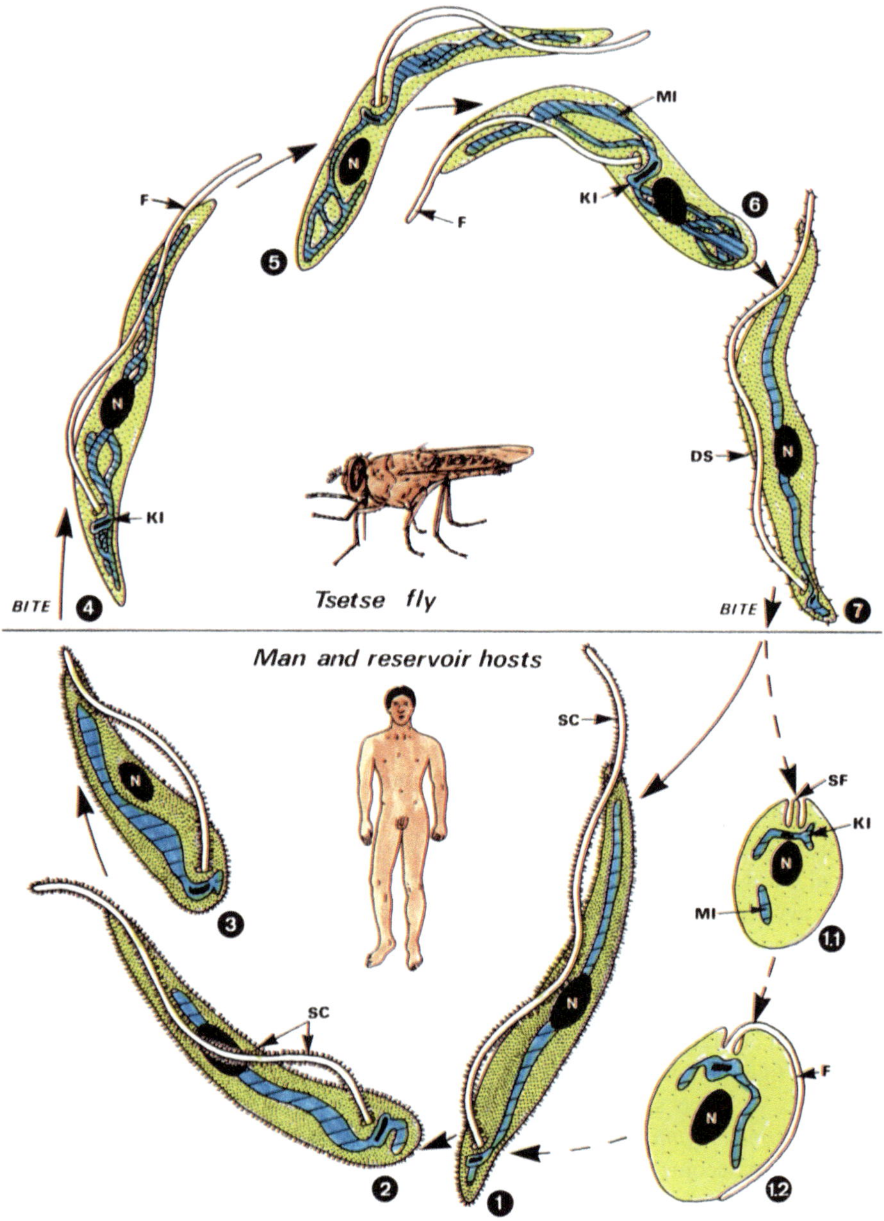

Fig. 4.20 Diagrammatic representation of the life cycle of *Trypanosoma brucei*. 1–3 Stages seen in blood (1.1–1.2 stages seen in brain). 4–7 stages in *Glossina* vectors. *F* flagellum; *KI* kinetoplast; *MI* mitochondrion; *N* nucleus; *SC* surface coat; *SF* short flagellum

variation had been definitively proven), contains 1000 of these VSG genes. They are located on all of the very heterogeneous chromosomes of these *Trypanosoma* species and contain about 10 % of the nuclear DNA. However, always only one of these VSG genes is expressed in a given individual stage, so that in a host's body at the same time large amounts of different parasitic stages are present, which cannot be killed at the same time by the host's defence system. The change of the activities of the individual gene occurs apparently after each cell division so that many of these parasites get the chance to survive. The components of a new VSG protein layer are apparently set free in the flagellar pocket by an exocytosis process. Inside this flagellar pocket, also processes of endocytosis occur at the same time. During the latter process, about 5 % of the own surface coat, which contains the host antibodies, are ingested per minute, thus diminishing considerably the effects of the antibodies. With respect to this great antigenic variability, it seems at least difficult if not impossible to develop a successful vaccine.

As soon as the blood stages have reached the intestinal tract of their vectors (*Glossina* species = tsetse flies), they discharge their VSG-surface layer and replace it by a layer of glycoprotein, which is named **procyclin** and **PARP** (=**procyclic acidic repetitive protein**). This protein does not undergo alterations and contains 400–500 remnants of amino acids, which are anchored (as VSGs) with the help of a GPI anchor (=glycosylphosphatidylinositol) in the cell membrane. The later occurring stages inside the salivary glands then express again the VSGs and thus may survive inside the blood of their vertebrate vector.

Flagellum

All stages in the life cycle of trypanosomes possess a single flagellum, which, however, has a stage-specific length and different starting sites (Figs. 4.17, 4.18 and 4.21). Also the so-called amastigote stages (Figs. 4.19, 4.20 and 4.30) possess a short flagellum, which, however, does not surpass the outer body line, since it is situated in a depression called flagellar pocket (Fig. 4.30b).

The flagellum is always anchored in the typical flagellar pocket with the help of a typical basal body (Fig. 4.18), which comprises 9 × 3 microtubules at its starting point (anchor) in the cytoplasm of the parasite. In the case of the members of the related bodonid groups, often remnants of a second flagellum are seen being also anchored in the cytoplasm with the help of a typical basal body system. The flagella of most species of the trypanosomatids are characterized by a so-called **paraxial rod** which is stretched along the typical set of 9 × 2 microtubules and consists of a network filaments which help to stabilize the flagellum (Fig. 4.18). This paraxial rod lacks in some species, e.g. *Crithidia oncopelti, Blastocrithidia culicis* and *Herpetomonas roitmani*. The flagellum is not used exclusively for movements but also helps to become attached at the inner surface of the mouthparts of vectors until they become ejected during bloodsucking.

Undulating Membrane

Trypo- and epimastigote stages show a so-called undulating membrane when studying them alive or stained with the help of a light microscope. Electron microscopical investigations, however, show that the flagellum is attached only at a few sites at the cell membrane with the help of so-called **desmosomes** (Figs. 4.17 and 4.18). Thus, movements of this attached flagellum have the same effects as the oars of a rowing boat.

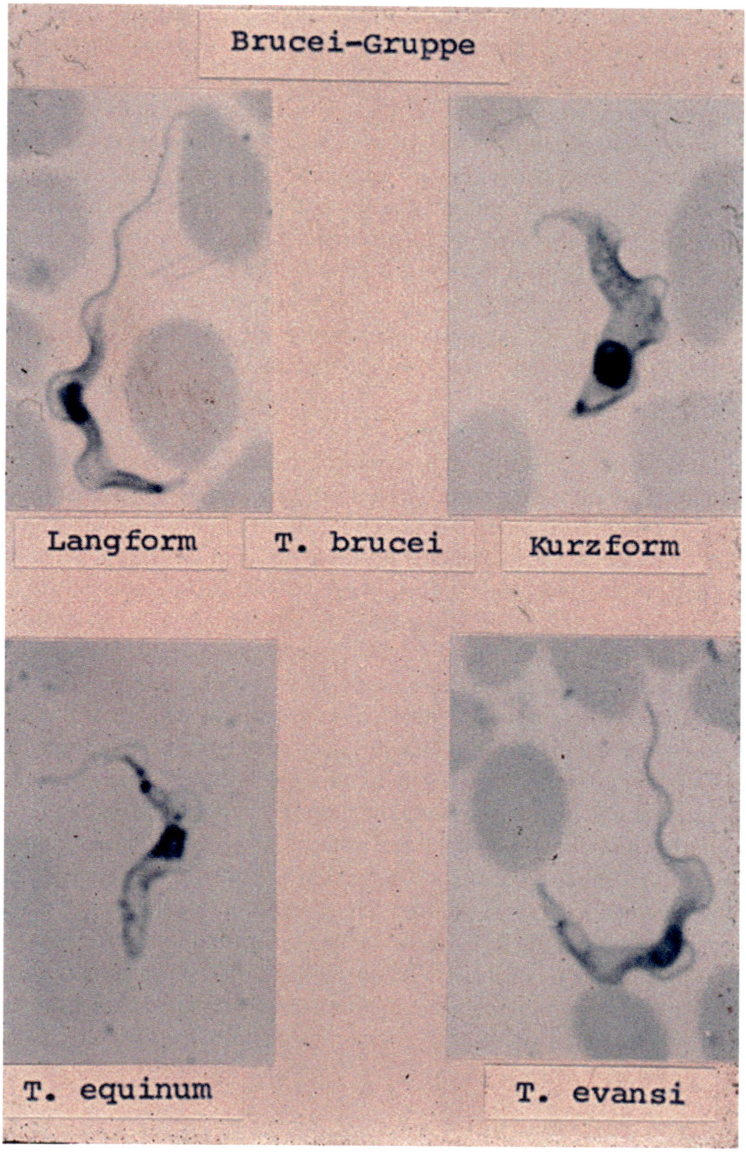

Fig. 4.21 Light micrograph of blood stages of stages of *Trypanosoma brucei* group (= Gruppe) of cattle. *Langform* = long form; *Kurzform* = short form

Kinetoplast

This term describes an enlarged region of the mitochondrion, which does not show the typical cristae formed by the inner membrane at other sites. In this region, which is situated close to the basal apparatus of the flagellum, the mitochondrion is filled by DNA strands in a species-specific arrangement. According to their size, these strands are described as **maxicircles** or **minicicles**. Besides the formation of peculiar proteins of the respiratory chain (maxicircles), their DNA steers the formation of the so-called **guide RNA** (minicircles) and thus the production mRNA, which is important for the maxicircles. Due to mutations as well as due to the activity of several compounds (acidines, phenanthrenes, diamidines), the kinetoplast is not visible in light microscopy and thus these stages are described as **akinetoplastic**. However, electron microscopic studies show that the enlargement of the mitochondrion persists but has lost there its DNA. Therefore, these stages are also described as dyskinetoplastic stages. They are able to survive in the blood of their vertebrate hosts, but cannot develop inside their vectors, since they do not possess the needed proteins of the respiratory chain.

Feeding

The members of the kinetoplastida ingest the materials, which are superfluently present in the body (intestine, blood vessels) of their hosts. Ingestion of food occurs by **phago-** or **pinocytosis** along the membrane of the flagellar pocket. The ingested molecules or particles are hydrolysed inside food vacuoles. Blood stages of these parasites have a glycolytic metabolism, while the stages inside insects or in tissue cultures proceed an **oxidative metabolism** whereby amino acids like proline are the main source of energy. The blood stages degrade sugars until pyruvate, which is finally excreted. During this process, they need within one hour an amount of glucose which corresponds to 50–100 % of their own dry weight. Polysaccharides are not stored inside their body, so that they need constant uptake of food. However, the species *Trypanosoma cruzi* (=agent of the Chagas disease) makes an exception. These stages are able to survive (with the help of fatty acids?) several days inside tissue cultures without constant uptake of glucose, since the blood stages possess all elements of the respiratory chain and since they are able to use the proteins which are stored in so-called **reservosomes**. The uptake of sugars occurs via the cell membrane by means of the so-called lightened diffusion (=cleared transport). The degradation of the sugars occurs inside organelles called **glycosomes**, which occur inside the cells in larger numbers, are membrane bound, appear electron dense and include all needed enzymes for glycolysis. In African trypanosomes, they have diameters of about 0.25 μm.

Peroxisomes

Peroxisomes are lacking in all trypanosomes. The purines that they need to build up the DNA and RNA are ingested with their food.

Development

As can be seen in Table 4.4, there are considerable differences in the pathogenicity of the different species of the Trypanosomatidae, which are apparently based on the effects of so-called **kinines**. According to this hypothesis, toxic substances were set free, when parasites are destroyed inside the blood of hosts. These substances have apparently deleterious effects on the walls of the blood vessels. In some cases, the

Table 4.4 Important species of the genus *Trypanosoma*

Species	Length (µm)	Vertebrate host	Disease	Symptoms	Distribution	Vector	Mode of transmission
T. brucei brucei	25–42	Equids, pigs, rodents, ruminants	Nagana	Fever, meningoencephalitis, paralysis	Tropical Africa	*Glossina* species	Bite
T. brucei gambiense	16–31	**Humans**, monkeys, dogs, pigs, antelopes	Sleeping sickness (less acute)	Swelling of neck lymph nodes, oedema, meningoencephalitis	West Africa	*Glossina* species, *G. palpalis*. *G. tachinoides*	Bite
T. brucei rhodesiense	20–30	**Humans** (rats are easily infected)	Sleeping sickness (acute form)	Fever, somnolence, see above	East Africa	*Glossina* species *G. morsitans*	Bite
T. congolense	9–18	Ruminants, predators	Nagana	Fever, raving madness, anaemia	Congo, Zululand	*Glossina* species	Bite
T. simiae	12–24	Sheep, goats, monkeys, pigs, camels	Lethal, hardly acute phase	Fever, raving madness, anaemia	East Africa	*Glossina* species	Bite
T. vivax	20–27	Ruminants, equids	Souma	Becomes chronic, self-healing	Tropical Africa	*Glossina* species	Bite
T. evansi	18–34	Equids, ruminants, dogs	Surra	Fever, oedema, anaemia	India, Africa. Sibiria, China, Australia, South and Central America	*Tabanus* and *Stomoxys* species	Mechanically during bite without development in the vector
T. equinum	20–30	Equids, cattle, capybaras	Mal de Caderas	Fever, anaemia	South and Central America	*Tabanus* species	Bite
T. equiperdum	18–28	Equids	Dourine	Genital schwelling, paralysis	Mediterranean region, India, China, Java, America	–	Mechanically during sexual act

(continued)

Table 4.4 (continued)

Species	Length (µm)	Vertebrate host	Disease	Symptoms	Distribution	Vector	Mode of transmission
T. cruzi	16–20	**Humans,** domestic and wild animals	Chagas disease	Oedema, myocarditis, affection of CNS	South America	Bugs: *Triatoma, Rhodnius*	Feces
T. rangeli	25–32	Rats, **humans**	**Apathogenic**	–	South America	Bugs: *Rhodnius*	Bite; feces doubtful
T. theileri	25–120	Cattle, antilopes	Apathogenic	–	Cosmopolitan	Tabanids	Feces
T. melophagium	25–70	Sheep	Apathogenic	–	Cosmopolitan	Sheep keds	Feces
T. lewisi	24–35	Rats	Apathogenic	–	Cosmopolitan	Rat fleas	Feces

sudden setting free of masses of such substances may induce a chain reaction leading to shock-like syndromes. Thus, the main damages in hosts due to trypanosomes are not based on extraction of important compounds (as it is the case in some worms, e.g. *Diphyllobothrium* species) but have their origin in the production and excretion of toxic substances by these flagellates. With respect to their vectors, the *Trypanosoma* species do not show severe signs of damage, if at all. Otherwise they would limit their chances to become transmitted to vertebrate hosts.

Reproduction
The reproduction of the trypanosomatids occurs mostly by binary longitudinal division, whereby the kinetoplast and the basal body of the flagellum are at first duplicated (Fig. 4.30a). Sexual processes have been shown in members of the *T. brucei* group, where also variable antigenetic structures have been shown. According to some studies, the trypanosomes run through their life cycle in a diploid state except for some specimens of the epimastigotes inside the intestine of the tsetse fly, which are haploid. Therefore, it can be postulated that there exist a preceeding **meiosis** and fusion of complete cells or at least fusions of nuclei.

4.3.1 Trypanosomes (Salivaria): e.g. Agents of Nagana

1. **Name**: Zulu language: *nagana* = stage of reduced awareness. Latin: *saliva* = saliva, spit; *vivax* = living. Sir David Bruce (1855–1931) = English researcher: Congo = African river; Rhodesia = ancient name for Zimbabwe (state in East Africa), named according to Cecil Rhodes (1853–1902), a rich English-South African businessman, who conquered as Prime Minister the region called later Rhodesia.
2. **Geographic distribution/epidemiology**: Mainly in Central, West, South and East Africa; the parasites may be distributed by animal transportation to other regions.
3. **Biology, morphology**: All species occur as trypomastigote stages extracellularly in the blood of their hosts (see Table 4.4) and are reproduced therein by longitudinal binary fission (Figs. 4.21 and 4.22). Their flagellum starts at the hind end inside a flagellar pocket and stretches along the surface to the anterior end where it becomes free (Figs. 4.17, 4.18 and 4.21). On light micrographs, it is seen to start close to the so-called kinetoplast.
   (a) ***Trypanosoma brucei brucei***:
   The stages of this species reach a length of 25–42 μm long; they are poly- and pleomorph. This species occur **only rarely** in ruminants.
   (b) ***Trypanosoma vivax vivax***
   The specimens reach a length of 18–27 μm and appear as monomorphic trypomastigotes in the blood of their hosts. This is the most common species in cattle.
   (c) ***Trypanosoma congolense congolense***

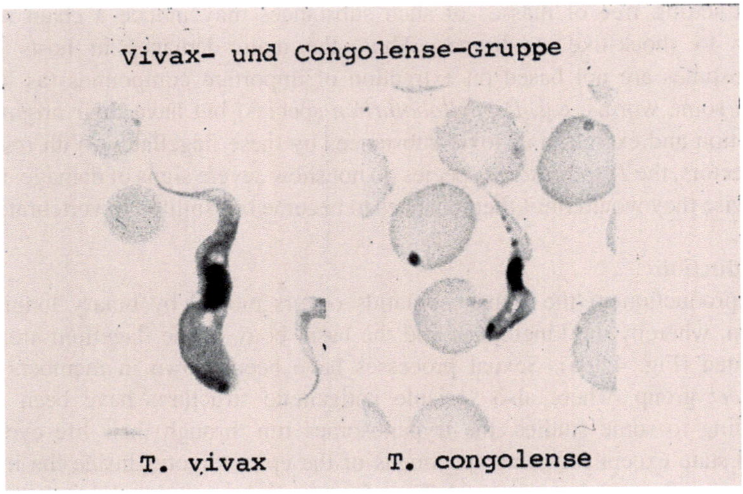

Fig. 4.22 Light micrographs of *T. vivax* and *T. congolense (Gruppe = group)*

The stages of this species reach a length of only 9–18 μm and the flagellum possesses only a very short free portion.

(d) ***Trypanosoma brucei gambiense*** (West Africa) and ***T.b. rhodesiense*** (East Africa). These species are very pathogenic in humans but lead to only low-grade diseases in cattle, which serve as **reservoir hosts**.

All these species are transmitted within the saliva during bites of tsetse flies (*Glossina* species). Both sexes of these flies suck blood and thereby they ingest trypomastigote stages. Inside their intestine, the **trypomastigote stages** are transformed (during binary fissions) into **epimastigotes**, which show the origin of their flagellum in front of the nucleus, which is situated in the middle of the parasitic stage (Fig. 4.20).

4. **Symptoms of disease (Trypanosomiasis, nagana)**: The severeness of the disease (called sleeping sickness) depends on the race of infected cattle. Some endemic cattle species are even **trypanotolerant**; i.e. they only show low-grade symptoms or even none. In the case of acute disease, the following symptoms are seen: high fever, anaemia, weakness, extreme loss of weight and eventual death. If infections are due to *T. congolense* and *T. b. brucei*, motility may be considerably disturbed due to brain lesions. In the case of small ruminants (e.g. sheep), the disease remains at a chronic status.
5. **Diagnosis**: Microscopical demonstration of the blood stages after enrichment methods and by the use of PCR test systems.
6. **Pathway of infection**: Percutaneously during bites of male or female tsetse flies.

7. **Prophylaxis**: In Africa: Elimination of tsetse flies in the surroundings of animal sites. Keeping away wild ruminants close to cattle, since they may be infected.
8. **Incubation period**: About 1 week.
9. **Prepatent period**: 1 week.
10. **Patency**: Several months, eventually until death.
11. **Therapy**: Drugs of choice are diminazen aceturate = Berenil®, Ganaseg® at a dose of 3.5 mg/kg bodyweight intramuscularly and ethidiumbromid (1 mg/kg bodyweight i.m.). In the case of existing resistances, the dose can be increased onto 7–8 mg/kg bodyweight. **Cave**: Diminazen is toxic for camels, donkeys and dogs. The latter can be treated with isomethamidium chloride (0.25–1 mg/kg bodyweight).

Further Reading

Connor RJ, van den Boscho P (2004) African animal trypanosomes. In: Coetzer JA, Tustin RC (eds) Infectious diseases of lifestock, vol 1. Oxford University Press, pp 251–296.

Deschamps JY et al (2016) Refractory hypoglycaemia in a dog infected with *Trypanosoma congolense*. Parasitology 23:1.

Kato CD et al (2016) The role of cytokines in the pathogenesis and staging of *Trypanosoma brucei rhodesiense* sleeping sickness. Allergy Asthma Clin Immunol 12:4.

Munday JC et al (2015) Transport proteins determine drug sensitivity and resistance in a protozoan parasite, *Trypanosoma brucei*. Front Pharmacol 6:32.

Sistrom M et al (2016) De novo genome assembly shows genome wide similarity between *Trypanosoma brucei brucei* and *T. b. rhodesiense*. PLoS One 11: e0147660.

Uzcanga GL et al (2016) Serodiagnosis of bovine trypanosomiasis cause by non tsetse transmitted *Trypanosoma vivax*. Vet Parasitol 218:31–42.

4.3.2 *Trypanosoma brucei evansi* (Agent of Surra)

1. **Name**: G.H. Evans (1835–1935); Sir David Bruce (1855–1931) = English researchers; Hindi: *surra* = indian word for bad, spoiled.
2. **Geographic distribution/epidemiology**: North Africa, India, Asia, Central and South America.
3. **Biology, morphology**: The trypomastigote stages inside the blood are monomorphous, reach a length of 18–34 μm and possess an anterior long free portion of the flagellum (Fig. 4.21). The transmission of the parasites occurs mechanically during the bite of tabanids and/or during blood licking of vampire bats in South and Central America. Thus, there is practically no reproduction or longer staying of the parasites inside these mechanical vectors.
4. **Symptoms of disease (Surra)**: Fever, oedemas, anaemia and others like in the case of Nagana.

5. **Diagnosis**: Microscopical demonstration of the blood stages (eventually after use of concentration methods); use of PCR-test systems.
6. **Pathway of infection**: Percutaneous due to mechanical injection during bites of tabanids and vampire bats with their contaminated mouthparts.
7. **Prophylaxis**: Outside of closed stables very difficult, since it is practically not possible to keep away tabanids and nightly active vampire bats.
8. **Incubation period**: 1 week.
9. **Prepatent period**: 1 week.
10. **Patency**: Several months.
11. **Therapy**: Berenil® 5–10 mg/kg body weight intramuscularly or phenanthridine derivates, e.g. Novidium® (homidium-chloride) 1 mg/kg bodyweight intramuscularly or Samorin (isometamidium) 0.24–0.5 mg/kg bodyweight intramuscularly.

Further Reading

Claes F et al (2005) *Trypanosoma equiperdum*: master of disguise or historical mistake? Trends Parasitol 21:17–24.

Hagos A et al (2010) Efficacy of Cymelarsan® and Diminasan® against *Trypanosoma equiperdum* in mice and horses. Vet Parasitol 171:200–206.

Lai DH et al (2008) Adaptions of *Trypanosoma brucei* to gradual loss of kinetoplast DNA: *T. equiperdum* and *T. evansi* are petite mutants of *T. brucei*. PNAS 105:1999–2004.

Njiru ZK et al (2004) Detection of *Trypanosoma evansi* in camels using PCR and CATT/T evansi tests in Kenya. Vet Parasitol 124:187–189.

4.3.3 Stercorarian Trypanosomes

1. **Name**: Latin: *sterx* = posterior end, hindback; *mel* = hony; *cruz* = cross. Greek: *korax* = hook; *phagein* = feeding, ingesting. Arnold Theiler (1867–1936): Swiss veterinarian working in South Africa; J. A. Lewis (~1937) American researcher, discoverer of rodent trypanosomes.
2. **Geographic distribution/epidemiology**: Worldwide; in Europe only a few species exist (Table 4.4).
3. **Biology, morphology**:
   (a) ***Trypanosoma theileri*** (Figs. 4.23 and 4.24)
   Species is considered to be apathogenic inside cattle. These very long, often bended stages (30–120 μm) are transmitted mechanically during bites of tabanids (genera *Tabanus, Haematopota*) and by ixodid ticks.
   (b) ***Trypanosoma melophagium***
   This species is found in sheep and does not induce symptoms of disease. The transmission occurs during bites of sheep louse flies (*Melophagus ovinus*).
   (c) ***Trypanosoma lewisi***

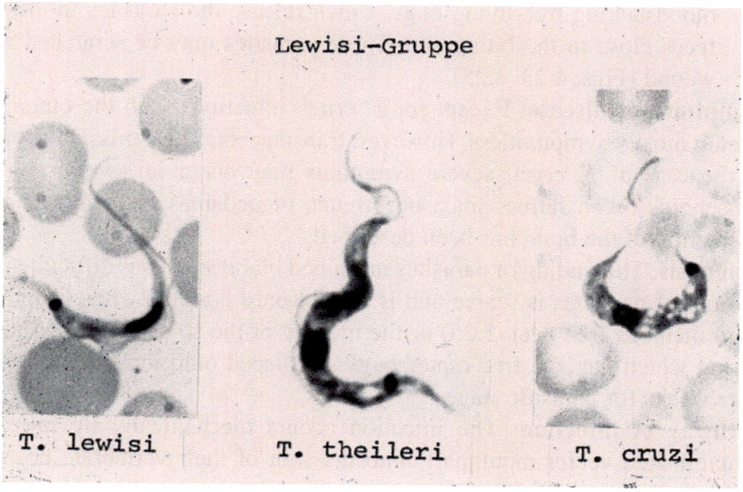

Fig. 4.23 Light micrographs of smear preparations of *Trypanosoma* stages of the *T. lewisi* group (= German: Gruppe)

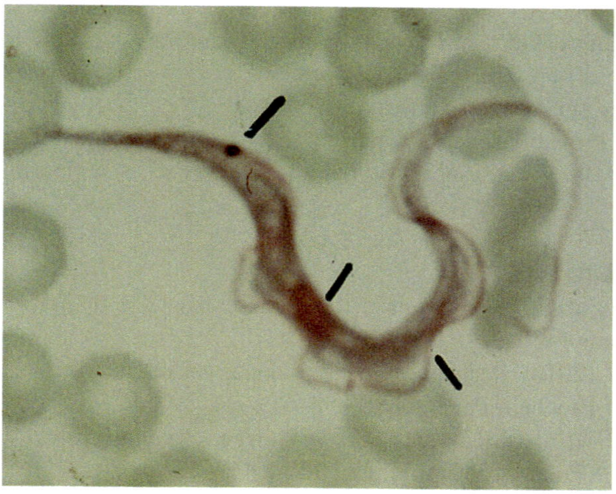

Fig. 4.24 Light micrograph of a Giemsa-stained *T. theileri* stage. The kinetoplast, the nucleus and an attachment site of the flagellum are marked

This worldwide occurring apathogenic species parasitizes in rats (e.g. *Rattus rattus*) and is the leading species close to the two species listed above. The stages measure 24–35 μm in length and are transmitted by rat fleas.

(d) ***Trypanosoma cruzi***

This species occurs in South America and infects a broad spectrum of hosts including wild and many domestic animals and humans. Vectors are bloodsucking free-living bugs, which release the parasites in droplets of feces close to the biting site. These parasites may be scratched into the wound (Figs. 4.23–4.25).

4. **Symptoms of disease**: Except for *T. cruzi*, infections with the other species remain mostly symptomless. However, transplacental transmission is possible. In the case of *T. cruzi*, severe symptoms may occur in several animals in households or on farms, since occurrence of oedemas, myocarditis or even infestation of the brain has been described.
5. **Diagnosis**: The finding of parasites in stained blood smears is difficult since the number of parasites is scarce and is mostly only possible when using enrichment methods (see Sect. 3.25) or the method of the so-called **xenodiagnoses**, during which parasite-free raptor bugs are placed onto the skin and checked after weeks for parasitic stages.
6. **Pathway of infection**: The infection occurs mechanically by injection of contaminated vector mouthparts into the skin of their vertebrate hosts or by rubbing parasite-containing feces of raptor bugs into wounds in the case of *T. cruzi*.
7. **Prophylaxis**: Except for *T. cruzi*, measurements are not needed; in the case of *T. cruzi*, stables and buildings should be kept free from raptor bugs, which might be infected.
8. **Incubation period**: Hard to define, since except for *T. cruzi* symptoms are low grade if at all present.
9. **Prepatent period**: Not known.
10. **Patency**: Probably several years.
11. **Therapy**: In general not affordable; in *T. cruzi* slaughtering of infected animals.

Further Reading

Cardoso MS et al (2016) Evasion of the immune response by *Trypanosoma cruzi*. Front Immunol 6:659.

Lun ZR et al (2016) Resistance to normal human serum reveals *Trypanosoma lewisi*. Mol Biochem Parasitol 199:58–61.

Martinkovic F et al (2012) *Trypanosoma* (*Megatrypanum*) *melophagium* in the sheep ked. J Eukaryot Microbiol 59:134–144.

4.3.4 *Trypanosoma brucei equiperdum*

1. **Name**: Greek: *trypanon* = drill; *soma* = body. Latin: *equus* = horse; *perdere* = loose. French/Arabic: *dourine*.
2. **Geographic distribution/epidemiology**: Worldwide; in Europe recently only in southern countries. The dourine of horses and donkeys is absent in Germany and Central Europe since 1953.

3. **Biology, morphology**: The trypomastigote stages measure about 25 μm in length and reproduce themselves by longitudinal binary fissions. They parasitize in different vessels of the sexual organs of male and female horses and donkeys. At a later stage, they may enter the lymph system and reach via blood also the brain. Morphologic diagnostic is difficult (Fig. 4.26c).
4. **Symptoms of disease (dourine)**: The infection occurs during mating activity of horses and donkeys by transmission of the flagellates, which live and reproduce themselves in the vagina and in the preputium. There they induce lesions and nodules, which at first are not noted for at least 1 week (often it takes much longer). Thus, the following symptoms are seen mostly only after several months. **Male animals**: The penis is permanently swollen and erected. Slimy material is discharged from the urethra and urinary urgency occurs rather often. Lymph nodes are permanently swollen along the urogenital system. **Female animals**: They show swelling of the vagina and permanent urinary urgency and are constantly "on heat" (Fig. 4.26b). Both sexes show during the progress of the infection whitish dots along the sexual organs (so-called toad spots) and urticarial and oedema along the breast and belly. In general, the disease, which has three phases, ends in a chronic phase but may also induce polyneuritis, paralysis disturbed movements, ptosis and a rather high mortality rates.
5. **Diagnosis**: The **dourine** is a disease, which must be announced at the relevant governmental authorities. It can be diagnosed by microscopical investigation of the urine, slime of urethra and vagina or even in blood smears. Most safe results are obtained by investigation with the help of KBR and ELISA methods.

   Differential diagnosis: Deckdruse or blister disease show similar general symptoms.
6. **Pathway of infection**: During horse mating, eventually mechanically by contaminated mouthparts of flies.
7. **Prophylaxis**: Repeated serological tests.
8. **Incubation period**: Ranging from 1 to 12 weeks.
9. **Prepatent period**: 5–7 days.
10. **Patency**: Eventually years.
11. **Therapy: Attention**: In many countries, this disease, which has to be announced to the veterinarian authorities, can only become treated under guidance and/or allowance of governmental veterinarian authorities and treatment is regulated by law. Infected animals have to become excluded from the herd. Suramin (Naganol®) or quinapyramin (trypacide®) or diminazene furat and homidium salts in the case of ruminants.

Further Reading

Li SQ et al (2009) Immunization with recombinant actin from *Trypanosoma evansi* induces protective immunity against *T. evansi*, *T. equiperdum* and *T. b. brucei* infection. Parasitol Res 104:429–435.

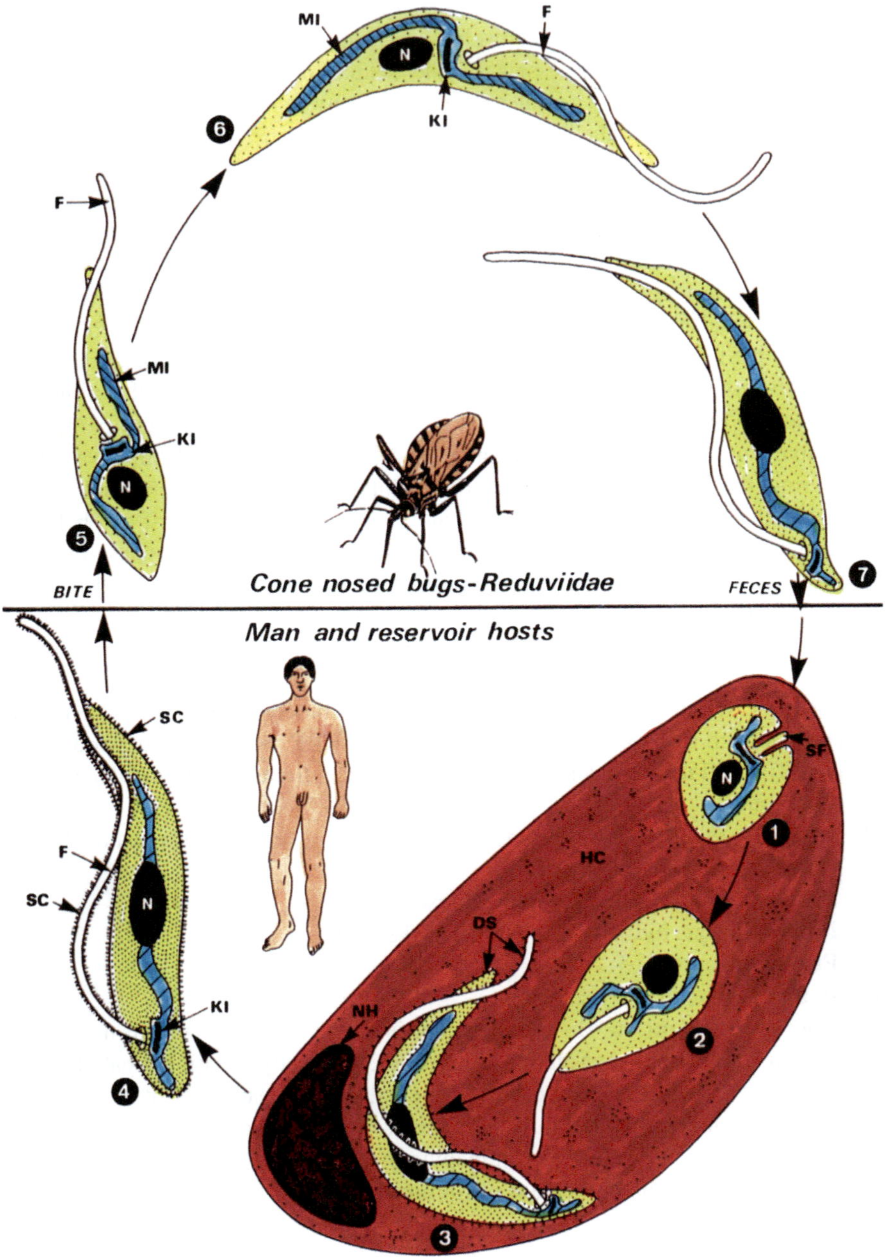

Fig. 4.25 Diagrammatic representation of the life cycle of *Trypanosoma cruzi* (agent of the Chagas disease). *DS* developing SC; *F* flagellum; *HC* host cell; *KI* kinetoplast; *MI* mitochondrion; *N* nucleus; *NH* host cell nucleus; *SC* surface coat; *SF* short flagellum; *V* vacuole; *ZY* cytoplasm

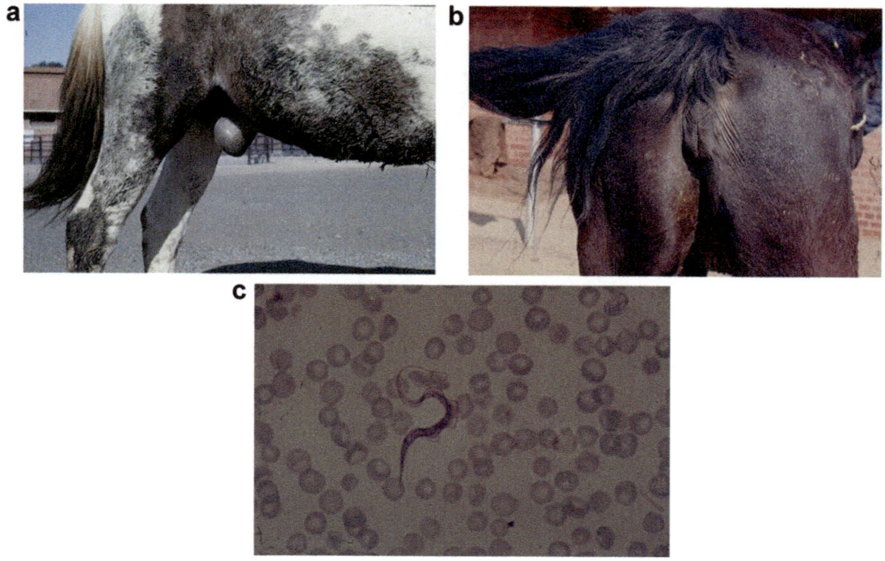

Fig. 4.26 (**a**) Magnification of a stallion with swollen sexual organs due to dourine. (**b**) Close-up photo of the genital opening of an infected mare. (**c**) Light micrograph of a stage of *T. brucei equiperdum*

Luciani M et al (2013) IgG antibodies from dourine infected horses identify a distinctive *Trypanosoma equiperdum* antigenic pattern of low molecular weight molecules. Vet Immunol Immunopathol 151:140–146.

Luckings AG et al (2004) Dourine. In: Coetzer JA, Tustin RC (eds) Infectious diseases of lifestock, vol 1. Oxford University Press.

4.3.5 Trypanosomes of Fish

4.3.5.1 *Trypanoplasma* Species

1. **Name**: Greek: *plasma* = motile mass. Latin: *cyprinus* = carp; *salmo* = salmon. Amédée Borrel (1867–1936) French microbiologist.
2. **Geographic distribution/epidemiology**: Worldwide, infecting fresh and saltwater fish.
3. **Biology, morphology**: These species belong to the order Bodonidae. *Trypanoplasma* species, e.g. *T. borreli* (carps), *T. cyprini* (carp) and *T. salmositica* (salmon), possess always two flagella and appear in the blood vessels of fish in two different shapes, which reproduce them by binary longitudinal divisions.
   (a) Slender, comma-shaped stages measuring 15 μm in length and 2–3 μm in width. The free ends of the flagella are at least 10–15 μm long. These stages are found during the first days of an infection (Fig. 4.27a, b).

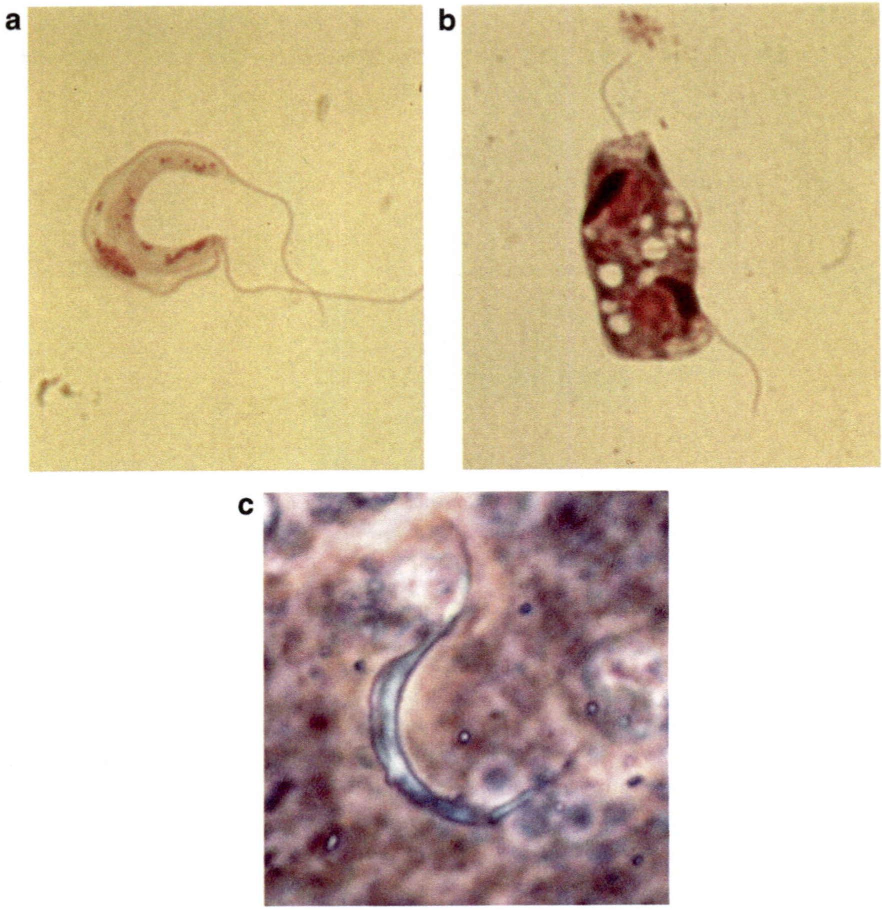

Fig. 4.27 Light micrographs of flagellate specimens of genera of the trypanoplasms of fish. (**a**) *Trypanosoma*. (**b**) *Trypanoplasma*. (**c**) *Cryptobia*

(b) Stumpy, polymorphous stages of 15–20 μm in size. Their free ends measure only 3–5 μm; these stages are seen mainly during chronic infections or after a period of coldness (Fig. 4.27c).
(c) Besides these rather defined stages, many types of intermediate forms can be observed. **Reproduction** occurs in any stage by binary fission.

Intermediate hosts and **vectors** are leeches, which suck blood at their host fish (e.g. specimens of the genuera *Piscicola* and *Hemiclepsis*). However, transmission experiments have shown that the species of the genus *Trypanoplasma* are very host specific.

4. **Symptoms of disease (sleeping sickness of fish, ichthyobodosis)**: Infected fish show clear signs of apathia, reduced feeding and loss of weight.

Physiological investigations indicated a considerable decrease of the haematocrit. Death cases occur mainly among young fish. In the case of **salmons**, exophthalmia, oedema, ascites and immunosuppression have been described as well as necrosis of liver and kidneys. In cases of fish being kept in warm water, mortality rates of up to 60 % have been reported.

5. **Diagnosis**: Microscopical demonstration of the agents of disease in Giemsa-stained blood smears (Figs. 4.27a–c).
6. **Pathway of infection**: Percutaneously during bloodsucking of leeches, which enter by vomiting fluids of the anterior portion of the intestine. **Attention**: Leeches remain lifelong infected!
7. **Prophylaxis**: A vaccination is available to protect salmons in smaller sea culture systems. Otherwise it is mostly reasonable to control the leeches in culturing ponds either by drying out or by treatment with Masoten® (metrifonate). This needs the following materials and conditions in order to 10,000 m^2 pond water:
   - 1.25 kg (at a water deepness of 50 cm),
   - 2.50 kg (at a water deepness of 1 m).

   In cases of a preventive pond treatment (=14 days before entering fish into the pond), also 2.5 kg or even 5 kg per 10,000 m^2 can be used. **Waiting times** before consumption of fish:
   - 0 days in the case as above described
   - 21 days in cases of bath of fish in higher concentrations (2.5 kg Masoten® for 100 kg) for 5–10 min in order to kill masses of parasites. See recommendations of the producer (Bayer, Leverkusen).
8. **Incubation period**: 14–17 days.
9. **Prepatent period**: Depending on the water temperature 5–60 days.
10. **Patency**: More than 100 days. Inapparent infections apparently remain persistent for long or even lifelong.
11. **Therapy**: Ornamental fish in aquaria: Flagellol® produced by Fa. Alpha-Biocare, Düsseldorf, distributed by Fa. Sera, Heinsberg, Germany.

Further Reading

Burreson EM (2007) Haemoflagellates of Oregon marine fishes: *Trypanosoma* and *Trypanoplasma*. J Parasitol 93:1442–1451.

Jurecka P et al (2009) Genetic resistance of carp to *Trypanosoma borreli*. Vet Immunol Immunopath 127:19–25.

Kovacevic N et al (2015). The analysis of the acute phase response during infection with *T. carassi*. Dev Comp Immunology 53:112–122.

Woo PT (ed) (2010) Fish diseases and disorders, vol 1: Protozoan and metazoan infections, 2nd edn. CAB International, Wallingford.

4.3.5.2 *Trypanosoma* Species of Fish

1. **Name**: See Sect. 4.3.5.1.
2. **Geographic distribution/epidemiology**: Worldwide.
3. **Biology, morphology**: Species of the genus *Trypanosoma* are also found in large numbers of different fish species belonging to the groups of fresh and saltwater fish. They appear twisted-comma shaped and reach depending on the species a size of 10–100 μm by 2–8 μm. They are found in many fish species mostly in small numbers, but their number may increase considerably in cases of overcrowded ponds. In the blood of fish, they appear in the trypomastigote stage showing a flagellum at each cell pole, which has a free portion of 10–15 μm in length (Fig. 4.27a). Vectors are apparently leeches. However, intense research is still lacking especially with respect of the situation inside leeches.
4. **Symptoms of disease**: The infections are mostly inapparent apart from the situation in overcrowded ponds, where masses of parasites are produced, if water temperature is considerably increased. Increase of parasites leads to weight losses in the fish population.
5. **Diagnosis**: Microscopical demonstration of the blood stages with the help of the Giemsa stain.
6. **Pathway of infection**: Percutaneously during sucking activity of leeches.
7. **Prophylaxis**: Drying of water ponds before entering a new fish population. Treatment of the pond water with the help of Masoten® (see *Trypanoplasma*).
8. **Incubation period**: Unknown due to lacking experimental investigations under different water temperature (at least for most species).
9. **Prepatent period**: Not fully understood, probably at least 1 week, depending on temperature.
10. **Patency**: Potentially lifelong.
11. **Therapy**: Unknown; mostly, however, not needed; keeping the ponds free from leeches is recommended.

Further Reading

Hayes PM et al (2014) Fish trypanosomes from South Africa. Parasit Vectors 7:50.

Lemos M et al (2015) Trypanosomes from Brazilian catfishes and leeches. Parasit Vectors 8:573.

Li M et al (2014) Glucocorticoid receptors in *Cryptobia salmositica*. Int J Parasitol 44:205–210.

Madison BN et al (2013) Duress without stress: *Cryptobia* infection in rainbow trout. J Endocrinol 218:287–297.

4.4 *Leishmania* Species

The *Leishmania* species are transmitted by bites of so-called sandflies (see Phlebotomidae, Fig. 4.28). The stages of the *Leishmania* species were 1903 described by the English scientists Leishman and Donovan and belong like the

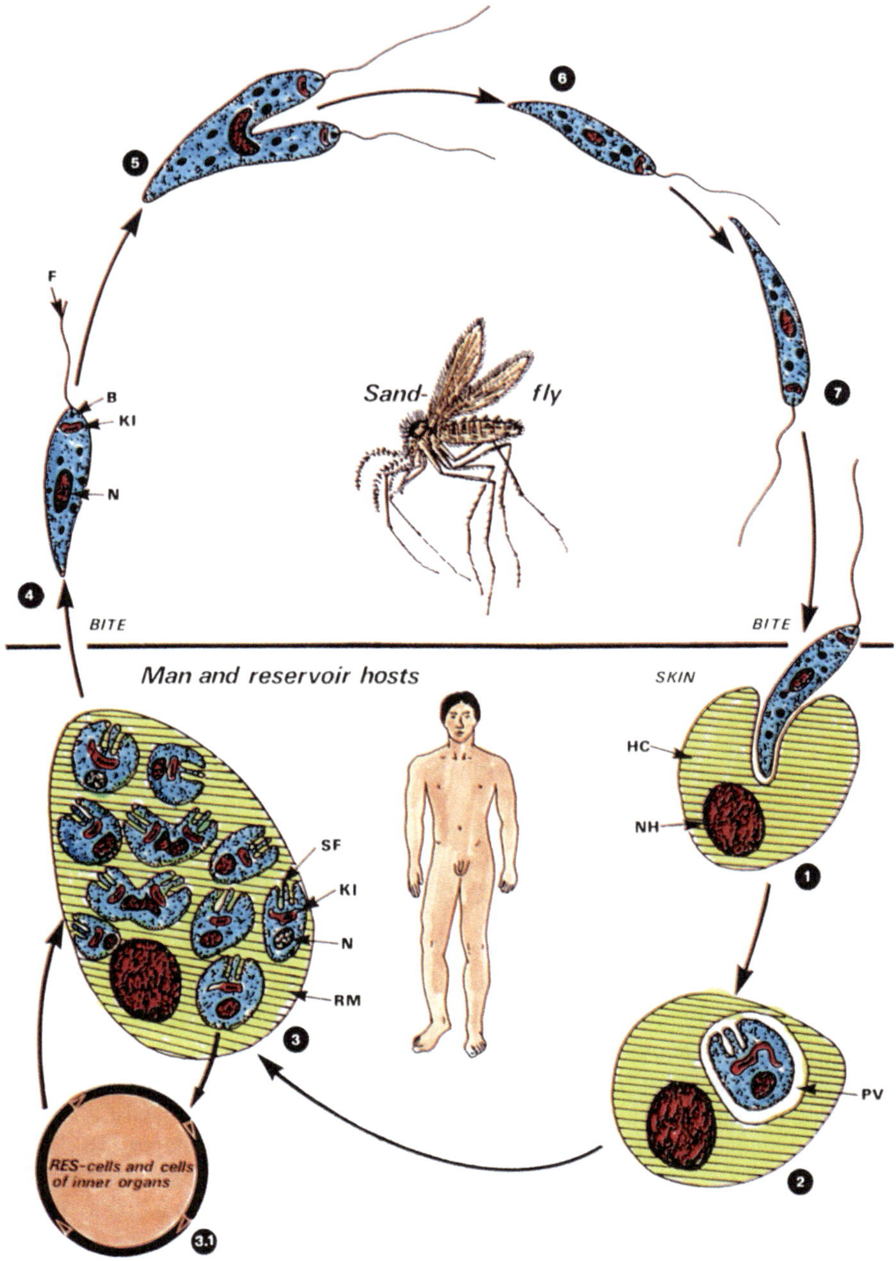

Fig. 4.28 Diagrammatic representation of the life cycle of the *Leishmania* species of humans and hosts like dogs. *B* basal body; *F* flagellum; *HC* host cell; *KI* kinetoplast; *N* nucleus; *NH* nucleus of host cell; *RM* remnants of host macrophage; *PV* parasitophorous vacuole

Table 4.5 Important *Leishmania* species of animals and humans

	Species	Disease	Geographical distribution	Reservoir host
Cutaneous leishmaniasis	*L. tropica minor*	"Dry" cutaneous leishmaniasis	Middle East, Mediterranean region, Asiea	Rodents, dogs
	L. tropica major	"Wet" cutaneous leishmaniasis, Oriental boil, Aleppo boil	Middle East, Mediterranean region, Asiea	Rodents, dogs
	L. aethiopica	Diffuse cutaneous leishmaniasis	Ethiopia, Kenya	Rock hyrax
	L. braziliensis braziliensis	Espundia, mucosal leishmaniasis	Mexiko → Brazil	Rodents, armadillos
	L. braziliensis peruvianae	Uta (like Oriental boil)	Peru	Dogs
	L. mexicana mexicana	Chiclero ulcer	Central America	Rodents
	L. mexicana amazonensis	Diffuse cutaneous leishmaniasis	Amazonas region	Rodents
	L. mexicana pifanoi	Diffuse cutaneous leishmaniasis	Venezuela	Rodents
Visceral leishmaniasis	*L. donovani donovani*	Kala Azar, Dum dum fever, visceral leishmaniasis	Near East, Mediterranean region, Africa, India, South America, former USSR	Dogs, foxes
	L. donovani infantum	Visceral leishmaniasis	Mediterranean region	Dogs
	L. donovani chagasi	Visceral leishmaniasis	South America	Dogs

agents of the sleeping sickness and of the Chagas disease to the group of so-called Kinetoplastida, which received their name from the Giemsa stainable, DNA-containing region of the single and long mitochondrion. The infection of humans and other hosts (Table 4.5) occurs in sandy and dry regions during bites of the bloodsucking species of sandflies of the genera *Phlebotomus* and *Lutzomyia*. During bloodsucking the insects inject so-called promastigote stages (10–12 μm long), which are ingested by macrophages whereby they are at first included into parasitophorous vacuoles but become later set free in the cytoplasm due to disruption of the membrane of the parasitophorous vacuole. These promastigotes then become transformed during repeated binary fissions into small **amastigotes**, which in reality are **micromastigotes** since their very short flagellum does not proceed the cell surface and thus is not visible in light microscopy but can be seen on electron micrographs (Fig. 4.30). Repeated divisions of these tiny, about 2–3 μm sized stages lead finally to the disruption of the host cell. The parasites then enter neighbour cells, which finally also become destroyed so that outer and inner lesions occur. The different *Leishmania* species cannot be differentiated from each other alone by morphological criteria. Thus physiological, immunological and molecular-biological methods are often needed to differentiate clearly between single species. However, species of the *Leishmania donovani* group enter also inner organs, while the species of the *Leishmania tropica* complex stay mainly in the skin region (see Table 4.5).

4.4.1 *Leishmania* Species of Dogs

1. **Name**: William B. Leishman (1865–1926), Charles Donovan (1863–1951), English researchers in the tropics. Latin: *infans* = child.
2. **Geographic distribution/epidemiology**: Dry regions; in Europe, North Africa: countries along the border of the Mediterranean Sea; worldwide in warm countries; up to 20 % of the dogs in Portugal and Spain are infected. In Germany, leishmaniasis is the second most imported disease of dogs; however, spreading does not yet occur since not enough sandflies are present there.
3. **Biology, morphology**:
   (a) ***L. donovani* group**
   These protozoan species are the agents of the visceral leishmaniasis, which is introduced due to destruction of cells of the reticulo-endothelial system (Fig. 4.29).
   (b) ***L. tropica* group, *L. infantum***
   These species are the agents of the skin leishmaniasis, which leads to the destruction of cells especially in the skin.
   The live cycle and details of the different species are shown in the Figs. 4.28, 4.29, 4.30 and 4.31. Vectors are so-called sandflies of the genera *Phlebotomus* and *Lutzomyia* (see chapter Insects).
4. **Symptoms of disease**:
   (a) **Visceral disease**:
   Swelling of the lymph nodes, spleen and liver; claw hypertrophia; alopecia; fever; diarrhoeas; anaemia; polymyositis; without treatment death.
   (b) **Cutaneous disease**:
   Skin ulcers, with brownish crusts; skin appears with scales and whitish knots.
5. **Diagnosis**: (**a**) Microscopical investigation of Giemsa-stained preparations obtained from skin crusts and by punction of bone mark or lymph nodes (Figs. 4.29, 4.30 and 4.31). (**b**) Use of serological methods (ELISA, IFAT) and PCR.
6. **Pathway of infection**: Percutaneously during bites of infected sandflies.
7. **Prophylaxis**: To leave dogs at home when travelling in endemic regions or to protect dogs with the help of repellents (e.g. Advantix®) **Attention**: Dogs may

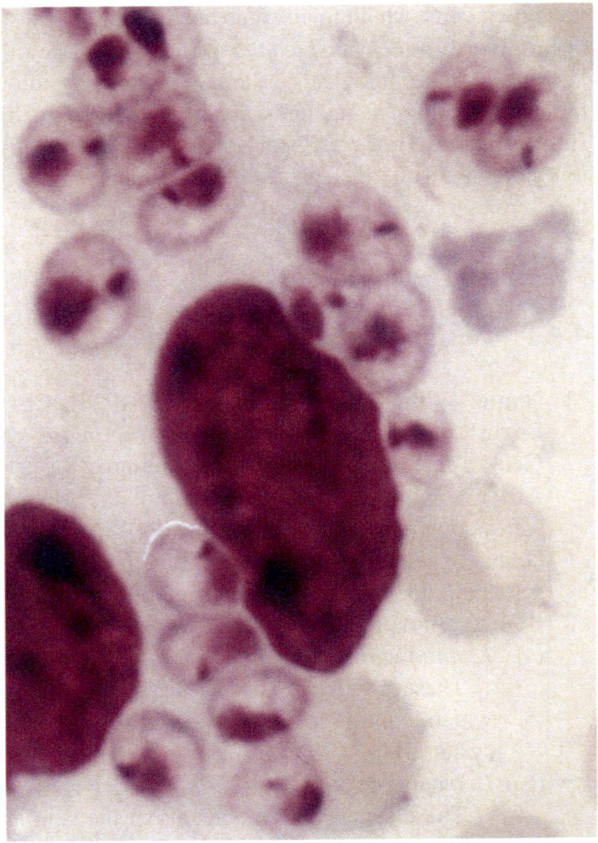

Fig. 4.29 Light micrograph of a Giemsa-stained smear preparation of a ruptured macrophage containing many amastigotes of *Leishmania donovani*. *K* Kinetoplast of the parasite; *N* nucleus; *NH* nucleus of the host cell

act as reservoir hosts for humans. Thus, importation of potentially infected dogs in regions without leishmaniasis may be dangerous.

8. **Incubation period**: Varying (some weeks up to several months).
9. **Prepatent period**: Parasites can be rather easily observed as soon as first symptoms of disease occur.
10. **Patency**: Years in the case of *L. donovani*; several months in the case of *L. tropica*.
11. **Therapy**: Chemotherapy of dogs does not lead to a complete healing. Treatment with allopurinol® (20 mg/kg bodyweight, orally, daily for 2–6 months) helps that symptoms disappear; however, recidives are always possible (still after 2 years treatment). Alternatives (without better prognosis) are glucantime® and amphotericin®.

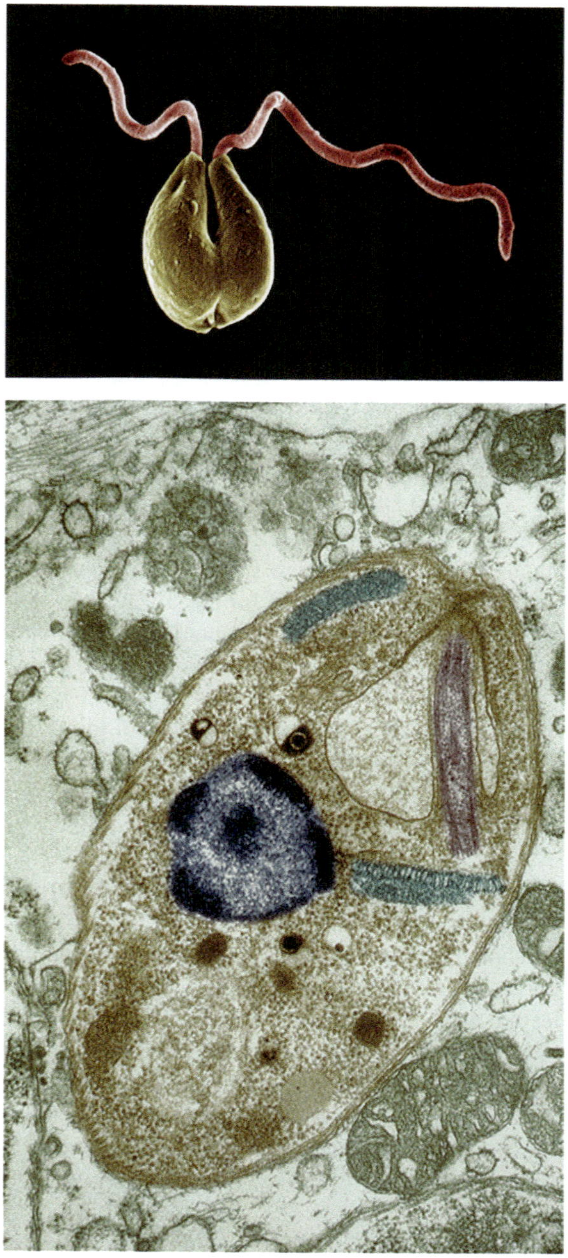

Fig. 4.30 (**a**) Scanning micrograph of a division stage of a promastigote stage of *Leishmania* sp. (**b**) Transmission electron micrograph of a section of an "amastigote" stage inside the cytoplasm of a degrading host cell showing the short flagellum in the flagellar pocket of *L. major*

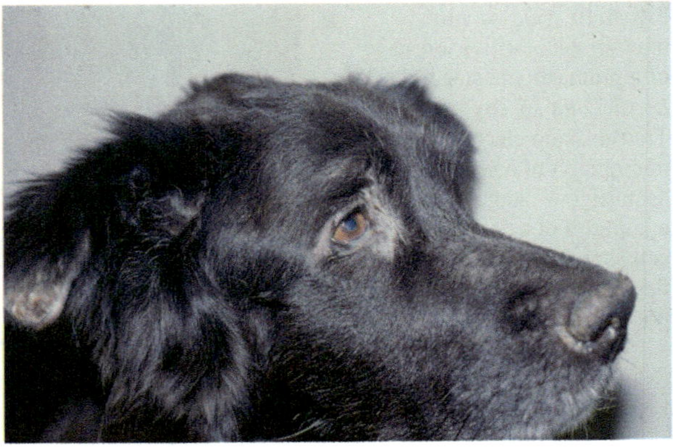

Fig. 4.31 Dog with crusts along the ears and above the eyes due to leishmaniasis

Further Reading

De Aruda MM et al (2016) Sensitivity and specificity of serological testing of canine *Leishmania* infection. Mem Inst Oswaldo Cruz 28:1–6.

Aslan M et al (2016) New insights into transmissibility of *Leishmania infantum* from dogs to sandflies. J Inf Dis 16:2–10.

Vida B et al (2016) Immunologic progression of canine leishmaniosis following vertical transmission in USA dogs. Vet Immunol Immunopathol 169:34–38.

Yasur-Landau D et al (2016) Allopurinol in *Leishmania infantum* from dogs with disease relapse. PLOS Negl Trop Dis 10:e0004341.

4.4.2 Leishmaniasis of Cats, Equids and Cattle

Leishmaniasis in cats was introduced in Europe by *L. infantum* and in Asia and Latin America by *L. venezuelensis* and *L. amazonensis*. Infected cats showed skin lesions and a generalized dermatitis. Diagnosis was done by PCR showing infection rates of 20–30 % (e.g. in Portugal).

Infections with *L. infantum* were documented in horses in South Europe. *L. braziliensis* was found in South America in horses and donkeys.

Ruminants are only infected in rare cases (e.g. *L. donovani* was shown in goats and in cattle), where, however, up to 10 cm large lesions had been observed.

Further Reading

Dantas-Torres F (2007) The role of dogs as reservoirs of *Leishmania* parasites. Vet Parasitol 149:139–146.

Martin-Sanchez J et al (2007) Infection by *Leishmania infantum* in cats. Vet Parasitol 145:267–273.

4.5 Amoebae

Amoebas are characterized by so-called **pseudopodia**, which are formed spontaneously as protrusions of the surface and which are used for movements and for uptake of food particles. These pseudopodia are protruded species specifically only in one direction (e.g. *Entamoeba* species, *Limax* species) or in various directions at the same time (e.g. *Acanthamoeba* species; Fig. 4.36a). Some amoebas are able to form single pseudopodia as well as several ones at the same time. In addition, flagellated stages may occur besides amoeboid stages (e.g. *Naegleria* species may have two flagella) (Fig. 4.36b). This is the reason why the so-called **Rhizopoda** were included into the flagellates and named **Sarcomastigophora**. Today, however, the *Entamoeba* species are considered as separate but very basic group of the protozoans. However, most of the amoebas are free living and only a few of them live as parasites (Table 4.6). Cysts are always formed by an exocytotic excretion of materials. **Reproduction** occurs by binary, often in equal fission.

4.5.1 *Entamoeba histolytica*

1. **Name**: Greek: *amoibos* = fluid, elastic, shape changing; *entos* = inside; *histos* = tissue; *lysis* = dissolution.
2. **Geographic distribution/epidemiology**: Common in warm countries, however, also introduced in European dog rearing facilities, rarely also in cats and in monkeys in Zoological gardens.
3. **Biology, morphology**: Infectious *E. histolytica* cysts in fresh feces contain four nuclei and reach diameters of 15–20 μm. The nucleolus is always situated centrally inside the nucleus thus being an important feature that distinguishes this species from other *Entamoeba* species (Figs. 4.32, 4.33 and 4.34). The 4-nucleated amoebas hatch in the lumen of the intestine; each nucleus divides once so that for a short period 8-nucleated stages can be seen (Fig. 4.32). This stage becomes divided into eight single nucleated amoeba (**minuta stages**), which start feeding intestinal fluids and particles. These uninucleated minuta stages divide repeatedly but stay mainly in the lumen of the intestine. Under not yet fully understood conditions, they grow up to become **magna stages** (Fig. 4.32), which enter the intestinal wall and reach via blood vessels liver and many other organs, where they initiate abscesses. These magna forms may reach diameters of up to 40 μm and contain—when observed by help of a light microscope—often vacuoles with included red blood cells. The minuta forms, which stay inside the intestine, excrete a protecting cyst wall as soon as they are transported with the intestinal contents into the colon region. Thus, they are found as uninucleated cysts in the feces.
4. **Symptoms of disease (Intestinal and extraintestinal amoebiasis)**: Low-grade infections lead to fluid feces which are excreted repeatedly several times per day and may considerably weaken dogs due to the considerable loss of water. In cases of severe infections (**ulcerative colitis**), fever, exiccosis due to

Table 4.6 Important amoebas

Species	Size (µm)	Host/Tissue	Most common number of nuclei in cysts	Pathogenicity
Entamoeba coli	20–45	**Humans**/colon	8	–
E. hartmanni	5–10	**Humans**/colon	4	–
E. histolytica		**Humans**/colon	4	+
L. h. Minuta form	10–18	Monkeys/**humans**/colon	4	?
L. h. Magna form	20–40	Intra- and extraintestinal abscesses	–	+
E. dispar	10–18	**Humans**/colon	4	?
E. polecki	10–20	**Humans**/pigs/colon	1	–
E. gingivalis	10–20	**Humans**/mouth	1	–
E. gallinarum	9–25	Gamefowl/caecum	8	–
E. invadens	9–38	Reptiles/colon	4	+
E. bovis	5–20	Ruminants/stomach	1	–
E. suis	5–25	Pigs/colon	1	–
E. equi	7–25	Horses/colon	1	–
E. cuniculi	10–15	Rabbits/intestine	1	–
Endolimax nana	6–15	**Humans**/colon	4	–
Jodamoeba bütschlii	8–20	**Humans**/pigs/colon	1	–
Naegleria gruberi	22	**Humans**/CNS	1	+
N. fowleri	20	**Humans**/CNS	1	+
Acanthamoeba sp.	40	**Humans**/CNS	1	+
Balamuthia mandrillaris	30	**Humans**/CNS, monkeys/CNS	1	+
Malpighamoeba melificae	10–15	Bees/intestine	2	+

severe water loss, bloody feces and formation of abscesses (in liver and other organs) occur, which often lead to death in untreated cases. However, it has been described that the disease may become chronic, so that it is often underestimated and people are surprised by the sudden death of their dog (Raether 1968).

5. **Diagnosis**: Microscopical demonstration of 2- or 4-nucleated cysts or amoebic stages with typical nuclei inside diarrhoeic feces of dogs (Figs. 4.33 and 4.34). Stages in the liver can be diagnosed with the help of methods such as IIFT, IHA, ELISA and computer tomography.
6. **Pathway of infection**: Oral by uptake of cysts within contaminated food or direct contact with cyst-containing feces when sniffing.
7. **Prophylaxis**: Especially in warm countries dogs should be kept away from other dogs and their feces.

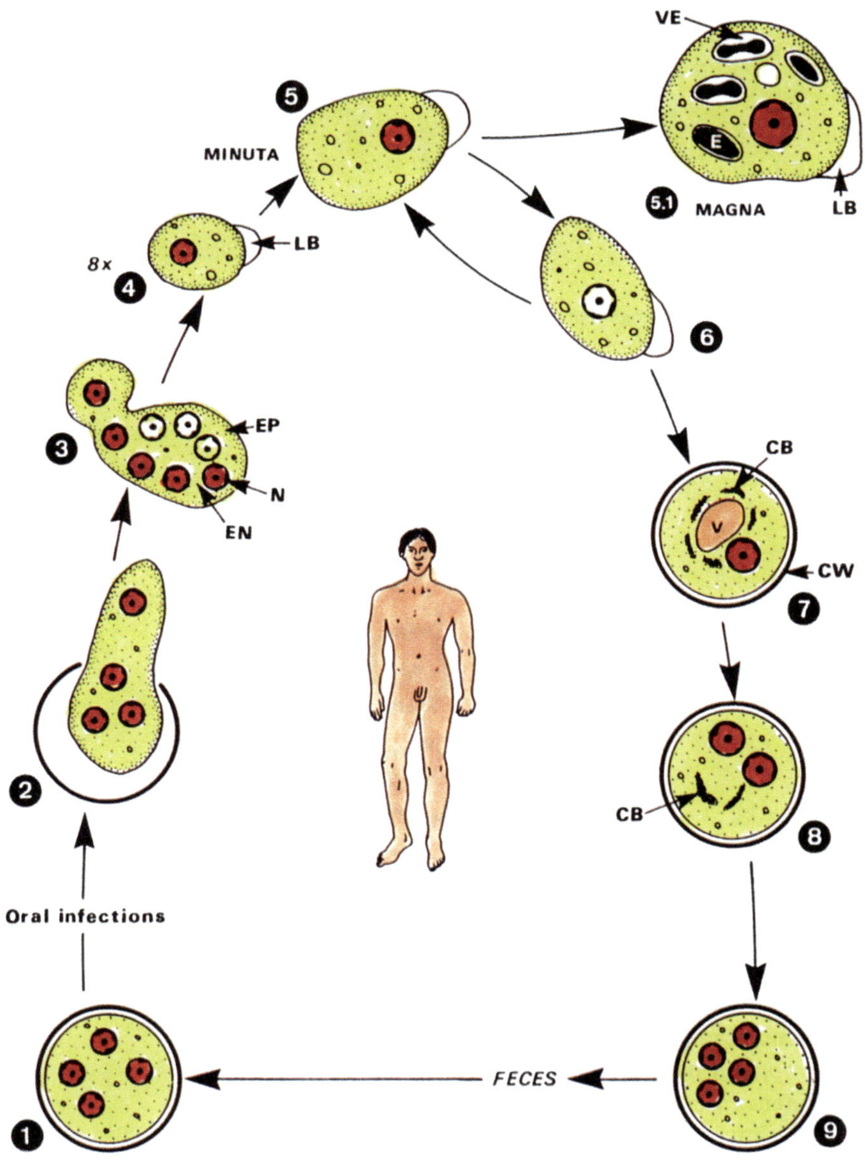

Fig. 4.32 Diagrammatic representation of the life cycle of *Entamoeba histolytica*, which runs in dogs and humans very similar as well as in other hosts. (**1**) 4-nucleated cyst, which is ingested orally by hosts. (**2–4**) After hatching from the cyst inside the intestine, binary fission of the nuclei occurs leading to 8-nucleated stages, which divide into 8 single nucleated cells (so-called **minuta stages**). These stages reproduce themselves constantly by repeated divisions. (**5**) Some of them grow up to **magna stages**, which enter the intestinal wall and ingest host cells (**5.1**), while others start to form cysts (**6–9**), which finally may become excreted within the feces of the host. *CB* chromidial bodies; *CW* cyst wall; *EN* endoplasm containing organelles and food vacuoles; *EP* ectoplasm; *LB* lobopodium = pseudopodium; *N* nucleus; *VE* vacuole containing erythrocytes

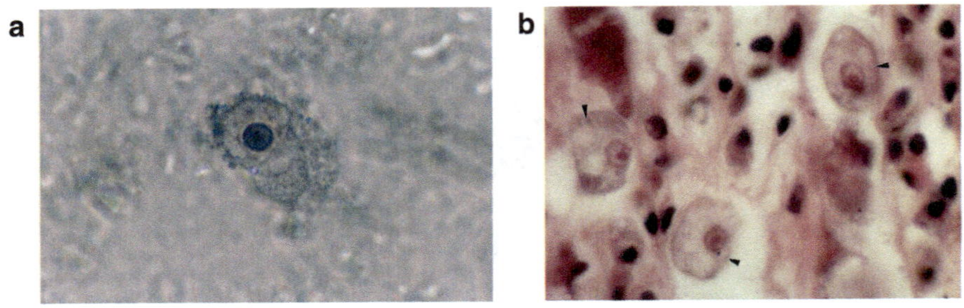

Fig. 4.33 Light micrographs of *E. hystolytica*. (**a**) Minuta form. (**b**) Magna forms inside the intestinal wall (*arrows*) showing the typical central nucleolus

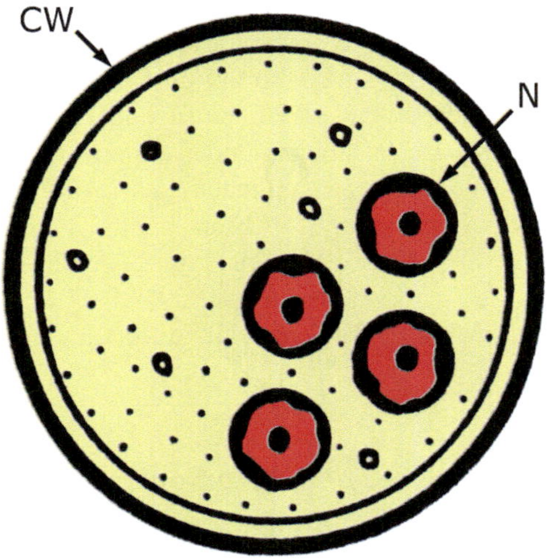

Fig. 4.34 Diagrammatic representation of a cyst of *E. histolytica*. Characteristic is the centrally situated nucleolus inside the nucleus. *CW* cyst wall; *N* nucleus

8. **Incubation period**: 2–10 days.
9. **Prepatent period**: 1 week–months
10. **Patency**: Eventually years.
11. **Therapy**: Use of products known from human medicine such as 5-nitroimidazoles in dosages such as those used in *Giardia* infections.

Further Reading

Alam MA et al (2015) *Entamoeba* infections in different populations of dogs in an endemic area of Lahore, Pakistan. Vet Parasitol 207:216–219.

Leelayoova S et al (2009) Prevalence of intestinal parasitic infections in military personnel and military dogs, Thailand. J Med Assoc Thai 92 Suppl 1:S53–S59.

Raether W (1968) Acute amoebic dysenteria in a dog. Kleintierpraxis 7:196–200.

Schär F et al (2014) The prevalence and diversity of intestinal parasitic infections in humans and domestic animals in a rural Cambodian village. Parasitol Int 63:597–603.

4.5.2 Amoebas of Reptiles and Frogs

1. **Name**: Latin: *invadere* = penetrating; *rana* = frog.
2. **Geographic distribution/epidemiology**: Worldwide. *Entamoeba histolytica* lead to an ulcerative enteritis especially if too many animals were kept in the same cage/biotope.
3. **Biology, morphology**: Among a large amount of described amoeba species in reptiles and frogs, mainly *Entamoeba invadens* (in snakes and carnivorous reptiles) and *E. ranarum* in frog larvae are of high importance. The pathogenic effects are increased especially if too many animals were kept in a cage with reduced hygienic conditions.
   - ***Entamoeba invadens***

     4-nucleated cyst reach a diameter of 10–20 μm. After oral uptake within food or from floor, so-called metacystic 8-nucleated stages are developed from the hatching 4-nucleated ones. These 8-nucleated stages have a size of up to 60 μm and give rise to 8 single nucleated stages, which reproduce themselves by binary fissions giving rise to 10–20 μm sized uninuclear amoebae. Similarly to *Entamoeba histolytica* (but without formation of magna stages), the stages of *E. invadens* may enter the intestinal wall and may become spread all over the body. Stages in the colon become encysted and are excreted as 4-nucleated cysts.
   - ***Entamoeba ranarum***

     This amoeba, which reaches as trophozoites a diameter of 10–30 μm and as 4-nucleated cyst 8–10 μm, is especially found in tadpoles.
4. **Symptoms of disease (Amoebiasis)**: Especially in captivity (=under stress), snakes, carnivorous lizards, etc., show high rates of morbidity and mortality, if they are infected with amoebas. Turtles, however, show rather low infection rates and low-grade symptoms. This is probably why they feed mainly plants, but it is supposed that they are symptomless vectors of such amoebas. The **entamoebiasis** appears as an **ulcerative gastritis** and as **haemorrhagic enteritis** and **colitis**. Infected animals do not feed, show loss of weight and convulsive cramps and are considerably thirsty. The feces appear bloody and slimy. In experimental infections, it was shown that death would occur within 2–11 weeks if no treatment would be given. In untreated cases—brought too late to veterinarians—death rates reached 100 %. In most cases, there occurred enormous liver swellings and thrombosis of many blood vessels.
5. **Diagnosis**: Demonstration of the cysts (4-nuclei = infectious cyst; 1–2 nuclei = non-mature cyst) looking like those of *Entamoeba histolytica* (Fig. 4.34). Fresh feces show also motile amoebas. In the case of turtles, only cultures of feces will give indications, since cyst numbers in feces are rather

low. For **differential diagnosis**, infections with salmonelles or nematodes have to be excluded.

6. **Pathway of infection**: Oral by uptake of 4-nucleated cysts.
7. **Prophylaxis: Amoebiasis** of snakes and other reptiles as well as of frogs can be avoided (or kept at least at a low level), if the following measurements are employed:
   (a) Keep new animals separately in quarantine for several weeks and clean cages and aquaria with hot steam or hot water.
   (b) Keep animals (especially snakes), which come from different climatic regions separately.
   (c) Avoid contacts between turtles and snakes and with other reptiles.
   (d) Use different equipment to clean the different cages or aquaria.
   (e) Keep animals at their appropriate temperatures; this will avoid stress.
   (f) Treat new imported animals preventively.
8. **Incubation period**: Very variable, often only a few days in the case of very young snakes or lizards.
9. **Prepatent period**: Mostly only a few days.
10. **Patency**: Possible lifelong in the case of turtles due to mostly inapparent infections; several weeks or months in cases of sensible and stressed animals.
11. **Therapy**:
    (a) Increasing the cage temperature at a level of 35 °C has led to an increasing healing effect during treatment. A sufficient amount of drinking water must always be offered.
    (b) Medical treatment with the help of the free-sale product Amoebol® (Fa. Sera, Germany; produced by Alpha-Biocare GmbH, Düsseldorf) or application of nitroimidazole.

Further Reading

Baseler LJ et al (2014) Pathology in practice. *E invadens* infection in a ball python. J Am Vet Med Assoc 245:501–503.

Brewer LA et al (2008) Analysis of commercial *Entamoeba histolytica* ELISA kits for the detection of *Entamoeba invadens* in reptiles. J Zoo Wildl Med 39:493–495.

Chia MY et al (2009) *Entamoeba invadens* myositis in a common water monitor lizard (*Varanus salvator*). Vet Pathol 46:673–676.

4.5.3 *Malpighamoeba mellificae*

1. **Name**: Marcello Malpighi (1628–1694) = Italian scientist. Honouring his person, the excretion organs of insects have been named: Malpighi tubules. Latin: *mellificere* = honey making.
2. **Geographic distribution/epidemiology**: Europe, the USA, New Zealand and probably other countries.

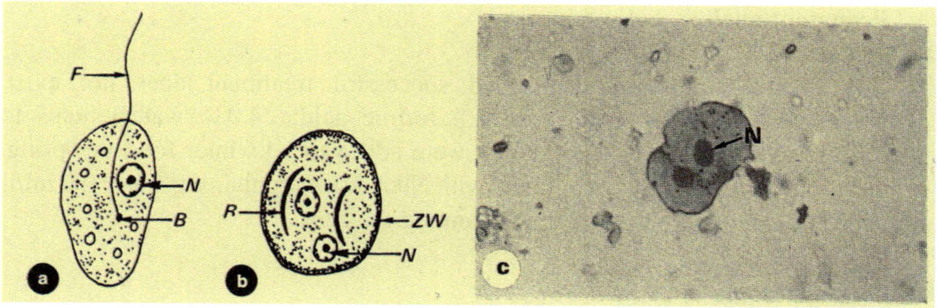

Fig. 4.35 *Malpighamoeba mellificae*. Diagrammatic representation of the trophozoites (**a**) and the cyst (**b**) and a light micrograph of the 2-nucleated cyst (**c**). *B* basal body; *F* flagellum; *N* nucleus; *R* rhyzostyl; *ZW* cyst wall

3. **Biology, morphology**: The motile stages of *M. mellificae* appear in two different shapes in the Malpighian tubules of bees: amoebic stages and uni-flagellated stages (Fig. 4.35). In addition, there occur after 3–4 weeks of repeated binary fissions so-called cysts, which include each two nuclei (Fig. 4.35b), measure about 5–15 μm in diameter and are excreted within the feces, thus leading to a quick infection of many of their adult members of bee stocks. The bee queens, however, are mostly rather seldom infected. The amoebic stages and especially the cysts may occur in such large numbers that they can block the excretion of feces. In addition, mixed infections with *Nosema apis* may increase the problem of defecation of the bee worker stages.
4. **Symptoms of disease (Amoebic flu of bees)**: Characteristic for this disease are the following observations:
   (1) Bees eject masses of fluid, greenish-greyish feces when starting their flight from the entrance of the stock.
   (2) The regions around the stock entrance are contaminated by greyish-whitish layers.
   (3) When opening the stock after winter, especially high numbers of dead bees can be found (especially when the winter had been rather cold). Even death of all hive members have been reported. **Self-healing** may occur in weeks of high summer temperatures and availability of sufficient food. It was reported that self-healing is much more common in cases of double infections with *Nosema apis* (apparently due to increased immune reactions).
5. **Diagnosis**: Microscopical determination of vegetative stages and cysts (Fig. 4.35) in the feces of sick appearing bees.
6. **Pathway of infection**: Oral uptake of cysts, which had been transported by foreign, infected bees into a stock (but only in cases when these foreign bees had excreted feces inside of the foreign hive).
7. **Prophylaxis**: Desinfection of the honeycombs by aeration with acetic acid. This kills cysts of *M. mellificae* and those of *Nosema apis*.
8. **Incubation period**: 2–3 weeks.

9. **Prepatent period**: 2–4 weeks.
10. **Patency**: Variable (often lifelong).
11. **Therapy**: A completely safe and successful treatment does not exist. Neotektin® (66.3 % acetic acid, 8.8 formic acid; 24.9 % water) helps to decrease the disease, if 20–30 drops were added to 1 l winter food. In spring, the honeycombs have to be sprayed with Neotektin® containing water (30 ml/l) and should decrease the death rates among the bees.

Further Reading

Ritter W (1994) Bee diseases. Ulmer, Stuttgart.
Zander E, Böttcher FK (1984) Diseases of bees. Ulmer, Stuttgart.

4.5.4 *Acanthamoeba* Species and Related Groups

There are several species of so-called free-living amoebae which are facultatively pathogenic. In animals, there are only rare cases of infections due to the species of the genera *Naegleria*, *Hartmannella*, *Balamuthia* (one single species in monkeys) and *Acanthamoeba* (Fig. 4.36a). Since these species are able to switch in seconds from the free-living amoeboid stage to a cyst, they are able to protect other agents of disease, e.g., from disinfecting solutions. For example, in our tests and those of other groups it turned out that their cyst stages may contain bacteria of the genera *Legionella*, *Pseudomonas* and *Chlamydia*. In the case of *Acanthamoeba*, 19 species have been described. Since many of them are found in humans (e.g. eyes, body cavities, etc.), they should also be present in animals (hidden in the brain) but apparently remained often undetected. The existence of clear species of *Acanthamoeba* becomes more and more doubtful. Researchers speak now of type isolates belonging to 19 different genotypes (T1–T19).

a

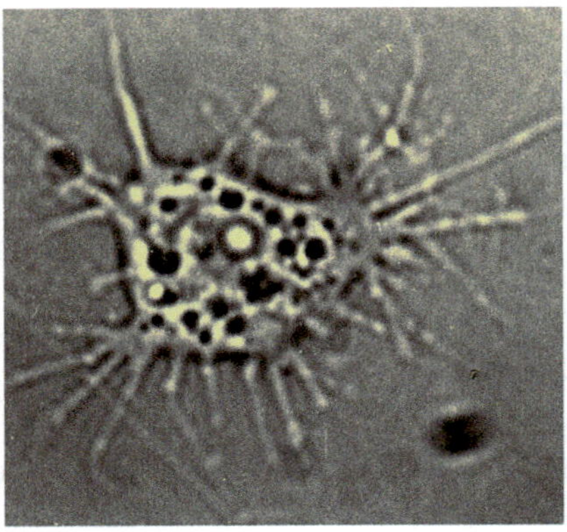

b

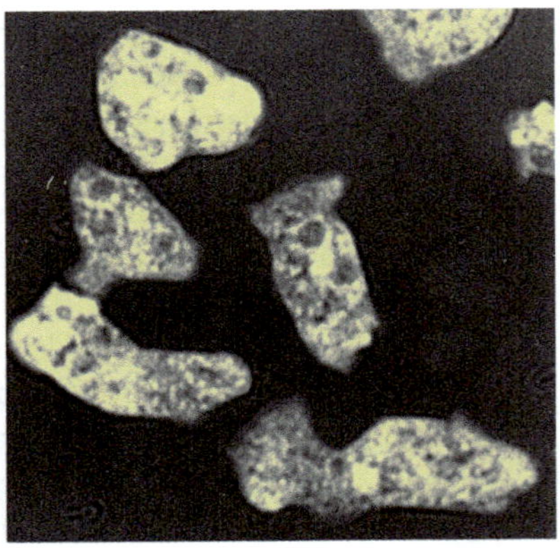

Fig. 4.36 Light micrograph of the amoebas of the genera *Naegleria* species (**b**) and *Acanthamoeba* (**a**)

Further Reading

Magnet A et al (2014) Novel *Acanthamoeba* 18s/rRNA gene sequence type from an environmental isolate. Parasitol Res 113:2845–2850.

Shoaib HM et al (2013) Evaluation of inhibitory potential of some selective methanolic plant extracts on biological characteristics of *Acanthamoeba castellanii* using human corneal epithelian cells in vitro. Parasitol Res 112:1179–1188.

4.6 Apicomplexa (Sporozoa)

This group of protozoan organisms is now according to many groups of nomenclaturally very active scientists listed among the animal phylum **Alveolata** covering the following units:

The subphylum Apicomplexa (former Sporozoa) is most important with respect to parasitic diseases of animals. Thus, the specimens of the following genera are the focus of this book. However, mostly economically important parasites are considered as well as those, which can be transmitted to humans.

Subphylum: Sporozoa (Apicomplexa)
1. Class: Perkinsea [today recognized as dinoflagellates (=free-living flagellates)]
2. Class: Sporozoea
1. Subclass: Gregarinia
Genus: *Monocystis* (parasites of evertebrates)
Genus: *Gregarina* (parasites of evertebrates)
2. Subclass: Coccidia
1. Order: Protococcida (Eucoccida)—without schizogony (parasites of evertebrates)
2. Order: Eucoccida (Schizococcida)—with schizogony
1. Suborder: Eimeriina
Genus: *Eimeria*
Genus: *Isospora*
Genus: *Cystoisospora*
Genus: *Toxoplasma*
Genus: *Sarcocystis*
Genus: *Besnoitia*
Genus: *Frenkelia* (syn. *Sarcocystis*)
Genus: *Hammondia*
Genus: *Neospora*
Genus: *Globidium*
Genus: *Cryptosporidium* (presumably belonging to the group of gregarines)
Genus: *Caryospora*
Genus: *Cyclospora*
2. Suborder: Haemosporina

(continued)

A. Haemosporidea
Genus: *Plasmodium*
Genus: *Leucocytozoon*
Genus: *Haemoproteus*
B. Piroplasmea
Genus: *Babesia*
Genus: *Theileria*
3. Suborder: Adeleina
Genus: *Karyolysus*
Genus: *Haemogregarina*
Genus: *Hepatozoon*
Supplement:
Genus: *Klossia*
Genus: *Adelina*

4.6.1 *Eimeria* Species

This genus was named in honour of the Swiss scientist **Theodor Eimer** (1843–1897), who worked as Professor in Tübingen (Germany). This genus contains about 800 species (their number is still increasing), which parasitize in the intestine (and some other organs) of plant feeding animals but are not found in humans. They belong in a narrow sense to the **Coccidia**, which got this name due to their ovoid shape.

The *Eimeria* species are very **host specific**; i.e. a species of chickens cannot be transmitted to other bird groups. The disease is called **coccidiosis**, which is one of the most important diseases of chickens, rabbits as well as many farm animals. They lead yearly to enormous losses especially if these animals were kept very close together.

The typical life cycle and its different stages are shown in Fig. 4.37. The infection always starts by oral uptake of infectious oocysts, which contain four sporocysts each being equipped with two slender, infectious (=cell penetrating) motile sporozoites (Figs. 4.38 and 4.39). Inside the intestine, the sporozoites are set free, enter the host cells and are included within a so-called parasitophorous vacuole. Inside these vacuoles, the parasites are rounded up, start feeding with the help of **micropores** (=ultracystostomes), obtain a spherical shape and start nuclear divisions, thus reaching the status of a multinucleated **schizont**. This schizont gives rise to long slender stages (**merozoites**), which leave the parasitophorous vacuole after disruption of the host cell and enter neighbouring cells. Due to repeated schizont formations and disruption, a lot of host cells are destroyed leading to the **diarrhoeas** which are typical for the **cocciciosis**. After a species-specific number of schizogonies, **male and female gamonts** are produced. The male gametes (microgametes) possess species specifically 2–3 flagella. They enter the female one (macrogamete), and this fusion gives rise to the formation of a

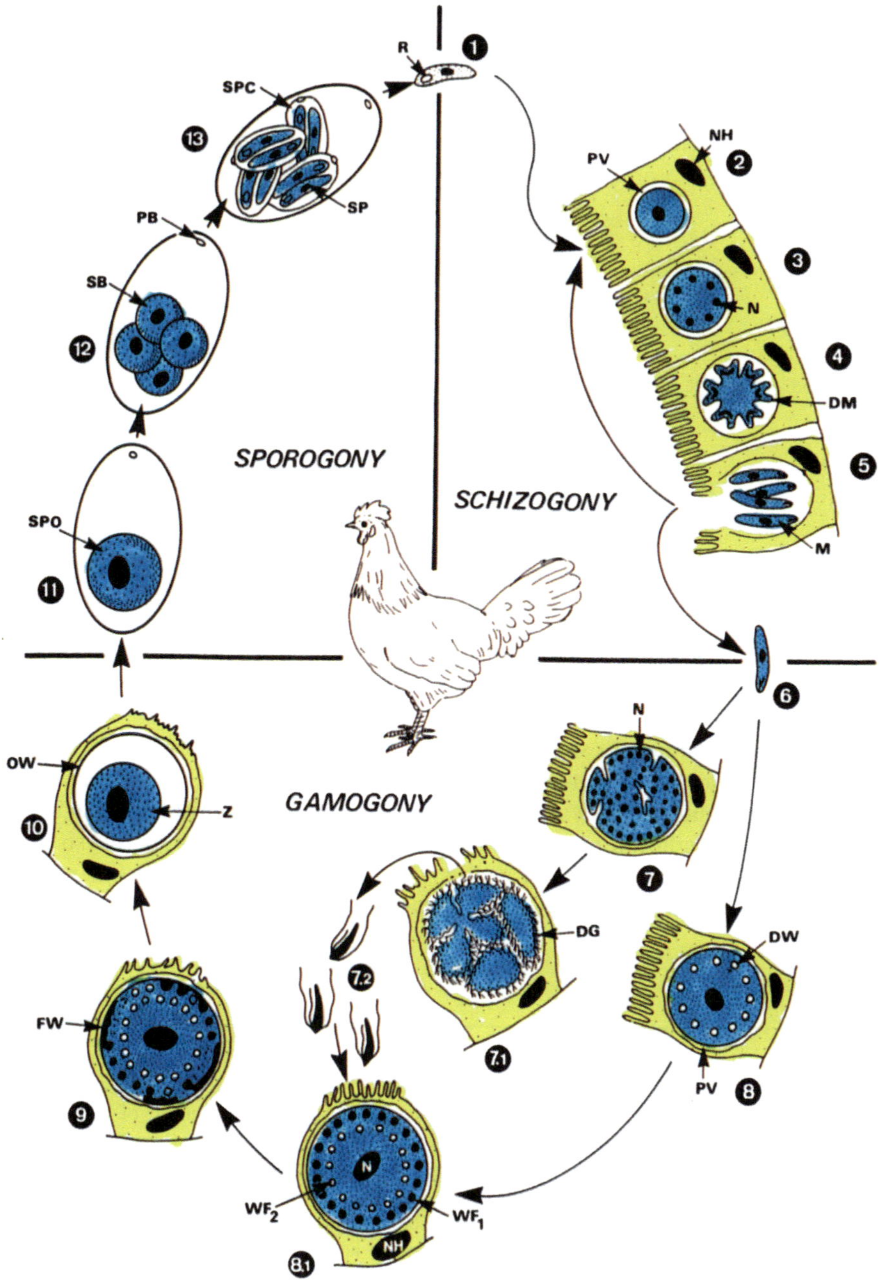

Fig. 4.37 Diagrammatic demonstration of the life cycle of chicken coccidian of the genus *Eimeria* (see Table 4.7). (**1**) After oral uptake of sporulated oocysts, the sporozoites hatch in the small intestine from the sporocysts. (**2–6**) After penetration, multinucleate schizonts are formed (**3**) inside a parasitophorous vacuole (PV). The schizonts produce motile merozoites (DM, M),

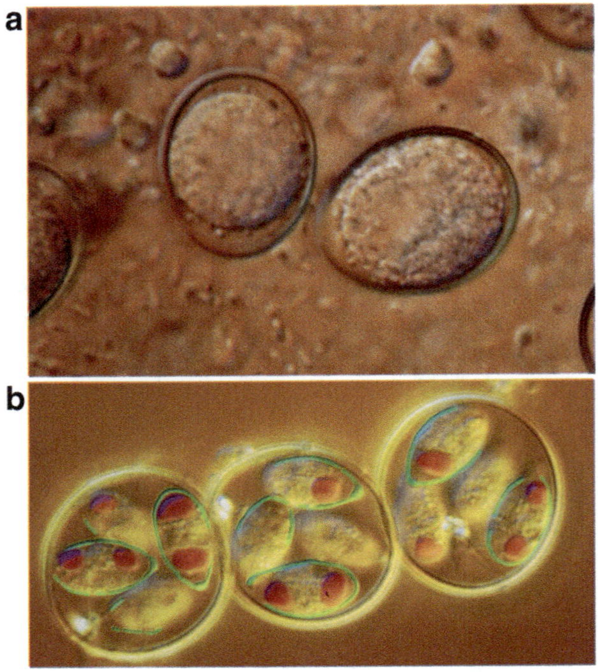

Fig. 4.38 Light micrographs of unsporulated (**a**) and sporulated (**b**) oocysts of the genus *Eimeria*. Inside the sporulated oocysts, each 4 sporocysts are visible

zygote, which excretes the substances of two wall-forming body types, which fuse at the surface and thus form the very resistant **oocyst wall**. As soon as these so-called **unsporulated oocysts** are excreted, oxygen gives the stimulus that inside the oocyst a **sporulation process** starts, which gives rise to four sporocysts each with two sporozoites. The oocyst wall is very stable and protects the inner infectious stages eventually for years even in chemical storage solutions.

Fig. 4.37 (continued) which may initiate another generation of schizonts in other intestinal cells (**2–5**) or become gamonts of different sex (**7**, **8**). (**7**) Formation of multinucleate microgamonts, which develop many flagellated microgametes (**7.1–7.2**). (**8**) Formation of uninucleate macrogamonts, which grow to be macrogametes (**8.1**) that are characterized by the occurrence of 2 types of wall-forming bodies (WF_1, WF_2). (**9**) After fertilization the young zygote forms the oocyst wall by consecutive fusion of both types of wall-forming bodies (FW). (**10**) Unsporulated oocysts are set free via feces (exceptions are reptile- and fish-parasitizing *Eimeria* spp.). (**11–13**) Sporulation (outside the host) is temperature dependent and leads to formation of 4 sporocysts, each containing 2 sporozoites (SP), which are released when the oocyst is ingested by the next host. *DG* developing microgametes; *DM* developing merozoite; *DW* developing wall-forming bodies; *FW* fusion of WF_1 to form the outer layer of OW; *M* merozoite; *N* nucleus; *NH* nucleus of host cell; *OW* oocyst wall; *PB* polar body (granule); *PV* parasitophorous vacuole; *R* refractile (=reserve) body; *SB* sporoblast; *SP* sporozoite; *SPC* sporocyst; *SPO* sporont; WF_1 wall-forming bodies I; WF_2 wall-forming bodies II; *Z* cytoplasm of zygote (=young oocyst)

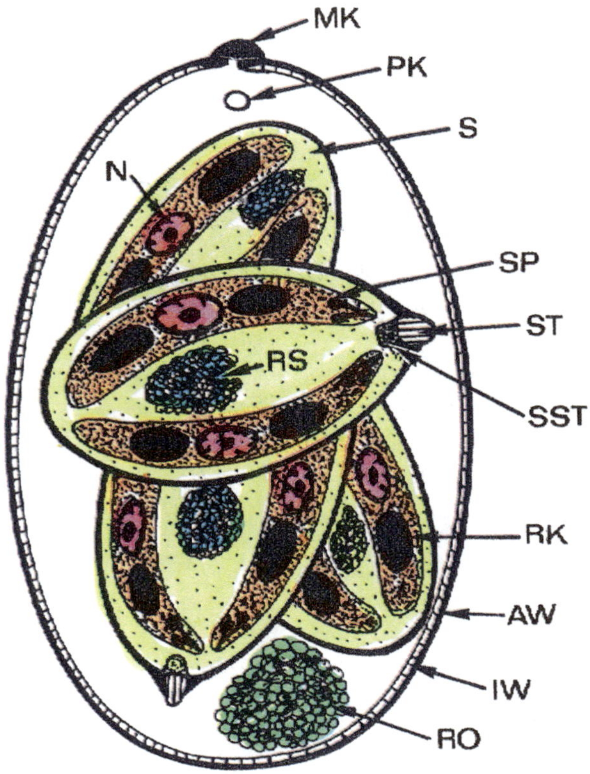

Fig. 4.39 Diagrammatic representation of an oocyst of an *Eimeria* species. *AW* Outer oocyst wall; *IW* Inner oocyst wall; *MK* Cover of micropyle; *N* Nucleus; *PK* Polar body; *RK* Refractile body; *RO* Residual body of oocyst; *RS* Residual body of sporocyst; *S* sporocyst; *SP* sporozoite; *SST* Substieda body; *ST* Stieda body

4.6.1.1 *Eimeria* Species of Ruminants

1. **Name**: The genus name honours the Swiss scientist Theodor Eimer (1843–1897). The species name takes in general relations to towns (e.g. Auburn, USA), shape (e.g. ellipsoids), names of scientists (e.g. Nina Kohl, Yakimow), hosts (*sus* = pig; *bovis* = cattle, etc.).
2. **Geographic distribution/epidemiology**: Worldwide; epidemics occur mainly in cases when animals are kept under bad health conditions or very close together. Prevalence rates may reach 80 % in some populations.
3. **Biology, morphology**: Cattle, sheep and goats harbour each at least ten different *Eimeria* species, which are morphologically and pathologically very different (Fig. 4.40). An exact species diagnosis is often very difficult without comparative material, but with respect to treatment mostly rough knowledge is sufficient. The following species have been described from the different hosts (among others) (Table 4.7):
   - **Cattle**: *E. bovis, E. auburnensins, E. zuerni, E. ellipsoidalis.*
   - ***Sheep***: *E. faurei, E. intricate, E. ovina, E. ovinoidalis.*
   - **Goat**: *E. arloingi, E. ninakohlyakomovi, E. christenseni.*

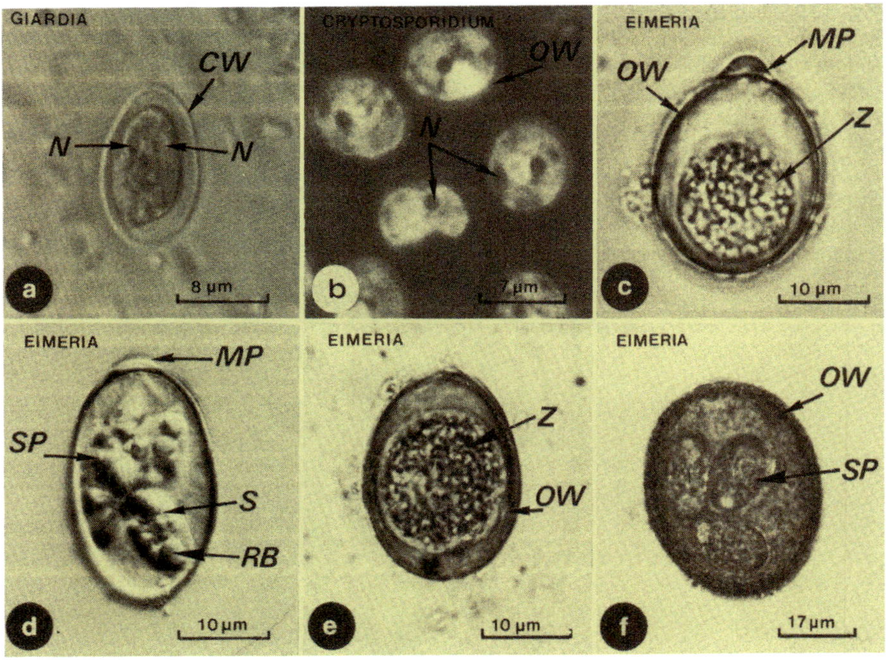

Fig. 4.40 Light micrographs of intestinal protozoans of ruminants. (**a**) Cyst of *Giardia* sp. (**b**) Oocysts of *Cryptosporidium* sp. (**c**) *Eimeria intricata*: unsporulated oocyst from sheep; (**d**) *E. ovina*: sporulated oocyst from sheep; (**e**) *E. bovis*: unsporulated oocyst from cattle; (**f**) *E. intricata*: sporulated oocyst from sheep with the thick oocyst wall. *CW* cyst wall; *MP* micropyle; *N* nucleus; *RB* residual body; *S* sporocyst; *SP* sporozoite; *Z* cytoplasm of the zygote (inside)

The following features are common:

(a) They are extremely host specific (e.g. those of cattle do not develop inside sheep) and the whole development occurs in a single host.
(b) The ovoid oocysts (with a spectrum of 17–56 µm in length) are excreted unsporulated (Fig. 4.38a).
(c) Sporulation time outside the body takes 1–4 days to produce four sporocyst containing each two sporozoites (Fig. 4.38b).
(d) The development inside the host cells occurs always inside a parasitophorous vacuole.
(e) The number of schizogonies inside a host is rather specific in each genus, while the size of the schizonts is species specific covering sizes from 15 µm up to 600 µm. Some of the big ones even got separate genus names (e.g. *Globidium*).
(f) After formation of oocysts by fusion of a macro- and a microgamete, the development is terminated inside the host and the inner differentiation of sporocysts and sporozoites starts, when the oocyst has been excreted and gets contact to oxygen.

Table 4.7 Important *Eimeria* species in important food animals for humans

Species	Host/location in tissue	Size of oocysts (μm)	Prepatency (days)	Pathogenicity
	Cattle			
E. bovis	Small intestine (posterior part)	23–34 × 17–23	15–21	+
E. auburnensis	Small intestine	36–42 × 19–26	17–20	+
E. zuernii	Small intestine	16–20 × 15–18	15–19	+
E. ellipsoidalis	Small intestine	18–26 × 13–18	8–13	+
	Sheep			+
E. faurei	Small intestine	22–33 × 19–24	12–15	+
E. intricata	Small intestine, caecum	40–56 × 30–42	20–27	+
E. ovina	Small intestine	23–36 × 16–24	19	+
E. ovinoidalis	Colon	17–25 × 13–20	10–15	
	Goats			
E. arloingi	Crypts of intestine	25–33 × 16–21	14–20	+
E. ninakohlyakimovae	Crypts of intestine	16–28 × 14–23	11–17	+
E. christenseni	Small intestine	34–41 × 23–38	14–23	+
	Pigs			
E. scabra	Small intestine	25–45 × 17–28	7–10	+
E. suis	Small intestine	13–20 × 11–15	10	+
	Horses/donkeys			
E. leuckarti	Small intestine	70–90 × 50–69	31–37	–
	Kaninchen			
E. intestinalis	Caecum, colon	23–32 × 15–20	10	+
E. perforans	Small intestine	16–28 × 12–16	4–6	+
E. magna	Small intestine	28–40 × 18–30	7–9	+
E. stiedai	Bile duct	26–40 × 16–25	12–16	+
	Rats			
E. contorta	Entire intestine	18–27 × 15–21	6	–
E. nieschulzi	Small intestine	16–26 × 13–21	7–8	+
	Mice			
E. falciparum	Caecum, colon	16–21 × 11–17	4–5	+
E. ferrisi	Caecum	17–20 × 14–16	4–5	–
	Chickens			
E. tenella	Caecum	23 × 19 (mean)	6	+
E. maxima	Small intestine	30 × 20 (mean)	5	+
E. necatrix	Small intestine	22 × 17 (mean)	6	+
E. praecox	Small intestine	21 × 17 (mean)	4	–
	Geese			
E. truncata	Kidneys	15–22 × 11–16	5	+

(continued)

Table 4.7 (continued)

Species	Host/location in tissue	Size of oocysts (μm)	Prepatency (days)	Pathogenicity
E. anseris	Small intestine	16–23 × 13–18	7	+
E. nocens	Colon	25–33 × 17–24	9	+
	Ducks			
E. danailovi	Small intestine	19–22 × 11–14	7	+
	Turkeys			
E. adenoeides	Colon, Caecum	25 × 17 (mean)	5	+
E. meleagrimitis	Small intestine	20 × 17 (mean)	5	+
	Pigeons			
E. labbeana	Small intestine	15–18 × 14–16	6	+
E. columbarum	Small intestine	19–21 × 17–20	6	−/+
	Fishes			
E. iroquoina	Intestine	8–10 × 11–14	16	+
E. columbarum	Small intestine	19–21 × 17–20	6	−/+
	Fishes (Gulf Kilifish)			
Calyptospora funduli	Liver	21 × 19	?	
	Fishes (Cyprinids)			
Goussia subepithelialis	Intestine	18–21	?	
G. carpelli	Intestine	8–14	12–19	
	Fishes (Salmonids)			
G. truttae	Intestine	10–12	?	
Epieimeria anguillae	Intestine	10–14	?	

Prepatency: Period of start of infection until moment of excretion of the first infectious stage = oocyst

+, present; −, not present

(g) Development inside host takes species-specific periods (e.g. 8–21 days) (=prepatent period).

(h) Severe and often deadly infections occur mainly in very young animals or in otherwise diseased animals.

4. **Symptoms of disease (Coccidiosis)**: The symptoms of disease are extremely severe especially in very young animals leading to heavy and often bloody diarrhoeas. Superinfections with bacteria are also common. In severe cases, additional fever, tetanic cramps and strabismus may occur besides bronchopneumonias. Death cases are common. Infections with *E. bovis* and *E. zuernii* may introduce **haemorrhagic typhlitis** and **colitis.** The species *E. alabamensis, E. auburnensis* and *E. ellipsoidalis* lead to **catarrhalic**

enteritis, while sheep may suffer from **haemorrhagic enteritis** due to infections with *E. ovoinoidalis* or **catarrhalic enteritis** due to *E. faurei*. However, survived infections lead to a species-specific, cell-induced immunity (however, the protection period is not fixed and may differ among single animals of the same species).

5. **Diagnosis**: Demonstration of the oocysts in the feces by help of flotation or S.A.F.C. and M.I.F.C. methods.
6. **Pathway of infection**: Oral by uptake of sporulated oocysts within contaminated food. **Note**: oocysts may survive for at least 1 year on meadows or in stables.
7. **Prophylaxis**: Regular, often repeated cleaning of stables from feces and use of methods to keep meadows clean. Desinfection products can be used in stables besides cleaning them with hot steam or hot water. The following anticoccidial products are registered: Lysococ®, P3-Incicoc®, Club-TGV®). If single food products are used, the addition of several chemical compounds may induce a certain prophylactic effect (e.g. addition of 10–40 mg monensin-Na/kg food for cattle or salinomycin-Na 10–30 mg/kg food of cattle, lambs and goats). **Important**: Use drugs exclusively after description of the veterinarian (see Table 4.8).
8. **Incubation period**: Species specific: 2–9 days.
9. **Prepatent period**: Species specific: 6–35 days.
10. **Patency**: Some weeks.
11. **Therapy**: Cattle, sheep and goats may be treated with sulfonamides or combinations (sulfonamide plus trimethoprim) in cases of clearly diagnosed coccidiosis. For other compounds, see Table 4.8. **Note**: Products have to be prescribed by the veterinarian and afford product-specific waiting periods after application.

Table 4.8 Anticoccidial drugs for ruminants[a]

Animal	Compound (producer)	Experimentally determined dose
Cattle	**Lasalocid**-Na (Roche)	1 mg/kg bodyweight daily for some weeks
	Monensin-Na (Elanco)	1 mg/kg bodyweight daily for 10 days
	Salinomycin-Na (Huvepharma)	2 mg/kg bodyweight daily for 3 weeks
	Meticlorpindol (Dow)	10 mg/kg bodyweight daily for 6 weeks
	DecoQuinate (May and Baker)	0.5–0.8 mg/kg bodyweight daily for 3–4 weeks
Sheep/goat	**Monensin** (Elanco)	1 mg/kg bodyweight daily for some days or 10–20 mg/kg in food
Sheep	**Meticlorpindol** (Dow)	250 mg/kg food for 65 days
Cattle/sheep	**Toltrazuril** (Bayer)	1 × 20 mg/kg bodyweight orally

[a] There are different registrations for European countries

Further Reading

Deplazes P et al (2013) Lehrbuch der Parasitologie für die Tiermedizin, 3rd edn. Enke, Stuttgart.

Kern G et al (2015) Analysis of risk factors for infections with gastrointestinal nematodes, *Eimeria* spp. and lungworms in German organic sheep farms. Berl Münch Tierärztl Wochenschr 128:233–239.

Pellerdy LP (1974) Coccidia and coccidiosis, 2nd edn. Paul Parey, Berlin.

Philippe P et al (2014) Comparative efficacy of diclazuril (Vecoxan®) and toltrazuril (Baycox bovis®) against natural infections of *Eimeria bovis* and *Eimeria zuernii* in French calves. Vet Parasitol 206:129–137.

Souza LE et al (2015) Epidemiology of *Eimeria* infections in sheep raised extensively in a semiarid region of Brazil. Rev Bras Parasitol Vet 24:410–415.

Sudhakara Reddy B et al (2015) Clinical coccidiosis in adult cattle. J Parasit Dis 39:557–559.

Yang R et al (2014) Longitudinal prevalence, oocyst shedding and molecular characterisation of *Eimeria* species in sheep across four states in Australia. Exp Parasitol 145:14–21.

4.6.1.2 *Eimeria* and *Isospora* Species Infecting Pigs

1. **Name**: Greek: *isos* = identical, similar; *sporos* = seed, spore. Latin: *scabra* = rough; *sus* = pig. Eimer = family name of Theodor Eimer (1843–1897), Professor in Tübingen, Germany.
2. **Geographic distribution/epidemiology**: Worldwide.
3. **Biology, morphology**: Pigs are infected by a series of *Eimeria* and *Isospora* species (Table 4.7), which can only diagnosed species specifically by specialists. However, exact diagnosis is not needed since control can be done by the available compounds. Some examples are presented here.
   (a) ***Eimeria scabra***

   Sporulated oocysts measure 25–45 × 17–28 μm; they have a rough, thick, brownish wall and contain a terminal polar body but no micropyle. The three generations and the male and female gametes are developed in vacuoles of the host's epithelial cells of the jejunum and ileum. Sporulated oocysts contain four sporocysts (with a Stieda body) containing each two sporozoites.

   (b) ***Eimeria suis***

   This also very pathogenic species has smaller oocysts measuring 13–20 × 11–15 μm.

   (c) ***Isospora suis***

   This species parasitizes in the small intestine. Its sporulated oocysts measure 17–22 × 17–19 μm, and it has a short prepatent period of only 5 days. The oocyst wall is smooth and colourless and does not contain a micropyle or a polar body. Sporulated oocysts contain two sporocysts without Stieda body containing each four sporozoites. Infection rates of piglets may reach 77 %.

4. **Symptoms of disease (Coccidiosis)**: Reduced growth and loss of weight. In cases of heavy infections, strong, intense, intermittent diarrhoeas occur, which appear watery, yellowish and foamy.
5. **Diagnosis**: Sporulation of oocysts occurs after 1–2 days in the case of *I. suis* and after 9–11 days in *E. scabra* after storage in 2.5 % $K_2Cr_2O_2$ = potassium bichromate.
6. **Pathway of infection**: Oral by uptake of sporulated oocysts.
7. **Prophylaxis**: Regular and intensive cleaning the stable by hot steam of disinfection products such as Lysococ®, P3-Incicoc®, Chevi75 or Club-TGV® anticoc.
8. **Incubation period**: 2–3 days
9. **Prepatent period**: *E. scabra*: 8–9 days; *I. suis*: 5–6 days.
10. **Patency**: About 1 week.
11. **Therapy**: Toltrazuril (20 mg/kg bodyweight).

Further Reading

Karamon J et al (2007) Prevalence of *Isospora suis* and *Eimeria* spp. in suckling piglets and sows in Poland. Vet Parasitol 147:171–175.

Koudela B et al (1990) Concurrent infection of enterocytes with *Eimeria scabra* and other enteropathogens in swine. Vet Parasitol 35:71–77.

Ruttkowski B et al (2001) PCR-based differentiation of three porcine *Eimeria* species and *Isospora suis*. Vet Parasitol 95:17–23.

Vitovec J et al (1987) Pathology and pathogenicity of *Eimeria scabra* (Henry, 1931) in experimentally infected pigs. Folia Parasitol (Praha) 34: 299–304.

4.6.1.3 Coccidia of Perissodactyla (*E. leuckarti*)

1. **Name**: Theodor Eimer (1843–1897), Friedrich Leuckart (1794–1843) and Rudolf Leuckart (1822–1898) were important German researchers in the field of parasites.
2. **Geographic distribution/epidemiology**: Worldwide.
3. **Biology, morphology**: Oocysts of the sporozoan parasite *Eimeria* (syn. *Globidium*) *leuckarti* are rather large reaching a size of 70–80 × 50–69 μm (Fig. 4.41a, b). They are characterized by an extreme thick outer cover and a colourless smooth inner one. The gamagony is proceeded intracellularly in cells of the lamina propria of the small intestine. The number of schizonts seems rather low, since they are found scarcely distributed in cells of the ileum measuring about 12.5 μm in diameter. The sporulation of the oocysts occurs very slowly outside of the body. It takes about 21 days at 25 °C and needs even 42 days at 15 °C.
4. **Symptoms of disease (Coccidiosis)**: In general, infections show low or even no symptoms of diseases.
5. **Diagnosis**: Demonstration of the typical oocysts in the feces with the help of the sedimentation method (Fig. 4.41). When using the flotation method, it is needed to increase the centrifugation period to 7–8 min, since these large oocysts have a very high specific weight.

a b c

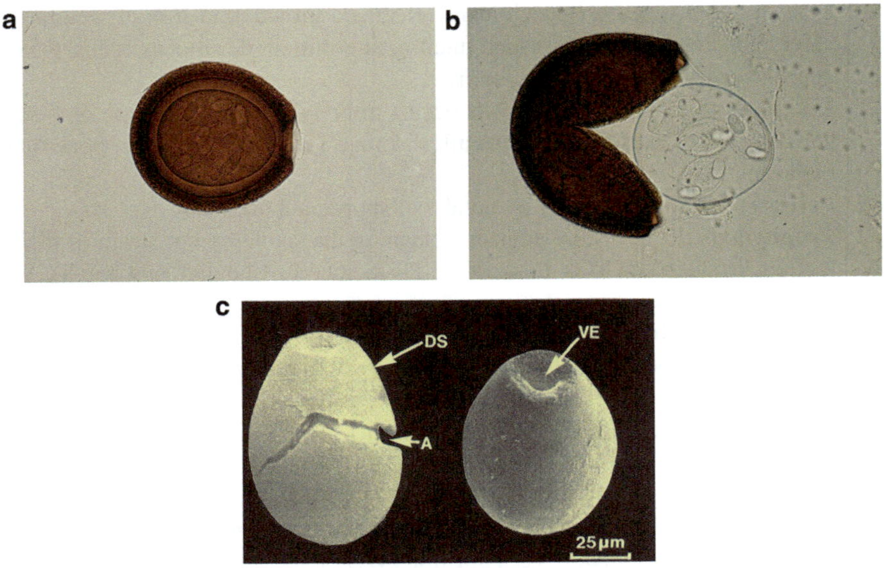

Fig. 4.41 Light micrographs of an *Eimeria leuckarti*. Unsporulated (**a**) and sporulated (**b**) oocyst, where the outer layer has been broken down at places. (**c**, **d**) Scanning electron micrographs of oocysts. *A* disrupted wall; *DS* thick wall of the oocyst; *VE* dorsal depression

6. **Pathway of infection**: Oral by uptake of sporulated oocysts.
7. **Prophylaxis**: Repeated cleaning of the stables with the help of hot steam or hot water and eventually by the use of disinfection compounds such as Lysococ®, P3-Incicoc®, Chevi75 or Club-TGV® anticoc.
8. **Incubation period**: 2–3 days (obtained in experimental infections).
9. **Prepatent period**: 31–37 days.
10. **Patency**: 3–18 days.
11. **Therapy**: Toltrazuril would solve the problem, but treatment is mostly not needed due to the inapparent symptoms of disease.

Further Reading

De Souza PN et al (2009) Natural infection by *Cryptosporidium* sp., *Giardia* sp. and *Eimeria leuckarti* in three groups of equines with different handlings in Rio de Janeiro, Brazil. Vet Parasitol 160:327–333.

Gülegen E et al (2016) *Eimeria leuckarti* infections in sport horses in Northwest Turkey. Berl Münch Tierärztl Wochenschr 129:52–55.

Lyons ET, Tolliver SC (2004) Prevalence of parasite eggs (*Strongyloides westeri*, *Parascaris equorum*, and strongyles) and oocysts (*Eimeria leuckarti*) in the feces of thoroughbred foals on 14 farms in central Kentucky in 2003. Parasitol Res 92:400–404.

4.6.1.4 *Eimeria* Species of Poultry

1. **Name**: The genus name honours the Swiss researcher Theodor Eimer (1843–1897), who worked in Tübingen (Germany).
2. **Geographic distribution/epidemiology**: Worldwide; especially common in large breeding and egg production units.
3. **Biology, morphology**: Among the economically used birds, infections by *Eimeria* are most important, since they induce yearly many millions of losses in Euro and US Dollars. Infections of birds with *Isospora* species are especially important among imported leisure bird species, where high death rates occur year by year due to infection with *Isospora* and *Eimeria* species. The latter (Table 4.7) are not only very host specific (e.g. the species of chickens cannot be transmitted to ducks or geese), but develop exclusively in cells of peculiar intestinal regions (Table 4.7). The following bird groups are parasitized by the following selected *Eimeria* species:

   (1) **Chickens**

   Up to nine different species are described. Six of them are very pathogenic: *E. acervolina* and *E. mitis* are found in the anterior region of the small intestine. *E. necatrix* and *E. maxima* settle in cells of the mid region of the small intestine, while *E. brunetti* is found in the ileum and rectum and *E. tenella* in the caeca. *E. mivati* and *E. hagani* are less pathogenic (infected hosts show reduced mortality). *E. praecox* is claimed even to be apathogenic. *E. mivati* is under discussion to be a variation of *E. acervulina*. The predominant sites of infection of the above-listed species are shown in Figs. 4.42, 4.43 and 4.44.

   (2) **Geese**

   Pathogenic species are *E. anseris* (small intestine and colon), *E. truncata* (in cells of the kidneys) and *E. kotlani* (in colon and cloaca cells).

   (3) **Ducks**

   E. kotlani, E. danailova and *Tyzzeria perniciosa* are all pathogenic parasitizing in the small intestine, but they occur rather seldom.

   (4) **Turkeys**

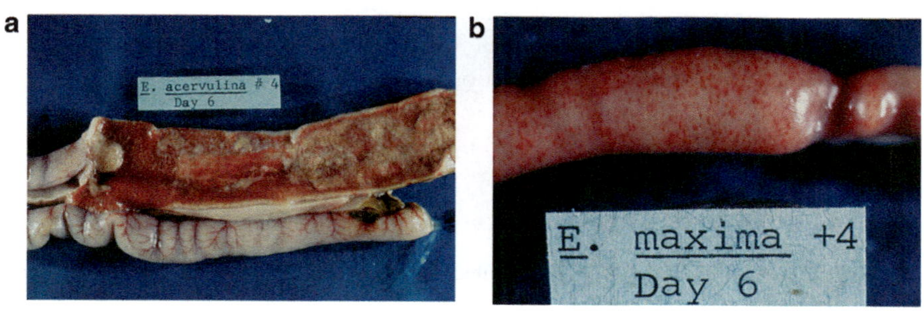

Fig. 4.42 Macrophotos of intestines of chickens parasitized by *Eimeria acervulina* (**a**) and *E. maxima* (**b**)

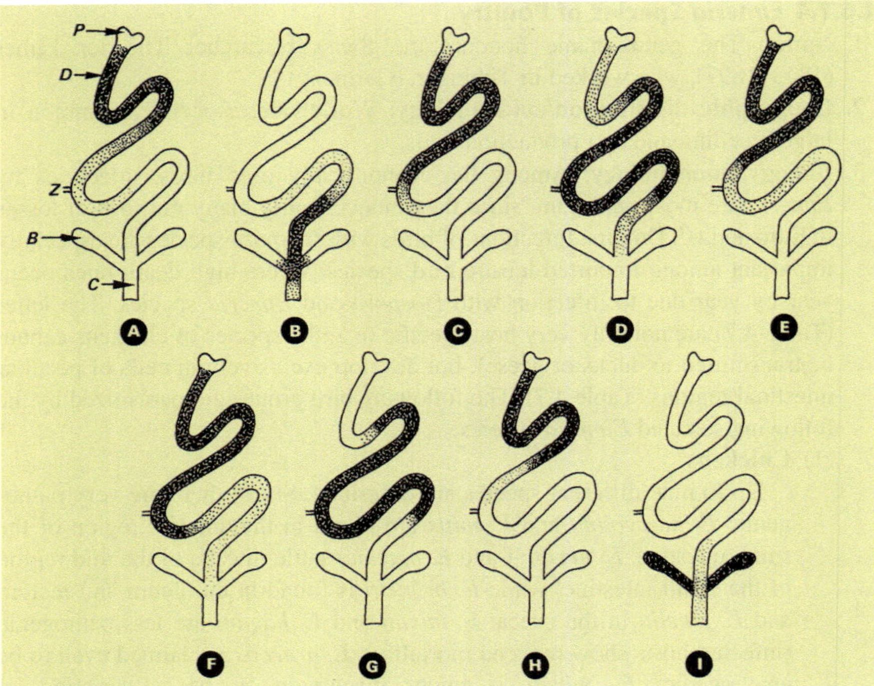

Fig. 4.43 Diagrammatic representation of the infection sites of the different *Eimeria* species of the domestic chickens according to various authors. (**a**) *E. acervulina*; (**b**) *E. brunetti*; (**c**) *E. hagani*; (**d**) *E. maxima*; (**e**) *E. mivati*; (**f**) *E. mitis*; (**g**) *E. necatrix*; (**h**) *E. praecox*; (**i**) *E. tenella*. *B* caecum; *C* colon; *D* duodenum; *P* pylorus; *Z* jejunum (shortened)

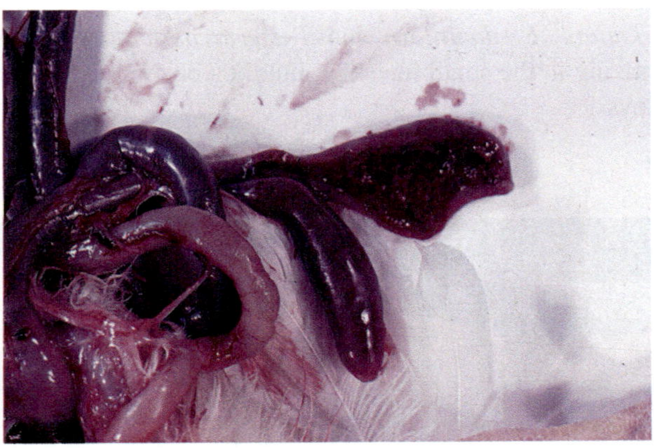

Fig. 4.44 Macrophoto of *Eimeria tenella*-infected caeca of a domestic chicken showing large haemorrhagies

The species *E. meleagrimitis* and *E. dispersa* (both in the small intestine) and *E. adenoeides* (ileum, caeca and colon) are of great importance if the birds are kept in masses.

(5) **Pigeons**

E. labbeana and *E. columbarum* are both pathogenic settling in epithelial cells of the small intestine.

(6) **Psittaciformes**

The species *E. dunsigni* parasitizes in the small intestine reaches a considerable high degree of pathogenicity but is not very common.

The specimens of the genera *Eimeria* and *Isospora* produce after a species-specific number of schizogonies and the phase of gamogony always unsporulated oocysts (Figs. 4.37 and 4.38a). Four sporocysts are formed (each containing two sporozoites) in the case of *Eimeria* but two sporocysts each with four sporozoites in the case of *Isospora* species (compare Fig. 4.46b). The size of the oocysts ranges (except for some species) in dimensions of 20–25 × 15–20 μm, and thus, higher microscopical magnifications are needed for exact species determination.

4. **Symptoms of disease (Coccidiosis)**: Acute coccidiosis is characterized by severe diarrhoeas based on several types of enteritis (catarrhalic up to haemorrhagic ones). The feces are a watery fluid, mixed with flakes of slime, contain blood and appear often greenish. The destruction of the epithelia and/or other organs induces considerable pathophysiological disturbances (e.g. increase of the acidity of the intestinal contents; loss of plasma proteins, blood and vitamins; loss of carbohydrates; dysfunction of the kidneys; hypothermia before death). In addition, secondary bacterial infections many increase these effects. Although all stages of birds can be infected, especially young and old animals are severely endangered, since their immunity status is only poorly developed. As a consequence of the enteritis, the total fitness is severely decreased, which leads to reduction of food uptake and loss of weight. In cases of haemorrhagic enteritis (Figs. 4.42, 4.43 and 4.44), the mortality rate is considerably increased. Reinfections are common and occur together with spreading of other agents of disease. Especially virus and bacterial infections are common.
5. **Diagnosis**: Coccidiosis is characterized by the occurrence of masses of unsporulated oocysts in the feces (Figs. 4.38, 4.39 and 4.40), which can be diagnosed microscopically with the help of flotation or by S.A.F. concentration methods. In cases of bloody and fluid feces, the use of concentration methods is mostly not needed, since then masses of oocysts occur. The exact species diagnosis can be obtained by special laboratories, but is mostly not needed, even if several parasitic species of parasites are obviously present, since the available medication covers practically all species. However, subclinical courses of coccidian infections might not be noted especially in cases of multifactorial diseases. Therefore, it is recommended—especially in cases of masses of animals—to proceed feces investigations at regular intervals.

6. **Pathway of infection**: Oral by uptake of sporulated oocysts (Figs. 4.38, 4.39 and 4.40) within food or drinking water. **Attention**: Sporulated oocysts are very resistant and remain infectious for at least 1 year!
7. **Prophylaxis**:
   (a) **Hygienic measurements**
   In the case of the present rearing conditions of birds, it is very difficult to avoid coccidial infections. In order to reduce the infection pressure, it is needed to discharge fecally contaminated soil or other floor material at regular intervals and to use disinfection products in empty stables such as P3-Incicoc®, Lysococ®, Chevi75, Club-TGV® or similar ones. Equipment and stables should also be cleaned by hot steam or hot water. This total cleaning can only be done, if the stables are used according to the principle: all in-all out, which means that the stables are completely empty before new birds are installed. **Important**: Also flies should be kept away from stables, since they are able to transmit coccidian oocysts as well as worm eggs.
   (b) **Chemoprophylaxis**
   Mass production of healthy birds for human consumption is only possible using regular treatment with coccidostatic chemical compounds within food (Table 4.9). However, it is needed that very young animals are able to develop a relevant immunity status, which can be obtained with the help of application of relevant chemical compounds. This protective immunity is developed since chemotherapy never kills all parasites especially before the background that daily slight reinfections occur. This fact makes it clear that from time to time the stables must be thoroughfully cleaned; otherwise the oocysts will accumulate on the floors.
   The efficacy of prophylactic treatments may become reduced by simultaneous MAREK infections or other infections. Thus, the goal of a chemoprophylaxis against coccidian infections can only be reached by the fact that the number of oocysts close to the birds is kept as low as possible, which affords the rotation of the pharmaceutical products used in a stable.
   (c) **Resistances against anticoccidial medicaments**
   Resistance is a growing problem in bird rearing systems. Application of polyether antibiotics apparently slows down the development of resistances against coccidial infections, since their targets are others than those of a typical coccidiocidal drugs. However, the long-lasting use has also allowed several resistances.
   (d) **Immunization**
   Immunization of birds against coccidian species is in principle possible by artificial infections with low doses of coccidians and simultaneous treatment with coccidiocidal compounds when given via drinking water. However, it is not clear how long such a so-called **premunity** is retained. Thus infections may break out, when the low dosage treatment has been stopped.
   Besides the non-attenuated alive vaccination product (Coccivac®) exist attenuated vaccines, which contain a species cocktail of oocysts of

Table 4.9 Food additions in bird food to avoid/suppress coccidiosis

Chemical term	Trade name	Producer/Distributor	Animal (max. age for treatment) (Behandlungshöchstalter)	Dosis ppm (waiting time)
METICLORPINDOL	Coyden®	Dow	Chickens; Guinea fowls	125 (5 days)
DECOQUINAT	Deccox®	Rhône-Poulenc (May u. Baker)	Chickens	20–40 (3 days)
MONENSIN-NATRIUM	Elancoban® (Coban®)	Elanco (Eli Lilly)	Chickens Young hens (16 weeks) Turkeys (16 weeks)	100–125 (3 days) 100–120 90–100 (3 days)
ROBENIDIN	Cycostat®	Cyanamid	Chickens, turkeys	30–36 (5 days)
LASALOCID-NATRIUM	Avatec®	Roche	Chickens, Young hens (16 weeks) Turkeys (12 weeks)	75–125 (5 days) 90–125 (5 days)
NARASIN	Monteban®	Elanco (Eli Lilly)	Chickens	60–70 (5 days)
SALINOMYCIN-NATRIUM	Sacox®	Huvepharma	Chickens	50–70 (5 days)
NICARBAZIN	Altek® 25	MSD	Chickens (4 weeks)	100–125 (9 days)
MADURAMICIN AMMONIUM	Cygro®	Cyanamid	Chickens	5 (7 days)
NARASIN/NICARBAZIN granulates (1:1)	Maxiban®	Elanco (Eli Lilly)	Chickens	80–100 (7 days)
DICLAZURIL	Clinacox®	Janssen	Chickens	1 (5 days)

Eimeria species of chickens (Paracox®, Livacox®). Their virulence is rather low, so that no real outbreaks of coccidiosis occur, but a relevant immunization is developed. However, such biotechnological vaccines are not free from risks.

(e) **Coccidiosis of doves and canary birds**

This disease is best controlled by daily cleaning of the floors and removal of feces, since the oocysts may survive there for months. Chemoprophylaxis is possible by application of the products via drinking water. **Attention**: Prophylactically used products containing sulfonamides or trimethoprim may lead to feather damages.

8. **Incubation period**: 3–5 days; however, low-grade infections do not show symptoms and might not be noted. Thus, repeated feces control is recommended from time to time.
9. **Prepatent period**: Species specific, but mostly 4–7 days.

Chickens: 4 days: *Eimeria praecox, E. acervulina, E. mitis, E. mivati*
5 days: *E. maxima, E. brunetti, E. tenella*
6 days: *E. necatrix*

Turkeys: 4 days: *E. dispersa*
5 days: *E. meleagrimitis, E. adenoeides*

Geese: 5 days: *E. truncata*
7 days: *E. anseris*
9 days: *E. nocens*

Ducks: 6 days: *Tyzzeria perniciosa*
8 days: *E. danilovi*

Doves: 6 days: *E. labbeana*
Canaries: 5 days: *Isospora canaria*
9 days: *I. serrini*

10. **Patency**: The acute patent period is mostly rather short, but reinfections lead to long-lasting infections. *Isospora* species may have patencies of up to 250 days.
11. **Therapy**: Intense causal treatment should only be done, if mortality rates are high and all birds show reduced growth and weight gain. One example is Baycox®, which is applied via drinking water covering the dose of 15–20 mg/kg body weight. Other compounds are listed in Table 4.9.

Further Reading

Chen RA et al (2016) Observation of the excretion pattern of a precocious line of *Eimeria necatrix* and the strengthening of immune homogeneity. Exp Parasitol 164:79–83.

Hoan TD et al (2016) Identification and immunogenicity of microneme protein 2 (EbMIC2) of *Eimeria brunetti*. Exp Parasitol 162:7–17.
Liu L et al (2016) Ultrastructural effects of acetamizuril on endogenous phases of *Eimeria tenella*. Parasitol Res 115:1245–1252.
Matsubayashi M et al (2016) Transcriptional profiles of virulent and precocious strains of *Eimeria tenella* at sporozoite stage; novel biological insight into attenuated asexual development. Infect Genet Evol 40:54–62.
Mehlhorn H (1972) Electron microscopy studies on the developmental stages of *Eimeria maxima* (Sporozoa, Coccidia). I. Ultrastructure of macrogametes. Parasitol Res 39:161–182.
Mehlhorn H (1972) Ultrastructural study of developmental stages of *Eimeria maxima* (Sporozoa, Coccidia). II. Fine structure of microgametes. Z Parasitenkd 40:151–163.
Mehlhorn H (1972) Electron microscopy studies on the developmental stages of *Eimeria maxima* from the domestic chicken. 3. Differentiation of microgametes with special reference to nuclear divisions. Parasitol Res 40:243–260.
Mehlhorn H et al (1984) The effects of sym. Triazinones on developmental stages of *Eimeria tenella*, *E. maxima* and *E. acervulina*: a light and electron microscopical study. Parasitol Res 70:173–182.
Moraes JC et al (2015) Prevalence of *Eimeria* spp. in broilers by multiplex PCR in the Southern region of Brazil on two hundred and fifty farms. Avian Dis 59:277–281.
Ogedengbe ME et al (2013) Sequencing the complete mitochondrial genome of *Eimeria mitis* strain USDA 50 (Apicomplexa: Eimeriidae) suggests conserved start positions for mtCOI- and mtCOIII-coding regions. Parasitol Res 112:4129–4136.
Pellerdy LP (1974) Coccidia and coccidiosis, 2nd edn. Paul Parey, Berlin.
Pop L et al (2015) Effects of artemisinin in broiler chickens challenged with *Eimeria acervulina*, *E. maxima* and *E. tenella* in battery trials. Vet Parasitol 214:264–271.
Répérant JM et al (2012) Pathogenicity of *Eimeria praecox* alone or associated with *Eimeria acervulina* in experimentally infected broiler chickens. Vet Parasitol 187:333–336.
Rochell SJ et al (2016) Effects of *Eimeria acervulina* infection severity on growth performance, apparent ileal amino acid digestibility, and plasma concentrations of amino acids, carotenoids, and α1-acid glycoprotein in broilers. Poult Sci. pii: pew035.

4.6.1.5 *Eimeria* Species of Rabbits

1. **Name**: The genus name honours the Swiss scientist Theodor Eimer (1843–1897) who worked in Tübingen, Germany. Species names describe features or locations of the species (e.g. Latin *perforans* = perforating; *intestinalis* = belonging to the intestine). Other species names honour scientists, e.g. Christian Stieda (1837–1918).
2. **Geographic distribution/epidemiology**: Worldwide.

3. **Biology, morphology**: In the case of animals belonging to the Lagomorpha and rodents, many very specific *Eimeria* species are described, which very often parasitize in the cells of the small intestine and which are very often highly pathogenic, so that the infections bring about severe economic losses in farms. As in the case of ruminants, infected rabbits excrete unsporulated oocysts, which reach a species-specific size of up to 40 µm in length. They develop within 2–7 days four sporocysts, each with two sporozoites (Fig. 4.45). **Rabbits** are parasitized by *Eimeria intestinalis, E. perforans* and *E. magna,* while *E. contorta* and *E. nieschulzi* are common in **rats** or *E. falciformis* and *E. ferrisi* in laboratory mice (*Mus musculus*).
4. **Symptoms of disease**:
   (a) The **intestinal coccidiosis** of rabbits leads to severe diarrhoeas accompanied by anaemia and weakness. Complications and even death may be introduced especially in cases of additional secondary bacterial infections.
   (b) The **liver coccidiosis** of rabbits due to infections with oocysts of *E. stiedai* (syn. *E. stiedae*) with considerable liver swellings leads to dysfunctions and death of the infected animal. When looking at histological sections, the liver shows yellowish whitish knots, which represent connective tissues replacing destroyed regions. The knots are even visible at the surface of the liver. The mortality rates are high especially when young animals are infected by several *Eimeria* species.
5. **Diagnosis**: Demonstration of the unsporulated oocysts in fresh feces with the help of fecal concentration method (Fig. 4.45).
6. **Pathway of infection**: Oral uptake of sporulated oocysts within contaminated food or drinking water.

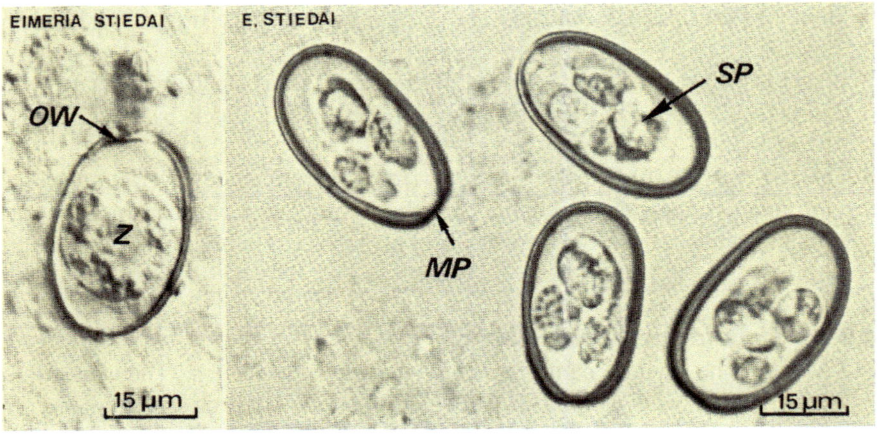

Fig. 4.45 Light micrographs of unsporulated and sporulated oocysts of *Eimeria stiedai* (*stiedae*) from rabbits. *Left*: unsporulated; *right*: sporulated ones. *MP* micropore; *OW* oocyst wall; *SP* sporocyst

7. **Prophylaxis**: Repeated removal of feces, cleaning floors with hot steam or hot water or addition of 10–15 ppm anticoccidia into the drinking water to decrease the infection pressure.
8. **Incubation period**: A few days. However, if only a few oocysts are ingested, the symptoms may be low grade or even hidden.
9. **Prepatent period**: Species specific: e.g. *E. intestinalis* = 9–11 days; *E. stiedai* = 12–14 days.
10. **Patency**: A few weeks (2–5) with increased oocyst shedding within the first 2 weeks.
11. **Therapy**: Toltrazuril (Baycox®): 25 ppm within drinking water for 2 days. Repetition after 5 days.

Further Reading

Al-Quraishy S (2012) Exogenous and endogenous stages of *Eimeria perforans* naturally infected domestic rabbit (*Oryctolagus cuniculus*) in Saudi Arabia: light microscopic study. Saudi J Biol Sci 19:31–34.

Ball SJ et al (2014) Observations on oocyst development of *Eimeria stiedai* in rabbits. Acta Parasitol 59:544–547.

Chen A et al (2014) Development of *Eimeria nieschulzi* (Coccidia, Apicomplexa) gamonts and oocysts in primary fetal rat cells. J Parasitol Res. 2013:Article ID 591520.

Heitlinger E et al (2014) The genome of *Eimeria falciformis*—reduction and specialization in a single host apicomplexan parasite. BMC Genomics 15:696.

Jing F et al (2012) Large-scale survey of the prevalence of *Eimeria* infections in domestic rabbits in China. Parasitol Res 110:1495–1500.

Pellerdy L (1972) Coccidia and coccidiosis, 2nd edn. Paul Parey, Berlin.

Tian SQ et al (2012) The complete mitochondrial genome sequence of *Eimeria magna* (Apicomplexa: Coccidia). Mitochondrial DNA 26:714–715.

Yan W et al (2013) Simultaneous identification of three highly pathogenic *Eimeria* species in rabbits using a multiplex PCR diagnostic assay based on ITS1-5.8S rRNA-ITS2 fragments. Vet Parasitol 193:284–288.

4.6.1.6 Coccidian Species of Fishes

1. **Name**: Most species names are related to the position of the parasites inside their hosts or give hints on their preferred hosts.
2. **Geographic distribution/epidemiology**: Worldwide, especially in ponds with huge numbers of cultivated fishes. Death rates may be enormous.
3. **Biology, morphology**: Coccidia belonging to several genera may be common parasites in fishes kept in artificial culture ponds and introduce severe economic losses. These parasites enter—species dependent—the epithelial cells of the intestine, kidneys, gonads, gills and/or swim bladder. In former times, all species had been summarized within the genus *Eimeria*. However, this changed during the last 20 years and today at least five important genera are known, where all sporulated oocysts possess four sporocysts each with two sporozoites.

(a) **Genus *Eimeria***:

Sporocysts are characterized by a single **stieda body**, which closes the sporocyst wall (Fig. 4.39).

(b) **Genus *Epieimeria***:

Sporocysts contain a stieda and a substieda body.

(c) **Genus *Goussia***:

Sporocysts do not possess a stieda body, but the sporocyst shell appears with two valves being separated by a longitudinal suture.

(d) **Genus *Crystallospora***:

The surface of the sporocyst shows 12 areas (=dodecaedric); the wall has no stieda body but possesses an equatorial suture.

(e) **Genus *Calyptospora***:

The sporocysts possess protrusions (=sporopodes) and the development is heteroxenous.

The infection of the hosts starts by the oral uptake of sporulated oocysts (=they contain four sporocysts each with two sporozoites) (Fig. 4.39). Inside the intestine, the oocyst wall is disrupted and the sporozoites hatch from the sporocysts. Depending on the species, the sporozoites enter specific host cells and start there the typical schizogony process (compare Fig. 4.37). However, in contrast to chicken Eimerians, the "fish Eimerians" develop their sporocysts and sporozoites already inside the host's body and thus they are excreted fully sporulated. With exception of the species of the genus *Calyptospora,* the life cycle of the fish Eimerian species belongs to the **monoxenous type** = only a single host type is infected, while some *Calyptospora* species have obligatorily two different hosts. The three phases (schizogony, gamogony and sporogony) of the life cycle of these *Calyptospora* species run in the tissues (liver/pancreas) of the final hosts (fish), while inside small crustaceans the ingested oocysts are either just stored or induce a schizogony leading to the production of merozoites, which are infectious for the final hosts when ingesting such infected crabs, etc. These species are defined as **intermediate hosts**. In contrast to the coccidians of chickens, the host specifity of fish coccidians is doubtful due to the fact that experimental infections are rare. With respect to a successful therapy, an exact species determination is mostly not needed.

Important species are:

- *Goussia subepithelialis*, which produces yellowish knots in the intestinal wall of carps reaching diameters of 2–4 mm. The fish show in general a slow growth. Oocysts measure 18–21 μm.
- *Goussia carpelli*, which lives in the small intestine of many carp species and reaches often a parasitaemia of 100 % in young animals. The spherical oocysts measure 8–14 μm in diameter. Their sporocysts contain a large residual body.
- *Goussia truttae*: This species is found in many salmonid fish species. The spherical oocysts are pale and measure 10–12 μm in diameter. The sporocysts do not possess a residual body.

- *Goussia clupearum*: This species is found in the liver and gonads of sardines and herrings.
- *Calyptospora* (syn. *Eimeria*) *funduli*: This species parasitizes in hepatocytes and cells of the ovary, intestine and swim bladder of *Fundulus* species. Intermediate hosts are *Palaeomonetes* and *Macrobrachium* species.
- *Eimeria anguillae:* This species parasitize inside the intestine of eels.
- *Eimeria acerinae*: This species is found inside the intestine of *Acerina* species.
- *Eimeria amurensis*: This species lives in cells of liver, kidney and gonads of *Pseudorasbora* species.
- *Eimeria ambassi*: This species induces abscesses in the muscles of barbe fishes.
- *Epieimeria anguillae*: This species occurs inside the intestine of eels.
- *Crystallospora cristalloides*: This species is found inside the intestine close to the pylorus of *Mortella* fishes.

In addition to the above-listed Eimeriid species, stages of the genus *Cryptosporidium* occur, reaching as cysts a diameter of about 5 μm. As in other hosts these specimens develop attached at the microvilli of the intestinal cells. It is not clear whether they can be transmitted to humans, which is the case in several *Cryptosporidium* species found in domestic/farm animals.

4. **Symptoms of disease (Coccidiosis)**: Several coccidians of fish introduce severe intestinal inflammations leading to yellowish-slimy feces. Fish lose weight, and their motility is reduced. The eyes are sunken in hollows and infected fishes are less resistant to secondary bacterial or fungal infections. Thus, death cases may be common especially in overcrowded ponds. Sections show necrosis in many regions of the intestine and adjacent organs.
5. **Diagnosis**: Documentation of oocysts in feces or in preparations of mucous layers.
6. **Pathway of infection**: Oral uptake of oocysts excreted by infected fish in ponds or aquaria.
7. **Prophylaxis**: In cases of ponds: young fish should be kept in separate ponds. Eventual treatment of empty ponds before introduction of water and fish. Do not enter small crustaceans in ponds or aquaria, since they may contain specimens of the genus *Calyptospora*.
8. **Incubation period**: Species specific: 2–14 days.
9. **Prepatent period**: Species-specific temperature dependent: 5–37 days or even months in moderate or cold climate.
10. **Patency**: A few weeks up to several months.
11. **Therapy**: In the case of rearing food fish, the use of coccidiostatica might be tried (Table 4.9). However, efficacy is not sure. Empty ponds can be treated with 1 kg/m^2 of calcium-nitrogen after drying and removal of mud. The chemical compounds should stay there for 8 days. The first filling of the pond has to be removed! **Attention**: chalk nitrogen is poisonous and thus use mouth-nose masks during work. In the case of fish in aquaria, the product Protazol®

produced by Fa. Alpha-Biocare GmbH (Düsseldorf) and marketed by Fa. Sera (Heinsberg, Germany) acts safely against protozoan parasites.

Further Reading

Ahmed MI et al (2010) Tissue responses of *Clarias gariepinus* (Catfish) to experimental *Eimeria* subepithelialis infection. Vet Parasitol 171:181–184.

Belova LM, Krylov MV (2000) Distribution of coccidians (Sporozoa: Coccidiida) in various systematic groups of fishes. Parazitologiia 34:522–533.

Dogga SK et al (2015) Phylogeny, morphology and metabolic and invasive capabilities of epicellular fish coccidium *Goussia janae*. Protist 166:659–676.

González-Kother P, González MT (2014) The first report of liver coccidian *Goussia cruciata* in jack mackerel, *Trachurus murphyi*, from the South Pacific and its relationship with host variables. Parasitol Res 113:3903–3907.

Lovy J, Friend SE (2015) Intestinal coccidiosis of anadromous and landlocked alewives, *Alosa pseudoharengus*, caused by *Goussia ameliae* n. sp. and *G. alosii* n. sp. (Apicomplexa: Eimeriidae). Int J Parasitol Parasit Wildl 4:159–170.

Molnar K, Baska F (1986) Light and electron microscopic studies on *Epieimeria anguillae* (Léger & Hollande, 1922), a coccidium parasitizing the European eel, *Anguilla anguilla* L. J Fish Dis 9:99–110.

Molnár K et al (2012) *Eimeria* that infect fish are diverse and are related to, but distinct from, those that infect terrestrial vertebrates. Infect Genet Evol 12:1810–1815.

Rosenthal BM et al (2016) Coccidian parasites of fish encompass profound phylogenetic diversity and gave rise to each of the major parasitic groups in terrestrial vertebrates. Infect Genet Evol 40:219–227.

Whipps CM et al (2012) Phylogeny of fish-infecting *Calyptospora* species (Apicomplexa: Eimeriorina). Parasitol Res 111:1331–1342.

4.6.2 *Isospora* and *Cystoisospora* Species

The species belonging to these genera are characterized by oocysts, which develop inside their oocyst wall two sporocysts each containing four sporozoites (Fig. 4.46b) after they had been excreted unsporulated within the feces (Fig. 4.46a; Table 4.10).

The life cycle contains in principle the same stages (schizonts, merozoites, gamonts, gametes and oocysts) as the members of the genus *Eimeria* (Fig. 4.46). In the case of the members of the genus *Cystoisospora*, a second **paratenic host** (=transportation host) is included in the life cycle (Fig. 4.47). These additional hosts (e.g. cattle, rodents, etc.) are infected by oral uptake of infectious (=fully sporulated) oocysts (Fig. 4.46b) from the feces of the regular final hosts. Then the sporozoites hatch from the sporocysts inside the intestine, pass the intestinal wall and enter, e.g., muscle cells of this host and survive therein for long inside a parasitophorous vacuole until the final host (e.g. a dog) ingests this piece of meat. Inside this final host, these "stored" sporozoites enter the intestinal cells and start

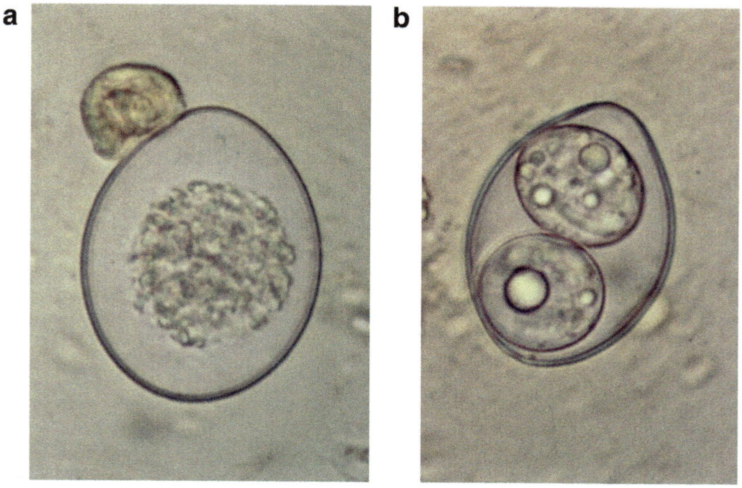

Fig. 4.46 Light micrographs of an unsporulated oocyst (**a**) and a sporulated oocyst (**b**) of *Cystoisospora felis*

Table 4.10 Important *Cystoisospora* and *Isospora* species

Species	Host/Site	Size of oocysts (μm)	Prepatent period (days)	Pathogenicity
Cystoisospora felis	Cats/small intestine	30–53 × 23–37	6–17	–
C. rivolta	Cats/duodenum	22–30 × 21–27	5	–
C. ohioensis	Dogs/small intestine	19–27 × 18–23	6	–
C. canis	Dogs/small intestine	36–44 × 29–36	9–11	–
C. burrowsi	Dogs/small intestine, caecum	21 × 8	6–9	+
C. rivolta	Dogs/small intestine, caecum	22–36 × 21–27	5–7	–
C. suis	Pigs/small intestine	17–22 × 17–19	5	+
I. belli	Humans/small intestine	20–33 × 10–19	9–10	+
I. erinacei	Hedgehog/small intestine	28–34 × 23–27	7–14	+
I. canaria	Canary/intestine	13 × 10	4–5	+
I. serini	Canary/liver, lung	12 × 10	9–10	+
I. lacacei	Sparrow/intestine	22–35 × 15–17	7–8	–

the schizogonic reproduction process (Fig. 4.47) as they would have done it, when this host had ingested oocysts with contaminated food. Thus, this type of life cycle of the *Cystoisospora* species enlarges the chances of a successful transmission.

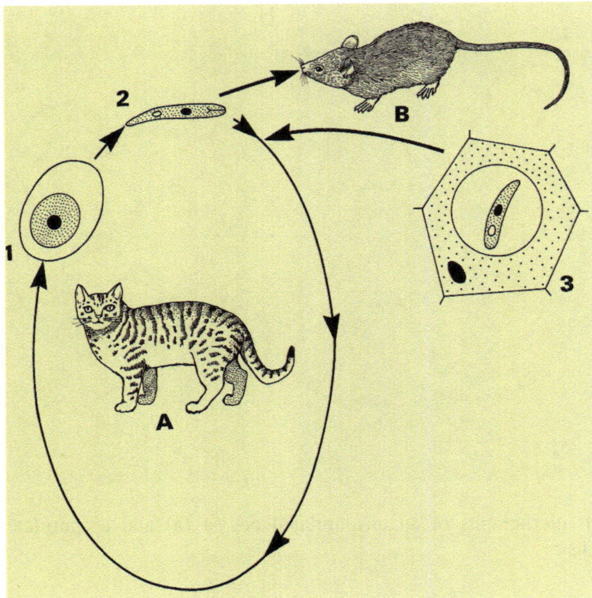

Fig. 4.47 Diagrammatic representation of the life cycle of *Cystoisospora felis* (**a**) which includes an intermediate host (**b**). (**a**) Schizogony, gamogony and sporogony run inside the intestinal epithelia of cats after ingestion of sporocysts containing sporozoites (**1**, **2**). (**b**) If a mouse ingests oocysts, the sporozoites (**2**) hatch inside their body, enter the intestinal wall and were transported to tissues, where they enter cells and stay there unchanged (**3**) until this mouse is ingested by a cat, where the "normal life cycle" starts again. These waiting stages in mice are also described as "**dormozoites**" = sleeping cells

4.6.2.1 *Isospora* and *Cystoisospora* Species of Carnivores

1. **Name**: Greek: *isos* = identical, similar; *sporos* = seed, persistent stages; *kystis* = bladder, capsule. Latin: *felis* = cat; *canis* = dog; *sus* = pig.
2. **Geographic distribution/epidemiology**: Worldwide; very common in young dogs and puppies as well as in cases of older dogs with a reduced immune status.
3. **Biology, morphology**: The *Cystoisospora* species are extremely species specific with respect to their final hosts (Table 4.10).
   (a) **Dogs**
   - Large oocysts (40 × 30 μm): *C. canis*;
   - Small oocysts (24 × 20 μm): *I. burrowsi; C. ohioensis* (21 × 18 μm).

   (b) **Cats**
   - Large oocysts (45 × 33 μm) (Fig. 4.46): *C. felis;*
   - Small oocysts (26 × 24 μm): *C. rivolta.*

These oocysts develop their sporocysts in a species-specific timing outside of the body (=1–4 days). It is needed to examine fresh feces; otherwise small stages might be kept for sarcosporidians.

After oral uptake of sporulated oocysts by dogs or cats, the sporozoites induce the life cycle comprising schizogony, gamogony and sporogony ending by excretion of unsporulated oocysts about 5–10 days after infection which again develop sporocysts and sporozoites. If such sporulated oocysts are ingested by non-specific hosts (e.g. cats, cattle), the sporozoites leave the intestine and enter body cells, wherein they stay unchanged as so-called **dormozoites** until this piece of meat is ingested by a dog in the case of *C. canis* or a cat in the case of *C. felis* (Fig. 4.47). The genus name *Cystoisospora* was created by the American scientist Jacob Frenkel (one of the discoverers of the life cycle of *Toxoplasma gondii*) in order to differentiate these *Isospora* species with facultative intermediate hosts from the *Isospora* species of birds, which are strictly monoxenous = one host species.

4. **Symptoms of disease (Coccidiosis)**: Low-grade infections remain mostly symptomless. In cases of high-grade infections with numerous oocysts, the animals show apathia and fluid, bloody feces for 1–7 days. Severe cases may lead to death due to massive destruction of the intestinal epithelia (haemorrhagic enteritis).
5. **Diagnosis**: Demonstration of the oocysts with the help of the flotation method or S.A.F.C.
6. **Pathway of infection**: Oral uptake of sporulated oocysts within fecally contaminated food or ingestion of sporozoites = dormozoites in, e.g., muscle cells of unspecific hosts (mice, cattle, etc.).
7. **Prophylaxis**: Removal of feces from rooms or cages of cats and dogs and disinfection of the floor by hot steam or hot water. Disinfection seems possible also by chemical compounds such as Lysococ® (4 %–30 min), Club-TGV anticoc, Chevi 75, P3-incicoc (5 %–2.5 h). **Important**: Raw meat should never be fed to dogs and cats (see also *Toxoplasma gondii*).
8. **Incubation period**: Variable (species specific): 3–4 days (e.g. *C. ohioensis*) or 6–8 days (e.g. *C. canis*).
9. **Prepatent period**: Species specific: 5 days (e.g. *C. rivolta*), up to 16 days (*C. felis*, Table 4.10).
10. **Patency**: Species specific: 4–12 days (*C. burrowsi*) up to 4 weeks (e.g. *C. canis, C. felis*).
11. **Therapy**: Drug of choice is toltrazuril (10 mg/kg bodyweight orally for 4–5 days).

Further Reading

Dubey JP (2014) Life cycle of *Cystoisospora felis* (Coccidia: Apicomplexa) in cats and mice. J Eukaryot Microbiol 61:637–643.

Dubey JP et al (2015) Experimental transmission of *Cystoisospora felis*-like coccidium from bobcat (*Lynx rufus*) to the domestic cat (*Felis catus*). Vet Parasitol 211:35–39.

Houk AE et al (2013) Experimentally induced clinical *Cystoisospora canis* coccidiosis in dogs with prior natural patent *Cystoisospora ohioensis*-like or *C. canis* infections. J Parasitol 99:892–895.

Houk AE, Lindsay DS (2013) *Cystoisospora canis* (Apicomplexa: Sarcocystidae): development of monozoic tissue cysts in human cells, demonstration of egress of zoites from tissue cysts, and demonstration of repeat monozoic tissue cyst formation by zoites. Vet Parasitol 197:455–461.

Matsubayashi M et al (2011) Phylogenetic identification of *Cystoisospora* spp. from dogs, cats, and raccoon dogs in Japan. Vet Parasitol 176:270–274.

Mitchell SM et al (2009) Development and ultrastructure of *Cystoisospora canis* Nemeséri, 1959 (syn, *Isospora canis*) monozoic cysts in two noncanine cell lines. J Parasitol 95:793–798.

4.6.2.2 *Isospora suis*

1. **Name**: Greek: *isos* = similar, identical; *sporos* = seed. Latin: sus = pig.
2. **Geographic distribution/epidemiology**: Worldwide; in piglets epidemics may occur with prevalence rates of up to 80 %. Due to the fact that sporulation of the oocyst takes only ~15 h at higher temperatures in stables and that the prepatent period is very short (5 days), infection spreads very rapidly in rearing facilities.
3. **Biology, morphology**: The oocysts containing two sporocysts each with four sporozoites measure 17–22 × 17–19 μm and have a colourless, smooth wall without protrusions. The development of the different stages runs like in the other *Isospora* species.
4. **Symptoms of disease**: Catarrhalic enteritis with high mortality rates may occur in very young piglets between days 5–20 after birth.
5. **Diagnosis**: Demonstration of unsporulated oocysts in feces (compare Fig. 4.46a).
6. **Pathway of infection**: Oral uptake of sporulated oocysts (compare Fig. 4.46b).
7. **Prophylaxis**: Quick removal of fresh feces from floors of stables; hot cleaning of the soil.
8. **Incubation period**: 3–5 days.
9. **Prepatent period**: 5–7 days.
10. **Patency**: 7–16 days.
11. **Therapy**: Toltrazuril, 20 mg/kg bodyweight.

Further Reading

Daugschies A et al (1999) Vorkommen von *Isospora suis*. Prakt Tierarzt 80:530–537.

Huang C et al (2016) Exploration of fluorescence-based real-time loop-mediated isothermal amplification (LAMP) assay for detection of *Isospora suis* oocysts. Exp Parasitol 165:1–6.

Langkjaer M, Roepstorff A (2008) Survival of *Isospora suis* oocysts under controlled environmental conditions. Vet Parasitol 152:186–193.

Mengel H et al (2012) Necrotic enteritis due to simultaneous infection with *Isospora suis* and clostridia in newborn piglets and its prevention by early treatment with toltrazuril. Parasitol Res 110:1347–1355.

Mundt HC et al (2007) Efficacy of various anticoccidials against experimental porcine neonatal isosporosis. Parasitol Res 100:401–411.

Schwarz L et al (2014) Superinfection of sows with *Cystoisospora suis* ante partum leads to a milder course of cystoisosporosis in suckling piglets. Vet Parasitol 204:158–168.

4.6.2.3 *Cyclospora* species

Cyclospora species (e.g. *C. viperae*) of snakes, *C. caryolytica* of moles, *C. cayetanensis* of humans and monkeys) develop spherical oocysts inside intestinal cells of their hosts. These oocysts measure 8–12 μm in diameter and develop outside of their hosts two 4–6 μm sized sporocysts, which contain two sporozoites each (Fig. 4.48). These stages are infectious, if orally ingested by their hosts. **Symptoms of disease** (severe diarrhoeas) occur mainly in stressed animals or in immunosuppressed humans. In humans, these oocysts had been first described as CLB bodies = *cyanobacteria-like bodies*. Healthy individuals often do not show symptoms of disease. Treatment in humans was successful when applying coxtrimoxazole (2 × 800 mg sulfamethoxazole/160 mg trimethoprim). Similar dosages/bodyweight could be used in animals.

Further Reading

Li N et al (2015) Identification and morphologic and molecular characterization of *Cyclospora macacae* n. sp. from rhesus monkeys in China. Parasitol Res 114:1811–1816.

Yamada M et al (2014) Intranuclear coccidiosis caused by *Cyclospora* spp. in calves. J Vet Diagn Invest 26:678–682.

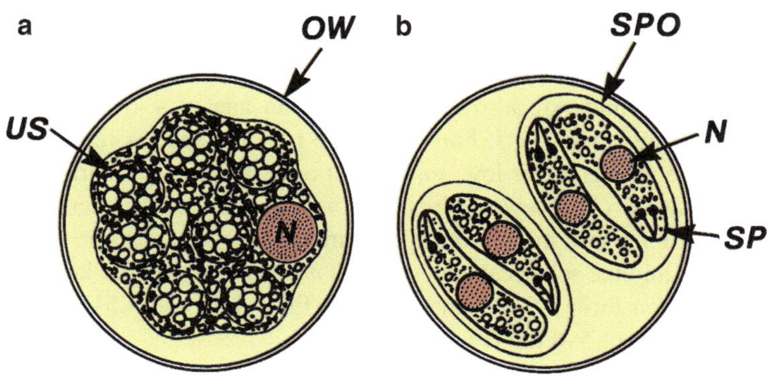

Fig. 4.48 Diagrammatic representations of an unsporulated (**a**) and a sporulated (**b**) oocysts of *Cyclospora* species. *N* nucleus; *OW* oocyst wall; *SP* sporozoites; *SPO* sporocyst; *US* undifferentiated oocyst cytoplasm

Zhao GH et al (2013) Molecular characterization of *Cyclospora*-like organisms from golden snub-nosed monkeys in Qinling Mountain in Shaanxi province, northwestern China. PLoS One 8(2):e58216.

4.6.3 *Cryptosporidium* species

The *Cryptosporidium* species have been found in the intestine of about 170 species of vertebrates including humans. In the narrow sense, they are no true coccidians, but are rather gregarines, since they do not develop intracellularly but are attached to the surface of intestinal cells with the help of a so-called **feeder organelle** (Fig. 4.49). Furthermore, there are no clearly defined species but so-called genotypes, which may occur in different hosts (Tables 4.11 and 4.12). As soon as an oocyst has been ingested by its host and reached the small intestine, the four sporozoites are released from the oocyst and become attached at the surface of an intestinal cell. Each sporozoite grows up to a spherical schizont, which forms an inner vacuole into which the developing merozoites protrude after repeated nuclear divisions. The schizogonic reproduction may be repeated several times so that the parasites may cover practically most of the intestinal cells (Fig. 4.52). The life cycle is documented in Fig. 4.49.

These parasites are described in a very broad spectrum of hosts (Tables 4.11, 4.12) but—except for the fact that all are **opportunistic organisms**, which occur in large crowds in diseased hosts or in generally immunocompromised organisms—many details of their interspecific relationships are still unknown.

Overview on *Cryptosporidium* Species

1. **Name**: Greek: *kryptein* = hiding; *spores* = seeds, spores.
2. **Geographic distribution/epidemiology**: Worldwide; epidemics among young or diseased hosts; humans can become infected by oocysts excreted by animals.
3. **Biology, morphology**: The different species/serotypes of the genus *Cryptosporidium* may infect humans as well as a broad spectrum of animals (Tables 4.11, 4.12). Especially young animals 3–30 days of age are extremely endangered, although epidemics are also common until an age of 4–6 months (seen in ruminants, pigs, dogs, cats, horses). While in some species (e.g. *C. parvum*) the host specifity is rather low, some other ones infect only a narrow spectrum of animals (e.g. *C. baileyi* of birds or *C. muris* of mice). Neither light microscopical nor electron microscopical features help to separate one species from the other. Only the spherical oocysts of *C. muris* are somewhat larger (7 μm in diameter) than those of the other species (Figs. 4.50 and 4.51), which reach only diameters of 5–6 μm (Fig. 4.52).

   The oocysts of the *Cryptosporidium* species do not contain sporocysts, but the early oocyst is transformed to the later sporocyst by transformation of the cyst wall as was seen in electron microscopical studies (Figs. 4.53, 4.54, 4.55 and 4.56). The final sporocyst is the infectious stage, which contains four

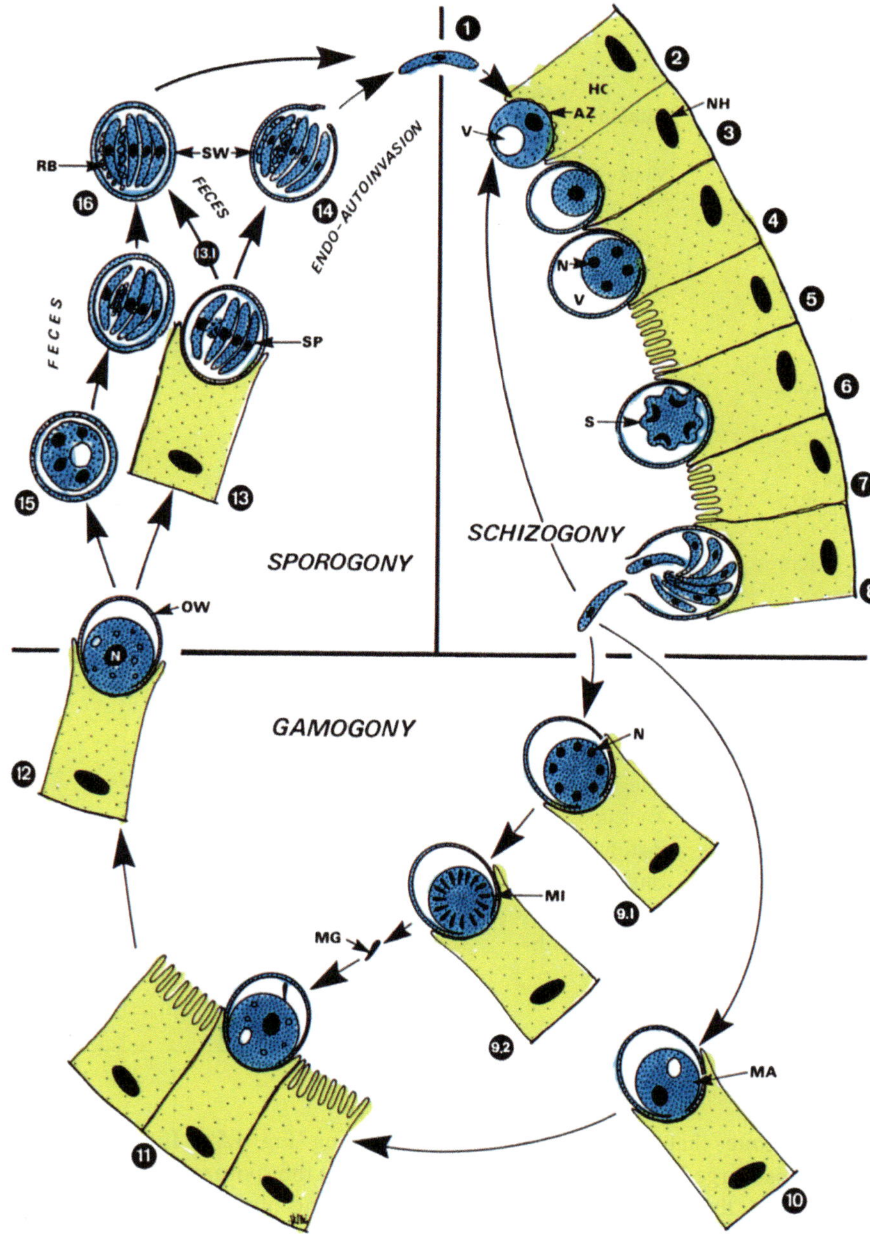

Fig. 4.49 Diagrammatic representation of the life cycle of *Cryptosporidium* species. (**1–8**) A sporozoite (excysted from an ingested oocyst) becomes attached at an intestinal cell (**1**) and develops into a schizont containing several merozoites (**8**), which may repeat schizont formation or introduce gamogony by growing up to a male (**9.1**) or female gamont (**10**). The male gamont produces microgametes (**9.2**), which fertilize the female gamete (**11**). Thereafter, a zygote is formed and becomes encysted, thus reaching the status of an oocyst (**12**). (**13–16**) Outside or inside

Table 4.11 *Cryptosporidium* species

Species	Main hosts	Size oocysts (μm)	Prepatent period (days)	Pathogenicity
Cryptosporidium parvum	**Humans**	5 × 4.5	2–3	+++
	Monkeys	?	2–3	+++
	Calves, lambs	2–3	2–3	+++
	Piglets, fowls	4–5	2–3	+++
	Cats, dogs, rabbits	4–6	2–3	+++
	Guinea pigs, mice, rats	2–3	2–3	+++
C. hominis	**Humans**, monkeys	5 × 4.5	2–3	+++
C. bailey	Young chicken	4.6 × 6.3	3–5	++
C. galli	Chicken	4.5 × 6	3–5	++
C. muris	Mice, rats, **humans**	7 × 5	4	++
C. canis	Dogs, **humans**	5 × 4.5	?	+
C. felis	Cats, **humans**	5 × 4.5	?	+
C. wrairi	Guinea pigs	5.2 × 4.5	?	+
C. molnari	Fish	4.7 × 4.4	?	+
C. andersoni	Cattle	5.5 × 7	?	++
C. bovis	Cattle, sheep	5 × 5	?	+
C. ryanae	Cattle	5 × 5	?	+
C. fayeri	Kangaroos	5 × 5	?	++
C. macropodium	Kangaroos	5 × 5	?	++
C. suis	Pigs, **humans**	5 × 5	3–5	+++
C. meleagridis	Birds, **humans**	5 × 4.6	3–5	+
C. crotali	Reptiles	2 × 5	?	+
C. serpentis	Snakes, lizards	5.8 × 6.8	2–7	+
C. varani	Warans	4 × 5	?	+
C. nasorum	Fish	2 × 5	?	+
C. scophthalmi	Turbots	4 × 5	?	+

sporozoites that remain invisible without staining (Fig. 4.50). If the oocysts are not discharged within the feces, they may release the sporozoites already inside the host, thus leading to a potential overcrowding of the intestine and to the severe symptoms of a strong diarrhoea. In those species, which parasitize the cells of the urogenital system, the oocysts can be diagnosed in the urine.

Fig. 4.49 (continued) the body each oocyst is transformed to an infectious sporocyst. *AZ* adhesion zone; *HC* host cell; *MA* macrogamete; *MG* microgamete; *MI* microgamont; *N* nucleus; *NH* nucleus of the host cell; *OW* oocyst wall; *RB* residual body; *S* schizont; *SP* sporozoites; *SW* sporocyst wall; *V* vacuole

Table 4.12 *Cryptosporidium* species—infected organs

Species	Localization
C. canis	Small intestine
C. felis	Small intestine
C. hominis	Small intestine
C. parvum	Small intestine
C. suis	Small intestine
S. wrairi	Small intestine
C. andersoni	Stomach
C. galli	Stomach
C. muris	Stomach
C. molnari	Stomach
C. serpentis	Stomach
C. baileyi	Whole intestine, cloaca, trachea
C. hominis, C. parvum, in HIV patients	Bile ducts, lungs, intestine

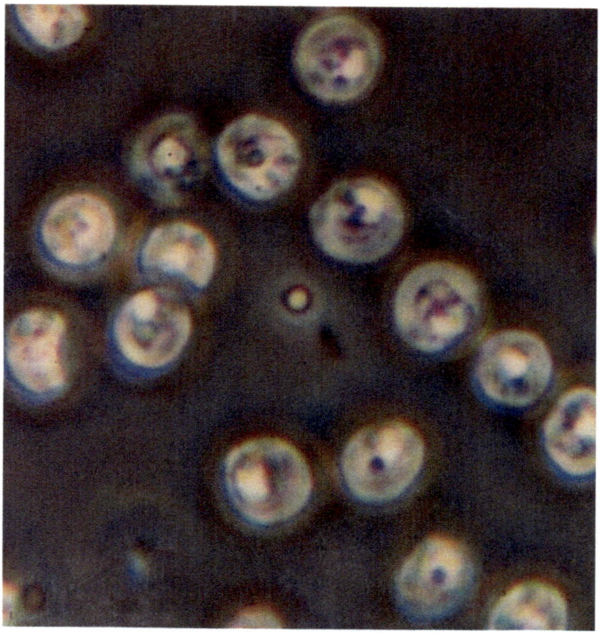

Fig. 4.50 Light micrograph of unstained oocysts of *Cryptosporidium* sp. in feces

New infections occur after oral uptake of oocysts/sporocysts in fecally contaminated food or drinking water (Fig. 4.49). The hatched sporozoites become attached at the microvilli of the small intestine (ileum, jejunum) without entering the host cells. After repeated formation of schizonts and merozoites (Figs. 4.51, 4.52 and 4.53), gamonts are formed (Fig. 4.55). After fertilization, again oocysts are formed (Fig. 4.56). According to recent

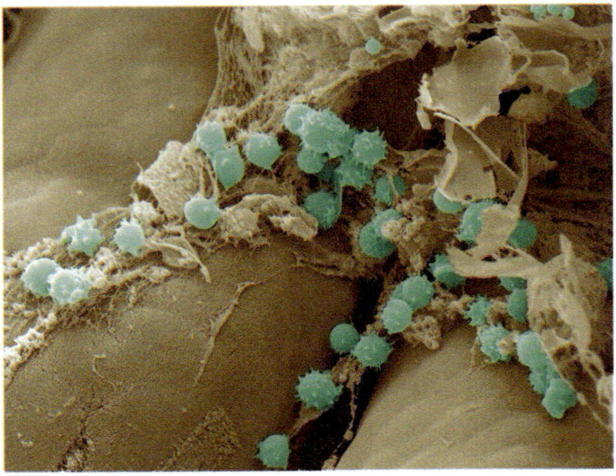

Fig. 4.51 Scanning electron micrograph of *Cryptosporidium* stages on damaged or even absent intestinal epithelium

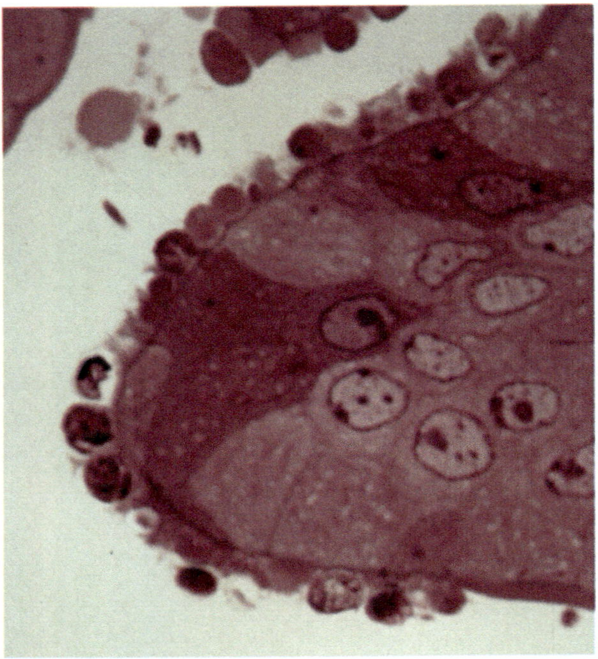

Fig. 4.52 Semithin section of an intestinal villus of a mouse being closely covered by developmental stages of *Cryptosporidium* sp

observations, it seems that there exist two types of oocysts. Type one produces its four sporozoites only after having reached the environment via feces, and thus, these oocysts may lead to infections of other animals. The second type of oocysts seems to produce the four sporozoites already inside the same host, which thus becomes overcrowded with parasitic stages. This type would lead to a constant **autoinfection**. Diseased animals may excrete up to ten million

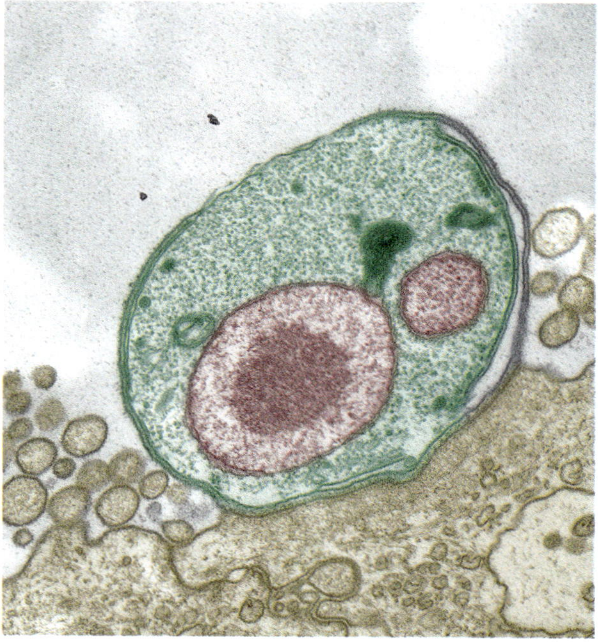

Fig. 4.53 Transmission electron micrograph of a young schizont of *Cryptosporidium parvum* on an intestinal cell

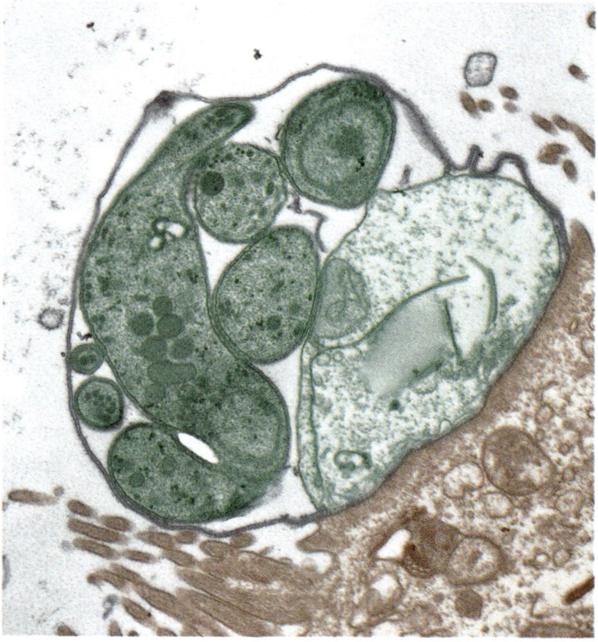

Fig. 4.54 Transmission electron micrograph of a mature *Cryptosporidium parvum* schizont on the surface of an intestinal host cell

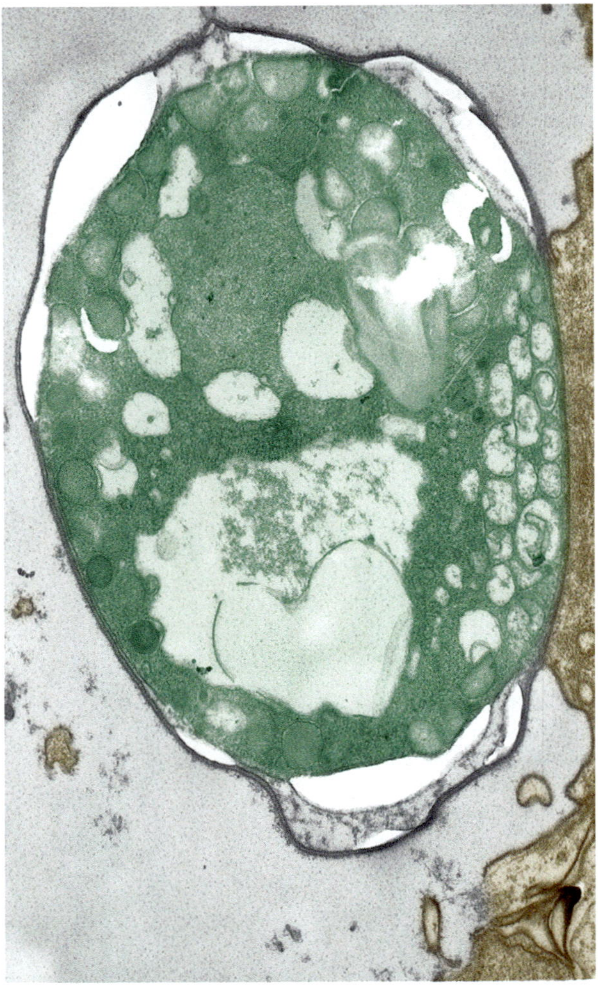

Fig. 4.55 Transmission electron micrograph of a young gamont of *C. parvum*

oocysts per gram feces, so that a single infected animal may endanger a whole herd of animals. Since *Cryptosporidium* infections are common but adult animals are only rarely infected, it can be concluded that low doses of this species may introduce immunity in older animals, which apparently lasts lifelong.

4. **Symptoms of disease**: Especially very young or diseased animals (with a weak immune system) are severely affected and may die due to permanent diarrhoeas. The feces appear then watery-yellowish and slimy. Maldigestion, malabsorption, dehydration, loss of weight and low-grade fever are very common symptoms. In severe cases, cramps, motility disturbances or even death may follow.

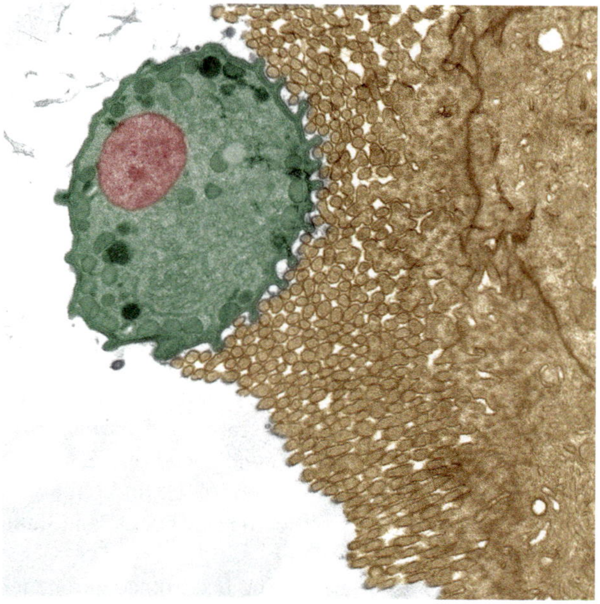

Fig. 4.56 Transmission electron micrograph of a section through a young oocyst of *C. parvum* attached at its host cell

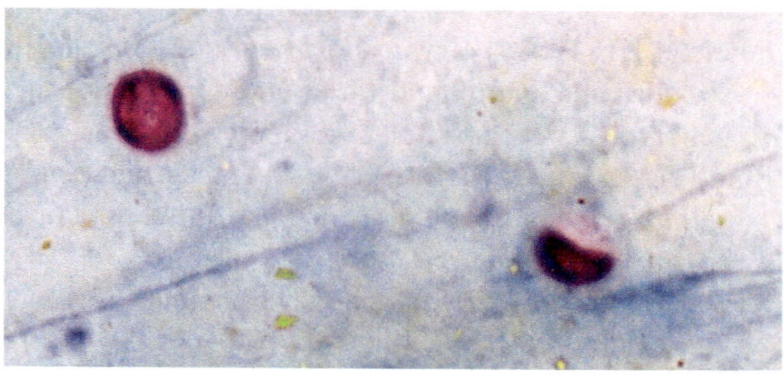

Fig. 4.57 Light micrograph of two oocysts of *C. parvum* stained according to Ziehl–Neelsen

5. **Diagnosis**: The best method is the microscopical demonstration of the oocysts inside feces or urine (Figs. 4.50 and 4.57).
6. **Pathway of infection**: Oral uptake of sporulated oocysts within fecally contaminated food. Oocysts remain infectious for at least 6 months, if a reasonable humidity is present.
7. **Prophylaxis**: Removal of the feces from stables; cleaning the floors with hot steam and/or anticoccidial products (see: *Eimeria*). **Attention**: Animal feces may contain oocysts being infectious for humans.
8. **Incubation period**: Variable, mostly 2–10 days.
9. **Prepatent period**: 2–4 days (experimentally obtained).
10. **Patency**: 12–14 days (experimentally obtained in previously uninfected animals).
11. **Therapy**: A fully effective treatment does not yet exist. Positive results had been obtained by application of **lasalocid** (15 mg/kg bodyweight 1× daily for

3 days) in calves and in lambs (5 mg/kg bodyweight daily for 3 days). Also the use of **halofuginone** (Halocur®, Fa. MSD) was successful in calves of 35–45 kg, if they received 8 ml on 7 consecutive days. Waiting time: 13 days. **Attention**: It is needed to give additionally electrolyte solutions (1.5–2.0 l) to allow rehydration.

Further Reading

Clode PL et al (2015) Life without a host cell: what is *Cryptosporidium*? Trends Parasitol 31:614–624.

Díaz P et al (2015) *Cryptosporidium* species and subtype analysis in diarrhoeic pre-weaned lambs and goat kids from north-western Spain. Parasitol Res 114:4099–4105.

Grinberg A, Widmer G (2016) *Cryptosporidium* within-host genetic diversity: systematic bibliographical search and narrative overview. Int J Parasitol. pii: S0020-7519(16)30002-9 (epub ahead of print).

Ibrahim MA et al (2016) Epidemiology and public health significance of *Cryptosporidium* isolated from cattle, buffaloes, and humans in Egypt. Parasitol Res. doi:10.1007/s00436-016-4996-3.

Kostopoulou D et al (2015) The occurrence and genetic characterization of *Cryptosporidium* and *Giardia* species in foals in Belgium, The Netherlands, Germany and Greece. Vet Parasitol 211:170–174.

Petersen HH (2014) *Cryptosporidium* and *Giardia* in Danish organic pig farms: seasonal and age-related variation in prevalence, infection intensity and species/genotypes. Vet Parasitol 214:29–39.

Ramo A et al (2016) Intra-species diversity and panmictic structure of *Cryptosporidium parvum* populations in cattle farms in Northern Spain. PLoS One 11(2): e0148811.

Ryan U, Hijjawi (2015) New developments in *Cryptosporidium* research. Int J Parasitol 45:367–373.

Yang R et al (2015) Genetic diversity of *Cryptosporidium* in fish at the 18S and actin loci and high levels of mixed infections. Vet Parasitol 214:255–263.

Yuan L et al (2014) Effects of different inoculation routes on the parasitic sites of *Cryptosporidium baileyi* infection in chickens. Exp Parasitol 145:152–156.

4.6.4 *Sarcocystis* Species

1. **Name**: Greek: *sarx* = meat; *kystis* = bladder.

The so-called Sarcosporidia, which at first had been called “psorospermes” = muscle stages, were at first described in the year 1843 by the Swiss scientist Friedrich Miescher (1811–1887) in the muscles of domestic mice, but they were found later in practically all mammals (including whales and humans) as well as in birds and reptiles (Mehlhorn and Heydorn 1978).

Fine structural identity of these cyst stages with other well-known coccidian species (especially with their merozoites and sporozoites) stimulated transmission experiments which finally showed that the *Sarcocystis* species have an obligatory two-host life cycle, which runs between animals acting as preys or predators. The schizogony—known from the classic Coccidia—occurs in the epithelial cells, e.g., of blood vessel or inner organs of preys (plant feeders or omnivores) as precursor of a phase inside muscle or brain cells, where so-called tissue cysts are developed, which may survive for long until a predator ingests them (Fig. 4.58). As soon as a carnivorous host has ingested such cyst-containing meat, the sexual cycle starts. The so-called cyst stages (=cyst merozoites) enter cells of the intestinal lamina propria and become therein either a male or a female gamont. While the female gamont just grows up, the microgamont gives rise to many flagellated microgametes, which penetrate singly into a macrogamete and thus lead to the formation of a zygote, which is finally transformed into an oocyst by fusion of so-called wall-forming bodies. Inside this oocyst, finally two sporocysts are formed each containing four sporozoites. The oocyst wall of the sarcosporidia is rather thin compared to that of *Eimeria* species. Therefore, **final hosts** (=predators = meat feeding animals or humans) excrete in their feces mostly single sporocysts (Fig. 4.59b), which are infectious for intermediate hosts (=plant feeders or omnivores—Fig. 4.58; Table 4.13). If such an intermediate host has ingested, such oocysts or single sporocysts schizonts are developed within 6 days inside endothelial cells of the liver, kidney, brain, etc. However, these schizonts are not found inside a parasitophorous vacuole, but they are situated immediately inside the cytoplasm (Fig. 4.58). These schizonts keep their 3-layered pellicle and also their nucleus is not divided into many smaller ones but grows up enormously. In a simultaneous division process, the nucleus and surrounding cytoplasm give rise to about 50–90 merozoites (Figs. 4.58 and 4.70). These merozoites enter neighbouring endothelial cells and give rise to secondary schizonts. During this phase, the intermediate hosts may show severe symptoms of disease and may even die (see Table 4.14) in case they are infected with a very virulent species.

Beginning at the 20th day after the infection, the merozoites of the second generation of merozoites enter cells of the brain and of the muscles, wherein they start to develop **tissue cysts** (Figs. 4.62, 4.63, 4.64, 4.65 and 4.66). These cysts are characterized by species-specific cyst protrusions, which can be used for species diagnosis (Figs. 4.60 and 4.67). At first, they are surrounded by a single membrane, under which, however, electron-dense material is deponed (Fig. 4.66). Depending on the species, this so-called primary cyst wall (20–100 nm thick) may develop species-specific protrusions (Fig. 4.67). The penetrated **merozoite** becomes spherical and is described as **metrocyte**, which starts the typical inner division (=**endodyogeny**; Fig. 4.68) (=neu). About 1 month after the infection, the young tissue cysts are filled with these metrocytes. During further repeated

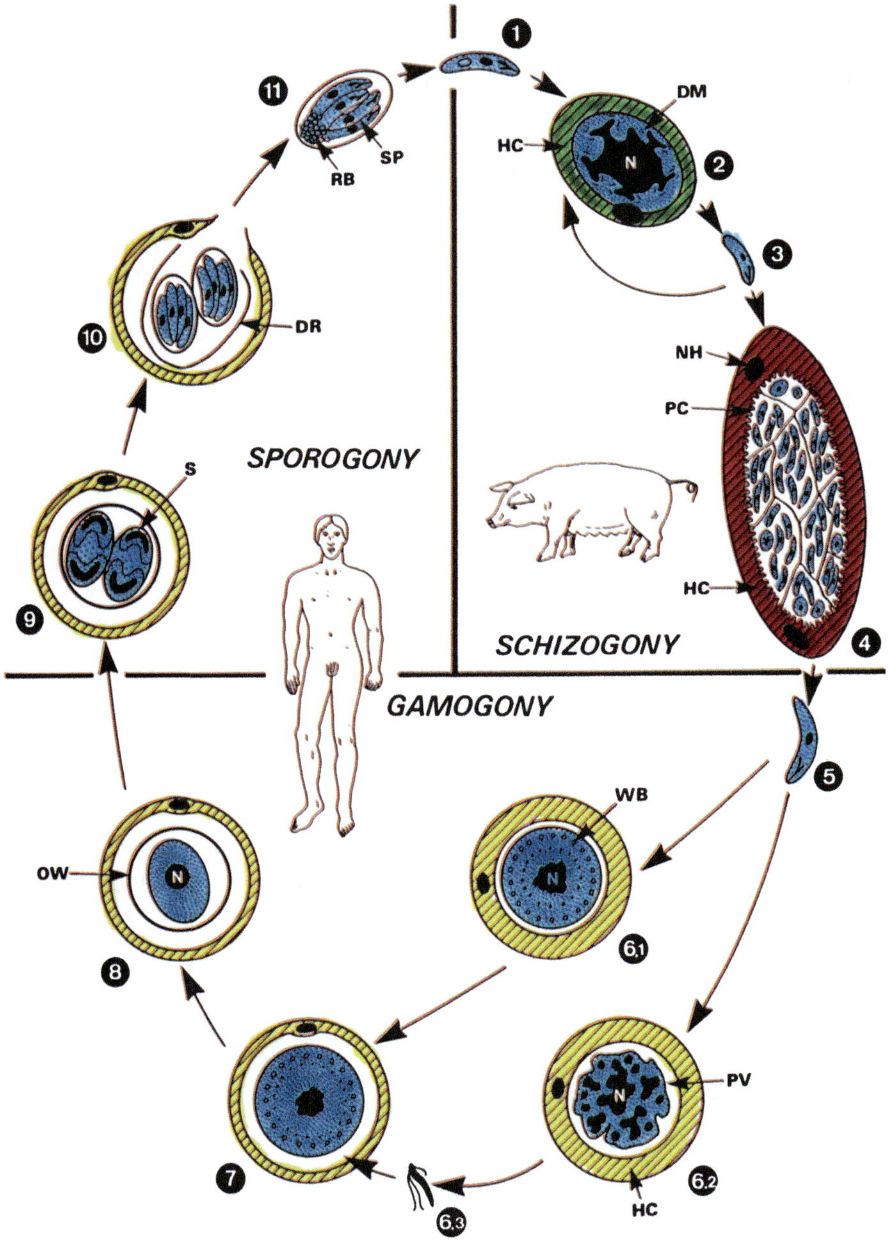

Fig. 4.58 Diagrammatic representation of the life cycle of *Sarcocystis suihominis*. (**1**) Motile sporozoites hatch from the ingested sporocysts inside the intestine of the intermediate host, i.e. swine. (**2**) Two generations of schizonts are formed (5–6 and 12–17 days after infection) inside endothelial cells of blood vessels, giving rise to 60–100 merozoites by endopolygeny. (**3**) Free motile merozoites; first-generation merozoites enter other endothelial cells and form schizonts, whereas merozoites of the second generation induce formation of tissue cysts. (**4**)

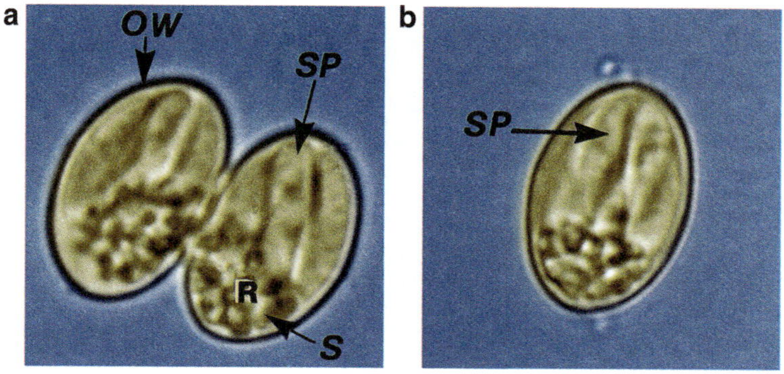

Fig. 4.59 Light micrographs of an oocyst (**a**) and a sporocyst (**b**) of *S. suihominis*. The thin wall of the oocyst is often already ruptured inside the intestine. The oocyst is already fully sporulated; i.e. the sporocysts contain each already four infectious sporozoites besides a residual body. *OW* oocyst wall; *R* residual body; *S* sporocyst; *SP* sporozoites

endodyogenies, the typical cyst merozoites are formed, which finally (beginning about 2 months after the infection) fill up the whole cyst interior, which is subdivided by amorphic material into so-called chambers (Fig. 4.74). The size of the tissue cysts differs among the species. Their size may reach (species specific) from a few μm up to several centimetres (Figs. 4.64, 4.65, 4.66, 4.67, 4.68, 4.69 and 4.75). In any way, the tissue cysts are covered by a so-called primary cyst wall, which separates the parasites from the surrounding host cell cytoplasm (Figs. 4.60 and 4.67).

As soon as the final host (carnivores, omnivores) have ingested raw, cyst-containing meat, the cyst merozoites are set free in the host's intestine. Then they enter the cells of the lamina propria. In the case that the number of such merozoites is large, this infection may induce severe diarrhoeas, leading to a large loss of water and a collapse of the infected animal. The penetrated cyst merozoites are transformed within 14 h into male or female gamonts (=micro-, macrogamonts). After fusion of the gametes, the oocyst is formed which are excreted fully sporulated within the feces

Fig. 4.58 (continued) Cyst formation inside typical cells (muscle fibres, brain cells); within these cysts, the parasites are reproduced by repeated endodyogeny leading to thousands of cyst merozoites which are situated inside chamber-like hollows. (**5**) When the final host man has eaten cyst containing raw or insufficiently cooked meat, the cyst merozoites are set free and enter cells of the lamina propria. (**6**) Formation of female (macrogametes; **6.3**) via gamonts (**6.1**, **6.2**) within 14 h after infection. (**7**) Fusion of gametes. (**8**) Formation of the oocyst wall around the zygote. (**9–11**) Formation of 2 sporocysts (containing 4 sporozoites each) inside the host cells. The smooth oocyst wall often becomes disrupted. Thus, fully sporulated oocysts are found in the feces (**11**). *DM* developing merozoites; *DR* disrupted oocysts wall; *HC* host cell; *N* nucleus; *NH* nucleus of the host cell; *OW* oocysts wall; *PC* primary cyst wall; *PV* parasitophorous vacuole; *RB* residual body; *S* sporocyst; *SP* sporozoite; *WB* oocysts wall-forming bodies (for further details, see Table 4.13)

Table 4.13 Important *Sarcocystis* species

Species—old name	Species—new name	Intermediate host	Final host	Old name for stages in final host	Pathogenicity
	Sarcocystis bovihominis	Cattle	**Humans**	*Isospora hominis*	–
S. cruzi	*S. bovicanis*	Cattle	Dogs	*I. bigemina*	+
S. hirsuta	*S. bovifelis*	Cattle	Cats	*I. bigemina*	–
S. fusiformis	*S. levinei*	Water buffaloes	Dogs	*I. bigemina*	–
	S. fusiformis	Water buffaloes	Cats	*I. bigemina*	–
S. miescheriana	*S. suihominis*	Pigs	**Humans**	*I. hominis*	+
	S. suicanis	Pigs	Dogs	*I. bigemina*	+
S. tenella	*S. ovicanis*	Sheep	Dogs	*I. bigemina*	+
	S. arieticanis	Sheep	Dogs	*I. bigemina*	+
	S. ovifelis	Sheep	Cats	*I. bigemina*	–
	S. medusiformis	Sheep	Cats	*I. bigemina*	–
S. moulei	*S. capracanis*	Goats	Dogs	*I. bigemina*	+
	S. hircicanis	Goats	Dogs	*I. bigemina*	–
	S. moulei	Goats	Cats	*I. bigemina*	–
S. gracilis	*S. S. gracilis*	Roe deer	Dogs	*I. bigemina*	–
S. bertrami	*S. equicanis*	Horses	Dogs	*I. bigemina*	–
	S. bertrami	Horses	Dogs	*I. bigemina*	–
	S. fayeri	Horses	Dogs	*I. bigemina*	–
S. sp.	*S. neurona*	Horses, Equids	Opossums	*I. sp.*	+
S. cameli	*S. camelis*	Camels	Dogs	*I. bigemina*	–
S. muris	*S. muris*	Mice	Cats	*I. bigemina*	+
S. sp.	*S. dispersa*	Mice	Owls (*Tyto*)	*I. sp.*	–
S. sp.	*S. cernae*	Mice (*Microtus*)	Falconids (*Falco*)	*I. sp.*	+
S. sp.	*S. murivipera*	Mice	Snakes	*I. sp.*	+
S. sp.	*S. singaporensis*	Rats (*Rattus*)	Snakes	*I. sp.*	+

S. cuniculi	*S. cuniculi*	Rabbit	Cats	*I. bigemina*	–
S. rileyi	*S. rileyi*	Duck	Dogs	*I. sp.*	–
S. horvathi	*S. horvathi*	Chicken	Dogs	*I. bigemina*	–
S. sp.	*S. falcatula*	Birds	Opossums	*I. sp.*	+
S. sp.	*S. podarcicolubris*	Lizard	Snakes	*I. sp.*	–
S. lindemanni	*S. -*	**Humans**	?	?	?
S. nesbiti	*S. -*	Monkeys, **Humans?**	?	?	?

+ = yes; – = no; ? = questionable

starting species specifically on days 4–11 after infection (Fig. 4.60). The final hosts excrete for about 6 weeks such oocysts (or even single sporocysts, since the very thin oocyst wall ruptures inside the feces prior to excretion).

Originally it was thought that each intermediate host (plant feeder, omnivores) is infected only by a single *Sarcocystis* species and that the differently looking tissue cysts have a different age. However, intense transmission experiments of the groups of Heydorn and Mehlhorn, Rommel and colleagues as well as Matuschka and colleagues or several American groups (Fayer, Frenkel, Dubey) showed that several *Sarcocystis* species may parasitize in the same final or intermediate host (see Table 4.13). In omnivorous lizards (e.g. *Gallotia galloti*), the same animal may be intermediate and final host at the same time for some *Sarcocystis* species. In the case that they feed tissue cysts in meat of smaller specimens of their own species, they excrete as final hosts oocysts. In the case they ingest oocyst-contaminated food, they become intermediate hosts and tissue cysts are formed inside their muscles.

With respect to economy, the *Sarcocystis* species of animals are of high importance due to the fact that they may induce during the schizogony severe symptoms of disease, which may even lead to death. Thus, farmers have to avoid any transmission among farm animals such as dogs/cats on one side and sheep, goats, cattle on the other.

Also with respect to human health, *Sarcocystis* species have a high impact, since humans may serve final and as well as intermediate host. In the case that humans are **final hosts**, they may suffer from severe diarrhoeas within 4–48 h after having ingested raw, cyst-containing meat. In the case that they are **intermediate hosts**, they develop muscle cysts (Fig. 4.61) (e.g. in the case of *S. lindemanni*, where the final host is not yet known).

Further Reading

Becker B et al (1979) Light and electron microscopic studies on the gamogony and sporogony of 5 *Sarcocystis* species. Zentralbl Bakteriol Orig A 244:394–404.

Frenkel JK et al (1979) Sarcocystinae: nomina dubia and available names. Parasitol Res 58:115–139.

Heydorn AO et al (1975) Proposal for a new nomenclature if the sarcosporidia. Parasitol Res 48:73–82.

Mackenstedt U et al (1990) DNA measurements and ploidy determination of different stages in the life cycle of *Sarcocystis muris*. Parasitol Res 76:662–668.

Mehlhorn H et al (1985) Announcement of the establishment of neohepantotypes of some important *Sarcocystis* species. Parasitol Res 71:689–692.

Mehlhorn H, Heydorn AO (1978) The Sarcosporidia: the fine structure and life cycle. Adv Parasitol 16:43–92.

Sommer I et al (1992) A comparison of the sporozoite and cyst merozoite surface proteins of *Sarcocystis* species. Parasitol Res. 78:396–403.

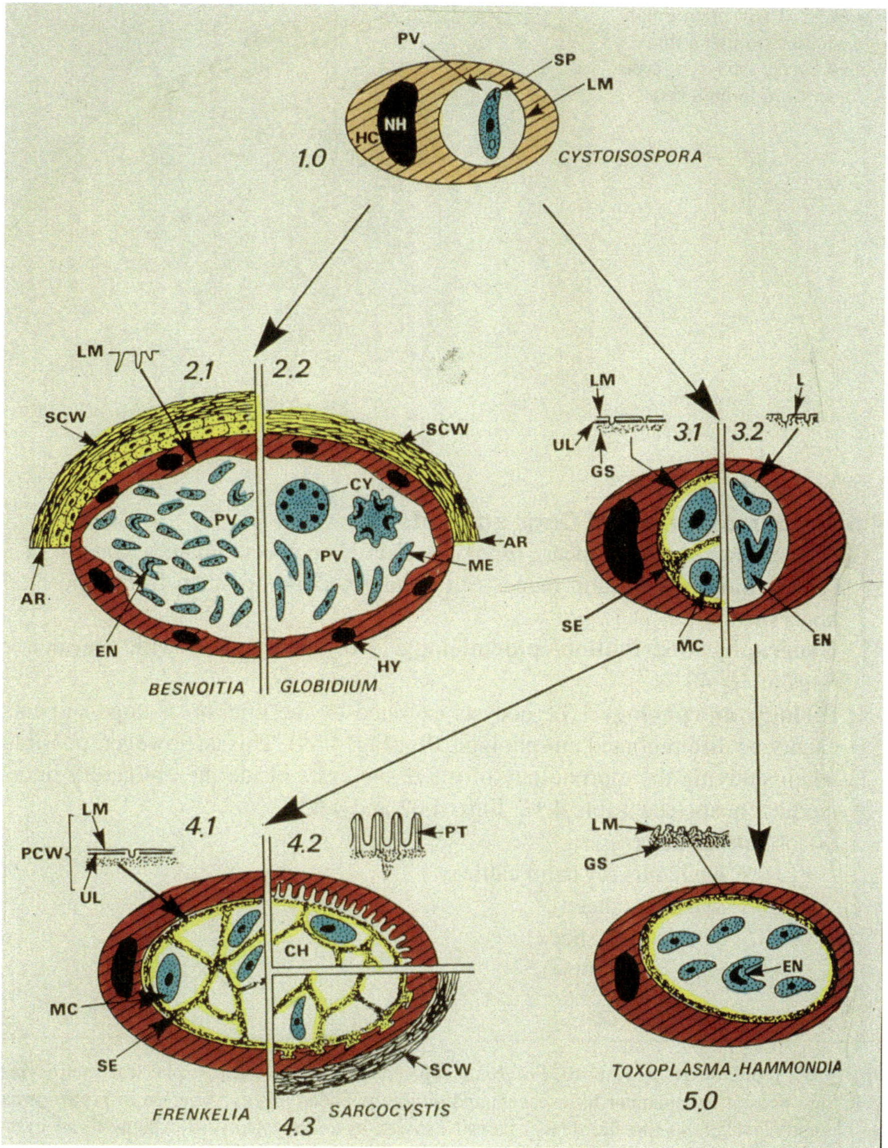

Fig. 4.60 Diagrammatic representation of cysts in different cyst-forming coccidia. (**1**) The simplest cyst formation. A parasite (sporozoites) is included in a parasitophorous vacuole (PV) which is bounded by a single cell membrane (LM). This is represented by the monozoic cysts of *Cystoisospora felis*, *C. rivolta* and *C. ohioensis* in transport (i.e. paratenic) hosts (such as mice). (**2**) In *Besnoita* spp. (**2.1**) and *Globidium* spp. (**2.2**) cysts, the original parasitophorous vacuole (PV) is enlarged and is filled by numerous parasites reproducing by endodyogeny (**2.1**) or schizogony (**2.2**). Even in old cysts, the PV is limited by a single unthickened cell membrane (LM). A secondary cyst wall (SCW) consisting of fibrillary material is always present; the host cell nuclei generally undergo hypertrophy and hyperplasia. (**3**) Young cysts of *Frenkelia* spp. and *Sarcocystis* spp. (**3.1**) and *Toxoplasma* spp. and *Hammondia* spp. (**3.2**) show the indicated features.

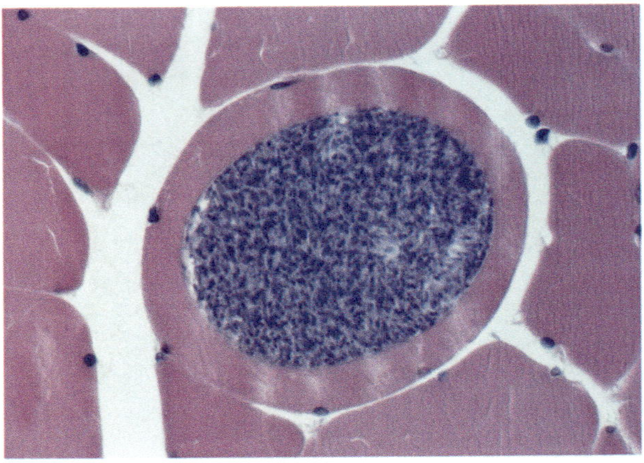

Fig. 4.61 Light micrograph of a section through a thin-walled *Sarcocystis* cyst (type: *S. lindemanni* in humans)

4.6.4.1 Sarcosporidia of Dogs and Cats

1. **Name**: Greek: *sarx* = meat; *sporos* = seeds. The species names consist of a composition of the Latin names of the intermediate hosts (in front) and final hosts (behind).
2. **Geographic distribution/epidemiology**: Worldwide; often with prevalence rates of 10–40 %.
3. **Biology, morphology**: The oocysts excreted by the final hosts dogs and cats cannot be differentiated morphologically (Fig. 4.59). This is, however, possible when studying the morphology of the tissue cysts inside the obligatory intermediate hosts (see Table 4.13; Figs. 4.62 and 4.63).
   (a) **Cysts in dogs**
      1. *S. bovicanis* (cysts in cattle).
      2. *S. ovicanis* (sheep).
      3. *S. arieticanis* (sheep).
      4. *S. equicanis* (horse).

Fig. 4.60 (continued) In cysts of *Frenkelia* spp. and *Sarcocystis* spp., spherical metrocytes (MC) are present (in chamber-like spaces) and divide by endodyogeny, whereas in *Toxoplasma gondii* and *Hammondia* spp. the slender parasites divide by endodyogeny. (**4**) Mature tissue cysts of *Frenkelia* and *Sarcocystis* are characterized by typical septa (SE) formed by the ground substances (GS). In *Frenkelia* spp. and some *Sarcocystis* spp. (**4.1**), the primary cyst wall (PCW) never forms long protrusions, whereas in other *Sarcocystis* spp. typical protrusions occur (**4.2**, **4.3**). With cysts of *S. ovifelis*, a secondary cyst wall (ECU) surrounds the parasitized muscle fibre (**4.3**). (**5**) The primary cyst wall of mature *T. gondii* and *Hammondia* spp. cysts remains smooth; the cysts are tightly filled with cyst merozoites (bradyzoites). Typical septa as well as metrocytes never oocur. *AR* artificially interrupted SCW; *CH* chamber-like space filled with parasites; *CY* cytomere; *EN* endodyogeny; *CS* ground substance; *HC* host cell; *HY* hypertrophic host cell nuclei; *LM* limiting single membrane of PV; *MC* metrocyte; *ME* merozoite; *N* nucleus; *NH* nucleus of host cell; *PCW* primary cyst wall; *PT* protrusion of PCW; *PV* parasitophorous vacuole; *SCW* secondary cyst wall; *SE* septum formed by GS; *SP* sporozoites; *UL* underlying dense material

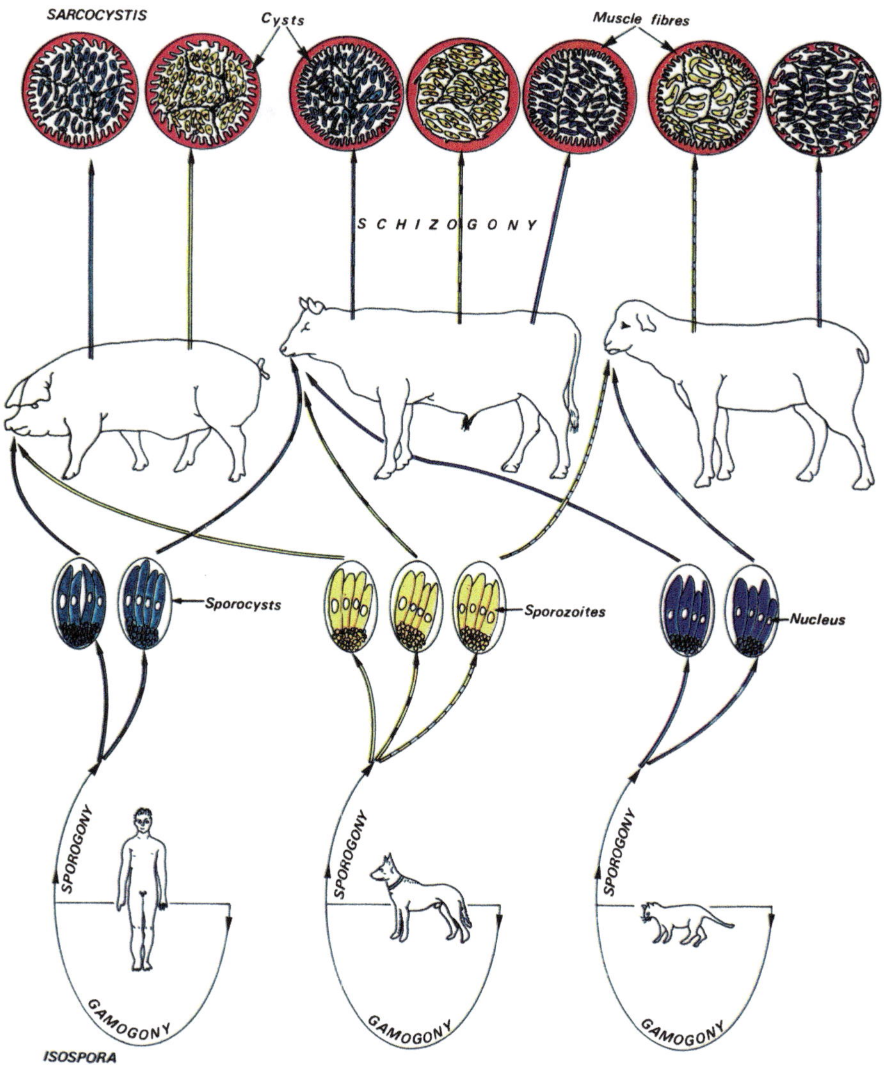

Fig. 4.62 Diagrammatic representation of the life cycles of the *Sarcocystis* species of humans and their most important domestic animals (according to Mehlhorn and Heydorn 1978). Pigs, cattle and sheep are **intermediate hosts**. The final hosts are infected by oral uptake of tissue cysts inside the muscles of intermediate hosts. Species in final hosts (below from *left* to *right*): *S. suihominis; S. bovihominis; S. suicanis; S. bovicanis; S. ovicanis; S. bovifelis; S. ovifelis*

5. *S. capracanis* (goat).
6. *S. hiricanis* (goat).
7. *S. suicanis* (pig).

(b) **Cysts in cats**

1. *S. bovifelis* (cattle).
2. *S. ovifelis* (sheep).

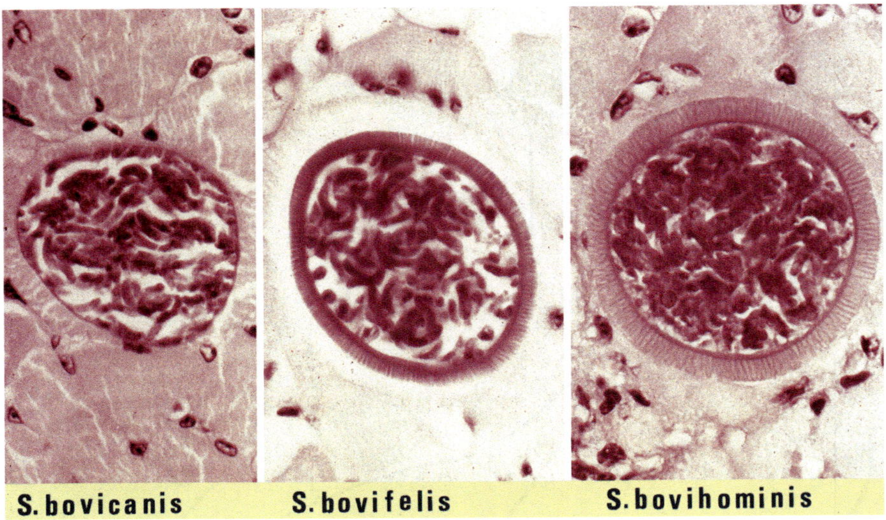

Fig. 4.63 Light micrographs of sections through *Sarcocystis* species in muscle of cattle

3. *S. muris* (mouse).
4. *S. cuniculi* (rabbits).
5. *S. cymruensis* (rats).

The development occurs in parasitophorous vacuoles inside cells of the lamina propria of the small intestine of the final hosts (Fig. 4.58).

In contrast to the *Cystoisospora* species and *Toxoplasma gondii* in the case of *Sarcocystis* species, no schizogonic process occurs but only the initiation of a gamogony followed by a sporogony. The sporulation of the oocysts is already finished in the intestine so that these final hosts excrete in their feces already sporulated oocysts (or even sporocysts if the thin oocyst wall has become disrupted). These oocysts measure about 12–16 × 8–10 µm and always contain two sporocysts each with four sporozoites (Fig. 4.59a). They are infectious only for their obligate intermediate hosts (see Table 4.11). After oral uptake of such oocysts, the sporozoites hatch from the sporocysts and enter endothelial cells of blood vessels, wherein two generations of schizonts occur (Fig. 4.58) being followed after about 1 month by the formation of tissue cysts inside muscle fibres (Figs. 4.63, 4.64, 4.65, 4.66, 4.67, 4.68 and 4.69). The stages (cyst merozoites) inside the tissue cysts reach infectivity about 2–3 months after infection and keep it for at least 1 year (probably much longer). If final hosts ingest such cysts within raw meat, the cycle is closed and the formation of oocysts starts again (Fig. 4.58).

4. **Symptoms of disease**: The development of the oocysts inside dogs and cats proceeds mostly symptomless. In rare cases, diarrhoeas may occur for 1–2 days. (Humans are much more sensitive and develop severe diarrhoeas, nausea or fever when infected with *S. suihominis*).

Fig. 4.64 Macrophoto of whitish cysts of *Sarcocystis ovifelis* in an oesophagus of sheep compared a German money coin

Fig. 4.65 Macrophoto of long stretched, whitish appearing cysts of a *Sarcocystis* species in a mouse

5. **Diagnosis**: Demonstration of the sporulated oocysts or tiny sporocysts (Fig. 4.59) in feces with the help of fecal concentration methods (e.g. flotation, S.A.F.C.).
6. **Pathway of infection**: Oral uptake of tissue cysts inside raw meat of the intermediate hosts.
7. **Prophylaxis**: Feeding only meat that was deep frozen or cooked. Cleaning of floors from feces of dogs and cats by hot water or hot steam.
8. **Incubation period**: 1–2 days.
9. **Prepatent period**: Species specific: dogs: 8–10 days; cats: 5–14 days.
10. **Patency**: Cats/dogs: 6–7 weeks.

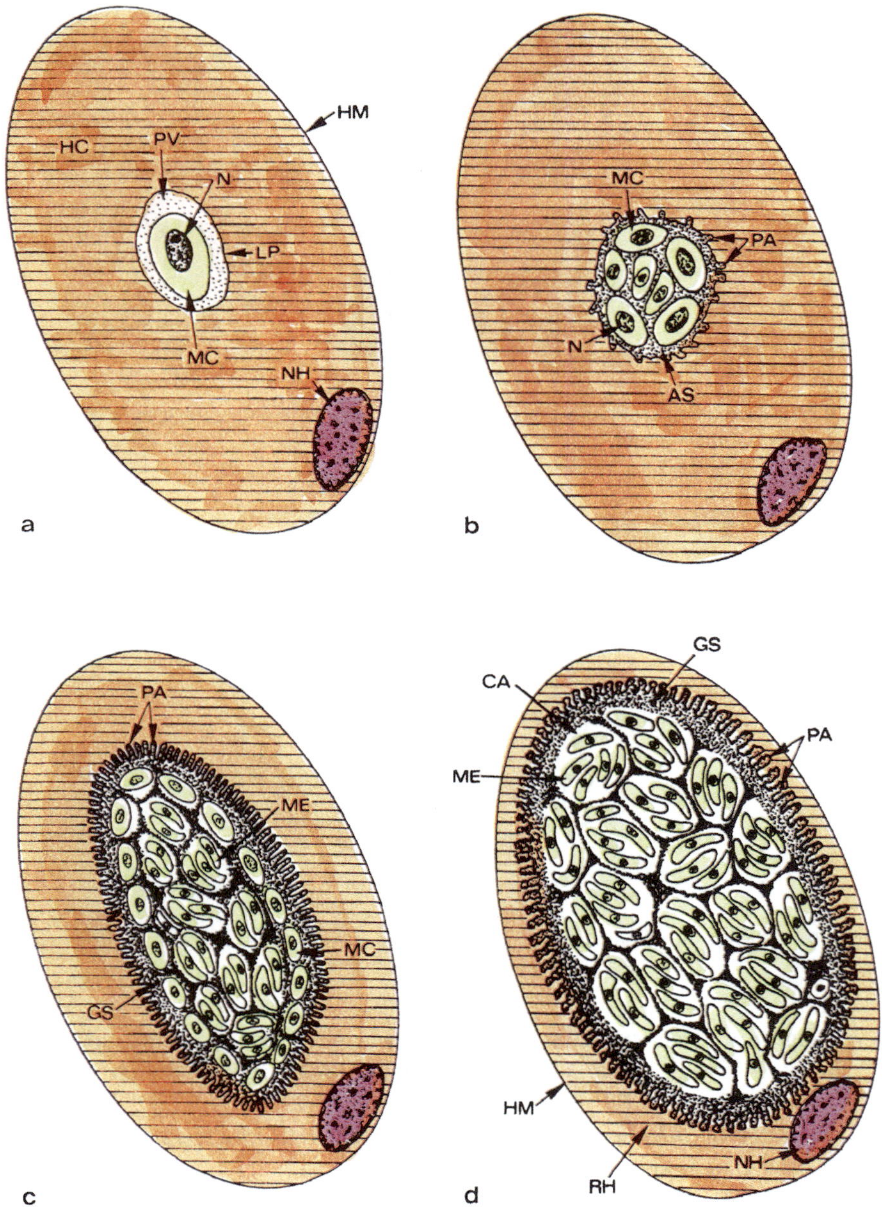

Fig. 4.66 Diagrammatic representation of the growth of a *Sarcocystis* cyst inside a muscle fibre after the penetration of second-generation merozoite. *AS* amorphous substance; *CA* chamber-like hollows; *GS* ground substance; *HC* host cell; *HM* host cell membrane; *LP* limiting membrane of parasitophorous vacuole; *MC* metrocyte; *ME* cyst merozoite; *N* nucleus; *NH* nucleus of host cell; *PA* protrusions; *PV* parasitophorous vacuole; *RH* remnants of host cell

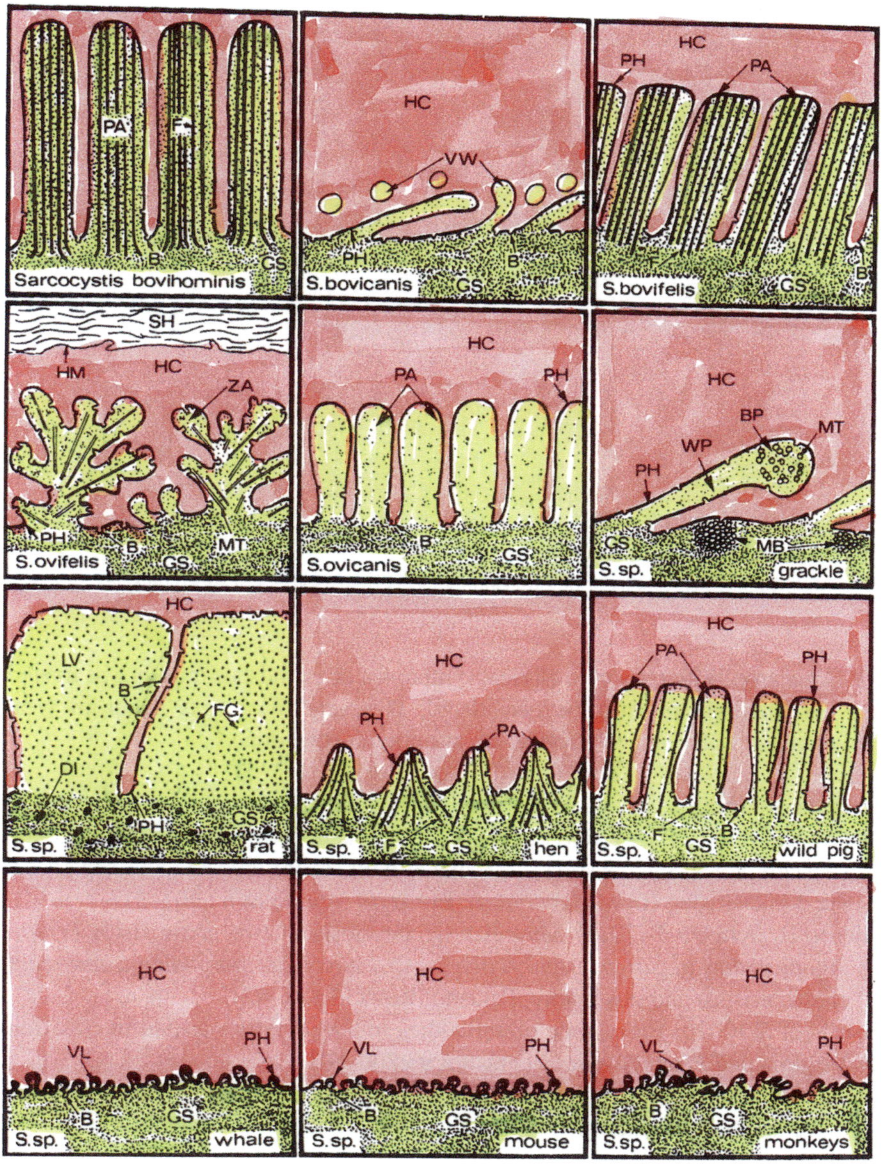

Fig. 4.67 Diagrammatic representation of the surfaces of species-specific *Sarcocystis* tissue cyst walls in muscle fibres. *B* sections through bubblelike protrusions or invaginations; *D* dense inclusions; *F* filaments; *FG* fine granules; *GD* ground substance of cyst; *HC* host cell; *HM* host cell membrane; *LV* large and broad protrusions; *MB* bundles of microtubules; *MT* microtubuli; *PA* palisade-like protrusions; *PH* primary cell wall; *SH* surrounding material; *VW* weak protrusions; *WP* weak palisades; *ZA* cauliflower-like protrusions

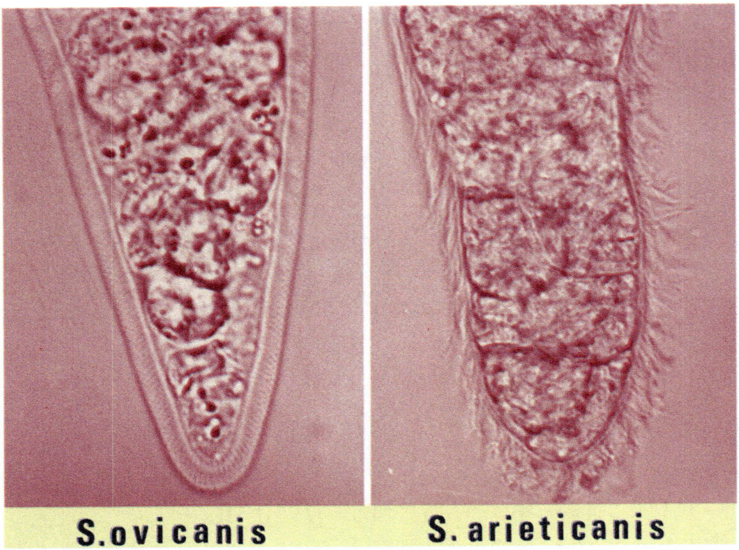

Fig. 4.68 Light micrographs of dog transmitted *Sarcocystis* species in sheep muscles

11. **Therapy**: With respect to clinical symptoms, dogs and cats must not be treated, but prophylaxis must be considered. If severe diarrhoeas occur in weak animals, water and electrolyte substitution must be done.

Further Reading

Dubey JP et al (2014) *Sarcocystis caninum* and *Sarcocystis svanai* n. spp. (Apicomplexa: Sarcocystidae) associated with severe myositis and hepatitis in the domestic dog (*Canis familiaris*). Eukaryot Microbiol 62:307–317.

Gjerde B, Schulze J (2014) Muscular sarcocystosis in two arctic foxes (*Vulpes lagopus*) due to *Sarcocystis arctica* n. sp.: sarcocyst morphology, molecular characteristics and phylogeny. Parasitol Res 113:811–821.

Hsu V et al (2010) Prevalence of antibodies to *Sarcocystis neurona* in cats from Virginia and Pennsylvania. J Parasitol 96:800–801.

Mehlhorn H, Heydorn AO (1978) The sarcosporidia (Protozoa, Sporozoa): life cycle and fine structure. Adv Parasitol 16:43–91.

Sykes JE et al (2011) Severe myositis associated with *Sarcocystis* spp. infection in 2 dogs. J Vet Intern Med 25:1277–1283.

4.6.4.2 Sarcosporidia of Ruminants

1. **Name**: Greek: *sarx* = meat; *sporos* = seed. Species names compose Latin names of intermediate and final hosts (e.g. *bovicanis* = Latin *bovis* = cattle; *canis* = dog).

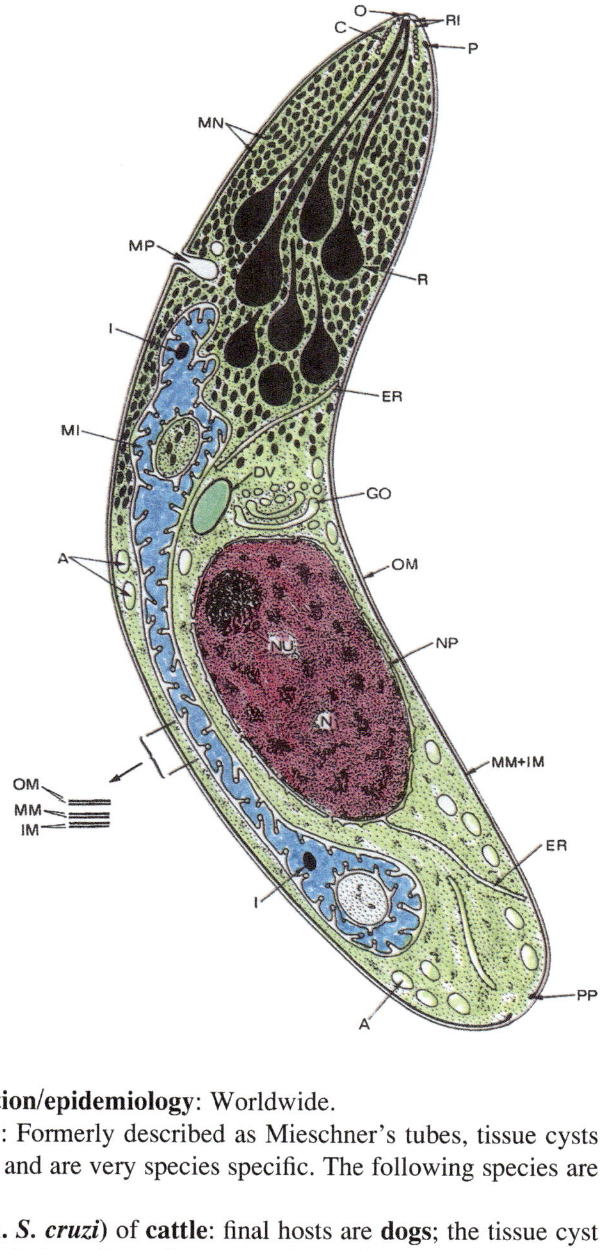

Fig. 4.69 Diagrammatic representation of the longitudinal section through a cystozoite (cyst merozoite) of *Sarcocystis* species, representing the basic appearance of the related stages of the genera *Toxoplasma, Besnoita, Frenkelia, Hammondia, Isospora* and *Eimeria*. *A* amylopectin; *C* conoid; *D* dense, spherical bodies; *DV* double-walled vesicle (now known as apicoplast being surrounded by 4 membranes); *ER* endoplasmic reticulum; *GO* Golgi apparatus; *I* dense inclusion; *IM* inner membrane; *MI* mitochondrion; *MM* middle membrane; *MN* micronemes; *MP* micropore; *N* nucleus; *NP* nuclear pore; *NU* nucleolus (karyosome); *O* opening of the conoid; *OM* outer membrane of pellicle; *P* anterior polar ring; *PP* posterior polar ring; *R* rhoptries; *RI* ring-like elements of the conoidal canopy

2. **Geographic distribution/epidemiology**: Worldwide.
3. **Biology, morphology**: Formerly described as Mieschner's tubes, tissue cysts occur in all ruminants and are very species specific. The following species are of importance:
   (a) ***S. bovicanis* (syn. *S. cruzi*)** of **cattle**: final hosts are **dogs**; the tissue cyst appears thin walled, since the surface protrusions (of about 3 cm in length) lay flattened along the cyst surface; pathogenic (Figs. 4.62 and 4.63).
   (b) ***S. bovifelis* (syn. *S. hirsuta*)** of **cattle**: final hosts are **cats**, tissue cysts are thick walled due to 5.5 μm long upright standing protrusions (Figs. 4.8 and 4.10).

(c) ***S. bovihominis* (syn. *S. hominis*)** of **cattle**: final hosts are **humans**; the tissue cysts are thick walled since the upright standing protrusions are up to 7–8 µm long (Figs. 4.62 and 4.63).
(d) ***S. ovifelis* (syn. *S. gigantea*)** of **sheep**: final hosts are **cats**; the cyst wall appears thin, cauliflower-like branched; protrusions measure 1–4.5 µm in length. (Figs. 4.62, 4.64, 4.66 and 4.67).
(e) ***S. ovicanis* (syn. *S. tenella*)** of **sheep**: final hosts are **dogs**; tissue cysts are short (4.5 µm) (Fig. 4.68).
(f) ***S. arieticanis*** of **sheep**: final hosts are **dogs**; cyst wall is rather thin; pathogenic (Figs. 4.68 and 4.70).
(g) ***S. medusiformis*** of **sheep**: final hosts are **cats**; cyst wall has lap-like protrusions; pathogenic.
(h) ***S. capracanis*** of **goats**: final hosts are **dogs**; cyst wall appears thick walled.
(i) ***S. hircicanis*** of **goats**: final hosts are **dogs**; cyst wall is rather thin.
(j) ***S. gracilis*** of **roe deer**: final hosts are **dogs**: they are microscopically tiny.

In all cases (**a–j**), the life cycle belongs to the obligatory type.

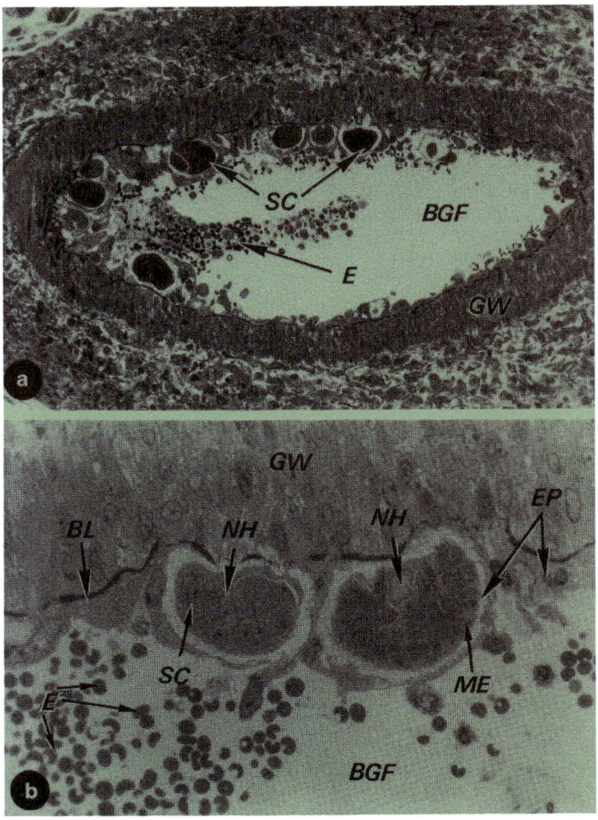

Fig. 4.70 (**a**, **b**) Light micrograph of *Sarcocystis arieticanis* schizonts in the walls of blood vessels of the omentum of sheep (intermediate hosts). The final hosts are dogs. *BGF* blood vessel; *BC* basal lamina; *E* erythrocyte; *EP* epithelial cell; *GW* wall of the blood vessel; *ME* merozoite; *NH* nucleus of the host; *SC* schizont

4. **Symptoms of disease (Sarcocystosis)**: Only those species which are transmitted by dogs are pathogenic for ruminants. During the acute phase of the disease (=persistence of endothelial schizonts), the following symptoms have been described: apathia, high fever, anaemia, icterus, haemorrhages, swellings of lymph nodes, liver and spleen, oedemas at ears and eventually death cases. As soon as cysts occur, symptoms become clearly reduced. After survival of a sarcocystosis, a specific immunity is produced by the infected animal; however, the length of the protection period is unknown, but in general symptoms are clearly reduced in cases of reinfections.
5. **Diagnosis**: During the acute phase (=first months after infection) the clear diagnosis is difficult but can be tried with the help of blood smears. Serological tests (IHAT, IFAT) are only successful after the acute phase of the disease.
6. **Pathway of infection**: Oral by uptake of oocysts and sporocysts in food contaminated with feces of final hosts.
7. **Prophylaxis**: Avoiding contact of ruminants with dogs and cats; do not feed raw meat to dogs and cats.
8. **Incubation period**: In the case of the intermediate hosts cattle/sheep 12–14 days (=length of the first phase of schizogony).
9. **Prepatent period**: Merozoites can be first seen about 12 days after infection.
10. **Patency**: Tissue cysts inside intermediate hosts (cattle) remain infectious for at least 1 year for final hosts.
11. **Therapy**: Therapy is still in an evaluation status. In **cattle**, **amprolium** (Amprolvet® 100 mg/kg bodyweight) showed convincing results when given daily for 30 days (acute symptoms were not severe). **Toltrazuril** (10 mg/kg bodyweight) was also successful if given orally on days 5, 10, 15, 20 and 25 after a postulated infection. The same promising result can be stated for **halofuginone-hydrobromide** (Stenerol®) at a dose of 0.66 mg/kg bodyweight on two consecutive days during the first schizogony in the hosts (goats, sheep, cattle). **Attention**: halofuginone may lead to diarrhoeas, loss of appetite and general intoxication if dosage is too high.

Further Reading

Akhlaghi M et al (2016) Molecular differentiation of bovine sarcocysts. Parasitol Res. doi:10.1007/s00436-016-5020-7.

Al Quraishy S et al (2014) *Sarcocystis arieticanis* (Apicomplexa: Sarcocystidae) infecting the heart muscles of the domestic sheep, *Ovis aries* (Artiodactyla: Bovidae), from K. S. A. on the basis of light and electron microscopic data. Parasitol Res 113:3823–3831.

Aryeetey M, Mehlhorn H, Heydorn AO (1980) EM studies on the development of *Sarcocystis capracanis* in goats. Zentralbl Bakteriol 247:534–550.

Gjerde B (2013) Phylogenetic relationships among *Sarcocystis* species in cervids, cattle and sheep inferred from the mitochondrial cytochrome c oxidase subunit I gene. Int J Parasitol 43:579–591.

Gjerde B (2016) The resurrection of a species: *Sarcocystis bovifelis* Heydorn et al., 1975 is distinct from the current *Sarcocystis hirsuta* in cattle and morphologically indistinguishable from *Sarcocystis sinensis* in water buffaloes. Parasitol Res 115:1–21.

Gjerde B et al (2016) Molecular differentiation of *Sarcocystis buffalonis* and *Sarcocystis levinei* in water buffaloes (*Bubalus bubalis*) from *Sarcocystis hirsuta* and *Sarcocystis cruzi* in cattle (*Bos taurus*). Parasitol Res. doi:10.1007/s00436-016-4998-1.

Heydorn AO, Mehlhorn H (1982) Fine structure of *Sarcocystis arieticanis*. Zentralbl Bakteriol Microbiol Hyg A 264:353–362.

Latif B et al (2013) High prevalence of muscular sarcocystosis in cattle and water buffaloes from Selangor, Malaysia. Trop Biomed 30:699–705.

Mehlhorn H, Heydorn AO (1978) The sarcosporidia (Protozoa, Sporozoa): life cycle and fine structure. Adv Parasitol 16:43–91.

Mehlhorn H et al (1975) Light and electron microscope studies on cysts of *Sarcocystis fusiformis* in calves. Zentralbl Bakteriol Orig A 232:373–391.

Moré G et al (2013) Development of a multiplex real time PCR to differentiate *Sarcocystis* spp. affecting cattle. Vet Parasitol 197:85–94.

Ng YH et al (2015) Modified use of methylene blue in the tissue compression technique to detect sarcocysts in meat-producing animals. Vet Parasitol 214:200–203.

Vangeel L et al (2013) Different *Sarcocystis* spp. are present in bovine eosinophilic myositis. Vet Parasitol 197:543–548.

4.6.4.3 Sarcosporidia of Horses

1. **Name**: German: *bertrami* = Bertram, family name of first describer of this species. Latin: *equicanis* = combination of host names *equus* = horse and *canis* = dog; *neurona* = these parasites enter the neurons of their host.
2. **Geographic distribution/epidemiology**: Worldwide (*S. bertrami*); New World (*S. neurona*).
3. **Biology, morphology**:
   (a) ***Sarcocystis bertrami* (syn. *S. fayeri, S. equinus, S. asinus*)**:

   The tissue cysts are macroscopically visible and may reach lengths of up to 15 mm. The protrusions of the primary cyst wall are hair-like. Dogs are worldwide final hosts; tissue cysts occur in practically all equids (e.g. horses, donkeys, mules, zebras): the cysts are of low-grade pathogenicity.

   (b) ***Sarcocystis equicanis***:

   These cysts are only light microscopically visible, since they measure only 600 × 50 µm. At first, they stand upright; later they become hair-like and reach a length of 5–11 µm. Final hosts are dogs, which excrete mostly sporocysts (15 × 10 µm) beginning 8 days after the infection. The patency of this excretion lasts up to 3 weeks. In horses, two schizogonies occur before cyst formation starts around the 30th day after the oral uptake of oocysts/sporocysts. The banana-shaped cystozoites reach infectivity about 60 days after the horse had ingested infectious oocysts (experimental data of Prof. Dr. A O Heydorn, Berlin).

(c) ***Sarcocystis neurona***:

This species was until now only described in South and North America. Horses are apparently only accidental hosts, which become infected by oral uptake of oocysts excreted by opossums (*Didelphis* species). Other intermediate hosts are many mammals (dogs, cats, racoons, monkeys) and several bird species, which are apparently the most important ones, since racoons feed rather often on birds. According to Dubey et al. (1991), only endodyogeny stages occur in the brain of horses, but no cysts. This underlines that horses are only accidental hosts.

4. **Symptoms of disease**: The species *S. bertrami* and *S. equicanis* induce mostly rather low-grade symptoms of disease (e.g. fever, apathia, anaemia, anorexia, loss of hair). Repeated infections remain mostly symptomless. *Sarcocystis neurona* may induce in horses the so-called **equine protozoal myelocephalitis (EPM)**, which may even lead to death. Similar symptoms are claimed for infections with *Neospora hughesi* of horses.
5. **Diagnosis**: Microscopical demonstration of tissue cysts (obtained by muscle biopsy) in squeezed unstained muscles or with the help of trypsin digestion of muscle portions.
6. **Pathway of infection**: Oral by uptake of oocysts/sporocysts excreted by final hosts such as dogs and racoons.
7. **Prophylaxis**: Avoid to feed dogs on raw horse meat.
8. **Incubation period**: Several months (at least two) in *S. equicanis* and a few weeks in *S. neurona*.
9. **Prepatent period**: Cysts are first seen in muscles about 30 days after infection.
10. **Patency**: Years.
11. **Therapy**: For treatment of infections with *Sarcocystis bertrami* and *S. equicanis* stages, see medicaments used in ruminants, but treatment is often not needed. However, for the treatment of EPM due to *Sarcocystis neurona* infections in the USA, the following recommendations are given:
    (a) **Ponazuril** (Marquis®): 5 g/kg bodyweight orally for 28 days or
    (b) **Nitazoxanide** (Navigator®): days 1–5: orally 25 mg/kg bodyweight, days 6–28: 50 mg/kg bodyweight.

    Note: Nitazoxanide may disturb the intestinal bacterial flora; thus, food additions (oils) should be made.

Further Reading

Aleman M et al (2016) *Sarcocystis fayeri* in skeletal muscle of horses with neuromuscular disease. Neuromuscul Disord 26:85–93.

Arias M et al (2012) Exposure to *Sarcocystis* spp. in horses from Spain determined by Western blot analysis using *Sarcocystis neurona* merozoites as heterologous antigen. Vet Parasitol 185:301–304.

Dubey JP (2004) Equine protozoal myeloencephalitis. In: Coetzer JA, Tustin RC (eds) Infections diseases of live stak, 2nd edn. Oxford University Press, Oxford.
Dubey JP et al (1991) *Sarcocystis neurona* n. sp.: the etiological agent of equine protozoal myeloencephalitis. J Parasitol 77:212–218.
Dubey JP et al (2015) An update on *Sarcocystis neurona* infections in animals and equine protozoal myeloencephalitis (EPM). Vet Parasitol 209:1–42.
Dubey JP et al (2016) Ultrastructure of *Sarcocystis bertrami* sarcocysts from a naturally infected donkey (*Equus asinus*) from Egypt. Parasitology 143:18–23.
Edwards GT (1984) Prevalence of equine *Sarcocystis* in British horses and comparison of two detection methods. Vet Rec 115:265–267.
Erber M, Geisel O (1981) Vorkommen und Entwicklung von 2 Sarkosporidienarten des Pferdes. Parasitol Res 65:283–291.
Pusterla N et al (2015) Daily feeding of diclazuril top dress pellets in foals reduces seroconversion to *Sarcocystis neurona*. Vet J 206:236–238.

4.6.4.4 *Sarcocystis* Species of Pigs

1. **Name**: Combinations of Latin names of final and intermediate hosts (*sus* = pig; *homo* = human; *canis* = dog).
2. **Geographic distribution/epidemiology**: Worldwide.
3. **Biology, morphology**: In pigs two *Sarcocystis* species occur:
   (a) ***Sarcocystis suihominis***: Final hosts are humans, who excrete oocysts and sporocysts within the feces (see Fig. 4.62). Tissue cysts reach in pigs a length of up to 5 mm and are found inside fibres of the skeletal muscles, muscle cells of the heart and brain. In light microscope, visible tissue cysts appear to be covered by a 1 mm thick wall, which in electron microscopy shows that there are large numbers of 14 μm long protrusions, which are lying flat along the surface of the cyst. The interior of the cyst is subdivided by amorphous material, thus forming chamber-like hollows, which are filled in young cysts by so-called large metrocytes and in older cysts by so-called cyst merozoites (cystozoites) like those in Figs. 4.73 and 4.74, which reach a length of up to 16 μm.
   (b) ***Sarcocystis suicanis* (syn. *S. miescheriana*)**: Final hosts are dogs, while the cysts inside the muscles of pigs reach a length of 1.5–2 mm and are found in heart muscle cells as well as in fibres of the skeletal muscles. The cyst wall reaches a thickness of up to 5 μm and is characterized by numerous palisade-like protrusions. The inner construction of the cyst does not differ from that of other *Sarcocystis* muscle cysts.

   In both species, two generations of schizonts precede cyst formation, which starts around the 30th day after infection by oral uptake of infectious oocysts within fecally contaminated food. About 60 days after infection of the pigs, the cysts are mature, i.e. contain the long banana-shaped cyst merozoites which will introduce the development of gamonts in the final hosts **dogs** or **humans** if they ingest cyst-containing meat.
4. **Symptoms of disease**: The severity of symptoms of disease of pigs depends on the amount of ingested oocysts/sporocysts within food that is contaminated by

feces of an infected dog. One million oocysts would lead to death of the infected piglet. An acute **sarcocystosis** during the repeated schizogonic reproduction inside endothelial cells is characterized by fever with two peaks (showing maxima on days 5 and 19 and on days 11–15 past infection). Especially during the second fever phase, the pigs show apathia and anaemia. Just prior to death, a significant cyanotic red coloration of the ears, of the tail and/or of other regions of the skin occurs (Figs. 4.71 and 4.72).

5. **Diagnosis**: Sarcocystosis in pigs during the first month can be diagnosed by examination of Giemsa-stained blood smears and after 2 months by obtaining muscle probes and digesting them followed by Giemsa staining of smear preparations (Fig. 4.73). Serological methods (IFAT, IHA, ELISA) may be used, too, but they are mostly only helpful if muscle cysts are already formed (Fig. 4.74).
6. **Pathway of infection**: Oral by uptake of oocysts or sporocysts excreted within feces of dogs (*S. suicanis*) or humans (*S. suihominis*).

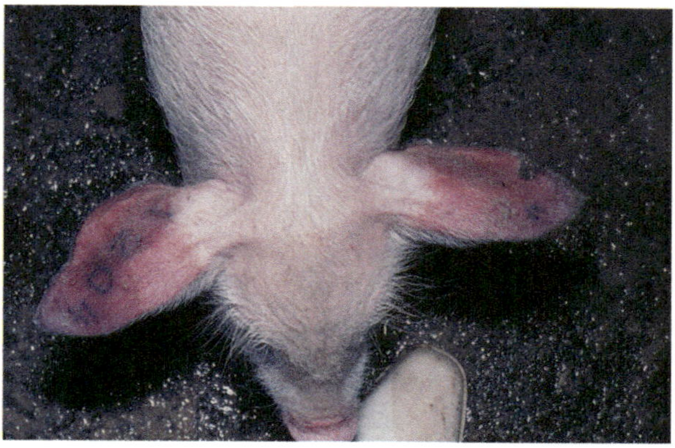

Fig. 4.71 This *Sarcocystis*-infected pig shows large haemorrhagies inside the ears due to numerous schizonts

Fig. 4.72 Macrophoto of two pigs of exact the same age. The *right* one had been experimentally infected with a rather low dosage of oocysts of *S. suihominis*, which led to a reduced growth

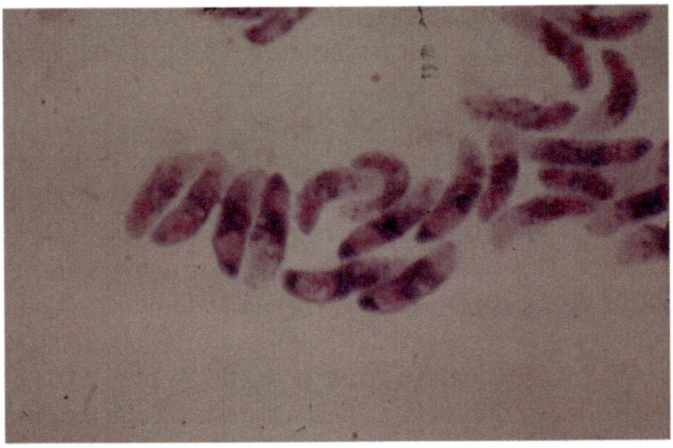

Fig. 4.73 Light micrograph of Giemsa-stained cystozoites of *Sarcocystis* species

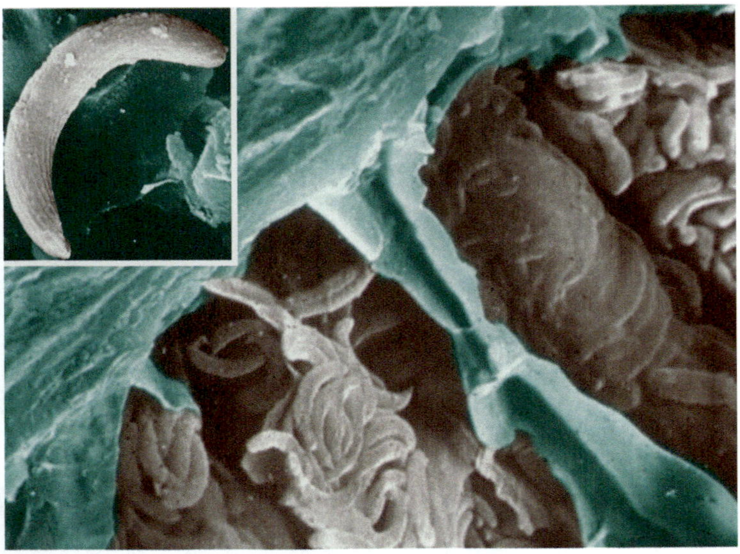

Fig. 4.74 Scanning electron micrograph of the periphery of a *Sarcocystis* cyst showing "chambers" containing cystozoites. *Inset*: Single banana-shaped cyst merozoite

7. **Prophylaxis**: Avoidance of contact of pigs with feces of dogs and of humans. Dogs should only be fed with well-done cooked meat or after deep freezing for at least 24 h. **Attention**: Also humans should not eat raw or undercooked meat of pigs.
8. **Incubation period**: In pigs: 5–6 days; humans <24 h.
9. **Prepatent period**: About 1 month (=appearance of first cysts in muscle fibres).
10. **Patency**: Tissue cysts remain infectious for at least 1 year after oral infection.
11. **Therapy**: Unknown, pigs can be experimentally treated with toltrazuril.

Further Reading

Chlabra MB, Samantaray S (2013) *Sarcocystis* and sarcocystosis in India: status and emerging perspectives. J Parasit Dis 37:1–10.

Coelho C et al (2015) Unraveling *Sarcocystis miescheriana* and *S. suihominis* infections in wild boar. Vet Parasitol 212:100–104.

Fayer R et al (2015) Human infections with *Sarcocystis* species. Clin Microbiol Rev 28:295–311.

Heydorn AO, Mehlhorn H (1978) Light and electron microscopic studies on *Sarcocystis suihominis*: the schizogony preceeding cyst formation. Zentralbl Bakteriol Orig A 240:123–134.

Mehlhorn H, Heydorn AO (1978) The Sarcosporidia: life cycle and fine structure. Adv Parasitol 16:43–103.

Mehlhorn H, Heydorn AO (1979) Electron microscopical study on gamagony of *Sarcocystis suihominis* in human tissue cultures. Parasitol Res 58:97–113.

Mehlhorn H et al (1985) Announcement of the establishment of neohepantotypes for some important *Sarcocystis* species. Parasitol Res 71:689–692.

Sommer I et al (1991) Biochemical and immunological characterization of major surface antigens of *Sarcocystis muris* and *S. suicanis* cyst merozoites. Parasitol Res 77:2004–2011.

Sommer I et al (1992) A comparison of sporozoites and cyst merozoite surface proteins of *Sarcocystis*. Parasitol Res 78:398–403.

Yan W et al (2013) Morphological and molecular characterization of *Sarcocystis miescheriana* of pigs. Parasitol Res 112:975–980.

4.6.4.5 Sarcosporidia of Birds

1. **Name**: Greek: *horvathi* = honouring the Hungarian discoverer Horvath; *falcatula* = named according to the suggested final hosts = falcons; *calchasi* = remembering the archaic Greek seer Kalchas who dreamed that a hawk hided itself and suddenly attacked a careless dove. Kalchas thus advised the Greek army which attacked the famous town Troja to leave back the wooden Trojan horse (with soldiers inside) and to hide themselves until the doors of the town were opened. This should be done by their soldiers during the next night, after the Trojan people have transported this "gift" into their town.
2. **Geographic distribution/epidemiology**: Worldwide.
3. **Biology, morphology**: In the muscles of domestic chickens at least two *Sarcocystis* species occur (*S. horvathi* and *S.* sp.). The tissue cysts of *S. horvathi* develop protrusions with a length of up to 3 μm. *S. falcatula* is found in the muscles of canary birds and doves, while opossums are claimed to act as final hosts. However, it seems that predator birds such as falcons and hawks may also act as final hosts. This is the case, e.g., in *Sarcocystis calchasi*, which leads in doves to a strongly reduced ability to fly. Further species are *S. columbae* and *S.* spec. from sparrow hawk. Thus, infected birds are an easy prey for predator birds, which thus become final hosts. Highly infected doves may even die in high ratios.

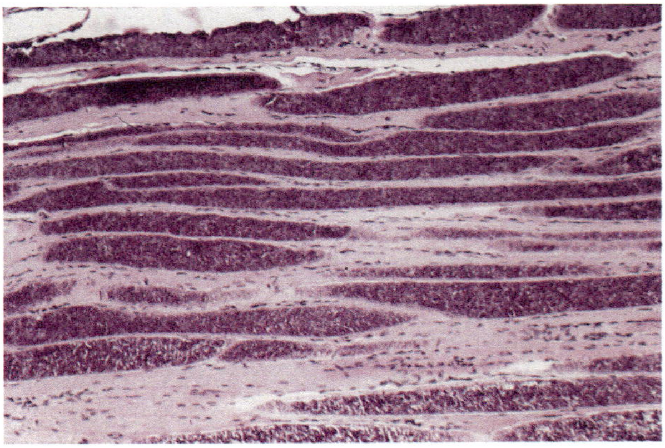

Fig. 4.75 Light micrograph of a section through several tissue cysts of *Sarcocystis calchasi* inside the muscle fibres of dying doves

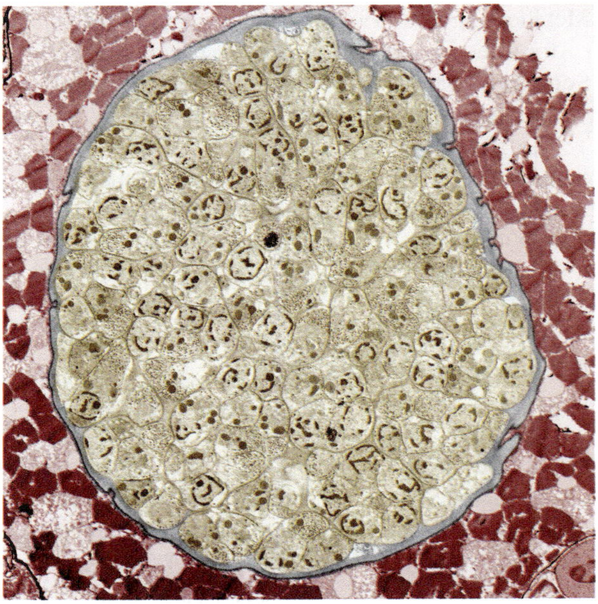

Fig. 4.76 Transmission electron micrograph of a section through a tissue cyst of *Sarcocystis calchasi* in the damaged muscles of a dove

4. **Symptoms of disease (Sarcocystosis)**: Low-grade infections remain inapparent apart from a slight weakness during flight activities. However, heavy infections may lead to death (Figs. 4.75 and 4.76).
5. **Diagnosis: Prey birds**: Microscopical demonstration of cysts and cyst merozoites in muscle probes of dead animals by digestion. However, symptoms of necrosis in muscles may have also other reasons. **Predator birds**: Demonstration of oocysts inside the feces by flotation method.
6. **Pathway of infection: Doves/prey birds**: Oral by uptake of oocysts within feces of raptor birds. **Predator birds**: Ingestion of tissue cysts within muscles of prey birds.

7. **Prophylaxis**: Avoidance of contamination of bird rearing facilities with feces of raptor/predator birds.
8. **Incubation period**: 2–8 weeks (after experimental infections). In cases of natural infections timing is not known, since either low-grade infections remain undetected and high-grade infections end with sudden death.
9. **Prepatent period: Prey birds**: Merozoites could be demonstrated about 10 days after an experimental infection of doves. Tissue cysts can be first seen after about 4 weeks. In **predator birds**: Timing of excretion of oocysts is unknown since experimental data are lacking.
10. **Patency**: In low-grade infections of prey birds, tissue cysts start to become damaged about 4 months after experimental infections. Length of the excretion period of oocysts by raptor birds is unknown.
11. **Therapy**: Unknown. However, infected doves might be treated with toltrazuril.

Further Reading

Gjerde B (2014) Molecular characterization of *Sarcocystis rileyi* from a common eider (*Somateria mollissima*) in Norway. Parasitol Res 113:3501–3504.

Olias P et al (2009) A novel *Sarcocystis* associated encephalitis and myositis in racing pigeons. Avian Pathol 38:121–128.

Olias P et al (2010a) *Sarcocystis*: species lethal for domestic pigeons. Emerg Inf Dis 16:497–499.

Olias P et al (2010b) *Sarcocystis calchasi* sp. nov. of the domestic pigeon and the Northern goshawk. Parasitol Res 106:577–585.

Olias P et al (2010c) *Sarcocystis calchasi* is distinct of *S. columbae* from the wood pigeon and *Sarcocystis* sp. from the sparrow hawk. Vet Parasitol 171:7–14.

Rimoldi G et al (2013) An outbreak of *Sarcocystis calchasi*: encephalitis in multiple psittacine species within an enclosed zoological area. J Vet Diagn Invest 25:775–781.

4.6.5 *Toxoplasma gondii*

1. **Name**: Greek: *toxon* = arch; *plasma* = cell contents; *gondii* = belonging to the gundi = African rodental mouse-like appearance, where the French scientists Nicolle and Manceaux detected 1907 (in Tunis) the bended infectious stages (today: tachy- and bradyzoites).
2. **Geographic distribution/epidemiology**: Worldwide. In practically all warm-blooded animals and in humans different stages of *T. gondii* may occur. Especially very young animals and animals with an immune suppression are highly endangered and may even die in cases they become infected for the first time.
3. **Biology, morphology: Final hosts** of this protozoan parasite are cats (e.g. domestic cats but also any other member of the widespread genera of the felids (Fig. 4.77), which excrete unsporulated oocysts (Fig. 4.78a) within their feces. These oocysts sporulate outside of the body of the final host and develop in their interior two sporocysts each containing 4 sporozoites

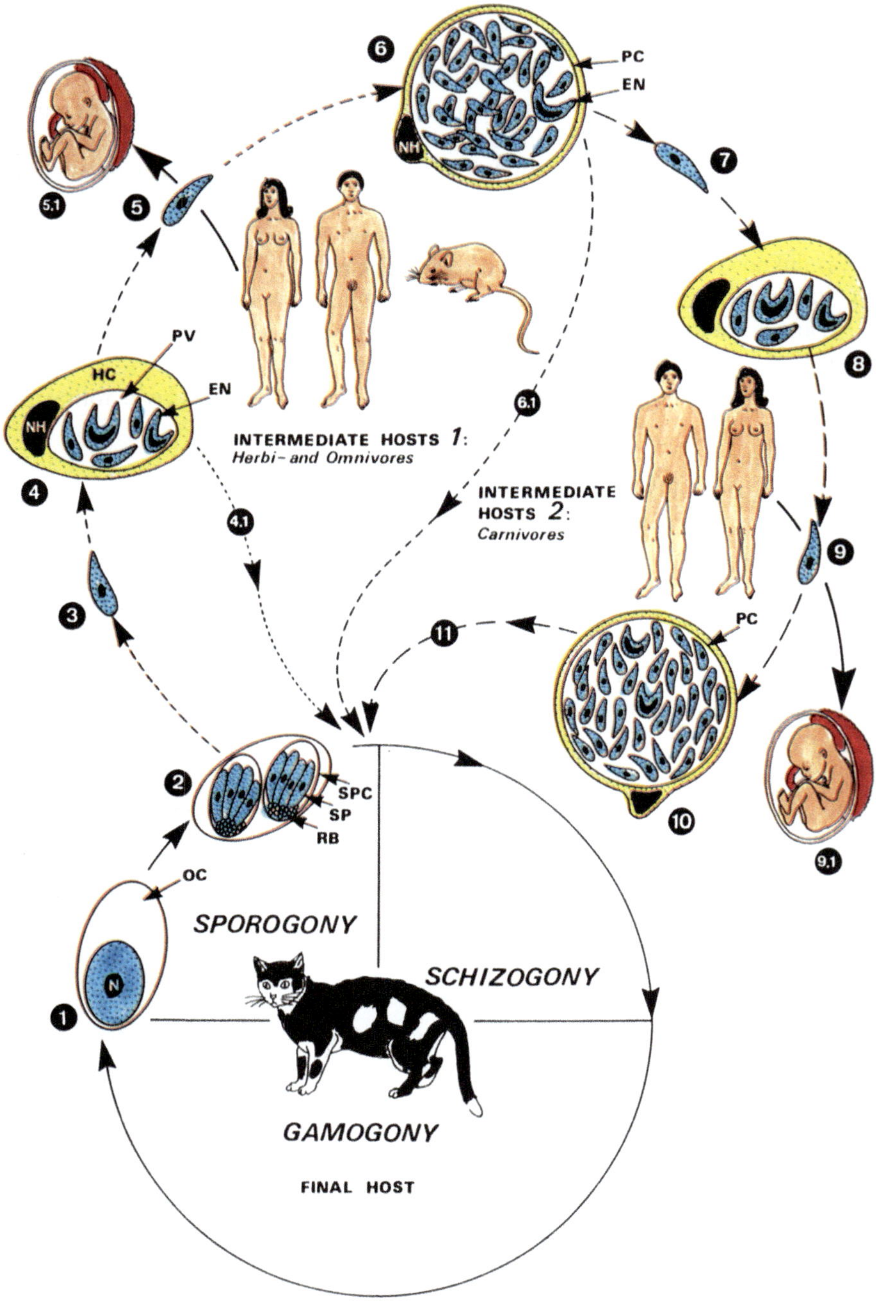

Fig. 4.77 Diagrammatic representation of the life cycle of *Toxoplasma gondii*. At the position of the human couple and the mouse, practically all warm-blooded vertebrates might be placed. The typical coccidian life cycle proceeds in the intestinal epithelium of felids (final host) which are infected by oral uptake of sporulated oocysts (**2**), ingestion of "pseudocysts" (**4.1**, **8**) or tissue cysts (**6.1**, **11**) with meat of various intermediate hosts (of 2 types). (**1**) Unsporulated oocysts are excreted with feces. (**2**) Sporulation (i.e. formation of sporocysts and sporozoites) occurs outside

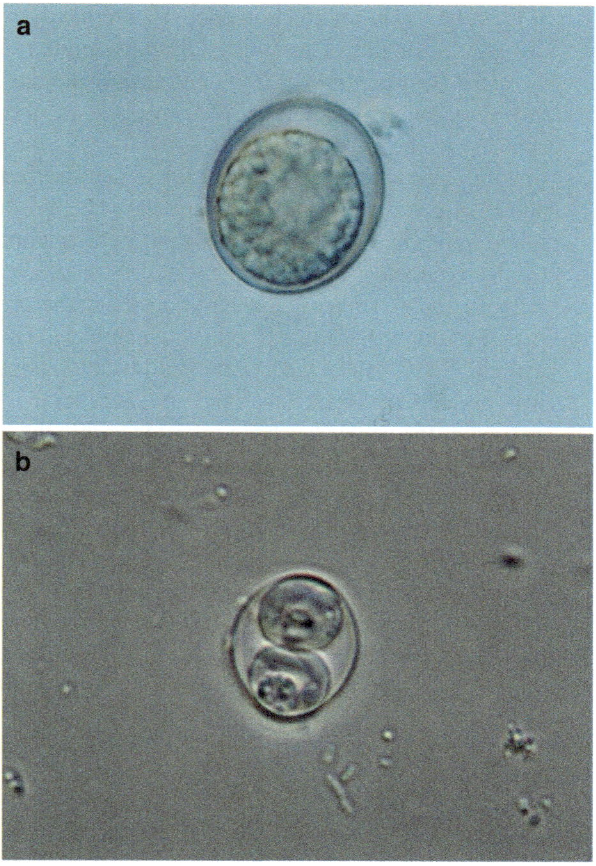

Fig. 4.78 *Toxoplasma gondii:* Light microscopic representations of (**a**) an unsporulated oocyst and (**b**) a sporulated cyst

Fig. 4.77 (continued) the final host. These stages may become spread by transport hosts such as flies and cockroaches. (**3**) After ingestion of oocysts by intermediate hosts of type 1, the sporozoites are set free inside its intestine and penetrate numerous types of extratestinal cells (i.e. cells of the RES). (**4**) Inside the host cell, the parasite reproduce by a typical binary fission (endodyogeny) leading to "pseudocysts" which are filled with merozoites (i.e. tachyzoites). (**4.1**) After ingestion of such pseudocysts, cats may become infected. (**5**) Free merozoite (tachyzoite) in blood or lymph fluid after bursting of a pseudocyst. (**5.1**) When the first infection occurs in pregnant women (or animals), these merozoites may pass into the placenta and infect the fetus, leading to severe damage. (**6**) Formation of tissue cysts, mainly inside brain and muscle cells. After several endodyogenies, these cysts (waiting stages) contain numerous cyst merozoites (bradyzoites, cystozoites) which are infectious for cats (**6.1**). (**7–10**) When carnivorous animals or man (intermediate hosts of type 2) ingest such tissue cysts with raw or insufficiently cooked meat, reproduction (see **3–6**) via pseudocysts is repeated, leading to the same tissue cysts (**10**) as in intermediate hosts of type 1. Diaplacental transmission (**9.1**) may also occur (see **5.1**), leading to congenital toxoplasmosis. (**11**) Cats may also become infected by ingestion of tissue cysts from type 2 intermediate hosts. Then they pass oocysts after 3–5 days, whereas the prepatent period is longer after an inoculation of pseudocysts (9–11 days) or oocysts (21–24 days). *EN* division by endodyogeny; *HC* host cell; *N* nucleus; *NH* nucleus of host cell; *OC* oocyst; *PC* primary cyst wall; *PV* parasitophorous vacuole; *RB* residual body; *SP* sporozoites; *SPC* sporocysts (for related species, see Coccidia/Table 5)

(Fig. 4.78b). These infectious oocysts are nearly spherical measuring in mean 10–14 × 8–12 μm. Thus, they are considerably smaller than the oocysts of *Isospora* (*Cystoisospora*) *felis*. The size of the infectious *Toxoplasma* oocysts, however, ranges in the level of the oocysts of the *Hammondia* and *Besnoitia* species (see below). If a cat ingests the oocysts, the released sporozoites enter at first the intestinal wall, stay there for some time, return back after about 12 days and then penetrate into epithelial cells. Therein they start at first the schizogonic reproduction being followed by the formation of female gametes (=macrogametes) and flagellated male gametes (=microgametes) (Fig. 4.70). After fusion of a macro- and a microgamete, the zygote becomes ensheathed by fusion of the wall-forming bodies, which had been produced inside the macrogamete. This ensheathed zygote is now called oocyst (Fig. 4.78a), and its stiff wall protects the inner development (sporulation), which, however, only starts as soon as the oocyst has been excreted within the feces. Exclusively outside of the body, each oocyst *Toxoplasma gondii* develops two sporocysts with 4 sporozoites (Fig. 4.78b). Thus, these oocysts belong to the *Isospora* type.

If such oocysts are ingested by intermediate hosts (e.g. pigs, cattle, sheep, but also humans), the sporozoites that were set free in the intestine enter the blood vessels and were ingested by defence cells such as macrophages being included therein in so-called parasitophorous vacuoles (Fig. 4.79). Therein they start a peculiar binary fission (**endodyogeny**); see Figs. 4.68 and 4.77). Finally, the therein produced **tachyzoites** (Figs. 4.79 and 4.80) enter muscle fibres and/or brain cells and start there again binary reproduction leading finally to the

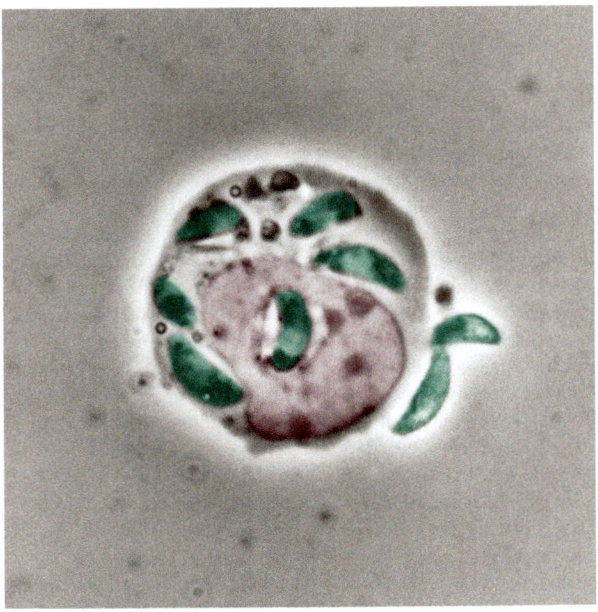

Fig. 4.79 *Toxoplasma gondii*: Light microscopic representation of a macrophage being infected by *T. gondii* tachyzoites, which lay inside the cell within so-called parasitophorous vacuoles. They will later become reproduced by endodyogeny and thus lead to rupture of this host cell

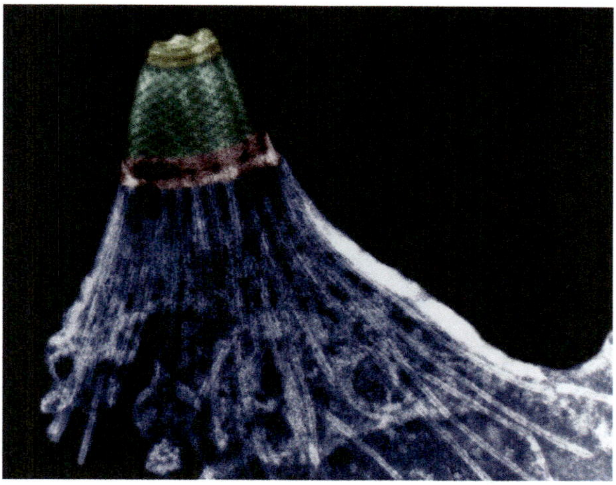

Fig. 4.80 Negative staining of the anterior pole of a tachyzoite of *T. gondii* showing the protruded conoid (consists of microtubules) and the two polar rings, where the subpellicular microtubules are attached. The typical, three-layered pellicle, which fully covers the surface of the parasite, has been removed during preparation

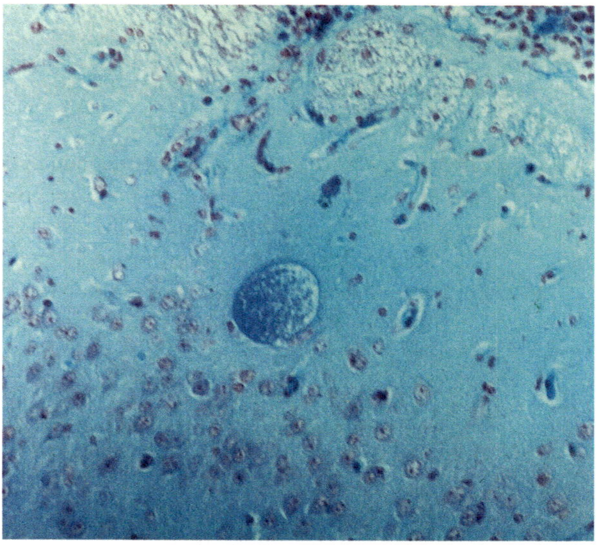

Fig. 4.81 Light micrograph showing a section through brain tissues containing a *Toxoplasma* tissue cyst

formation of **tissue cysts** which contain the slowly reproducing **bradyzoites** (Figs. 4.81 and 4.82).

These bradyzoites in cysts and the previously occurring tachyzoites are infectious for cats, if they feed raw or undercooked meat of intermediate hosts. However, also predator animals and humans may become infected, if they ingest such stages (Fig. 4.77). This explains why *Toxoplasma gondii* is probably the most ubiquitous parasite on earth.

4. **Symptoms of disease**:
   (a) **Cats**:
   In the case of a low-grade infection, no clinical symptoms occur. In cases of high-grade infections, cats show phases of diarrhoea, which stop

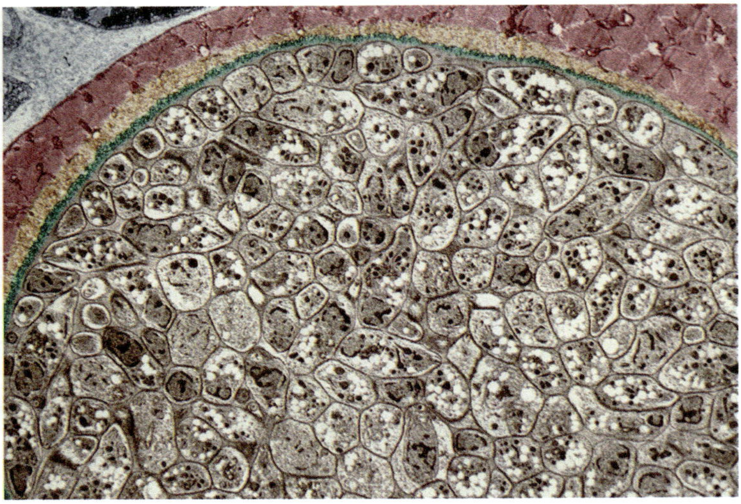

Fig. 4.82 Transmission electron micrograph of a section of a *Toxoplasma gondii* tissue cyst inside a muscle fibre. Note that the cyst fills nearly the whole muscle fibre, which shows degenerated material (*yellow*) close to the smooth cyst limitation. The cyst interior is filled by numerous cross-sectioned bradyzoites, which are infectious for cats if they ingest this piece of meat in a raw status

often without treatment and may introduce a rather solid immunity. Old cats or immunosuppressed cats, however, may also become severely ill, showing an acute toxoplasmosis which is due to the infection of numerous macrophages and different tissues (e.g. muscles or legs and heart). Also the brain of such cats may become infected. Thus, such cats may even die from an own generalized toxoplasmosis, but it is much more common in infected intermediate hosts. In these cases, mostly hepatitis, myocarditis, pneumonia and encephalitis are diagnosed. In cases of prenatal infections, often abortions of cat puppies may occur, too.

(b) **Dogs**:

Dogs are apparently bad hosts for *T. gondii*, since mainly only young dogs show severe symptoms of disease. However, toxoplasmosis might be combined with infections such as pneumonia or diarrhoeas. Also cases of ataxia have been reported in heavily infected dogs.

(c) **Cattle**:

These animals—although being infected—show in general no or only low-grade infections.

(d) **Sheep and goats**:

These hosts are highly susceptible and reach seroprevalence rates of up to 100 %. Pregnant animals show often cases of abortus. Low infections introduce in general few and low-grade symptoms of disease.

(e) **Pigs**:

Toxoplasmosis affects practically only very young animals, which then show apathia, fever, pneumonia, cyanosis and/or paresis of the hind legs.

However, infectious cysts remain lifelong present (especially, since pigs are often slaughtered already at an age of 1 year).

(f) **Rodents, rabbits, minks, birds**:

These animals may all be infected; however, the grade of symptoms of disease is low. In cases that those animals were kept close together, local epidemics may occur among young animals.

5. **Diagnosis**:
   (A) **Final hosts cats**:

   Demonstration of the typical unsporulated oocysts in fresh feces (Fig. 4.78a) or sporulated ones (Fig. 4.78b) in stools, which have been situated for several days outside of the body.

   (B) **Intermediate hosts**:

   Microscopical demonstration of cysts in muscle probes of infected animals. Also a broad spectrum of serological tests give indications; however, it remains often doubtful whether an infection is acute or old. **Note**: In some countries (e.g. Germany), toxoplasmosis of dogs and cats has to be announced to the veterinarian authorities!
6. **Pathway of infection of cats**: (**1**) Oral uptake of sporulated oocysts within feces of other cats or (**2**) ingestion of mice containing muscle cysts of *T. gondii*,
7. **Prophylaxis**:
   (1) Do not feed raw meat to cats.
   (2) Prohibit entrance of cats into stables or farmed animals.
   (3) In households and farms: Clean regularly defecation places of cats.
   (4) Sheep might be protected by vaccines (e.g. Ovilis Toxovax®).
   (5) Humans: **Important**: Pregnant women without antibodies against *T. gondii* should avoid contact with cats and their feces and they should not eat raw meat. Furthermore they should be tested at monthly intervals for rising antibodies.
8. **Incubation period**: Variable depending on the amount of ingested oocysts; mostly: a few up to 7 days.
9. **Prepatent period**: Cats excrete oocysts:
   (a) In case of oral uptake of sporulated oocysts; 21–24 days;
   (b) In case of feeding pseudocysts (infected macrophages) within meat of an infected intermediate host: 9–11 days.
   (c) In case of feeding of mature muscle cysts: 3–5 days.
10. **Patency**: 1–15 days.
11. **Therapy**: Application of **toltrazuril** (Baycox®); a permanent application of 5–10 mg/kg bodyweight suppresses the excretion of oocysts by cats and thus can be used in households with a pregnant woman without antibodies against *T. gondii* (Rommel et al. 2006).

    In cases of an acute toxoplasmosis of dogs and cats, the application of **clindamycin** or **sulfadiazine** plus **trimethoprim** stops severe symptoms.

    Attention: **Pyrimethamin** (Doraprim®) cannot be used, since it induces teratogenic effects in doses needed to eliminate the *Toxoplasma* stages.

Further Reading

Bacci C et al (2015) Detection of *Toxoplasma gondii* in free-range, organic pigs in Italy. Int J Food Microbiol 202:54–56.

Castans P et al (2016) Experimental ovine toxoplasmosis. Vet Res 47:43.

Cornelissen JBWJ et al (2014) An experimental *Toxoplasma gondii* dose response challenge model to study therapeutic or vaccine efficacy in cats. PLoS One 9: e104740.

Djokic V et al (2016) *Toxoplasma gondii* infection in pork produced in France. Parasitology 143:537–567.

Dubey JP et al (2014) Isolation and genetic characterization of *T. gondii* from black bears (*Ursus americanus*), bobcats (*Lynx rufus*) and feral cats (*Felis catus*). J Eukaryot Microbiol 62:410–415.

Lopes WDZ et al (2013) Sexual transmission of *Toxoplasma gondii* in sheep. Vet Parasitol 195:47–56.

Munoz J, Mayer DCG (2016) *Toxoplasma gondii* and *Giardia duodenalis* infections in domestic dogs in New York City public parks. Vet J. doi:10. 1016/j.tvjl.2016.02.015.

Nicolle J, Manceaux G (1909) Sur un protozoaire nouveau du gondi. Arch Inst Pasteur 2:97–101.

Rommel M et al (2006) Textbook of veterinarian parasitology, 6th edn. Paul Parey, Berlin.

Sarvi S et al (2015) Cattle toxoplasmosis. Asian J Trop Med 2015:120–126.

Verheist D et al (2014) Seroprevalence of *T. gondii* in domestic sheep. Vet Parasitol 205:57–61.

Zhang XX et al (2015) Seroprevalence and risk factors of *T. gondii* infection in stray dogs. Parasitol Res 114:4725–4729.

4.6.6 *Hammondia* Species

1. **Name**: The genus name honours the American parasitologist Datus M. Hammond (1911–1974) who initiated together with German scientists (Erich Scholtyseck, Heinz Mehlhorn) and members of the groups of Bill Chobotar and Ron Fayer basic research on economically important coccidian parasites. One of the species names honours my friend Prof. Dr. Alfred-Otto Heydorn (Berlin), who received the Robert-Koch Research price and discovered together with Prof. Dr. Michael Rommel (Hannover) and my group numerous coccidian life cycles.
2. **Geographic distribution/epidemiology**: Worldwide; in Germany up to 1 % of dogs and cats act as final hosts.
3. **Biology, morphology**: The *Hammondia* species are coccidians belonging to the Protozoa and parasitize in the intestinal cells of dogs and related species or in cats and relatives as final hosts, but use a broad spectrum of intermediate hosts. The tissue cyst can morphologically not become differentiated from those of *Toxoplasma gondii* (Figs. 4.81 and 4.82). However, they can be

differentiated with the help of serological test systems and by the fact that their life cycle is an obligatory one.

(a) ***Hammondia heydorni***:

Final hosts (excreting oocysts) are dogs, foxes and coyotes. Intermediate hosts are cattle, sheep, goats, roe deer, elks and also guinea pigs. Also dogs may act as intermediate hosts. The oocysts excreted by the final hosts had been named at first as "small stages of *Isospora bigemina* of dogs", measuring 10–14 × 9–13 μm. They are colourless and possess only a rather thin oocyst wall. In humid surroundings and room temperatures, they sporulate, i.e. they produce inside two sporocysts each with four infectious sporozoites. Intermediate hosts become infected by oral uptake of such sporulated oocysts. After hatching in the intestine, the sporozoites enter the intestinal wall and reach via bloodstream muscles, where they enter and start the formation of thin-walled tissue cysts looking like those of *T. gondii* (Fig. 4.81). The stages (cystozoites) inside the muscle cyst are infectious for the final hosts, which excrete fresh oocyst about 7–17 days after infection. **Attention**: In cases when dogs ingest such sporulated oocysts, they may also develop tissue cysts as it is the case in typical intermediate hosts.

(b) ***Hammondia hammondi***:

Final hosts are cats, while mice, rats, pigs, guinea pigs, hamsters, dogs and monkeys of the genus *Saguinus* may become intermediate hosts. The oocysts measure 11–13 × 11 μm and have a rather thick wall (of up to 0.5 μm in diameter). They appear pale and colourless. After a phase of different endodyogenies during the first 11 days after the infection of intermediate hosts, the formation of tissue cysts starts in skeletal muscle fibres (rarely also in cells of heart and brain). The tissue cysts in muscle fibres reach a size of 0.4 × 0.1 mm, have like *Toxoplasma gondii* no chamber-like subdivisions and contain exclusively cyst merozoites, measuring 7 × 2 μm. Metrocytes never occur. The final host does never serve as additional intermediate host!

4. **Symptoms of disease**: Final hosts show practically no symptoms of disease, while intermediate hosts may suffer severely from infections with *H. hammondi*, such as myositis and myocarditis leading to a high mortality rate. Intermediate hosts of *H. heydorni* show only slight symptoms of disease.
5. **Diagnosis**: Microscopical demonstration of the oocysts in the feces of the final hosts with the help of the flotation method.
6. **Pathway of infection**: Obligatory oral uptake of tissue cysts within raw meat of intermediate hosts.
7. **Prophylaxis**: Avoiding to feed raw meat to final hosts.
8. **Incubation period**: None in final hosts, since practically no symptoms occur. However, intermediate hosts of *H. hammondi* may show severe symptoms of disease starting within the first 2 weeks.
9. **Prepatent period**: Final hosts: 7–17 days *H. heydorni*; 5–13 days *H. hammondi*.
10. **Patency**: 1–29 days (in experimental infections).
11. **Therapy**: Unknown; substance used in *Toxoplasma* infections should be tested.

Further Reading
Dubey J et al (2014) Life cycle of *Hammondia hammondi* in cats. J Eukaryot Microbiol 62:346–352.
Frenkel JK, Smith DD (2003) Determination of the genera of cyst forming coccidians. Parasitol Res 91:384–389.
Heydorn AO, Mehlhorn H (2001) Further remarks on *Hammondia hammondi* and the taxonomic importance of obligate heteroxeny. Parasitol Res 87:573–577.
Mehlhorn H, Heydorn AO (2003) Electron microscopy study of in-vitro cultures of *Hammondia heydorni* tachyzoites after passages through dogs, mice, rats, guinea pigs and jirds. Parasitol Res 90:512–520.
Schares G et al (2008) Characterization of a repetitive DNA fragment in *Hammondia hammondi* and its utility for the species differentiation of *H. hammondi* from *Toxoplasma gondii*. Mol Cell Probes 22:244–251.
Walzer KA et al (2014) *Hammondia hammondi* harbours functional orthologs of the host-modulation effectors GRA15 and ROP16 but is distinguished from *Toxoplasma gondii* by a unique transcriptional profile. Eukaryot Cell 13:1507–1518.

4.6.7 *Neospora* Species

1. **Name**: Greek: *neos* = new; *sporos* = seed. Latin: *canis* = dog. Hughes = American protozoologist. This genus is very probably identical with *Hammondia* (Heydorn and Mehlhorn 2002).
2. **Geographic distribution/epidemiology**: Worldwide, although Dubey et al. (2002) believed that this genus species exists separately besides *Hammondia heydorni*. On overcrowded meadows, mass infections of cattle may occur.
3. **Biology, morphology**: Nomenclature of this genus is under discussion. Fact is that the groups Gottstein (Bern), Heydorn (Berlin) and Mehlhorn (Düsseldorf) showed the identity of *Hammondia hammondi* with the by Dubey newly named species *Neospora caninum* (Müller et al. 2001).
   (a) ***Neospora caninum* (syn. *Hammondia heydorni*)**:
   Final hosts are dogs and coyotes; intermediate hosts (with tissue cysts) are cattle (but also carnivores). Oocysts measure 9–11 μm (Fig. 4.83).
   (b) ***Neospora hughesi***:
   Final hosts are unknown or not sufficiently proven; intermediate hosts (with tissue cysts) are horses. Oocysts (Fig. 4.83) are identical to those of *H. heydorni*, measuring 9–11 μm in diameter.
4. **Symptoms of disease**:
   (a) ***Neospora caninum* (syn. *Hammondia heydorni*)**:
   After first infections, abortions may occur in pregnant **cattle**. In some countries, this is the reason of 90 % of cattle abortion. Also retention of fertility occurs. Newborn animals may seem healthy but develop later often encephalomyelitis. **Dogs**: Intrauterine infected baby dogs may show

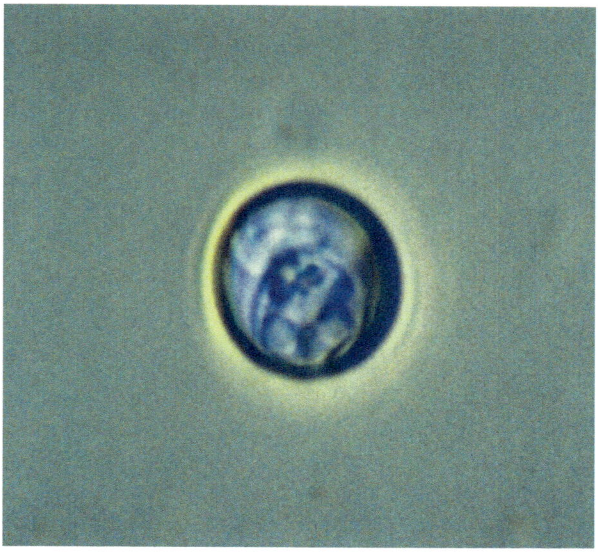

Fig. 4.83 Light micrograph of a *Neospora caninum* (syn. *Hammondia heydorni*) oocyst from feces of dogs

a polyradiculitis-myositis syndrome with paralysis and weakness of muscles. **Other animals**: No severe symptoms have been reported.

(b) ***Neospora hughesi***:

Disturbances of motility are described in intermediate hosts (horses) due to cysts in the brain.

5. **Diagnosis**:
   (a) **Dogs and other final hosts**:

   Fecal demonstration of oocysts or PCR demonstration of tissue parasites.

   (b) **Intermediate hosts**:

   Demonstration of antibodies.
6. **Pathway of infection**:
   (a) **Intermediate hosts**: Oral uptake of oocysts within fecally contaminated food.

   (b) **Final hosts**: Uptake stages in raw meat of intermediate hosts (but also very probably by uptake of oocysts).
7. **Prophylaxis**: Dogs should not be fed with raw meat of potential intermediate hosts. The intermediate hosts could probably be protected by a vaccine (available in the USA).
8. **Incubation period**: **Dogs**: about 4–5 days; **typical intermediate hosts**: it takes at least 14 days until symptoms occur, if inner organs are infected.
9. **Prepatent period**: 4–5 days.
10. **Patency: Dogs**: weeks; **cattle**: moth to years.
11. **Therapy**: Treatment with toltrazuril (dogs, other animals); ponazuril (horses).

Further Reading

Darius AK, Mehlhorn H, Heydorn AO (2004) Effects of toltrazuril and ponazuril on the fine structure and multiplication of the NC-1 strain of *Neospora caninum* (syn. *Hammondia heydorni*) in cell cultures. Parasitol Res 92:453–458.

Dubey JP et al (2002) *Neospora caninum* and *Hammondia heydorni* are separate species. Trends Parasitol 18:66–69.

Gondim LF et al (2004) Coyotes (*Canis latrans*) are definitive hosts of *Neospora caninum*. Int J Parasitol 34:159–161.

Hemphill A, Gottstein B (2006) *Neospora caninum* and neosporosis—recent achievements in hosts and parasite cell biology and treatment. Acta Parasitol 51:15–25.

Heydorn AO, Mehlhorn H (2002) A re-evaluation of *Neospora* and *Hammondia* spp. Trends Parasitol 18:246–247.

Müller N et al (2001) Comparative molecular investigation of Nc5-PCR amplicons from *Neospora caninum* NC-1 and *Hammondia heydorni*. Parasitol Res 87:883–885.

Sun W-W et al (2015) Herd-level prevalence and associated risk factors for *Toxoplasma gondii*, *Neospora caninum*, *Chlamydia abortus* and bovine viral diarrhoea virus in China. Parasitol Res 114:4211–4218.

4.6.8 *Besnoitia* Species

1. **Name**: The species name has its origin in the name of the French veterinarian, Professor at the University Toulouse: C. Besnoit. Species names honour other scientists or describe names of hosts (see Table 4.14): e.g. Latin: *capra* = goat; *tarand* = reindeer; *oryctofelisi* = comes from *oryctolagus* = rabbit; *felis* = cat.
2. **Geographic distribution/epidemiology**: Worldwide; in Europe *B. besnoiti* has been described in Bavaria, in Switzerland and in southern countries (France, Spain, Portugal).

Table 4.14 Important *Besnoita* species

Species	Intermediate host (tissue cysts)	Final host	Oocyst size (μm)	Pathogenicity
B. besnoiti	Cattle, rodents	Predator birds?	?	+
B. caprae	Goats	?	?	+
B. darligi	Opossums, lizards	Cats	11–13 × 10–13	–
B. jellisoni	Rodents (e.g. mice)	?	?	+
B. wallacei	Rodents (e.g. rats)	Cats	16–19 × 10–13	–
B. oryctofelisi	Rabbits	Cats	12 × 11	–
B. tarandi	Reindeer	?	?	+
B. benetti	Horses, donkeys	?	?	–

+ = yes; – = no; ? = questionable

3. **Biology, morphology**:

Besnoitia besnoiti is described in cattle and goats infesting the epidermis, subepidermal tissues, eyes and many mucous layers of the body, where rather large tissue cysts (up to 6 mm) occur (Figs. 4.84, 4.85 and 4.86). *B. neotomofelis* is described in the Southern Plains woodrat (*Neotoma micropus*); other species occur worldwide and are listed in Table 4.14. The life cycle is not yet clear. Most cases are described in cattle, where 10–20 % of the infected animals showed clear tissue cysts measuring 2–5 mm in diameter. They represented considerably enlarged host cells which were surrounded by a thick layer of amorphous material, which apparently has its origin in damaged surrounding cells and newly included collagen. This secondary layer may also become

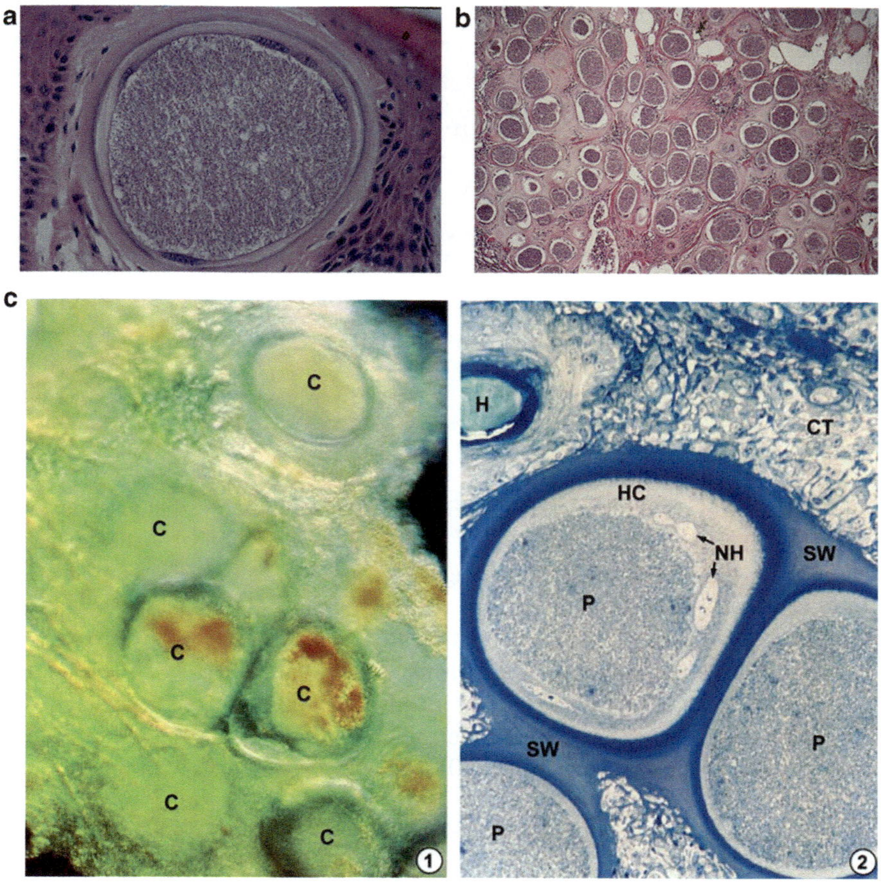

Fig. 4.84 (**a–c**) *Besnoitia besnoiti*: in host tissues: (**a**) Cyst at higher magnification showing clearly that the host cell contains the parasites in a vacuole and that the host cell is surrounded by an amorphous layer thus forming a wall. (**b**) Tangential section through the skin showing how close the tissue cysts are laying side by side. (**c1**) Surface section of a host showing closely arranged cysts (C). (**c2**) Semithin section of two cysts of *B. besnoiti* showing clearly host cells and surrounding material. *CT* connective tissues; *H* host tissues; *HC* host cell; *NH* nuclei of host cells; *P* parasites; *SW* secondary cyst wall

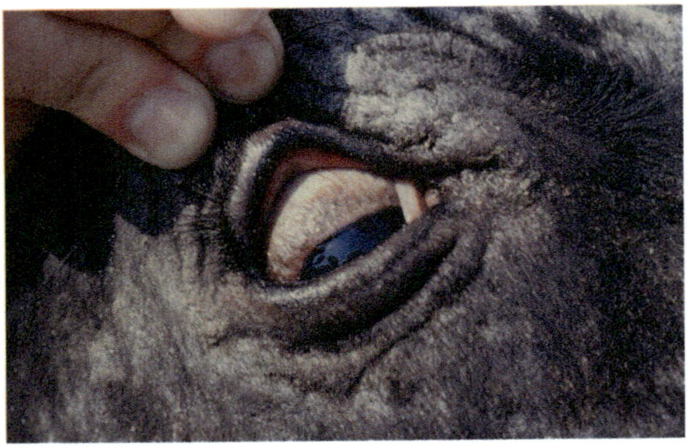

Fig. 4.85 *B. besnoiti* cysts inside an eye of an infected cow

Fig. 4.86 Macrophoto showing the hairless and horny skin of a *B. besnoiti*-infected cow

calcified. Inside the tissue cysts (Figs. 4.84a, b), numerous (thousands) cyst merozoites are found, which are produced by the typical banana-shaped endodyogeny (Fig. 4.68). These stages measure 9 × 25 µm. The complete life cycle is not yet clear although some Russian authors claim that cats should be final hosts (at least for one not yet clearly defined species). However, since many cysts are situated in the epidermis (e.g. of cattle; Figs. 4.85 and 4.86) a transmission pathway via licking or bloodsucking flies seems reasonable. Also during body contacts cyst merozoites might be transferred from one animal to the other. The paper Olias et al. (2011) shows a world map with the different species.

4. **Symptoms of disease (*B. besnoiti*)**:
   (a) **Acute phase**:
   Before the formation of tissue cysts, swellings of lymph nodes, oedemas in different organs, disturbances of motility, fever (41–42 °C), reduced food uptake and high rates of morbidity and mortality have been described. When puncturing lymph nodes many dividing endodyogeny stages can be observed.

(b) **Chronical phase**:

During formation of the rather large tissue cysts (Figs. 4.84, 4.85 and 4.86) skin eczemas, loss of hair, induration of the skin (cutis, subcutis), loss of weight and fever occur; bulls may become sterile after an orchitis.

The general three clinical phases may occur singly or following each other in an affected herd:

- Animals with symptoms: 10–20%.
- Animals showing only scleroconjunctival symptoms: 10–20%.
- Animals without apparent symptoms but showing seropositivity: rest.

5. **Diagnosis**: Besnoitosis can be diagnosed cytologically by microscopical inspection of cyst contents or lymph node punctures, if cysts and skin swellings are not clearly visible by inspection with naked eye. Hidden infections can be diagnosed with the help of serological methods such as ELISA, IFA, Westernblot or by PCR.
6. **Pathway of infection**: Probably orally by oral uptake of still unknown oocysts. However, direct transmission by body contacts seems reasonable, too, as well as transmission of cyst merozoites with the help of mouthparts of licking or bloodsucking insects.
7. **Prophylaxis**: Difficult, since the life cycle is not fully known. However, evidently infected animals should be immediately separated from the herd.
8. **Incubation period**: Probably 2 weeks as obtained in experimental transmissions.
9. **Prepatent period**: Depending on the grade of infection, parasite stages may be observed in blood smear preparations.
10. **Patency**: Cysts remain apparently lifelong infectious.
11. **Therapy**: Unknown; however, in the case of goats the parenteral application of antimony solution (1 % 0.6 ml/kg bodyweight) led to healing of an acute infection.

Further Reading

Basso W et al (2011) Exploring the life cycle of *Besnoitia besnoiti*: experimental infection of putative definitive and intermediate host species. Vet Parasitol 178:223–234.

D'Haese J et al (1977) Comparative electron microscope study of pellicular structures in coccidian (*Sarcocystis*, *Besnoitia* and *Eimeria*). Int J Parasitol 7:505–518.

Dubey JP, Jabsley MJ (2010) *Besnoitia neotomofelis* n. sp. (Protozoa, Apicomplexa from the Southern plains woodrat (*Neotoma micropus*). Parasitology 137:1731–1747

Kiehl E et al. (2010) Molecular biological comparison of different *Besnoitia* species and stages from different countries. Parasitol Res 106:889–894.

Lesser M et al (2012) First cases of besnoitiosis in cattle in Switzerland. Schweiz Arch Tierheilkd 154:469–474.

Mehlhorn H et al (2009) Another Africa disease in Central Europe: besnoitiosis of cattle. I. Light and electron microscopical study. Parasitol Res 104:861–868.
Olias P, Shade B, Mehlhorn H (2011) Molecular pathology, taxonomy and epidemiology of *Besnoitia* species. Infect Genet Evol 11:1564–1576.
Scholtyseck E, Mehlhorn H (1974) Fine structure of cyst and cyst wall of *Sarcocystis tenella, Besnoitia jellisoni, Frenkelia* sp. and *Toxoplasma gondii*. J Protozool 21:284–294.

4.6.9 Further Species with Tissue Cysts

1. **Morphology/life cycle**

   There are some more protozoan species, which also produce tissue cysts such as had been shown in Sects. 4.6.4–4.6.8. Most important are species of the genera *Frenkelia, Caryospora* and *Globidium*. Only some significant species are selected here:

   (a) **Genus *Frenkelia***: (named honouring the US-American Jacob (Jack) Frenkel (1924–2013), who was one of the discoverer of the *Toxoplasma gondii* life cycle). *Frenkelia clethrionomys buteonis* is a common species forming cyst in mice of the genus *Clethrionomys* (Fig. 4.87). These cysts are found mainly in the brain and are characterized by a lobulated shape and a very smooth limiting cover. Due to the presence of these cysts, the mice are less quick and less alert, so that they become a rather easy prey of the buzzard (*Buteo* species). Molecular biological methods have shown that these *Frenkelia* species are very similar to those of the genus *Sarcocystis*, so that some authors consider *Frenkelia* as a sister genus.

   (b) **Genus *Caryospora***: The typical coccidian life cycle of the members of this genus (i.e. schizogony, gamagony, sporogony) occurs inside the intestinal cells of the final host of type 1 (e.g. in the rattle snake in the case of *C. bigenetica*). This final host excretes within its feces unsporulated ovoid oocysts with a diameter of 15 μm. As soon as these oocysts are set free, a

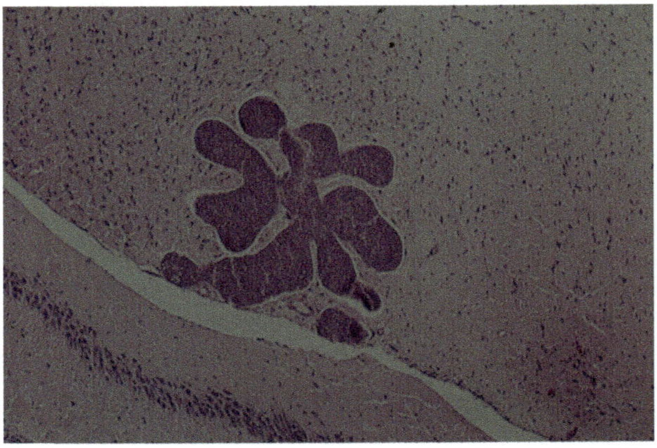

Fig. 4.87 *Frenkelia clethrionomys buteonis*. Light micrograph of a cyst in the brain of peculiar mouse species (*Clethrionomys* sp.). Note the lobulated surface

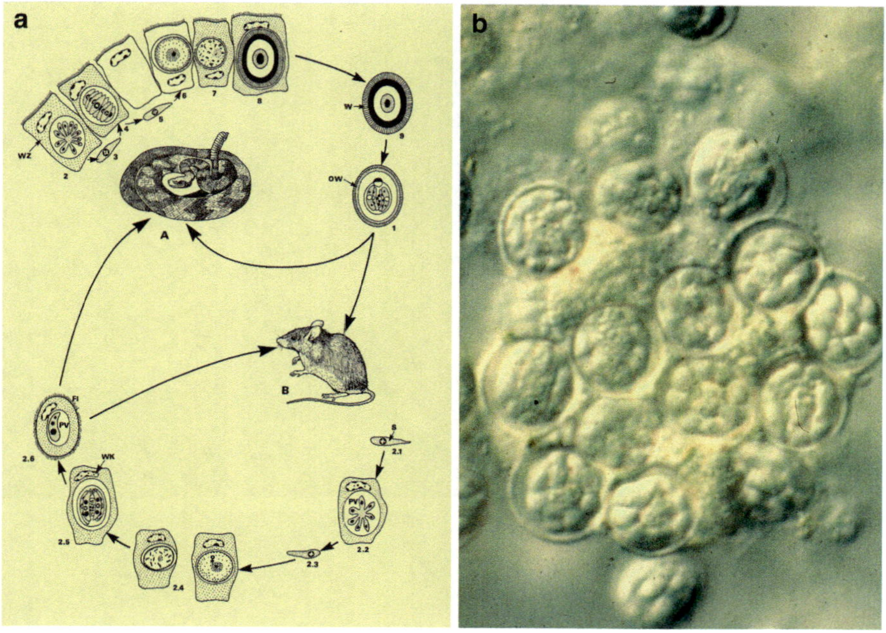

Fig. 4.88 (**a**) Diagrammatic representation of the life cycle of *Caryospora bigenetica* (according to Sundermann 1988). (**a**) First type of final host (rattlesnake). (**1**) The sporulated oocyst contains a single sporocysts with 8 sporozoites. (**2–4**) After oral infection with oocysts, 2 generations of schizonts are formed inside the intestinal epithelium. (**5, 6**) Male and female gamonts are formed and later gametes occur. The male gamont (G) forms numerous gametes. (**7, 8**) After fertilization, a thick-walled oocyst is formed inside the host cell and becomes free within the feces. While sporulating on the ground, a single sporocysts with 8 sporozoites is formed inside each oocyst. (**b**) Second type of final host (mice, cotton rats, dogs). (**2.1–2.4**) Repetition of the schizogonic and gamogonic development (described in **1–8**). The sporulation of oocysts may occur in skin regions (**2.4**). (**2.5**) Sporozoites that were set free from their sporocysts while still inside the skin of their hosts enter other host cells and induce formation of so-called caryocysts, i.e. cells being surrounded by fibrillary material. Those caryocysts are infectious for both host types (if taken up orally). Furthermore, these caryocysts are infectious for hosts of the second type during close skin contacts. *CA* caryocyst; *FI* filamentous material; *HC* host cell; *ME* merozoite; *OO* oocyst; *OW* oocyst wall; *PV* parasitophorous vacuole; *S* sporozoites

single sporocyst containing eight sporozoites is formed inside the oocyst (Figs. 4.87 and 4.88). If **final hosts of type 2** (rodents, dogs) ingest such sporulated oocysts, the sporozoites wander into the skin and produce there (inside cells) at first schizonts and merozoites, then gamonts and gametes and finally again infectious oocysts occur, which may infect snakes as soon as they ingest such infected mice. Thus, the species of this genus have two types of final hosts which stay in relations as **predator** and **prey**. Infected mouse cells are also described as caryocysts and thus are name-giving for the genus. Transmission of the stages of these parasites may also occur just by body contacts. It was noted that infected dogs show symptoms of

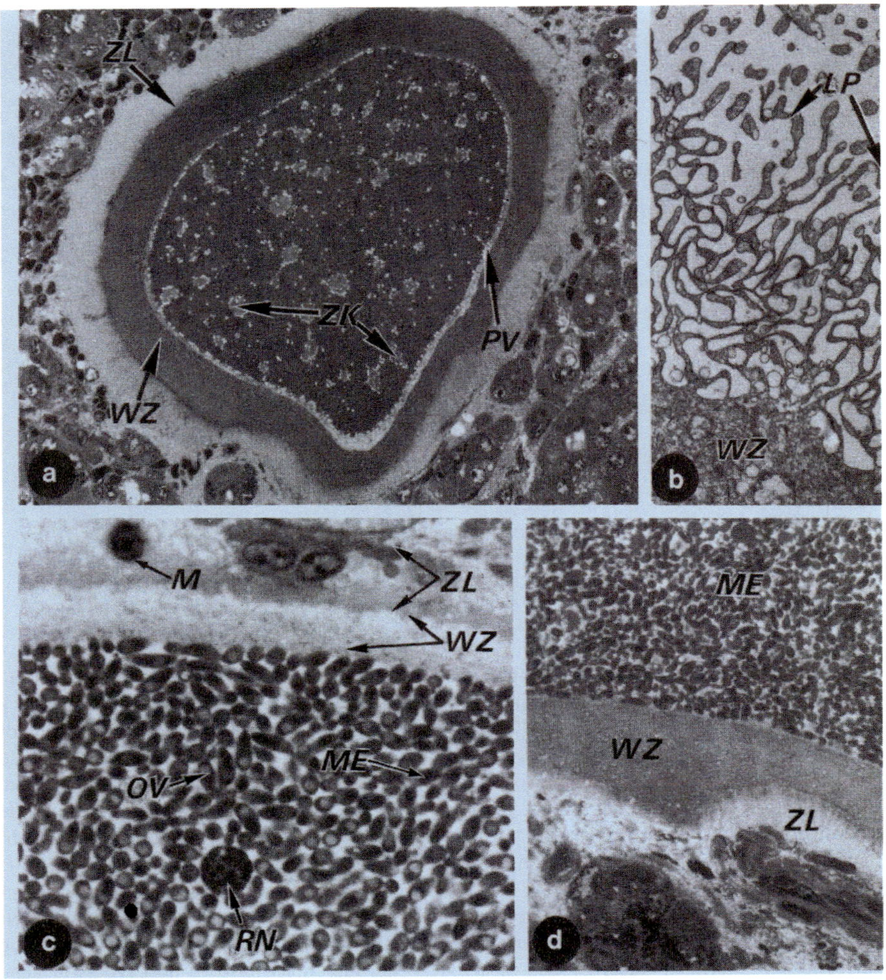

Fig. 4.89 Light micrograph of sections through so-called *Globidium* stages (=giant schizonts?) obtained from the abomasum of sheep. (**a**) Cross section through a young cyst in a host cell during the phase of nuclear divisions. (**b**) The surface of those "cysts" is considerably enlarged by protrusions (LP). (**c**, **d**) Sections through cyst stages containing differentiated = infectious merozoites (ME). One cyst stage (**c**) shows merozoite with clear inner structures and thus belongs probably to another species. *LP* long outer protrusion of the host cell; *M* monocyte; *ME* merozoite; *OV* ovoid, clear inclusion; *PV* parasitophorous vacuole; *RN* remnant nuclei; *WZ* host cell; *ZK* centres of nuclear divisions in a schizont; *ZL* zone of long protrusions of the host cell

immune suppression, which may lead to death in the case of further infections.

(c) **Genus *Globidium***: These species develop large (1–2 mm) tissue cysts, which start from parasitophorous vacuoles inside cells of the abomasum of ruminants. These whitish appearing "globules" contain thousands of cyst merozoites (Figs. 4.89a–d). Morphological and electron microscopical studies showed that there exist apparently several species. Some authors claim that all these schizonts belong to not yet defined *Eimeria* species. However, their ultrastructure seems too different from that of the *Eimeria* merozoites.

2. **Therapy**: First studies showed that the anticoccidial compound toltrazuril is effective against the different stages of the genera *Frenkelia, Caryospora* and *Globidium.*

Further Reading

A. Caryospora

Alfaleh FA et al (2013) *Caryospora cherrughi* sp. n. infecting *Falco cherrug* in Saudi Arabia. Parasitol Res 112:971–974.

Sundermann CA et al (1988) Complete development of *Caryospora bigenetica* (Apicomplexa: Eimeriidae) in vitro. J Protozool 35:465–469.

Sundermann CA, Lindsay DS (1989) Ultrastructure of in vivo-produced caryocysts containing the coccidian *Caryospora bigenetica* (Apicomplexa: Eimeriidae). J Protozool 36:81–86.

Viana LA et al (2015) *Caryospora bigenetica* in South America: new hosts and distribution records. Braz J Vet Parasitol 24:101–104.

Yang R et al (2014) A new *Caryospora* coccidian species from the laughing kookaburra (*Dacelo novaequineae*). Exp Parasitol 145:68–73.

B. *Frenkelia*

Dubey JP et al (2000) *Frenkelia microti* infection in a chinchilla (*Chinchilla laniger*) in the USA. J Parasitol 86:1149–1150.

Laakone J, Henttonen H (2000) Ultrastructure of *Frenkelia* sp. from a Norwegian lemming. J Wildl Dis 36:362–366.

Scholtyseck E, Mehlhorn H (1974) Fine structure of the cyst and cyst wall of *Sarcocystis tenella, Besnoita jellisoni, Frenkelia* sp. and *Toxoplasma gondii.* J Parasitol Zool 21:284–294.

Tichet-Calvet E et al (2004) *Frenkelia* parasites in a small mammal community. Parasite 11:301–310.

C. *Globidium*

Mehlhorn H et al (1984) Two types of *Globidium* cysts of goats. Parasitol Res 70:731–737.

Sénaud J, Heydorn AO, Mehlhorn H (1984) Three new types of *Globidium* of sheep: An in-vivo and in-vitro investigation. Parasitol Res 70:721–729.

4.6.10 *Babesia* Species

Babesia and *Theileria* species (Tables 4.15 and 4.16) are described together with some other species as so-called **piroplasms**. This trivial name is based on the pear-like shape of some of their developmental stages (Latin: *pirum* = pear). They are transmitted by ixodid ticks from vertebrate hosts to vertebrate hosts. The life cycles

Table 4.15 Important *Babesia* species

Species	Vector tick	Stage of tick	Vertebrate hosts	Size inside erythrocytes (μm)	Geographical distribution
Large *Babesia species*					
B. bigemina	*Boophilus* spp.	Nymphs, adults	Cattle, water buffaloes, wild ruminants	5 × 2	South Europe, America, Africa, Asia, Australia
B. bovis	*Boophilus* spp., *Ixodes* spp., *Rhipicephalus bursa*	Larves	Cattle, water buffaloes, wild ruminants	2.5 × 1.5	Europe, America, Africa, Asia, Australia
B. divergens	*Ixodes ricinus*	Larves	Cattle, wild ruminants, **humans**	1.5 × 0.5	Europe
B. major	*Haemaphysalis punctata*	Adults	Cattle	3 × 1.5	Europe, North-West Africa
B. motasi	*Haemaphysalis* spp., *Rhipicephalus bursa*	Adults	Sheep, goats	4 × 2.5	South Europe, Near East, South Russia, Africa, Asia
B. ovis	*Rhipicephalus bursa*	Adults	Sheep, goats	2 × 1	South Europe, Near East, South Russia, Africa, Asia
B. caballi[b]	*Hyalomma* spp., *Dermacentor* spp. *Rhipicephalus* spp.	Adults	Horses, donkeys, zebras	4 × 2.5	Europe, Asia, Africa, America, Australia
B. canis[a]	*Rhipicephalus sanguineus*, *Haemaphysalis leachi*, *Dermacentor reticulatus*	Nymphs, adults	Dogs, canids, foxes	5 × 2.5	Europe, Africa, Australia
B. trautmanni	*Rhipicephalus* spp.	?	Pigs	4 × 2.5	South Europe, Africa
B. herpailuri	?	?	Wild cats	3 × 2.2	South America
B. pantherae	?	?	Wild cats, lions	2.5 × 1.5	Africa

Small *Babesia* species with doubtful systematical position					
B. gibsoni (syn. *Microbabesia*)	*Haemaphysalis bispinosa*, *Rhipicephalus sanguineus*	All stages	Dogs, foxes, other canids	1.2–2.1	Asia, Africa, India, Japan,
B. microti group	*Dermacentor* spp., *Rhipicephalus* spp., *Ixodes* spp.	Larves, nymphs	Rodens, **humans**	1.5–2	Europe, North America
B. felis (syn. *Achromaticus felis*)	*Haemaphysalis leachi*	?	Cats incl. lions, wild cats	1.5–2	Africa
	B. rodhaini = *B. quadrigemina* (syn. *Achromaticus*)	?	?	Mice	1.5–2
Europe					

[a]*Babesia canis* is now subdivided into three subspecies *B. canis canis* (Europe, transmitted by *Dermacentor reticulatus*), *B. canis vogeli* (transmitted in the tropics and subtropics by *Rhipicephalus sanguineus*) and *B. canis rossi* (very pathogenic in South Africa, transmitted by *Haemaphysalis leachi*)
[b]*B. equi* of horses is now placed into the genus *Theileria*

Table 4.16 Important *Theileria* species

Species	Vector tick	Vertebrate host	Disease	Endemic region
Theileria parva parva	*Rhipicephalus appendiculatus*, *R.* spp.	Cattle, cape buffalo	East coast fever	Africa
T. parva lawrencei	*Rhipicephalus appendiculatus*, *R.* spp.	Cattle, cape buffalo	Corridor disease	Africa
T. annulata	*Hyalomma* sp.	Cattle, water buffalo	Mediterranean coast fever	Africa, Asia
T. mutans	*Amblyomma* sp.	Cattle, cape buffalo	Mild theleriosis, pseudo coast fever	Africa
T. velifera	*Amblyomma* sp.	Cattle, cape buffalo	–	Africa
T. taurotragi (syn. *Cytauxzoon*)	*R. appendiculatus*, *R.* spp.	Cattle, antelope	Mild theileriosis, pseudo coast fever	Africa
T. sergenti (syn. *T. orientalis*)	*Haemaphysalis* spp.	Cattle, buffalo	Oriental theileriosis	Asia, Europe, Australia, North Africa
T. orientalis	*Haemaphysalis longicornis*	Cattle	Asiatic theileriosis	Japan, South and East Asia
T. hirci	*Hyalomma* spp.	Sheep, goats	Theileriosis	Africa, Europe
T. ovis	*Rhipicephalus* spp., *Hyalomma* spp.	Sheep	–	Africa
T. separata	*Rhipicephalus evertsi*, *R.* spp.	Sheep	–	Africa
T. equi (syn. *Babesia*)	*Dermacentor* spp., *Hyalomma* spp., *Rhipicephalus* spp.	Horses, mules, donkeys	Horse theileriosis	South Europe, Africa, Asia, America

of the species belonging to the genera *Babesia* and *Theileria* are very similar to those of the *Plasmodium* and related species (=agents of malaria) except for the fact that the latter are transmitted during the bloodsucking process of female mosquitoes (Figs. 4.80, 4.99 and 4.105; Table 4.17).

4.6.10.1 *Babesia* Species of Dogs

1. **Name**: Victor Babès (1854–1926), Romanian veterinarian, who discovered 1888 these parasites in the blood of cattle. Latin: *canis* = dog; *caballus* = horse; *ovis* = sheep; *bovis* = cattle; *divergere* = to be different; other subspecies

names are based on family names of discoverers (e.g. Vogel, Rossi, Conradi, Rodhain, Trautmann, etc.).

2. **Geographic distribution/epidemiology**: Worldwide; in Europe, however, mainly in South Europe, and some few autochthonic cases in Germany.
3. **Biology, morphology**: The former species *Babesia canis* is now divided into three subspecies: *B. canis canis* (Europe), *B. canis vogeli* (Tropics) and *B. canis rossi* (South Africa). The stages of the life cycle are diagrammatically shown in Fig. 4.90. The stages inside erythrocytes are called merozoites although they are formed during repeated binary fissions (Figs. 4.90 and 4.91). They appear pear shaped and reach a length of about 6.5 μm. Vectors (and final hosts) are ixodid ticks (e.g. members of the genera *Rhipicephalus, Haemaphysalis* and *Dermacentor*) (Table 4.15). Inside the ticks the gamogony occurs in the intestine and the sporogony in the salivary glands. The infectious sporozoites are transmitted during bloodsucking of the ticks (Fig. 4.90). Ookinetes of the *Babesia* enter also the ovaries of female ticks. Thus already tick larvae are possibly infected and thus may transmit these agents of disease (=so-called **transovarian transmission**). Besides *B. canis* also stages of *B. gibsoni, B. conradae* may infect dogs in different regions (Table 4.15).
4. **Symptoms of disease**: Acute infected animals show apathia, fever, enlargement of the spleen, anaemia and icterus. Even in cases of a chronically progressing babesiosis the anaemia persists (Fig. 4.92). In some cases also cerebral dysfunctions, which led to paralysis as well as to epileptic attacks. In case of *B. canis rossi* disease may progress extraordinarily, since death cases already occurred 24 h after an infection.
5. **Diagnosis**: Intensive anamnesis and microscopical demonstration of the intraerythrocytic stages (Fig. 4.91) after obtaining blood and colouring the blood smear by Giemsa stain. Since blood smears show too few parasites, serological investigations will help (e.g. IFAT is done in special laboratories).
6. **Pathway of infection**: Bites of vector ticks which belong among others to the following species:
   (a) In **Europe**: *Dermacentor reticulatus, Rhipicephalus sanguineus*;
   (b) **Subtropics, Egypt**: *Rhipicephalus sanguineus*;
   (c) **Africa (South to the Sahara) and India**: *Haemaphysalis leachi*.
7. **Prophylaxis**: Tourists should leave their dogs at home. Protection by
   - **Collars** with insecticides;
   - Use of **repellents**: Advantix®, Exspot®, Anticks® (produced by Fa. Alpha-Biocare of the author);
   - **Vaccination** trials: Use of Nobivac®Prio; Pirodog®. However, protection is not total;
   - Prevention by **advance treatment**: Imidocarb (1×6 mg/kg bodyweight subcutaneously, twice in an interval of 14 days), starting 4 weeks before departure in a *Babesia* endemic country. Protection is reached for about 4 weeks.

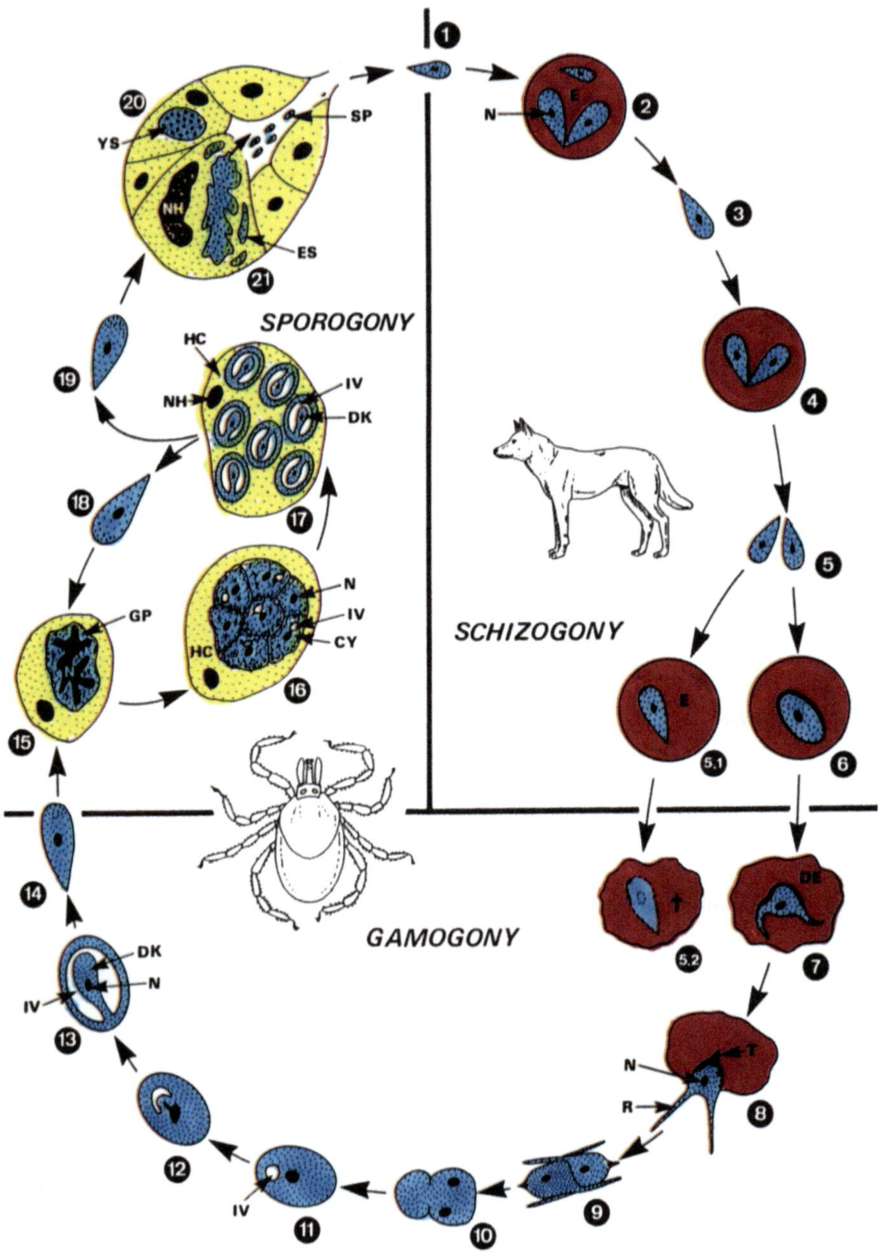

Fig. 4.90 Diagrammatic representation of the life cycle of *Babesia canis*. (**1**) Sporozoites in the saliva of feeding tick. (**2–5**) Asexual reproduction in erythrocytes of vertebrate host (dog) by binary fission, producing merozoites (**5**) which enter other erythrocytes. When merozoites are ingested by a tick (**5.1**), they become digested inside the gut (**5.2**). (**6**) Some merozoites become ovoid gamonts. (**7, 8**) After ingestion into the tick's intestinal cells, the ovoid gamonts form

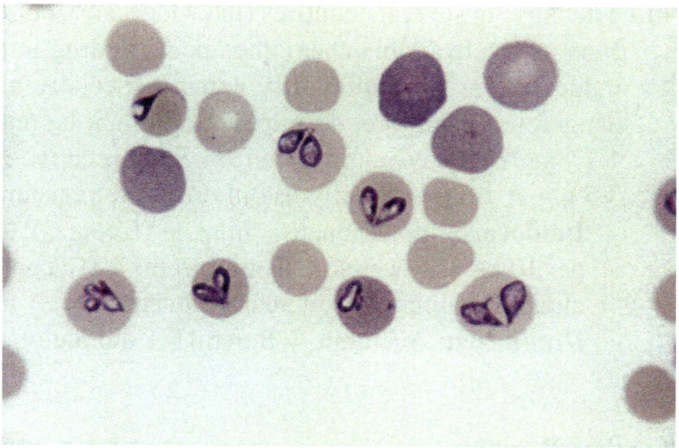

Fig. 4.91 Light micrograph of a smear preparation showing red blood cells parasitized by dividing stages of *Babesia canis canis*

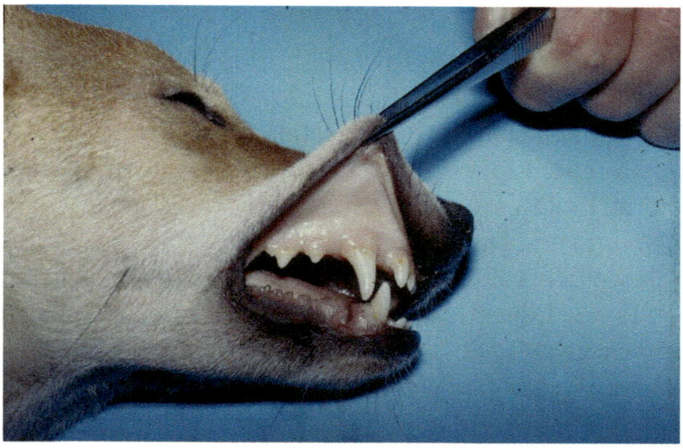

Fig. 4.92 Dog suffering from anaemia due to an infection with *B. canis canis*

8. **Incubation period**: Variable, mostly 2–3 weeks.
9. **Prepatent period**: 2 weeks.
10. **Patency**: In cases of a strong immune system and low-grade infections, parasites can be seen for years with the help of the blood examination (thick droplet method). This phenomenon is based on a so-called **premunity**.

Fig. 4.90 (continued) protrusions and thus appear as ray bodies (**8**). Fusion of 2 uninucleate ray bodies (gametes). (**10**) Formation of a zygote. (**11–14**) Formation of a single kinete from a zygote inside the inner vacuole. The kinet leaves the intestinal cell and enters cells of various organs (including the eggs) of the vector ticks. (**15–18**) Formation of several kinetes (sporokinetes). This process is repeated (**15–18**) and also proceeds in eggs of ticks. The infection is thus transmitted to the next generation of ticks (i.e. transovarial transmission). (**19.21**) Some of the kinetes penetrate cells of the salivary glands, where a large multinuclear sporont (YS, ES) is formed (inside hypertrophic host cells), finally giving rise to thousands of small sporozoites (SP), which are injected during the feeding act (i.e. transstadial transmission). *CY* cytomere (uninucleate); *DE* digested erythrocyte; *DK* developing kinete; *E* erythrocyte; *ES* enlarged sporont (forming sporozoites); *GP* growing parasite (polymorphic stage); *HC* nucleus of host cell; *IV* inner vacuole; *N* nucleus; *NH* nucleus of host cell; *R* raylike protrusion; *SP* sporozoites; *T* thorn-like apical structure; *YS* young sporont

11. **Therapy**: In several countries (like Germany), no relevant drugs are registered. Most of the available chemotherapeutical drug initiate severe side effects and induce rather high **mortality rates**: Especially use of diminazene might be dangerous. Therefore, dose reduction might be reasonable: e.g. application of 1.17 mg/kg bodyweight given on 3 consecutive days seems more safe than 3.5 mg at 1 day. Causally useful drugs are registered:
    - **Imidocarb** (dipropionate = Imizole®Coopers): 6 mg/kg bodyweight = 0.5 ml per 10 kg bodyweight (taken from the 12 % commercial solution). However, clinical healing will not achieve **premunity**.
    - **Diminazen**: Aceturate = Berenil® from Sanof/Aventis.

Further Reading

Ayoob AL et al (2010) Clinical management of canine babesiosis. J Vet Emerg Crit Care 20:77–89.

Baneth G et al (2015) Reclassification of *Theileria annae* as *Babesia vulpes* sp. nov. Parasit Vectors 8:207.

Bartley PM et al (2016) Detection of *Babesia annae* DNA in lung exudate samples from Red foxes (*Vulpes vulpes*) in Great Britain. Parasit Vectors 9:84.

Barutzki D et al (2007) The babesiosis of dogs—an autochthonic disease in Germany. Dtsch Tierärztebl 55:284–293.

Clancey N et al (2010) *Babesia* (*Theileria*) *annae* in a red fox (*Vulpes vulpes*) from Prince Edward Island, Canada. J Wildl Dis 46:615–621.

Di Cicco MF et al (2012) Re-emergence of *Babesia conradae* and effective treatment of infected dogs with atovaquone and azithromycin. Vet Parasitol 187:23–27.

Hamel D et al (2012) Epidemiological aspects on vector-borne infections in stray and pet dogs from Romania and Hungary with focus on *Babesia* spp. Pasrasitol Res 110:1537–1545.

Kjemtrup AM et al (2006) *Babesia conradae*, sp. Nov., a small canine *Babesia* identified in California. Vet Parasitol 138:103–111.

Schein E et al (1979) Electron microscopical studies on the development of *Babesia canis* (Sporozoa) in the salivary glands of the vector tick *Dermacentor reticulatus*. Acta Trop 36:229–241.

Schetters TP et al (2009) Systemic inflammatory responses in dogs experimentally infected with *Babesia canis*. Vet Parasitol 162:7–15.

Solano-Gallego L, Baneth G (2011) Babesiosis in dogs and cats. Expanding parasitological and clinical spectra. Vet Parasitol 181:48–60.

Sunaga F et al (2014) Protective efficacy of *Babesia gibsoni* culture-derived exoantigens against the challenge infection in dogs. Parasitol Res 113:1681–1686.

Yamasaki M et al (2014) In vitro and in vivo safety and efficacy studies of amphotericin B on *Babesia gibsoni*. Vet Parasitol 205:424–433.

4.6.10.2 *Babesia* Species of Ruminants

1. **Name**: Victor Babès (1854–1926): Romanian veterinarian, who detected these parasites in cattle in the year 1888. Latin: *divergere* = being different/starting divisions; *maior* = larger; *bos* = cattle; *bigeminus* = two times, double.
2. **Geographic distribution/epidemiology**: Worldwide; in Europe rather few cases.
3. **Biology, morphology**: The sequence of the life cycles of the listed species runs like that of *B. canis* depicted in Fig. 4.90.
   (a) **Cattle**
   - ***Babesia divergens*** (agent of meadow red disease, bloody urine disease). The parasites are mainly situated at the periphery of the erythrocyte. These stages measure about 1.7 × 1 μm (Figs. 4.93 and 4.95). Inside the vector ticks (see Table 4.15), these parasites are constantly transmitted into the eggs, so that many of the tick larvae are already infected when hatching from the egg outside the body.
   - ***B. major*** is only low-grade pathogenic. The erythrocytic stages reach a size of 3.5 × 1.5 μm. These stages are centrally placed inside the red blood cells and appear typically pear shaped. Vectors are *Haemaphysalis punctata* ticks. In Germany they occur mainly in Northern regions.
   - ***B. bovis*** appears in Europe mainly in southern countries. Inside the erythrocytes, the stages reach a size of 2.4 × 1.5 μm and their shape is described as signet-ring-like or ovoid (Fig. 4.93). They induce a disease named **epidemic haemoglobinuria**. Vector and final host are ticks of

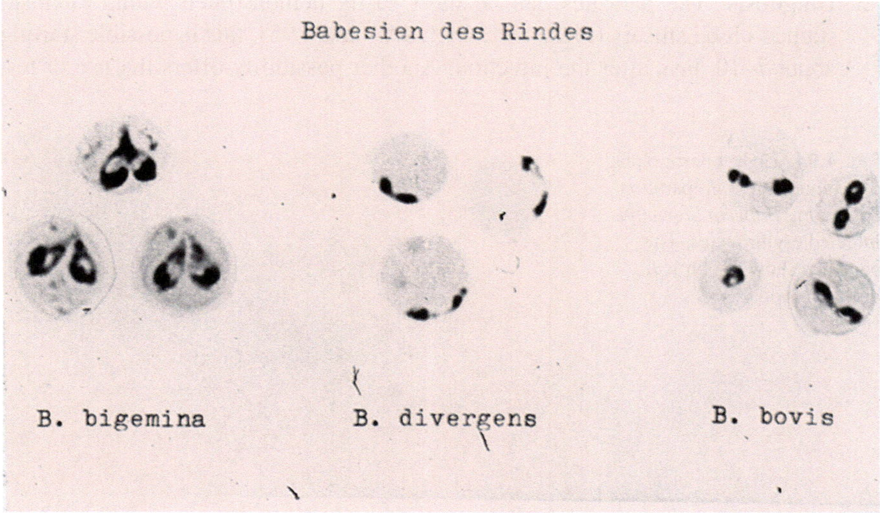

Fig. 4.93 Light micrographs of *Babesia* stages of cattle (German = Rind). Note the typical position inside the red blood cells

the genera *Boophilus, Rhipicephalus* and *Ixodes*. The transovarial transmission of the parasites onto the larva 1 occurs regularly.

- ***B. bigemina*** stages are the agents of the so-called **Texas fever** (or red water fever), which is also found in rare cases in Europe. The erythrocytic stages are spherical, ovoid or pear shaped and reach a size of 4–5 × 2–3 μm just filling the whole erythrocyte when divided (Figs. 4.93, 4.94 and 4.98). Vectors are ticks of the genus *Boophilus*.

(b) **Sheep**

The species of the sheep are only found in rather rare cases and are rather low-grade pathogenic. The most important species of sheep is *Babesia motasi*. Their stages inside measure 2–3.8 × 1.8 μm and produce rather few division stages (=thus leading to low-grade infections). In Central Europe, they are transmitted by *Ixodes ricinus*; in other countries species like *Rhipicephalus bursa, Dermacentor marginatus* and *Haemaphysalis punctata* act as vectors (Fig. 4.95).

4. **Symptoms of disease (Babesiosis; babesiasis):** In the cases of an infection with pathogenic species (*B. bovis, B. divergens, B. bigemina*), older animals show significant clinical symptoms starting about 8 days after the infection. These are fever for 1 week (40–41.8 °C), icterus and intestinal problems (diarrhoeas, paresis). Later haemoglobinuria (=dark to black urine; Fig. 4.96), anaemia, inability to stand up and coma follow. Exitus is rather common in severe cases. If the disease has been survived, the parasites remain present in the bloodstream, so that new ticks may become infected during bloodsucking. If there are double infestations with another parasite (e.g. *Fasciola hepatica*), babesiosis can become reinforced—a fact which occurs in Central Europe very often during the months May until September.
5. **Diagnosis**: The parasites can be most easily demonstrated within Giemsa-stained blood smears (Figs. 4.91, 4.93, 4.94 and 4.95); this is possible starting about 7–10 days after the infection. Another possibility offers the use of test

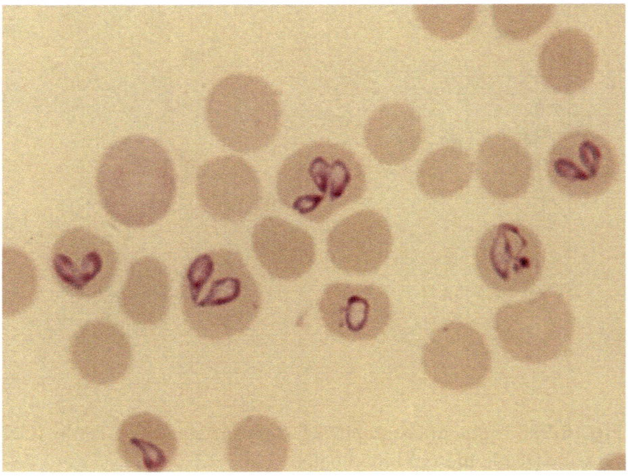

Fig. 4.94 Light micrographs of a blood smear preparation containing *Babesia bigemina*-infected erythrocytes. The parasites show the typical binary fissions

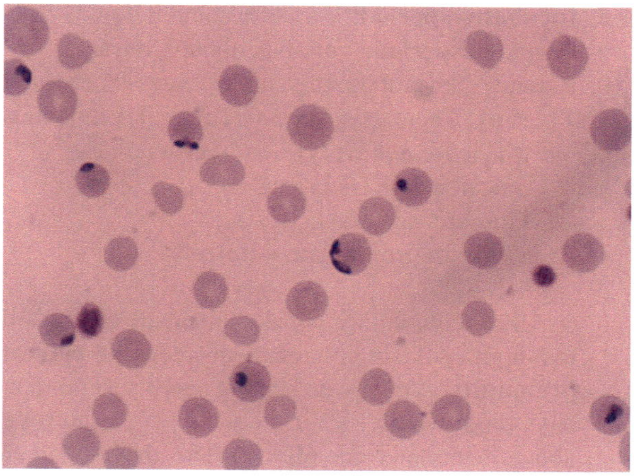

Fig. 4.95 Light micrograph of a blood smear preparation containing stages of *B. divergens* in red blood cells being often situated closely at the inner surface of the host cell

Fig. 4.96 Macrophoto of tubes with normal urine of cattle (*left*) and of cattle infected with *Babesia* stages. Note the intense haemoglobinuria

sticks that announce even traces of blood in the urine. Serological methods (ELISA, IFAT, IHA) are useful to detect low-grade infections of apparently diseased animals.

6. **Pathway of infection**: The transmission of the *Babesia* stages (sporozoites) occurs by bloodsucking of ticks. All stages (larvae, nymphs and adults) may be infected (=containing finally sporozoites in their saliva).
7. **Prophylaxis**: Most effective methods to protect cattle from invasion of ticks are those which bring acaricides onto the fur. Really acting vaccines are not available.

8. **Incubation period**: Variable, about 8 days.
9. **Prepatent period**: Variable, about 8 days.
10. **Patency**: Several months, up to years, depending on the species.
11. **Therapy**: There exist several chemical compounds, which are used worldwide, but which are not everywhere registered.
   - Amicarbalid = Diapram®
   - Diminazenaceturat = Berenil®
   - Imidocarb proprionate = Imizol®

   All of them induce considerable side effects—especially if the dose used is too high. All single treatments depress clinical symptoms and lead to a **premunity**. A complete healing needs several treatments at intervals of several days (and mostly higher dosages, which, however, are dangerous due to the above-mentioned side effects). The above-cited chemical compounds are efficacious against *B. bovis* (syn. *B. argentina, B. berbera, B. colchica*), *B. bigemina, B. major* and (with some restrictions) also against *B. divergens* (syn. *B. occidentalis, B. caucasica, B. carelica*).

Further Reading

Adjou Moumouni PF et al (2015) Molecular detection and characterization of *Babesia bovis*, *Babesia bigemina*, *Theileria* species and *Anaplasma marginale* isolated from cattle in Kenya. Parasit Vectors 8:496.

Erster O et al (2015) Molecular detection of *Babesia ovis* in sheep and ticks using the gene encoding *B. ovis* surface protein D (BoSPD). Vet Parasitol 214:282–288.

Esmaeilnejad B et al (2014) Evaluation of antioxidant status, oxidative stress and serum trace mineral levels associated with *Babesia ovis* parasitemia in sheep. Vet Parasitol 205:38–45.

Jaramillo Ortiz JM et al (2014) Vaccine strategies against *Babesia bovis* based on prime-boost immunizations in mice with modified vaccinia Ankara vector and recombinant proteins. Vaccine 32:4625–4632.

Liu J et al (2014) A PCR method targeting internal transcribed spacers: the simultaneous detection of *Babesia bigemina* and *Babesia bovis* in cattle. Acta Parasitol 59:132–138.

Hornok S et al (2014) Re-emergence of bovine piroplasmosis in Hungary: has the etiological role of *Babesia divergens* been taken over by *B. major* and *Theileria buffeli*? Parasit Vectors 7:434.

Maeda H et al (2013) Inhibitory effect of cyclophilin A from the hard tick *Haemaphysalis longicornis* on the growth of *Babesia bovis* and *Babesia bigemina*. Parasitol Res 112:2207–2213.

Mehlhorn H, Schein E (1984) The piroplasms: Life cycle and sexual stages. Adv Parasitol 23:37–104.

Niu Q et al (2013) Sequence and organization of the rhoptry-associated-protein-1 (rap-1) locus for the sheep hemoprotozoan *Babesia* sp. BQ1 Lintan (*B. motasi* phylogenetic group). Vet Parasitol 198:24–38.

Rossouw I et al (2015) Morphological and molecular descriptors of the developmental cycle of *Babesia divergens* parasites in human erythrocytes. PLoS Negl Trop Dis 9(5):e0003711.
Wang JM et al (2013) A sero-epidemiological survey of Chinese *Babesia motasi* for small ruminants in China. Parasitol Res 112:2387–2391.

4.6.10.3 *Babesia* Species of Horses

1. **Name**: Victor Babès (1854–1926), Romanian veterinarian. Latin: *equus*, *caballus* = horse.
2. **Geographic distribution/epidemiology**: Worldwide, in Europe mainly in Southern countries, parasitizing besides in horses in many related species (e.g. Donkeys, mules) (Fig. 4.97).
3. **Biology, morphology**:
   (a) ***Babesia equi*** (now transferred to ***Theileria equi***)
   This is the most common species parasitizing horses. There are stages present inside lymphocytes. Especially characteristic are division stages appearing as so-called **Maltese-cross stages** (Fig. 4.98; *B. equi*), which give rise directly to four erythrocytic merozoites. These stages are small, reaching a length of about 2 μm. Vectors and final hosts are ticks of the genera *Hyalomma* and also *Rhipicephalus evertsi*. Inside these ticks, the eggs are not infected. Therefore, there is **no transovarial transmission**.
   (b) ***Babesia caballi***
   The merozoites (which occur exclusively in erythrocytes) are pear shaped and reach a length of 3 μm (Fig. 4.98; *B. caballi*). They are reproduced by binary fission. Vectors and final hosts are ticks of the genera *Dermacentor, Hyalomma* and *Rhipicephalus*. Here the parasites are found always in the tick eggs and thus the larvae are already infected (=**transovarial transmission**).
4. **Symptoms of disease (Babesiosis)**: *B. equi* (syn. *Theileria*) is significantly more pathogenic than *B. caballi* reaching a parasitaemia of up to 80 % while infection rates in *B. equi* range around 10 %. Symptoms are high fever, apathia, haemorrhagies, haemolytic anaemia, icterus, splenomegaly and liver swelling. In *Theileria* (*Babesia*) *equi* infections, the urine of infected animals shows haemoglobinuria (=dark red urine) much more often than it is the case in *B. caballi* infections. Acute infections may lead to death within a few days especially in the case of old horses (induced by lung oedema). After survival of an infection in general, **premunity** occurs, which, however, does not cover the other species = there is **no cross immunity**.

   Chronic infections have been reported especially for *B. caballi* lasting for several months up to 4 years. They are accompanied by loss of weight, weakness of the hind legs and reduced fitness.

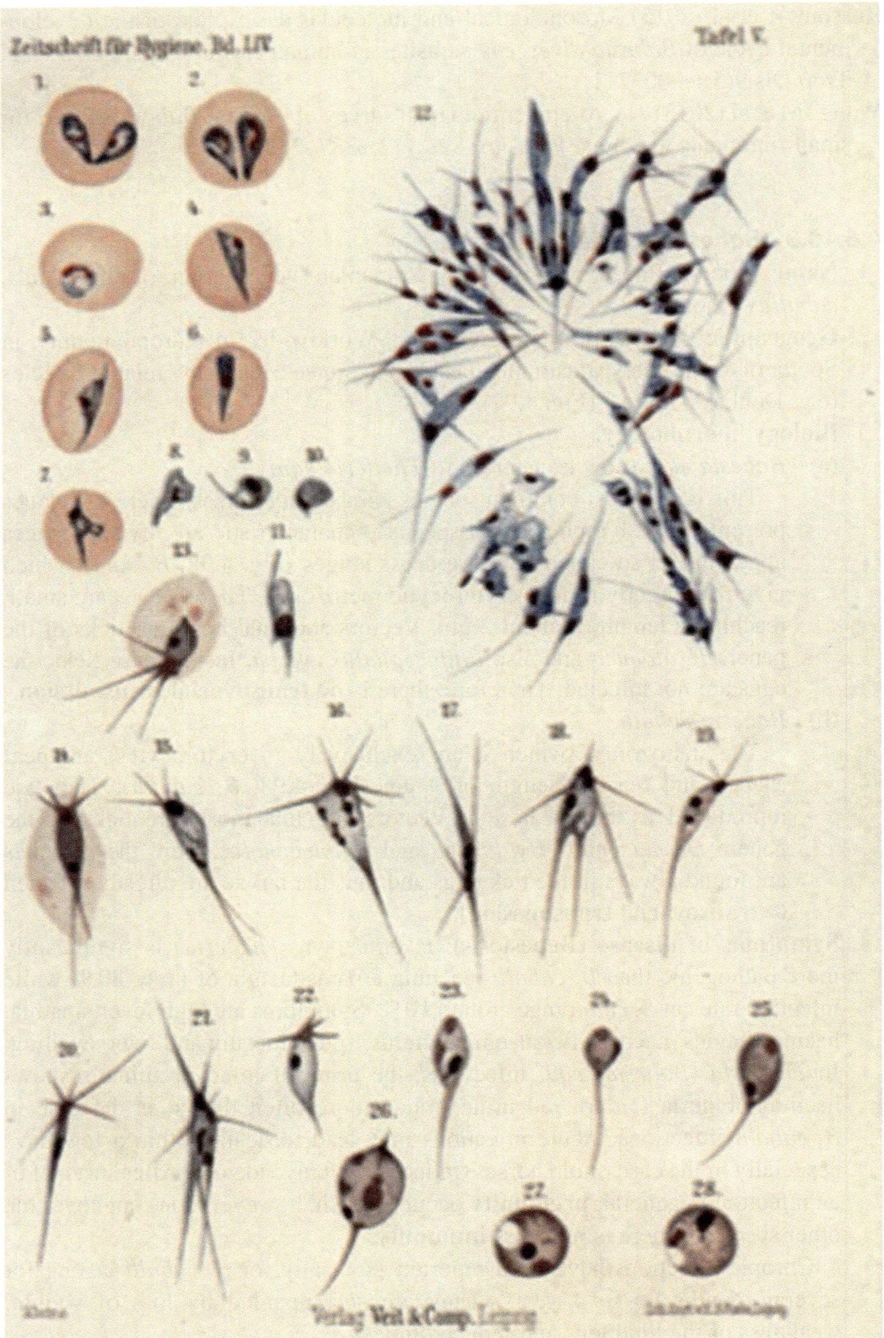

Fig. 4.97 Diagrammatic representation of the *Babesia* stages of cattle (**1–7**) and in ticks (**8–28**) drawn 1905 by Robert Koch and co-workers during an Africa expedition

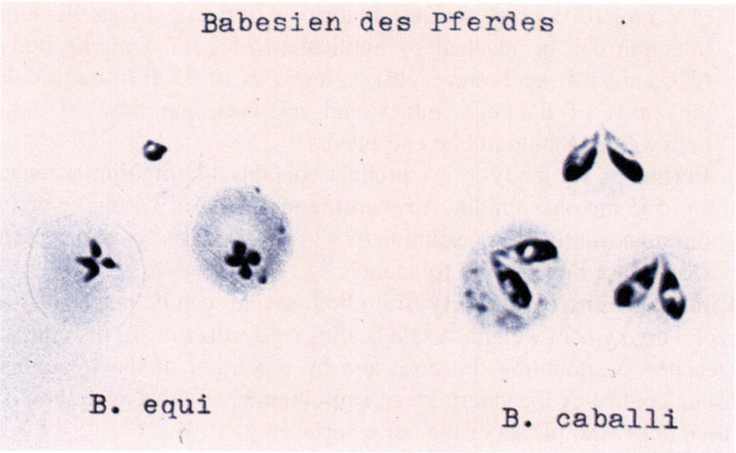

Fig. 4.98 Light micrograph of the stages of *Babesia equi* (=*Theileria equi*) and *B. caballi* inside red blood cells of equids (= German: Pferde)

5. **Diagnosis**: During fever attacks the parasites can be diagnosed with the help of Giemsa stain blood smears (Fig. 4.98). In acute infections, this can be done even in the first 14 days; later it might become difficult and serological tests should then be used (KBR, IFAT, ELISA, SELISA), which are offered by special laboratories. Such investigations are obligatory for horses prior to being imported into the USA and several other countries like Australia and Canada.
6. **Pathway of infection**: Percutaneous during bites of vector ticks. The sporozoites are included in the saliva.
7. **Prophylaxis**: Daily check for attached ticks and—if present—to remove them without squeezing them. Application of acaricides prior to placing the horses onto the meadow. Application of **imidocarb** (imizol®): 1 × 2.4 mg/kg bodyweight = 2 ml/100 kg bodyweight offers protection for about 1 month. However, these products are not available in several countries. Washing the horses with the non-poisonous product Mite-Stop® (Fa. Alpha-Biocare, Neuss, Germany) protects for about 4 days from tick attacks; same protection occurs with the help of the TaonX insect-protection product (also produced by Alpha-Biocare).
8. **Incubation period**: 2–10 days.
9. **Prepatent period**: 2–7 days.
10. **Patency**: Eventually several years (especially in infections with *B. caballi*).
11. **Therapy**: Infections with *B. caballi* can more easily controlled by application of chemotherapeutical products than *B. equi* (syn. *T. equi*). The following compounds/products show significant effects:
    (a) **Imidocarb** = imizol® and imixol®: dosis: 2.4 mg/kg bodyweight = 2 ml/ 100 kg bodyweight given intramuscularly leads to clinical healing (**premunity**). **Sterilization** of an infection can be reached in the case of *Theileria* (*Babesia*) *equi* when giving 4 × 4.8 mg/kg bodyweight

(4 × 4 ml/100 kg bodyweight). In the case of *B. caballi*, sterilization of the infection can be reached by application of 2 × 2.4 mg/kg bodyweight (2 × 2 ml/100 kg bodyweight) at intervals of 72 h intramuscularly. In the cases of donkeys, mules and relatives, the dose of 2.4 mg/kg bodyweight should not be exceeded.

(b) **Berenil®**: 1 g ready-to-use product contains 445 mg diminazenaceturate and 555 mg phenamidin. **A recommended dose** is 3.5 mg/kg bodyweight intramuscularly or 7 % solution (5 ml per 100 kg bodyweight). **Attention**: Do not use this product for camels.

Clincal healing (premunity) from both species can be reached by application of 4 mg/kg bodyweight. A safe healing (=sterilization of the infection) can be reached by doubling the dose and by repetition of the treatment under constant control by the veterinarian. **Importance**: Divided dose should always be used at several places of the horse surface.

Further Reading

Bartolomé Del Pino LE et al (2016) *Babesia caballi* and *Theileria equi* infections in horses in Central-Southern Italy: sero-molecular survey and associated risk factors. Ticks Tick Borne Dis 7:462–469.

García-Bocanegra I et al (2014) Seroprevalence and risk factors associated with *Babesia caballi* and *Theileria equi* infection in equids. Vet J 195:172–178.

Hussain MH et al (2014) Seroprevalence of *Babesia caballi* and *Theileria equi*. Vet Parasitol 202:248–256.

Mehlhorn H, Schein E (1998) Redescription of *Babesia equi* Laveran, 1901 as *Theileria equi* Mehlhorn, Schein 1998. Parasitol Res 84:467–475.

Piantedosi D et al (2014) Seroprevalence and risk factors associated with *Babesia caballi* and *Theileria equi* infections in donkeys from Southern Italy. Vet J 202:578–582.

Ribeiro AJ et al (2013) Prevalence of *Theileria equi*, *Babesia caballi* and *Anaplasma phagocytophilum* in horses from the north of Portugal. Parasitol Res 112:2611–2617.

Wise LN et al (2013) Review of equine piroplasmosis. J Vet Intern Med 27:1334–1346.

4.6.10.4 Piroplasms of Cats

In felids, a series of piroplasms have been described to occur inside blood and/or lymph system. Since some of them are first reproduced inside lymphocytes, they belong rather to the genus *Theileria*. The following two species have some importance:

(a) ***Babesia felis***:

This species occurs in cats in the regions of the Mediterranean countries, India, the USA and South Africa (leopards, lions, house cats). The stages in the erythrocytes are only 1–2.5 μm long. Vectors are ticks of the genus

Haemaphysalis. **Symptoms of disease** are anaemia, icterus and general weakness and even in several cases death may occur. **Control**: Imported cats should be treated with products used against dog babesiosis.

(b) ***Cytauxzoon felis***:

This species occurs mainly in North America, and its specimens reach a length of about 2 µm. Vectors are ticks of the genus *Dermacentor* (especially *D. variabilis*).

(c) ***Babesia herpailuri* (syn. *B. pantherae*)**:

This species was described in South America inside the jaguarondi and in South Africa in leopards and house cats.

Further Reading

Ayoob AL et al (2010) Feline babesiosis. J Vet Emerg Crit Care 20:90–97.

Clarke LL, Rissi DR (2015) Neuropathology of natural *Cytauxzoon felis* infection in domestic cats. Vet Pathol 52:1167–1171.

Hamel D et al (2012) Epidemiological aspects on vector-borne infections in stray and pet dogs from Romania and Hungary with focus on *Babesia* spp. Pasrasitol Res 110:1537–1545.

Sherrill MK, Cohn LA (2015) Cytauxzoonosis: Diagnosis and treatment of an emerging disease. J Feline Med Surg 17:940–948.

Solano-Gallego L, Banetti G (2011) Babesiosis in dogs and cats—expanding parasitological and clinical spectra. Vet Parasitol 181:48–60.

4.6.10.5 Piroplasms of Mice and Man

Rodents may harbour several *Babesia* species. At least one of which (*Babesia microti*) can be transmitted to humans and mostly introduces a chronic and latent babesiosis. Characteristic are typical tetrade blood stages, which look very similar to those of *B. equi* (Fig. 4.98). Rodents at home do not need treatment; infections of humans occur by bites of ticks of the genus *Ixodes* (see Table 6.2).

Further Reading

Dunn JM et al (2014) *Borrelia burgdorferi* promotes the establishment of *Babesia microti* in the northeastern United States. PLoS One 9:e115494.

Eshoo MW et al (2014) Broad-range survey of tick-borne pathogens in Southern Germany reveals a high prevalence of *Babesia microti* and a diversity of other tick-borne pathogens. Vector Borne Zoonotic Dis 14:584–591.

Farkas R et al (2015) First report on *Babesia* cf. *microti* infection of red foxes (*Vulpes vulpes*) from Hungary. Parasit Vectors 8:55.

Yabsley MJ, Shock BC (2012) Natural history of zoonotic *Babesia*: role of wildlife reservoirs. Int J Parasitol Parasit Wildl 2:18–31.

4.6.11 *Theileria* Species

1. **Name**: The genus name honours the Swiss/South African veterinarian Arnold Theiler (1867–1936). Species names derive from Latin: *mutare* = changing;

parvus = small; *annulata* = ring-like; *ovis* = sheep; *hircus* = goat; *velum ferrare*/*gerere* = sail bearing; *orientalis* = coming from the orient; *separate* = divided, separate; *tauritragi* = on cattle. Lawrence = name of an English discoverer.

2. **Geographic distribution/epidemiology**: Worldwide, in Europe, however, only rare cases, while these species are of extreme high importance in Africa due to their high pathogenicity.
3. **Biology, morphology**:

   A. **Cattle species** (Table 4.16)

   - ***T. mutans***: agent of the so-called "mild theileriosis" and of the "pseudo-coast fever", both occurring in Africa. The intra-erythrocytic stages are mostly spherical with diameters of about 1.5 μm. Division stages are called "maltese cross stages", since four merozoites are formed at the same time (Fig. 4.98). Vectors are in West Europe: *Haemaphysalis punctata* (therefore occur often mixed infections with *Babesia major*); in Africa *Amblyomma* species act as vectors (and final hosts) (see Fig. 4.99).
   - ***Theileria parva parva***: agent of the so-called "East-coast fever". The stages are extremely pathogenic and lead to mortality rates of up to 90 %. Vectors are *Rhipicephalus appendiculatus* ticks (Figs. 4.99, 4.100, 4.101, 4.102, 4.103, 4.104 and 4.105).
   - ***T. parva lawrencei***: agent of the so-called "Corridor disease" in Africa south of the Sahara. Vectors are ticks of the species *Rhipicephalus appendiculatus*.
   - ***T. annulata***: agent of the so-called "Mediterranean coast fever" in North Africa, Near and Middle East and Central Asia. Vectors are *Hyalomma* species (Figs 4.101 and 4.102).

   B. **Species in sheep/goats** (Table 4.16)

   - ***T. ovis***: This species apparently includes several low-grade infectious species (strains) parasitizing smaller ruminants. The erythrocytic stages measure only 1.3 μm in length, but they transform the red blood cells to a polymorphic appearance. Vectors in Europe are members of the genera *Rhipicephalus*, *Haemaphysalis* and also *Ixodes ricinus*. Co-infection with *T. motasi* may occur.
   - ***T. hirci***: This pathogenic species occurs in Europe only in southern countries. The erythrocytic stages reach a length of up to 2 μm. This species possesses schizonts in the lymphocytes which are up to 20 μm in diameter, while those of the other *Theileria* species are smaller, but become more often reproduced while the host cells start new divisions.

   C. **Species in horses**

   See *Babesia equi*, which is old name, but which is still used in the literature (Fig. 4.105).

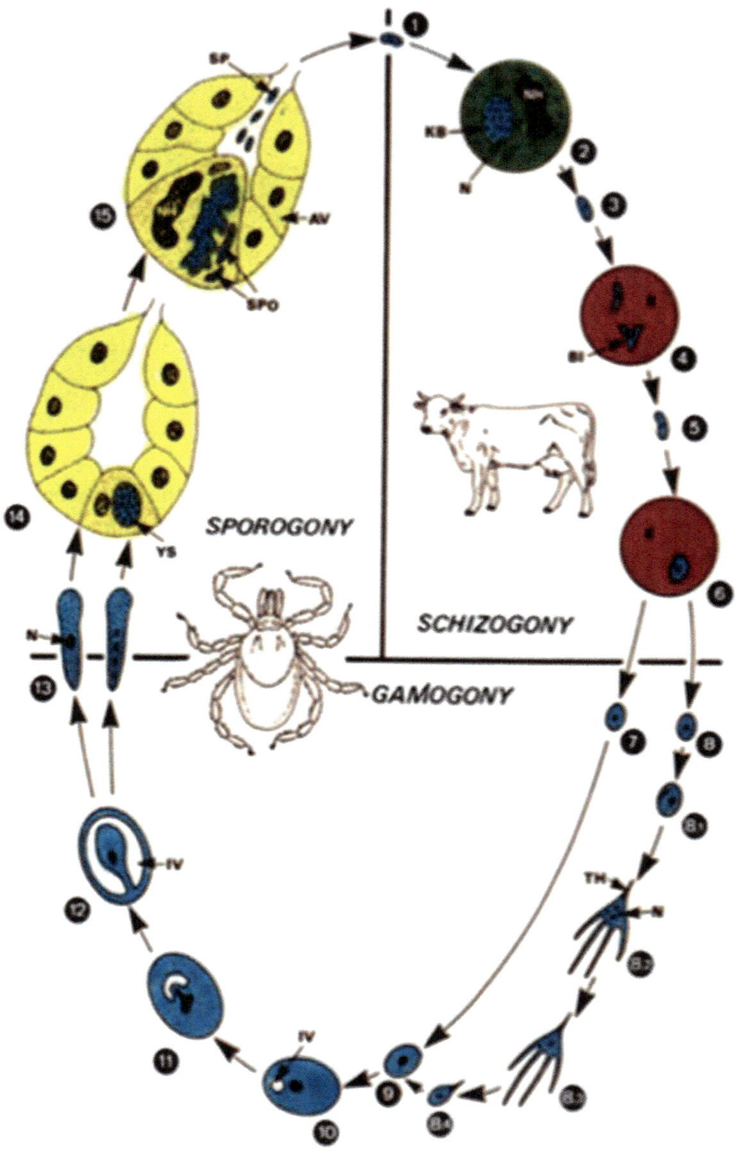

Fig. 4.99 Diagrammatic representation of the life cycle of *Theileria* spp. (for hosts, see Table 4.16). (**1**) Sporozoites are injected during blood meal of ixodid ticks. (**2**) Schizont (Koch's body) inside the cytoplasm of newly dividing lymphocyte, eventually forming merozoites. (**3**) Free motile merozoites enter erythrocytes. (**4**) Binary fission inside erythrocyte (at a low rate). (**5**) A few free merozoites enter other erythrocytes. (**6**) Formation of spherical or ovoid stages (i.e. gamonts). (**7**, **8**) Gamonts free in blood masses inside tick gut. (**8.1–8.4**) Formation of 4-nucleate microgamonts (**8.2**) which give rise by fission to uninucleate microgametes (**8.3–8.4**). The latter fuse with the macrogamete (**9**). (**9**) Macrogamete. (**10**) Zygote. (**11–13**) Formation of motile kinete from ovoid immobile zygote inside intestinal cells of the tick. Note that the developing kinete

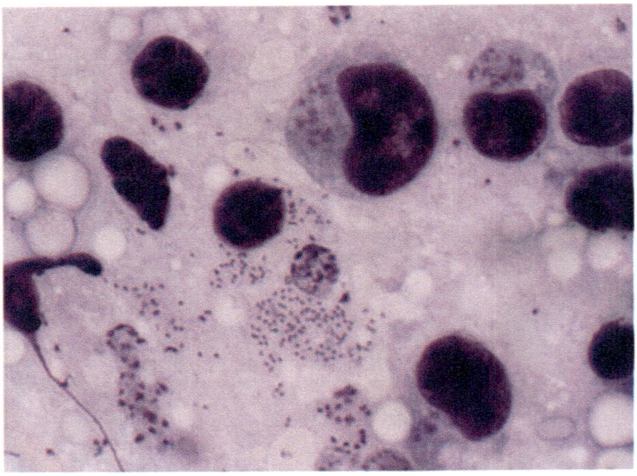

Fig. 4.100 Light micrograph of a lymph node punction of a cattle infected with *T. parva*. The lymphocytes contain *Theileria* schizonts. Many merozoites are visible after bursting of cells

4. **Symptoms of disease (Theileriosis)**: In the case of the pathogenic species, fever starts on day 7 p.i. reaching 41 °C. During this phase, numerous schizonts are formed inside the lymphocytes. Then follows anaemia, icterus, lymph node and spleen swellings, rarely haemoglobinuria (in the case of kidney infarcts, mortality rates may reach 50 % in the case of older animals). It was noted that *Theileria* infections lead to a severe immunosuppression.
5. **Diagnosis**: Demonstration of the Koch's bodies (spheres) in lymphocytes and, starting from day 10 p.i, the red blood cell stages (Figs. 4.101, 4.104 and 4.105).
6. **Pathway of infection**: Percutaneous during tick bites.
7. **Prophylaxis**: Regular tick control.
8. **Incubation period**: 5–7 days in pathogenic species.
9. **Prepatent period**: From day 5 p.i. schizonts occur in lymph nodes.
10. **Patency**: In case of survival: eventually years.
11. **Therapy**:
    (a) **Parvaquone** = Clexon®: Solution i.m. injection.
    (b) **Buparvaquone** = Butalex: solution i.m. injection.
    (c) **Halofuginone** = Halofuginone-lactat = Terit® oral application.
    (d) **Special for horses**: (*T. equi* = imidocarb).
    Attention: Treatment may lead to strong side effects. Thus, doses beyond 4 mg/kg bodyweight are dangerous.

Fig. 4.99 (continued) protrudes into an enlarging vacuole (IV) within the zygote. In *T. parva* kinetes (**13**), division of the nucleus may start before they leave the intestinal cells. (**14**) After moult of the ixodid tick and attachment to a new host, kinetes enter the cytoplasm of cells of salivary glands and give rise to young sporonts which grow and initiate repeated nuclear divisions. (**15**) Parasitism leads to considerable enlargement of the host cell and its nucleus. Inside the giant host cell, the sporont forms thousands of sporozoites. The latter become transmitted during the next blood meal. *AV* alveolar cell of salivary glands; *BI* binary fission; *E* erythrocyte; *IV* inner vacuole; *KB* Koch's body (=schizont); *N* nucleus; *NH* nucleus of host cell; *SP* sporozoites; *SPO* sporont; *TH* thornlike structure; *YS* young sporont

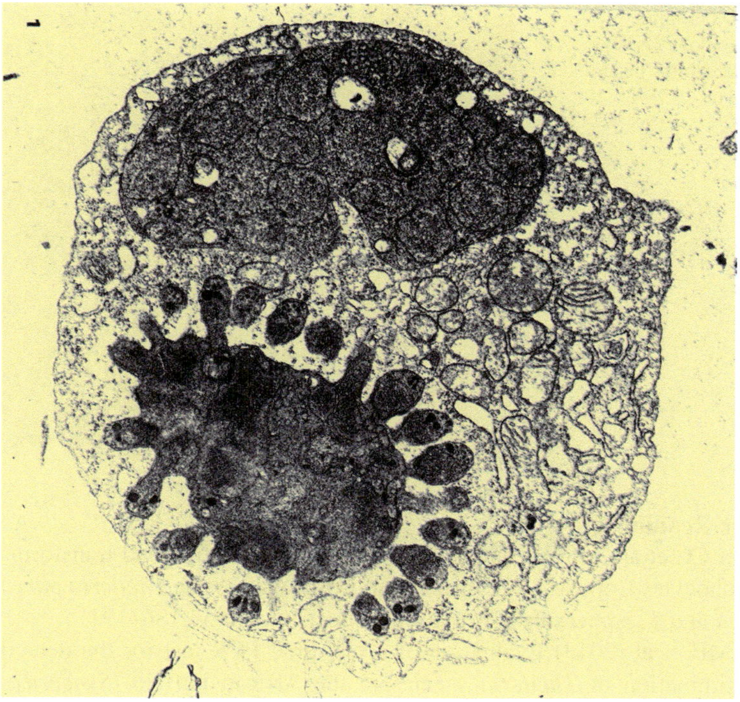

Fig. 4.101 Transmission electron micrograph through a lymphocyte infected by two schizonts of *T. parva*, one in the stage nuclear reproduction and the other already producing merozoites

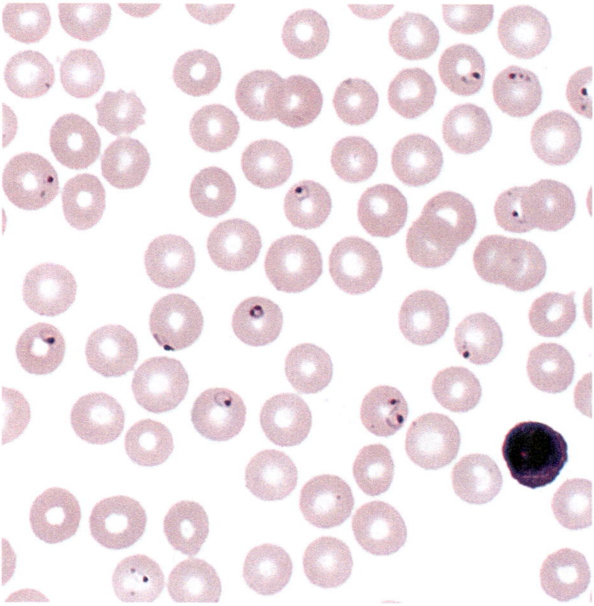

Fig. 4.102 Giemsa-stained blood smear showing erythrocytes containing merozoites of *T. annulata*

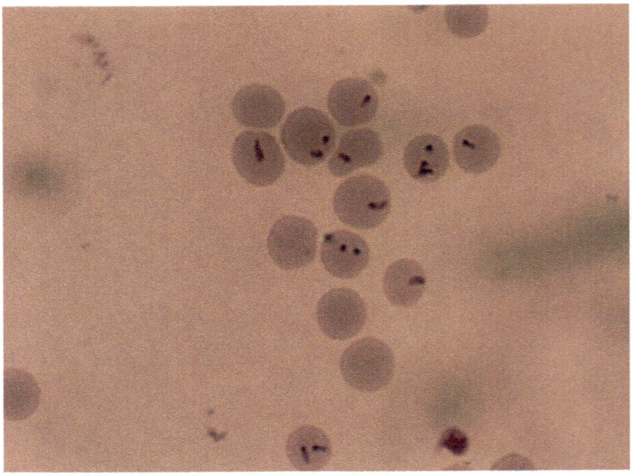

Fig. 4.103 Light micrograph of *Theileria parva* merozoites inside erythrocytes

Further Reading

Baldwin CL et al (1986) Comparative analysis of infection and transformation of lymphocytes from African buffalo and Boran cattle with *Theileria parva* subsp. *parva* and *T. parva* subsp. *lawrencei*. Infect Immun 53:186–191.

Chaisi ME et al (2013) Evaluation of a real-time PCR test for the detection and discrimination of *Theileria* species in the African buffalo (*Syncerus caffer*). PLoS One 8:e75827.

Chauhan HC et al (2015) Comparison of molecular and microscopic technique for detection of *Theileria annulata* from the field cases of cattle. Vet World 8:1370–1374.

Maxie MG et al (1982) A comparative study of the disease in cattle caused by *Theileria parva* or *T. lawrencei*: II. Hematology, clinical chemistry, coagulation studies and complement. Vet Parasitol 10:1–19.

Mehlhorn H et al (1975) Electron microscope studies on development stages of *Theileria annulata* (Dschunkowsky, Luhs, 1904) in the intestine and haemolymph of *Hyalomma anatolicum excavatum* (Koch, 1844). Parasitol Res 48:137–150.

Mehlhorn H et al (1978) Electron microscopic studies on the development of kinetes of *Theileria parva* Theiler, 1904 in the gut of the vector ticks *Rhipicephalus appendiculatus* Neumann, 1901. Acta Trop 35:123–136.

Mehlhorn H et al (1979) Electron-microscopic studies on *Theileria ovis* Rodhain, 1916: development of kinetics in the gut of the vector tick, *Rhipicephalus evertsi evertsi* Neumann, 1897, and their transformation within cells of the salivary glands. J Protozool 26:377–385.

Mehlhorn H et al (1981) Electron microscopical study on the effect of halofuginone on *Theileria parva*. Tropenmed Parasitol 32:231–233.

Mehlhorn H, Schein E (1977) Electron microscopic studies of the development of kinetes in *Theileria annulata* Dschunkowsky & Luhs, 1904 (Sporozoa, Piroplasmea). J Protozool 24:249–257.

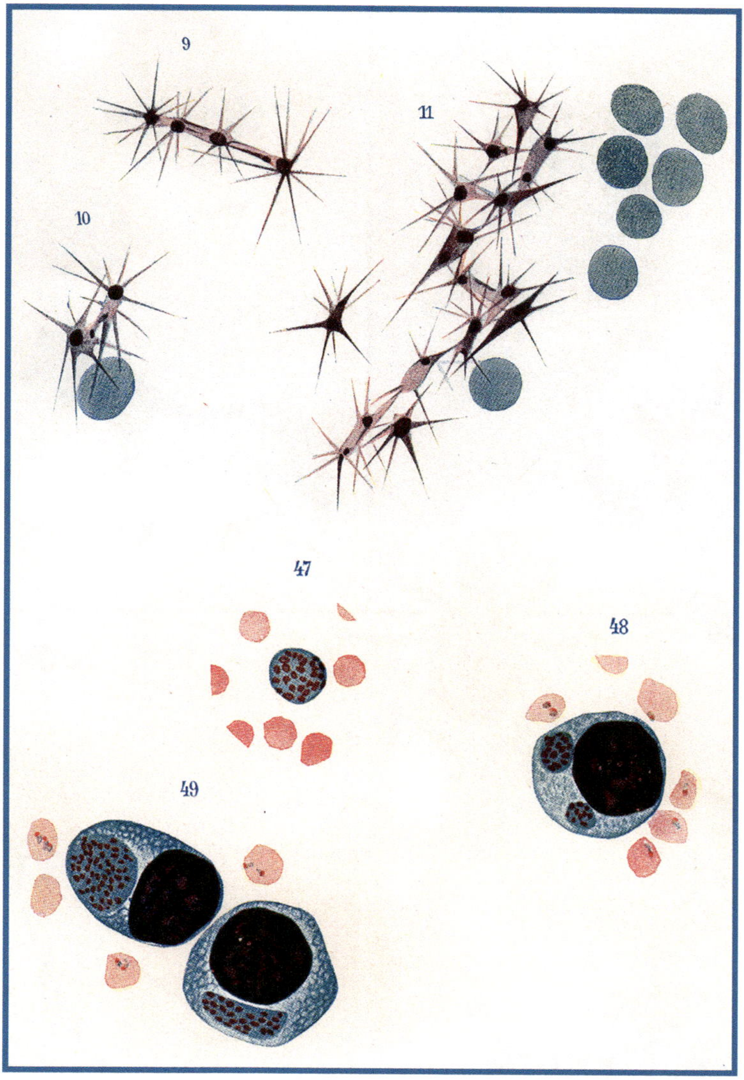

Fig. 4.104 Table published by the German Nobel Prize winner (1905) Robert Koch (1843–1910) and co-worker Friedrich Karl Kleine showing stages of the agents of East Coast fever of cattle

Mehlhorn H, Schein E (1984) The piroplasms: life cycle and sexual stages. Adv Parasitol 23:37–104.
Morrison WI et al (2015) Understanding the basis of parasite strain-restricted immunity to *Theileria parva*. Annu Rev Anim Biosci 3:397–418.
Shayan P et al (2016) Identification and characterization of *Theileria ovis* surface protein (ToSp) resembled TaSp in *Theileria annulata*. Parasitol Res. doi:10.1007/s00436-016-4929-1.

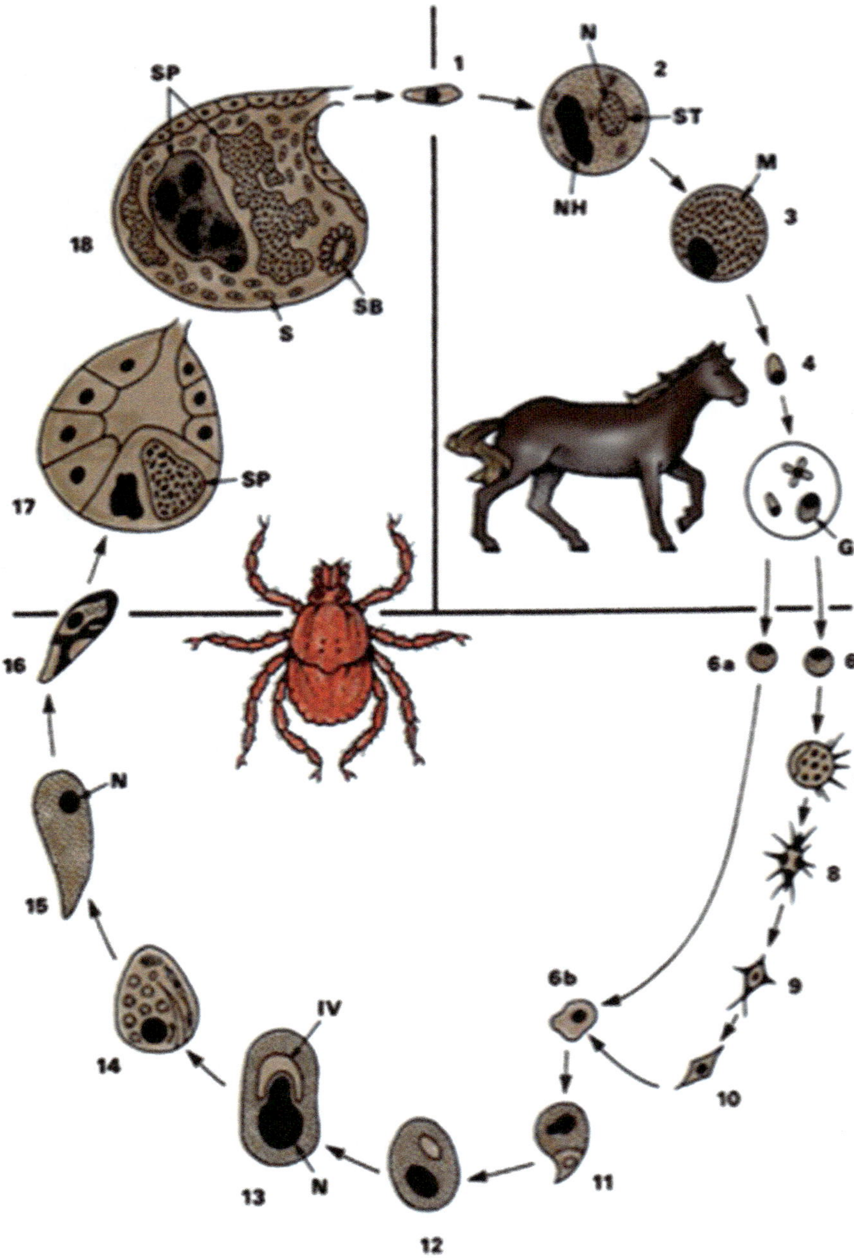

Fig. 4.105 Diagrammatic representation of the life cycle of *Theileria equi* (syn. *Babesia equi*). (**1**) Sporozoite injected with tick (nymph, adult female) saliva. (**2**) Young schizont in a lymphocyte (macroschizont, Koch's body). (**3**) Late schizont in a lymphocyte during the formation of merozoites (microschizont). (**4**) Free merozoite. (**5**) Reproduction inside erythrocytes—note the occurrence of Maltese cross-like dividing stages and the presence of spherical stages (gamonts).

4.6.12 Agents of Malaria or Malaria-Like Diseases

Within this group of blood parasites are collected some species of the genera *Plasmodium* (of birds), *Leucocytozoon* (of birds), *Haemoproteus* (of birds and reptiles), *Hepatozoon* (of dogs, mice, reptiles), *Hepatocystis* (of monkeys), etc. (Table 4.17).

4.6.12.1 *Plasmodium* Species

1. **Name**: Greek: *plasma* = fluid, but structured material. Latin: *juxtanucleare* = nuclei side by side; *gallinaceum* = belonging to chicken; *relictum* = remnant; *praecox* = early.
2. **Geographic distribution/epidemiology**: Worldwide.
3. **Biology, morphology**: In Europe only a few species exist, which have no importance for farm animals. Since the pathogenic agents of **chicken malaria** (*P. juxtanucleare, P. gallinaceum*) are lacking, only *P. relictum* (syn. *P. praecox*) introduces severe diseases in sparrows, doves, ducks, swans but also in penguins in zoological gardens. Especially the latter are highly endangered by death due to malaria. Vectors are mainly species of the genera *Anopheles* and *Culex*, but also *Culiseta* females are apparently able to transmit sporozoites to the different bird species. At first, schizonts are produced in endothelial cells of the liver and lung, which give rise to merozoites that enter again endothelial cells and become again schizonts. Then produced merozoites enter erythrocytes already 5–7 days after the infection and start there the intraerythrocytic schizogony, which leads to the destruction of the parasitized erythrocytes. The intraerythrocytic **schizonts** appear spherical up to polymorphous. Since they are rather large, the nucleus of the erythrocytes is always pressed aside (Figs. 4.106, 4.107 and 4.111a). About 8–32 merozoites are produced within 12–36 h (depending on the species) inside infected erythrocytes. The final rupture of the infected cell finally sets free the merozoites but also remnants of the digested haemoglobin. This so-called pigment may induce fever and weak animals may die in cases of massive

---

Fig. 4.105 (continued) (**6**) After engorgement of ticks the ovoid/spherical gamonts undergo further development within the blood masses inside the intestine (mostly inside host cells). (**7–10**) By divisions, some raylike microgametes (**10**) are produced by microgamonts (**7, 8**). (**11**) Fusion (syngamy) of gametes. (**12–16**) Inside the zygote (**12**) a slender, motile, club-shaped kinete is developed, which leaves the intestinal cells and enters via haemolymph the salivary gland cells of the ticks after their molt (larva-nymph or nymph-adult female) and their attachment to another host. (**17**) Penetrated kinetes grow up inside the salivary gland cells and give rise to multinucleated sporonts. (**18**) The multinucleated sporonts are divided into numerous small sporoblasts (SB) which form sporozoites by a budding process at their periphery; during the next sucking period the sporozoites are injected (with the saliva) to the new host. *G* gamont; *IN* inner vacuole; *M* microschizont; *N* nucleus; *NH* nucleus of the host cell; *S* sporozoites; *SB* sporoblast; *SP* sporonts

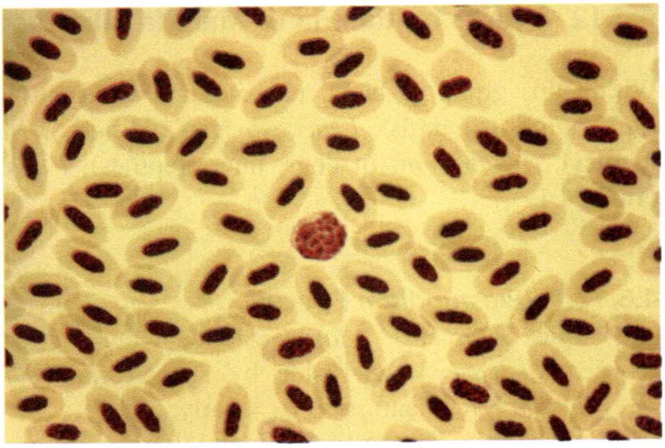

Fig. 4.106 Light micrograph of a Giemsa-stained blood smear preparation showing in the center a schizont of *Plasmodium relictum* which has been released from its host cell

Table 4.17 Agents of animal malaria and related diseases

Species	Hosts	Habitat	Vector
Hepatozoon muris (syn. *H. perniciosum*)	Rats, mice	Liver cells, leucocytes	Rat mites
H. erhardovae	Voles (Clethrionomys sp.)	Lung, leucocytes	Rat fleas
H. canis *H. americanum* *H. felis*	Dogs, felids	Liver, spleen, leukocytes	Ticks (*Rhipicephalus sanguineus, Ixodes* species)
H. aegypti	Snakes (*Spalerosophis diadema*)	Lung, erythrocytes	Mosquitoes (*Culex* species)
Plasmodium gallinaceum	Chickens	Erythrocytes	Mosquitoes (*Culex* species)
P. relictum (syn. *P. praecox*)	Sparrows, penguins	Erythrocytes	Mosquitoes (*Culex* species)
Haemoproteus columbae	Doves	Erythrocytes	Midges (*Culicoides* species)
H. nethionis	Ducks	Erythrocytes	Louse flies
Leucocytozoon simondi	Geese, ducks	Leukocytes	Black flies, midges
L. smithi	Turkeys	Leukocytes	Midges
Hepatocystis kochi, *H. simiae*	African monkeys	Liver, erythrocytes	Midges
Karyolysus species	Reptiles	Erythrocytes	Bloodsucking mites
Haemogregarina species	Turtles	Erythrocytes	Leeches
Schellackia species	Reptiles	Erythrocytes, intestinal cells	Mechanical transmission, mites

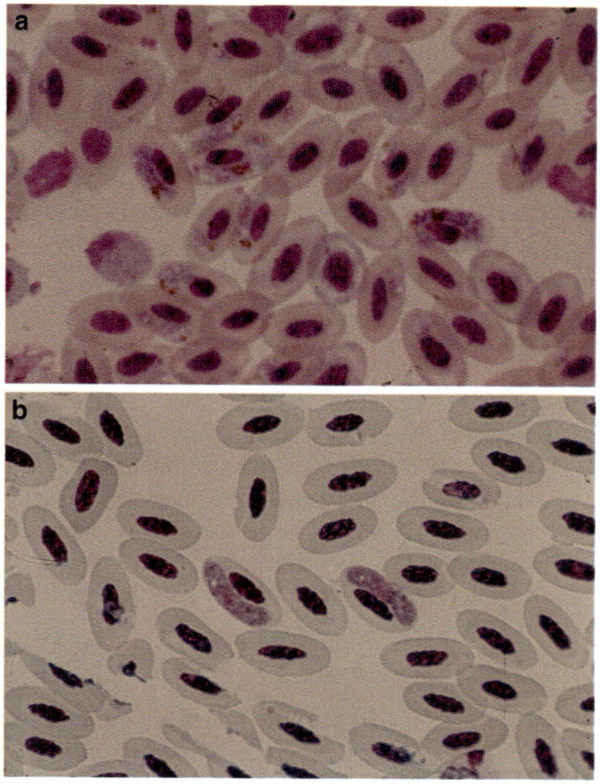

Fig. 4.107 (**a**, **b**) Light micrographs of banana-shaped *Plasmodium* stages inside the nucleated erythrocytes

infection. The rupture of the parasitized erythrocytes is in contrast to human malaria agents not synchronized (with the exception of *Plasmodium relictum*). After several repeated schizogonic processes, the gamogony starts whereby sexually determined merozoites grow up either to a large male or female gamont. If the vector insect has ingested such infected erythrocytes, these gamonts are released in the intestine of the mosquito. While female gamonts just increase to become a macrogamete, the male gamont becomes divided and gives finally rise to 4–8 motile male gametes. After fertilization, the zygote becomes motile and thus is called **ookinete**, which penetrates the intestinal epidermis, becomes situated in a space between the intestinal cells and the basal layer of the intestine and grows up to become an **oocyst**, wherein the **sporozoites** are produced during an asexual process. As soon as the oocyst wall is ruptured, the sporozoites pass the body cavity of the mosquito and enter the salivary gland cells. During the next bloodsucking, the sporozoites may be injected to the new host.

4. **Symptoms of disease (Malaria)**: In Europe, birds may become infected during summer, while in the tropics the cycle runs all the year around. Sparrows show very low-grade symptoms of disease—if at all. Thus, they act apparently as **reservoir hosts**. Inside young or less adapted bird species (especially in

penguins), however, severe symptoms may occur (high fever, high-grade enlargements of the liver and spleen, anaemia, extreme weakness, apathia, coma and high rates of mortality).

5. **Diagnosis**: Demonstration of parasitic stages inside **Giemsa-stained** blood smear preparations or using the so-called **thick droplet method**, which is very helpful in cases, where only few erythrocytes are infected.
6. **Pathway of infection**: Percutaneous during bites of vector insects (mosquitoes).
7. **Prophylaxis**: Control of mosquitoes close to houses and rams; removal of water, wherein mosquitoes may breed.
8. **Incubation period**: About 1 week.
9. **Prepatent period**: About 1 week.
10. **Patency**: Depending on the species eventually years.
11. **Therapy**: There are several compounds/products, which, however, are not always very efficacious. These products have to become entered into the drinking water of farmed birds in endemic regions. Among the antimalarial compounds, the following are most commonly used: chloroquine and primaquine but also artesunate, quinacrine, pyrimethamine, chlorguanil or trimethoprim. However, toxicity may be high. Thus, at first low dosages should be entered.

Further Reading

Kumnuan R et al (2013) In vivo transmission blocking activities of artesunate on the avian malaria parasite *Plasmodium gallinaceum*. Vet Parasitol 197:447–454.

Matos de Macchi B et al (2013) Chickens treated with a nitric oxide inhibitor became more resistant to *Plasmodium gallinaceum* infection. Vet Res 44:8

Lauron EJ et al (2015) De novo assembly and transcriptome analysis of *Plasmodium gallinaceum*. Malar J 14:296.

4.6.12.2 *Hepatozoon* Species

1. **Name**: Greek: *heaps* = liver; *zoon* = animal. Latin: *canis* = dog; *mus* = mouse; *sus* = pig.
2. **Geographic distribution/epidemiology**: Worldwide, in Europe mainly around the Mediterranean Sea. Hosts are dogs, red foxes, hyenas and jackals but also cats. Other parasitic species are also described inside reptiles and in frogs.
3. **Biology, morphology**: *Hepatozoon canis* and *H. americanum* have a typical coccidian life cycle. The dog and the above-listed canids and felids are obligate **intermediate hosts**. Inside their endothelial cells (i.e. liver, spleen, bone mark), the asexual reproduction (**schizogony**) occurs. The produced merozoites enter leucocytes and are transformed into gamonts (Fig. 4.108). **Final Hosts** and **vectors** are ticks (*Rhipicephalus sanguineus*, *Amblyomma maculatum*, *Dermacentor* species and *Ixodes* species), which ingest the 10×5 µm measuring gamonts during their blood meal at an infected intermediate host. Inside of the intestine of the ticks, the **gamogony** occurs. The finally formed zygote

Fig. 4.108 Light micrograph of a gamont (H) of *Hepatozoon canis* in a ruptured leucocyte. *E* erythrocyte

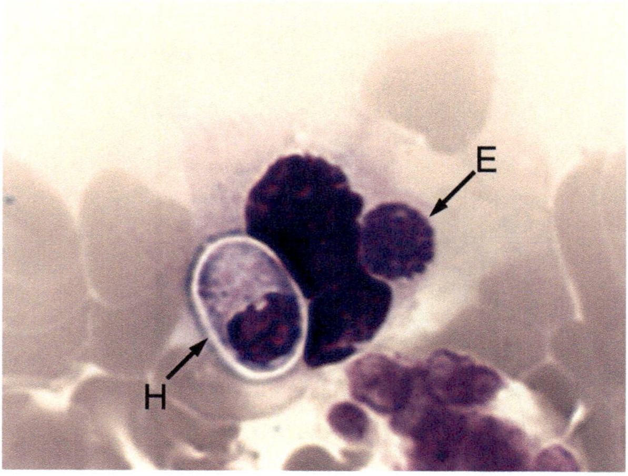

(=motile ookinete) leaves the intestine of the tick and penetrates into the body cavity, where the **sporogony** occurs giving finally rise to oocysts and sporocysts. The latter contain 16 motile **sporozoites**. The cycle is finished as soon as the dog or other intermediate hosts ingest the infected tick (Table 4.17).

4. **Symptoms of disease (Hepatozoonosis)**: Leading symptoms are swellings of the lymph nodes, fever, loss of weight, apathia, rough hair and multiple lesions (necrosis) inside all infected organs. The blood picture shows hyperglobulinaemia, hypoalbuminaemia and increased values of creatine kinase and alkaline phosphatase. The intensity of symptoms may vary, but death cases may occur, too (especially due to infections with *H. americanum*).
5. **Diagnosis**: The diagnosis is difficult, since the parasites inside the leucocytes occur mainly starting from day 28 after infection, while unspecific clinical symptoms are seen very early after the infection. Thus, the disease can only be diagnosed during the first 4 weeks p.i. by punction of bone mark or liver, which in low-grade infections would be too dangerous with respect to side effects. Serological tests (IFAT, ELISA) and PCR show the parasites very soon after infection.
6. **Pathway of infection**:
   (a) ***Hepatozoon canis, H. americanum***: Oral by ingestion of infected ticks.
   (b) ***H. muris***: Ingestion of infected bloodsucking mites.
   (c) ***H. erhardovae***: Ingestion of infected fleas.
   (d) ***H. aegypti***: This species lives not only in the organs of canids or felids but also in reptiles (Fig. 4.109). However, the pathway of infection is similar, since the agents of disease are transmitted when reptiles ingest infected *Culex* mosquitoes or they become infected when these mosquitoes suck blood.
7. **Prophylaxis**: Use of repellents or acaricidal and insecticidal products sprayed onto the hair of dogs and cats (e.g. Anticks®). Removal of attached ticks

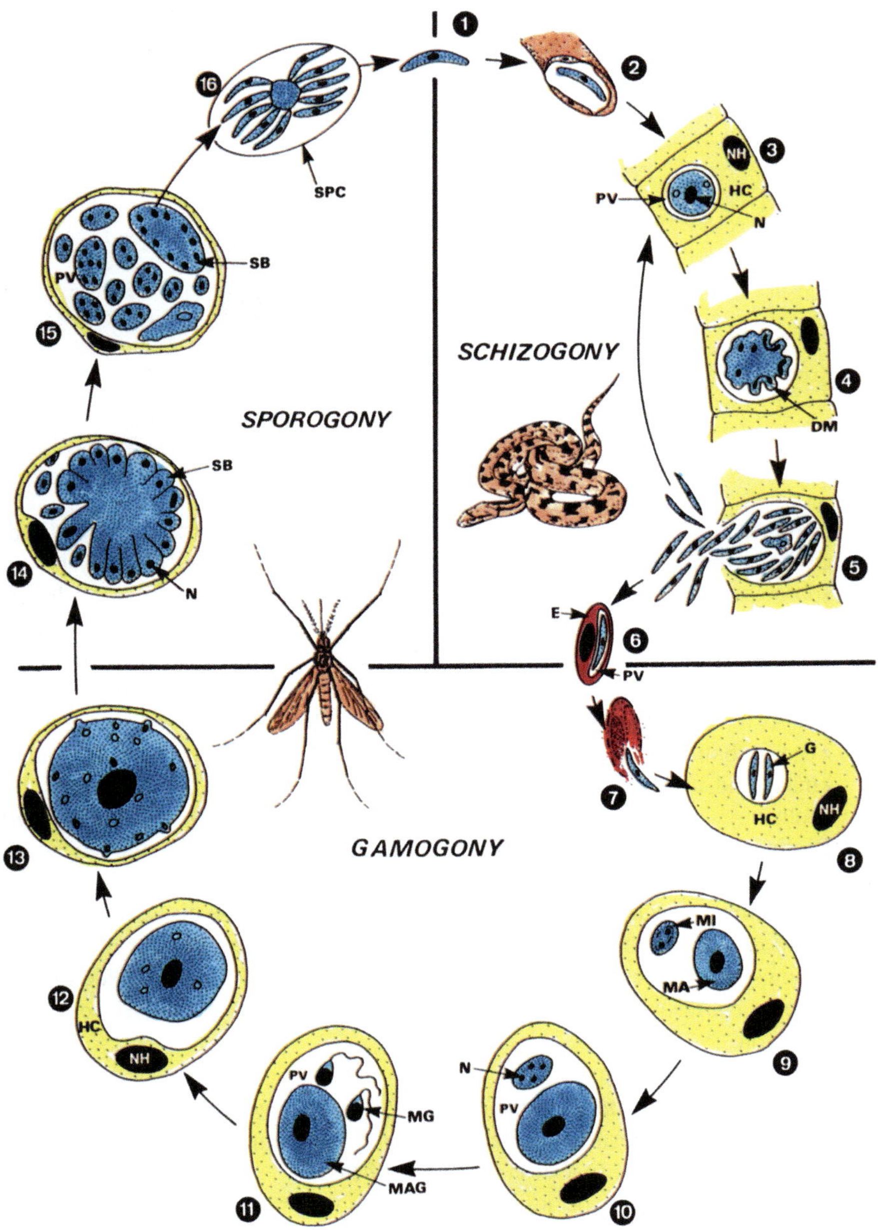

Fig. 4.109 Diagrammatic representation of the life cycle of *Hepatozoon aegypti* inside snakes (*Spalerosophis diadema;* Colubridae) and mosquitoes (*Culex pipiens*). (**1**, **2**) Sporozoites are injected during bite of the female mosquito and enter the lung capillaries of the snake. (**3–5**) After penetration into endothelial cells (**3**), they grow to be schizonts (**4**) which form merozoites (**5**). (**6**) The free merozoites may enter other endothelial cells, where they repeat schizont formation (**3**) or penetrate into erythrocytes and become gamonts of different sex. (**7**, **8**) After

without squeezing them. **Chemoprophylaxis**: imidocarb 1 × 6 mg/kg bodyweight subcutaneously.

8. **Incubation period**: 2–4 weeks.
9. **Prepatent period**: 4–6 weeks.
10. **Patency**: Potentially several years.
11. **Therapy**: There does not exist a tested treatment scheme. However, in many cases the subcutaneous application of 5–6 mg/kg bodyweight **imidocarb** at intervals of 14 days leads to sufficient effects (treatment must be done until gamonts have been disappeared from blood).

Further Reading

Allen KE et al (2011) Novel *Hepatozoon* in vertebrates from the Southern United States. J Parasitol 97:648–653.

Baneth G et al (2015) Canine vector borne co-infections: *Ehrlichia canis* and *Hepatozoon canis* in the same host monocytes. Vet Parasitol 208:30–34.

Johnson EM et al (2007) Field survey of rodents for *Hepatozoon* infections in an endemic focus of American hepatozoonosis. Vet Parasitol 150:27–32.

Karadjian G et al (2015) Systematic revision of the adelaid haemogregarines. Parasitology 22:31.

Moco TC et al (2012) Morphological, morphometric, and molecular characterization of *Hepatozoon* spp. from naturally infected *Caudisoma durissa terrifica* snakes. Parasitol Res 110:1393–1401.

Najm NA et al (2014) *Hepatozoon canis* in German red foxes and their ticks. Parasitol Res 113:2679–2685.

4.6.12.3 *Leucocytozoon* Species

1. **Name**: Greek: *leukos* = white; *kytos* = cell; *zoon* = animal.
2. **Geographic distribution/epidemiology**: Worldwide.
3. **Biology, morphology**: *Leucocytozoon* species (e.g. *L. simondi* of geese and ducks; *L. smithi* of turkeys; see Table 4.17) occur in a rather large number of birds (Figs. 4.110 and 4.111c, d). After the bite of the vector insects (members

Fig. 4.109 (continued) sucking by a mosquito, the gamonts are set free (**7**), migrate into the haemocoel and penetrate into host cells, where they associate in pairs within a parasitophorous vacuole (**8**). (**9**) On day 2 after infection, differentiation to mico- and macrogamonts occurs. (**10**) Nucleus of microgamonts divides. (**11**) Formation of the non-flagellated microgametes on day 3 after infection. (**12**, **13**) Fertilization on day 4 and growth of the young oocyst during the following days. (**14**) Formation of sporoblasts on days 8–10. (**15**) Formation of 15–75 sporoblasts, the nuclei of which divide several times. Then the sporoblast forms a smooth wall and thus becomes a sporocysts. (**16**) These sporocysts become disrupted, leading to distribution of sporozoites inside the vector. The main method of infection is the bite, but in experiments oral infection of snakes was also possible. *DM* developing merozoite; *E* erythrocyte; *G* gamonts (of different sex); *HC* host cell; *MA* macrogamont; *MI* microgamont; *MG* microgamete; *N* nucleus; *NH* nucleus of the host cell; *PV* parasitophorous vacuole; *SB* sporoblast; *SPC* sporocysts

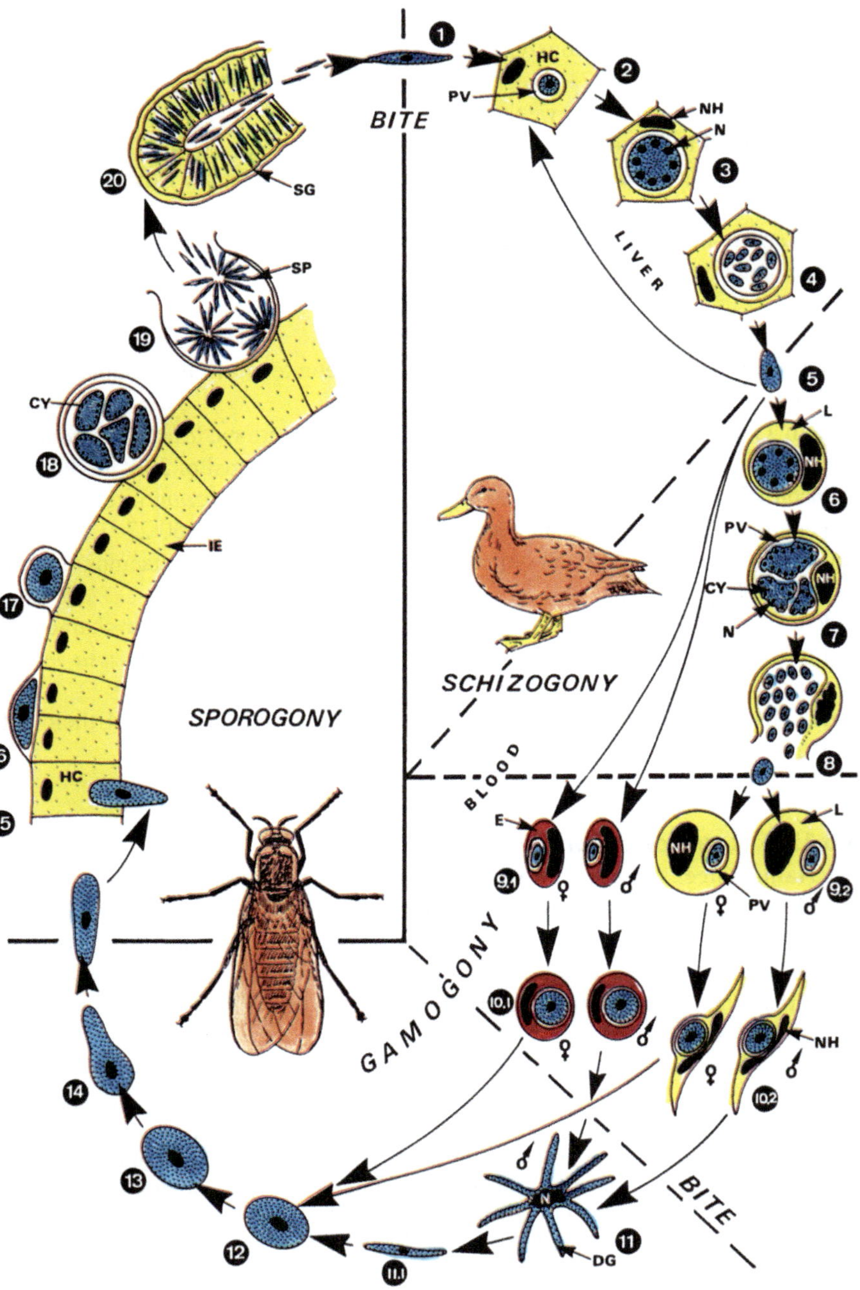

Fig. 4.110 ***Leucocytozoon simondi***. Diagrammatic representation of the life cycle in its vertebrate hosts (domestic and wild ducks and geese) and in its vector (*Simulium* sp., blackflies). (**1–5**) Sporozoites injected by the *Simulium* fly are carried by the bloodstream to the liver, where they enter Kupffer cells, and form the multinucleate first-generation schizonts. The latter give rise to small merozoites (**5**) which may reinfest other hepatic cells (**2**) or invade lymphoid cells (**6–8**) or

of the group of black flies and midges), the sporozoites enter the cells of various tissues and start to grow up to schizonts. The **schizonts** of the first generation occur in the liver and are rather small (10–18 μm). Those of the second generation occur on days 4–6 after infection and are found in the lung, liver, heart, brain, kidneys, intestine and lymph nodes. Since they reach diameters of up to 160 μm, they are described as megaloschizonts. Starting about 1 week after infection (but mostly after 10–12 days p.i.), **gamonts** can be found in leucocytes inside the peripheral blood vessels. Since the parasites grow up considerably (up to 20 μm), the parasitized host cells appear spindle shaped (Fig. 4.110: 10.2). Other gamonts appear spherical. The gamonts are ingested during the bites of the vector insects. After formation of the gametes inside the stomach system of, e.g., simuliid vectors, large ookinetes are formed, which reach a length of up to 30 μm and are transformed into 10–13 μm sized oocysts when being attached at the stomach wall. In contrast to *Plasmodium* species, *Leucocytozoon* species produce rather few sporozoites, which finally enter the salivary glands (Fig. 4.110: 18–20).

4. **Symptoms of disease**: *L. simondi* infections lead in ducks and geese to disturbances during motions which appear staggering. The head and neck are twisted and the head shakes considerably. Infected animals are often very weak and show a strong anaemia (leading to a whitish-bluish appearance of the mucous layers, a pale beak and whitish skin between the toes). In addition, often symptoms of icterus and of spleno- or hepatomegaly may occur. In the case of young animals, high rates of mortality may occur due to bleedings and inner vascular haemolysis. Besides ducks and geese, also parrots may become severely infected, showing similar symptoms of disease and also considerable mortality rates. *L. smithi* and *L.* (*akiba*) *caulleryi* in chickens leads to similar symptoms and mortality rates.

Fig. 4.110 (continued) erythrocytes (**9.1**). (**6–8**) After invasion of lymphoid cells or macrophages 4–6 days after infection, large schizonts (=megaloschizonts) of 60–150 μm diameter are formed, which via cytomers (**7**) produce numerous merozoites (**8**). (**9–12**) Having entered lymphoid cells, the majority of merozoites probably develop into gamonts (**9.2**), but it is thought that some may initiate further asexual reproduction. During the formation of the finally elongate or ovoid gamonts (20 × 5 μm), the host cells become distorted and appear elongated spindle shaped (**10.2**). Occasionally, spherical gamonts appear (**10.1**) which are thought to originate from hepatic merozoites (**5**) that have penetrated erythrocytes instead of lymphoid cells. However, there is no evidence that these differ functionally from the elongate forms. When the vector has sucked blood, the formation of gametes (**11**, **12**) is initiated inside the gut, leading, after fertilization, to an extracellular zygote (**13**). (**13–17**) The immobile zygote is transformed into a motile ookinete, which enters the intestinal wall (**15**), migrates through the cytoplasm of a gut cell and begins its transformation into an oocyst, situated between basal membrane and epithelial cells of the gut (**17**). (**18–20**) Formation of multinucleate sporoblasts (**18**) which give rise to numerous sporozoites (**19**, SP). The latter are released into the body cavity and migrate to the salivary glands (**20**). These slender sporozoites are finally injected into the next host. *CY* cytomere; *DG* developing microgamete; *E* erythrocyte; *HC* host cell; *IE* intestinal epithelium; *L* lymphoid cell/macrophage; *N* nucleus; *NH* nucleus of host cell; *PV* parasitophorous vacuole; *SG* salivary gland; *SP* sporozoite

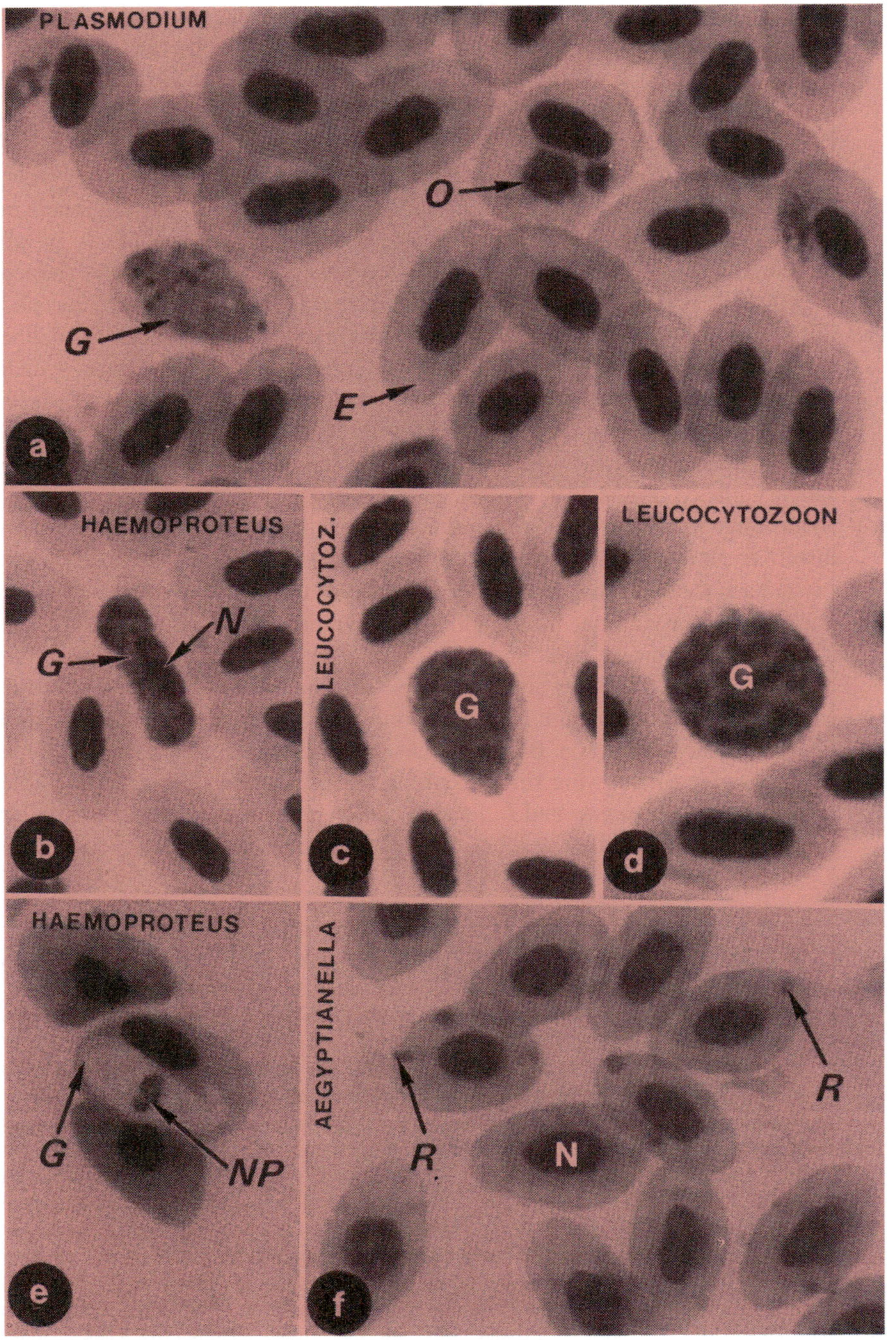

Fig. 4.111 Light micrograph of blood parasites of birds on farms. (**a**) *Plasmodium relictum* (syn. *P. praecox*). (**b**, **e**) *Haemoproteus columbae*; (**c**, **d**) *Leucocytozoon simondi*. (**f**) *Aegyptianella pullorum* (Rickettsiales) *E* erythrocyte; *G* gamont; *N* nucleus; *NP* nucleus of the parasite; *O* trophozoite; *R* rickettsial stage

5. **Diagnosis**: Demonstration of infected leucocytes (Fig. 4.110 and 4.111c, d) within Giemsa-stained blood smears or in organ dabbing preparations.
6. **Pathway of infection**: Percutaneously during bites of infected black flies and/or *Culicoides* midges.
7. **Prophylaxis**: Under normal outdoor rearing conditions, the birds cannot become effectively protected. Indoor stables, however, can be protected by contact insecticides sprayed at the walls.
8. **Incubation period**: About 1 week.
9. **Prepatent period**: Free merozoites can be diagnosed in the blood beginning at the 4–6th day after infection. The typical gamonts appear around day 10 p.i.
10. **Patency**: Practically lifelong, although gamonts seem to disappear from time to time from the peripheral blood, while the schizogony runs on.
11. **Therapy**: There is no peculiar registered therapeutical product. However, according to literature reports, several products act therapeutically or prophylactically. **Pyrimethamine** (0.5–1 ppm) and different sulfonamides in combination with pyrimethamine as well as **furazolidone** (100–150 ppm) can be added **prophylactically** into the food of endangered birds. **Therapy** can be done with the same products by several times repeated feeding of the above-cited substances. Some papers claimed that the combination of **halofuginone** (3 ppm) and **furazolidone** (116 ppm) offers protection.

Further Reading

Imura T et al (2014) The apicoplast genome of *Leucocytozoon caulleryi*. Parasitol Res 113:823–828.

Omori S et al (2008) Two extra chromosomal genomes of *Leucocytozoon caulleryi*. Parasitol Res 103:953–957.

Reeves AB et al (2015) Genetic diversity and host specificity varies across three genera (*Plasmodium, Haemoproteus, Leucocytozoon*) of blood parasites in ducks. PLoS One 10(2):e0116661.

Shutler D et al (2010) Relationships between circulating leucocytes and *Leucocytozoon simondi*. Comp Biochem Physiol A 156:46–49.

Steele EJ, Noblet GP (2001) Gametogenesis, fertilization and ookinete differentiation of *Leucocytozoon smithi*. J Eukaryot Microbiol 48:118–125.

4.6.12.4 *Haemoproteus* Species

1. **Name**: Greek: *haima* = blood; *proteus* = named after the Greek god of the sea, who could change his shape. Latin: *Columba* = dove.
2. **Geographic distribution/epidemiology**: Worldwide.
3. **Biology, morphology**: *Haemoproteus* species (e.g. *H. columbae* of **doves**, *H. nettionis* of **ducks** and **geese**) run their schizogonies inside the endothelial cells of inner organs (lung, spleen, liver), while inside the erythrocytes the sausage-like or dumbbell-like gamonts occur, which surround the host cell nucleus. The gamonts are first seen about 4 weeks after the infection (Fig. 4.111b, e). Vectors are often so-called louse flies (e.g. *Pseudolynchia canariensis* = syn. *Lynchia maura*). However, in the case of *H. columbae*,

midges (genus *Culicoides*) act as vectors injecting the sporozoites, which are produced in oocysts that are located inside the intestinal wall. After rupture of the oocyst wall, they are set free and enter the salivary gland cells.

4. **Symptoms of disease (Pseudomalaria)**: In general, infections lead to rather smooth symptoms which may appear as reduced food uptake, restlessness and low-grade anaemia.
5. **Diagnosis**: Microscopical demonstration of gamonts in Giemsa-stained blood smear preparations (Figs. 4.111b, e). However, diagnosis is only successful if done starting on day 28–34 p.i.
6. **Pathway of infection**: Percutaneously during bites of vector insects.
7. **Prophylaxis**: Vector control inside stables.
8. **Incubation period**: Mostly it takes several weeks until first slight symptoms may be observed.
9. **Prepatent period**: 4–5 weeks: noted by first appearance of gamonts in blood smears.
10. **Patency**: Often several years.
11. **Therapy**: In general, treatment is not needed; severe infections in young animals can be treated using imidocarb and doxycycline.

Further Reading

Bukauskaite D et al (2015) Biting midges (*Culicoides*) transmit *Haemoproteus* parasites of owls: evidence from sporogony and molecular phylogeny. Parasit Vectors 8:303.

Coral AA et al (2015) In vitro development of *Haemoproteus columbae* with perspectives for genomic studies of avian haemosporidian parasites. Exp Parasitol 157:163–169.

Valkunias G (2005) Avian malaria parasites and other haemosporidia. CRC, Boca Raton.

Valkunia G et al (2015) The evidence for rapid gametocyte viability changes in the course of parasitaemia in *Haemoproteus* parasites. Parasitol Res 114:2903–2909.

4.6.12.5 *Hepatocystis* Species

These protozoans (*H. kochi; H. simiae*) parasitize in African monkeys being transmitted by midges (Ceratopogonidae). They build giant schizonts inside liver cells and gamonts in erythrocytes and disturb physiological processes (Table 4.17).

Further Reading

Ayouba A et al (2012) Ubiquitous *Hepatocystis* infections in monkeys. Int J Parasitol 42:709–713.

Thurber MI et al (2013) Co-infection and cross species transmission of divergent *Hepatocystis* lineages in a wild African primate community. Int J Parasitol 43:613–619.

4.6.12.6 Further Species Inside Red Blood Cells

In water dwellings, reptiles and in fishes as well as in lizards many species occur which are grouped into the genera *Lunkesterella, Haemogregarina, Karyolysus, Schellackia*, etc., which are either transmitted mechanically by mites (*Schellackia* = syn. *Lainsonia*) or during bloodsucking acts, e.g., of leeches. The parasitaemia is in general very high and animals kept inside closed systems (aquaria, terraria) are especially high endangered. Their life cycles are in general only poorly understood.

Further Reading

Curtis LM et al (2013) *Gnathia aureamaculosa*, a likely definitive host of *Haemogregorina balistapi* and potential vector for *Haemogrearina bigemina* between fishes of the Great Barrier Reef, Australia. Int J Parasitol 43:361–370.

Dvorakova N et al (2015) Haemogregarines of freshwater turtles from South East Asia with a description of *Haemogregarina sacaliae* and a redescription of *Haemogregarina pellegrini*. Parasitology 142:816–826.

Haklova-Kocikova B et al (2014) Morphological and molecular characterization of *Karyolyrus*. Parasit Vectors 7:555.

Megia-Palma R et al (2013) Phylogenetic analysis based on 18SrRNA gene sequences of *Schellackia* parasites. Parasitology 140:1149–1157.

Soares P et al (2014) *Haemogregarina* spp. in a wild population from *Podocnemis unifilis* in the Brazilian Amazonia. Parasitol Res 113:4499–4503.

4.7 Ciliates and Flagellates

The members of this group belong to either the Ciliophora (bearing rows of cilia) or Flagellata (bearing species-specific amounts of flagella). Some of them are parasites, but most of them are free living and belong to the giant amounts of the worldwide existing plankton in fresh or saltwater. The most important features are the cilia and flagella which are both constructed according to the same plan: nine pairs of peripheral and one pair of central microtubules. In contrast to most members of the flagellates, which possess only a single nucleus, the ciliates are equipped by a large macronucleus and a tiny micronucleus. While flagellates are reproduced by binary longitudinal divisions, the ciliates are divided by cross division. However, ciliates have developed a system that allows exchange of chromosomes. This occurs during the so-called conjugation, which starts by the common adhesion of two partners. While the macronucleus disappears, the partners exchange a portion of their micronucleus, which fuses with the portion left behind.

Besides many very important free-living ciliates, some of them are important parasites especially of fishes but also of higher vertebrates (Table 4.18).

Table 4.18 Important ciliate parasites

Species	Main hosts	Habitat	Size (μm)
Balantidium coli	Humans, pigs, monkeys	Colon	50–100 × 45
Balantidium sp.	*Sheep*	Intestine	45 × 35
Buxtonella sulcata	Ruminants	Caecum	50–130 × 60
Nyctotherus sp.	Humans	Intestine	26 × 10
N. cordiformis	Frogs	Colon	60–200
N. ovalis	Sheep	Colon	90–185
Ichthyophthirius multifiliis	Fishes	Skin	50–500
Cryptocaryon irritans	Fishes	Skin	30–300
Chilodonella cyprini	Fishes	Skin	40 × 15
Trichodina spp.	Fishes	Skin	80 × 80
Epistylis spp.	Fishes	Skin	30 × 10
Carchesium spp.	Fishes	Skin	30 × 10
Glossatella spp.	Fishes	Skin	35 × 9
Apiosoma pisciola	Fishes	Skin	80 × 38
Enchelys parasitica	Salmons	Skin	60 × 20

4.7.1 *Balantidium coli*

1. **Name**: Greek: *balantion* = sack; *kolon* = colon, terminal portion of the intestine. Latin: *cilium* = cilium, motile bristle.
2. **Geographic distribution/epidemiology**: Worldwide, high infection rates, if pigs are kept too close together.
3. **Biology, morphology**: This ciliate protozoan (Fig. 4.112) parasitizes in the lumen of colon and caecum of **pigs, cattle, horses, donkeys** and **monkeys** (but occasionally also in humans). It grows up reaching a size of 100 μm in mean (ranging from 50 to 150 μm). In its interior the typical micro- and macronuclei lay closely side by side. Reproduction occurs by repeated binary fissions along the median central axis (Figs. 4.112 and 4.113). During a so-called **conjugation** process, two individuals of this species fuse for a short time in their central region and exchange a so-called wandering nucleus, which represents the half of a divided micronucleus. Inside the colon of the host, the vegetative stages excrete a cyst wall. The resulting spherical cyst stage (Fig. 4.114) has a diameter ranging between 50 and 100 μm and becomes excreted within the feces. The remaining vegetative stages inside the intestine are able to enter into the intestinal wall and thus may lead to significant damages. Food is taken up at a special place called cytopharynx.
4. **Symptoms of disease (Balantidium flu)**: Characteristic are so-called *Balantidium* diarrhoeas (=slimy-fluid feces) in addition to typhlitis and colitis. In general, the symptoms become only severe, when the host suffers from an immune suppression due to another parasitic infection or in cases of virus-induced diseases and due to other immune suppressions. Many infections, however, may remain completely symptomless.

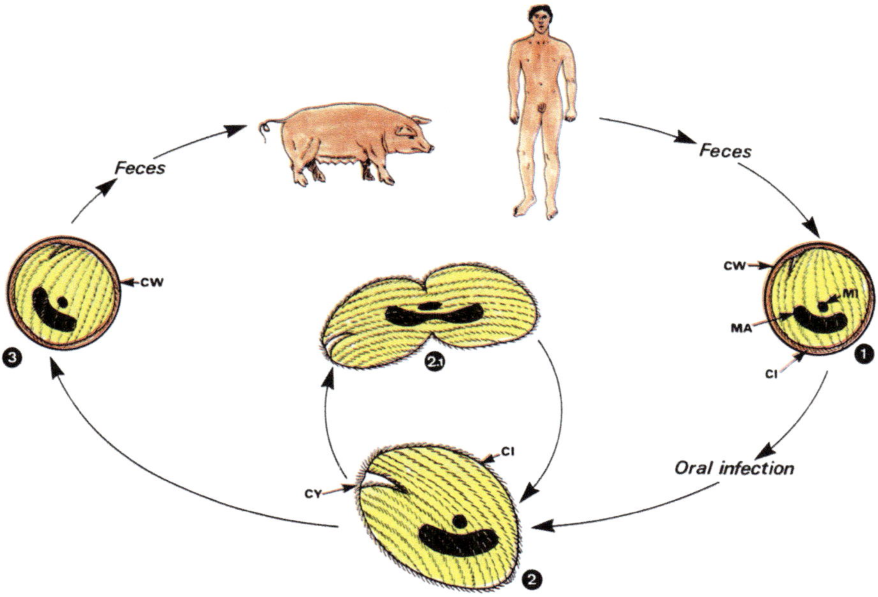

Fig. 4.112 Diagrammatic representation of the life cycle of *Balantidium coli*. The infection occurs by oral uptake of cyst stages within pig feces. (**1**) Cyst stage in pig feces. (**2**, **2.1**) Trophozoites are reproduced by binary cross sections inside the intestinal tube. (**3**) Inside the colon, the trophozoites excrete a wall and thus cysts are formed. *CI* cilia standing in rows; *CW* cyst wall; *CY* cytopharynx; *MA* macronucleus; *MI* micronucleus

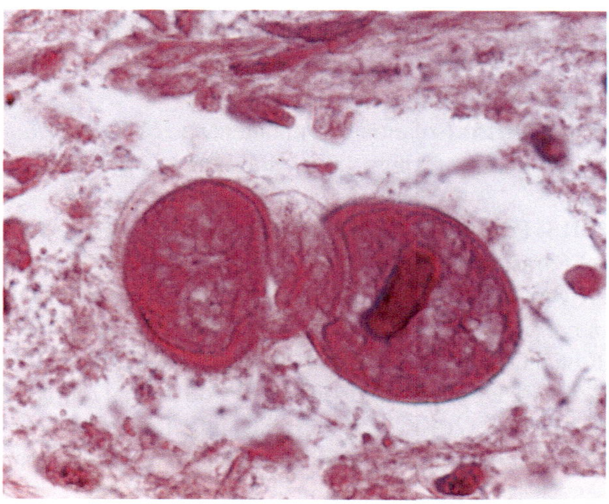

Fig. 4.113 Light micrograph of a dividing trophozoite of *Balantidium coli* inside the intestinal wall of a pig

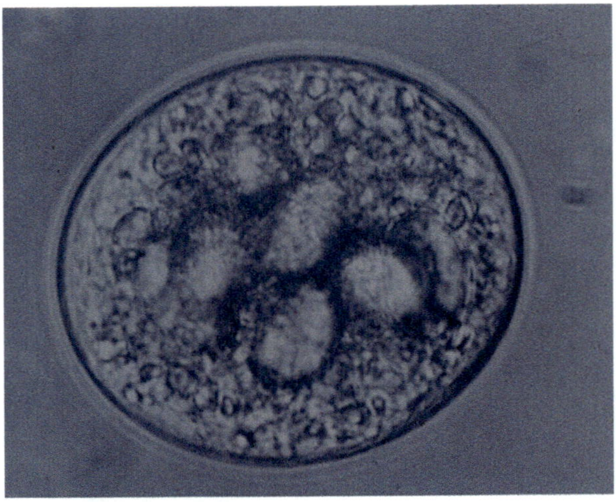

Fig. 4.114 Light micrograph of a cyst stage of *Balantidium coli*

5. **Diagnosis**: Microscopical demonstration of cysts and the motile amoebas in fresh warm feces (Figs. 4.113 and 4.114).
6. **Pathway of infection**: Oral uptake of cysts inside fecally contaminated food.
7. **Prophylaxis**: Regular and intense cleaning of the floor of pig, cattle and horse stables by hot steam or the use of effective means for chemical disinfection.
8. **Incubation period**: Days up to weeks.
9. **Prepatent period**: Days up to weeks.
10. **Patency**: Eventually years.
11. **Therapy**: Literature reports claim healing after treatment with **doxycycline** and with **trimethoprime** combined with a sulfonamide.

Further Reading

Da Silva Barbosi A (2015) Isolation and maintenance of *Balantidium coli* cultured from fecal samples of pigs and non-human primates. Vet Parasitol 2010:240–245.

Khan A et al (2013) Prevalence, hematology, and treatment of balantidiasis among donkeys. Vet Parasitol 196:203–205.

Nilles-Bije L, Rivera WL (2010) Ultrastructural and molecular characterization of *Balantidium coli* isolated in the Philippines. Parasitol Res 106:387–394.

Pomajbikova K et al (2013) Novel insights into the genetic diversity of *Balantidium* and *Balantidium*-like cyst forming ciliates. PLoS Negl Trop Dis 3:e2140.

Ponce-Gordo ML et al (2014) New insights into the molecular phylogeny of *Balantidium* based on the analysis of new sequences of species from fish hosts. Parasitol Res 113:4327–4333.

4.7.2 *Buxtonella sulcata*

1. **Biology, morphology**: *B. sulcata* is a flattened cosmopolitan ciliate which occurs in high prevalence rates (~45–50 %) in the intestine of cattle and buffaloes. It measures about 250–300 × 130–150 µm. Like *Balantidium coli*, it forms spherical cysts (~200 µm), which are excreted within the feces and may infect other animals (Fig. 4.115). Infections may lead to severe diarrhoeas (often in co-infection with *Balantidium coli*). Food uptake occurs by help of a "cell mouth" (vestibule), which starts at the anterior pole and ends with a deepening at the posterior pole.
2. **Therapy**: Application of **dimetridazole** (e.g. Entryl®): 4 days oral application of 250 mg/kg bodyweight daily.

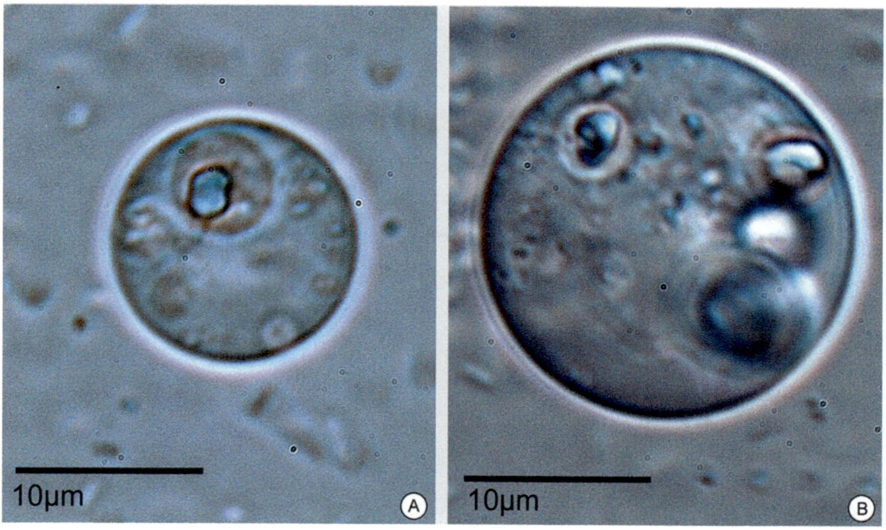

Fig. 4.115 Light micrographs of two cysts (**a**, **b**) of *Buxtonella sulcata* obtained from cattle feces

Further Reading

Grim JN et al (2015) Light microscopic morphometrics, ultrastructure and molecular phylogeny of the putative pycnotrichid ciliate, *Buxtonella sulcata*. Eur J Parasitol 51:425–436.

Hasheminasab SS et al (2015) *Buxtonella* spp. like infection in cattle. Ann Parasitol 61:247–251.

4.7.3 *Ichthyophthirius multifiliis* and Related Ciliates

1. **Name**: Greek: *ichthys* = fish; *phtheir* = louse; *trichodes* = hairy; *chilodon* = lip. Latin: *multum* = much; *filia* = daughter; *filius* = son.

2. **Biology, morphology**:
   (a) ***Ichthyophthirius multifiliis*** (agent of the so-called white dots disease). This disease is due to the occurrence of large numbers of parasitic stages (trophozoites) inside the skin and gills of freshwater fish (Figs. 4.116, 4.117, 4.118 and 4.119). The infection of fish occurs by entering of 15–40 μm sized **swarmers (tomits, theronts)**, which need to find a host within 96 h after leaving the cyst on the floor of their watery biotope (Fig. 4.116). Inside the skin, the swarmers grow up to 100–1000 μm sized spherical trophozoites (Figs. 4.117, 4.118 and 4.119). These stages introduce the typical white-dotted appearance of the fish skin (Fig. 4.119a). Having reached a suitable size, the trophozoites leave the skin and swim to the ground of their watery biotope, where they become encysted. Inside these cysts, they produce by repeated binary fissions up to 1024 swarmers, which finally leave the cyst and infect other fishes. In **saltwater** fish, the

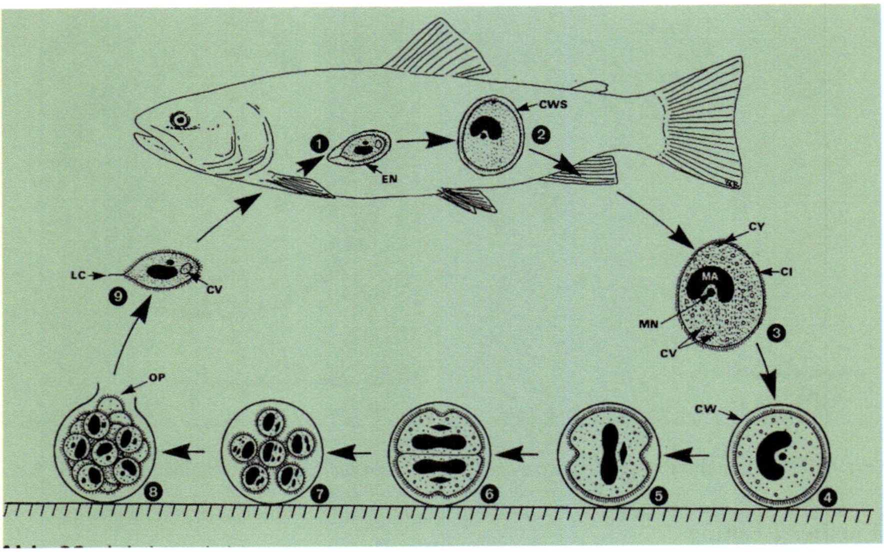

Fig. 4.116 Diagrammatic representation of the life cycle of *Ichthyophthirius multifiliis* parasitizing many species of freshwater fish. (**1**) Swarmer penetrates the skin of a fish and becomes encysted (EN) by host tissues. (**2**) The encysted swarmer grows to be a trophozoite, reaching a diameter of up to 1 mm; the skin then presents with a large greyish pustule at these places. (**3**) Upon rupture of the pustules, the trophozoites, which have numerous contractile vacuoles, are liberated and swim about feebly. Upon coming to rest on the bottom of the pond, the trophozoite secretes a thick-walled gelatinous cyst wall about itself. (**4–8**) Within an hour of encystation, the mother trophozoites starts reproduction by simple transverse division, being repeated until up to 1024, 30–50 μm long, pear-shaped swarmers (with a single contractile vacuole) are formed. (**9**) After rupture of the cyst, the liberated swarmer attaches to the skin of fish within 1 day (unattached ones die during the second day). *CI* cilia; *CV* contractile vacuole; *CW* cyst wall of cysts on the bottom of the pond; *CWS* cyst wall in skin; *CY* cytostome; *EN* encysting swarmer; *LC* long terminal cilium; *MA* macronucleus; *MI* micronucleus; *OP* opening rupture of CW; *SW* swarmer

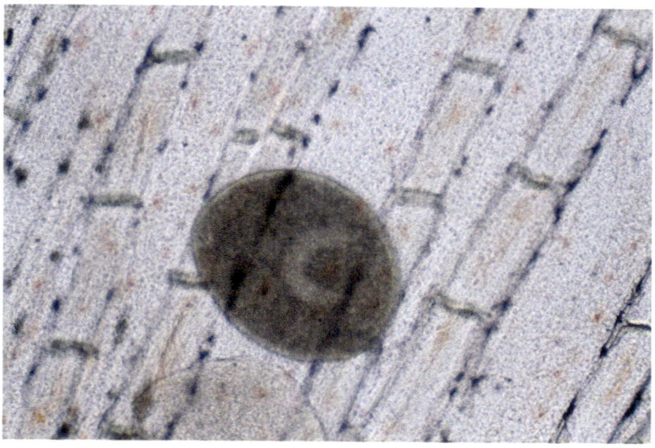

Fig. 4.117 Light micrograph of a growing trophozoites of *Ichthyophthirius multifiliis* below the scales of a fish

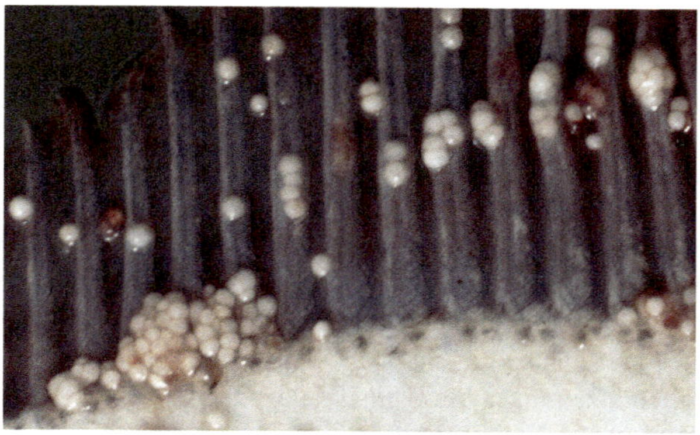

Fig. 4.118 Macrophoto of *Ichthyophthirius multifiliis* in the back fin of an ornamental fish

a b

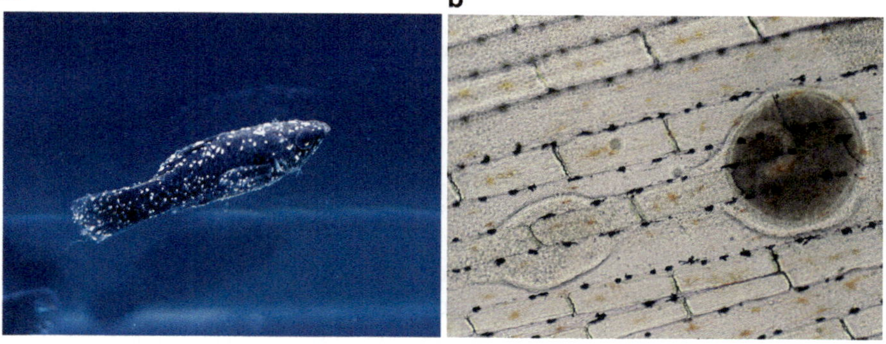

Fig. 4.119 (**a**, **b**) *Ichthyophthirius multifiliis*. (**a**) Black molly fish with whitish dots. (**b**) Light micrograph of a portion of the fish skin parasitized by a trophozoites

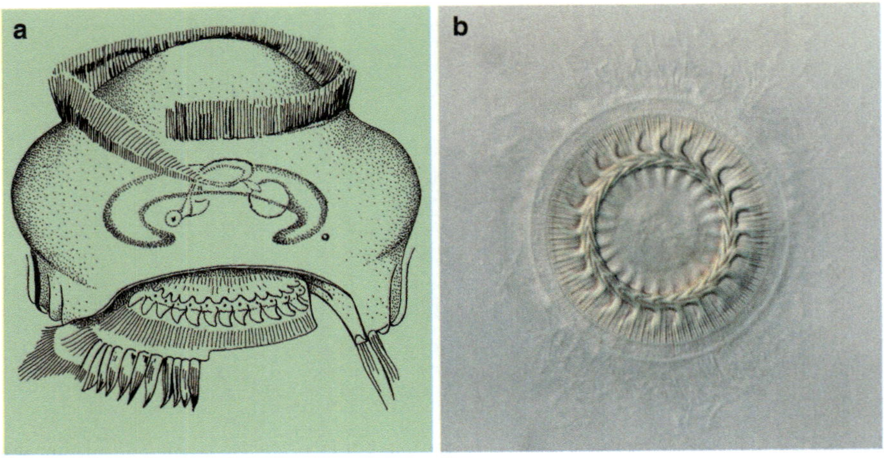

Fig. 4.120 *Trichodina* sp. (**a**) Diagrammatic representation of a trophozoites. (**b**) Light micrograph of the ventral side of a trophozoites, whereby the parasite becomes attached at the surface of fishes

closely related species *Cryptocaryon irritans* induces identical symptoms as *I. multifiliis* in freshwater fish.

Symptoms of disease: Infected fish show breathing problems in cases of heavy infections of the gills. Further symptoms are loss of weight, loss of motility and increased mortality rates especially in very young fishes.

(b) ***Trichodina* species and relatives**: These ciliates are dorso-ventrally flattened (Fig. 4.120) and possess adoral rows of cilia and teeth. These parasites are attached at the skin of weakened fishes and even enter the skin. Their size ranges between 70 and 90 μm.

Symptoms of disease: Large inflamed (reddish) areas, gloomy skin with nodules.

(c) ***Chilodonella* species**: These stages reach mostly a size of 35 × 15 μm and parasitize on the surface of the gills and skin of freshwater fishes and introduce a darkening of the skin. *C. cyprini* of carps (skin darkener) is very large (60–70 μm), has a heart-like shape and destroys the skin especially of young fishes.

(d) **Peritrich ciliates**: This is a group of skin parasites, which induce rather few damages at healthy skin, but enlarge lesions in diseased skin. This group contains species of the genera *Apiosoma, Glossatella, Carchesium* and *Epistylis*. In all cases, the fishes are attacked by the swarmers of these species.

(e) **Flagellates**: The specimens of the genus *Ichthyobodo* (syn. *Costia*) reach a size of 10–20 μm and possess two differently sized flagella (10 and 17 μm; Fig. 4.121). They become attached at the host's skin and become anchored there with the help of a rhizome-like protrusion. This induces damages of the surface and thus gives rise to inflammations.

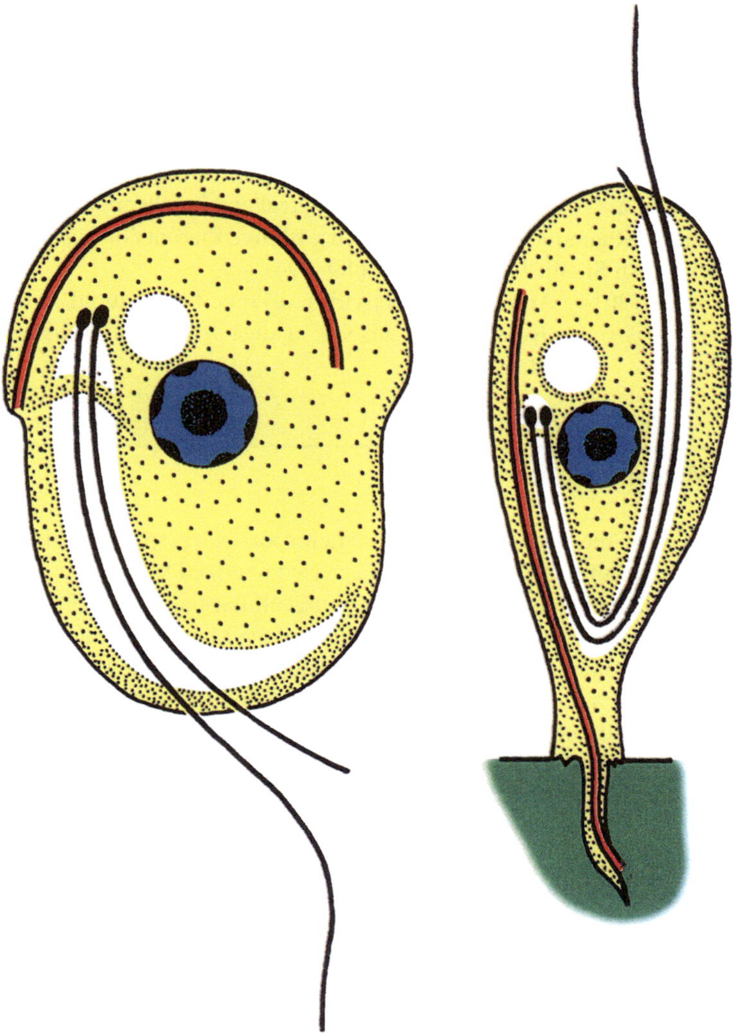

Fig. 4.121 *Ichthyobodo* sp. (syn. *Costia* sp.). Diagrammatic representation of a free swimming stage (*left*) and an attached one

3. **Control of the parasitic ciliates and flagellates**
   (1) **Empty fish ponds** can be disinfected rather easily by using solutions of formalin and/or potassium permanganate. However, all these products have to be used with absolute caution, since they are skin harming and toxic.
   (2) **Infected fish skin**:
      (a) **Food fish**: Long-time bath in water containing 2–4 mg/10 l malachite oxalate for 6 days in the case of carps and 10 days for trouts.

(b) **Ornamental fish**: Medicinal bath using the Ciliol®/Protazol® of Fa. - Alpha-Biocare GmbH, Düsseldorf, distributed by Fa. Sera, Heinsberg (Germany).

Further Reading

Mehlhorn B et al (2005) Health for ornamental fishes. Springer, Heidelberg.

Tancredo KR et al (2015) Haemato-immunological and biochemical parameters of silver catfish Rhamdia quelen immunized with live theronts of *Ichthyophthirius multifiliis*. Fish Shellfish Immunol 45:689–694.

Yao JY et al (2014) Antiparasitic activities of specific bacterial extracellular products of *Streptomyces griseus* SDX-4 against *Ichthyophthirius multifiliis*. Parasitol Res 113:3111–3117.

4.8 Microsporidia

The members of the protozoan phylum Microspora developed so-called **spores**. They are the transmittable stages, which are characterized by a single protrudable tubular polar filament and their obligatory intracellular parasitism (Fig. 4.122 and 4.123). The spores are excreted within the feces of their hosts, and thus, they may be ingested within contaminated food by other hosts. As soon as they have reached the target organ inside a host, the polar filament is extruded and injected into a host cell. Then the amoeboid, membrane-bound cytoplasm of the parasite creeps through the tube into the cytoplasm of the host cell and is placed within a parasitophorous vacuole. Therein the parasitic stages ingest food and grow up to reach the stage of a **meront (schizont)**. In some species was observed that two nuclei, which are the result of an autogamic process, fuse in an apparently primitive sexual process. In other species belonging to the genera *Amblyospora, Edhazardia, Culicospora, Thelohania*, etc. Clear sexual processes have been described showing a change of the ploidy status and in addition host changes. For example, 2-nucleated sporonts inside larvae of mosquitoes produce via a meiosis 8 haploid spores (**octospores**). Later in adult mosquitoes further spores are produced, which enter the eggs of the mosquitoes and thus infect the next host generation. Many microsporidian species are specialized on cells of vertebrates including humans and farm animals, if these hosts suffer from immune suppression or are diseased due to infections with viruses, bacteria or parasites (Table 4.19).

4.8.1 *Encephalitozoon cuniculi*

1. **Name**: Greek: *encephalon* = brain; *zoon* = animal; Latin: cuniculus = rabbit.
2. **Geographic distribution/epidemiology**: Worldwide, up to 80 % of the rabbits are infected, but also other animals and immune-suppressed humans may be infected.

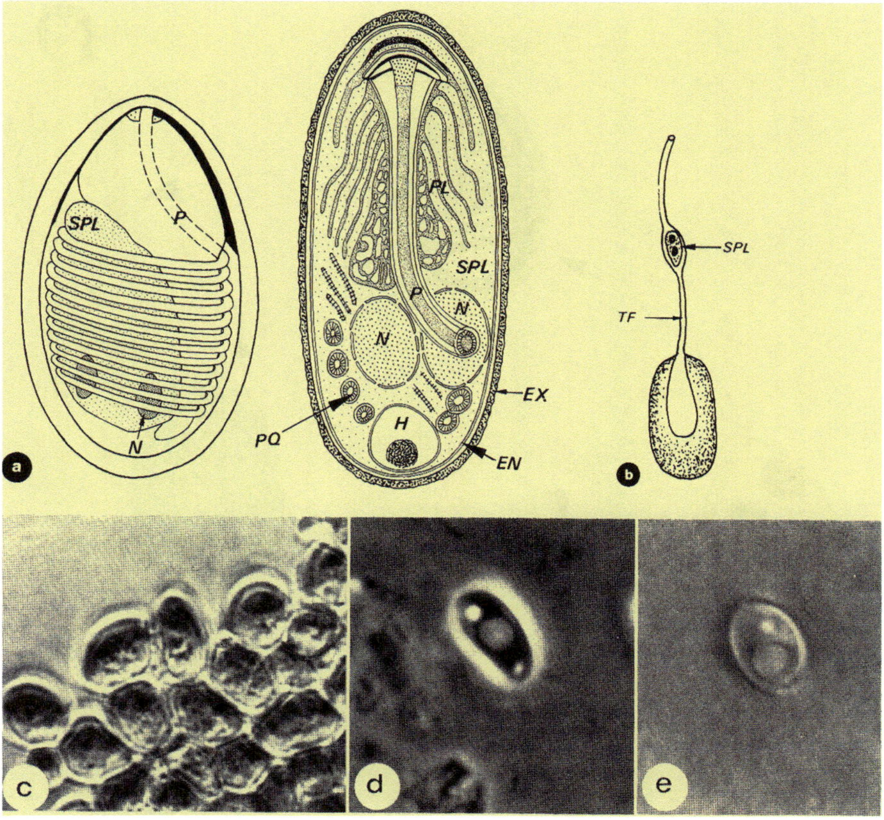

Fig. 4.122 (**a**, **b**) Diagrammatic representations of stage of *Nosema apis*. (**c–e**) Light micrographs of *Nosema apis* stages. *DZ* intestinal cells; *E* developmental stages; *EN* endospore; *EX* exospore; *H* hind vacuole; *N* nucleus; *NW* nucleus of the intestinal cell; *P* polar filament; *PL* polaroplast; *PQ* polar filament cross sectioned; *SP* spore; *SPL* sporoplasm; *SW* wall of shell; *TF* tubular polar structure

3. **Biology, morphology**: This microsporidian protozoan is characterized by the formation of ellipsoid, uninuclear spores, which measure 2.5 × 1.4 μm and contain each a polar capsule, which includes a hollow, tubular polar and about 24 μm long filament being arranged in 5 windings (Fig. 4.123). In case that such a spore is ingested by a rabbit, the polar filament becomes extruded inside the intestine and is injected into a cell. The cytoplasm creeps through the tube into the cell and is enclosed in a parasitophorous vacuole. After a period of repeated binary fissions, during which the host cell is destroyed, the parasites are drifted with the help of the bloodstream to many organs. Especially inside the kidneys, they are found in high numbers. At first, the meronts are reproduced in parasitophorous vacuoles inside the cells. These **meronts** reach a size of 6 μm in diameter and contain 1–2 nuclei. The formation of spores starts with the division of a two-nucleated stage, thus producing two so-called

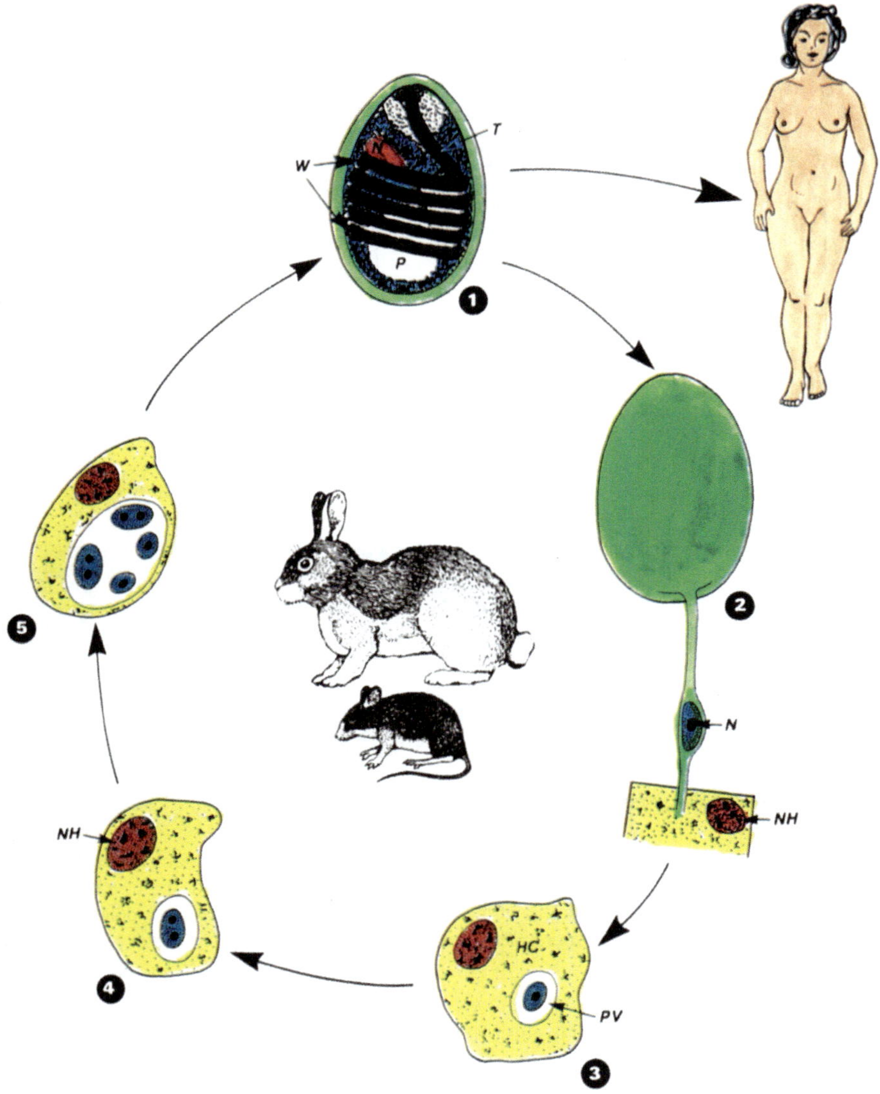

Fig. 4.123 Diagrammatic representations of the life cycle of *Encephalitozoon cuniculi*, which may parasitize within a variety of hosts including immune depressive humans. (**1**) The infection of AIDS patients occurs via oral uptake of spores that derive from urine of animals (via contaminated food or via touching of furs). The mature uninuclear spore is characterized by 5 windings of the polar tube (**1**) and the occurrence of a posterior vacuole (P). (**2, 3**) In human intestine, the spore extrudes the polar tube being injected into a host cell. The uninuclear sporoplasm creeps through the tube in the cytoplasm of the host cell, where it is included within a parasitophorous vacuole. (**4, 5**) Reproduction by repeated binary fissions. The last binary fission (**5**) leads to 2 uninuclear sporoblasts, which each grow up and differentiate into an infectious cyst. The latter are freed when the host cell is used up and bursts. Thus, these spores may become distributed in the whole body set free in human stool. *HC* host cell; *N* nucleus; *NH* nucleus of host cell; *P* posterior vacuole; *W* windings of the polar tube; *T* polar tube

Table 4.19 Important species of microsporidia

Species	Hosts	Habitat	Size (µm)	Occurrence
Glugea anomala	Fish (*Gasterosteus* spp.)	Connective tissue	2 × 6	Arctic
G. fennica	*Fish* (*Lota lota*)	Cutis, fins	2.6 × 7	Finland
G. truttae	Bees (*Apis mellifica*)	Yolk sack	1.5 × 5	Europe
Nosema apis	**Humans** (AIDS)	Intestine	5 × 9	Worldwide
Microsporidium ceylonensis	**Humans** (AIDS)	Cornea	3 × 5	Worldwide
M. africanum	**Humans** (AIDS)	Cornea	3 × 5	Worldwide
M. ocularum	**Humans** (AIDS)	Cornea	3 × 5	USA
Vittaforma cornea	**Humans** (AIDS)	Cornea	3 × 4	Europe
Brachiola vesicularum	**Humans** (AIDS)	Muscles	2 × 2.9	USA
B. connori	**Humans** (AIDS)	All organs	2 × 4	USA
Pleistophora typicalis	Fish	Skeletal muscles	2.3 × 4.4	Europe
P. anguillarum	Fish (eel)	Skeletal muscles	3 × 5	Japan, Taiwan
P. danilewski	Reptiles, frogs	Skeletal muscles	4 × 2	Europe
Trachipleistophora hominis	**Humans** (AIDS)	Muscles	2.4 × 4	Europe
T. anthropophthera	**Humans** (AIDS)	Many organs	2 × 3.7	Europe, USA
Loma branchilis	Fish (*Gadus* sp.)	Gills	2.3 × 4.8	Boreo-Arctic
Thelohania californica	Mosquitoes (*Culex* sp.)	Intestine	3 × 6	America
T. baueri	Fish (fresh water)	Ovaries	2.7 × 5.4	Gulf of Finland
Ichthyosporidium giganteum	Fish (salt water)	Connective tissues	4 × 7	Atlantic coast
Microsporidium cotti	Fish (salt water)	Testes	9 × 3	Atlantic coast
M. schuetzi	Frogs (*Rana* sp.)	Ovaries	7 × 2	USA
Encephalitozoon lacertae	Lizards (*Podarcis* sp.)	Intestine	3.5 × 1.5	France
E. cuniculi	Rabbits, rats, mice, guinea pigs, hamster, sheep, goats, dogs, foxes. Felids monkeys, **humans**	Intestine, other organs	2.5 × 1.5	Worldwide
E. helleri	**Humans** (AIDS)	Many organs	2.4 × 1.5	Worldwide
E. intestinalis	**Humans** (AIDS)	Many organs	2.0 × 1.5	Worldwide
Enterocytozoon bieneusi	**Humans** (AIDS), pigs, dogs, monkeys	Intestine, lung	1.5 × 0.5	Worldwide

sporoblasts. This process takes about 2–5 days. The infectious spores are then excreted mainly within the urine but also in feces. Besides rabbits also mice, rats, guinea pigs, hamsters, goats, sheep, pigs, horses, dogs, foxes, cats and many primates and humans may become infected. Especially immune-suppressed humans are highly endangered.

4. **Symptoms of disease (Encephalitozoonosis)**: In the case of rabbits, only rather mild symptoms occur but remain chronically. In severe cases of all potential hosts, symptoms of nephritis and encephalitis may occur accompanied by paralysis. In case of a generalization, infections of several organs mortality rates are high. In immune-suppressed humans, infections of brain and eyes are rather common and many cases of encephalitis are fatal.
5. **Diagnosis**: Microscopical demonstration of spores inside the urine or in coloured tissue sections of biopsy material. For tests like IFAT, ELISA and Carbon immunoassay Fa. Teichmann, Uppsala, Sweden, offers defined antigenic material.
6. **Pathway of infection**: Oral or nasal uptake of spores from urine of infected animals. Fetus may apparently become infected on a transplacental pathway.
7. **Prophylaxis**: Disinfection of rabbit stables; immune suppressive persons should avoid contacts especially with rabbits due to their high infection rates.
8. **Incubation period**: A few days up to weeks; in the case of transplacental transmission 1 day.
9. **Prepatent period**: 2–5 days.
10. **Patency**: In the case of rabbits: months up to years.
11. **Therapy**: Chemotherapy: unknown; antibiotics like axytetrazycline reduce spore formation on 50 %. **Toltrazuril** acts in general on microsporidia as well as **fenbendazole**. However, apparently full clearance does not occur.

Further Reading

Abu-Akkada S et al (2015) First detection of an ignored parasite, *Encephalitozoon cuniculi*, in different hosts in Egypt. Parasitol Res 114:843–850.

Da Fonseca Ferreira-da-Silva M et al (2014) Identification of the microsporidian *Encephalitozoon cuniculi* as a new target of the IFN-inducible IRG resistance system. PLOS Pathog 10:e100449.

Künzel F et al (2014) Comparison of an indirect fluorescent antibody test with Western blot for the detection of serum antibodies against *Encephalitozoon cuniculi* in cats. Parasitol Res 113:4457–4462.

Rodriguez-Tovar LE (2016) *Encephalitozoon cuniculi*: grading the histological lesions in brain, kidney, and liver during primo infection outbreak in rabbits. J Pathog. Article ID 5768428.

4.8.2 Microsporidia of Fishes

1. **Name**: Greek: *micros* = small; *sporos* = seed, spore.
2. **Geographic distribution/epidemiology**: Worldwide.

3. **Biology, morphology**: Microsporidia occur in most of the fresh and saltwater fish species. They produce uninuclear spores of 10 μm diameter at the maximum. Inside the spore wall at the apical pole, a tubular polar filament arises winding several times (compare Fig. 4.122a). While intestinal cells are parasitized only by the stages of a few species, other organs (e.g. muscles) may harbour a broad spectrum of microsporidia. **Examples**: In the intestine, the following stages may be seen:
   (a) Spores of *Glugea hertwigi* are found in the intestine of stints reaching a size of 5.5 × 2.3 μm.
   (b) Spores of *Glugea stephani* reach a size of only 3.5 × 23 μm in the intestine of flat fishes.
4. **Symptoms of disease (Bump disease)**: Leading symptoms are bump-like swellings of the body surface due to inner tumour-like swollen knots containing amounts of parasites and degenerated host tissues. In general, a hyperplasia of the connective tissues is prominent as well as granular wounds and inflamed hydrops. Chronic infections remain often low grade, but acute cases show high mortality rates.
5. **Diagnosis**: Microscopical demonstration of spores (compare Fig. 4.122) in stained smear preparations obtained from infected organs.
6. **Pathway of infection**: Oral by uptake of free spores in water or by feeding spore-bearing intermediate hosts (e.g. small crustaceans).
7. **Prophylaxis**: Disinfection of the rearing ponds prior to bringing in new fishes; regularly repeated health control of fishes kept in ponds.
8. **Incubation period**: Variable, mostly some days.
9. **Prepatent period**: About 1 week.
10. **Patency**: Lifelong.
11. **Therapy**:
   (a) A specific registered chemotherapeutic is not yet registered. The very good efficacy of **triazinones**, however, was already shown (Mehlhorn et al. 1988).
   (b) **Disinfection of the empty pond** is a reliable method to minimize spores on the pond's floor. This can be done by specialists using **calcium** products. However, after treatment the floor of the pond has to be cleaned from any chemical remnants, since otherwise toxic effects may hit the later therein incubated fishes.
   (c) **Disinfection of breeding vessels** by formalin (40–80 ml 40 % formalin per litre water for 24 h) is efficacious but dangerous if these vessels are not thoroughfully cleaned by freshwater.
   (d) **Killing of microsporidian spores** in a fish-free pond can be done using 1 mg/m^2 chalk nitrogen. Treatment should be done in autumn, when fishes have been taken out and in spring before fishes become entered. **Attention**: Chalk nitrogen is highly toxic. Thus, masks have to be worn during treatment and water has to be tested before new fishes are entered.

Further Reading

Brown AV et al (2010) Description of 5 new *Loma* species. J Eukaryot Microbiol 57:529–553.

Ferguson JA et al (2007) Spores of two fish microsporidia (*Pseudoloma neutrophila* and *Glugea anomala*) are highly resistant to chlorine. Dis Aquat Organ 76:2005–2014

Franzen C et al (2006) Transfer of the members of the genus *Brachiola* to the genus *Anncaliia* based on ultrastructural and molecular data. J Eukaryot Microbiol 53:26–35.

Markness JE et al (2013) Demonstrated efficacy of a pilot heterologous whole spore vaccine against microsporidial gill disease in rainbow trout. Clin Vaccine Immunol 20:1483–1484.

Khan RA (2005) Prevalence and influence of *Loma branchialis* (Microspora) on growth and mortality in Atlantic cod (*Gadus morhua*). J Parasitol 91:1230–1232.

Mehlhorn H et al (1988) Toltrazuril effective against a broad spectrum of protozoan parasites. Parasitol Res 75:64–66.

4.8.3 *Nosema* Species of Bees

1. **Name**: Greek: *nosos* = disease. Latin: *apis* = bee.
2. **Geographic distribution/epidemiology**: Worldwide; Infections may become epidemic dimensions with high mortality rates among honeybees.
3. **Biology, morphology**: There exist two species of the genus *Nosema* in bees: *N. ceranae* and *N. apis*. The spores of these species reach a size of 9 × 5 μm and contain a large coiled polar filament, which becomes extruded as soon as these spores have been ingested by a bee (Figs. 4.122 and 4.124). The extruded tubular polar filament is injected into an epithelial cell of the bee and the sporoplasm creeps into the host cell. After several asexual divisions (**merogony**), which give rise to 6–8 μm sized, two-nucleated meronts start the development of large plasmodia which introduce the production of numerous spores (**sporogony**), which are excreted within the feces after the host cell has been disrupted. At the end of the yearly activity phase of the bees, the reproduction of the *Nosema* stages stops. *Nosema*-infected bees are often co-infected by stages of the amoeba *Malpighamoeba mellificae*, which leads to similar signs of disease as the *Nosema* species.
4. **Symptoms of disease (Nosematosis, bee flu, bee spring disease)**: Infected bees excrete masses of yellowish fluid feces, which lead to a drying out effect of the bees and finally to death. This disease hits mainly the young bees in springtime. The loss of the young bees leads to lack of food for growing bees; as a consequence the number of bees in a hive remains small and many of them are too weak to fly. Since the bees excrete their feces outside their stock, the farmer does mostly not note the persistent infection until the bee number is extremely reduced.

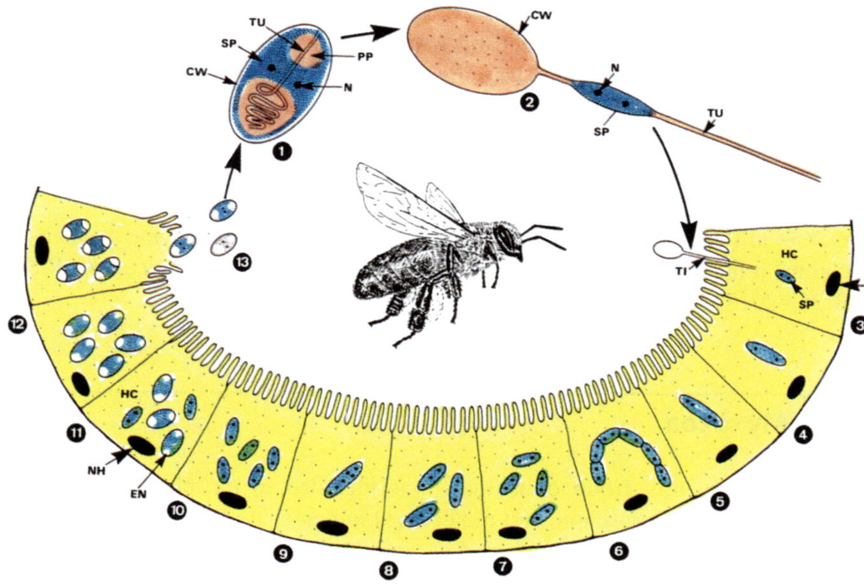

Fig. 4.124 Diagrammatic representation of the developmental cycle of *Nosema apis* inside bees. (**1**, **2**) Infectious spore with and without extruded polar filament, which is injected into host cells. (**3–12**) Asexual reproduction process inside the host cell cytoplasm. *CW* cyst wall; *EN* encystation; *HC* host cell; *N* nucleus; *NH* nucleus of the host cell; *PP* polaroplast; *SP* sporoplasm; *T* injected polar tubule; *TU* polar tube

5. **Diagnosis**: Microscopical demonstration of the typical spores in the feces of a bee or inside the organs of dead bees (Figs. 4.122 and 4.124).
6. **Pathway of infection**: Oral by uptake of spores within contaminated food.
7. **Prophylaxis**: Isolation of infected stocks from non-infected ones and discharge of dead bees from the inside of a stock. **Attention**: Spores remain infectious for years (e.g. in honey at least 3 months!). Therefore, honeycombs should be removed from parasitized hives and replaced by clean ones. Use of steam containing acidic acid to clean the honeycombs.
8. **Incubation period**: Days in springtime, months in autumn.
9. **Prepatent period**: Spores can be demonstrated very shortly after an infection of a stock. Timing depends on the amount of infected bees.
10. **Patency**: Lifelong of the bees due to a quick follow-up of the stages of the microsporidia.
11. **Therapy**: In many countries, there exist no registered products. In some countries, compounds like **fumagillin** or Hg compounds are available. Furthermore, **triazinones** have shown high activity; however, they must be applied in low dosages. Furthermore, it is needed to avoid any weakening of the stocks, e.g., by avoiding infestation with *Varroa* mites.

Further Reading

Charbonneau LR et al (2015) Effects of *Nosema apis*, *N. ceranae* and co-infections on honey bee (*Apis mellifera*) learning and memory. Sci Report 6:22626.

Kurze C et al (2016) *Nosema* sp. infections cause no energetic stress in tolerant honey bees. Parasitol Res. doi:10.1007/s00436-016-4988-3.

Maside X et al (2015) Population genetics of *Nosema apis* and *N. ceranae*: one host (*Apis mellifera*) and two different histories. PLoS One e0145604.

Peng Y et al (2015) Seminal fluid of honey bees contains multiple mechanisms to combat infections of the sexually transmitted pathogen *Nosema apis*. Proc R Soc B 263:20151785.

4.9 Myxozoa

In former times, this group of animals was included in the phylum Protozoa; however, electron microscopic and molecular biological studies showed that they are true Metazoa. The Myxozoa of our days comprise a group of parasites which are transmitted by **spores**, which have their origin in multicellular precursors, possess 1–6 polar capsules and form often several covers. The polar capsules may be—depending on the different species—distributed at both poles (e.g. genus *Myxidium*) or restricted on one pole (Figs. 4.125 and 4.126). Inside the host's intestine, the polar filaments are set free and help to become attached at the intestinal wall cells. Then the sporoplasm is set free and enters species-specific tissues, where again spores are produced, which often only are set free after the death of the host or have to become ingested by raptor fish. The development of the Myxozoa stages occurs inside of intercellular spaces (e.g. inside cartilage or other connective tissues). However, recent publications showed that also an intracellular parasitism is not as rare as formerly believed. This can be observed in species of the genera of the order Multivalvulida (e.g. *Kudoa, Unicapsula*), which are found in oocytes,

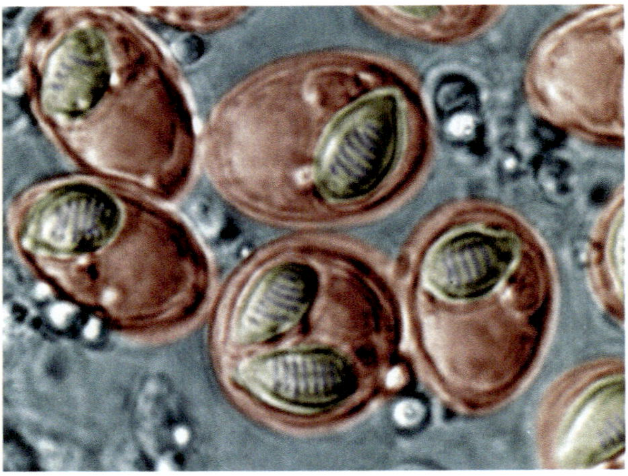

Fig. 4.125 Light microscopical aspect of Myxozoa cysts possessing two polar capsules at the same pole. In some cases, the second capsule is situated in a deeper position

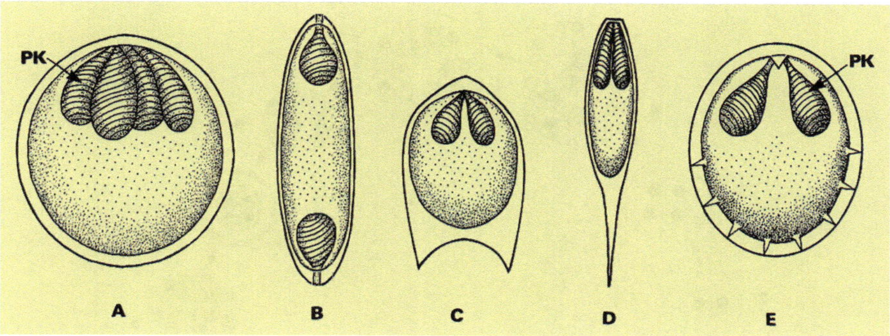

Fig. 4.126 Diagrammatic representation of the spores of different Myxozoa species. (**a**) *Chloromyxum* sp.; (**b**) *Myxidium* sp.; (**c**) *Hoferellus cyprini*; (**d**) *Henneguya psorospermica*; (**e**) *Myxobolus pfeifferi* (according to Reichenow). *PK* polar capsule

wherein also stages of *Myxobolus cyprini* may occur. The approximate 1350 species of the Myxozoa parasitize mainly in fishes and may lead to enormous economic losses, since the three diseases **swim bladder infection** of carps, **PKD** = proliferative kidney disease and the **whirling disease** of salmonids are rather common.

The life cycles of the members of the class **Myxozoa** are very complex and many details are still unknown. Thus, the known and confirmed data are:

(1) The transmission of the specimens occurs on a direct way by oral uptake of free spores or spores in prey fishes.
(2) The sporoplasm of the Myxozoa species leaves the cyst wall and creeps actively into the final sites of parasitism.

However, there are also exceptions from the two above-listed common features:

(a) ***Myxobolus cerebralis*** has an obligate host change between tubificid annelids and fishes. The stages inside the tubificids are known since long and had been kept for a separate species (Fig. 4.127). They are exclusively infectious for salmonids.
(b) In the case of ***Tetracapsuloides bryosalmonis***, several extrasporogonic developmental stages occur inside the wall of the swim bladder of carps. The same is seen in the case of the proliferative kidney disease (PKD) of salmonid fishes leading to a real flooding of infected tissues, where masses of spores become developed.

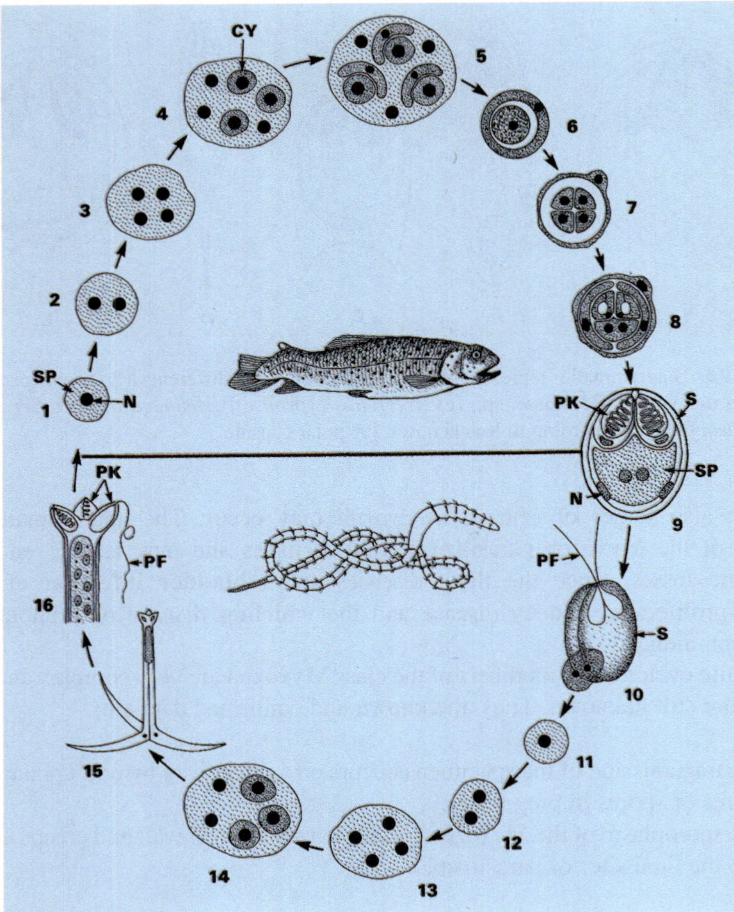

Fig. 4.127 Diagrammatic representation of the life cycle of *Myxobolus* (syn. *Myxosoma*) *cerebralis*. (**1–3**) After ingestion of the tubificid free-living worm containing the stages, which had been described as *Triactinomyxon ignotum*, the latter release inside the intestine of the trout the 1-nucleated sporoplasms. (**4–9**) During nuclear and plasma divisions, finally many spores are formed in the cartilage, which are set free, when fish died or are ingested by a raptor fish. (**10–16**) If water annelids (family Tubificidae) ingest such spores, finally many stages are produced, which had formerly been named as *Triactinomyxon ignotum* and contain many sporoplasms. *CY* cytomere; *N* nucleus; *P* pericyte; *PF* polar tube (filament); *PK* polar capsule; *S* sporogenic cell; *SP* sporoplasm

As above-mentioned Myxozoa are true multicellular organisms, and thus, they are kept by many researchers for "true metazoans", which had undergone a dedifferentiation while having adapted a parasitic lifestyle. Today it is believed that the Myxozoa are relatives of the Cnidaria within the phylum Coelenterata, since the latter possess also extrudible polar filaments (=so-called cnids). Due to this, the Myxozoa represent a separate phylum among the Metazoa.

System
Phylum: MYXOZOA
Class: Myxosporea
Order: Bivalvulida
Suborder: Bipolarina
Genus: *Myxidium*
Genus: *Sphaeromyxa*
Suborder: Eurosporina
Genus: *Ceratomyxa*
Genus: *Tetracapsuloides*
Genus: *Sphaerospora*
Suborder: Platysporina
Genus: *Myxosoma*
Genus: *Myxobolus*

Further Reading

Bartosova P et al (2015) Morphology and phylogeny of two new species of *Sphaeromyxa* from marine fish. Parasitology 142:660–674.

Holzer AS et al (2013) Who's who in renal sphaerosporids. Parasitology 140:46–60.

Kumar G et al (2015) Interaction of *Tetracapsuloides bryosalmonae*, the causative agent of proliferative kidney disease, with host proteins in the kidney of *Salmo trutta*. Parasitol Res 114:1721–1727.

Manrique WG et al (2016) Ultrastructural description of *Myxobolus cuneus*. Parasitol Res. doi:10.1007/s00436-016-5026-1.

Matthews PD et al (2015) Ultrastructure and ssrRNA sequencing of *Myxidium* amazonense. Parasitol Res 114:4675–4683.

Matthews PD et al (2016) Morphological and ultrastructural aspects of *Myxobolus niger*. Acta Trop 158:214–219.

Morsy K et al (2016) *Hennequya collaris* parasite of the greenband parrotfish. Parasitol Res. doi:10.1007/s00436-016-4968-7.

Rangel LF et al (2016) *Tetractinomyxon* stages genetically consistent with *Sphaerospora dicentrarchi*. Parasitology. doi:10:1017/s0031182016000512.

4.9.1 *Myxobolus* (syn. *Myxosoma*) *cerebralis* in Fishes

1. **Name**: Greek: *myxa* = slime; *ballein* = excrete. Latin: *cerebrum* = brain.
2. **Geographic distribution/epidemiology**: Worldwide in rearing systems of trouts—epidemics may occur, if the fishes are kept in very closely filled ponds. This species is widely distributed especially in Europe and North America.
3. **Biology, morphology**: This parasite induces extremely high mortality rates in young fishes. The infectious stages are ovoid spores, which measure 7–9.5 × 7–10 μm. Two polar capsules are situated at the anterior pole; each contains an extrudable polar tube (filament), which is coiled in 5 windings

(Fig. 4.128). Each of these two polar tubes is situated in a separate cell. The 2-nucleated sporoplasm and the two scale-formation cells are further inclusions inside the outer spore wall. This feature obtained with the help of electron microscopy proves that these stages are no true protozoans, but probably reduced metazoans. At least for *Myxobolus cerebralis*, it is proven that the life cycle includes two hosts: **fish** and **annelid worms** of the group of *Tubifex* and relatives (Fig. 4.127). Inside these *Tubifex* worm typical Myxozoa spores are produced which possess three long shell protrusions. These spores are known for long under the name *Triactinomyxon ignotum* and have been systematically been placed into the group of Ascetosporidia. These *Triactinomyxon* spores are infectious for the fish host, induce there the spore formation of the *Myxosoma* type. These spores are produced inside the cartilage of the fish host. At first large plasmodia occur, which give rise by division to smaller daughter plasmodia. Then by a schizogonic process single nucleated cells occur. One of them surrounds as so-called **pericyte** another one, which thus become the **sporogonic cell**. The total of these two cells is called **pansporoplast**, within which finally two of the multicellular *Myxobolus* spores are formed. Within each of the original plasmodium stages, many of these spores are finally produced, the wall of which is very strong due to the included chitin (Fig. 4.127). The whole development takes about 8–12 months.

4. **Symptoms of disease (Whirling disease)**: The first symptoms of this disease can be noted about 5–7 weeks after the infection: **Turning swimming**: it looks like as if the fish tries to bite into its tail fin. At the same time melanization (getting black) of the scales of posterior body region occurs. The backbones become weakened so that the body gets a C-like shape. Destruction of the spinal cord and the cartilage leads to disturbances in movements and paresis. The reduced motility and fitness lead to a strongly reduced food uptake and thus to weakness, which makes the fishes susceptible for other agents of disease. The development of the spores inside the cartilage takes about 4 months so that the symptoms of disease grow up rather slowly. However, in any case high mortality rates are reached. Spores are set free from dead fish or within feces of raptor fish.
5. **Therapy**: Treatment of *Myxobolus cerebralis* infections is still in an experimental state. The use of systematically acting compounds such as **acetarsol** and **furazolidone** brought promising results. The activity of **triazinones** on sporoplasms and even spores of *M. cerebralis* was proven by Mehlhorn et al. (1988), when fish were placed in water containing the medicament. Since the above-cited chemicals registered are not or not in many countries registered, the available means of control is the disinfection of the empty pond (twice a year) using calcium-nitrogen.

Further Reading

Hedrick RP et al (2012) Invasion and initial replication of ultraviolet, irradiated waterborne infective stages of *Myxobolus cerebralis* results in immunity to whirling disease in rainbow trout. Int J Parasitol 42:657–660.

Mehlhorn H et al (1988) Toltrazuril against a broad spectrum of protozoan parasites. Parasitol Res 75:64–68.

Murcia S et al (2015) *Myxobolus cerebralis* infection risk in native cutthroat trout *Oncorhynchus clarkii*. J Fish Dis 38:637–652.

4.9.2 *Tetracapsuloides bryosalmonae* of Salmonids

1. **Name**: Greek: *tetra* = four; *bryon* = moss. Latin: *capsula* = capsule; *capsuloides* = capsule-like: *salmonae* = belonging to the salmon fish.
2. **Biology, morphology**: This parasite was formerly called **PKX**. It is characterized by four polar capsules (like Fig. 4.126a). The disease due to this myxosporidian parasite is called **proliferative kidney disease (PKD)** and is characterized by granulomes and inflammations of the kidney, which lead to apathia of the infected animals, swellings of the abdominal trunk, anaemic gills, ataxias, etc. Severe cases lead to death of the infected fishes. The infection rates of fish in rearing ponds and/or in free fresh or saltwater environments increase considerably with rising temperatures.
3. **Diagnosis**: The infectious spores can be proven by microscopical analysis of tissue probes of weak fishes.
4. **Prophylaxis**: Cleaning of empty ponds or ponds after removal of the fishes by chemicals (see Microsporidia of fishes).
5. **Therapy**: Medical treatment is still unknown.

Further Reading

Jencic V et al (2014) A survey of *Tetracapsuloides bryosalmonae* infections. J Fish Dis 37:711–717.

Kumar G et al (2015) Interaction of *Tetracapsuloides bryosalmonae*, the causative agent of proliferative kidney disease, with host proteins in the kidney of *Salmo trutta*. Parasitol Res 114:1721–1727.

Mehlhorn B et al (2006) Fish parasites. Springer, Heidelberg (in German).

4.9.3 *Sphaerospora renicola*

1. **Name**: Greek: *sphaera* = ball; *sporos* = seed. Latin: *ren, renis* = kidney.
2. **Biology, morphology**: The infectious spores of this species are developed inside the wall of the swim bladder of fishes as well as in the kidneys. They reach a size of 6–8 μm and are mainly set free in urine or after death of an infected fish and inside the feces of a predator, which had ingested fish tissues.
3. **Symptoms of disease**: Damage of kidney and swim bladder inflammation (**SBI**). The disease occurs mainly in ponds filled with high numbers of fishes, where often up to 100 % of the animals are infected. Young fish are especially sensible to infections, while fishes at an age of 2–3 years seem to be rather immune.
4. **Pathway of infection**: Ingestion of spores within the water or feeding infected fish.

5. **Therapy**: There are no special methods or medicaments available to treat a persistent infection. However, pond disinfection twice a year and rearing only low numbers of fish in a pond will decrease the infection risk.

Further reading

Eszterbayer E, Szekely SC (2004) Molecular phylogeny of the kidney parasite *Sphaerospora renicola* from common carp (*Cyprinus carpio*) and *Sphaerospora* ep. From goldfish. Acta Vet Hung 52:469–478.

Holzer AS et al (2013) Who's who in renal sphaerosporids. Parasitology 140:46–60.

4.10 *Blastocystis* Species (Blastocystosis)

1. **Name**: Greek: *blastos* = germ; *kystos* = cyst; the name is based on the first descriptions of Alexeieff (1911) and Brumpt (1912). Alexeieff took the name from the fungus *Blastomyces* and Brumpt added the species name *hominis*, although these organisms occur practically in all vertebrates.
2. **Geographic distribution/epidemiology**: Worldwide; always connected with more or less intensive diarrhoeas.
3. **Biology, morphology**: The different *Blastocystis* isolates are spherical anaerobic unicellular protozoans, which occur inside the intestinal fluid and feces of humans and of many animals and belong to the most encountered microorganisms in fecal samples of humans and animals. Although their cover is rather thin, they look like cysts and reach sizes ranging from 5 to 200 μm (Fig. 4.128a–c). Since a large central vacuole occupies the central portion of these organisms, the cytoplasm is pushed to the peripheral rim (Fig. 4.128a, b). This cytoplasm contains several nuclei, mitochondrion like organelles and Golgi apparatus. The central vacuole has a storage function and contains various substances such as carbohydrates, lipids and basic proteins. The wall appears in different shapes, which have been described as vacuolated, multivacuolar, avacuolar, granular, amoeboid and cyst forms. The latter are apparently excreted within the feces of the host and should be the infectious stages (being mainly seen in feces), while the thin-walled ones remain in the intestine and repeat permanently a reproduction by binary fissions (Fig. 4.128c). Also reproduction by budding was seen: then three or more protrusions occur along a dividing intestinal stage. In all cases (stages inside the intestine or inside the feces), several nuclei were seen when using DAPI staining. In total, 10 subtypes of the genus *Blastocystis* have been descried in the literature.
4. **Symptoms of disease**: Immune competent animals show practically no symptoms of disease. However, immune incompetent animals, animals infected additionally with other parasites or humans may suffer severely from diarrhoeas due to this parasite. Large animals and humans may excrete 10 l fluid feces per day. However, immune competent animals may not show any symptoms for years.

5. **Diagnosis**: Microscopical inspections of the fluid feces show the typical cyst-like stages (Fig. 4.128).
6. **Pathway of infection**: Oral uptake of cysts in fecally contaminated food.
7. **Prophylaxis**: Separate infected animals from others and clean regularly stable floor with hot water/steam.
8. **Incubation period**: In cases of immune suppressive and very young hosts: a few days. In cases of immune competent hosts, often practically no symptoms are noted.
9. **Prepatent period**: Unknown, however, immune instable animals will become infected in a few days.
10. **Patency**: Practically lifelong due to repeated self-infections.
11. **Therapy**: Unknown; however, some authors succeeded in diminishing diarrhoeas by application of metronidazole.

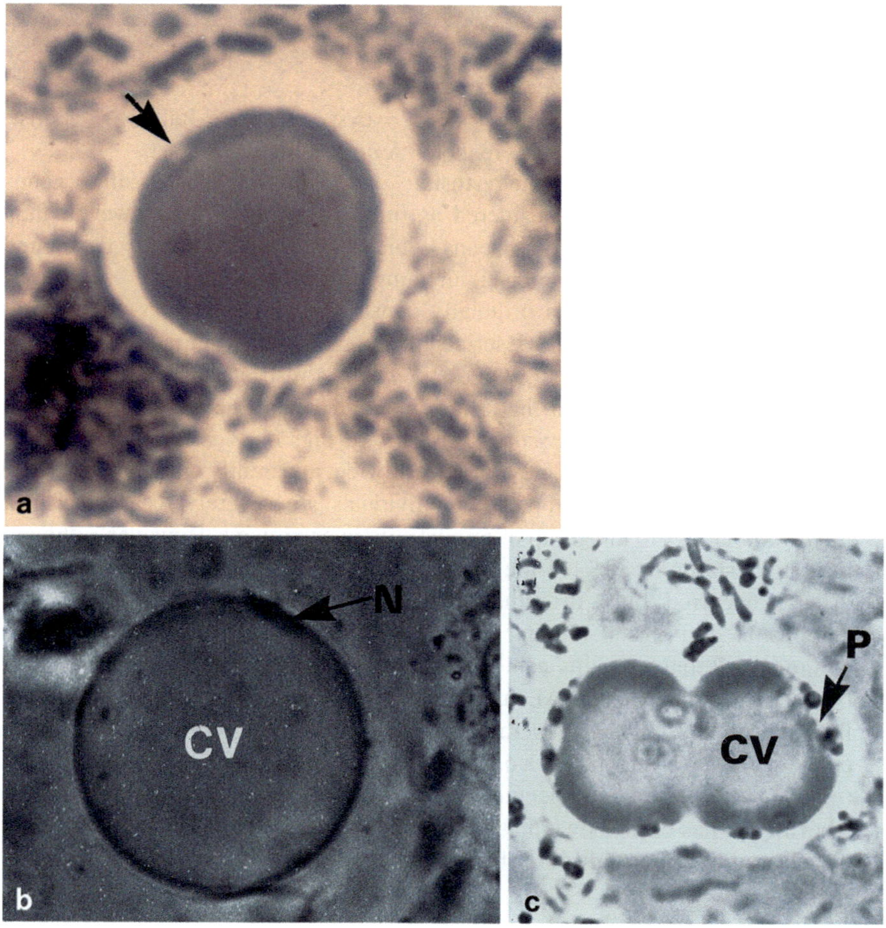

Fig. 4.128 (**a–c**) *Blastocystis* species: light micrographs of single cyst stages (**a**, **b**) and in binary division (**c**). *CV* central vacuole; *N* nucleus; *P* peripheral cytoplasm

Further Reading

Alexeieff A (1911) Sur la nature des formations dites kystes de *Trichomonas intestinalis*. C R Soc Biol 71:296–298.

Brumpt E (1912) *Blastocystis hominis* et forms voisines. Bull Soc Pathol Exot 5:725–730.

Mehlhorn H et al (2012) *Blastocystis*: pathogen or passenger? Parasitology research monographs, vol 4. Springer, Heidelberg.

Tan TC et al (2009) Genetic variability of *Blastocystis* isolates. Parasitol Res 105:1283–1286.

Yoshikawa H et al (2011) Evaluation of DNA extraction kits for molecular diagnosis of human *Blastocystis* subtypes from fecal samples. Parasitol Res 109: 1045–1050.

Zierdt CH (1973) Studies on *Blastocystis hominis*. J Protozool 20:114–121.

4.11 *Pneumocystis* Species

1. **Name**: Greek: *pneuma* = air; *kystos* = cyst. Carini = Italian scientist; Jirovec = Czech scientist; Jack Frenkel = American parasitologist.
2. **Life cycle/disease (Pneumocystosis)**: The specimens of the genus *Pneumocystis* are widely considered as fungi (Ascomycetes). However, they show also signs of typical protozoans, which have been apparently developed due to their parasitic lifestyle inside the lung of mammalians.

   Pneumocystis carinii was originally considered as the species parasitizing humans, although Jirovec had described the human species at first, which is now called *Pneumocystis jirovecii*, while *P. carinii* is restricted to animals. In the latter case, it has turned out that there apparently exist subspecies in the main animal groups such as *P. carinii f. sp. suis* of pigs or *P.c. murina*. Although *Pneumocystis carinii* stages are found worldwide, apparently only very young or old animals or immune-suppressed individuals are severely diseased. Thus, *P. carinii* belongs to the group of **opportunistic organisms**. However, although in many tests up to 80 % of the animals showed this peculiar parasite, severe disease occurred only in rather few animals. The life cycle stages are demonstrated in Figs. 4.129 and 4.130.
3. **Diagnosis**: Reliable results can be obtained by microscopical investigation of bronchoalveolar lavage (BALF), while some investigations just of the mouth fluid of animals failed in several cases. In recent times, Weissenbacher-Lang et al. (2016) have established a quantitative real-time polymerase chain reaction (qPCR) to detect porcine pneumocystosis.
4. **Therapy**: It has been shown that application of **cotrimazole** plus **prednisolone** or application of **pentamidine** showed curative effects.

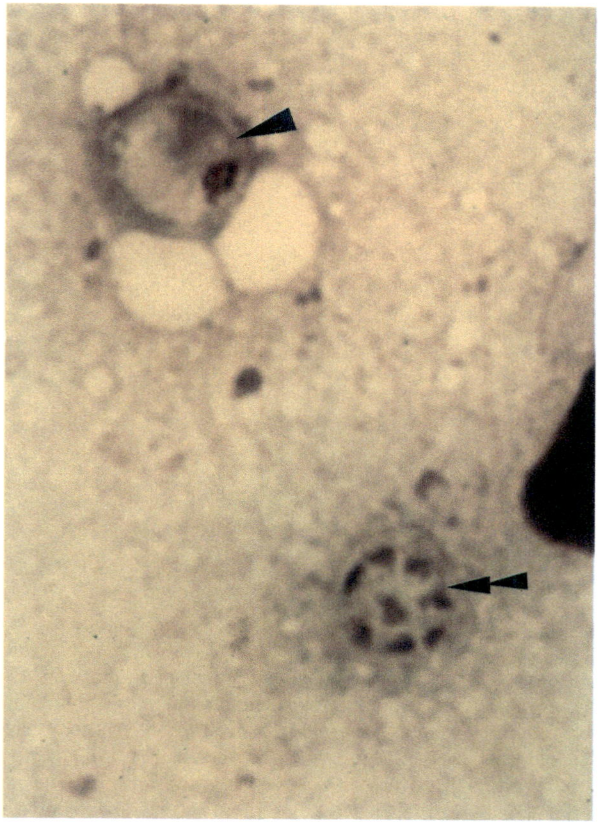

Fig. 4.129 Light micrograph of a smear preparation showing each a single nucleated and a multinucleated stage (*arrows*) of *Pneumocystis jiroveci*

Further Reading

Li WJ et al (2015) Diagnosis of *Pneumocystis* pneumonia using serum (1–3) –β-D-Glucan: a bivariate meta-analysis and systematic review. J Thorac Dis 7:2214–2225.

Montesinos I et al (2015) Comparison of 2 real-time PCR assays for diagnosis of *Pneumocystis jurovecii* pneumonia. Dian Microbiol Inf Dis 82:143–147.

Ruan S et al (2016) Treatment with interleukin-7-restores host defense against *Pneumocystis* in CD4+ T-lymphocyte-depleted mice. Infect Immun 84:108–116.

Weissenbacher-Lang C et al (2016) Establishment of a quantitative real-time PCR for the detection of *Pneumocystis carinii f. sp. suis* in bronchoalveolar lavage samples from pigs. J Vet Diagn Invest. doi:10.1177/1040638716641158.

Yiannakis EP et al (2016) Systematic review of outbreaks of *Pneumocystis jiroveci* pneumonia. J Hosp Inf 93:1–8.

Yoshikawa H et al (1987) Freeze fracture studies on *Pneumocystis carinii*. Parasitol Res 73:132–138.

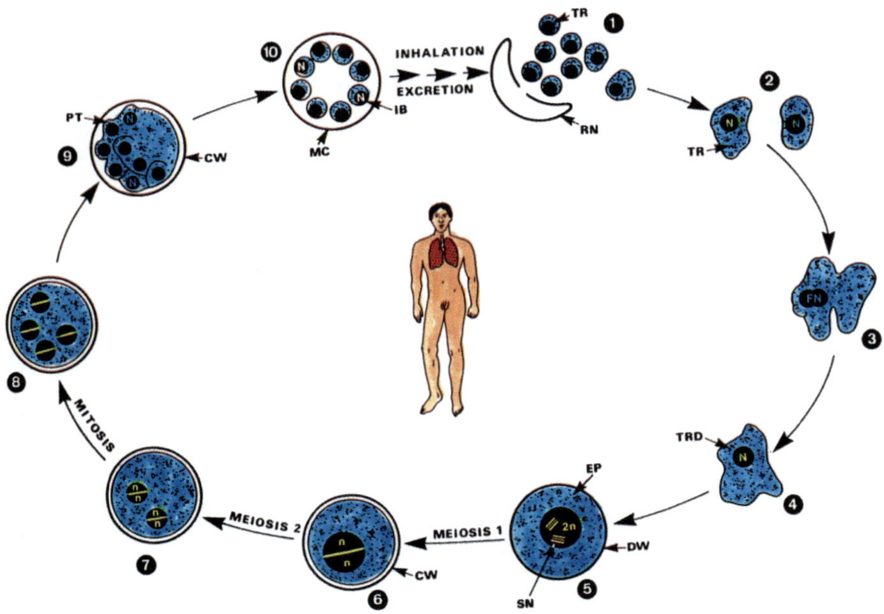

Fig. 4.130 Diagrammatic representation of the life cycle of *Pneumocystis carinii* of animals and *P. jirovecii* of humans. This diagram is based on results of the Yoshikawa group (Nara, Japan). (**1**) Hosts inhale cysts, which are excreted in nasal or bronchial fluids. Inside the lungs, the cyst wall breaks up and releases 8 haploid trophozoites, which become attached at the surface of cells of the lung alveoli. (**2–4**) Always two of the trophozoites fuse, thus forming a diploid zygote (=diploid trophozoites), which may proceed binary divisions. (**5**) The zygotes excrete a cyst wall. (**6–8**) Formation of 8 nuclei (1 step meiosis). (**9**) Formation of 8 intracystic body (=young trophozoites). (**10**) Mature cyst with 8 intracystic bodies. This cyst may be released from the lung included in slime. It can lead to **autoinfections** (e.g. common in HIV patients) or patients with any other immune suppression, which may lead to death. *CW* cyst wall; *DW* developing cyst stage; *EP* early cyst stage; *FN* fusion of the nuclei; *IB* intracystic body; *MC* mature cyst; *N* nucleus; *PT* protrusion = developing trophozoites; *RN* remnants of the cyst wall; *SN* synaptonemal complex; *TR* trophozoites; *TRD* diploid trophozoites

Further Reading (Joint List for This Chapter)

Abeywardena H et al (2015) A perspective on *Cryptosporidium* and *Giardia*. Adv Parasitol 88:243–301

Abu-Akkada S et al (2015) First detection of an ignored parasite, *Encephalitozoon cuniculi*, in different hosts in Egypt. Parasitol Res 114:843–850

Adjou Moumouni PF et al (2015) Molecular detection and characterization of *Babesia bovis*, *Babesia bigemina*, *Theileria* species and *Anaplasma marginale* isolated from cattle in Kenya. Parasit Vectors 8:496

Adriana G et al (2016) *Giardia duodenalis* genotypes in domestic and wild animals. Vet Parasitol 217:71–73

Ahmed MI et al (2010) Tissue responses of *Clarias gariepinus* (Catfish) to experimental *Eimeria* subepithelialis infection. Vet Parasitol 171:181–184

Akhlaghi M et al (2016) Molecular differentiation of bovine sarcocysts. Parasitol Res. doi:10.1007/s00436-016-5020-7

Al Quraishy S et al (2014) *Sarcocystis arieticanis* (Apicomplexa: Sarcocystidae) infecting the heart muscles of the domestic sheep, *Ovis aries* (Artiodactyla: Bovidae), from K. S. A. on the basis of light and electron microscopic data. Parasitol Res 113:3823–3831

Alam MA et al (2015) *Entamoeba* infections in different populations of dogs in an endemic area of Lahore, Pakistan. Vet Parasitol 207:216–219

Aleman M et al (2016) *Sarcocystis fayeri* in skeletal muscle of horses with neuromuscular disease. Neuromuscul Disord 26:85–93

Alexeieff A (1911) Sur la nature des formations dites kystes de *Trichomonas intestinalis*. C R Soc Biol 71:296–298

Alfaleh FA et al (2013) *Caryospora cherrughi* sp. n. infecting *Falco cherrug* in Saudi Arabia. Parasitol Res 112:971–974

Allen KE et al (2011) Novel *Hepatozoon* in vertebrates from the Southern United States. J Parasitol 97:648–653

Al-Quraishy S (2012) Exogenous and endogenous stages of *Eimeria perforans* naturally infected domestic rabbit (*Oryctolagus cuniculus*) in Saudi Arabia: light microscopic study. Saudi J Biol Sci 19:31–34

Amini A et al (2014) Trichomonads in birds—a review. Parasitology 141:733–747

Arias M et al (2012) Exposure to *Sarcocystis* spp. in horses from Spain determined by Western blot analysis using *Sarcocystis neurona* merozoites as heterologous antigen. Vet Parasitol 185:301–304

Aryeetey M, Mehlhorn H, Heydorn AO (1980) EM studies on the development of *Sarcocystis capracanis* in goats. Zentralbl Bakteriol 247:534–550

Aslan M et al (2016) New insights into transmissibility of *Leishmania infantum* from dogs to sandflies. J Inf Dis 16:2–10

Ayoob AL et al (2010a) Clinical management of canine babesiosis. J Vet Emerg Crit Care 20:77–89

Ayoob AL et al (2010b) Feline babesiosis. J Vet Emerg Crit Care 20:90–97

Ayouba A et al (2012) Ubiquitous *Hepatocystis* infections in monkeys. Int J Parasitol 42:709–713

Bacci C et al (2015) Detection of *Toxoplasma gondii* in free-range, organic pigs in Italy. Int J Food Microbiol 202:54–56

Baldwin CL et al (1986) Comparative analysis of infection and transformation of lymphocytes from African buffalo and Boran cattle with *Theileria parva* subsp. *parva* and *T. parva* subsp. *lawrencei*. Infect Immun 53:186–191

Ball SJ et al (2014) Observations on oocyst development of *Eimeria stiedai* in rabbits. Acta Parasitol 59:544–547

Baneth G et al (2015a) Reclassification of *Theileria annae* as *Babesia vulpes* sp. nov. Parasit Vectors 8:207

Baneth G et al (2015b) Canine vector borne co-infections: *Ehrlichia canis* and *Hepatozoon canis* in the same host monocytes. Vet Parasitol 208:30–34

Bartley PM et al (2016) Detection of *Babesia annae* DNA in lung exudate samples from Red foxes (*Vulpes vulpes*) in Great Britain. Parasit Vectors 9:84

Bartolomé Del Pino LE et al (2016) *Babesia caballi* and *Theileria equi* infections in horses in Central-Southern Italy: sero-molecular survey and associated risk factors. Ticks Tick Borne Dis 7:462–469

Bartosova P et al (2015) Morphology and phylogeny of two new species of *Sphaeromyxa* from marine fish. Parasitology 142:660–674

Barutzki D et al (2007) The babesiosis of dogs—an autochthonic disease in Germany. Dtsch Tierärztebl 55:284–293

Baseler LJ et al (2014) Pathology in practice. *E. invadens* infection in a ball python. J Am Vet Med Assoc 245:501–503
Basso W et al (2011) Exploring the life cycle of *Besnoita besnoiti*: experimental infection of putative definitive and intermediate host species. Vet Parasitol 178:223–234
Becker B et al (1979) Light and electron microscopic studies on the gamogony and sporogony of 5 *Sarcocystis* species. Zentralbl Bakteriol Orig A 244:394–404
Belova LM, Krylov MV (2000) Distribution of coccidians (Sporozoa: Coccidiida) in various systematic groups of fishes. Parazitologiia 34:522–533
Benchimol M et al (2016) The fungal metabolite gliotoxin inhibits proteasome proteolytic activity and induces an irreversible pseudocystic transformation and cell death in *Tritrichomonas foetus*. Parasitol Res 115:3057–3069
Bouzid M et al (2015) The prevalence of *Giardia* infection in dogs and cats. Vet Parasitol 207:181–202
Brewer LA et al (2008) Analysis of commercial *Entamoeba histolytica* ELISA kits for the detection of *Entamoeba invadens* in reptiles. J Zoo Wildl Med 39:493–495
Brown AV et al (2010) Description of 5 new *Loma* species. J Eukaryot Microbiol 57:529–553
Brumpt E (1912) *Blastocystis hominis* et forms voisines. Bull Soc Pathol Exot 5:725–730
Bukauskaite D et al (2015) Biting midges (*Culicoides*) transmit *Haemoproteus* parasites of owls: evidence from sporogony and molecular phylogeny. Parasit Vectors 8:303
Burreson EM (2007) Haemoflagellates of Oregon marine fishes: *Trypanosoma* and *Trypanoplasma*. J Parasitol 93:1442–1451
Cardoso MS et al (2016) Evasion of the immune response by *Trypanosoma cruzi*. Front Immunol 6:659
Castans P et al (2016) Experimental ovine toxoplasmosis. Vet Res 47:43
Chaisi ME et al (2013) Evaluation of a real-time PCR test for the detection and discrimination of *Theileria* species in the African buffalo (*Syncerus caffer*). PLoS One 8:e75827
Charbonneau LR et al (2015) Effects of *Nosema apis*, *N. ceranae* and co-infections on honey bee (*Apis mellifera*) learning and memory. Sci Report 6:22626
Chauhan HC et al (2015) Comparison of molecular and microscopic technique for detection of *Theileria annulata* from the field cases of cattle. Vet World 8:1370–1374
Chen A et al (2014) Development of *Eimeria nieschulzi* (Coccidia, Apicomplexa) gamonts and oocysts in primary fetal rat cells. J Parasitol Res. 2013:Article ID 591520.
Chen RA et al (2016) Observation of the excretion pattern of a precocious line of *Eimeria necatrix* and the strengthening of immune homogeneity. Exp Parasitol 164:79–83
Chia MY et al (2009) *Entamoeba invadens* myositis in a common water monitor lizard (*Varanus salvator*). Vet Pathol 46:673–676
Chlabra MB, Samantaray S (2013) *Sarcocystis* and sarcocystosis in India: status and emerging perspectives. J Parasit Dis 37:1–10
Claes F et al (2005) *Trypanosoma equiperdum*: master of disguise or historical mistake? Trends Parasitol 21:17–24
Clancey N et al (2010) *Babesia* (*Theileria*) *annae* in a red fox (*Vulpes vulpes*) from Prince Edward Island, Canada. J Wildl Dis 46:615–621
Clarke LL, Rissi DR (2015) Neuropathology of natural *Cytauxzoon felis* infection in domestic cats. Vet Pathol 52:1167–1171
Clode PL et al (2015) Life without a host cell: what is *Cryptosporidium*? Trends Parasitol 31:614–624
Coelho C et al (2015) Unraveling *Sarcocystis miescheriana* and *S. suihominis* infections in wild boar. Vet Parasitol 212:100–104
Connor RJ, van den Boscho P (2004) African animal trypanosomes. In: Coetzer JA, Tustin RC (eds) Infectious diseases of lifestock, vol 1. Oxford University Press, Oxford, pp 251–296
Coral AA et al (2015) In vitro development of *Haemoproteus columbae* with perspectives for genomic studies of avian haemosporidian parasites. Exp Parasitol 157:163–169

Cornelissen JBWJ et al (2014) An experimental *Toxoplasma gondii* dose response challenge model to study therapeutic or vaccine efficacy in cats. PLoS One 9:e104740

Curtis LM et al (2013) *Gnathia aureamaculosa*, a likely definitive host of *Haemogregorina balistapi* and potential vector for *Haemogrearina bigemina* between fishes of the Great Barrier Reef, Australia. Int J Parasitol 43:361–370

D'Haese J et al (1977) Comparative electron microscope study of pellicular structures in coccidian (*Sarcocystis*, *Besnoitia* and *Eimeria*). Int J Parasitol 7:505–518

Da Fonseca Ferreira-da-Silva M et al (2014) Identification of the microsporidian *Encephalitozoon cuniculi* as a new target of the IFN-inducible IRG resistance system. PLoS Pathog 10:e100449

Da Silva Barbosi A (2015) Isolation and maintenance of *Balantidium coli* cultured from fecal samples of pigs and non-human primates. Vet Parasitol 2010:240–245

Dado D et al (2012) Prevalence and genotype of *Giardia duodenalis* from dogs in Spain: possible zoonotic transmission and public health importance. Parasitol Res 111:2419–2422

Damron GW (1976) Gastrointestinal trichomonads in horses: occurrence and identification. Am J Vet Res 37:25–28

Dantas-Torres F (2007) The role of dogs as reservoirs of *Leishmania* parasites. Vet Parasitol 149:139–146

Darius AK, Mehlhorn H, Heydorn AO (2004) Effects of toltrazuril and ponazuril on the fine structure and multiplication of the NC-1 strain of *Neospora caninum* (syn. *Hammondia heydorni*) in cell cultures. Parasitol Res 92:453–458

Daugschies A et al (1999) Vorkommen von *Isospora suis*. Prakt Tierarzt 80:530–537

De Aruda MM et al (2016) Sensitivity and specificity of serological testing of canine *Leishmania* infection. Mem Inst Oswaldo Cruz 28:1–6

De Souza PN et al (2009) Natural infection by *Cryptosporidium* sp., *Giardia* sp. and *Eimeria leuckarti* in three groups of equines with different handlings in Rio de Janeiro, Brazil. Vet Parasitol 160:327–333

Deplazes P et al (2013a) Lehrbuch der Parasitologie für die Tiermedizin. Enke, Stuttgart

Deplazes P et al (2013b) Lehrbuch der Parasitologie für die Tiermedizin, 3rd edn. Enke, Stuttgart

Deschamps JY et al (2016) Refractory hypoglycaemia in a dog infected with *Trypanosoma congolense*. Parasitology 23:1

Dezfoulian O et al (2010) *Hexamita* infection associated with diarrhoea and stunting in native turkey poults. Trop Biomed 27:504–508

Di Cicco MF et al (2012) Re-emergence of *Babesia conradae* and effective treatment of infected dogs with atovaquone and azithromycin. Vet Parasitol 187:23–27

Díaz P et al (2015) *Cryptosporidium* species and subtype analysis in diarrhoeic pre-weaned lambs and goat kids from north-western Spain. Parasitol Res 114:4099–4105

Djokic V et al (2016) *Toxoplasma gondii* infection in pork produced in France. Parasitology 143:537–567

Dogga SK et al (2015) Phylogeny, morphology and metabolic and invasive capabilities of epicellular fish coccidium *Goussia janae*. Protist 166:659–676

Duarte M et al (2009) Cytotoxicity of solubilisation vehicles for *Trichomonas gallinae* and *Tritrichomonas foetus*. Vet Parasitol 166:167–170

Dubey JP (2004) Equine protozoal myeloencephalitis. In: Coetzer JA, Tustin RC (eds) Infectious diseases of livestock, 2nd edn. Oxford University Press, Oxford

Dubey JP (2014) Life cycle of *Cystoisospora felis* (Coccidia: Apicomplexa) in cats and mice. J Eukaryot Microbiol 61:637–643

Dubey JP, Jabsley MJ (2010) *Besnoita neotomofelis* n. sp. (Protozoa, Apicomplexa from the Southern plains woodrat (*Neotoma micropus*). Parasitology 137:1731–1747

Dubey JP et al (1991) *Sarcocystis neurona* n. sp.: the etiological agent of equine protozoal myeloencephalitis. J Parasitol 77:212–218

Dubey JP et al (2000) *Frenkelia microti* infection in a chinchilla (*Chinchilla laniger*) in the USA. J Parasitol 86:1149–1150

Dubey JP et al (2002) *Neospora caninum* and *Hammondia heydorni* are separate species. Trends Parasitol 18:66–69
Dubey JP et al (2014a) *Sarcocystis caninum* and *Sarcocystis svanai* n. spp. (Apicomplexa: Sarcocystidae) associated with severe myositis and hepatitis in the domestic dog (*Canis familiaris*). Eukaryot Microbiol 62:307–317
Dubey JP et al (2014b) Isolation and genetic characterization of *T. gondii* from black bears (*Ursus americanus*), bobcats (*Lynx rufus*) and feral cats (*Felis catus*). J Eukaryot Microbiol 62:410–415
Dubey J et al (2014c) Life cycle of *Hammondia hammondi* in cats. J Eukaryot Microbiol 62:346–352
Dubey JP et al (2015a) Experimental transmission of *Cystoisospora felis*-like coccidium from bobcat (*Lynx rufus*) to the domestic cat (*Felis catus*). Vet Parasitol 211:35–39
Dubey JP et al (2015b) An update on *Sarcocystis neurona* infections in animals and equine protozoal myeloencephalitis (EPM). Vet Parasitol 209:1–42
Dubey JP et al (2016) Ultrastructure of *Sarcocystis bertrami* sarcocysts from a naturally infected donkey (*Equus asinus*) from Egypt. Parasitology 143:18–23
Dunn JM et al (2014) *Borrelia burgdorferi* promotes the establishment of *Babesia microti* in the northeastern United States. PLoS One 9:e115494
Dvorakova N et al (2015) Haemogregarines of freshwater turtles from South East Asia with a description of *Haemogregarina sacaliae* and a redescription of *Haemogregarina pellegrini*. Parasitology 142:816–826
Edwards GT (1984) Prevalence of equine *Sarcocystis* in British horses and comparison of two detection methods. Vet Rec 115:265–267
Ehsan AM et al (2015) Assessment of zoonotic transmission of Giardia and Cryptosporidium between cattle and humans. PLoS One 10:e0118
Erber M, Geisel O (1981) Vorkommen und Entwicklung von 2 Sarkosporidienarten des Pferdes. Parasitol Res 65:283–291
Erster O et al (2015) Molecular detection of *Babesia ovis* in sheep and ticks using the gene encoding *B. ovis* surface protein D (BoSPD). Vet Parasitol 214:282–288
Eshoo MW et al (2014) Broad-range survey of tick-borne pathogens in Southern Germany reveals a high prevalence of *Babesia microti* and a diversity of other tick-borne pathogens. Vector Borne Zoonotic Dis 14:584–591
Esmaeilnejad B et al (2014) Evaluation of antioxidant status, oxidative stress and serum trace mineral levels associated with *Babesia ovis* parasitemia in sheep. Vet Parasitol 205:38–45
Eszterbayer E, Szekely SC (2004) Molecular phylogeny of the kidney parasite *Sphaerospora renicola* from common carp (*Cyprinus carpio*) and *Sphaerospora* sp. From goldfish. Acta Vet Hung 52:469–478
Farkas R et al (2015) First report on *Babesia* cf. *microti* infection of red foxes (*Vulpes vulpes*) from Hungary. Parasit Vectors 8:55
Fayer R et al (2015) Human infections with *Sarcocystis* species. Clin Microbiol Rev 28:295–311
Ferguson JA et al (2007) Spores of two fish microsporidia (*Pseudoloma neutrophila* and *Glugea anomala*) are highly resistant to chlorine. Dis Aquat Organ 76:2005–2014
Flechter R et al (2012) Control of *Giardia* infections with rodinazole and intensive hygiene management in dog kennel. Vet Parasitol 187:93–98
Franzen C et al (2006) Transfer of the members of the genus *Brachiola* to the genus *Anncaliia* based on ultrastructural and molecular data. J Eukaryot Microbiol 53:26–35
Frenkel JK, Smith DD (2003) Determination of the genera of cyst forming coccidians. Parasitol Res 91:384–389
Frenkel JK et al (1979) Sarcocystinae: nomina dubia and available names. Parasitol Res 58:115–139
Frey CF et al (2009) Intestinal *Trichomonas foetus* infection in cats. Parasitol Res 104:783–788
García-Bocanegra I et al (2014) Seroprevalence and risk factors associated with *Babesia caballi* and *Theileria equi* infection in equids. Vet J 195:172–178

Geurden T, Olson M (2011) *Giardia* in pets and farm animals and their zoonotic potential. In: Luja H, Svard S (eds) *Giardia*: a model organism. Springer, Vienna, pp 71–85
Gjerde B (2013) Phylogenetic relationships among *Sarcocystis* species in cervids, cattle and sheep inferred from the mitochondrial cytochrome c oxidase subunit I gene. Int J Parasitol 43:579–591
Gjerde B (2014) Molecular characterization of *Sarcocystis rileyi* from a common eider (*Somateria mollissima*) in Norway. Parasitol Res 113:3501–3504
Gjerde B (2016) The resurrection of a species: *Sarcocystis bovifelis* Heydorn et al., 1975 is distinct from the current *Sarcocystis hirsuta* in cattle and morphologically indistinguishable from *Sarcocystis sinensis* in water buffaloes. Parasitol Res 115:1–21
Gjerde B, Schulze J (2014) Muscular sarcocystosis in two arctic foxes (*Vulpes lagopus*) due to *Sarcocystis arctica* n. sp.: sarcocyst morphology, molecular characteristics and phylogeny. Parasitol Res 113:811–821
Gjerde B et al (2016) Molecular differentiation of *Sarcocystis buffalonis* and *Sarcocystis levinei* in water buffaloes (*Bubalus bubalis*) from *Sarcocystis hirsuta* and *Sarcocystis cruzi* in cattle (*Bos taurus*). Parasitol Res. doi:10.1007/s00436-016-4998-1
Gondim LF et al (2004) Coyotes (*Canis latrans*) are definitive hosts of *Neospora caninum*. Int J Parasitol 34:159–161
González-Kother P, González MT (2014) The first report of liver coccidian *Goussia cruciata* in jack mackerel, *Trachurus murphyi*, from the South Pacific and its relationship with host variables. Parasitol Res 113:3903–3907
Grabensteiner E et al (2006) Broad dissemination of *Histomonas meleagridis*. Parasitol Int 55:317–322
Grim JN et al (2015) Light microscopic morphometrics, ultrastructure and molecular phylogeny of the putative pycnotrichid ciliate, *Buxtonella sulcata*. Eur J Parasitol 51:425–436
Grinberg A, Widmer G (2016) Cryptosporidium within-host genetic diversity: systematic bibliographical search and narrative overview. Int J Parasitol 46:465–471
Gülegen E et al (2016) *Eimeria leuckarti* infections in sport horses in Northwest Turkey. Berl Munch Tierarztl Wochenschr 129:52–55
Hagos A et al (2010) Efficacy of Cymelarsan® and Diminasan® against *Trypanosoma equiperdum* in mice and horses. Vet Parasitol 171:200–206
Haklova-Kocikova B et al (2014) Morphological and molecular characterization of *Karyolyrus*. Parasit Vectors 7:555
Hamel D et al (2012) Epidemiological aspects on vector-borne infections in stray and pet dogs from Romania and Hungary with focus on *Babesia* spp. Pasrasitol Res 110:1537–1545
Hasheminasab SS et al (2015) *Buxtonella* spp. like infection in cattle. Ann Parasitol 61:247–251
Hauck R, Hafez HM (2013) Experimental infections with the protozoan parasite *Histomonas meleagridis*: a review. Parasitol Res 112:19–34
Hayes PM et al (2014) Fish trypanosomes from South Africa. Parasit Vectors 7:50
Hedrick RP et al (2012) Invasion and initial replication of ultraviolet, irradiated waterborne infective stages of *Myxobolus cerebralis* results in immunity to whirling disease in rainbow trout. Int J Parasitol 42:657–660
Heitlinger E et al (2014) The genome of *Eimeria falciformis*—reduction and specialization in a single host apicomplexan parasite. BMC Genomics 15:696
Hemphill A, Gottstein B (2006) *Neospora caninum* and neosporosis—recent achievements in hosts and parasite cell biology and treatment. Acta Parasitol 51:15–25
Hess M et al (2006) Clonal cultures of *Histomonas meleagridis*, *Tetratrichomonas gallinarum* and *Blastocystis* sp. Parasitology 133:547–554
Hess M et al (2015) *Histomonas meleagridis*—new insights into an old pathogen. Vet Parasitol 208:67–76
Heydorn AO, Mehlhorn H (1978) Light and electron microscopic studies on *Sarcocystis suihominis*: the schizogony preceeding cyst formation. Zentralbl Bakteriol Orig A 240:123–134

Heydorn AO, Mehlhorn H (1982) Fine structure of *Sarcocystis arieticanis*. Zentralbl Bakteriol Microbiol Hyg A 264:353–362
Heydorn AO, Mehlhorn H (2001) Further remarks on *Hammondia hammondi* and the taxonomic importance of obligate heteroxeny. Parasitol Res 87:573–577
Heydorn AO, Mehlhorn H (2002) A re-evaluation of *Neospora* and *Hammondia* spp. Trends Parasitol 18:246–247
Heydorn AO et al (1975) Proposal for a new nomenclature if the sarcosporidia. Parasitol Res 48:73–82
Hoan TD et al (2016) Identification and immunogenicity of microneme protein 2 (EbMIC2) of *Eimeria brunetti*. Exp Parasitol 162:7–17
Holzer AS et al (2013) Who's who in renal sphaerosporids. Parasitology 140:46–60
Hornok S et al (2014) Re-emergence of bovine piroplasmosis in Hungary: has the etiological role of *Babesia divergens* been taken over by *B. major* and *Theileria buffeli*? Parasit Vectors 7:434
Houk AE, Lindsay DS (2013) *Cystoisospora canis* (Apicomplexa: Sarcocystidae): development of monozoic tissue cysts in human cells, demonstration of egress of zoites from tissue cysts, and demonstration of repeat monozoic tissue cyst formation by zoites. Vet Parasitol 197:455–461
Houk AE et al (2013) Experimentally induced clinical *Cystoisospora canis* coccidiosis in dogs with prior natural patent *Cystoisospora ohioensis*-like or *C. canis* infections. J Parasitol 99:892–895
Hsu V et al (2010) Prevalence of antibodies to *Sarcocystis neurona* in cats from Virginia and Pennsylvania. J Parasitol 96:800–801
Huang C et al (2016) Exploration of fluorescence-based real-time loop-mediated isothermal amplification (LAMP) assay for detection of *Isospora suis* oocysts. Exp Parasitol 165:1–6
Hussain MH et al (2014) Seroprevalence of *Babesia caballi* and *Theileria equi*. Vet Parasitol 202:248–256
Ibrahim MA et al (2016) Epidemiology and public health significance of *Cryptosporidium* isolated from cattle, buffaloes, and humans in Egypt. Parasitol Res. doi:10.1007/s00436-016-4996-3
Imura T et al (2014) The apicoplast genome of *Leucocytozoon caulleryi*. Parasitol Res 113:823–828
Jaramillo Ortiz JM et al (2014) Vaccine strategies against *Babesia bovis* based on prime-boost immunizations in mice with modified vaccinia Ankara vector and recombinant proteins. Vaccine 32:4625–4632
Jencic V et al (2014) A survey of *Tetracapsuloides bryosalmonae* infections. J Fish Dis 37:711–717
Jing F et al (2012) Large-scale survey of the prevalence of *Eimeria* infections in domestic rabbits in China. Parasitol Res 110:1495–1500
Johnson EM et al (2007) Field survey of rodents for *Hepatozoon* infections in an endemic focus of American hepatozoonosis. Vet Parasitol 150:27–32
Jurecka P et al (2009) Genetic resistance of carp to *Trypanosoma borreli*. Vet Immunol Immunopathol 127:19–25
Kakandelwa C et al (2016) Prevalence of *Giardia* in dairy cattle. Vet Parasitol 215:114–116
Karadjian G et al (2015) Systematic revision of the adelaid haemogregarines. Parasitology 22:31
Karamon J et al (2007) Prevalence of *Isospora suis* and *Eimeria* spp. in suckling piglets and sows in Poland. Vet Parasitol 147:171–175
Kato CD et al (2016) The role of cytokines in the pathogenesis and staging of *Trypanosoma brucei rhodesiense* sleeping sickness. Allergy Asthma Clin Immunol 12:4
Keeling PJ, Brugerolle G (2006) Evidence from SSUrRNA phylogeny that *Octomitus* is a sister lineage to *Giardia*. Protist 157:2005–2012
Kern G et al (2015) Analysis of risk factors for infections with gastrointestinal nematodes, *Eimeria* spp. and lungworms in German organic sheep farms. Berl Munch Tierarztl Wochenschr 128:233–239
Khan RA (2005) Prevalence and influence of *Loma branchialis* (Microspora) on growth and mortality in Atlantic cod (*Gadus morhua*). J Parasitol 91:1230–1232

Khan A et al (2013) Prevalence, hematology, and treatment of balantidiasis among donkeys. Vet Parasitol 196:203–205

Kiehl E et al (2010) Molecular biological comparison of different *Besnoita* species and stages from different countries. Parasitol Res 106:889–894

Kjemtrup AM et al (2006) *Babesia conradae*, sp. Nov., a small canine *Babesia* identified in California. Vet Parasitol 138:103–111

Kolisko M et al (2008) Molecular phylogeny of diplomonads and enteromonads. BMC Evol Biol 8:205

Kostopoulou D et al (2015) The occurrence and genetic characterization of *Cryptosporidium* and *Giardia* species in foals in Belgium, The Netherlands, Germany and Greece. Vet Parasitol 211:170–174

Koudela B et al (1990) Concurrent infection of enterocytes with *Eimeria scabra* and other enteropathogens in swine. Vet Parasitol 35:71–77

Kovacevic N et al (2015) The analysis of the acute phase response during infection with *T. carassi*. Dev Comp Immunol 53:112–122

Kumar G et al (2015) Interaction of *Tetracapsuloides bryosalmonae*, the causative agent of proliferative kidney disease, with host proteins in the kidney of *Salmo trutta*. Parasitol Res 114:1721–1727

Kumnuan R et al (2013) In vivo transmission blocking activities of artesunate on the avian malaria parasite *Plasmodium gallinaceum*. Vet Parasitol 197:447–454

Künzel F et al (2014) Comparison of an indirect fluorescent antibody test with Western blot for the detection of serum antibodies against *Encephalitozoon cuniculi* in cats. Parasitol Res 113:4457–4462

Kurze C et al (2016) *Nosema* sp. infections cause no energetic stress in tolerant honey bees. Parasitol Res. doi:10.1007/s00436-016-4988-3.

Laakone J, Henttonen H (2000) Ultrastructure of *Frenkelia* sp. from a Norwegian lemming. J Wildl Dis 36:362–366

Lai DH et al (2008) Adaptations of *Trypanosoma brucei* to gradual loss of kinetoplast DNA: *T. equiperdum* and *T. evansi* are petite mutants of *T. brucei*. Proc Natl Acad Sci USA 105:1999–2004

Langkjaer M, Roepstorff A (2008) Survival of *Isospora suis* oocysts under controlled environmental conditions. Vet Parasitol 152:186–193

Latif B et al (2013) High prevalence of muscular sarcocystosis in cattle and water buffaloes from Selangor, Malaysia. Trop Biomed 30:699–705

Lauron EJ et al (2015) De novo assembly and transcriptome analysis of *Plasmodium gallinaceum*. Malar J 14:296

Leelayoova S et al (2009) Prevalence of intestinal parasitic infections in military personnel and military dogs, Thailand. J Med Assoc Thai 92 Suppl 1:S53–S59

Lemos M et al (2015) Trypanosomes from Brazilian catfishes and leeches. Parasit Vectors 8:573

Leonhard S et al (2007) The molecular characterization of *Giardia* from dogs. Vet Parasitol 150:33–38

Lesser M et al (2012) First cases of besnoitosis in cattle in Switzerland. Schweiz Arch Tierheilkd 154:469–474

Levi MG et al (2015) *Spironucleus meleagridis*. Vet Parasitol 208:169–173

Levine ND (1973) Protozoan parasites of domestic animals and man, 2nd edn. Burgess, Minneapolis

Li SQ et al (2009) Immunization with recombinant actin from *Trypanosoma evansi* induces protective immunity against *T. evansi*, *T. equiperdum* and *T. b. brucei* infection. Parasitol Res 104:429–435

Li M et al (2014) Glucocorticoid receptors in *Cryptobia salmositica*. Int J Parasitol 44:205–210

Li N et al (2015a) Identification and morphologic and molecular characterization of *Cyclospora macacae* n. sp. from rhesus monkeys in China. Parasitol Res 114:1811–1816

Li WJ et al (2015b) Diagnosis of *Pneumocystis* pneumonia using serum (1–3) –β-D-Glucan: a bivariate meta-analysis and systematic review. J Thorac Dis 7:2214–2225
Liebhardt D et al (2014) A single strain of *Tetratrichomonas* causes fatal typhlohepatitis in red legged pariges (*Alectoris rufa*). Avian Pathol 43:473–480
Liebhart D et al (2013) Vaccination against histomoniasis prevents a stop in egg production. Avian Pathol 42:79–84
Liu J et al (2014) A PCR method targeting internal transcribed spacers: the simultaneous detection of *Babesia bigemina* and *Babesia bovis* in cattle. Acta Parasitol 59:132–138
Liu L et al (2016) Ultrastructural effects of acetamizuril on endogenous phases of *Eimeria tenella*. Parasitol Res 115:1245–1252
Lloyd D, Williams CF (2014) Comparative biochemistry of *Giardia*, *Hexamita* and *Spironucleus*. Mol Biochem Parasitol 197:43–44
Lloyd D et al (2015) Motility of the diplomonadid fish parasite *Spironucleus vortens*. Microbiology 161:213–218
Lopes WDZ et al (2013) Sexual transmission of *Toxoplasma gondii* in sheep. Vet Parasitol 195:47–56
Lovy J, Friend SE (2015) Intestinal coccidiosis of anadromous and landlocked alewives, *Alosa pseudoharengus*, caused by *Goussia ameliae* n. sp. and *G. alosii* n. sp. (Apicomplexa: Eimeriidae). Int J Parasitol Parasit Wildl 4:159–170
Luciani M et al (2013) IgG antibodies from dourine infected horses identify a distinctive *Trypanosoma equiperdum* antigenic pattern of low molecular weight molecules. Vet Immunol Immunopathol 151:140–146
Luckings AG et al (2004) Dourine. In: Coetzer JA, Tustin RC (eds) Infectious diseases of lifestock, vol 1. Oxford University Press, Oxford
Lun ZR et al (2016) Resistance to normal human serum reveals *Trypanosoma lewisi*. Mol Biochem Parasitol 199:58–61
Lyons ET, Tolliver SC (2004) Prevalence of parasite eggs (*Strongyloides westeri*, *Parascaris equorum*, and strongyles) and oocysts (*Eimeria leuckarti*) in the feces of Thoroughbred foals on 14 farms in central Kentucky in 2003. Parasitol Res 92:400–404
Mackenstedt U et al (1990) DNA measurements and ploidy determination of different stages in the life cycle of *Sarcocystis muris*. Parasitol Res 76:662–668
Madison BN et al (2013) Duress without stress: *Cryptobia* infection in rainbow trout. J Endocrinol 218:287–297
Maeda H et al (2013) Inhibitory effect of cyclophilin A from the hard tick *Haemaphysalis longicornis* on the growth of *Babesia bovis* and *Babesia bigemina*. Parasitol Res 112:2207–2213
Magnet A et al (2014) Novel *Acanthamoeba* 18s/rRNA gene sequence type from an environmental isolate. Parasitol Res 113:2845–2850
Manrique WG et al (2016) Ultrastructural description of *Myxobolus cuneus*. Parasitol Res. doi:10.1007/s00436-016-5026-1
Markness JE et al (2013) Demonstrated efficacy of a pilot heterologous whole spore vaccine against microsporidial gill disease in rainbow trout. Clin Vaccine Immunol 20:1483–1484
Marming K (2010) Update on the diagnosis and management of *Tritrichomonas foetus* infections in cats. Top Comp Anim Med 25:145–148
Martinkovic F et al (2012) *Trypanosoma* (*Megatrypanum*) *melophagium* in the sheep ked. J Eukaryot Microbiol 59:134–144
Martin-Sanchez J et al (2007) Infection by *Leishmania infantum* in cats. Vet Parasitol 145:267–273
Maside X et al (2015) Population genetics of *Nosema apis* and *N. ceranae*: one host (*Apis mellifera*) and two different histories. PLoS One 10(12):e0145609
Matos de Macchi B et al (2013) Chickens treated with a nitric oxide inhibitor became more resistant to *Plasmodium gallinaceum* infection. Vet Res 44:8

Matsubayashi M et al (2011) Phylogenetic identification of *Cystoisospora* spp. from dogs, cats, and raccoon dogs in Japan. Vet Parasitol 176:270–274

Matsubayashi M et al (2016) Transcriptional profiles of virulent and precocious strains of *Eimeria tenella* at sporozoite stage; novel biological insight into attenuated asexual development. Infect Genet Evol 40:54–62

Matthews PD et al (2015) Ultrastructure and ssrRNA sequencing of *Myxidium* amazonense. Parasitol Res 114:4675–4683

Matthews PD et al (2016) Morphological and ultrastructural aspects of *Myxobolus niger*. Acta Trop 158:214–219

Maxie MG et al (1982) A comparative study of the disease in cattle caused by *Theileria parva* or *T. lawrencei*: II. Hematology, clinical chemistry, coagulation studies and complement. Vet Parasitol 10:1–19

McDougald LR (2005) Blackhead disease (histomoniasis) in poultry: a critical review. Avian Dis 49:462–476

Megia-Palma R et al (2013) Phylogenetic analysis based on 18SrRNA gene sequences of *Schellackia* parasites. Parasitology 140:1149–1157

Mehlhorn H (1972a) Electron microscopy studies on the developmental stages of *Eimeria maxima* (Sporozoa, Coccidia). I. Ultrastructure of macrogametes. Parasitol Res 39:161–182

Mehlhorn H (1972b) Ultrastructural study of developmental stages of *Eimeria maxima* (Sporozoa, Coccidia). II Fine structure of microgametes. Z Parasitenkd 40:151–163

Mehlhorn H (1972c) Electron microscopy studies on the developmental stages of *Eimeria maxima* from the domestic chicken. 3. Differentiation of microgametes with special reference to nuclear divisions. Parasitol Res 40:243–260

Mehlhorn H, Heydorn AO (1978a) The Sarcosporidia: the fine structure and life cycle. Adv Parasitol 16:43–92

Mehlhorn H, Heydorn AO (1978b) The sarcosporidia (Protozoa, Sporozoa): life cycle and fine structure. Adv Parasitol 16:43–91

Mehlhorn H, Heydorn AO (1978c) The Sarcosporidia: life cycle and fine structure. Adv Parasitol 16:43–103

Mehlhorn H, Heydorn AO (1979) Electron microscopical study on gamagony of *Sarcocystis suihominis* in human tissue cultures. Parasitol Res 58:97–113

Mehlhorn H, Heydorn AO (2003) Electron microscopy study of in-vitro cultures of *Hammondia heydorni* tachyzoites after passages through dogs, mice, rats, guinea pigs and jirds. Parasitol Res 90:512–520

Mehlhorn H, Schein E (1977) Electron microscopic studies of the development of kinetes in *Theileria annulata* Dschunkowsky & Luhs, 1904 (Sporozoa, Piroplasmea). J Protozool 24:249–257

Mehlhorn H, Schein E (1984) The piroplasms: life cycle and sexual stages. Adv Parasitol 23:37–104

Mehlhorn H, Schein E (1998) Redescription of *Babesia equi* Laveran, 1901 as *Theileria equi* Mehlhorn, Schein 1998. Parasitol Res 84:467–475

Mehlhorn H et al (1975a) Light and electron microscope studies on cysts of *Sarcocystis fusiformis* in calves. Zentralbl Bakteriol Orig A 232:373–391

Mehlhorn H et al (1975b) Electron microscope studies on development stages of *Theileria annulata* (Dschunkowsky, Luhs, 1904) in the intestine and haemolymph of *Hyalomma anatolicum excavatum* (Koch, 1844). Parasitol Res 48:137–150

Mehlhorn H et al (1978) Electron microscopic studies on the development of kinetes of *Theileria parva* Theiler, 1904 in the gut of the vector ticks *Rhipicephalus appendiculatus* Neumann, 1901. Acta Trop 35:123–136

Mehlhorn H et al (1979) Electron-microscopic studies on *Theileria ovis* Rodhain, 1916: development of kinetics in the gut of the vector tick, *Rhipicephalus evertsi evertsi* Neumann, 1897, and their transformation within cells of the salivary glands. J Protozool 26:377–385

Mehlhorn H et al (1981) Electron microscopical study on the effect of halofuginone on *Theileria parva*. Tropenmed Parasitol 32:231–233
Mehlhorn H et al (1984a) The effects of sym. Triazinones on developmental stages of *Eimeria tenella*, *E. maxima* and *E. acervulina*: a light and electron microscopical study. Parasitol Res 70:173–182
Mehlhorn H et al (1984b) Two types of *Globidium* cysts of goats. Parasitol Res 70:731–737
Mehlhorn H et al (1985) Announcement of the establishment of neohepantotypes of some important *Sarcocystis* species. Parasitol Res 71:689–692
Mehlhorn H et al (1988a) Toltrazuril effective against a broad spectrum of protozoan parasites. Parasitol Res 75:64–66
Mehlhorn H et al (1988b) Toltrazuril against a broad spectrum of protozoan parasites. Parasitol Res 75:64–68
Mehlhorn B et al (2005) Health for ornamental fishes. Springer, Heidelberg
Mehlhorn B et al (2006) Fish parasites. Springer, Heidelberg (in German)
Mehlhorn H et al (2009) Another Africa disease in Central Europe: besnoitiosis of cattle. I. Light and electron microscopical study. Parasitol Res 104:861–868
Mehlhorn H et al (2012) *Blastocystis*: pathogen or passenger? Parasitology research monographs, vol 4. Springer, Heidelberg
Mengel H et al (2012) Necrotic enteritis due to simultaneous infection with *Isospora suis* and clostridia in newborn piglets and its prevention by early treatment with toltrazuril. Parasitol Res 110:1347–1355
Michi A et al (2016) A review of sexually transmitted bovine trichomoniasis. Theriogenology 85:781–791
Mitchell SM et al (2009) Development and ultrastructure of *Cystoisospora canis* Nemeséri, 1959 (syn, *Isospora canis*) monozoic cysts in two noncanine cell lines. J Parasitol 95:793–798
Moco TC et al (2012) Morphological, morphometric, and molecular characterization of *Hepatozoon* spp. from naturally infected *Caudisoma durissa terrifica* snakes. Parasitol Res 110:1393–1401
Molnar K, Baska F (1986) Light and electron microscopic studies on *Epieimeria anguillae* (Léger & Hollande, 1922), a coccidium parasitizing the European eel, *Anguilla anguilla* L. J Fish Dis 9:99–110
Molnár K et al (2012) *Eimeria* that infect fish are diverse and are related to, but distinct from, those that infect terrestrial vertebrates. Infect Genet Evol 12:1810–1815
Montesinos I et al (2015) Comparison of 2 real-time PCR assays for diagnosis of *Pneumocystis jurovecii* pneumonia. Diagn Microbiol Inf Dis 82:143–147
Moraes JC et al (2015) Prevalence of *Eimeria* spp. in broilers by multiplex PCR in the Southern region of Brazil on two hundred and fifty farms. Avian Dis 59:277–281
Moré G et al (2013) Development of a multiplex real time PCR to differentiate *Sarcocystis* spp. affecting cattle. Vet Parasitol 197:85–94
Morrison WI et al (2015) Understanding the basis of parasite strain-restricted immunity to *Theileria parva*. Annu Rev Anim Biosci 3:397–418
Morsy K et al (2016) *Hennequya collaris* parasite of the greenband parrotfish. Parasitol Res. doi:10.1007/s00436-016-4968-7
Müller N et al (2001) Comparative molecular investigation of Nc5-PCR amplicons from *Neospora caninum* NC-1 and *Hammondia heydorni*. Parasitol Res 87:883–885
Munday JC et al (2015) Transport proteins determine drug sensitivity and resistance in a protozoan parasite, *Trypanosoma brucei*. Front Pharmacol 6:32
Mundt HC et al (2007) Efficacy of various anticoccidials against experimental porcine neonatal isosporosis. Parasitol Res 100:401–411
Munoz J, Mayer DCG (2016) *Toxoplasma gondii* and *Giardia duodenalis* infections in domestic dogs in New York City public parks. Vet J. doi:10.1016/j.tvjl.2016.02.015
Murcia S et al (2015) *Myxobolus cerebralis* infection risk in native cutthroat trout *Oncorhynchus clarkii*. J Fish Dis 38:637–652

Najm NA et al (2014) *Hepatozoon canis* in German red foxes and their ticks. Parasitol Res 113:2679–2685

Ng YH et al (2015) Modified use of methylene blue in the tissue compression technique to detect sarcocysts in meat-producing animals. Vet Parasitol 214:200–203

Nicolle J, Manceaux G (1909) Sur un protozoaire nouveau du gondi. Arch Inst Pasteur 2:97–101

Nilles-Bije L, Rivera WL (2010) Ultrastructural and molecular characterization of *Balantidium coli* isolated in the Philippines. Parasitol Res 106:387–394

Niu Q et al (2013) Sequence and organization of the rhoptry-associated-protein-1 (rap-1) locus for the sheep hemoprotozoan *Babesia* sp. BQ1 Lintan (*B. motasi* phylogenetic group). Vet Parasitol 198:24–38

Njiru ZK et al (2004) Detection of *Trypanosoma evansi* in camels using PCR and CATT/T evansi tests in Kenya. Vet Parasitol 124:187–189

Ogedengbe ME et al (2013) Sequencing the complete mitochondrial genome of *Eimeria mitis* strain USDA 50 (Apicomplexa: Eimeriidae) suggests conserved start positions for mtCOI- and mtCOIII-coding regions. Parasitol Res 112:4129–4136

Olias P et al (2009) A novel *Sarcocystis* associated encephalitis and myositis in racing pigeons. Avian Pathol 38:121–128

Olias P et al (2010a) *Sarcocystis*: species lethal for domestic pigeons. Emerg Inf Dis 16:497–499

Olias P et al (2010b) *Sarcocystis calchasi* sp. nov. of the domestic pigeon and the Northern goshawk. Parasitol Res 106:577–585

Olias P et al (2010c) *Sarcocystis calchasi* is distinct of *S. columbae* from the wood pigeon and *Sarcocystis* sp. from the sparrow hawk. Vet Parasitol 171:7–14

Olias P, Shade B, Mehlhorn H (2011) Molecular pathology, taxonomy and epidemiology of *Besnoita* species. Infect Genet Evol 11:1564–1576

Omori S et al (2008) Two extra chromosomal genomes of *Leucocytozoon caulleryi*. Parasitol Res 103:953–957

Pallant L et al (2015) The epidemiology of infections with *Giardia* species and genotypes in well cared dogs and cats. Parasit Vectors 8:2

Pellerdy L (1972) Coccidia and coccidiosis, 2nd edn. Paul Parey, Berlin

Pellerdy LP (1974) Coccidia and coccidiosis, 2nd edn. Paul Parey, Berlin

Peng Y et al (2015) Seminal fluid of honey bees contains multiple mechanisms to combat infections of the sexually transmitted pathogen *Nosema apis*. Proc R Soc B 263:20151785

Petersen HH (2014) *Cryptosporidium* and *Giardia* in Danish organic pig farms: seasonal and age-related variation in prevalence, infection intensity and species/genotypes. Vet Parasitol 214:29–39

Pham AND et al (2013) Establishing monoeukaryotic *Histomonas meleagridis* cultures from in vivo infection contaminated with *Tetratrichomonas gallinarum* and *Blastocystis* spp. Parasitology 140:1266–1274

Philippe P et al (2014) Comparative efficacy of diclazuril (Vecoxan®) and toltrazuril (Baycox bovis®) against natural infections of *Eimeria bovis* and *Eimeria zuernii* in French calves. Vet Parasitol 206:129–137

Piantedosi D et al (2014) Seroprevalence and risk factors associated with *Babesia caballi* and *Theileria equi* infections in donkeys from Southern Italy. Vet J 202:578–582

Pomajbikova K et al (2013) Novel insights into the genetic diversity of *Balantidium* and *Balantidium*-like cyst forming ciliates. PLoS Negl Trop Dis 3:e2140

Ponce-Gordo ML et al (2014) New insights into the molecular phylogeny of *Balantidium* based on the analysis of new sequences of species from fish hosts. Parasitol Res 113:4327–4333

Pop L et al (2015) Effects of artemisinin in broiler chickens challenged with *Eimeria acervulina*, *E. maxima* and *E. tenella* in battery trials. Vet Parasitol 214:264–271

Poyton SL et al (2004) Ultrastructure of *Spironucleus salmonis* and *Hexamita salmonis*. Dis Aquat Org 60:49–64

Pusterla N et al (2015) Daily feeding of diclazuril top dress pellets in foals reduces seroconversion to *Sarcocystis neurona*. Vet J 206:236–238

Raether W (1968) Acute amoebic dysenteria in a dog. Kleintierpraxis 7:196–200
Ramo A et al (2016) Intra-species diversity and panmictic structure of *Cryptosporidium parvum* populations in cattle farms in Northern Spain. PLoS One 11(2):e0148811
Rangel LF et al (2016) *Tetractinomyxon* stages genetically consistent with *Sphaerospora dicentrarchi*. Parasitology. doi:10:1017/s0031182016000512
Reeves AB et al (2015) Genetic diversity and host specificity varies across three genera (*Plasmodium, Haemoproteus, Leucocytozoon*) of blood parasites in ducks. PLoS One 10(2):e0116661
Répérant JM et al (2012) Pathogenicity of *Eimeria praecox* alone or associated with *Eimeria acervulina* in experimentally infected broiler chickens. Vet Parasitol 187:333–336
Ribeiro AJ et al (2013) Prevalence of *Theileria equi, Babesia caballi* and *Anaplasma phagocytophilum* in horses from the north of Portugal. Parasitol Res 112:2611–2617
Richter B et al (2010) First report of typhlitis/typhlohepatitis caused by *Tetratrichomonas gallinarum*. Avian Pathol 39:499–503
Rimoldi G et al (2013) An outbreak of *Sarcocystis calchasi*: encephalitis in multiple psittacine species within an enclosed zoological area. J Vet Diagn Invest 25:775–781
Ritter W (1994) Bee diseases. Ulmer, Stuttgart
Rochell SJ et al (2016) Effects of *Eimeria acervulina* infection severity on growth performance, apparent ileal amino acid digestibility, and plasma concentrations of amino acids, carotenoids, and α1-acid glycoprotein in broilers. Poult Sci 95:1573–1581
Rodriguez-Tovar LE (2016) *Encephalitozoon cuniculi*: grading the histological lesions in brain, kidney, and liver during primo infection outbreak in rabbits. J Pathog. Article ID 5768428
Rommel M et al (2006) Textbook of veterinarian parasitology, 6th edn. Paul Parey, Berlin
Rosenthal BM et al (2016) Coccidian parasites of fish encompass profound phylogenetic diversity and gave rise to each of the major parasitic groups in terrestrial vertebrates. Infect Genet Evol 40:219–227
Rossouw I et al (2015) Morphological and molecular descriptors of the developmental cycle of *Babesia divergens* parasites in human erythrocytes. PLoS Negl Trop Dis 9(5):e0003711
Ruan S et al (2016) Treatment with interleukin-7-restores host defense against *Pneumocystis* in CD4+ T-lymphocyte-depleted mice. Infect Immun 84:108–116
Ruttkowski B et al (2001) PCR-based differentiation of three porcine *Eimeria* species and *Isospora suis*. Vet Parasitol 95:17–23
Ryan U, Hijjawi (2015) New developments in *Cryptosporidium* research. Int J Parasitol 45:367–373
Sarvi S et al (2015) Cattle toxoplasmosis. Asian J Trop Med 2015:120–126
Schär F et al (2014) The prevalence and diversity of intestinal parasitic infections in humans and domestic animals in a rural Cambodian village. Parasitol Int 63:597–603
Schares G et al (2008) Characterization of a repetitive DNA fragment in *Hammondia hammondi* and its utility for the species differentiation of *H. hammondi* from *Toxoplasma gondii*. Mol Cell Probes 22:244–251
Schein E et al (1979) Electron microscopical studies on the development of *Babesia canis* (Sporozoa) in the salivary glands of the vector tick *Dermacentor reticulatus*. Acta Trop 36:229–241
Schetters TP et al (2009) Systemic inflammatory responses in dogs experimentally infected with *Babesia canis*. Vet Parasitol 162:7–15
Scholtyseck E, Mehlhorn H (1974a) Fine structure of cyst and cyst wall of *Sarcocystis tenella, Besnoitia jellisoni, Frenkelia* sp. and *Toxoplasma gondii*. J Protozool 21:284–294
Scholtyseck E, Mehlhorn H (1974b) Fine structure of the cyst and cyst wall of *Sarcocystis tenella, Besnoita jellisoni, Frenkelia* sp. and *Toxoplasma gondii*. J Parasitol Zool 21:284–294
Schwarz L et al (2014) Superinfection of sows with *Cystoisospora suis* ante partum leads to a milder course of cystoisosporosis in suckling piglets. Vet Parasitol 204:158–168
Sénaud J, Heydorn AO, Mehlhorn H (1984) Three new types of *Globidium* of sheep: An in-vivo and in-vitro investigation. Parasitol Res 70:721–729

Shayan P et al (2016) Identification and characterization of *Theileria ovis* surface protein (ToSp) resembled TaSp in *Theileria annulata*. Parasitol Res. doi:10.1007/s00436-016-4929-1

Sherrill MK, Cohn LA (2015) Cytauxzoonosis: diagnosis and treatment of an emerging disease. J Feline Med Surg 17:940–948

Shoaib HM et al (2013) Evaluation of inhibitory potential of some selective methanolic plant extracts on biological characteristics of *Acanthamoeba castellanii* using human corneal epithelian cells in vitro. Parasitol Res 112:1179–1188

Shutler D et al (2010) Relationships between circulating leucocytes and *Leucocytozoon simondi*. Comp Biochem Physiol A 156:46–49

Sistrom M et al (2016) De novo genome assembly shows genome wide similarity between *Trypanosoma brucei brucei* and *T. b. rhodesiense*. PLoS One 11:e0147660

Slapeta J et al (2015) Differences in the fecal microbiome of non-diarrhoeic clinically healthy dogs and cats associated with *Giardia duodenalis* infection. Int J Parasitol 45:585–594

Soares P et al (2014) *Haemogregarina* spp. in a wild population from *Podocnemis unifilis* in the Brazilian Amazonia. Parasitol Res 113:4499–4503

Solano-Gallego L, Baneth G (2011) Babesiosis in dogs and cats. Expanding parasitological and clinical spectra. Vet Parasitol 181:48–60

Sommer I et al (1991) Biochemical and immunological characterization of major surface antigens of *Sarcocystis muris* and *S. suicanis* cyst merozoites. Parasitol Res 77:2004–2011

Sommer I et al (1992a) A comparison of the sporozoite and cyst merozoite surface proteins of *Sarcocystis* species. Parasitol Res 78:396–403

Sommer I et al (1992b) A comparison of sporozoites and cyst merozoite surface proteins of *Sarcocystis*. Parasitol Res 78:398–403

Souza LE et al (2015) Epidemiology of *Eimeria* infections in sheep raised extensively in a semiarid region of Brazil. Rev Bras Parasitol Vet 24:410–415

Steele EJ, Noblet GP (2001) Gametogenesis, fertilization and ookinete differentiation of *Leucocytozoon smithi*. J Eukaryot Microbiol 48:118–125

Sudhakara Reddy B et al (2015) Clinical coccidiosis in adult cattle. J Parasit Dis 39:557–559

Sun W-W et al (2015) Herd-level prevalence and associated risk factors for *Toxoplasma gondii*, *Neospora caninum*, *Chlamydia abortus* and bovine viral diarrhoea virus in China. Parasitol Res 114:4211–4218

Sunaga F et al (2014) Protective efficacy of *Babesia gibsoni* culture-derived exoantigens against the challenge infection in dogs. Parasitol Res 113:1681–1686

Sundermann CA, Lindsay DS (1989) Ultrastructure of in vivo-produced caryocysts containing the coccidian *Caryospora bigenetica* (Apicomplexa: Eimeriidae). J Protozool 36:81–86

Sundermann CA et al (1988) Complete development of *Caryospora bigenetica* (Apicomplexa: Eimeriidae) in vitro. J Protozool 35:465–469

Sykes JE et al (2011) Severe myositis associated with *Sarcocystis* spp. infection in 2 dogs. J Vet Intern Med 25:1277–1283

Tan TC et al (2009) Genetic variability of *Blastocystis* isolates. Parasitol Res 105:1283–1286

Tancredo KR et al (2015) Haemato-immunological and biochemical parameters of silver catfish Rhamdia quelen immunized with live theronts of *Ichthyophthirius multifiliis*. Fish Shellfish Immunol 45:689–694

Thompson RC, Monis PT (2011) Taxonomy of *Giardia* species. In: Lujan HD, Svard S (eds) *Giardia*: a model organism. Springer, New York

Thurber MI et al (2013) Co-infection and cross species transmission of divergent *Hepatocystis* lineages in a wild African primate community. Int J Parasitol 43:613–619

Tian SQ et al (2012) The complete mitochondrial genome sequence of *Eimeria magna* (Apicomplexa: Coccidia). Mitochondrial DNA 26:714–715

Tichet-Calvet E et al (2004) *Frenkelia* parasites in a small mammal community. Parasite 11:301–310

Tysnes KR et al (2014) Subclinical *Giardia* in dogs: a veterinary conundrum relevant to human infection. Trends Parasitol 30:520–528

Uzcanga GL et al (2016) Serodiagnosis of bovine trypanosomiasis cause by non tsetse transmitted *Trypanosoma vivax*. Vet Parasitol 218:31–42
Valkunia G et al (2015) The evidence for rapid gametocyte viability changes in the course of parasitaemia in *Haemoproteus* parasites. Parasitol Res 114:2903–2909
Valkunias G (2005) Avian malaria parasites and other haemosporidia. CRC, Boca Raton
Vangeel L et al (2013) Different *Sarcocystis* spp. are present in bovine eosinophilic myositis. Vet Parasitol 197:543–548
Verheist D et al (2014) Seroprevalence of *T. gondii* in domestic sheep. Vet Parasitol 205:57–61
Viana LA et al (2015) *Caryospora bigenetica* in South America: new hosts and distribution records. Braz J Vet Parasitol 24:101–104
Vida B et al (2016) Immunologic progression of canine leishmaniosis following vertical transmission in USA dogs. Vet Immunol Immunopathol 169:34–38
Vitovec J et al (1987) Pathology and pathogenicity of *Eimeria scabra* (Henry, 1931) in experimentally infected pigs. Folia Parasitol (Praha) 34:299–304
Walzer KA et al (2014) *Hammondia hammondi* harbours functional orthologs of the host-modulation effectors GRA15 and ROP16 but is distinguished from *Toxoplasma gondii* by a unique transcriptional profile. Eukaryot Cell 13:1507–1518
Wang JM et al (2013) A sero-epidemiological survey of Chinese *Babesia motasi* for small ruminants in China. Parasitol Res 112:2387–2391
Weissenbacher-Lang C et al (2016) Establishment of a quantitative real-time PCR for the detection of *Pneumocystis carinii f. sp. suis* in bronchoalveolar lavage samples from pigs. J Vet Diagn Invest. doi:10.1177/1040638716641158
Whipps CM et al (2012) Phylogeny of fish-infecting *Calyptospora* species (Apicomplexa: Eimeriorina). Parasitol Res 111:1331–1342
Wise LN et al (2013) Review of equine piroplasmosis. J Vet Intern Med 27:1334–1346
Woo PT (ed) (2010) Fish diseases and disorders. vol 1: Protozoan and metazoan infections, 2nd edn. CAB International, Wallingford
Yabsley MJ, Shock BC (2012) Natural history of zoonotic *Babesia*: role of wildlife reservoirs. Int J Parasitol Parasit Wildl 2:18–31
Yamada M et al (2014) Intranuclear coccidiosis caused by *Cyclospora* spp. in calves. J Vet Diagn Invest 26:678–682
Yamasaki M et al (2014) In vitro and in vivo safety and efficacy studies of amphotericin B on *Babesia gibsoni*. Vet Parasitol 205:424–433
Yan W et al (2013a) Simultaneous identification of three highly pathogenic *Eimeria* species in rabbits using a multiplex PCR diagnostic assay based on ITS1-5.8S rRNA-ITS2 fragments. Vet Parasitol 193:284–288
Yan W et al (2013b) Morphological and molecular characterization of *Sarcocystis miescheriana* of pigs. Parasitol Res 112:975–980
Yang R et al (2014a) Longitudinal prevalence, oocyst shedding and molecular characterisation of *Eimeria* species in sheep across four states in Australia. Exp Parasitol 145:14–21
Yang R et al (2014b) A new *Caryospora* coccidian species from the laughing kookaburra (*Dacelo novaequineae*). Exp Parasitol 145:68–73
Yang R et al (2015) Genetic diversity of *Cryptosporidium* in fish at the 18S and actin loci and high levels of mixed infections. Vet Parasitol 214:255–263
Yao C (2013) Diagnosis of *Trichomonas foetus*. J Med Microbiol 62:1–9
Yao C, Köster LS (2015) *Trichomonas foetus* infection. Vet Res 46:35
Yao JY et al (2014) Antiparasitic activities of specific bacterial extracellular products of *Streptomyces griseus* SDX-4 against *Ichthyophthirius multifiliis*. Parasitol Res 113:3111–3117
Yasur-Landau D et al (2016) Allopurinol in *Leishmania infantum* from dogs with disease relapse. PLoS Negl Trop Dis 10:e0004341
Yiannakis EP et al (2016) Systematic review of outbreaks of *Pneumocystis jiroveci* pneumonia. J Hosp Inf 93:1–8

Yoshikawa H et al (1987) Freeze fracture studies on *Pneumocystis carinii*. Parasitol Res 73:132–138

Yoshikawa H et al (2011) Evaluation of DNA extraction kits for molecular diagnosis of human *Blastocystis* subtypes from fecal samples. Parasitol Res 109:1045–1050

Yuan L et al (2014) Effects of different inoculation routes on the parasitic sites of *Cryptosporidium baileyi* infection in chickens. Exp Parasitol 145:152–156

Zander E, Böttcher FK (1984) Diseases of bees. Ulmer, Stuttgart

Zaragatzki E et al (2010) Light and transmission electron microscopic studies on the encystation of *Histomonas meleagridis*. Parasitol Res 106:977–983

Zhang XX et al (2015) Seroprevalence and risk factors of *T. gondii* infection in stray dogs. Parasitol Res 114:4725–4729

Zhao GH et al (2013) Molecular characterization of *Cyclospora*-like organisms from golden snub-nosed monkeys in Qinling Mountain in Shaanxi province, northwestern China. PLoS One 8(2): e58216

Zierdt CH (1973) Studies on *Blastocystis hominis*. J Protozool 20:114–121

Zimre-Grabensteiner E et al (2011) Genetically different clonal isolates of *Trichomonas gallinae*. Parasitol Int 60:213–215

Worms (Helminths) 5

The last 50 years brought considerable progress in the treatment of parasites in humans and animals, although important deficits are still present in diseases due to parasites, which occur in a few of hosts, or in cases, where originally efficacious compounds have been hit by growing resistances of the parasites. Thus, especially farm animals are highly endangered by parasitosis, since the animals are living closely together in stables and/or on meadows, which increases repeated infections of large numbers of hosts. With respect to agents of disease, the following worm groups are of importance: **Trematodes** (sucking worms), **Cestodes** (tape worms), **Nematodes** (thread worms, round worms, whipworms, filariae, etc.), **Pentastomida** (tongue worms), **Acanthocephala** (thorny headed worms) and **Hirudinea** (leeches).

5.1 Platyhelminthes

This phylum comprises the following groups although details are under constant discussion.

1. **Class**: Turbellaria (free-living worms)
2. **Class**: Trematodes (sucking worms, flukes)
Subclass: Aspidobothrea
Subclass: Monogenea
Subclass: Digenea
3. **Class**: Cestodes (tapeworms)
Subclass: Cestodaria
Subclass: Eucestoda

The second and third class contain mainly parasitic species. The system is under constant change. Thus, only the most important groups are listed here.

© Springer International Publishing Switzerland 2016

H. Mehlhorn, *Animal Parasites*, DOI 10.1007/978-3-319-46403-9_5

5.1.1 Trematodes

These "flukes" or "sucking worms" belong together with the "tapeworms" to the phylum Platyhelminthes and are characterized by their dorso-ventrally flattened body. Due to this structure, the food uptake, which is especially important for the intestine-less tapeworms, is considerably facilitated. Trematodes, however, possess an often bifurcated, blind ending intestine, which has its opening in the mouth which is surrounded by the apically situated oral sucker. Characteristic for the trematodes is the fact that they are covered by a non-cellular layer called **tegument** (=neodermis), which is able to take up food and thus supports the parasitic lifestyle of these animals. The worms are mostly **hermaphrodites** (except for the important group of the schistosomes and related species), which harbour both male and female sexual organs. Details are shown below in the selected species of each group.

5.1.1.1 Aspidobothrea

Name: Greek: *aspis* = shield; *bottryon* = small pit.

This group comprises rather few species, which are characterized by a very large sucker, which is called **opisthaptor** or **Baer's disc**, which covers nearly fully the whole ventral side and is provided by tiny hooks (Fig. 5.1a). The Aspidobothrea are practically exclusive endoparasites of poikilothermic water animals but are also found as ectocommensals on mussels, snails and crustaceans. Especially the species of the genus *Aspidogaster* are important parasites in cultures of marine organisms. The development of the Aspidobothrea is direct without any change of generations. However, some species proceed their larval development in a series of hosts following each other (Table 5.1). Due to their low importance as agents of disease, no further information is given here.

5.1.1.2 Monogenea

Name: Greek: *monos* = single; *genesis* = reproduction.

Most of the monogenean species live as ectoparasites on the skin and gills of poikilothermic aquatic vertebrates such as fishes, reptiles and amphibia. Exceptions are some endoparasitic species, which live inside the bladder or oesophagus of the above-listed hosts (Table 5.1). To become anchored inside or on the host, the monogeneans have developed species specifically 1–3 anterior oral suckers, which surround the mouth being called **prohaptors**. A posterior sucker (called **opisthaptor**) has the same attaching function. Depending on the structure of this second holdfast organ, it is differentiated between monogeneans called **Monoopisthocotylea** and those described as **Polyopisthocotylea**. In the case of the first group, the sucker is large and not subdivided but contains 1–3 pairs of large, central hooks and is surrounded by 12–16 small hooklets (Figs. 5.1 and 5.2). In the case of the Polyopisthocotylea, the opisthaptor is subdivided into several smaller suckers (Fig. 5.2). However, some species may possess additional larger hooks. The existence of such hooks is recently considered as a very ancient

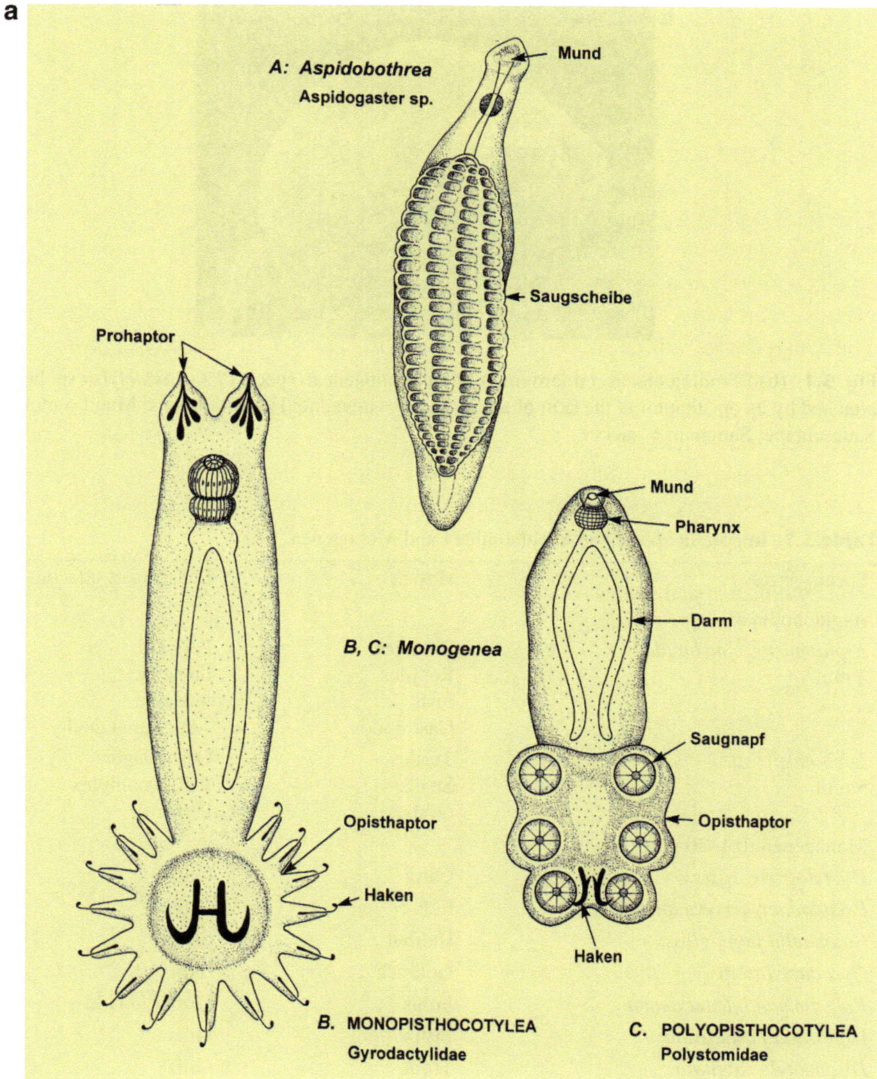

Fig. 5.1 (**a**) Diagrammatic representation of specimens of the Aspidobothrea and of the Monogenea

criterion as these of the tapeworms. Thus, he classified both Monogenea and Cestodes (tapeworms) are included in the new taxon **Cercomeromorphae**. The **ontogenesis** of the Monogenea runs always directly with a single reproduction phase on a single host. The **eggs** of many monogenean species are operculated = they possess a cover which is removed while the larva is hatching. However, the eggs are often covered by filament-like structures, which help to anchor the eggs at plants or on the soil. The larva (**oncomiracidium**) which hatches

b
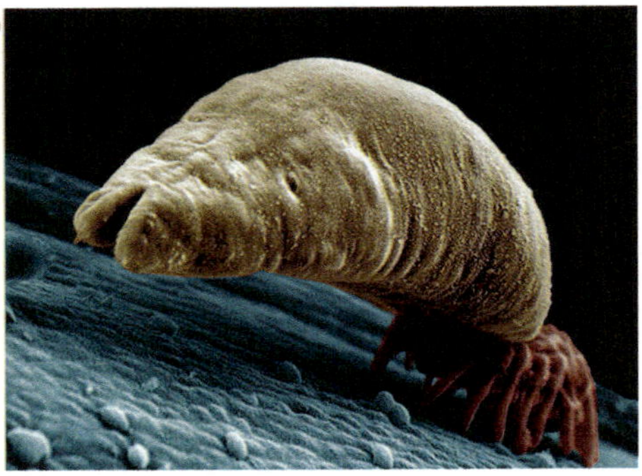

Fig. 5.1 (**b**) Scanning electron micrograph of the monogenean species *Pseudodactylus* sp. being attached by its opisthaptor at the skin of a fish. Darm = intestine; Haken = hooks; Mund = mouth; Saugscheibe; Saugnapf = sucker

Table 5.1 Important species of Aspidobothrea and Monogenea

Species/size	Hosts	Location of attachment
Aspidobothrea		
Aspidogaster conchicola 3 mm	Bivalves Reptiles Fish Gastropods	Pericard Intestine Intestine Abdominal cavity
Lophotaspis sp. 5 mm	Turtles Snails Oysters	Oesophagus Pallial complex Pericard
Monogenea (0.1–20 mm)		
Dactylogyrus vastator	Carp	Gills
Pseudodactylogyrus anguillae	Eels	Gills
Entobdella hippoglossi	Halibut	Skin
Gyrodactylus sp.	Goldfish	Skin, gills
Polystomum integerrimum	Frogs	Urinary bladder
Diplozoon paradoxum	Fish	Gills
Discocotyle sagittata	Trout	Gills

from an egg is covered by cilia and possesses a so-called **eyespot**, by which it might sense light (Fig. 5.3). This larval stage is already provided with the opisthaptor showing its species-specific structure. Within 24 h, the swimming oncomiracidium larva has to become attached at a fish in order to survive.

Monogeneans on fishes

1. **Name**: Greek: *monos* = single, unique; *diplos* = double; *zoon* = animal; *poly*s = many, multiple; *stoma* = mouth. Latin: *integer* = intact, clean; *paradoxus* = contradictory.

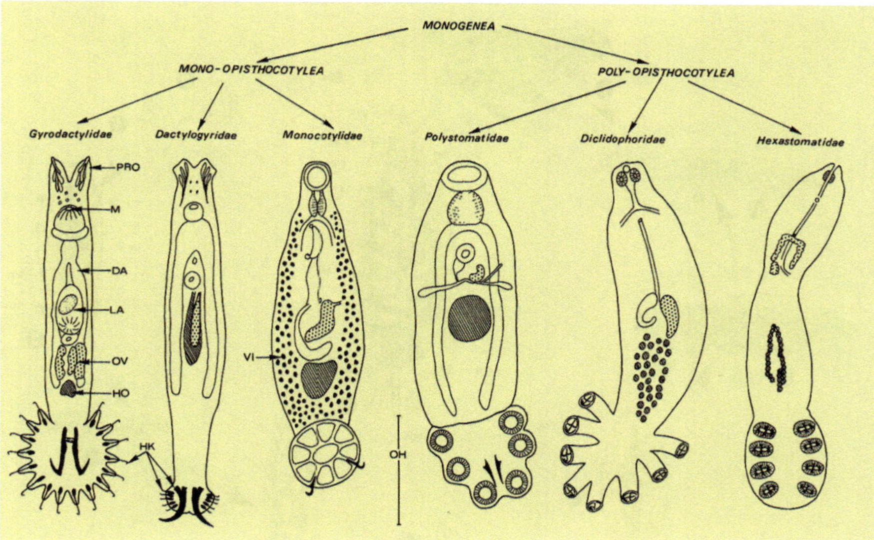

Fig. 5.2 Diagrammatic representation of some species of the Monogenea. DA = intestine; HK = hooks; HO = testes; LA = larva; M = mouth; OH = opisthaptor; OV = ovary; PRO = prohaptor; VI = vitellarium

2. **Geographic distribution/epidemiology**: Worldwide in fresh and saltwater. In cases of aquaria mass development may occur killing sensible fishes.
3. **Biology, morphology**: Practically all fish species may become infected; several species, which occur in a large variety of different shapes along the skin and on the gills (Fig. 5.2). Some examples are selected and described below:
   (a) ***Polystomum integerrimum***

      This parasite was already described in the year 1791 (!) at days, where microscopy was at its beginnings. This species, which reaches a length of 10 mm, differs from others, since it lives mainly inside the bladder of its hosts (frogs) (Fig. 5.3). Fecundity of this parasite is coordinated with that of its host. This is needed, since many frogs enter water only during the phase of egg laying. Thus, the larvae of this monogenean worm have only then the chance to enter their hosts. The operculated eggs of these worms are then excreted within the urine of infected frogs (see life cycle in Fig. 5.3). The short-cut development depicted in Fig. 5.3) is rather an exception for the worldwide distributed species *Polystomum integerrimum* but practically the rule for the American species *P. nearcticum*.

   (b) ***Diplozoon paradoxum***

      The specimens reach a length of about 4 mm and parasitize on the gills of fishes. The peculiarity of this species is based on the fact that always two specimens are fused lifelong in a cross-over status (Fig. 5.4). The

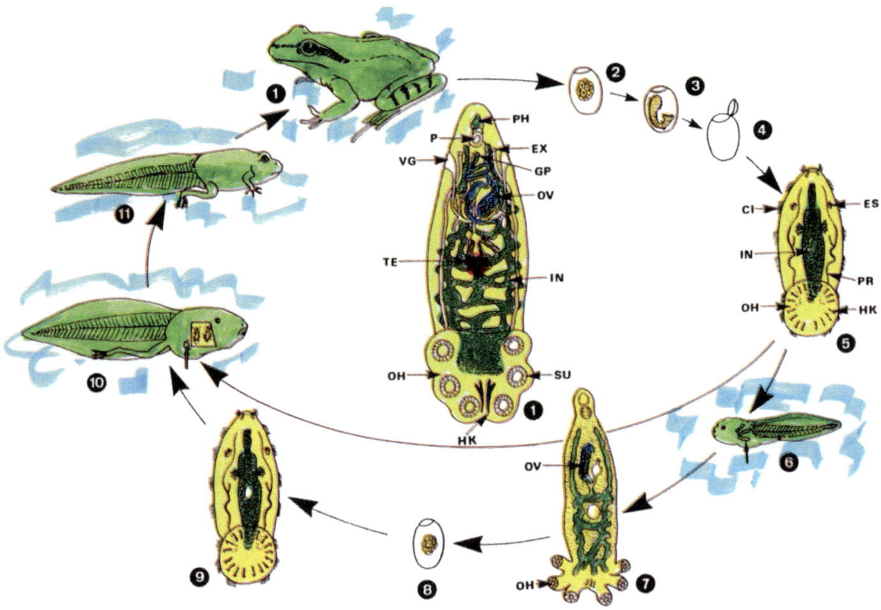

Fig. 5.3 Diagrammatic representation of the life cycle of *Polystomum integerrimum*. **1** Adult fluke inside urinary bladder/cloaca: when frogs become mature, flukes also reach maturity (apparently stimulated by the sexual hormones of the host). **2–5** Operculated eggs are set free with feces; in water each egg develops a larva with hooks (**oncomiracidium**) which leaves the egg and may initiate two different developmental cycles (6–9 or 10, 11). **6–9** When the oncomiracidia has become attached to tadpoles with outer gills (6; *arrow*), they grow into gill (or branchial) flukes (7) which are thought to represent neotenic forms. These branchial flukes produce a few eggs (8) which give rise to new oncomiracidia (9). **10, 11** When the oncomiracidium enters inner gills of tadpoles via spiracle (19; *arrow*), the development to the final bladder generation is initiated. When the tadpole undergoes metamorphosis, the worm passes out of the branchial chamber, migrates down the host's intestine and may become established in the host's bladder where it reaches sexual maturity within 3 years (in the frog). CI = cilia; ES = eye spot; EX = excretory pore; GP = genital pore (uterus + vas deferens of testis); HK = hooks; IN = intestine; OH = opisthaptor (caudal disk); OV = ovary; P = pharynx; PH = prohaptor (mouth); PR = protonephridia; SU = sucker; TE = multilobed testis with sperms; VG = vagina

fusion persists lifelong for about 5 years and always in springtime a new reproduction phase is initiated, when a so-called **diporpa larva** hatches from the eggs, which had been excreted 10 days before being provided with long filaments used for attachment at plants or at the surface of animals (Fig. 5.4).

4. **Symptoms of disease**: Monogeneans may kill their hosts in overcrowded ponds, since bacteria may enter the wounds at the attachment sites. Death occurs often during mixed infections with nematodes, which weaken the fishes additionally.

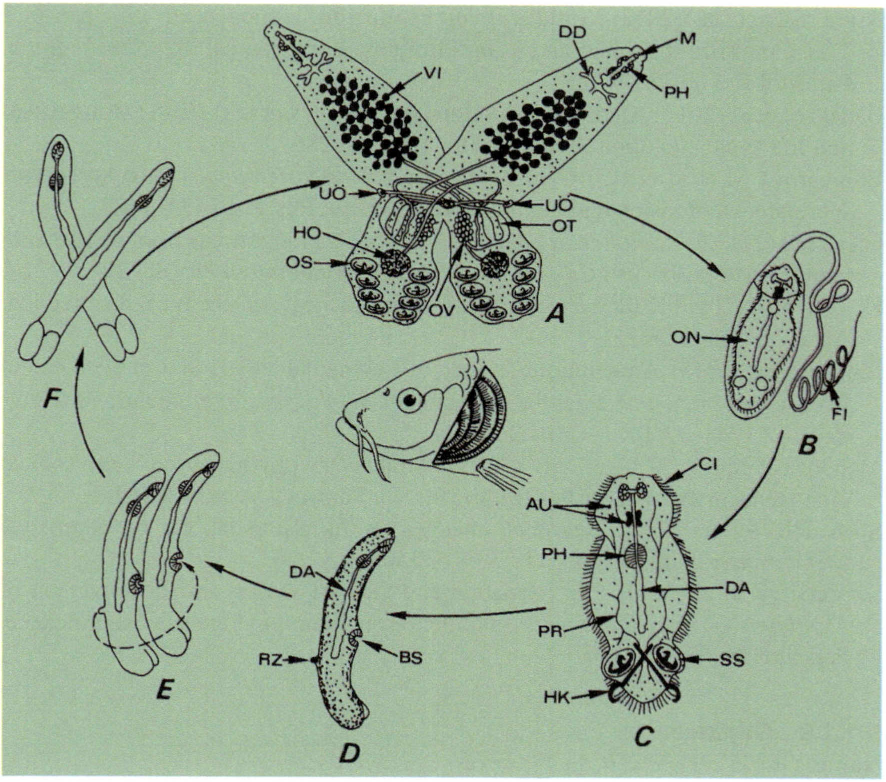

Fig. 5.4 Diagrammatic representation of the life cycle of *Diplozoon paradoxum* on the gills of cyprinid fish. **1** Adults on the gills of fish: **2** Egg with an oncomiracidium larva. **3** Free oncomiracidium. **4** After attachment to the gills of a host, the oncomiracidium is transformed into the **diporpa larva. 5**, **6** Fusion of two diporpas on the host; each diporpa attaches its sucker (VS) at the dorsal papilla (DP) of the other. This process stimulates their maturation and cross-fertilization. The bloodsucking adults can live for years in this form of permanent copulation. CI = cilia; DP = dorsal papilla; EY = eyes; FI = filament; HK = hook; IB = intestinal branch; IN = intestine; M = mouth; OH = opisthaptors with suckers; ON = oncomiracidium; OT = ootype; OV = ovary; PH = pharynx; PR = protonephridium; SU = sucker (clamps); TE = testis; UO = uterus opening; VI = vitellarium (vitelline gland); VS = ventral sucker

5. **Therapy**: In the case of ornamental fishes: addition of the **praziquantel** containing compound **Tremazol**® produced by Fa. Alpha-Biocare GmbH (Düsseldorf) marketed internationally by Fa. Sera, Heinsberg, Germany, to food.

Further Reading

Alves PV et al (2015) A redescription of *Lobatostoma kemostoma* from the Florida pompano fish *Trachinatus carolinus*. J Helminthol 89:335–344.

Arafa SZ et al (2012) Surface features of the monogenean gill parasites *Pseudodactylus anguillae* and *P. bini* from the European eel *Anguilla Anguilla*. Parasitol Res 110:127–133.

Badets M et al (2013) Alternative development in *Polystoma gallieni* (Monogenea) and life cycle evolution. Exp Parasitol 135:283–286.

Bombarova M et al (2015) Cytogenetics of *Aspidogaster limacoides*: karyotype, spermatocyte division and genome size. Parasitol Res 114:1473–1483.

Giese EG et al (2015) *Rohdella amazonica* n. sp. (Aspidogastrea) from the Amazonian banded puffer fish *Colomesus psittacus*. J Helminthol 89:288–293.

Koski P et al (2016) Killing *Cyrodactylus salaris* by heat and chemical disinfection. Acta Vet Scand 58:21.

Ling F et al (2015) Anthelmintic efficacy of cinnamaldehyde and cinnamic acid from cortex cinnamon essential oil against *Dactylogyrus intermedius*. Parasitology 142:1744–1750.

Mendoza-Palmero CA et al (2015) Molecular phylogeny of neotropical monogeneans from cat fishes (Siluriformes). Parasites Vectors 8:164.

Shah HB et al (2013) Seasonal changes in infra-populations of *Diplozoon kashmirensis*. Parasitol Res 112:3347–3356.

Swiderski Z et al (2011) Egg formation and the early embryonic development of *Aspidogaster limacoides* with comments on their phylogenetic significance. Parasitol Int 60:371–380.

5.1.1.3 Digenea

Name: Greek: *digenesis* = double sexes.

This group of the trematodes contains species, which run through several generations during their life cycle and use obligatorily at least two different hosts. The **final hosts** harbour the adult worms, which in most cases are **hermaphrodites** (i.e. both sexes occur inside an individual). In the **intermediate hosts,** reproduction occurs asexually—if at all, since there exist some species that use the intermediate host just as "**transport host**", wherein larval stages are stored until the final host ingests this so-called **paratenic host**. Some important species of the Digenea are listed in Table 5.2. Figs. 5.5 and 5.7 show the basic organization of the Digenea, while in Fig. 5.6 the different developmental stages within the developmental cycles of important species are compared.

The group name **Digenea** (=two generations) was chosen, since their life includes both a change of hosts and an alternation of different generations (Fig. 5.6). According to morphological criteria, the following groups are differentiated:

1. **Gasterostomes**: the intestine is not branched but appears sack-like and does not open at the apical pole but more ventrally.
2. **Monostomes**: these worms possess only one sucker; mostly the ventral one is reduced.

Table 5.2 Important Digenea (examples)

Family/species	Final hosts/tissues	Worm length (mm)	Egg size (μm)	First intermediate host (snails)[a]	Second intermediate host[b]	Prepatency in final host (weeks)
Family Diplostomatidae						
Alaria canis	Dog, fox; Small intestine	3–4	70 × 130	*Helisoma* sp.	Tadpoles[c]	5
Family Schistosomatidae						
Schistosoma mansoni	**Humans**, lab animals; liver and intestinal veins	6–10♂ 7–14♀	50 × 150	*Biomphalaria* sp.	–	5–7
S. haematobium	**Humans**, monkeys; veins of the urogenital system	10–15♂ 20 ♀	50 × 150	*Bulinus* sp., *Physopsis* sp.	–	10–12
S. japonicum	**Humans**, domestic animals; liver and intestinal veins	12–20♂ 28♀	55 × 90	*Oncomelania* sp.	–	3–10
S. intercalatum	**Humans**, rodents, ungulates; intestinal veins	♂♀	60 × 160	*Bulinus* sp.	–	5–7
Family Echinostomatidae						
Echinostoma ilocanum, E. lindoensis	**Humans**, dogs; small intestine	2.5–6.5	65 × 95	*Gyraulus* sp., *Hippeutis* sp.	Snails (*Pila*), bivalves (*Corbicula*)	3
E. revolutum	Birds, mammals; rectum, caecum	10–22	65 × 110	*Helisoma* sp.	Tadpoles, snails (*Physa* sp.)	3
Family Fasciolidae						
Fasciola hepatica	Ruminants, **humans**; bile ducts	20–30	70 × 140	*Lymnaea* sp.	On water plants	8–13
F. gigantica	Ruminants; bile ducts	25–75	90 × 140	*Lymnaea* sp.	On water plants	9–13
Fasciolopsis buski	**Humans**, pigs; small intestine	30–75	80 × 135	*Segmentina* sp., *Hippeutis sp.*, *Planorbis* sp.	On water plants and their fruits	9–13

(continued)

Table 5.2 (continued)

Family/species	Final hosts/tissues	Worm length (mm)	Egg size (µm)	First intermediate host (snails)[a]	Second intermediate host[b]	Prepatency in final host (weeks)
Family Paramphistomatidae						
Paramphistomum microbothrium	Ruminants; rumen	3–12	70 × 160	*Bulinus* sp., *Stagnicola* sp.	On water plants	13–15
P. cervi	Ruminants; rumen	5–12	85 × 140	*Bulinus* sp., *Planorbis* sp.	On water plants	9–16
Watsonius watsoni	**Humans**; small intestine	8–10	75 × 125	*Bulinus sp.*, *Stagnicola* sp.	On water plants	10
Family Dicrocoeliidae						
Dicrocoelium dendriticum	Sheep, cattle, dogs, **humans**; bile ducts	6–10	25 × 40	*Helicella* sp., *Zebrina* sp.	Ants	6–10
Family Prosthogonimidae						
Prosthogonimus pellucidus	Chicken, ducks, geese; cloaca, oviduct	8–12	15 × 25	*Bithynia* sp.	Dragonflies: adults + larva	1–3
Family Troglotrematidae						
Paragonimus westermani	**Humans**, predators; lung many animals	7–12	60 × 90	*Semisulcospira* sp., *Brolia* sp., *Hua* sp., *Thiara* sp.	Crabs, *Eriocheir* sp.; pigs = paratenic hosts	8–12
P. kellicotti	**Humans**, predators; lung	9–16	55 × 85	*Pomatiopsis* sp.	Crabs	22–24
Nanophyetus salmincola	Dogs, foxes; small intestine	1–2.5	45 × 80	*Oxytrema* sp.	Fishes	1–15
Family Opisthorchiidae						
Opisthorchis (=Clonorchis) sinensis	**Humans**, fish feeding mammals; bile ducts	10–25	15 × 30	*Parafossarulus* sp. *Semisulcospira* sp. *Bulinus* sp.	Fishes (cyprinids, salmonids)	2–2.5
O. felineus (=O. tenuicollis)	Fish feeding mammals; bile ducts	7–12	11 × 30	*Bithynia* sp.	Fishes (cyprinids, salmonids)	2–3
Metorchis conjunctus	Dogs, cats, **humans**; gall bladder	1–6.5	15 × 25	*Amnicola sp.*	Fishes (*Catostomus*)	4–5

Family Heterophyidae						
Metagonimus yokogawai	Fish feeding mammals, **humans**; small intestine	1–2	16 × 28	*Semisulcospira sp.*	Frehswater fish	1–2
Heterophyes heterophyes	Fish feeding mammals, **humans**; small intestine	1–2	14 × 24	*Pirenella sp. Cerithidia sp.*	Brackish water fishes	1–2

[a]As first intermediate host numerous other gastropod species in the same biotope can possibly be infested

[b]Neither in true second intermediate hosts nor in cysts on water plants proliferation occurs, but differentiation of organs of metacercariae occurs

[c]So-called mesocercariae are parasitizing here. Third intermediate hosts are snakes and rats (containing metacercariae), which are ingested by the final host

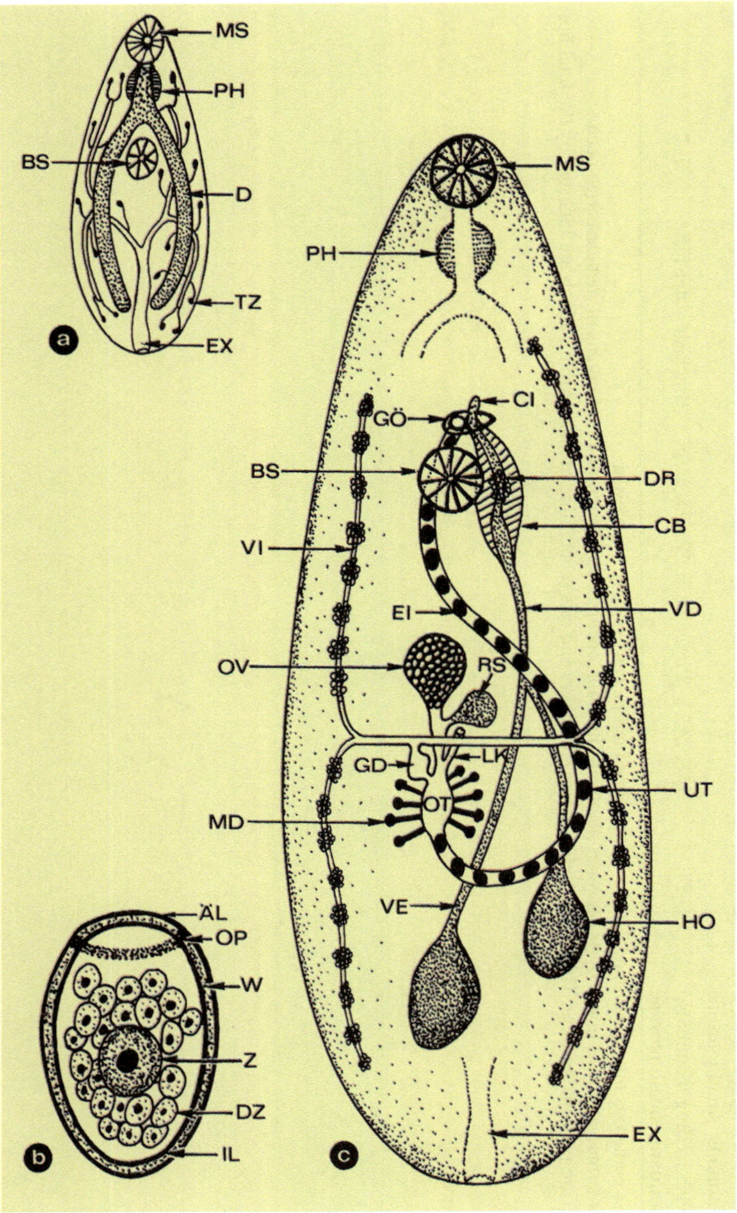

Fig. 5.5 Diagrammatic representation of a digenean trematode showing its organization. (**a**) Intestine and excretion systems. (**b**) Operculated egg (=it possesses a cover). (**c**) Detailed representation of the sexual organs. AL = outer layer of lipoproteins; BS = ventral sucker; CB = cirrus sack; CI = cirrus; D = intestine; DR = "prostate" gland; DZ = yolk cell; EI = egg; EX = excretion channel; GD = paired yolk channel; GÖ = genital opening;: IL = inner lipoprotein layer; HO = testes; LK = Laurer's channel; MD = Mehlis' gland; MS = oral sucker; OP = opening site of the egg; OT = ootype; OV = ovarium; PH = pharynx; RS = receptaculum geminis; TZ = terminal cell; UT = uterus; VD = vas deferens; VE = vas efferens; VI = vitellarium; W = sclerotized wall; Z = zygote

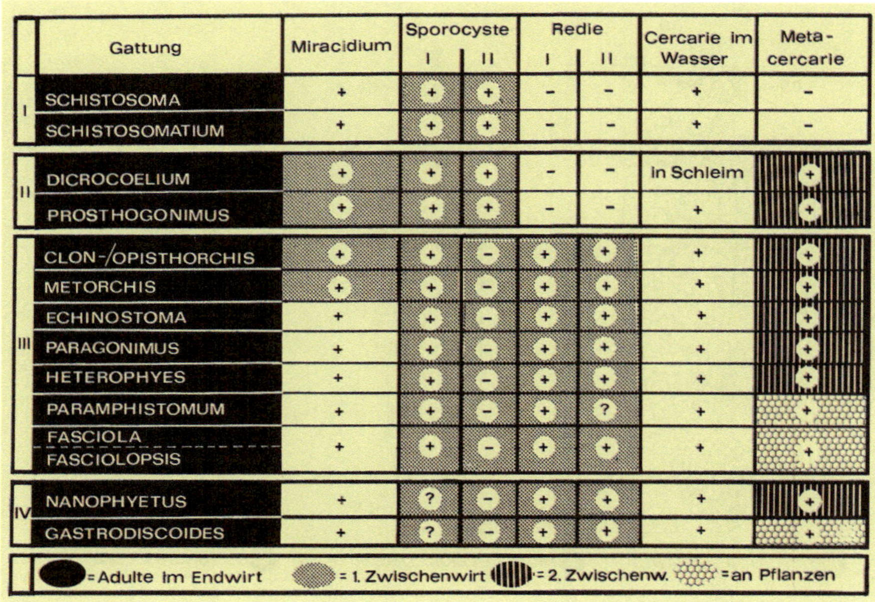

Fig. 5.6 Survey on the presence or absence (±) of specific developmental stages occurring in the life cycles of digenean trematodes. **Translation**: *Gattung* = genus; *Cercarie im Wasser* = cercariae in water; *Adulte im Endwirt* = adults in final host; *Zwischenwirt* = intermediate host; *an Pflanzen* = attached at plans

3. **Distomes**: two suckers are present; the ventral one is species specifically situated between the anterior and posterior pole.
4. **Amphistomes**: the ventral sucker occurs exclusively at the posterior pole.
5. **Echinostomes**: the anterior = oral sucker is surrounded by rows of hooklets.
6. **Holostomes**: these species possess besides oral and ventral suckers an additional one = the so-called **tribocytic holdfast** system.

   The specimens of these six groups are all hermaphrodites while the seventh group develops males and females:
7. **Schistosomes**: The flattened leaf-like male forms a longitudinal groove (so-called **canalis gynaecophorous**, wherein it bears lifelong (!) the female, which has a circular diameter.

Many species of these trematodes are important parasites infecting humans and their domestic animals and may introduce severe diseases. Some of them are listed in the following (shortened) survey:

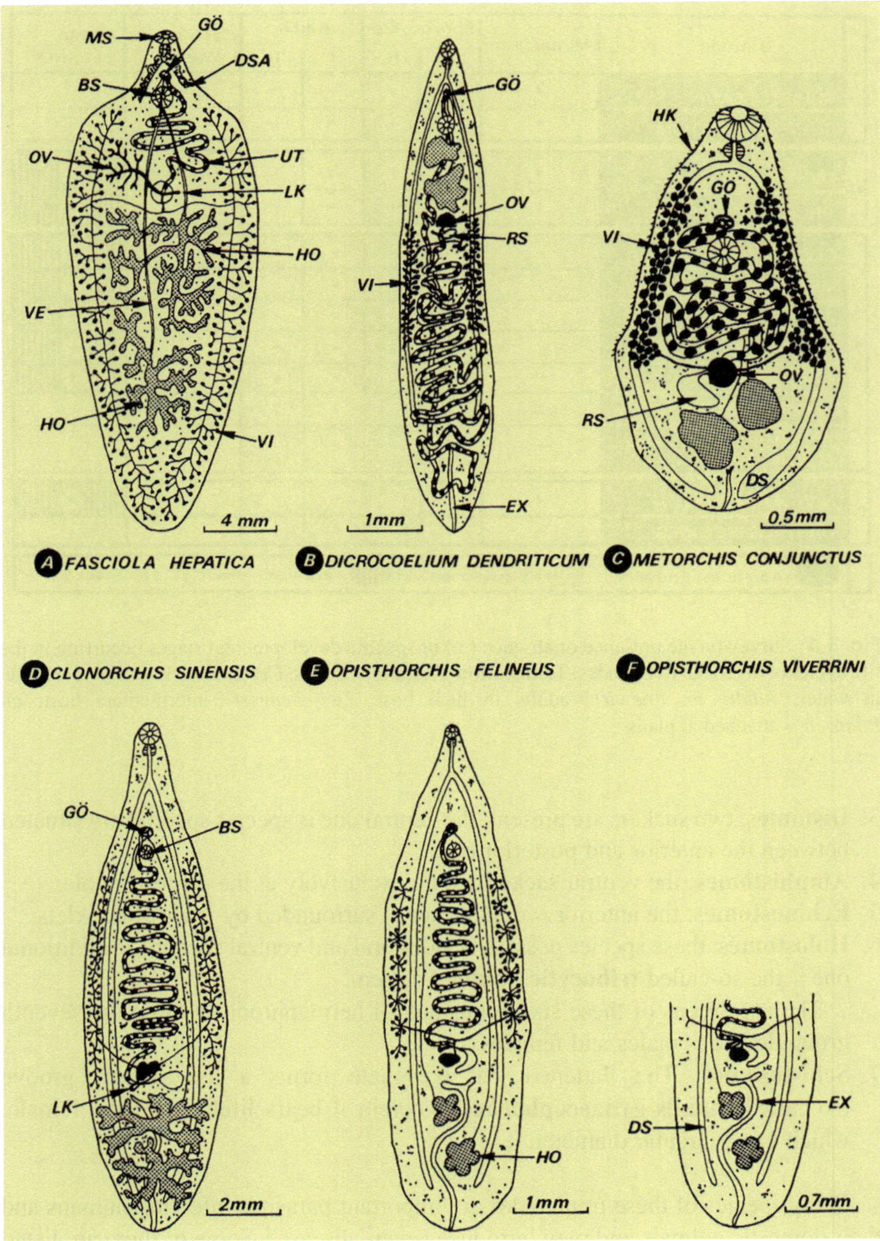

Fig. 5.7 Diagrammatic representation of the morphology of some important digenean trematodes. BS = ventral sucker; DSA = intestinal branches cut off in drawing; DS = intestinal branches; EX = excretion channel; GÖ = genital opening; HK = tegumental hooks; HO = testes; LK = Laurer's channel; MS = oral sucker; OV = ovary; RS = receptaculum seminis; VE = vas deferens; VI = vitellarium

System

Subclass: **Digenea**

Superorder: Anepitheliocystidia (the embryonic excretion bladder is retained)

  Order: Strigeata (cercariae with furcated tail, miracidia possess 2 pairs of protonephridia)

    Family: Strigeidae
      Diplostomatidae
      Schistosomatidae
      Spirochidae
      Cyclocoeliidae
      Bucephalidae u.a.

  Order: Echinostomata (cercariae with straight tail; miracidia with one pair of protonephridia),

    Family: Fasciolidae
      Gastrodiscidae
      Paramphistomatidae
      and others

Superorder: Epitheliocystidia (the excretion bladder is newly formed, the cercarial tail straight)

  Order: Plagiochiata (some species possess an oral stylet)

    Family: Dicrocoeliidae,
      Plagiorchiidae
      Prosthogonimidae
      Troglotrematidae u.a.

  Order: Opisthorchiata (cercariae have always excretion channels; no oral stylet)

    Family: Opisthorchiidae,
      Heterophyidae u.a.

A. *Fasciola hepatica* (Large liver fluke)

1. **Name**: Greek: *hepar* = liver. Latin: *fasciola* = small tape.
2. **Geographic distribution/epidemiology**: Worldwide.
3. **Biology, morphology**: The brownish-yellow appearing adult flukes of *Fasciola hepatica* reach a length of 3–5 cm, are hermaphroditic and live inside the bile ducts of their hosts (Figs. 5.8–5.11, 5.14 h). After copulation, these digenetic flukes produce several thousands of typical, operculated eggs, which are daily excreted within the feces or which may be stored inside the gallbladder for several weeks (Figs. 5.8, 5.9 and 5.10). These ovoid, thin-walled eggs measure 130–145 × 70–90 μm, appear golden yellow and contain—when excreted by the host—only the fertilized egg cell and a large number of yolk cells (Fig. 5.11). Some strains of *F. hepatica* are claimed to produce eggs reaching a length of up to 180 μm. This large fluke is common in many domestic and wild ruminants but may infect also other herbivorous mammals, e.g. rats and humans. The anus-less intestine of this flukes has two strands, which are laterally enormously branched (Figs. 5.9 and 5.10), thus forming so-called **diverticula**. The surface tegument of these worms is covered with numerous scales, which help the worm to remain fixed inside the bile ducts

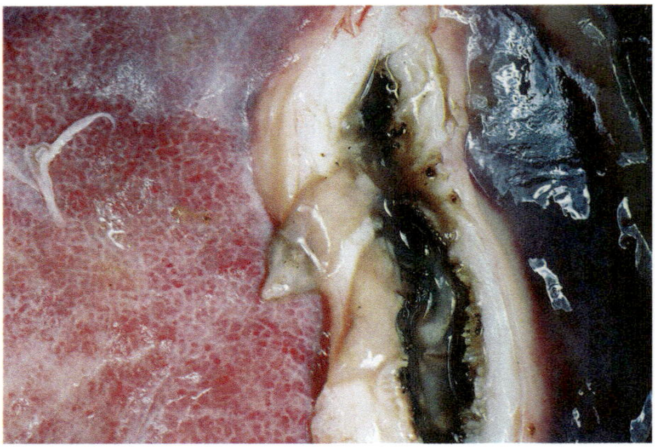

Fig. 5.8 Section of an opened bilary duct with extruding adult *Fasciola hepatica* flukes

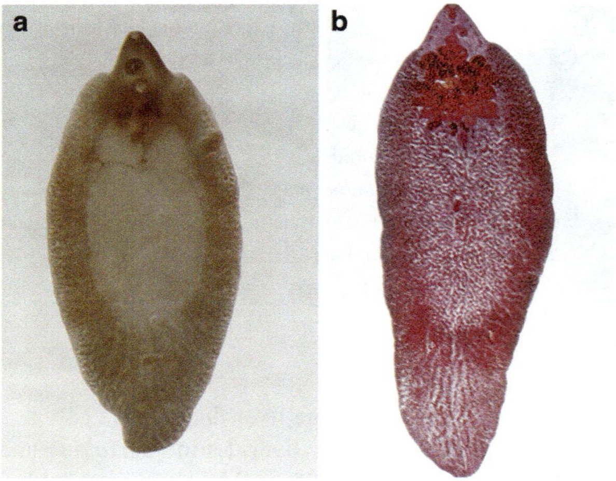

Fig. 5.9 Light micrographs of (**a**) an unstained *Fasciola hepatica* fluke and (**b**) another carmine red-stained specimen

(Figs. 5.8 and 5.10b). Outside of the host, the so-called **miracidium** larva is developed within 2–4 weeks at European temperatures. This miracidium larva reaches a length of 130 μm and is provided with a large number of rows of cilia, which help to move inside water after hatching as soon as the operculum has been blasted off. With the help of a light receptor (ocellum), the miracidium may find a suitable first intermediate host (e.g. snails of the genus *Galba* (syn. *Lymnaea*). The host must be entered within 20–30 h after hatching, since the stored glycogen reserves are exhausted. The penetration into the skin of the snails occurs with the help of proteolytic enzymes. Inside the snail, the miracidium releases the cilia and becomes a long stretched tube, which reaches a length of 0.5 mm and is now called **sporocyst**. Further sporocysts are formed from germ cells inside the sporocysts of the first generation. At least, two generations of **rediae** are developed which reach a size of 1.5–2 mm and

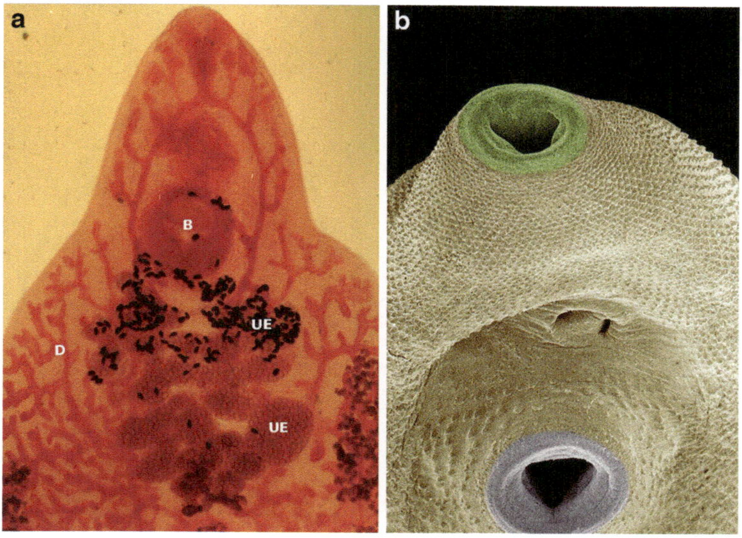

Fig. 5.10 *Fasciola hepatica* (**a**) The carmine red-stained anterior end shows the typical lateral intestinal diverticula which reach until the anterior tip of the worm. (**b**) Scanning electron micrograph of the anterior end showing the large anterior and ventral suckers as well as the two openings of the male and genital system. The surface of this fluke species is covered by numerous scales fortifying the whole body. B = ventral suckers; D = intestinal diverticula; UE = uterus containing black stained eggs

possess already an oral sucker, a pharynx and a non-branched, cylindrical intestine (often several generations of rediae may be produced inside the snail). Finally, ovoid **cercariae** are formed inside the rediae measuring 400 × 200 μm and being provided with a 1 mm long tail which allows to swim in the water when the cercariae are released from the snail, which occurs about 5–8 weeks after the snail had been infected. These cercariae swim in the freshwater biotope and become attached within a few minutes at plants and throw off their tail. Very quickly, a cyst wall (consisting of mucoproteins) is formed around this spherical stage, which is now called **metacercaria** and has diameters of 200–300 μm.

If a final host ingests such metacercariae, they hatch inside its intestine and penetrate within 24 h into the intestinal wall and reach within 24 h the peritoneal cavity, from where many of them enter the liver within 4–7 days. Then it takes another week until the larvae have reached their final destination: the bile ducts. There it takes at least two further months until the young worms reach maturity and start to produce eggs (after copulation) during their remnant lifespan of 8–12 months.

Related species such as the so-called giant liver fluke *Fasciola gigantica*, which reaches a size of 2.5–7.5 cm × 1 cm and is found in Africa, Asia and Near East, or the large American liver fluke *Fascioloides magna* (6–9 cm × 2.5 cm) in North America have similar life cycles—as well as the

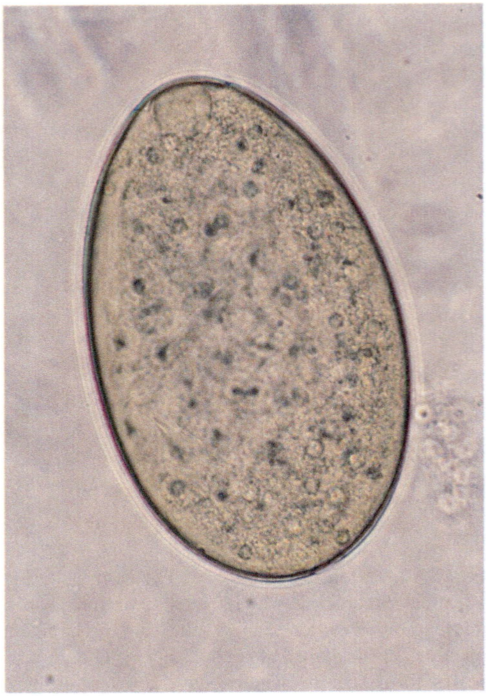

Fig. 5.11 Light micrograph of an unstained and not yet embryonated egg of *Fasciola hepatica*

small elk fluke *Parafasciolopsis fasciolaemorpha* which reaches a size of 3–7 mm × 2 mm.

4. **Symptoms of disease (Fascioliasis)**: The fascioliasis shows several phases
   1. Wandering of the juvenile stages from the intestine into the peritoneal cavity (starts about 7 days p.i. and leads to lesions and peritonitis).
   2. Migration of juveniles inside the liver parenchyma for up to 6–7 weeks leading to destruction of tissues and inflammations.
   3. Settling inside the bile ducts starts from the 6–7th week after infection and leads to destruction of the bile ducts, fibrosis, cirrhosis, reduced food uptake and loss of weight.

   Depending on the amount of liver flukes inside a host, the symptoms of disease may range from slight to severe and even death may occur.

   General symptoms: weakness, oedemas, inappetence, cachexia, loss of weight, anaemia, eosinophilia and loss of hair.
5. **Diagnosis**: Demonstration of the large, unembryonated eggs in fresh feces with the help of the sedimentation method (Figs. 5.11 and 5.14e, h).
6. **Pathway of infection**: Oral by uptake of metacercariae attached at water plants inside or close to water ponds on meadows.
7. **Prophylaxis**: Repeated chemotherapeutical treatment of cattle or other ruminants in order to reduce infection pressure on meadows. Offering drinking

water in special vessels and encircling water ponds on meadows by fences. Decrease of snail population in ponds with the help of molluscicides.

8. **Incubation period**: 2–6 weeks (migrating phase may lead to liver damages and related symptoms). In cases of ingestion of many metacercariae, symptoms may start earlier (already after 5 days).
9. **Prepatent period**: 2–3 months.
10. **Patency**: At least 2 years, however, in general repeated infections will occur on meadows with many infectious stages.
11. **Therapy**: Drug of choice is **triclabendazole** (sheep: 10 mg/kg bodyweight; cattle: 12 mg/kg bodyweight, orally), since both juvenile stages and adult worms are killed. Note: Waiting times are obligatory before slaughtering.

Further reading

Ashrafi K et al (2014) Fascioliasis: a worldwide parasitic disease of importance in travel medicine. Travel Med Inf Dis 12:636–649.

Hanna REB et al (2016) *Fasciola hepatica*: a light and electron microscope study of the ovary and of the development of oocytes within eggs in the uterus provides an insight into reproductive strategy. Vet Parasitol 221:93–103.

Hanna REB (2015) *Fasciola hepatica*: Histology of the reproductive organs and differential effects of triclabendazole on drug-sensitive and drug-resistant fluke isolates. Pathogens 4:431–456.

Kelley JM et al (2016) Current thread of triclabendazole resistance in *Fasciola hepatica*. Trends Parasitol. doi:10.1016/j.pt.2016.03.002.

Kravlova-Hromadova I et al (2011) Multiple origins of European populations of the giant liver fluke *Fascioloides magna*. Int J Parasitol 41:373–383.

Toet H et al (2014) Liver fluke vaccines in ruminants. Int J Parasitol 44:915–927.

B. *Dicrocoelium dendriticum* (syn. *D. lanceolatum*) (Small liver fluke, lancet fluke)

1. **Name**: Greek: *dicroos* = double; *koilia* = body cavity, hollow. Latin: *dendriticus* = furcated.
2. **Geographic distribution/epidemiology**: Worldwide, especially in Europe, America, Asia, North Africa.
3. **Biology, morphology**: The hermaphroditic small liver fluke has a size of 8–12 mm × 1.5–2.5 mm and is found in the bile ducts and gallbladder of house and wild ruminants, horses, camels, rabbits, hares and occasionally also in humans (Figs. 5.12 and 5.13). The intestine is furcated without lateral protrusions. The pair of testes is subdivided into lobes and is situated behind the ventral sucker. First intermediate hosts are terrestrial snails of the genera *Zebrina, Helicella*, etc., which feed the feces of ruminants containing the eggs of this fluke. The larva (miracidium) hatches from the egg inside the snail and with the help of an apical pike they pass the intestinal wall and reach the body cavity, where they grow up into sack-like mother sporocysts, which

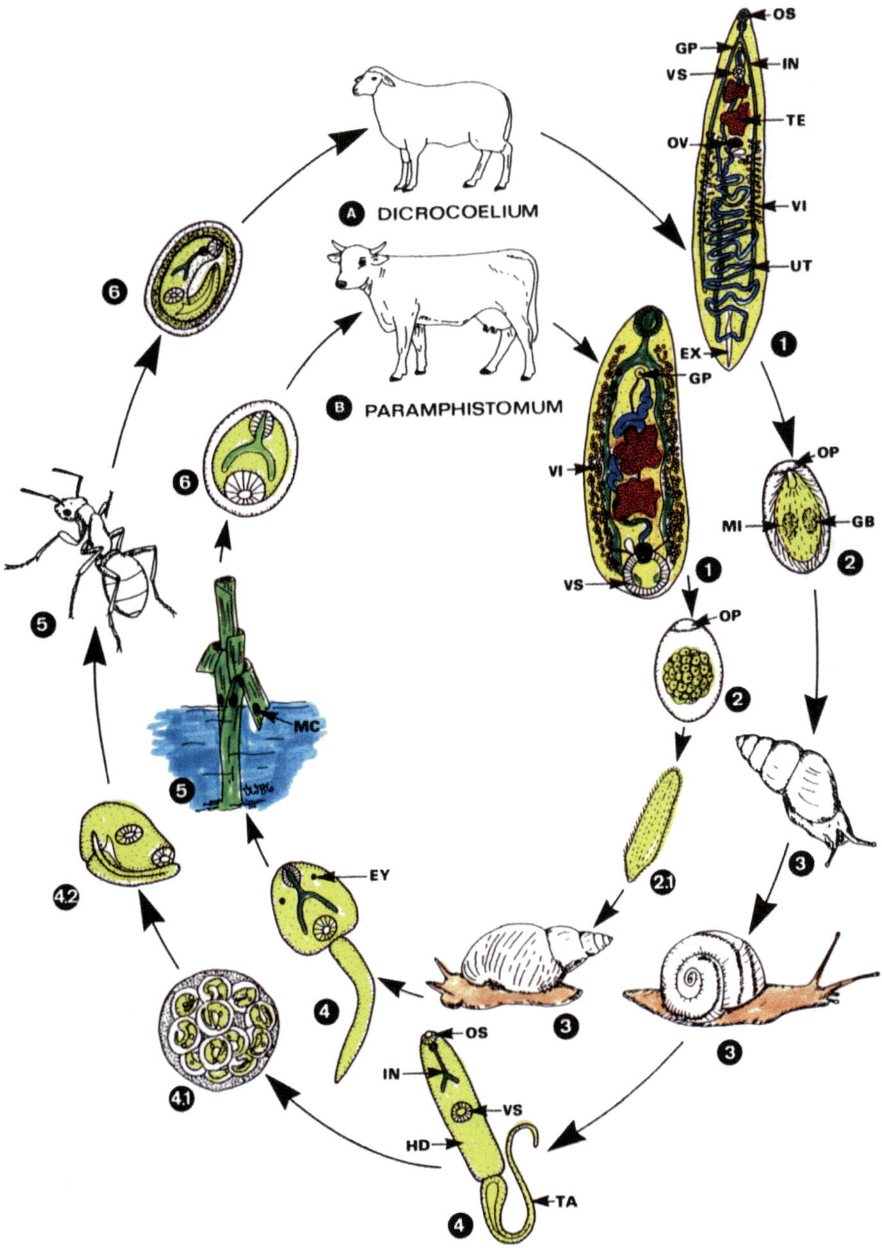

Fig. 5.12 Diagrammatic representation of the life cycles of the flukes *Dicrocoelium dendriticum* (**A**) and *Paramphistomum cervi* (**B**) in sheep and cattle (final hosts: Digenea/Table 5.3). **1** Adult worms in the bile ducts (**a**) or rumen (**b**). **2** Eggs are excreted in feces fully embryonated in (**a**) but not in (**b**). **2.1** In *P. cervi* the finally formed miracidium hatches from the egg and enters a water snail, whereas in *D. dendriticum* land-living snails swallow the eggs containing the miracidium. **3–4** Intermediate hosts for *P. cervi* are water snails of the genera *Bulinus, Planorbis, Stagnicola* and

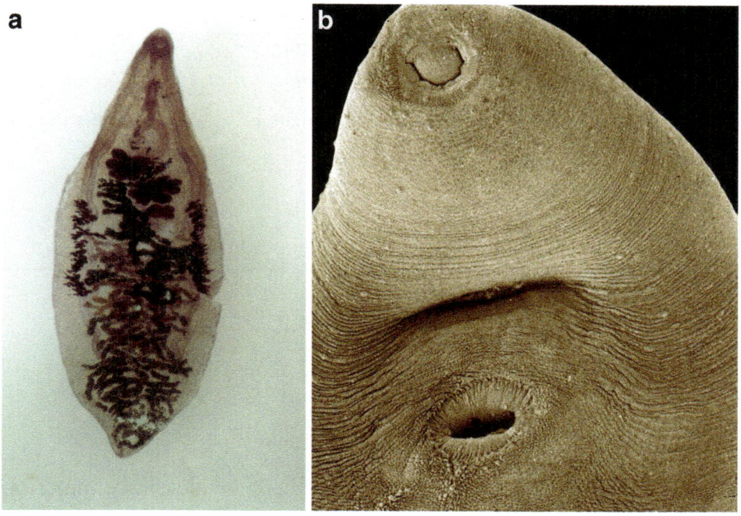

Fig. 5.13 *Dicrocoelium dendriticum* (syn. *D. lanceolatum*). (**a**) Light micrograph of the stained whole body. (**b**) Scanning electron micrograph showing the oral and ventral sucker, while the openings of the female and sexual organs are hidden there by a lateral folding. Note that the surface (=tegument) of this small liver fluke does not show scales like *Fasciola hepatica* (Fig. 5.10)

give rise (from germinal cells) to daughter sporocysts, which finally release via a "birth opening" about 10–40 tailed cercariae (Fig. 5.12). The development until the formation of the cercariae is rather slow and may take 3–5 months. The cercariae leave the daughter sporocysts and wander into the breathing

Fig. 5.12 (continued) Anisus, whereas in *D. dendriticum* terrestrial snails of the genera *Zebrina* or *Helicella* are involved. Development in the snails proceeds via two generations of sporocysts in the snail (**3**) or are excreted by the snails within slime balls (**4.1**). **5–6** In *D. dendriticum* ants become second intermediate hosts when feeding slime balls. Most of the cercariae encyst in the haemocoel (**6**) as metacercariae and can then infect the final host, where 1 or 2 cercariae enter the subesophageal ganglion, encyst there and cause an alteration of the ant's behaviour. When the temperature drops in the evening hours, the infected ants climb to the tips of grass (or other plants) and grasp them firmly with their mandibles, while uninfected ants return to their nests. The infected ants remain attached until the next morning, when they warm up and resume normal behaviour. These attached ants may be swallowed by plant-eating mammals. In *P. cervi* the free-swimming cercariae (with two eye spots) encyst on herbage and other objects (**6**), thus becoming metacercariae. Upon being swallowed along with forage, excystment of the cercariae of both species occurs in the duodenum. From there they enter the bile duct (*D. dendriticum*) or return (via the intestinal wall) into the abomasum (*P. cervi*). And from there they pass to the rumen, where they attach among the villi. EY = eye spot; EX = excretory bladder; GB = germ balls; GP = genital pore; HD = head; IN = intestine; MC = metacercaria; MI = miracidium; OP = operculum; OS = oral sucker; OV = ovary; TA = tail; TE = testis; UT = uterus; VI = vitellarium; VS = ventral sucker

Table 5.3 Anthelmintics against infestation with *Fasciola* for ruminants

Chemical short term	Recommended dose (mg/kg body weight)		Effect on *Fasciola* stages		Waiting time (days)	
	Cattle	Sheep	Immature	Adult	Edible tissues	Milk
Albendazole	7.5 orally	–	–	3	21–28	5
Netobimin	7.5 orally	–	–	3	10–20	5
Closantel	10.0 orally	–	–	1	28	b
Triclabendazole	12.0 orally	10.0 orally	1	1	14	b

Legend: **1** Efficacy 95–100 %
2 Efficacy ca. 80 %
3 Efficacy ca. 60 %
[a]Not in the first months of pregnancy
[b]Not for dams during lactation time

cavity of the snail and become enclosed there inside slimy material, finally appearing as "slime balls" (of 10 mm in diameter), which are set free and glue at plants. These slime balls may each include 3000–5000 cercariae, so that ants are intensively infected, if they ingest such slime balls. Inside the ants (second intermediate host), one or two of the ingested cercariae enter the ganglion of the ant, while the rest pass into the abdominal cavity, where they excrete a 20 μm thick wall and thus become **metacercariae**. These metacercariae measure finally about 400 × 250 μm. The "**brainworm**" (=the metacercaria in the suboesophageal ganglion) changes the behaviour of the ants. In the evening hours, when temperatures go down, the infected ants do not enter their stock but remain (fixed by their mandibles) at plants and thus may be ingested next morning by feeding ruminants.

Related species are *Dicrocoelium hospes* (the African lancet liver fluke), which reaches a size of 5–7 mm × 1 mm and uses ants of the genera *Dorylus* and *Camponotus* as second intermediate host. *Dicrocoelium chinensis* occurs in East Asia. Related species are *Eurytrema* species, which occur in the pancreas of cattle (*E. pancreaticum*) in Asia and South America, reaching a size of 10–16 mm × 5 mm and *E. procyonis*, which lives in the pancreas of racoons in the USA. These worms, however, are tiny and measure only 2 × 1 mm. *E. coelomaticum* may induce pancreatitis, fibrosis and other clinical symptoms in sheep and cattle.

4. **Symptoms of disease (Dicrocoeliosis)**: Low-grade infections remain mostly symptomless. However, retardation of development may occur due to reduced liver activity. Heavy infections may lead to **liver cirrhosis** after phases of cholangitis and formation of granulomas and abscesses. In sheep also cases of anaemia had been described. Liver can mostly no longer be used for human consumption.
5. **Diagnosis**: Demonstration of the tiny (40 × 25 μm sized), operculated eggs (Fig. 5.14d, g) with the help of concentration methods (M.I.F.C., S.A.F.C.).

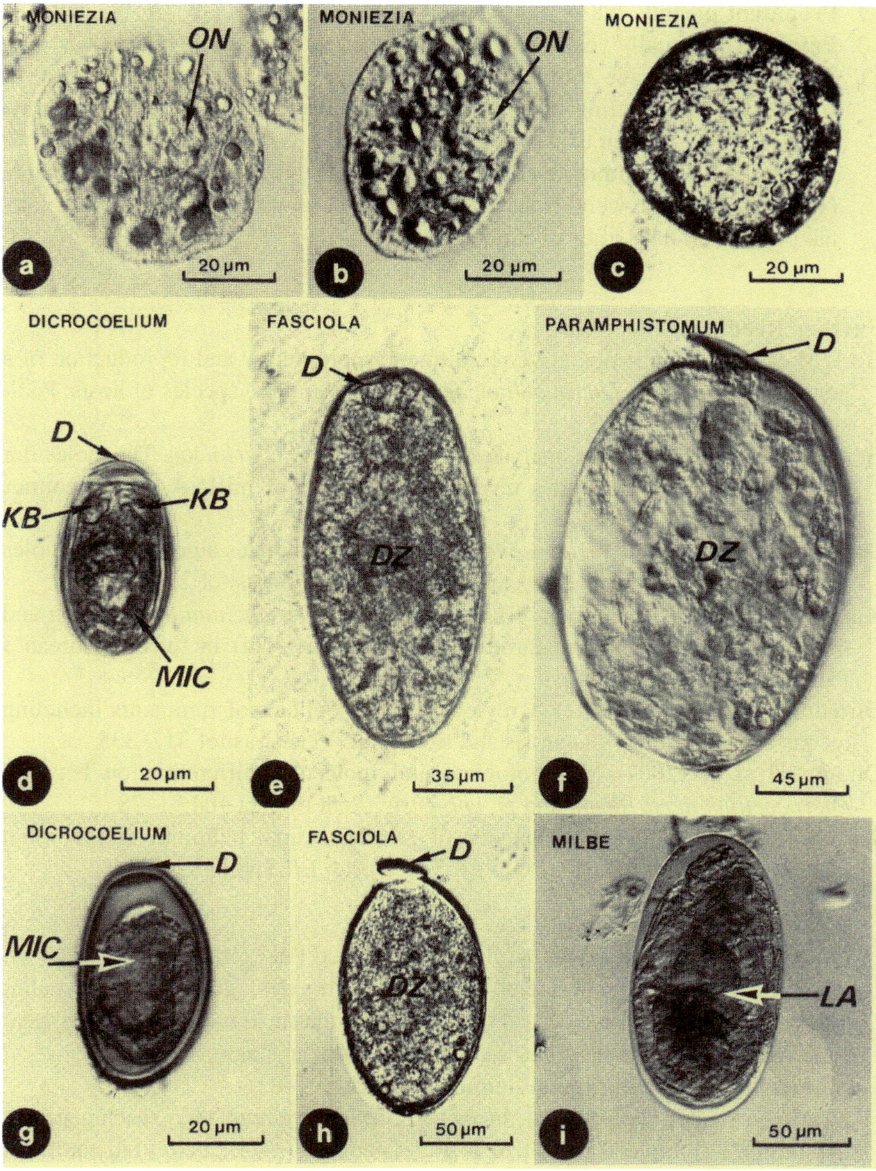

Fig. 5.14 Table showing the eggs of flat worms and a mite (Milbe). D = cover (=operculum); DZ = yolk cells; KB = germ balls; LA = larva; MIC = miracidium; ON = oncosphaera

6. **Pathway of infection**: Cattle: Oral uptake of metacercariae containing ants fixed at plants.
7. **Prophylaxis**: Very difficult if animals were kept on meadows in the nature.
8. **Incubation period**: Weeks up to months.

9. **Prepatent period**: 7–8 weeks.
10. **Patency**: Eventually years, especially in cases of repeated infections.
11. **Therapy**: Low-grade infections mostly do not need treatment. Efficacy of some compounds acting against *Fasciola* is lower in *Dicrocoelium*. Reports showed efficacy when using **tiabendazole** (250 mg/kg bodyweight, orally, no waiting time), **albendazole** (1 × 10–15 mg/kg bodyweight) and **mebendazole** (15–20 mg/kg bodyweight). Liver regeneration, however, is retarded and may take up to 100 days after treatment.

Further Reading

Beck M et al (2015) Comparative recruitment, morphology and reproduction of a generalist trematode, *Dicrocoelium dendriticum*, in three species of hosts. Parasitology 142:1297–1305.

Bernal D et al (2014) Surface analysis of *Dicrocoelium dendriticum*. The molecular characterization of exosomes reveals the presence of miRNAs. J Proteomics 105:232–241.

Castellano PP et al (2015) *Dicrocoelium dendriticum* induces autophagic vacuoles accumulation in human hepatocarcinoma cells. Vet Parasitol 212:175–180.

Köse M et al (2015) Infections of larval stages of *Dicrocoelium dendriticum* and *Brachylaima* sp. in brown garden snail (*Helix aspersa*) in turkey. Korean J Parasitol 53:647–651.

Otronto D, Traversa D (2002) A review of dicrocoeliosis of ruminants including recent advances in the diagnosis and treatment. Vet Parasitol 317–335.

Otranto D et al (2007) Morphological and molecular differentiation between *Dicrocoelium dendriticum* and *D. chinensis*. Acta tropica 104:91–98.

Trambow SR et al (2015) Prevalence of gastrointestinal helminth infections in ovine populations of Kashmir valley. Vet World, EISSN: 2231–0916.

C. *Paramphistomum cervi* (Rumen fluke) and related flukes

1. **Name**: Greek: *para* = besides; *amphi* = on both sides; *stoma* = mouth. Latin: *cervus* = stag. The name describes the fact that the suckers at both poles appear like mouths and that the fluke was first described in stags.
2. **Geographic distribution/epidemiology**: Worldwide.
3. **Biology, morphology**: Adult flukes of *Paramphistomum cervi* reach a size of 13 mm × 3–5 mm and parasitize as do related species (see below) in cattle but also in other ruminants. They are characterized by a small anterior and a large posterior sucker (Figs. 5.12 and 5.15). The eggs are very large (~up to 180 μm × 100 μm) and are excreted within the feces (Fig. 5.14f). Inside the egg, a miracidium larva is developed in temperature-dependent time (10–20 days). This miracidium enters water snails after having hatched inside freshwater. Common first intermediate hosts are snails of the genera *Planorbis* and *Anisius* and related species. Via sporocysts and rediae finally cercariae are produced, which are set free and become attached at plants at the border of the

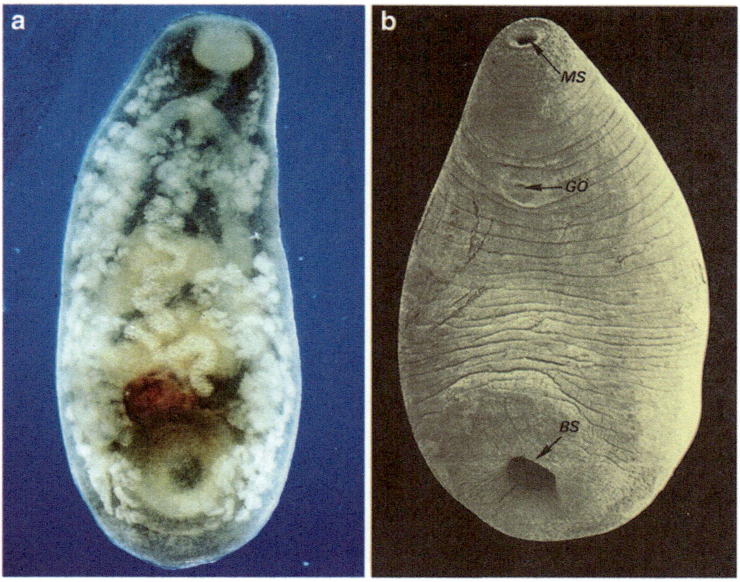

Fig. 5.15 Light micrograph (**a**) and scanning electron micrograph (**b**) of an adult worm of *Paramphistomum cervi*. BS = Ventral sucker; GÖ = genital openings; MS = oral sucker

water and excrete an outer sheath, thus reaching the stage of a metacercaria. If these stages are ingested by another final host, they start development into an adult worm (Fig. 5.12). This stage is reached about 10 weeks after the ingestion.

Further related species occur: *P. daubneyi* (syn. *Calicophoron daubneyi*) (4.5–9.5 mm × 2.5–6.5 mm), which is found in Europe and Africa in cattle, buffaloes and sheep; *P. ichikawai*, which reaches a size of 5–7 mm × 2–3 mm in cattle and sheep in Central Europe, in Russia and in Australia; and (besides others) *Explanatum* (syn. *Gigantocotyle*) *explanatum*, which reaches a size of 8–12 mm × 3–6 mm and occurs in Asia, in Africa and in the Caribbean. Hosts are many ruminants.

4. **Symptoms of disease (Paramphistomatosis)**: The infection with many worms inside the intestine may lead to many severe symptoms, since the mucous layer of the intestine may be destroyed leading to enteritis, diarrhoeas, catarrhalic symptoms and anorexia. In the case of heavy infections of the rumen, digestion problems occur leading to loss of weight and general fitness. Young and weak animals may lose weight and die. Immunity is weak, if at all present.
5. **Diagnosis**: Microscopical demonstration of the rather large operculated eggs, measuring 180 µm × 100 µm (Fig. 5.14f) by using the sedimentation method.
6. **Pathway of infection**: Oral by uptake of metacercariae attached at plants inside and around water ponds.

7. **Prophylaxis**: Snail control by molluscicids in water ponds and offering other means for drinking water on meadows.
8. **Incubation period**: Two weeks.
9. **Prepatent period**: 12–15 weeks.
10. **Patency**: About 1–1.5 years.
11. **Therapy**: Treatment is difficult. Even high doses of albendazole (15–20 mg/kg bodyweight) showed treatment failures in some cases.

Further Reading

Choudhary V et al (2015) Transcriptome analysis of the adult rumen fluke *Paramphistomum cervi* following next generation sequencing. Gene 570:64–70.

Khedri J et al (2015) Prevalence and intensity of *Paramphistostomum* spp. Iran J Parasitol 10:268–272.

Saowakon N et al (2013) *Paramphistomum cervi*: The in-vitro effect of plumbagin on motility, survival and tegument structure. Exp Parasitol 133:179–186.

Titi A et al (2014) Ruminal paramphistomosis in cattle from Northeastern Algeria: prevalence, parasite burdens and species identification. Parasite 21:50

Zintl A et al (2014) Bovine paramphistomes in Ireland. Vet Parasitol 204:199–208.

D. Trematodes parasitizing in dogs and cats

1. **Name**: Greek: *opisthen* = behind; *orchis* = testis. Latin: *tenuis* = thin; *collum* = neck; *felis* = cat. The term *Opisthorchis* refers to the placement of the testes in the hind region of the worm. Greek: *heteros* = the other; *phyle* = origin.
2. **Geographic distribution/epidemiology**: Worldwide; common in Asia, Europe; up to 15 % of cats may be infected in Asia.
3. **Biology, morphology**: Most common are flukes of the species *Opisthorchis felineus* (syn. *tenuicollis*) and *O. viverrini*, which may occur in a broad spectrum of at least accidentally fish-eating hosts (cats, dogs, foxes, pigs and humans. These flukes, which reach a length of 8–12 mm, parasitize inside the bile ducts of their hosts (Fig. 5.16). A significant diagnostic criterion is the fact that their two testes have only slightly depressed lobes (Fig. 5.16a). The anterior testis has four lobes and the posterior one five.

   Heterophyes species, which are very small (~2 mm), are common in the small intestine of dogs and cats; e.g. *H. heterophyes* had been found in Asia and in South European countries in up to 16 % of cats and dogs (Figs. 5.18 and 5.19). Their surface is closely covered with scales. In the case of *Opisthorchis* species as well as in *Heterophyes* species, water snails serve as first intermediate hosts and fresh and brackish water fishes as second intermediate hosts. In the latter, the encysted metacercariae are situated in muscles and various other organs.

   In South-East Asia (Korea, China) also the species *Clonorchis sinensis* (10–30 mm × 2–5 mm) is common in dogs and cats as well as *Metorchis bilis* (2.5–4.5 × 1.5 mm) in Europe and North America and *Pseudoamphistomum truncatum* in Europe, Russia and India.

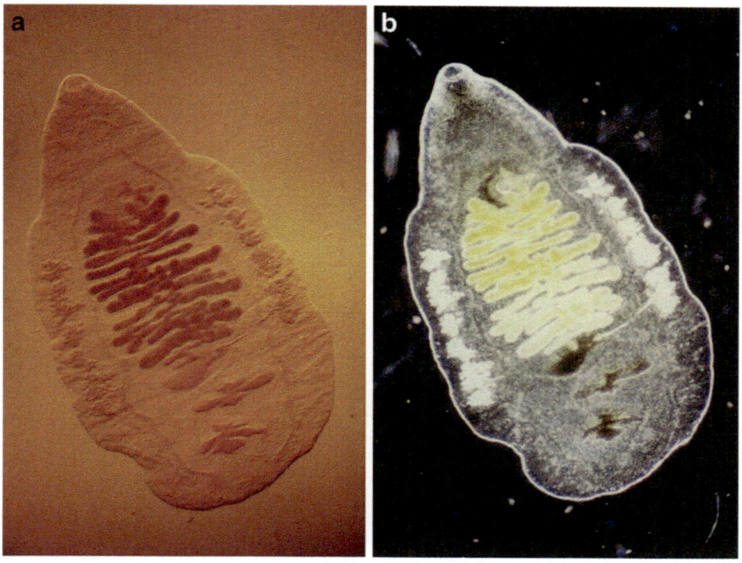

Fig. 5.16 Light micrographs of unfixed adult worms of *Opisthorchis tenuicollis* (syn. *felineus*) using different techniques. Note the shape of the two testes in terminal region of the worms

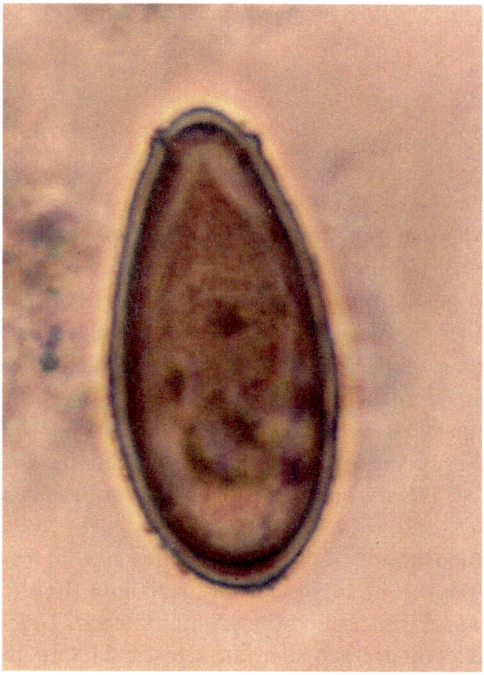

Fig. 5.17 Light micrograph of an egg of *Opisthorchis felineus*

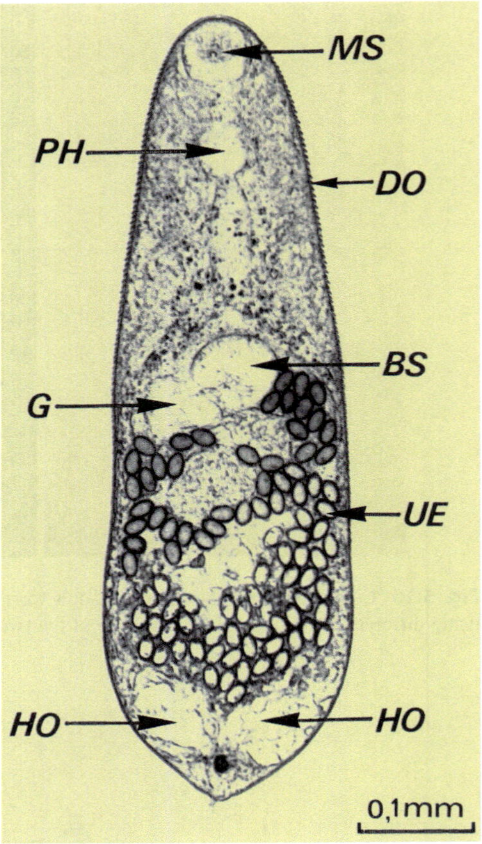

Fig. 5.18 Light micrograph of an adult worm of *Heterophyes heterophyes*. BS = ventral sucker; DO = thorns/scales; G = gonocotyl; HO = testis; MS = oral sucker; PH = pharynx; UE = uterus with eggs

4. **Symptoms of disease (Opisthorchiasis, heterophyiasis)**:
   (a) ***Opisthorchis tenuicollis***: Depending on the amount of ingested metacercariae, the symptoms of disease may be absent or severe, showing intense inflammations of the bile ducts, enlargement of the liver and proliferations of the bile duct epithelia, which may induce carcinoma (also common in humans!). **Clinical symptoms** are vomiting, loss of appetite, intestinal disorders, icterus, anaemia, oedemas and/or ascites.
   (b) ***Heterophyes* species**: In cases of heavy infections: vomiting, disturbances in digestion, bloody diarrhoeas and hosts are easily susceptible to other infectious diseases.
5. **Diagnosis**: Demonstration of the rather small eggs with the help of the flotation method. They possess in all species an operculum (Figs. 5.17 and 5.20). The eggs of *O. tenuicollis* measure 30 × 15 μm; those of *Heterophyes* species are somewhat smaller and reach a mean size of 24 × 14 μm. However, other species of this group of parasites may reach larger sizes (up to 100 μm in length).

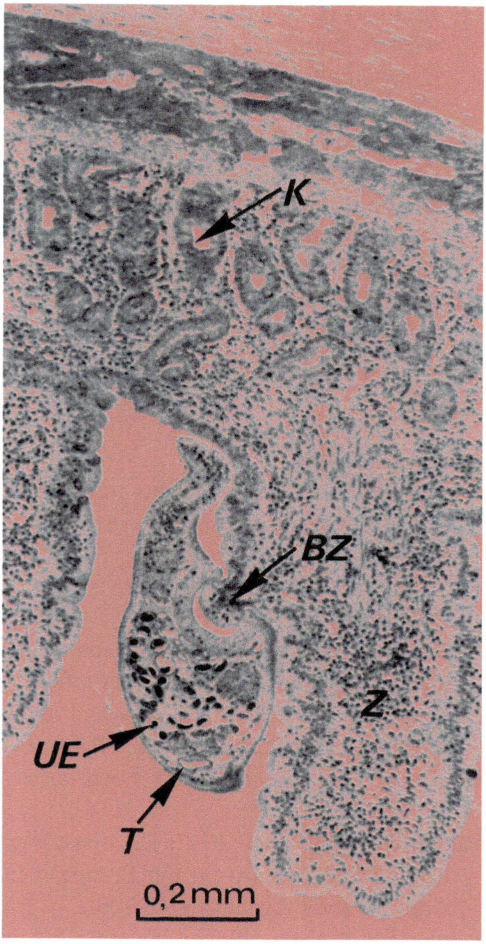

Fig. 5.19 Light micrograph of a section through the intestine of a cat showing an attached *Heterophyes* fluke. BZ = ventral sucker; K = crypts of the intestinal wall; T = hind end; UE = uterus filled with eggs; Z = protrusion of the intestinal wall

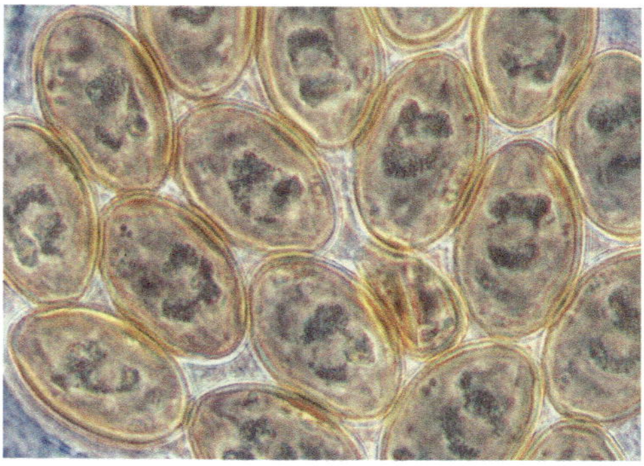

Fig. 5.20 Light micrograph of *Heterophyes* eggs (they already contain a miracidium larva)

6. **Pathway of infection**: Cats, dogs and humans are infected by ingesting raw fish meat containing 1–2 mm sized metacercariae of this species. In the case of *O. tenuicollis*, freshwater fishes are infected, while in the case of *Heterophyes* species, especially fishes are infected which live in brackish or saltwater.
7. **Prophylaxis**: Do not feed raw fish to home dogs or cats and avoid also own consumption of such fish meat.
8. **Incubation period**: About one week.
9. **Prepatent period**: 3–4 weeks in the case of *O. tenuicollis* and 7–9 days in the case of *Heterophyes* species.
10. **Patency**: Years in the case of *O. tenuicollis* but only 1–3 months in the case of *Heterophyes* species.
11. **Therapy**: Oral uptake of praziquantel (3 days each 25 mg/kg bodyweight).

Further Reading

Ashour DS et al (2014) Insights into regulatory molecules of intestinal epithelial cell turnover during experimental infection by *Heterophyes heterophyes*. Exp Parasitol 143:48–54.

Dao TM et al (2014) *Opisthorchis viverrini*-like liver fluke in birds from Vietnam. J Helminthol 88:441–446.

Elshazly AT et al (2008) Comparison of three immunodiagnostic tests for experimental *Heterophyes heterophyes* infection in dogs. Vet Parasitol 151:196–202.

El-Seifi MA et al (2012) ELISA and some biochemical tests of Heterophyidae infection in laboratory animals. Parasitol Res 110:679–687.

Keiser J, Utzinger J (2009) Food-borne trematodiases. Clin Microbiol Rev 22:466–483.

Pakharukova MY et al (2015) The first comprehensive study of praziquantel effects in vivo and in vitro on the European liver fluke *Opisthorchis felineus*. Int J Antimicrobiol Agents 46:94–100.

Petney TN et al (2013) The zoonotic fish borne liver flukes *Clonorchis sinensis*, *Opisthorchis felineus* and *O. viverrini*. Int J Parasitol 43:1031–1046.

Pomaznoy M et al (2013) Adult *Opisthorchis felineus* major protein fractions: comparison with liver flukes *O. viverrini* and *Clonorchis sinensis*. Exp Parasitol 135:297–306.

Schuster RK et al (2007) Liver flukes in dogs and treatment with praziquantel. Vet Parasitol 150:362–365.

Thaenkhan U et al (2012) Families Opisthorchiidae and Heterophyidae: Are they distinct? Parasitol Int 61:90–93.

Traversa A et al (2012) A large outbreak of *Opisthorchis felineus* in Italy. Clin Microbiol Inf Dis 31:1089–1093.

E. *Schistosoma* species

1. **Name**: Greek: *schizein* = cutting; *soma* = body. Latin: *bos, bovis* = cattle; *japonicus* = Japanese; *indicus* = indian; *spindalis* = spindle-like; *nasalis* = belonging to the nose; Orient = eastern regions. Theodor

Bilharz = name of the German discoverer of the worms, which got the genus name *Schistosoma* after the First World War (~1919). Before this time, the genus was named *Bilharzia*.

2. **Geographic distribution/epidemiology**: Worldwide, but restricted to warm-humid countries.
3. **Biology, morphology**: These flukes, which belong to the trematodes, differ from most other trematode genera, since they include male and female worms, which, however, live lifelong as couples (Figs. 5.21a, b), whereby the leaf-like male worm bears the female within a body fold. These couples live depending on the species in veins of the nose, the intestine or the urogenital system of their hosts which include ruminants, camels, rodents, antelopes, carnivores, goats, racoons and in many other species besides occasionally in humans. The surface of the males include so-called tegumental hooks, which help to stay in peculiar regions of the blood vessel system without being driven away. The eggs (Figs. 5.22a–d) are provided with species-specific protrusions, which help them to pass from the interior of blood vessels to the outside (=species specific into the lumina of the intestine, bladder or the nose tubes. During their passing to the outside, the miracidium larva is developed inside the eggs, which do not possess an operculum (cover) but are ruptured as soon as they reach water after being excreted from the host's body. Inside water, the hatched larva (miracidium) swims around and has to enter an intermediate host (many water snails), where two generations of sporocysts are formed before the motile cercariae are set free. These cercariae (Fig. 5.22e) have a furcated tail in contrast to those of other trematodes. Tail forward the cercariae swim in freshwater. In case they reach a final host (obligatorily needed within 24 h), they enter anterior end forward into the skin of the host, releasing the tail outside of the host's body. There they stay as so-called **schistosomulum** for some days. Many of these stages are killed by defence system of the host. However, a good bunch of them enters finally the blood vessel system and is transported by the bloodstream to

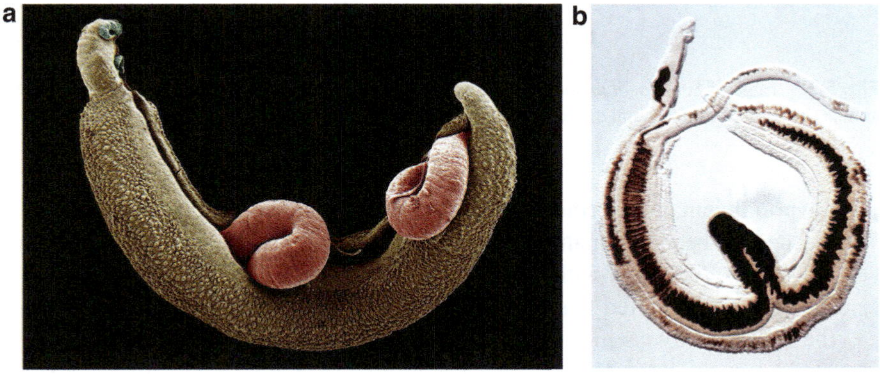

Fig. 5.21 Adult worms of *Schistosoma mansoni*. (**a**) Scanning electron micrograph. (**b**) Light micrograph showing that the intestine of the female is filled with host blood

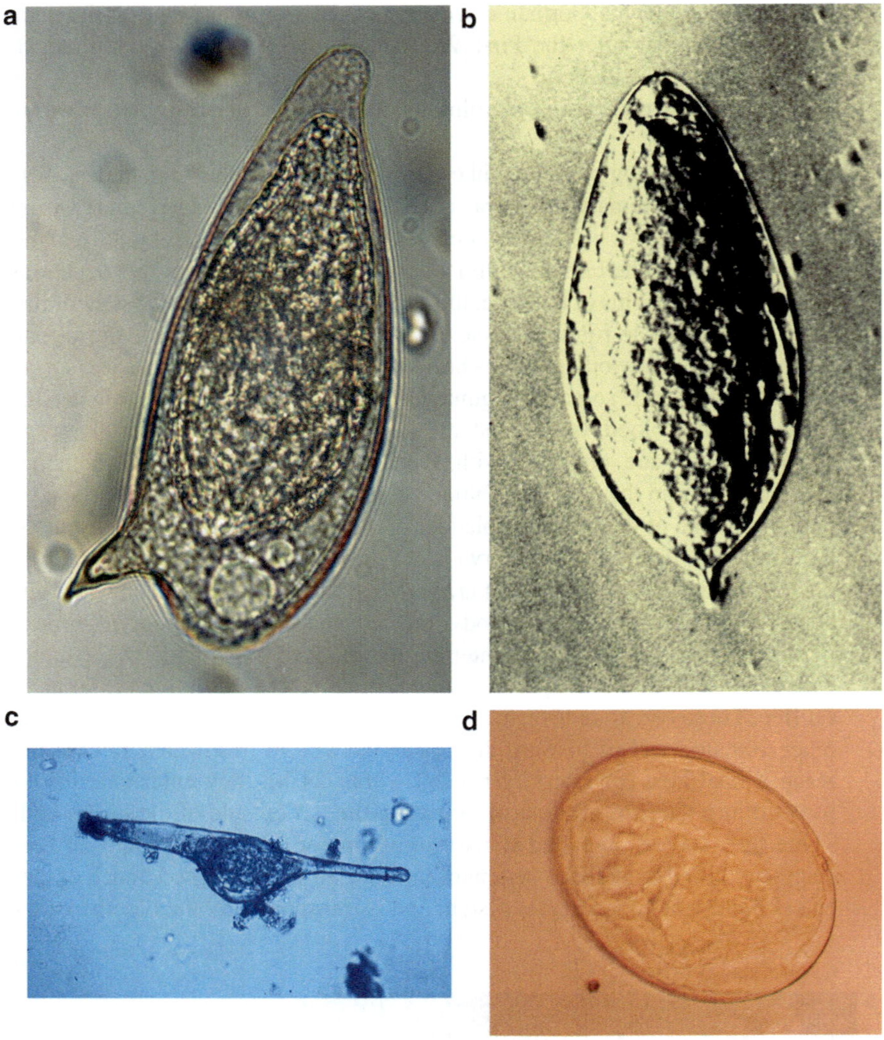

Fig. 5.22 Eggs of *Schistosoma* species containing each a miracidium larva. (**a**) Type with lateral spine; (**b**) Type with terminal spine; (**c**) Spindle-like type; (**d**) Type with a mini-spine

the portal vein, where couple formation occurs at a very early stage. As soon as the couple becomes mature, the females excrete species specifically shaped eggs (Figs. 5.22a–d), within which a miracidium larva is developed. Many of the eggs are clustered in small vessel of liver, etc., but several are able to pass the wall of body cavities. Thus, they may reach the lumen of the intestine, the bladder or the nose hollow, from where they have the chance to reach water, wherein the miracidium larva may hatch and is prepared to enter a new host.

Animals might become infected by the following species:

(a) ***Schistosoma bovis***: This species lives in ruminants, monkeys and humans in Africa and Iran; ♂ = 14 mm, ♀ = up to 25 mm; eggs (170 × 70 μm) possessing a terminal thorn. Ten percent of human infections in these regions are due to this species. Intermediate hosts are *Bulinus* and *Planorbis* species besides some others.
(b) ***Schistosoma japonicum***: This species lives in dogs and humans in East Asia; ♀ = until 26 mm, ♂ 15 mm; eggs (85 × 25 μm) with a tiny lateral thorn. Intermediate hosts are *Oncomelania* snails.
(c) ***Schistosoma matheei***: This species occurs in ruminants and monkeys in many regions of Africa. ♂ = 14 mm; ♀ = 25 mm; the eggs (170 × 70 μm) bear a terminal thorn; *Bulinus* snails are intermediate hosts.
(d) ***Schistosoma nasale***: This species occurs on the Indian subcontinent and parasitizes in the veins of the nose of buffaloes and cattle reaching infection rates of up to 50 %. The eggs appear boomerang-like and measure 350–380 μm × 50–80 μm (Fig. 5.22c). Intermediate hosts are snails of the genus *Indoplanorbis*.
(e) ***Schistosoma spindale***: This species occurs in ruminants and dogs in India and in the Far East. The up to 16 mm long adults live mainly in the mesenterial veins, but are found everywhere in the blood vessels. The eggs measure 300 × 80 μm, appear spindle like and possess a terminal spike (Fig. 5.22c). In cases of infections with large numbers, many fatal cases had occurred. Intermediate hosts are snails of the genera *Planorbis, Indoplanorbis* and *Lymnaea*.
(f) ***Schistosoma curassoni***: This species occurs in Africa, and the females have a length of 18–26 mm. Hosts are cattle and sheep; first eggs appear 40 days after infection within the feces.
(g) ***Heterobilharzia americana***: This species occurs in racoons, opossums, dogs, deer, etc., in the USA. The females reach a length of 9–18 mm and stay in the mesenterial veins. Eggs appear mostly about 70 days after infection. The snail *Lymnaea cubensis* is one of the intermediate hosts.

4. **Symptoms of disease (Schistosomiasis, bilharziasis)**: According to the site of parasitism, blood in urine, feces or the nose. Infected animals show loss of weight, decreasing fitness, rough fur, anaemia (clearly visible in the eyes), granulomatous enlargements and tumours in liver, spleen, etc. In the case of high-grade infections animals may die.
5. **Diagnosis**: Microscopical demonstration of the rather large eggs after using the sedimentation method. Portions of feces, urine or nasal mucus are filled up with water. Then the sediment is several times resuspended and finally investigated by light microscopy.
6. **Pathway of infection**: Percutaneous: the cercariae enter the skin when the final host has entered freshwater ponds with infected snails.
7. **Prophylaxis**: Snail control in water ponds in contact with farm animals (e.g. use of the antisnail product Bayluscid®. Vaccination of cattle using attenuated cercariae.

8. **Incubation period**: **(1) Dermatitis** due to penetrating cercariae is visible within 2 days. Reactions on wandering larvae become noticeable after 1 week, but general severe symptoms do not occur (species specific) before 4–6 weeks after infection.
9. **Prepatent period**:
   Examples:

S. bovis	44 days
S. mattheei	42 days
S. japonicum	40 days
S. nasale	77–116 days
S. spindale	44–76 days

10. **Patency**: Years.
11. **Therapy**:
    (a) Snail control: Use of molluscicides like Bayluscid® in ponds on meadows.
    (b) Chemotherapy: Drug of choice is **praziquantel** (3 × 10–35 mg/kg bodyweight). Since praziquantel is rather expensive, prophylactic methods must be used, too.

Further Reading

Becker B, Mehlhorn H (1980) Light and electron microscopic studies on the effects of praziquantel on *Schistosoma mansoni, Dicrocoelium dendriticum* and *Fasciola hepatica*. Parasitol Res 63:113–128.

Bian CR et al (2015) Serological prevalence of *Schistosoma japonicum* in motile populations in previously endemic and now non-endemic regions in China. PLoS One. doi:10.1371/journal.pone.0128896.

Higon M et al (2011) Screening trematodes for novel intervention targets: a proteomic and immunological comparison of *Schistosoma haematobium, S. bovis* and *E. caproni*. Parasitology 138:1607–1619.

Latchumikanthan A et al (2014) Occurrence of *Schistosoma nasale* infection in bullocks of Puducherry. J Parasitol Dis 38:238–240.

Li YS et al (2014) The *Schistosoma japonicum* self-cure phenomenon in water buffaloes. Int J Parasitol 44:167–171.

Moné H et al (2015) Introgressive hybridization of *Schistosoma haematobium* by *S. bovis* at the origin of the first case report of schistosomiasis in Corsica, France. Parasitol Res 114:4127–4133.

Murthy GSS et al (2013) Evaluation of a polyclonal antibody based on sandwich ELISA for the detection of faecal antigens in *Schistosoma spindale* infections in bovines. J Parasitol Dis 37:47–51.

Sato Y et al (2008) Mitochondrial DNA sequence and gene order of the Sri Lankan *Schistosoma nasale* is affiliated to the African/Indian group. Parasitol Int 57:460–464.

Southgate VR et al (1995) Mating behaviours in mixed infections of *Schistosoma haematobium* and *S. mattheei*. Parasitol Res 81:651–656.
Vicente B et al (2014) A *Fasciola hepatica* derived fatty acid binding protein induces protection against schistosomiasis caused by *Schistosoma bovis*. Exp Parasitol 145:145–151.
Vimalraj PG, Latchumikanthan A (2015) *Schistosoma spindale* infection in a captive jackal (*Canis aureus*). J Parasitol Dis 39:120–121.
Weyher A et al (2010) Molecular identification of *Schistosoma mattheei* from feces of Kinda (*Papio cynocephalus kindae*) and Grayfoot baboon (*Papio ursinus griseipes*) in Zambia. J Parasitol 96:184–190.

F. Schistosomatid trematodes of birds

1. **Name**: Greek: *thrix, trichos* = tiny hair; *ornis, ornithos* = bird. Theodor Bilharz (1825–1862), a German physician, who in Cairo (Egypt) at first described the worms which are called today *Schistosoma* species. The genus *Diplostomum* was later renamed *Bilharzia*.
2. **Geographic distribution/epidemiology**: Worldwide.
3. **Biology, morphology**:
   (a) ***Trichobilharzia* species** occur in blood vessels close to the intestine in **ducks** and other **water birds**. The female worms reach a size of 3–5 mm, while the males are slightly larger. The latter bear the females in a species-specific hook containing ventral fold called **canalis gynaecophorous**. Inside the uterus of the females, only a single egg is located, which reaches a size of 140–210 μm × 50–70 μm and appears spindle like. As soon as it passes into the intestinal lumen, it already contains the miracidium, which may enter a broad spectrum of water snakes, when the host's feces have entered water (compare Fig. 5.22c).
   (b) ***Bilharziella polonica***: This species lives in the mesenterial blood vessels of **ducks**. The males reach a length of 4 mm, while the female are smaller (~2 mm). Both have a flattened body shape. The 400 μm long eggs excreted by the female have a typical shape being provided with a very long filament-like structure at one of the poles (Fig. 5.25). They contain already a miracidium when they are deponed into water within the duck's feces. Intermediate hosts are snails of the genera *Lymnaea* and *Planorbis* (Fig. 5.25b).
   (c) ***Ornithobilharzia* species**: These species parasitize in **seagulls** and several other **water birds** and occasionally in **geese**. Related species are found in Asia also in cattle, horses and camels. The species inside birds reach a length of up to 8 mm. Their ovoid eggs measure 60–70 μm × 50 μm and contain already a miracidium when deponed in feces. The rather thick but smooth eggshell is provided with a terminal thorn (Fig. 5.25a). Intermediate hosts are several water snails.
4. **Symptoms of disease (Schistosomatidosis of birds)**: Most of the infections remain harmless. However, heavy infections lead to bloody feces, anaemia and

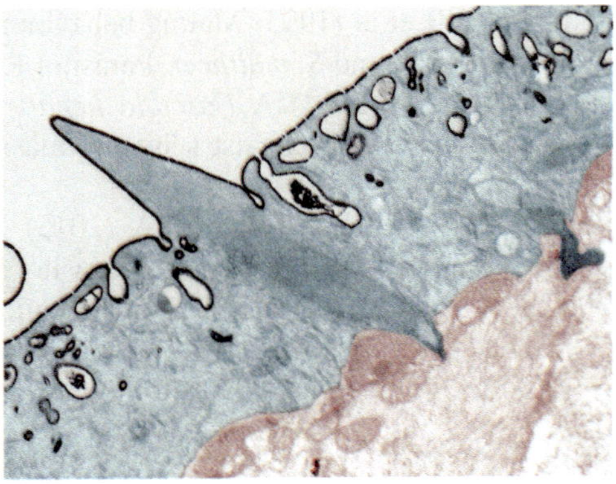

Fig. 5.23 Transmission electron micrograph of a section of the tegument of *S. mansoni* showing a section of a tegumental thorn and the dense surface coat, which protects from the antibodies of the host

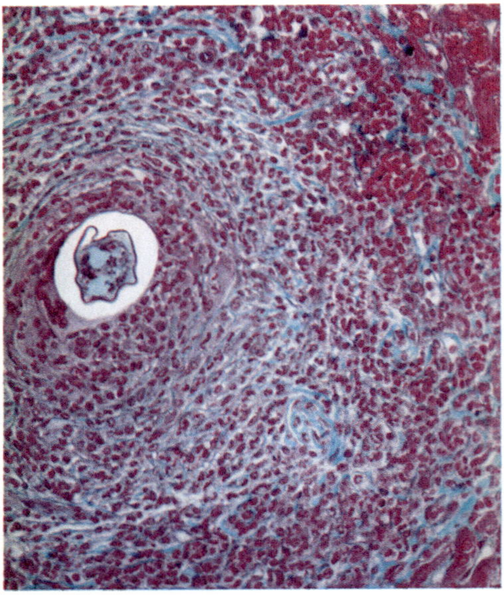

Fig. 5.24 Section through an egg in the liver surrounded by a granulome containing large amounts of collagen fibres

massive disturbances of the functions of the liver and intestine, where granulomas occur like those in other species (Fig. 5.24). In these cases, mortality rates are high. If **humans** are infected (e.g. when bathing in natural lakes), cercariae enter the skin as they do in birds, but they are killed by the defence system of the skin. Then the remnants of these cercariae induce a peculiar inflammation called "**bathing dermatitis**".

5. **Diagnosis**: Microscopical proof of the typical eggs (Fig. 5.25) in the feces with the help of the sedimentation method.
6. **Pathway of infection**: Percutaneously by penetrating cercariae during water contact (lakes, springs, rivers, etc.).

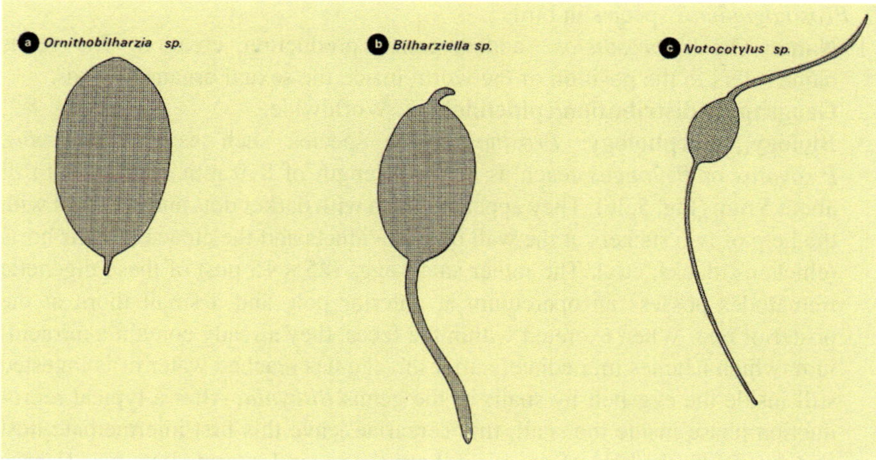

Fig. 5.25 Diagrammatic representation of the eggs of some schistosomatid trematodes of birds, the cercariae of which enter the skin of animals and humans leading to a so-called "bathing dermatitis" in humans, since there no further development occurs in contrast to birds

7. **Prophylaxis**: Practically impossible with respect to free-living birds. **Humans**: Using the repellent spray Viticks (Fa. Alpha-Biocare, Neuss, Germany), which protects at least for 1 h when bathing in lakes.
8. **Incubation period**: Birds: 2 weeks; **humans**: in the case of bathing dermatitis: 24 h.
9. **Prepatent period**: 3–4 weeks until eggs appear first in feces.
10. **Patency**: At least 4–5 months.
11. **Therapy**: None in free-living birds. Cultured ducks and geese might be treated with praziquantel, if symptoms become visible. Humans: use of desensitizing creams.

Further Reading

Boyce K et al (2012) Identification of a new species of digenean *Notocotylus malhamensis* from the bank vole *Myodes glareolus*. Parasitology 139:1630–1639.

Mahdavi SA et al (2013) Survey of migratory birds (Anatidae: *Anas* sp.) for schistosome parasites. Iran J Parasitol 8:333–336.

Santoro M et al (2011) Helminth community structure of the Mediterranean gull (*Ichthyaetus melanocephalus*) in Southern Italy. J Protozool 97:364–366.

Skala V et al (2014) Insights into the development of *Notocotylus attenuatus*. Parasitol Int 63:94–99.

Tolstenkov OO et al (2012) The neuro-muscular system in fresh water furcocercaria from Belarus: I. Schistosomatidae. Parasitol Res 110:185–193.

Yakhchali M et al (2013) Detection of infection with larval stages of *Ornithobilharzia turkestanicum* using PCR. Iran J Parastiol 8:627–633.

G. *Prosthogonimus* species in birds

1. **Name**: Greek: *prosthetos* = added; *gone* = production, creation. The genus name refers to the position of the worm inside the sexual organs of birds.
2. **Geographic distribution/epidemiology**: Worldwide.
3. **Biology, morphology**: *Prosthogonimus* species such as *P. pellucidus, P. ovatus* or *P. longus* reach as adults a length of 8–9 mm with a width of about 5 mm (Fig. 5.26). They appear whitish with darker dots and are fixed with the help of two suckers at the wall of the oviducts and the cloaca of their hosts (chickens, ducks, etc.). The rather small eggs (25 × 15 μm) of these digenetic trematodes possess an operculum at anterior pole and a small thorn at the posterior one. When excreted within the feces, they already contain a miracidium, which hatches immediately after the egg has reached water or is ingested still inside the eggshell by snails of the genus *Bithynia*. After a typical reproduction phase inside the snail, tiny cercariae leave this first intermediate host and enter with the help of an apical thorn the second intermediate host (larvae of dragon flies; e.g. stages of *Libellula* or *Cordulia*, etc.). Inside their body cavity, they are transformed into metacercariae and become infectious. If the final host (many birds) ingests such dragon fly larvae, the adult flukes develop inside the oviducts and cloaca of their final host and start egg production after copulation with other flukes (Fig. 5.26).
4. **Symptoms of disease (so-called dragonfly disease)**: The eggshells of the birds become extremely weak, so that they are squeezed and the contents flow away. Birds stop egg laying and show apathia, weakness and erection of the body (so-called penguin position): Also death cases may occur due to peritonitis initiated by secondary infections.
5. **Diagnosis**: Demonstration of the worm eggs inside the feces with the help of methods like M.I.F.C. or S.A.F.C.
6. **Pathway of infection**: Oral by uptake of metacercaria inside larvae of dragon flies or when adult dragon flies are dropped to earth during rain.
7. **Prophylaxis**: Hardly possible, if birds are kept close to ponds or lakes.
8. **Incubation period**: 1–2 weeks.
9. **Prepatent period**: In case of chickens: 1–2 weeks; in ducks: 3 weeks.
10. **Patency**: Chickens: 1–2 months; ducks: up to 5 months.
11. **Therapy**: A defined therapy is not known; Anthelmintics such as albendazole and fenbendazole should work. The same is claimed for praziquantel, which can be used (after introductory tests) in costful birds.

Further Reading

Heneberg P et al Integrative taxonomy of Central European parasitic flatworms of the family Prosthogonimidae. Lühe, 1909 (Tremadoda: Plagiorchiidae). Parasitol Int 64:264–273.

Leok CS et al (2002) Morphology of the oviduct fluke *Prosthogonimus ovatus* from Indonesian native chickens. J Vet Med Sci 64:1129–1131.

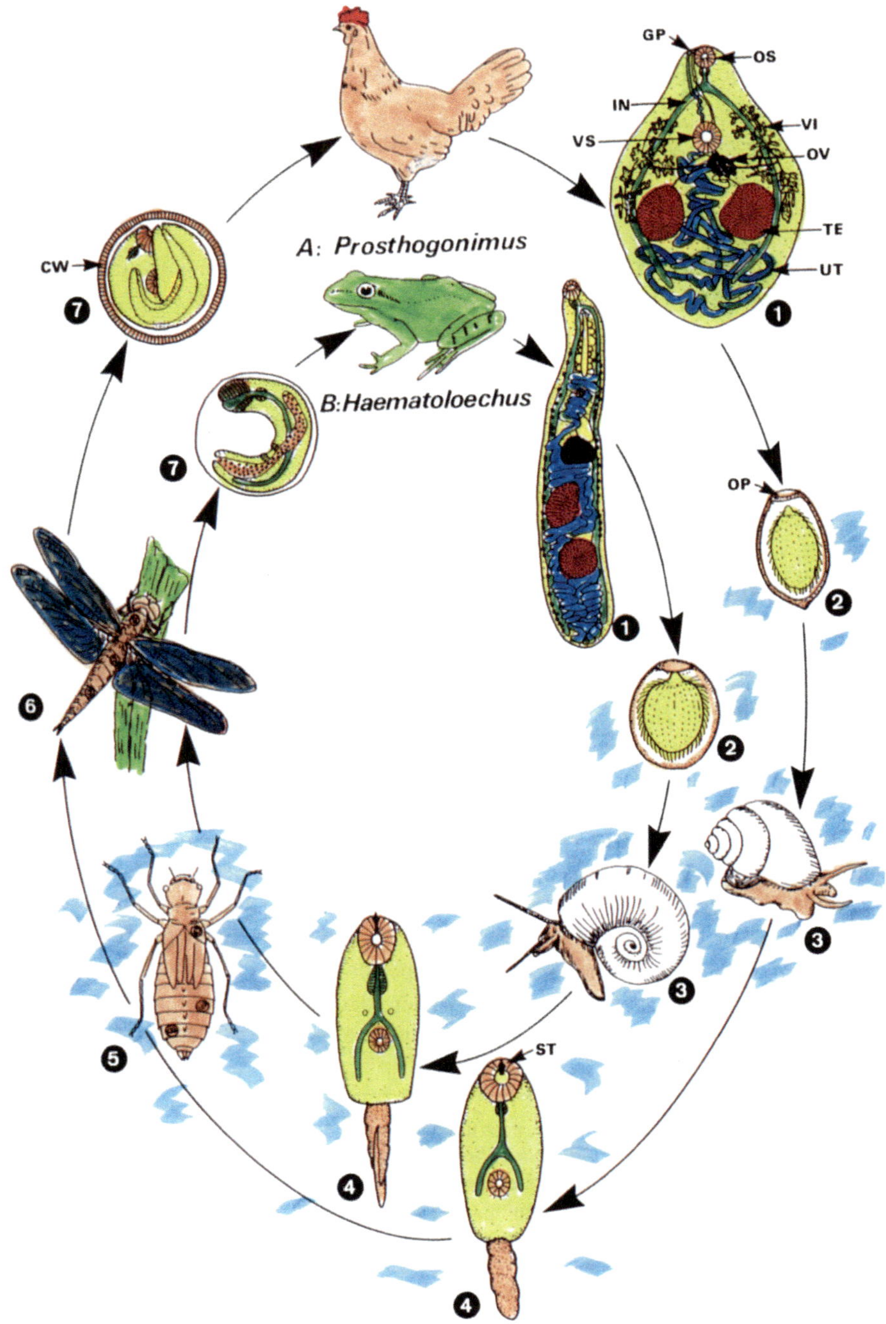

Fig. 5.26 Diagrammatical demonstration of the life cycles of trematodes with xiphidiocercariae. (**A**) *Prosthogonimus macrorchis* (7–8 × 5–6 mm) parasitizes in the oviduct, in the bursa fabricii and in the hindgut of chickens, ducks and their relatives and can decrease or even prevent egg laying. (**B**) *Haematoloechus* spp. (8 × 1.5 mm) are parasitic in the lungs of frogs and toads. **1** Adult worms. **2–4** Eggs are commonly passed with feces. They contain a fully developed miracidium,

Monteiro CM et al (2011) Helminth parasitism in the Neotropical cormorant (*Phalacrocorax brasilianus*) in Southern Brazil: effect of host size, weight, sex and maturity state. Parasitol Res 109:849–855.

H. Further digenetic flukes in birds

1. **Name**: Greek: *stoma* = mouth; *para* = side by side; *orchis* = testes. Latin: *echinatus* = thorny; *revolutus* = rolled back, behind situated.
2. **Geographic distribution/epidemiology**: Worldwide, especially common in mass breed farms.
3. **Biology, morphology**: In the feces of birds, the eggs of the following species may be found:
   (a) **Echinostomatid trematodes**: The adults reach a length of 2 cm and are characterized by a collar containing characteristic numbers of spines. This collar surrounds the mouth (Figs. 5.27 and 5.28). One of the most important species is *Echinostomum revolutum*, where the collar shows 37 spines. This species occurs in the caecum and colon of many **waterbirds**, while most other *Echinostoma* species are found in the small intestine. The operculated eggs, which are found in the feces, measure 90–-130 μm × 60–70 μm (Fig. 5.29). They are excreted in an unembryonated status, but within 18–30 days (depending on the temperature) the miracidium larva is developed if they are entered into water. First intermediate hosts are snails of the genera *Lymnaea* and *Planorbis*, which finally excrete cercariae that enter other water snails, tadpoles or mussels, which thus become the second intermediate host, wherein the infectious **metacercariae** are developed that may infect again other final hosts.
   (b) ***Echinoparyphium revolutum*** (45 collar spines, measuring 5 mm in length and producing eggs of a size of 110 μm × 80 μm) and ***Hypoderaeum conoideum*** (47–53 collar spines, 12–15 mm long and producing eggs of a size of 100 μm × 65 μm) parasitize also in the intestine and show a similar development as the *Echinostoma* species.
   (c) **Strigeid trematodes** reach only a length of 0.2–0.9 mm (exceptionally up to 3 mm). They appear spherical and are divided by depression into two

Fig. 5.26 (continued) which, however, does not hatch in water. When they are swallowed by their intermediate hosts (*H.* sp.—*Planorbula, Planorbis, Lymnaea* spp.; *P. macrorchis*—*Amnicola* spp.), daughter sporocysts produce numerous short-tailed cercariae; their oral sucker is provided with a stylet. **5** When the feebly swimming cercariae pass by the posterior ends of the naiads of dragonflies, they may become sucked into the "anal lung", from where they penetrate the thin cuticle and encyst nearby (as metacercariae; 7). **6, 7** When the naiad metamorphoses into a teneral and finally into an adult, the metacercariae remain encapsulated in the abdomen. Infections of final hosts occur when they swallow infected juvenile or adult dragonflies. Inside the final hosts, the young worms reach the sites of final location by creeping (*H.* spp., up the oesophagus and down the trachea; *P. macrorchis*, from cloaca to the different places). CW = cyst wall; GP = genital pore; IN = intestine; OP = operculum; OS = oral sucker; O = ovary; ST = stylet of OS; TE = testis; UT = uterus with eggs; VI = vitellarium; VS = ventral sucker

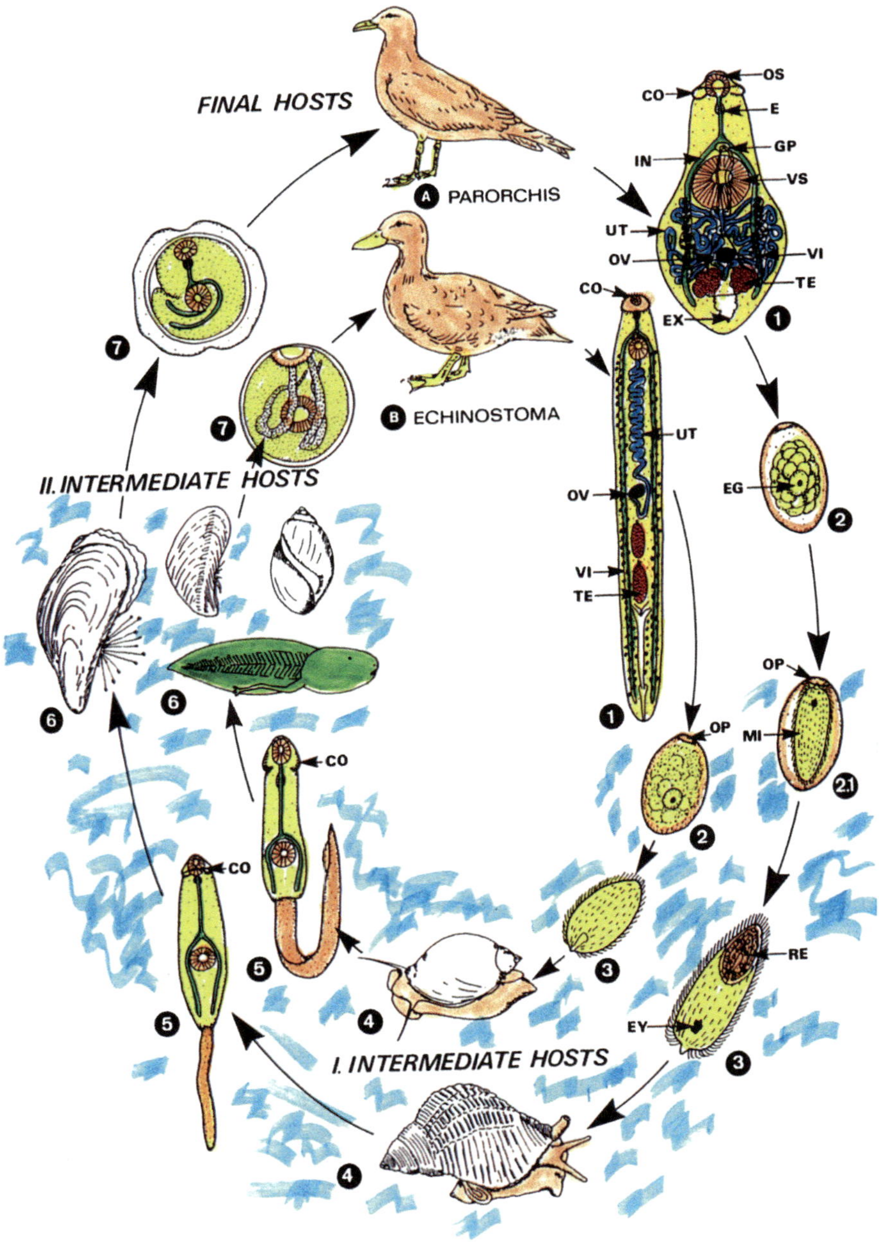

Fig. 5.27 Diagrammatic representation of the life cycle of the echinostomatid trematodes, *Parorchis acanthus* (**A**) and *Echinostoma revolutum* (**B**). **1** Adult flukes; *P. acanthus* (10 mm) lives in the bursa fabricii or rectum of herring gulls; *E. revolutum* (4–22 mm) is found in the rectum and/or caeca of ducks, geese and occasionally humans. Adults are characterized by their collar head with a characteristic number of spines. **2–4** The operculated eggs are passed unembryonated (*Echinostoma*) or fully embryonated (*Parorchis*) and containing a miracidium. The hatched

differently sized halves. They parasitize mainly in the intestines of **ducks** and **doves**. Important and common species are *Apatemon gracilis, Cotylurus cornutus* and *Parastrigea robusta*. Their ovoid eggs are excreted in an unembryonated status. They possess an operculum at the smaller side and reach a size of 90–110 μm × 60–70 μm. After the phase of embryonation in water, the operculum is blasted off, the miracidium hatches and enters finally water snails (e.g. *Lymnaea* species and related ones), which thus become the first intermediate host. Therein cercariae are formed, which finally enter second intermediate hosts (e.g. leeches, amphibia, mussels, water snails), where they are transformed to the infectious metacercariae. These are again infectious for the final hosts.

(d) ***Notocotylus attennatus***: This species lives in the caeca and rectum of many birds (also in farmed birds!) and reaches a size of 5 × 1.4 mm. Their ventral side is covered by spines. A ventral sucker is lacking. The 20 μm × 10 μm sized eggs are characterized by the possession of terminal filaments of a length of up to 0.2 mm (!) (Fig. 5.25). Mussels serve as first intermediate hosts, at the surface of which the metacercariae become attached.

4. **Symptoms of disease (Trematodiasis)**: Low-grade infections are rather harmless. However, mass infections lead in all cases to intestinal catarrhs, loss of weight and bloody diarrhoeas. Fatal cases are common in high-grade infections especially in young birds. **Humans** may also get infected by such trematodes. Their cercariae may enter human skin but are killed therein, and thus, symptoms of the typical **cercarial dermatitis** occur.
5. **Diagnosis**: Microscopical demonstration of the eggs with the help of the sedimentation method.
6. **Pathway of infection**:
   (a) Schistosomatid trematodes: percutaneously by entering of cercariae.
   (b) Other trematodes (listed here) are transmitted orally during ingestion of second intermediate hosts containing metacercariae.

Fig. 5.27 (continued) miracidium, which in *Parorchis* already includes a well-developed mother redia, finally penetrates the first intermediate host, water snails of the genera *Nucella* (*Parorchis*, seawater), *Physa* or *Heliosoma* (*Echinostoma*, freshwater). **5** Within the intermediate hosts, reproduction occurs via sporocysts (not present in *Parorchis*) and 2 generations of rediae, which finally give rise to free-swimming long-tailed cercariae, which are provided with a spiny collar (CO). **6–7** The cercariae of *E. revolutum* enter a variety of second intermediate hosts (snails, fingernail clams, tadpoles) and encyst inside soft parts as metacercariae (7), whereas cercariae from *P. acanthus* attach to the surface of snail shells and/or gills and excrete their metacercarial cyst wall (7). The final host becomes infected when ingesting metacercariae, the excystment of which takes place in the duodenum. From there, the young worms wander to their final sites. After a prepatent period of about 20 days, eggs are found in feces. E = erythrocyte; EG = egg cell; EY = eye spot; EX = excretory bladder; GP = genital pore; IN = intestine; MI = miracidium; OP = operculum; OS = oral sucker; OV = ovary; TE = testes; UT = uterus; VI = vitellarium; VS = ventral sucker

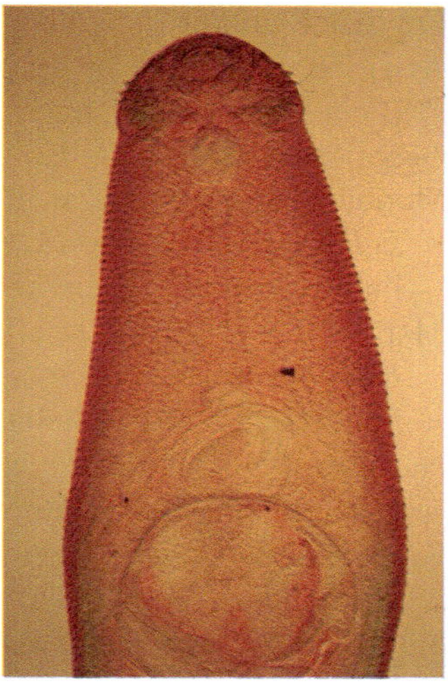

Fig. 5.28 Light micrograph of the anterior end of *Echinostoma revolutum*. The anterior collar region shows the typical spines. The anterior sucker is much smaller than the ventral one, which is depicted at lower region of the micrograph

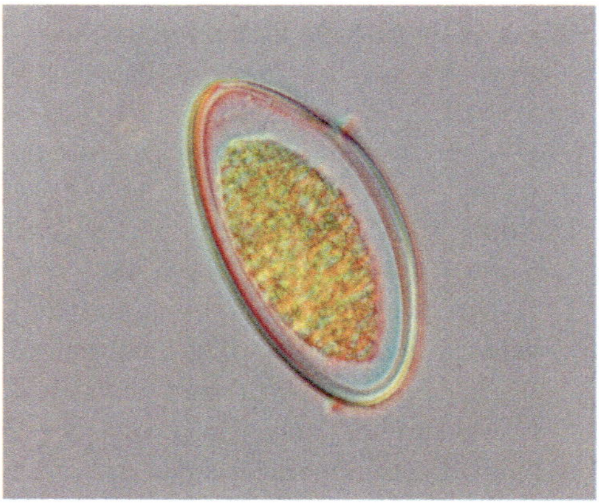

Fig. 5.29 Light micrograph of the egg of *Echinostoma revolutum*, which contains only a quantity of cells but not yet the miracidium

7. **Prophylaxis**: Nearly impossible, since infections occur outside of stables, where wild birds may "infect" water ponds.
8. **Incubation period**: Mostly unknowns: weeks.
9. **Prepatent period**: Species specific: e.g. *Echinostomum revolutum*: 15–19 days; *Echinoparyphium* sp. 7–12 days; *Apatemon gracilis*: 4 days; *Cotylurus cornutus*: 4–7 days; *Trichobilharzia cameroni*: 3–4 weeks.
10. **Patency**: Species specific: 6–10 months in *E. revolutum*; 13 days in *A. gracilis*; one month in *C. cornutus* and 4 months in *T. cameroni*.
11. **Therapy**: The literature shows test results for **triclabendazole** and **flubendazole** (Flubenol®: 5 days each 10–50 mg/kg bodyweight per os). Other compounds like benzimidazole or praziquantel will be effective, too.

Further Reading

Azizi H et al (2015) Experimental life cycle of *Hypoderaeum conoideum* (Echinostomatidae). Iran J Parasitol 10:102–104.

Georgieva S et al (2014) *Echinostoma revolutum.* Parasites Vectos 7:520.

Morley NJ et al (2010) The effects of host size and temperature on the emergence of *Echinoparyphium recurvatum* cercariae from *Lymnaea peregra* under natural light conditions. J Helminthol 84:307–316.

Morley NJ et al (2010) The effects of host size and temperature on the emergence of *Echinoparyphium recurvatum* cercariae from *Lymnaea peregra* under natural light conditions. J Helminthol 84:317–326.

Nagataki M et al (2015) Mitochondrial DNA sequences of 37 collar-spined echinostomes. Inf Genetics Evol 35:56–62.

Rondeland D et al (2015) Larval trematode infections ion *Lymnaea glabra* populations. Parasite 22:38.

Saijuntha W et al (2011) Genetic characterization of *Echinostoma revolutum* and *Echinoparyphium recurvatum.* Parasitol Res 108:751–755.

Sereno-Uribe AL et al (2015) Morphological and molecular analyses of larval and adult stages of *Echinoparyphium recurvatum.* J Helminthol 84:458–464.

Yank X et al (2015) Mitrochondrial genome of *Hypoderaeum conoideum*-comparison with selected trematodes. Parasites Vectors 8:97.

I. Digenetic trematodes parasitizing in fishes

1. **Name**: Greek: *stoma* = mouth; *azygos* = single; *sphaera* = ball-like, sphere; *bunos* = hill; *derma* = skin. Latin: *crepidula* = small bowl.
2. **Geographic distribution/epidemiology**: Worldwide.
3. **Biology, morphology**: The number of digenetic trematodes which parasitize as adults in the intestinal tract of fishes is very large (=fishes are **final hosts**). However, fishes may also serve as intermediate hosts in cases they are bearing the infectious **metacercariae** of other trematodes. For these parasites, the fishes act as **intermediate hosts**.

   The dioecious adult flukes are attached with the help of their two suckers at the intestinal wall and excrete operculated eggs, which already contain the miracidium larva, which after hatching invades water snails or mussels, which then become the first intermediate hosts inside which asexual reproduction lead via sporocysts (eventually rediae) finally to infectious cercariae. These stages enter finally second intermediate hosts such as small crustaceans (Entomostraca), water insect stages, etc. If these hosts are ingested by fishes, the metacercariae develop into adult worms inside species-specific regions of the intestinal tract.

   Important species are:

   (a) ***Crepidostomum farionis***. This fluke lives inside the intestine of fresh water fishes (e.g. salmons, trouts) and reaches a size of 2–6 mm × 1.5 mm. Diagnostic criteria are pigmental dots along the pharynx, a ventral sucker being situated in the anterior half of the body and testes laying behind each other in the terminal half of the body. The eggs measure 65–-85 μm × 40–45 μm. First intermediate hosts are probably mussels, while larvae of ephemerid flies act as secondary intermediate host.

(b) The related species of the genus ***Sanguinicola*** live in the kidneys.

(c) ***Azygia lucii***: This species parasitizes in the stomach and throat of fresh-water fishes such as perch, trout, pike, etc., reaches a size of 5.5 cm × 1–5-mm and has a cylindrical outer appearance. The two testes are located behind each other in the hind body of its host. Eggs release a cilia-free miracidium. Intermediate hosts are water snails such as e.g. *Lymnaea palustris*.

(d) ***Sphaerostoma* species**: These species parasitize a broad spectrum of hosts (e.g. carps, eels, perches, pikes) and reach a length of 4.2 mm. Their ventral sucker is situated in the centre of their body and is double sized compared to the oral one. The two testes are situated behind each other, and the ovary is located between both testes. The eggs measure 76 μm × 60 μm in mean; two intermediate hosts are involved: (**a**) water snails (e.g. *Bulimus* species) and (**b**) leeches of the genus *Herpobdella*. The different *Sphaerostoma* species ingest mainly the intestinal contents of their host and thus are much less damaging than other species which feed at intestinal organs of their hosts.

(e) ***Bunodera* species**: These species parasitize in freshwater fish such as perches, pikes and trouts and reach a length of 4.5 mm. Their head may be protruded with the help of a mobile neck region. The testes are situated at the very end of their body, while the ovary is located just before. Their eggs are rather large (100 μm × 50 μm). Intermediate hosts are mussels; accidental hosts may be *Cyclops* crustaceans and related species.

4. **Symptoms of disease**: The pathogenicity of the above-listed trematodes is rather variable depending not only on the amount of parasites included in the tissues. Infected fishes show retardation in growth, loss of weight, necrosis, intestinal inflammations and weakness that enhances secondary bacterial or viral infections.
5. **Diagnosis**: Microscopical demonstration of the operculated eggs within the feces. Sections show the adult worms attached at the intestinal wall.
6. **Pathway of infection**: Oral by ingestion of infected intermediate hosts.
7. **Prophylaxis**: Disinfection measurements in fish ponds and avoiding too large numbers of fishes inside ponds.
8. **Incubation period**: Species specific: ranging from days to weeks.
9. **Prepatent period**: Species specific, mostly several weeks.
10. **Patency**: Lifelong in adult fishes.
11. **Therapy**: Application in ponds: R.P. blue grain concentrate (Verman®) (Chlor-tetracycline together with di-n-butylcinnoxide) plus vitamins: 10 g/kg mixed food for 1 day. Repetition after 5 days. Waiting for fish meat: 30 days. For use in smaller ponds: praziquantel (1 × 5 mg/kg fish within dry food). For ornamental fishes: Tremazol® (Fa. Alpha-Biocare, Neuss, Germany) sold via Fa. Sera, Heinsberg, Germany) used as medicinal bath.

Further Reading

Evans NA (1977) The site preferences of two digeneans (*Asymphylodora kubanicum* and *Sphaerostoma bramae* in the intestine of the roach. J Helminthol 51:197–204.

Kuhn JA et al (2015) Parasite communities of two three-spined stickleback populations. Parasitol Res 114:1327–1339.

Poddubnaya LG et al (2012) Ultrastructure of the vitellarium in the digeneans *Phyllodistomum angulatum* and *Azygia lucii*. Acta Parasitol 57:235–246.

Quilichini Y et al (2007) Ultrastructural study of spermiogenesis and the spermatozoon of *Crepidostomum metoecus*: a parasite of *Salmo trutta*. J Parasitol 93:458–468.

5.2 Tapeworms (Cestodes)

The systematic position of the intestine-less, extremely flattened cestodes, which got their descriptive name from the Greek term "*kestos*" = belt, band, is under constant discussion. Most systems, however, accept the existence of two main groups, which are characterized by the number of hooks in their larval stages. The **Cestodaria**, which are non-important with respect to economy and human/ animal diseases, possess 10 hooks and thus are described as **decacanth**. On the other hand, the larvae of the **Eucestoda** possess only six hooks and are thus described as **hexacanth**.

The **Cestodaria**, which include some groups of the Monogenea, are unsegmented, dioecious worm-like appearing stages without the scolex, which is the typical holdfast organ of tapeworms.

A. **Traditional system** (extract)
Phylum: PLATYHELMINTHES
Class: Cestoda (extract)
1. Subclass Cestodaria: (decacanth larvae, 10 hooks)
Order: Amphilinidea
Order: Gyrocotylidea
2. Subclass Eucestoda: (hexacanth larvae, six hooks) (Table 5.4)
e.g. Order Caryophyllidea
Order: Pseudophyllidea
Family: Diphyllobothridae
Order: Proteocephalea
Order: Cyclophyllidea
Family: Dioecocestidae
Family: Hymenolepididae
Family: Taeniidae
Family: Mesocestoididae
Family: Dilepididae
Family: Davaineidae

(continued)

Family: Anoplocephalidae
Family: Dipylididae
Phylogenetic System (extract)
Phylum: PLATYHELMINTHES
Cercomeromorphae
Monogenea
Cestoda (=tapeworms in large sense)
Gyrocotylidae
Nephroposticophora
Amphilinidea
Cestoidea (=tapeworms in narrow sense)
Caryophyllidea
Eucestoda (Table 5.4)
(This group contains several families like those in the traditional system)

The **Eucestoda** (Table 5.4)—apart from the Caryophyllidea, which are considered as larvae with mature sexual organs (**neoteny, progenesis**) or as a complete different group—show the following regions of their tape-like body:

- **Scolex** (=head) (Fig. 5.30),
- **Proliferation zone** and
- **Strobila** (=chain of segments) (Figs. 5.31 and 5.32).

Scolex (Head)

The scolex is in relation to the size of the whole adult worm very small reaching mostly just 1–2 mm in length. However, the scolex of the Eucestoda is equipped by several holdfast systems, which allow the firm attachment at the intestinal wall of their hosts and are also means used for species and/or genus diagnosis. There may occur the following structures:

(a) **Bothria**: They occur as pairs of longitudinal, slit-like grooves being equipped by rather weak muscles (Fig. 5.30).
(b) **Acetabula**: This term describes circularly arranged suckers, which occur in species-specific numbers at the surface of adult tapeworms. For example, the members of the Cyclophyllidea possess four suckers, which are symmetrically arranged around the scolex. These suckers are provided with strong muscles. In the case of the Proteocephala, a fifth sucker is additionally present at the top of the scolex.
(c) **Rostrum (Rostellum)**: Among the Cyclophyllidea, some species possess a scolex that may contain a protrudable region (Fig. 5.38). This portion is armed in some species with a crown of hooks, which help to attach the worm at the intestinal wall (e.g. *Taenia solium*, *T. pisiformis*, *Dipylidium caninum*, *Hymenolepis* (*Rodentolepis*) *nana*, *Echinococcus* species, *Multiceps multiceps*, etc.) In other species such a crown of hooks or even single hooks are absent (e.g. *Taenia saginata*, *Hymenolepis diminuta*, *Mesocestoides*

Table 5.4 Important species of Eucestodes (selection)

Order/species	Length of the adult worm (m)	Egg size (µm) (mean)	Final host (s)	Prepatency (weeks)	Intermediate host (IH) /tissues	Stage in intermediate host/name of species in intermediate host
Order Pseudophyllidea *Diphyllobothrium latum*	Up to 20	50 × 70	**Humans**, cats, dogs	3	1. **IH**: Crustaceans (Copepoda); abdominal cavity 2. **IH**: Fishes; muscles 3. **IH**: Raptor fishes	1. IH: Procercoid 2. IH: Plerocercoid (=Sparganum) 3. IH: Plerocercoid (=Sparganum)
Spirometra erinacei europaei	1	35 × 60	Cats, dogs, **humans**	2–4	1. **IH**: Crustaceans (Copepoda); abdominal cavity 2. **IH**: Frogs, snakes; muscles	1. IH: Procercoid 2. IH: Plerocercoid (=Sparganum) 3. IH: Plerocercoid (=Sparganum)
Order Proteocephala *Proteocephalus amblоplitis*	0.3	40	Perches, raptor fishes	4	1. **IH**: Crustaceans (Copepoda); 2. **IH**: Fishes; muscles	1. IH: Procercoid 2. IH: Plerocercoid (=Sparganum) 3. IH: Plerocercoid (=Sparganum)
Order Cyclophyllidea Family Taeniidae *Taenia solium*	2–7	35–40	**Humans**	5–12	Pigs, **humans**; several tissues	Cysticercus; *C. cellulosae*
Taenia saginata	8–12	35–40	**Humans**	10–12	Cattle; several tissues	Cysticercus; *C. bovis* (*C. inermis*)
Taenia (=Hydatigera) taeniaeformis	0.6	35	Cats, dogs	7	Rats, mice; several tissues	Strobilocercus; *Cysticercus fasciolaris*
Taenia hydatigena	1	20	Dogs	11–12	Ruminants; ommentum of the intestine	Cysticercus; *C. tenuicollis*
Taenia ovis	1	30	Dogs, foxes	6–7	Sheep; muscles	Cysticercus; *C. ovis*
Taenia pisiformis	0.5–2	35	Dogs, cats	6	Rodents; ommentum	Cysticercus; *C. pisiformis*
Taenia (=Multiceps) multiceps	0.4–1	33	Dogs, foxes	6	Sheep, **humans**; brain	Coenurus; *C. cerebralis*
M. serialis	0.2–0.7	35	Dogs, foxes	1–2	Hare, rabbits; connective tissues	Coenurus; C. serialis

Echinococcus granulosus	2.5–6 mm	35	Dogs, wolves, foxes	6–9	Ruminants, **humans**: liver, lung	Hydatid; *Echinococcus hydatidosus* = *cysticus*
E. multilocularis	1.4–4 mm	35	Foxes, cats, dogs	4–6	Mice, **humans**; liver and others	Alveolar cyst; *Echinococcus alveolaris*
E. oligarthrus	Up to 2.9 mm	35	Wild cats		Rodents, **humans**	Unicystic echinococcosis
E. vogeli	4–6 mm	35	Forest dogs		Rodents, **humans**	Polycystic echinococcosis
Family Mesocestoididae *Mesocestoides lineatus*	0.3–2	40	Foxes, dogs, cats	2–3	1. **IH**: Mites (suspected); abdominal cavity 2. **IH**: Rodents, birds; abdominal cavity	1. **IH**: Cysticercoid 2. **IH**: Tetrahyridium
Family Dipyliidae *Dipylidium caninum*	0.2–0.5	50	Dogs, foxes, cats	2–2.5	Dog flea; abdominal cavity	Cysticercoid
Fam. Hymenolepididae *Vampirolepis* (syn. *Hymenolepis*) *nana*	20–40 mm	40–50	**Humans**, rodents	4	Insects; abdominal cavity, but also direct development	Cysticercoid
Hymenolepis carioca	30–80 mm	60–70	Gamefowl	2–3	Insects; abdominal cavity	Cysticercoid
Family Anoplocephalidae *Moniezia expansa*	4–10	50	Sheep, goats, cattle	4–6	Moss mites (oribatids); abdominal cavity	Cysticercoid
Anoplocephala perfoliata	25–80 mm	50	Horses	4–6	Moss mites (oribatids); abdominal cavity	Cysticercoid
Family Davaineidae *Davainea proglottina*	1–4 mm	30	Chicken	2	Slugs; several tissues	Cysticercoid
Raillietina tetragona	25 mm	35	Chicken	6	Insects; abdominal cavity	Cysticercoid

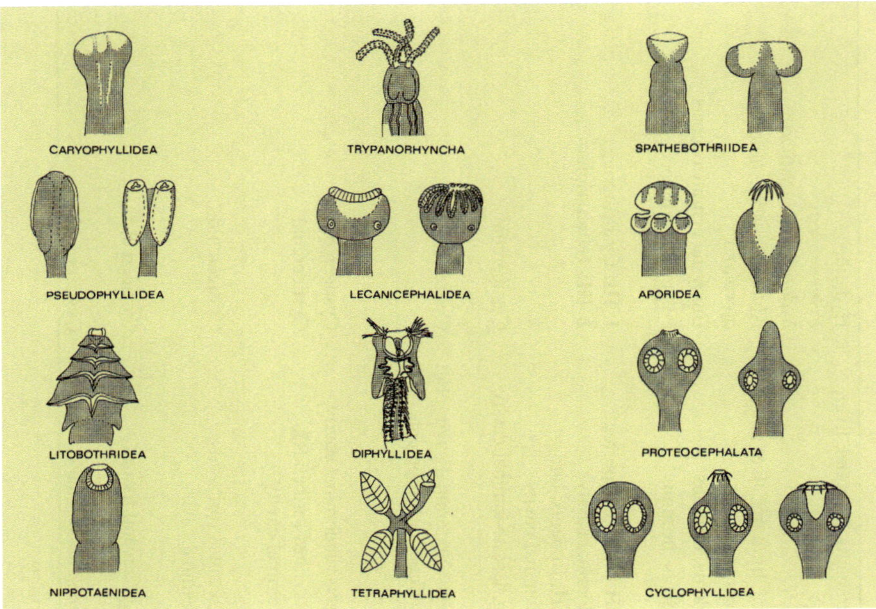

Fig. 5.30 Diagrammatic representation of the scolices of different groups of tapeworms

lineatus; Figs. 4.85b, 5.30, 5.40 and 5.51). The scolices of many other—not considered here—species possess many other structures and systems, which have been developed to solve the problem of the fixation of the worm at the intestinal wall.

Proliferation zone

This zone has often only a length of several millimetres in very long cestodes. However, this region is very important, since this is the production site of the cellular basis of the growing strobila. It is characterized by high-speed cell divisions and by cell differentiation processes. This enormous activity can be blocked by application of the anthelmintic compound praziquantel, since tapeworms treated with low doses of praziquantel showed exclusively damages. However, many processes need further examination.

Strobila

The body of the Eucestoda appears with so-called **proglottids**, which support the impression that the worm consists of a chain of separate units. However, the fact is only that the tapeworms have developed repeated groups of female and male sexual organs arranged in groups behind each other in a tape-like body, which forms a small lateral fold below each set of sexual organs (Fig. 5.31a–c). The shape of these special regions (called proglottids) may vary considerably in the groups of the cestodes (Figs. 5.31, 5.32, 5.36, 5.38, 5.40, 5.41, 5.44 and 5.48). In no case, a separating wall is formed between the proglottids. These morphological differences

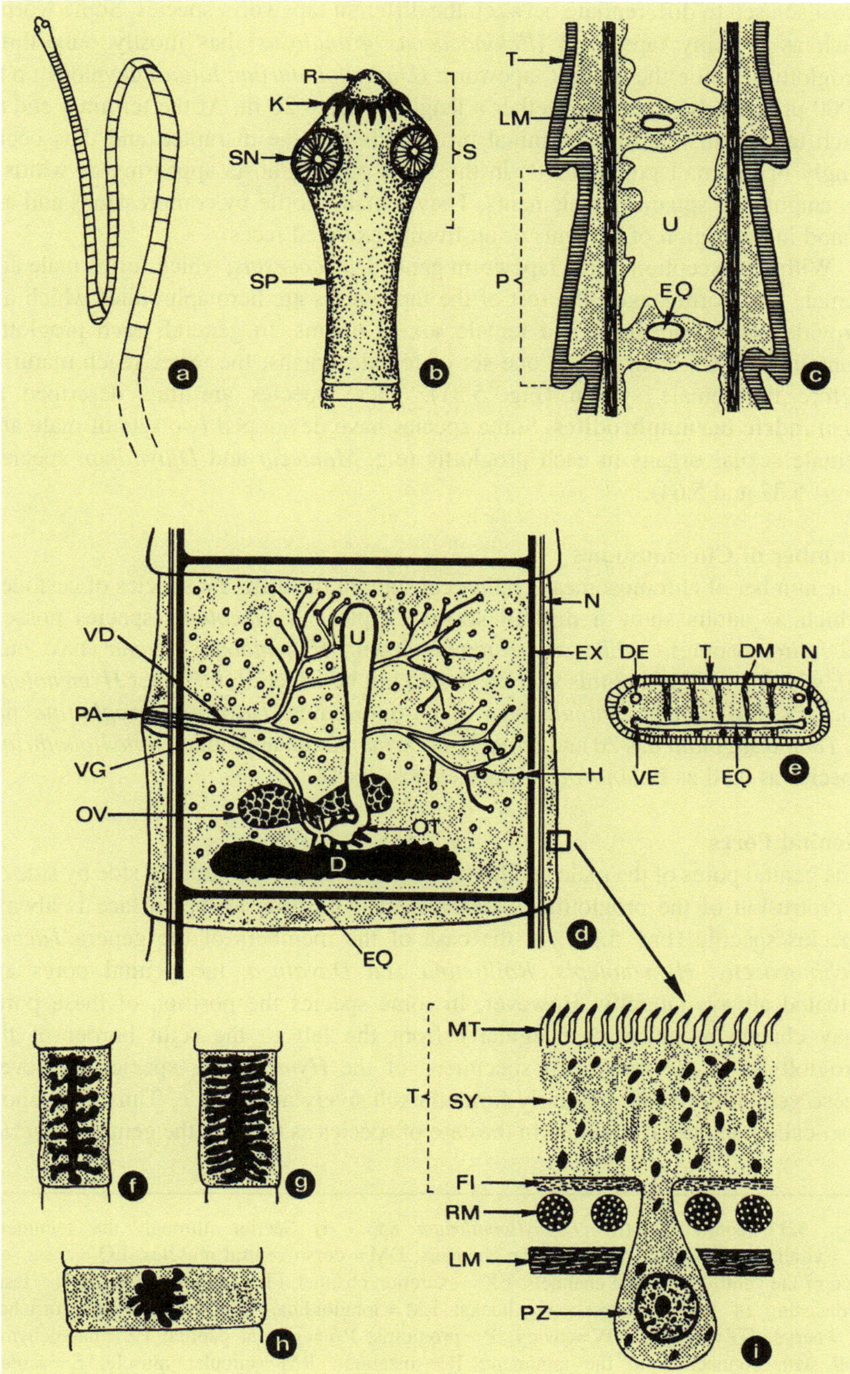

Fig. 5.31 Diagrammatic representation of the organization of cestodes (tapeworms). (**a**) Strobila; (**b**) Anterior pole; (**c**) Longitudinal section through gravid proglottids; (**e**) Dorsal aspect of a proglottis; (**f–h**) Uterus status of gravid proglottids: (**f**) = *Taenia solium*; (**g**) = *Taenia saginata*;

are also used to differentiate between the different tapeworm species. Some worms such as the tiny tapeworm (*Echinococcus granulosus*) has mostly only three proglottids, while the longest tapeworm (*Diphyllobothrium latum*) develops up to 4000 proglottids reaching together a length of up to 20 m. At the terminal end of each tapeworm, the most terminal proglottids become disrupted and thus occur singly or in small groups (2–3) in the feces of their hosts appearing as whitish, rectangular or square-like elements. They remain motile by compressions and are found in the region of the anus or on freshly excreted feces.

With the exception of the tapeworm genus *Dioecocestus*, which forms male and female tapeworm stages, the rest of the tapeworms are hermaphrodites which are provided with both male and female sexual organs. In general, each proglottis contains one set of male and one set of female organs; the testes reach maturity before the female system (Fig. 5.31). These species are thus described as **protandric hermaphrodites**. Some species have developed two sets of male and female sexual organs in each proglottis (e.g. *Moniezia* and *Dipylidium* species; Figs. 5.37 and 5.64).

Number of Chromosomes

The number of chromosomes is rather constant in the different species of cestodes, which as adults show a **diploid status**. Typical *Hymenolepis* species possess 12 chromosomes, while *Rodentolepis* (syn. *Hymenolepis*) *nana* has only 10, which is a sign that this species should not be added to the other *Hymenolepis* species. *Dipylidium caninum* has 10 chromosomes, *Davainea proglottina* has 8, *Taenia saginata* has 20 and *Diphyllobothrium latum* and other *Diphyllobothrium* species as well as *Echinococcus* species possess 18.

Genital Pores

The genital pores of the male and female sexual systems are situated side by side on a protrusion of the proglottids. However, their position at the surface is always species specific (Fig. 5.32). In the case of the members of the genera *Taenia, Echinococcus, Hymenolepis, Raillietina* and *Davainea*, the genital pores are situated always laterally. However, in some species the position of these pores may change regularly or irregularly from the left to the right border of the proglottids. In the individual specimens of the *Hymenolepis* species, however, these genital pores are regularly formed exclusively at one side. Thus, they show a so-called unilateral position. In the case of species as those of the genus *Moniezia*,

---

Fig. 5.31 (continued) (**h**) = *Diphyllobothrium* spp.; (**i**) Section through the tegument. D = vitellarium; DE = dorsal excretion channels; DM = dorso-ventral muscles; EQ = cross section of the ventral excretion channels; EX = excretion channel; FI = filamentous layer; H = testis (consisting of many spheres); K = hooks; LM = longitudinal muscles; MT = microtriches; N = nerve; OT = ootype; OV = ovary; P = proglottis; PA = genital papilla; PZ = parenchymal cell with connection to the tegument; R = rostellum; RM = circular muscle; S = scolex; SN = sucker; SP = zone of cell divisions; SY = syncytial layer (=no cell walls are present); T = tegument; U = uterus; VE = ventral excretion channel; VD = vas deferens; VG = vagina

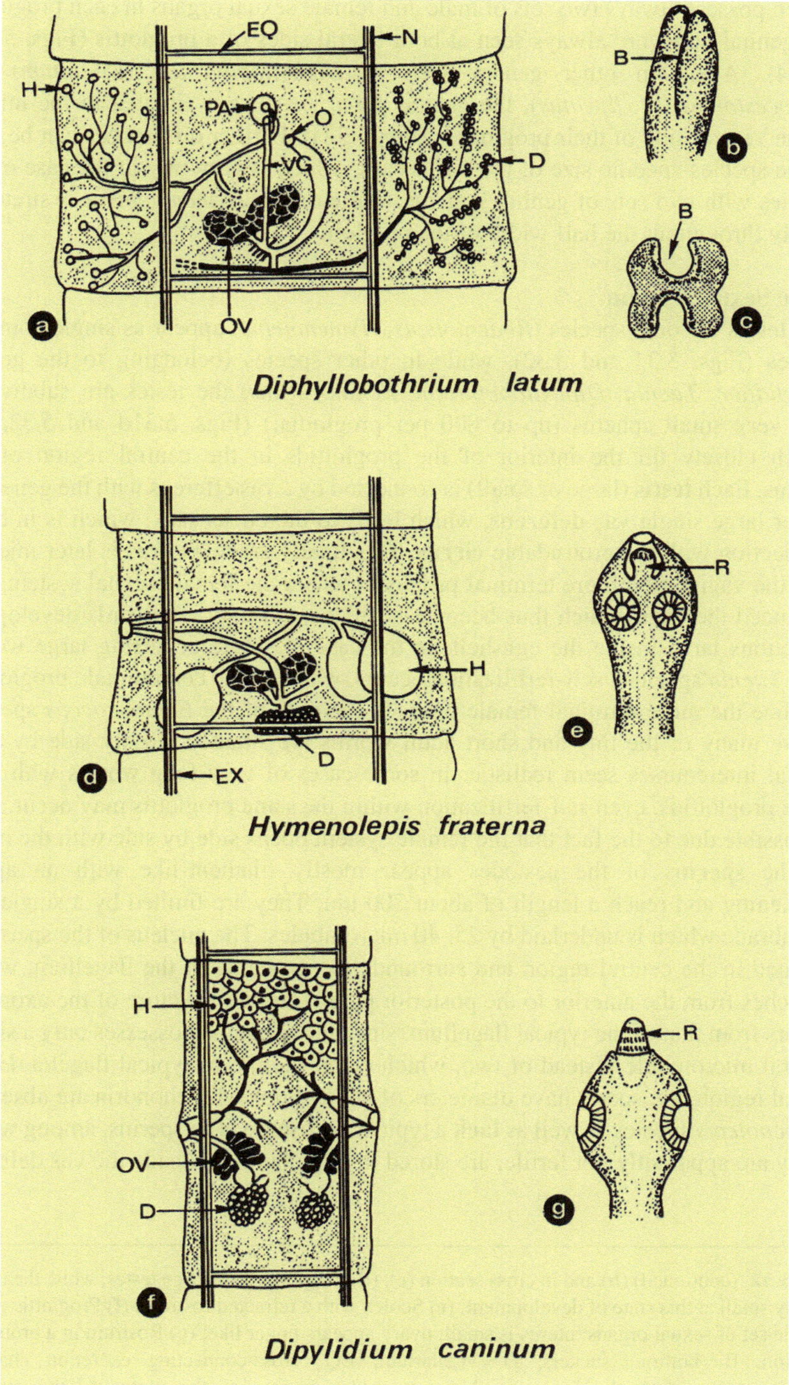

Fig. 5.32 Diagrammatic representation of proglottids and scolices of tapeworms. (**a**) Proglottis with a central sexual papillum and the opening of the uterus. The vitellarium and the testes are drawn only on one side, but they are present in both lateral regions. (**b**, **c**) Scolices in surface view

which possess always two sets of male and female sexual organs in each proglottis, the genital pores are always seen at both lateral sides of a proglottis (Figs. 5.32f; f. 64). Again in other genera such as *Diphyllobothrium* (*D. latum*) and *Mesocestoides* (*M. lineatus*), the genital pores are always situated in the middle of the ventral side of their proglottids (Fig. 5.32a). Further differences can be seen in the species-specific size of the cirrus sack, which is very tiny in the case of the species with two sets of genital systems, while e.g. in *Hymenolepis* it is stretched nearly throughout the half width of the proglottis.

Male Sexual System

The testes of some species (*Rodentolepis, Hymenolepis*) appear as single compact globes (Figs. 5.31 and 5.32), while in other species (belonging to the genera *Dipylidium, Taenia, Diphyllobothrium, Echinococcus*) the testes are subdivided into very small spheres (up to 800 per proglottis!) (Figs. 5.31d and 5.32a, f), which closely fill the interior of the proglottids in the central region of the worms. Each **testis** (large or small) is connected by a vas efferens with the generally rather large single **vas deferens**, which leads to the cirrus sack, which is in close connection with the protrudable **cirrus**. This finally erected cirrus is later injected into the vagina of a more terminal proglottis where the female sexual systems has produced the eggs, which thus become fertilized and may afterwards develop the infectious larva inside the eggshell. In the case of single occurring large worms (e.g. *Taenia* species), self-fertilization occurs, whereby the central male proglottids fertilize the more terminal female ones. In the case of the *Echinococcus* species, where many of the tiny and short adult worms are situated closely side by side, sexual intercourses seem realistic. In some cases of very long worms with very large proglottids, even self-fertilization within the same proglottis may occur. This is possible due to the fact that the female system opens side by side with the male.

The **sperms** of the cestodes appear mostly filament-like with an apical thickening and reach a length of about 200 µm. They are limited by a single cell membrane which is underlaid by 25–40 microtubules. The nucleus of the sperms is situated in the central region and surrounds the axoneme of the flagellum, which stretches from the anterior to the posterior pole. The fine structure of the axoneme differs from that of the typical flagellum, since the axoneme possesses only a single central microtubule instead of two, which occur in cilia or typical flagella. In the apical region, the sperm have diameters of about 4 µm. Mitochondria are absent in *Hymenolepis* species as well as **lack** a typical **acrosome**. The sperms, among which many are apparently not fertile, are stored in large bunches inside the vas deferens

Fig. 5.32 (continued) (**b**) and in cross section (**c**). (**d**) Proglottis with large testes, while the uterus is very small at this state of development; (**e**) Scolex with a retracted rostrum; (**f**) Proglottis with a double set of sexual organs; uterus is small; ovary appears finger like; (**g**) Rostrum in a protruded position. B = bothria = suckers; D = vitellarium; EQ = cross-connecting excretion channel; EX = excretion channel; H = testes; N = nerve; O = opening of the uterus to the outside; OV = ovary; P = proglottis; PA = sexual papilla; R = rostrum; VG = vagina

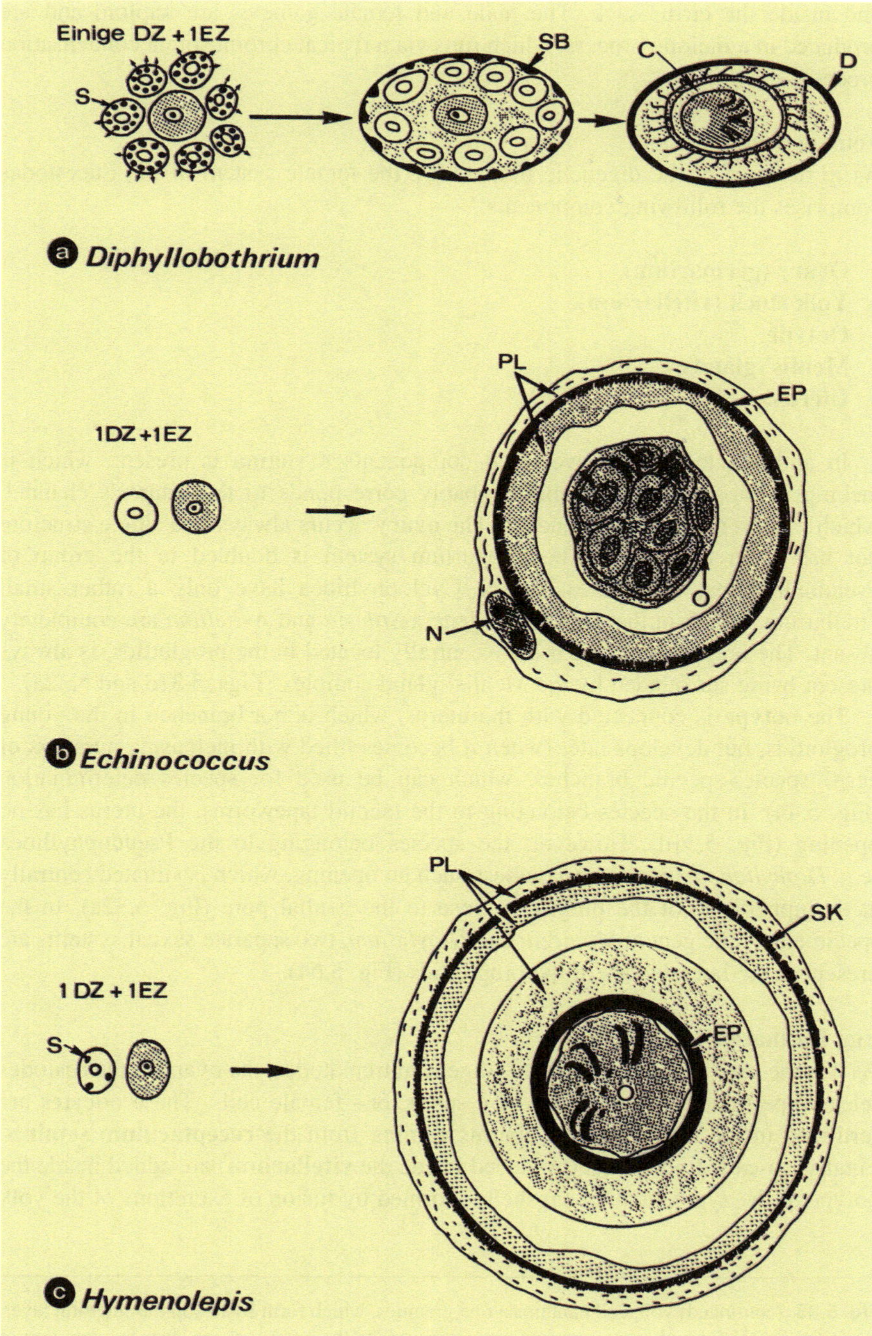

Fig. 5.33 Diagrammatic representation of different types of egg formation in tapeworms (as seen in electron microscopy). (**a**) The eggshell is formed by excretions of numerous (18) yolk cells. (**b**) The layer, which is described as embryophore, is apparently produced by a single yolk cell, which, however, does not contain visible granules. There is no sclerotization of the surface layer. (**c**) The

and inside the cirrus sack. The male and female gametes are haploid and are produced in a meiotic process, which runs via a typical chromosomal condensation process.

Female Sexual System

As in the case of the digenetic trematodes, the female system of the Eucestodes comprises the following components:

- **Ovary (germarium),**
- **Yolk stock (vitellarium),**
- **Ootype,**
- **Mehlis' glands,**
- **Uterus.**

In addition to these above-listed components a **vagina** is present, which is lacking in most trematodes, but probably corresponds to the Laurer's channel, which is present in some digeneans. The **ovary** occurs always as a single structure but has often two lobes. The vitellarium system is doubled in the group of Pseudophyllidea, while most of the Cyclophyllidea have only a rather small vitellarium, which in the case of the genera *Stilesia* and *Avitellina* are completely absent. The ootype, which is mostly centrally located in the proglottids, is always present being surrounded by the Mehlis' gland complex (Figs. 5.31d and 5.32a).

The **ootype** is connected with the uterus, which is not branched in the young proglottids, but develops later (when it becomes filled with increasing numbers of eggs) species-specific branches, which can be used for species determination (Fig. 5.44). In the species belonging to the taeniid tapeworms, the uterus has no opening (Fig. 5.31d). However, the species belonging to the Pseudophyllidea (e.g. *Diphyllobothrium latum*) possess such an opening, which is situated centrally at the upper side of the proglottis close to the genital pore (Fig. 5.32a). In the specimens of the genera *Moniezia* and *Dipylidium*, two separate sexual systems are present at the lateral sides of the proglottids (Fig. 5.64).

Egg Formation

As it is the case in the species of the digenean trematodes, the ovary of the cestodes releases periodically—regulated by a sphincter—female cells. These **oocytes** are fertilized inside the **ootype** by sperms coming from the **receptaculum seminis**. Finally, so-called yolk cells (produced inside the **vitellarium**) are added inside the ootype to the zygote and an eggshell is formed by fusion of excretions of the yolk

---

Fig. 5.33 (continued) yolk cell contains a few granules, which form by fusion the sclerotin layer, while the finally formed larva (oncosphaera) gives rise to the embryophore layer by excretion of fluid compounds. C = coracidium with oncosphaera; D = cover; DZ = yolk cell; EP = embryophore; EZ = egg cell; N = nucleus; O = oncosphaera; PL = cytoplasmatic material; S = secretion droplets; SB = shell formation; SK = sclerotized shell

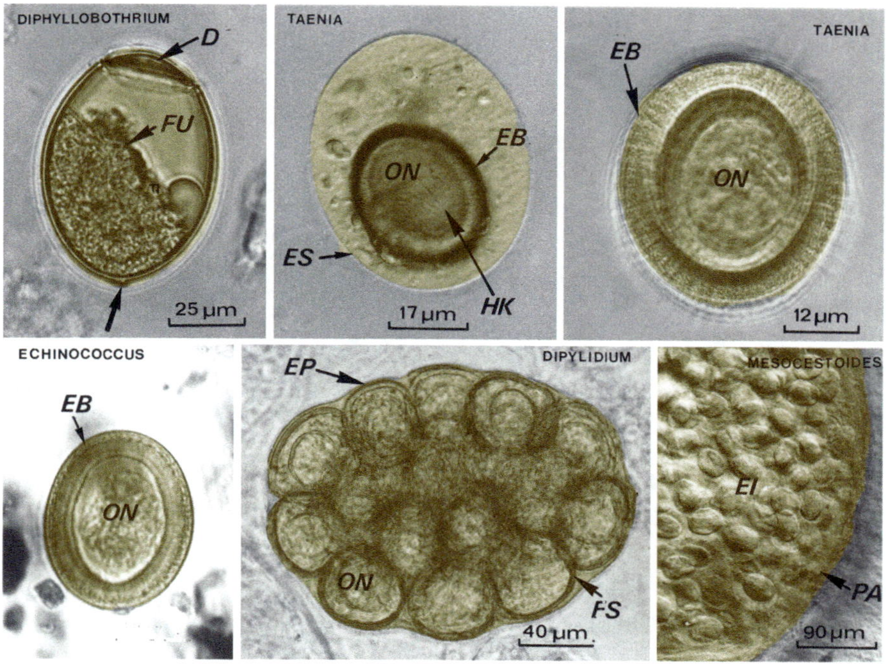

Fig. 5.34 Light micrographs of the eggs of different tapeworms. D = cover (=operculum); EB = embryophore; EI = egg; EP = egg package; ES = eggshell; FS = outer layer of the eggshell; FU = cell formation; HK = hooks; ON = oncosphaera; PA = paruterine organ

cells (Figs. 5.31, 5.32 and 5.33). The eggshell is rather thick in the case of *Diphyllobothrium latum*, since there 18–20 yolk cells excrete the material (Fig. 5.33a). However, in other species, where only one or even no yolk cell is added (as in members of the family Thysanomidae), the eggshell remains smooth and sclerotization does not occur (Fig. 5.33b, c). The fertilized eggs are released from the ootype and thus are pushed into the enlarging uterus. The further development varies considerably within the different group of tapeworms. Thus, the following two groups simplify the different processes which occur in the different species:

- **Thick-walled, sclerotized "eggs" with an operculum**:

  These so-called operculated eggs occur in the species belonging to the Pseudophyllidea. They start their embryonic development only as soon as they have reached an outside-body situation after being excreted within the feces of their hosts.
- **Non- or low-graded sclerotized eggs**:

  These eggs, which do not possess an **operculum,** belong e.g. to the species of the genera *Taenia, Dipylidium* and *Stilesia*. In these species, the development of the larva (oncosphaera) starts already inside the uterus. Thus, the growing larvae

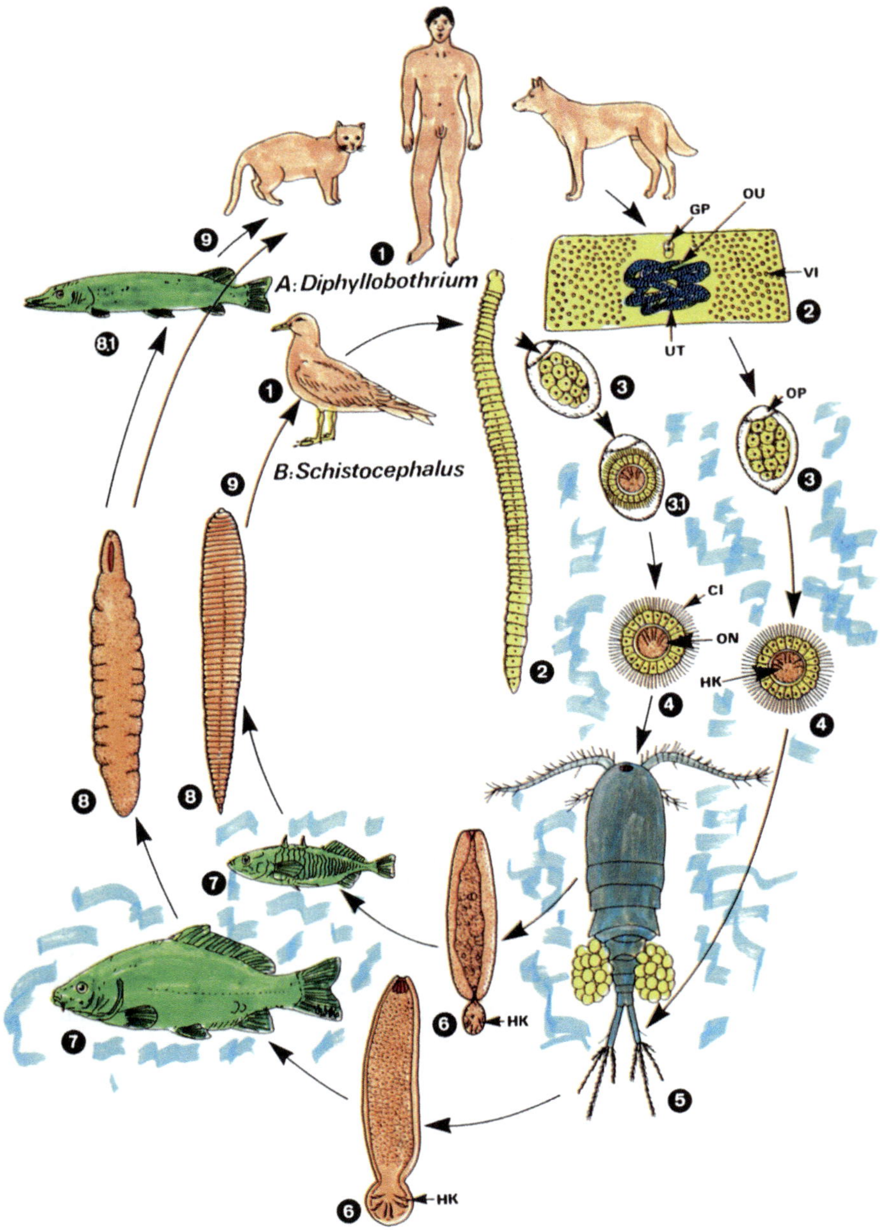

Fig. 5.35 Diagrammatic representation of the life cycle of pseudophyllidean cestodes. (**A**) *Diphyllobothrium latum* inhabits the intestine of human, cats, dogs and other fish-eating animals (**final hosts**), being attached to the intestinal wall with two longitudinal bothria. (**B**) *Schistocephalus solidus* occurs in the intestine of a wide range of fish-eating birds. *Ligula intestinalis* has a very similar life cycle. **1** Final hosts. **2** Adults. *D. latum* reaches a maximum size of 24 m; its mature proglottids are broader than long; coils of the gravid uterus form a centrally located rosette. *S. solidus* is lanceolate shaped with a size of 5–8 × 1 cm, bothria-like apical indentations are of poor adhesive power. **3** The operculated eggs are excreted unembryonated; completion of development to coracidium larva (**3.1**) takes one to several weeks depending on the

mainly contribute to the later occurring, rather thin eggshell, which is described as **embryophore** (Figs. 5.33 and 5.34).

In each of the proglottids, species-specific numbers of eggs are formed. Thus, the proglottids of the *Echinococcus* species contain about 200 eggs, while in *Taenia* species up to 100,000 eggs may be produced. They are in reality—as shown above—no true eggs but represent a conglomerate of cells developing finally the young larva. The uterus of the group of the last proglottids (=**gravid proglottids**) is fully filled with the species-specific eggs and thus appear in a species-specific shape (Fig. 5.44). Such terminal proglottids are released (**apolytically**) singly or in groups from the **strobila** (=chain of all proglottids). For example, in the case of the *Taenia* species about 3–10 proglottids become released at the same time and are excreted within the feces of the host. In the case of *Diphyllobothrium* spp., the eggs are already released still inside the intestine, so that the proglottids seen in the feces are empty and the typical eggs are mixed within the host's feces. This type of excretion of small bands of empty proglottids is described as "**pseudo-apolytic**". In other species, the proglottids

Fig. 5.35 (continued) water temperature. **4** Free coracidium larva containing the oncosphaera which is endowed with six hooks. **5–6** Having ingested free coracidia, several species of copepods are suitable intermediate hosts within which development of second-stage larvae (procercoid, 6) occurs. **7–8** As second intermediate hosts, brackish and freshwater fish become infected by ingesting infected copepods. Inside the intestine, the procercoid is released and eventually bores its way into the body cavity and muscles where it grows rapidly into a plerocercoid (sparganum). In *D. latum* the plerocercoids remain mainly undifferentiated, whereas in *S. solidus* the plerocercoids show the main features of the adults (i.e. division into 62–92 proglottids and the presence of genital anlagen; however, they are not yet fertile). Unlike *D. latum*, the progenetic plerocercoids of *S. solidus* are extremely specific in their host, developing only in the body cavity of the marine and freshwater forms of the 3-spined stickleback (*Gasterosteus aculeatus*). **8.1** In *D. latum* plerocercoids may become accumulated without further development in the muscles (not encysted) of carnivorous fish (paratenic host). **9** Infections of final hosts occur by ingestion of raw meat of fish containing plerocercoids. Having reached the intestine, the plerocercoids of *D. latum* grow rapidly and become adult worms in 5–6 weeks, whereas *S. solidus* plerocercoids mature rapidly (within 36–48 h) and release eggs. Humans, who accidently eat meat of fish containing plerocercoids of other nonhuman pseudophyllideans, may also become infected; however, plerocercoids do not mature there, but creep around inside the human body, leading to a disease called **sparganosis**. CI = cilia; GP = genital pore; HK = hooks of ON; ON = oncosphaera; OP = operculum; OU = opening of the uterus; UT = uterus; VI = vitellarium. Diagrammatic representation of the developmental cycles of the so-called fish tapeworms *Diphyllobothrium latum* and *Schistocephalus solidus*. **1** Adult worms live in the intestines of their hosts—a broad spectrum of fish-eating mammals. **2–6** The terminal proglottids (**2**) release operculated eggs, which contain a ciliated larva or the anlagen of a larva (depending on the species (**3**). This larva is called **coracidium** (**4**). If such larvae are ingested by the first intermediate host (=copepod crustaceans) (**5**), the so-called **procercoid** larva is developed (**6**). Such procercoids are infectious for fishes (=second intermediate hosts), within which the transformation to the **plerocercoid larva** occurs. These larvae are infectious for the final host as soon as they ingest raw, larva-containing fish meat. Raptor fish (**9**), which ingest other infected fishes, may serve as so-called **staple hosts**, where often large numbers of infectious plerocercoids have been observed. CI = cilia; GP = genital pores; HK = hooks; ON = oncosphaera larva; OP = operculum, cover of the egg; OU = opening of the uterus; UT = uterus; VI = vitellarium are already released **in the intestine** from the strobila just after fertilization. Such phenomena are described as **euapolytic** or **hyperapolytic**.

It is noteworthy that those eggs, which leave the uterus via an opening, are mostly thick walled and thus protected in the new surroundings. This protection outside of the host's body is given in the other cases by the fact that the whole proglottis is excreted and that wall of embryophores is rather thick. In some cases (e.g. *Dipylidium caninum* or *Raillietina carioca*), several eggs glue together and are enclosed in a common capsule so that "egg packages" are formed (Fig. 5.34). A similar function has apparently the so-called **paruterine organ**, which occurs in *Mesocestoides* species. This "organ" is formed by fortification of the whole uterus (Fig. 5.48) enclosing all eggs. Most of the tapeworm eggs are very resistant against influences outside of the body of their hosts. In tests, it was proven that eggs of *Echinococcus* species may survive temperatures reaching from minus 40 °C to plus 40 °C. Thus, storage of dead but potentially *E. multilocularis* containing foxes should be done at minus 80 °C or at 60 °C plus.

5.2.1 Tapeworms of Carnivores

Carnivores such as dogs and cats are (like humans) final hosts of tapeworms. Inside their intestine live the adult worms, which release periodically or even permanently proglottids, which contain often already fully infectious larvae (Fig. 5.34). In case that these eggs are ingested (within fecally contaminated food) by **intermediate hosts**, the infectious larva is developed inside their body. If these larvae are ingested (e.g. in raw meat) by final hosts, the larva grows up to a hermaphrodite adult worm. The male sexual organs reach maturity inside the proglottids of the first third of the body and fertilize the female stages (eggs) being situated in the proglottids just behind the midregion of the tapeworm, while in the terminal proglottids formation of the infectious larvae occurs. These eggs are stored inside the uterus and are only set free as soon the mature proglottids have been excreted within the feces of the final host and have set free the eggs after rupture of the proglottids, wherein they had been developed.

5.2.1.1 Fish Tapeworms Inside Dogs and Cats

1. **Name**: Greek: *dis* = two; *phyllon* = leaf; *bothrium* = slit, groove; *speira* = spiral; *metra* = uterus; Latin: *latum* = large, wide.
2. **Geographic distribution/epidemiology**: Worldwide in regions with freshwater lakes or brackish waters.
3. **Biology, morphology**: ***Diphyllobothrium latum*** is only rarely found in dogs and cats but nevertheless represents the most common fish tapeworm of carnivores. Even more rare are infections with ***Spirometra erinacei***, which looks very similar to *D. latum* (Fig. 5.35). This worm reaches in dogs a length of up to 3 m and possesses a scolex with two longitudinal grooves (bothria) and its eggs measure 67–75 μm × 45–54 μm (Fig. 5.34). The (when excreted)

empty proglottids are broader than long and measure 20 mm × 5 mm. The eggs are from the operculated type. They are unembryonated when excreted, but develop quickly the coracidium larva, which possesses cilia along its surface that enables them to swim. **First intermediate hosts** are small crustaceans (copepods, *Cyclops*), **second intermediate hosts** are fishes, within which the infectious plerocercoid stage is developed. If dogs and cats are fed with raw or undercooked fish, they become infected and act as final hosts bearing the adult worm (Fig. 5.35).

4. **Symptoms of disease**: Infections with single worms remain often symptomless for a long time. However, large worms or multiple infections may introduce disturbances of digestion and rarely also anaemia. In other hosts than dogs or cats (e.g. in humans or pigs), the ingested plerocercoids do not become mature but are stored as **spargana** in muscles and may introduce inflammations as well as (rarely) vitamin B12 deficiency reactions.
5. **Diagnosis**: Demonstration of eggs within feces, since the uterus has an opening and releases still inside the host's intestine most of the eggs. Several empty proglottids are always found in the feces clutching together (Fig. 5.36).
6. **Pathway of infection**: Oral uptake of larvae (plerocercoids) in raw fish meat.
7. **Prophylaxis**: Do not feed raw fish meat to cats or dogs.
8. **Incubation period**: 3–6 weeks.
9. **Prepatent period**: 1–6 weeks (depending on the species).
10. **Patency**: Years.
11. **Therapy**: Application of **praziquantel** (1 × 5 mg/kg bodyweight) as oral application.

Further Reading

Kavana NJ et al (2014) The life cycle of *Spirometra* species from Peninsular Malaysia. Trop Biomed 31:487–495.

Scioscia NP et al (2014) The Pampas fox (*Lycalopex gymnocercus*) as a new definitive host for *Spirometra erinacei*. Actas tropica 133:78–82.

Usmanova NM, Kazakou VI (2010) The DL1 repeats in the genome of *Diphyllobothrium latum*. Parasitol Res 107:449–452.

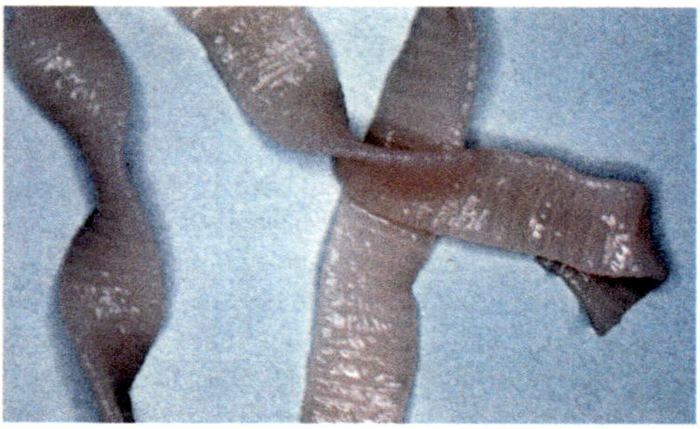

Fig. 5.36 Macrophoto of a tape consisting of terminal proglottids of *Diphyllobothrium latum* excreted by a host after treatment with praziquantel

Wang F et al (2014) *Spirometra* severely infecting wild-caught snakes from food market in Guangzhou and Shenzhen. Scientists World J 2014: ID874014.

Wicht B et al (2010) Inter- and intraspecific characterization of tapeworms of the genus *Diphyllobothrium* from Switzerland using nuclear and mitochondrial DNA targets. Parasitol Int 59:35–39.

Yoneva A et al (2014) First study of vitellogenesis of the broad fish tapeworm *Diphyllobothrium latum*. Parasitol Int 63:747–753.

5.2.1.2 *Dipylidium caninum* (Cucumber Seed Tapeworm)

1. **Name**: Greek: *dis* = two; *pyle* = pore. Latin: *canis* = dog.
2. **Geographic distribution/epidemiology**: Worldwide.
3. **Biology, morphology**: This tapeworm, which is the most common one in dogs and which also infects humans, may reach lengths of 20–50 cm. The scolex has a width of only 0.5 mm and possesses 4 suckers and a rostrum with 3–4 crowns of hooks. In each proglottids, two sets of male and female sexual organs are formed (Figs. 5.37 and 5.38). This can be only hardly recognized in terminal proglottids, since they are completely filled with the 200 × 120 μm sized egg packages each containing 8–30 eggs which measure 34–40 μm in diameter

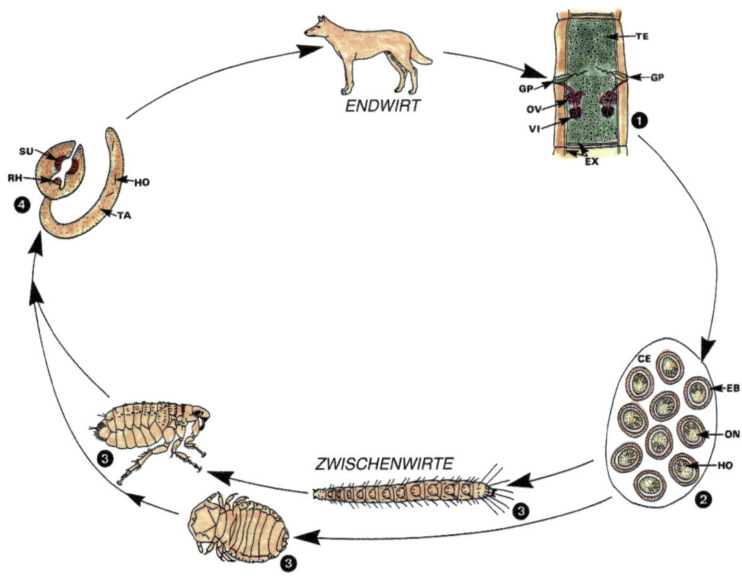

Fig. 5.37 Diagrammatic representation of the developmental cycle of one of the tapeworms of dogs (*Dipylidium caninum*). **1–3** The dog excretes proglottids (**1**), which contain egg packages (**2**). If larvae of mallophages or fleas (**3**) ingest such egg packages, the cysticercoid larvae finally occur also inside adult fleas and mallophages. If these are ingested by the same dog a reinfection occurs, while infections of other dogs occur as soon as the fleas or mallophages have entered another host. CE = egg package; EB = embryophore; ES = eggshell; EX = excretion vessel; GP = genital pore; HO = hook; ON = oncosphere; OV = ovary; RH = hook at the rostellum; SU = sucker; TA = tail of cysticercoid larva; RE = testes; VI = vitellarium; Endwirt = final host; Zwischenwirt = intermediate host

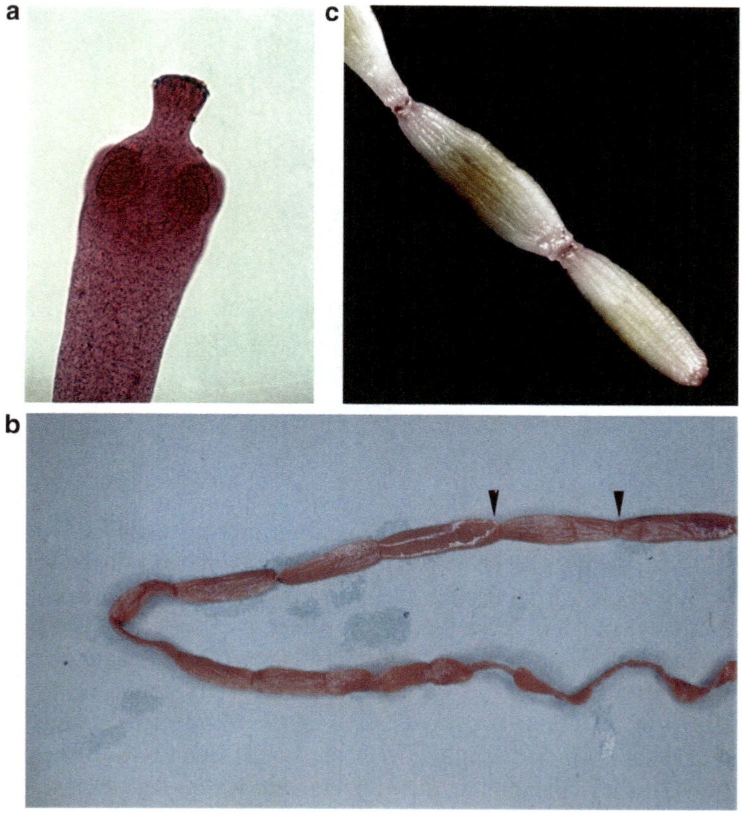

Fig. 5.38 Light micrographs of *Dipylidium caninum*. (**a**) Anterior end with protruded rostrum. (**b**) Ricegrain-like terminal proglottids obtained from dog feces. (**c**) Portions of the strobila being excreted after treatment with praziquantel

(Fig. 5.34). The released gravid proglottids appear similar to cucumber seeds and are known to creep around for a rather long period on the discharged feces (or in the hair around the dog's anus). During this activity, the surface of the proglottids is dried and the egg packages and single eggs containing the oncosphaera larva are set free. If detritus feeding insects (e.g. flea larvae, biting lice) ingest such eggs, they become the first intermediate host, within which the oncosphaera larva is transformed into the **cysticercoid** larva, which reaches infectivity as soon as this intermediate host has reached its adult stage.

4. **Symptoms of disease**: The itching, which occurs as soon as many motile proglottids creep around the anus, leads to an often seen reaction of dogs: they rub their anus along the soil, which looks like sledging (Fig. 5.39). However, in many cases symptoms remain low grade or unspecific (stress, loss of weight, digestion disturbances, hair disorders, etc.). Extremely heavy infections with numerous adult worms may lead to blocking of the intestine and death. However, these cases are extremely rare.

Fig. 5.39 Diagrammatic representation of a dog sliding its anus on the soil in order to decrease itching

5. **Diagnosis**: Demonstration of the typical cucumber seed-like proglottids in fresh feces, which can also be recognized by their reddish-brownish colour (Fig. 5.38b, c). During drying, these proglottids shrink and appear ricegrain like. If these proglottids are placed into water, the single egg packages can be identified with the help of the light microscope. They measure 200 × 120 μm (Figs. 5.34 and 5.37).
6. **Pathway of infection**: **Dogs, cats**: Oral uptake of infected intermediate hosts (fleas, biting lice) during nibbling in the fur. **Humans**: Mostly children are infected, if they contaminate their fingers with worm eggs during stroking of pet animals and have afterwards hand-mouth contacts.
7. **Prophylaxis**: Repeated treatment of pet animals with anti-worm drugs, use of anti-ectoparasite compounds and keeping clean the resting places of the domestic animals.
8. **Incubation period**: Variable: 43–6 weeks.
9. **Prepatent period**: 2–3 weeks.
10. **Patency**: About 1 year; however, reinfections are common, so that the worm load persists in many cases.
11. **Therapy**: Chemotherapy with praziquantel accompanied by application of anti-flea compounds on pet animals in the surroundings of human dwellings.

Further Reading

Beugnet F et al (2013) Preventive efficacy of Frontline® Combo and Certifect® against *Dipylidium caninum* infestation of cats and dogs using a natural flea infestation model. Parasite 20:7.

Beugnet F et al (2014) Occurrence of *Dipylidium caninum* from client-owned cats and dogs in Europe using a new PCR detection assay. Vet Parasitol 205:300–306.

Fourie JJ et al (2013) Prophylactic treatment of flea-infested dogs with an imidacloprid/flumethrin collar (Seresto®, Bayer) to preempt infection with *Dipylidium caninum*. Parasitol Res 112:S36–S46.

5.2.1.3 *Taenia* Species

1. **Name**: Greek: *taina* = tape; *hydatis* = bladder containing water. Latin: *pisum* = pea; *forma* = structure, shape; *multiceps* = with many heads; *crassiceps* = fatty headed.
2. **Geographic distribution/epidemiology**: Worldwide.
3. **Biology, morphology**:
   (a) ***Taenia pisiformis*** (Figs. 5.40, 5.41 and 5.42).
   The adult worms reach a length of 2 m. The gravid terminal proglottids measure 8–10 mm × 4–5 mm. The egg containing uterus shows a median strand with 8–14 lateral ones (Fig. 5.44). **Intermediate hosts** are rabbits, hares and several rodents, within which occurs the pea-sized, whitish cysticercus larva (Fig. 5.43).

Fig. 5.40 Scanning electron micrograph of the scolex and the anterior developing proglottids of the tapeworm *Taenia pisiformis*

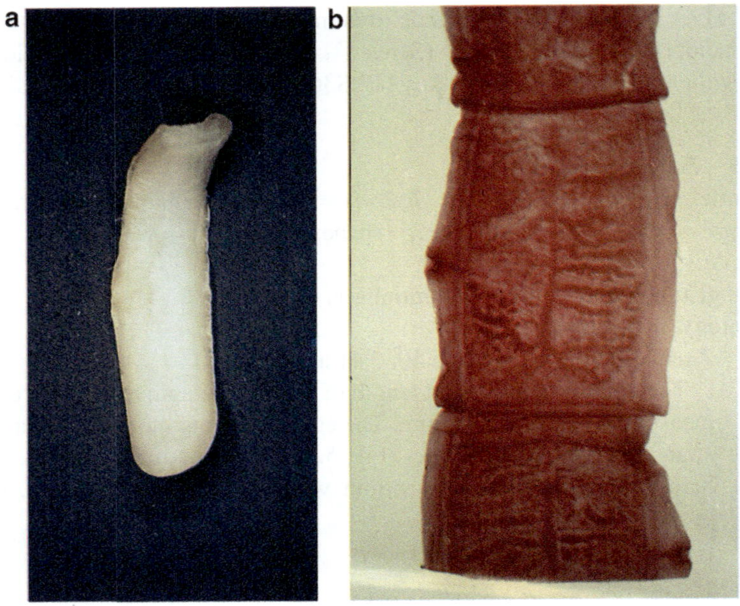

Fig. 5.41 (**a**) Macrophoto of a typical *Taenia* proglottis found in feces. (**b**) Micrograph of three ruthenium red-stained *Taenia* proglottids showing the egg-filled uteri with their typical lateral branches

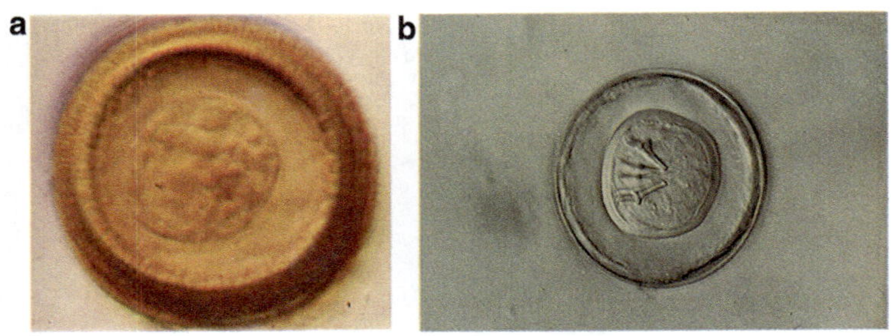

Fig. 5.42 Light micrographs of tapeworm eggs without the thin outer egg capsule. (**a**) Example of the genus *Taenia*. The outer thick-walled layer is the so-called embryophore; inside can be seen the so-called 6-hook larva = oncosphaera. (**b**) Example of an egg of the species *Hymenolepis diminuta*, where the six hooks of the oncosphaera larva can be clearly seen

(b) ***Taenia ovis***:

This tapeworm reaches a length of about 120 cm; the gravid proglottids reach a size of 8×4 mm; the uterus shows 20–25 lateral branches (Fig. 5.44). **Intermediate hosts** are sheep and goats, where the thin-walled cysticercus larva occurs inside skeletal muscles and in the heart (Fig. 5.45).

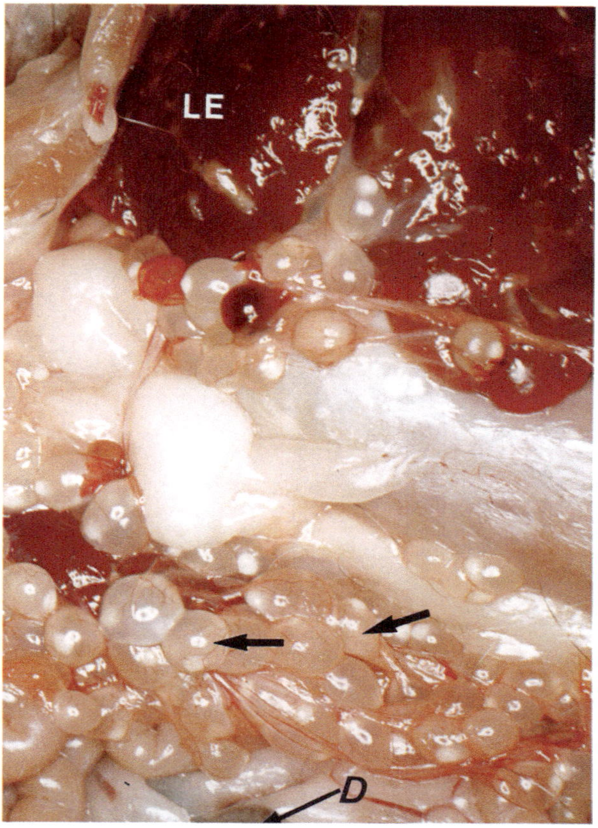

Fig. 5.43 Macrophoto of the ommentum region of a rodent containing large numbers of the pea-sized, bladder-like cysticerci of *Taenia pisiformis* (*arrows*). D = intestine; LE = liver

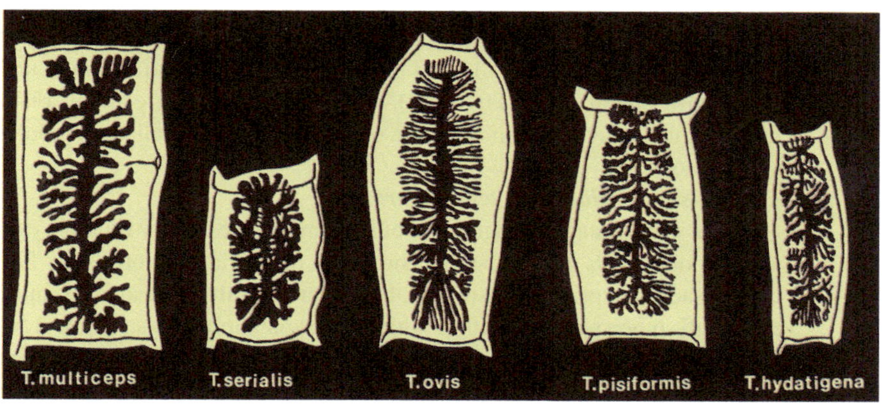

Fig. 5.44 Diagrammatic representation of the terminal proglottids of some *Taenia* species of carnivores, showing the typical branching patterns of the uterus

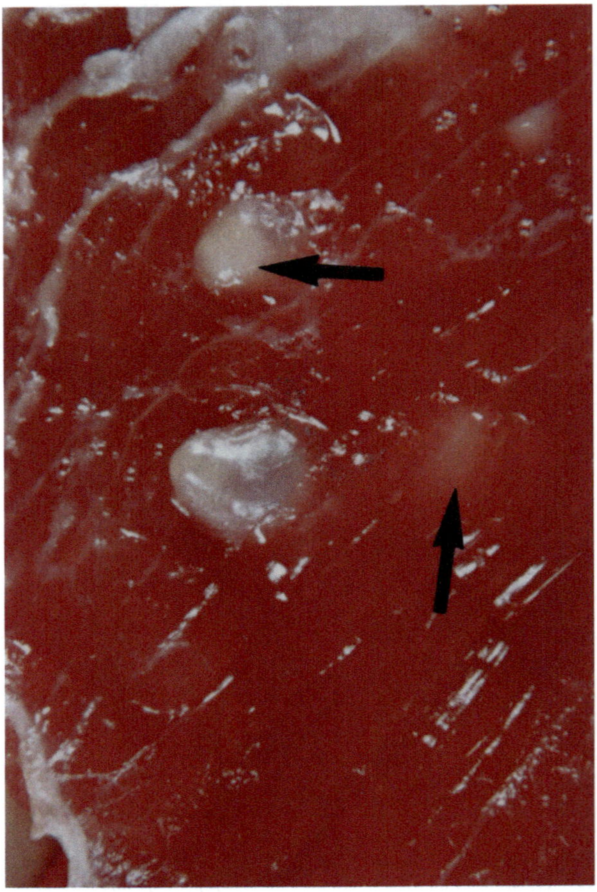

Fig. 5.45 Macrophoto of **cysticerci** of *Taenia ovis* in the muscles of a sheep (*arrows*)

(c) ***Taenia hydatigena***:

This tapeworm reaches a length of about 1 m. The uterus of the gravid proglottids, which measure 12 × 6 mm, shows a short median strand and 6–10 pairs lateral ones (Fig. 5.44). Intermediate hosts are many plant feeders, in the liver and mesentery of which the thin-walled cysticerci are developed.

(d) ***Taenia* (syn. *Hydatigera*) *taeniaeformis***:

These worms reach a length of about 60 cm and are mainly found in cats and do not possess a neck region. The uterus of gravid proglottids has 5–9 lateral, sack-like protrusions (Fig. 5.44). **Intermediate hosts** are rats, mice and many other rodents, within which the tapeworm-like strobilocercus larva is located reaching a length of up to 3 cm and ending in terminal bladder (Fig. 5.46).

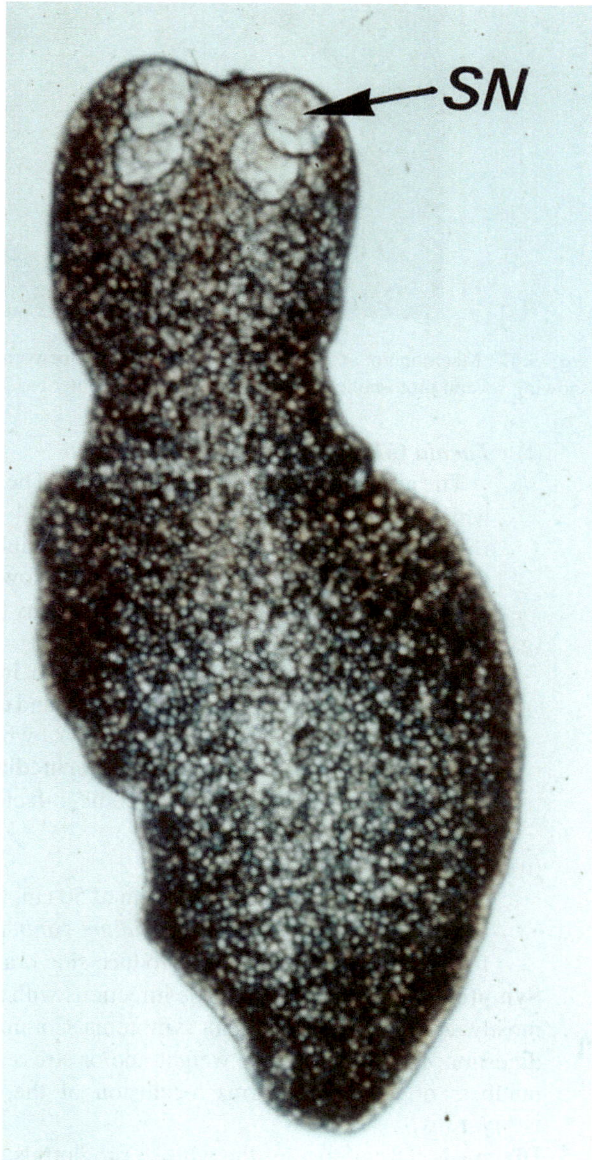

Fig. 5.46 Macrophoto of a so-called strobilocercus of the tapeworm *Taenia taeniaeformis*. SN = sucker

(e) ***Taenia krabbei* (syn. *T. cervi*)**:

The adult worm reaches a length of 2–5 m. Its terminal proglottids measure 12–15 mm × 5–7 mm and contain an uterus with 10–12 pairs of lateral strands which are further branched. **Intermediate hosts** are deers bearing the cysticercus larvae inside their muscles.

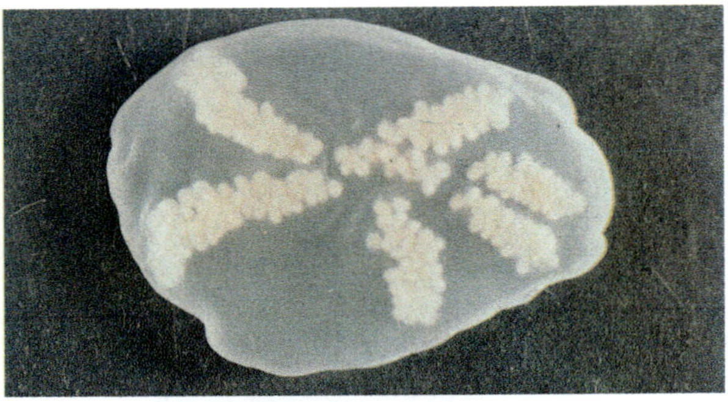

Fig. 5.47 Macrophoto of a coenurus larva of the tapeworm *Taenia (Multiceps) multiceps* showing several protoscolices at the inner side of the outer capsule

(f) ***Taenia (Multiceps) multiceps***:
The adult worms reach a length of 1 m. The uterus of gravid proglottids, which measure 8–12 mm × 5 mm, forms 14–26 pairs of lateral branches (Fig. 5.44). **Intermediate hosts** are mainly sheep. The neurotropic ***Coenurus cerebralis*** larva (Fig. 5.47), however, develops also in the brain of many other animals and in **humans (!)**.

(g) ***Taenia serialis***:
The adult worms measure up to 70 cm in length. Gravid proglottids have a size range of 8–16 mm × 3–5 mm and contain an uterus with 20–26 pairs of lateral branches (Fig. 5.44), which, however, often have interconnections with each other. **Intermediate hosts** are many rodents. However, also humans may become infected. The *Coenurus* larva is always settling in the muscles.

(h) ***Taeni crassiceps***:
The adult worm reaches a length of 50 cm, thus being the second largest tapeworm of dogs behind *Dipylidium caninum*. The cysticercus larva is found in mice and is able to produce side branches at its terminal pole.

4. **Symptoms of disease**: Low-grade infections with a single worm or a few ones mostly do not lead to significant symptoms. Common signs are disturbances of digestion, low-grade loss of weight and/or stress symptoms. In cases of high numbers of adult tapeworms, occlusion of the intestine may occur (*ileus verminosus*).
5. **Diagnosis**: Occurrence of the whitish proglottids, which can be seen with the naked eye in the feces. Eggs obtained by enrichment methods show often even their thin capsule (Fig. 5.34), which, however, is often destroyed during preparation, so that they then appear to be limited by the radially striated wall of the embryophore (Fig. 5.42a). Inside the embryophore always the fully developed and infectious **oncosphaera** larva (=six-hook larva) is located (e.g. Fig. 5.42b). The diameter of the eggs in this state of development is species specific but ranges at a level of 30–35 μm.

6. **Pathway of infection**: **Final hosts**: Oral uptake of the infectious larvae (e.g. cysticercus, strobilocercus, coenurus) within raw meat of intermediate hosts, which may become infected by oral uptake of worm eggs in fecally contaminated food.
7. **Prophylaxis**: Do not feed raw meat to dogs, cats, etc. (final hosts).
8. **Incubation period**: Species specific; mostly several weeks.
9. **Prepatent period**:

Taenia pisiformis	~50 days
T. ovis	60–130 days
T. hydatigena	50–70 days
T. taeniaeformis	36–80 days
T. (Multiceps) muilticeps	45–47 days
T. crassiceps	31–42 days

10. **Patency**: Years (often 2–5).
11. **Therapy**: **Praziquantel** is a broad band spectrum antihelminthic against tapeworms (1 × 5 mg/kg bodyweight—oral/subcutaneous). Further products (with a less large spectrum) are **bunamidin-hydrochloride** (e.g. Scoluban® 25–30 mg/kg bodyweight orally—maximum 600 mg/per animal) or **nitroscanate** (Lopatol®) (50 mg/kg bodyweight, orally).

    Benzimidazole carbamates eliminate by a single cure both tapeworms and trematodes: **Fenbendazole** (Panacur®) 3 days each 50 mg/kg bodyweight orally or **mebendazole** (Telmin®) 5 days, dose depending on the bodyweight. The same efficacy is given in the case of Drontal plus® (**praziquantel + pyrantel + febantel**) when given 1 tablet per 10 kg bodyweight.

Further Reading

Ash A et al (2015) Controlling *Taenia solium*: impact of a triple dose albendazole-regime. Acta Trop. doi:10.1016/j.actatropica.2015.05.018.

Dupuy C et al (2014) Prevalence of *Taenia saginata* cysticercosis in French cattle in 2010. Vet Parasitol 203:65–72.

Eichenberger RM et al (2013) Multi-test analysis and model-based estimation of the prevalence of *Taenia saginata* in naturally infected dairy cows. Int J Parasitol 43:853–859.

McFadden AMJ et al (2011) Investigation of an outbreak of *Taenia saginata*. Vet Parasitol 176:177–184.

Fogaca RC et al (2014) Phage-displayed peptides as capture antigens in an innovative assay for *Taenia saginata* infected cattle. Appl Microbiol Biotechnol 98:8887–8894.

Krecek RC et al (2008) Prevalence of *Taenia solium* cysticercosis. Vet Parasitol 154:38–47.

Lightowlers MW (2010) Eradication of *Taenia solium* cysticercosis. Int J Parasitol 40:1183–1192.

Lopez WBZ et al (2014) Historic of therapeutic efficacy of albendazole sulphoxide administered in different routes, dosages and treatment schemes against *Taenia saginata*. Exp Parasitol 137:14–20.
Okello AL et al (2016) Assessing the impact of a joint human-porcine intervention package for *Taenia solium* control. Acta Trop. 159:185–191.
Trevisan C et al (2016) Severe seizures in pigs naturally infected with *Taenia solium*. Vet Parasitol 220:67–71.

5.2.1.4 *Mesocestoides* Species

1. **Name**: Greek: *mesos* = central; *kestos* = tape, belt.
2. **Geographic distribution/epidemiology**: Worldwide.
3. **Biology, morphology**: Some species with not fully determined life cycle reach a length of about 40 cm and a width of 2 mm (Fig. 5.48). They are found as adults in the intestine of cats, dogs (*M. litteratus*) and more often in foxes (*M. lineatus*). They are clearly determined by the occurrence of the so-called paruterine organ (Fig. 5.48) in the terminal proglottids. Their scolex shows four suckers but no hooks. **First intermediate hosts** are mites (which is not completely clear). **Second intermediate hosts** are field mice and other rodents, within which the infectious **tetrathyridium larva** is developed reaching a size of 0.5–2 cm in diameter. In non-suitable hosts such as badges and hedgehogs but also in suitable hosts such as foxes, dogs and cats, the intestinal wall can be permeated. The larva thus may enter the peritoneal cavity, where they are encapsulated but remain infectious.
4. **Symptoms of disease**: Mostly no symptoms or very low-grade ones occur as in other tapeworms.
5. **Diagnosis**: Demonstration of the typical gravid proglottids in feces (Fig. 5.48). The eggs (Fig. 5.34) measure 40–60 μm × 35–43 μm.
6. **Pathway of infection**: Oral uptake of tetrathyridea larvae (within rodents).

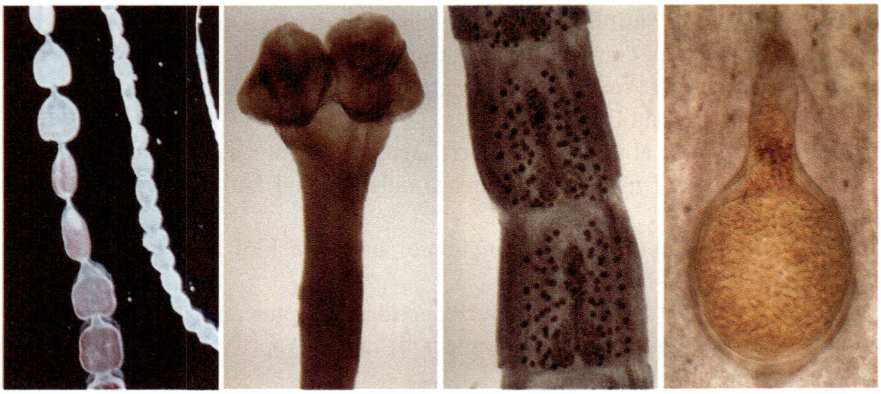

Fig. 5.48 *Mesocestoides* sp.: Light micrographs from left: strobila showing the typically shaped proglottids; unarmed scolex; proglottids (coloured) in the midregion of the body; one of the final proglottids containing the egg containing **paruterine organ**

7. **Prophylaxis**: Practically not possible; however, regular deworming keeps worm load small.
8. **Incubation period**: Variable.
9. **Prepatent period**: Depending on the host species: 2–3 weeks.
10. **Patency**: In the case of cats and dogs: months up to years.
11. **Therapy**: Use of praziquantel (1 × 5 mg/kg bodyweight).

Further Reading

Skirnisson K et al (2016) Occurrence of *Mesocestoides canislagopodis* in mammals. Parasitol Res. doi:10.1007/s00436-016-5006-5.

Szell Z et al (2015) Environmental determinants of the spatial distribution of *Mesocestoides* sp. Vet Parasitol 212:427–430.

Zalesny G, Hildebrand J (2012) Molecular identification of *Mesocestoides* spp. Parasitol 110:1055–1061.

5.2.1.5 *Echinococcus* Species

1. **Name**: Greek: *echinos* = hedge hock; *coccos* = nucleus. Latin: *granulus* = with grains; *multilocularis* = with many tiny chambers/hollows.
2. **Geographic distribution/epidemiology**: See Table 5.5.
3. **Biology, morphology**: As can be seen in Table 5.5, more and more *Echinococcus* species are described in recent times. However, they induce mostly very similar symptoms, so that the two most common species are described in the following:

Table 5.5 Species of the genus *Echinococcus*

Species, strain, genotypes (g)	Final hosts	Intemediate hosts
E. canadensis **G 8, G 10**	Dogs, wolves	Cervids, **humans**
E. equinus **G 4**	Dogs	Horses, donkeys, zebras
E. felidis	Lions	Wildlife
E. granulosus **G 1, G 2, G 3**	Dogs, canids	Sheep, cattle, goat, camel, pig, **humans**
E. intermedius **G 6, G 7, G 9**	Dogs	Pigs, goats, camels, **humans**
E. multilocularis	Foxes, dogs, cats	Rodents, pigs, horses, **humans**
E. oligoarthus	Wild cats	Rodents, **humans**
E. ortleppi **G 5**	Dogs	Cattle, sheep, goats, **humans**
E. shiquicus	Tibet fox	Pika (Family Ochotonidae)
E. vogeli	South American forest dog	Rodents, pakas, **humans**

G = genotypes (as far as known); results collected from papers of the groups of Eckert, Deplazes and Thompson

(a) ***Echinococcus granulosus***:

This species prefers as **final host** apparently the dog, while cats and foxes are less often infected (Figs. 5.49 and 5.50). The adult worm reaches a length of 2.5–6 mm, whereby the last (=gravid) proglottis is considerably larger than all others before (Fig. 5.52). The pores of the genital open just before or just behind the middle of the proglottis. The uterus of the excreted terminal proglottis shows lateral protrusions. The scolex of this rather tiny tapeworm is attached with the help of four suckers and a crown of hooks at the intestinal wall and enters deeply into the intestinal villi (Figs. 5.53 and 5.54). The hooks of the scolex are of different size: there are small ones measuring 19–35 μm and large ones 25–40 μm. All together 30–42 hooks may occur. In most cases, large crowds of worms are present being anchored very close to each other, so that fertilizations may occur by neighbouring worms. The eggs of *E. granulosus* belong to the *Taenia* type (Fig. 5.42a) and contain already prior to the discharge the **oncosphaera** larva, which is provided with six characteristic hooks. The strains of *E. granulosus* use a very broad spectrum of **intermediate hosts**; e.g. pigs, ruminants and horses are common hosts and humans may also become infected. There is a broad spectrum of animals, which are believed to harbour separate strains or even separate species (see Table 5.5) (Eckert et al. 1993). Also the so-called hydatid larval stages have slightly different shapes in these various intermediate hosts. Inside these hydatids—the name was given since they are filled with a fluid—other bladder-like compartments of which are formed along the inner layer, wherein the protoscolices (=heads) of the later tapeworms are produced asexually (Figs. 5.58 and 5.59). Each of these protoscolices grows up to a fertile worm, if a final host (dogs, cats, foxes) ingests such pieces of the hydatid. Thus, this life cycle contains a sexual generation in the final host and asexual generations in the intermediate hosts (Fig. 5.49). This type of life cycle would described as **metagenesis** (Fig. 5.55).

(b) ***Echinococcus multilocularis***:

This worm is mainly found in foxes and only rarely in dogs and cats. Adult worms measure only 1.5–3 mm and thus are considerably smaller than *E. granulosus* (Fig. 5.50). The adult specimens—although shorter—have more proglottids than *E. granulosus*. The last proglottis is often larger than all others together. The genital pores of the proglottids are situated clearly before the midst of the proglottis. The uterus of the gravid proglottis mostly does not show lateral protrusions, but appears sack like (Fig. 5.49). The scolex (Fig. 5.51) possesses a crown of 26–36 hooks, where the large ones (25–29 μm) as well as the small ones (19–24 μm) are smaller than those of *E. granulosus* worms. The eggs of this species can only hardly (if at all) be differentiated from those of *E. granulosus* or from those of *Taenia* species. **Intermediate hosts** are mainly rodents (field mice, rats). However, also humans may become infected and would develop the multilocular cyst inside its tissues, if they ingest an egg

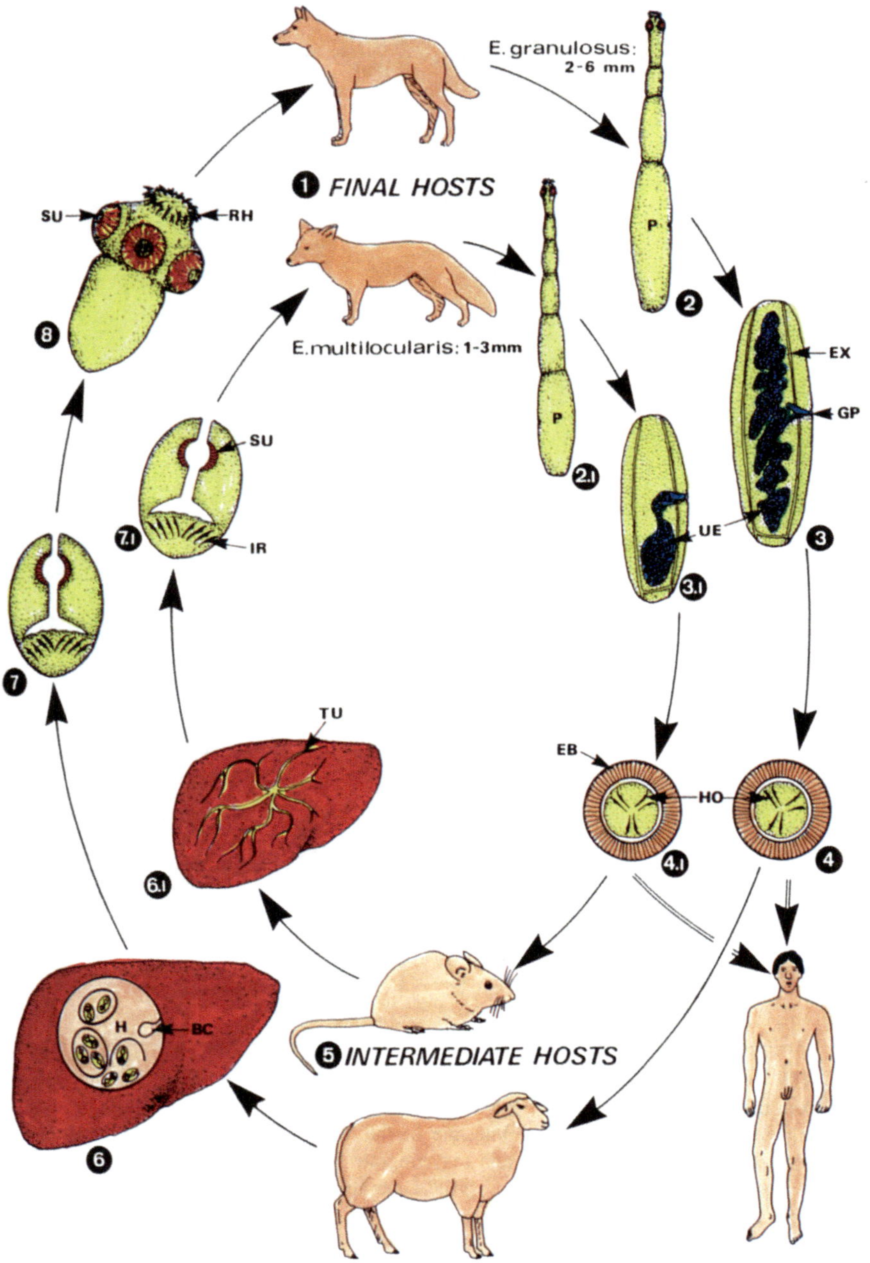

Fig. 5.49 Diagrammatic representation of the life cycles of *Echinococcus granulosus* (1–8) and *E. multilocularis* (1.1–7.1). **1**, **1.1** Final hosts may be dog, cat or fox with clear, species-specific preference. **2–3.1** Adult worms, which live in the small intestine of the final host, may be differentiated according to the size of the terminal proglottids (P), shape of uterus (UE) and size of rostellar hooks. **4**, **4.1** Eggs containing an infectious oncosphaera larva are released from the

Fig. 5.50 Scanning electron micrograph of adult worms of *Echinococcus granulosus* (*left*) and *E. multilocularis*. Note the different sizes

Fig. 5.49 (continued) detached drying proglottid in the feces of the host; eggs are indistinguishable from those of *Taenia* spp. **5**, **5.1** Eggs are orally ingested by intermediate hosts or man with contaminated food. **6**, **6.1** Inside the intestine of the intermediate hosts (including man), the oncosphaera hatches, enters the wall and may migrate (via blood) to many organs. Cysts are formed mostly in the liver and lung; in *E. granulosus* large unilocular hydatids occur, which are filled with fluid (containing thousands of protoscolices), whereas in *E. multilocularis* a tubular system infiltrates the whole organ (giving rise to alveolar aspects in sections). **7–8.1** In brood capsules of both cyst types, protoscolices are formed, which may become evaginated (8) even inside their cysts. Evaginated or not, protoscolices are fully capable to infect final hosts when these animals ingest infected organs of intermediate hosts. BC = brood capsule; EB = embryophore of the egg; EX = excretory vessels; GP = genital pore; H = hydatid; HO = hooks of oncosphaera; IR = invaginated rostellar hooks; P = proglottid; RH = rostellar hooks; SU = sucker; TU = tubular system; UE = uterus containing eggs

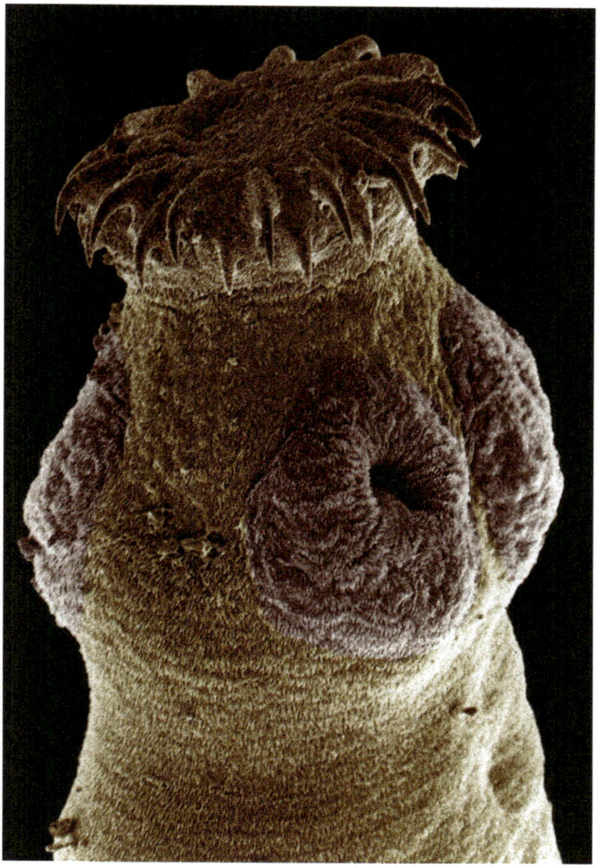

Fig. 5.51 Scanning electron micrograph of the scolex region of *Echinococcus multilocularis*

containing the typical **oncosphaera** larva (Fig. 5.49). These multilocular cysts consist of widespread tubules which do not contain a fluid but grow by solid protrusions of 10–20 μm in diameter entering the tissues of many organs (liver etc.; Figs. 5.57a, b and 5.58). Inside these tubular strands, protoscolices are formed, which infect final hosts, if they ingest such pieces of meat (Figs. 5.49 and 5.58). During surgery of humans, such fine strands with multitasking cells are set free. They become distributed via lymph and bloodstream and start to produce new cysts in other organs. Thus, surgery of humans has to be done very cautiously; otherwise single cells being set free from these multilocular cyst will act like cancer cells and start permanent reproduction.

4. **Symptoms of disease**: The final hosts (dogs, cats, foxes) mainly do not show symptoms of disease when infected with the *Echinococcus* species, so that infections often remain undetected for long. This endangers intermediate hosts, which live together in close neighbourhood with these final hosts. Only rarely occur haemorrhagic or catarrhalic symptoms in **final hosts**, so that infections

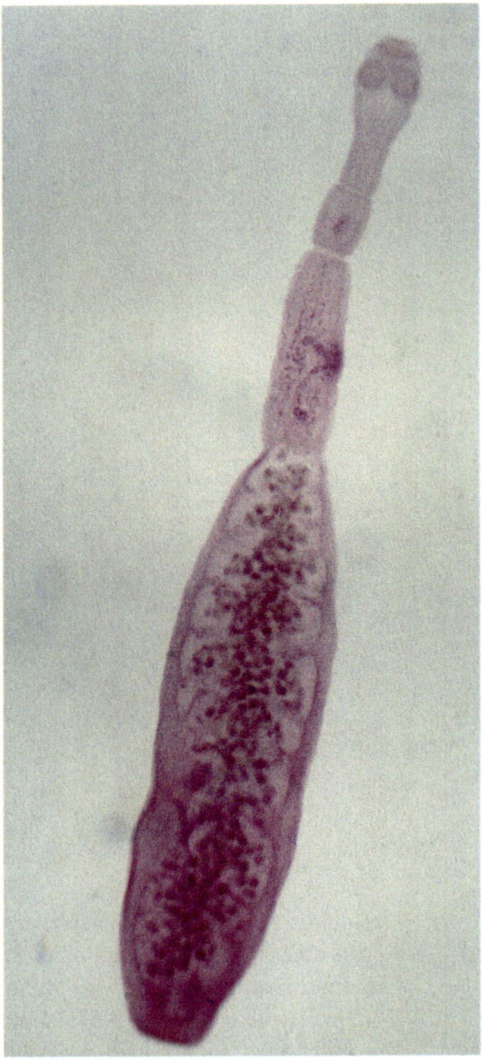

Fig. 5.52 Light micrograph of an adult (stained) worm of *Echinococcus granulosus* showing the large egg-filled proglottis

remain undetected. **Intermediate hosts**, however, show organ-specific symptoms due to the presence of growing cysts, which are formed by both *Echinococcus* species.

5. **Diagnosis**: Infected final hosts excrete in their feces the tiny whitish appearing, 1–2 mm long proglottids, which can easily be missed, if there are only worms present in the intestine. However, in cases of mass infections the feces may appear like covered by white "dust" due to large amounts of proglottids, which, however, shrink quickly in drying feces (Fig. 5.42a). **Important**: Proglottids do not appear during each defecation! However, fresh proglottids show internal structures, which give hints to which strains of *Echinococcus granulosus* the excreted proglottids belong (Fig. 5.60).

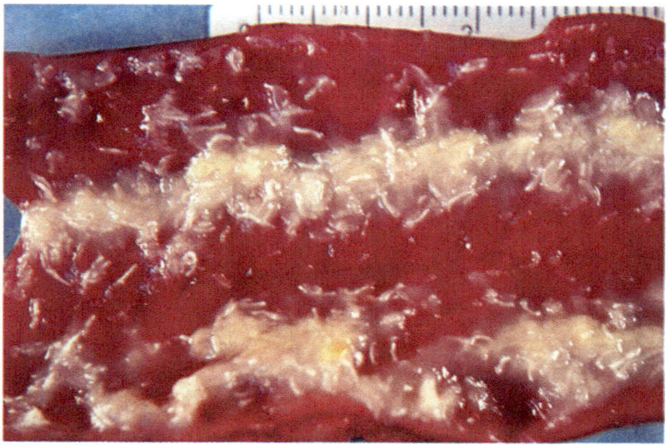

Fig. 5.53 Macrophoto of the internal aspect of the intestine of a dog containing many of the whitish appearing adults of *Echinococcus granulosus*

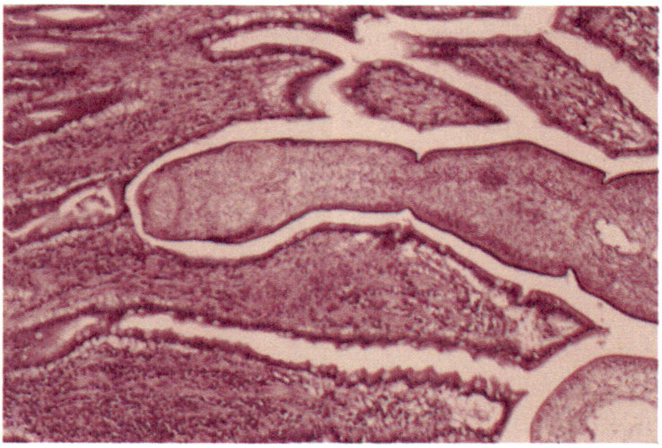

Fig. 5.54 Microscopical aspect of a section of the intestine of a fox showing the anterior end of *Echinococcus multilocularis* sticking between the intestinal villi

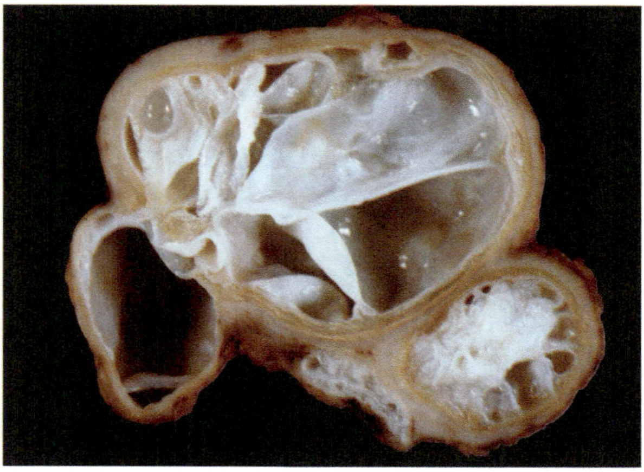

Fig. 5.55 Macrophoto of three attached hydatids of *Echinococcus granulosus* removed from the uterus of a women. Note the chamber-like compartments, which had been filled with fluid

Fig. 5.56 Macrophoto of a sheep liver being infected with numerous hydatids of the tapeworm *Echinococcus granulosus*

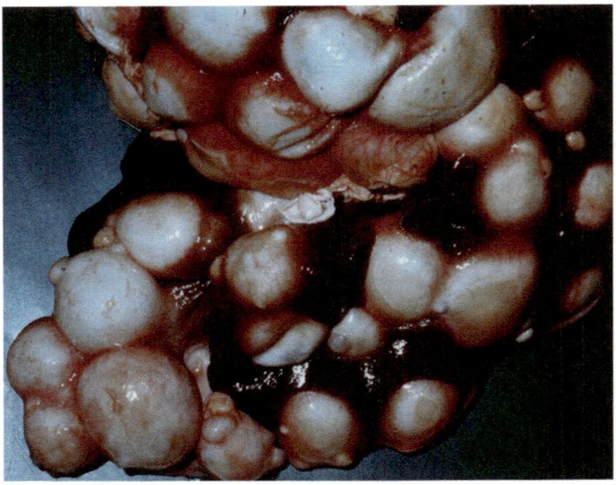

Fig. 5.57 Macrophoto of a human liver containing the strands of a so-called **alveococcus** cyst of the fox tapeworm *Echinococcus multilocularis*

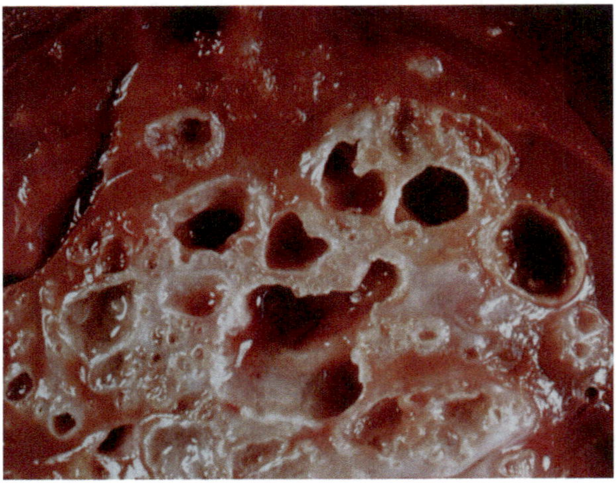

Fig. 5.58 Light micrograph of a semithin section through the periphery of a cyst of *Echinococcus multilocularis* showing two infectious protoscolices

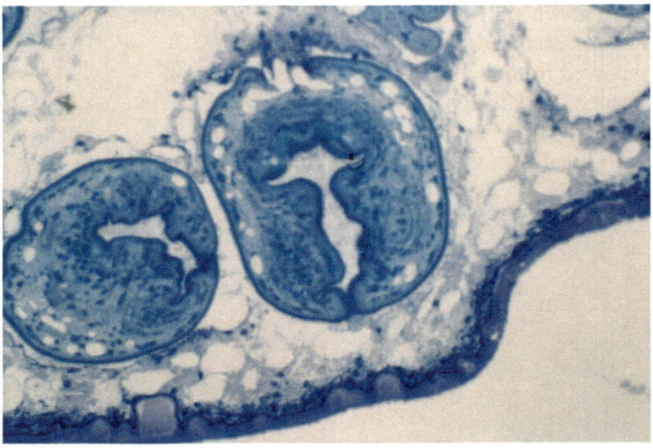

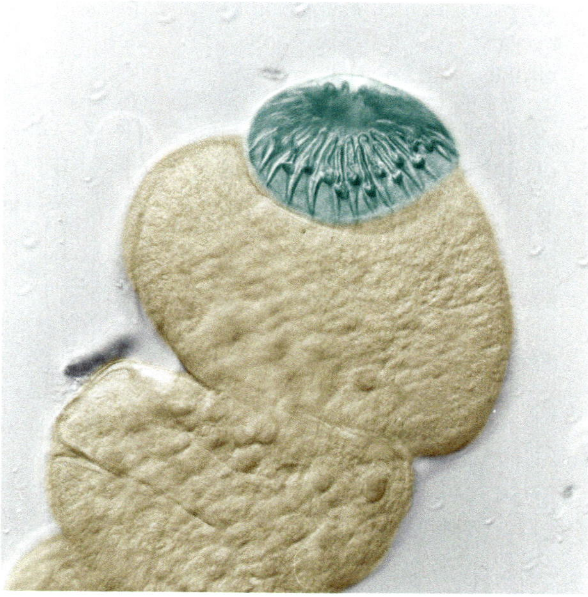

Fig. 5.59 Scanning electron micrograph of a young protruding protoscolex of *Echinococcus multilocularis*

6. **Pathway of infection**: **Final hosts**: Oral uptake of protoscolices within raw or undercooked organs of infected intermediate hosts. **Intermediate hosts**: Oral uptake of eggs excreted within the feces of final hosts (cats, dogs, foxes).
7. **Prophylaxis**: Do not feed raw meat to dogs and cats and avoid contact to their feces. Regular deworming of final hosts helps to keep away infections from humans and animals around humans. Keep farm animals away from fox feces.
8. **Incubation period**: **Final hosts**: 4–5 weeks; **intermediate hosts**: eventually years.
9. **Prepatent period**: **Final hosts**: Depending on the age and type of the host: in the case of *E. granulosus*: at least 35–42 days; in *E. multilocularis*: about 35 days.
10. **Patency**: *E. granulosus*: at least 6–7 months, rarely up to 2 years; *E. multilocularis*: rarely more than 5–6 months.
11. **Therapy**: **Final hosts (dogs, cats)**: Application of praziquantel (1×5 mg/kg bodyweight, orally or subcutaneously). **Importance**: Hot cleaning of sleeping places of infected domestic dogs/cats, washing their fur after treatment, since eggs may stick herein. In regions with many infected mice, treatment should be repeated every two months. **Intermediate hosts**: Humans: Cysts due to *E. granulosus* can be surgically removed; however, cysts due to *E. multilocularis* must be treated practically lifelong with **albendazole** or **mebendazole**, since these products do not kill the growing cysts but only stop their enlargement. **Note:** Recently Swiss veterinarians showed that dogs can also become intermediate hosts of *E. multilocularis* (www.escap.ch).

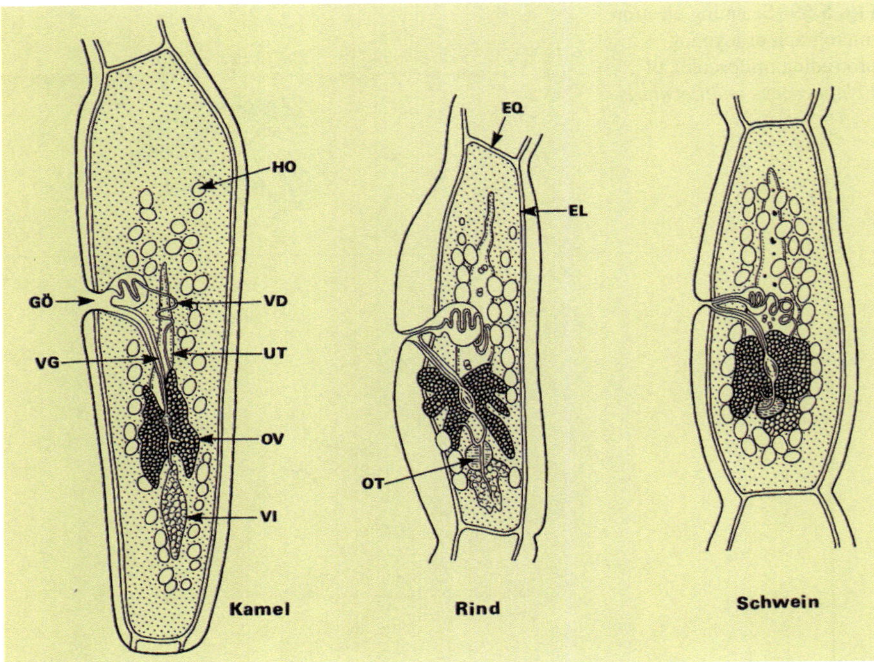

Fig. 5.60 *Echinococcus granulosus*: diagrammatic representation of the morphological investigation of defined strains of *Echinococcus granulosus* (Eckert et al. 1993) showing different proglottids. EL = longitudinal excretion channel; EQ = cross-running excretion channel; GÖ = genital opening; HO = testis; OT = ootype with Mehlis' glands; OV = ovary (=germarium); UT = uterus; VD = vas deferens; VG = vagina; VI = vitellarium; German: Kamel = camel; Rind = cattle; Schwein = pig

Further Reading

Ahmed ME et al (2013) First report on circulation of *Echinococcus ortleppi* in humped camel. BMC Vet Res 9:127.

Boulana B et al (2013) First report of *Echinococcus shiquicus* in dogs (China). Acta Trop 127:21–24.

Bonfana B et al (2015) *Echinococcus equinus* and *E. granulosus* from the UK. Int J Parasitol 95:161–166.

Carmena D, Cardona GA (2014) *Echinococcus* in wild carnivorous species. Vet Parastiol 202:69–94.

Chaignat V et al (2015) The brown hare as novel intermediate host for *E. multilocularis* in Europe. 114:3167–3169.

Chaligannis I et al (2015) *Echinococcus granulosus* infection dynamics in livestock in Greece. Parasitol Res 114:3167–3169.

Craig P et al (2015) *Echinococcus granulosus* epidemiology and state-of-the-art of diagnostics in animals. Vet Parasitol 213:132–148.

Coucher MA et al (2016) Cystic echinococcosis in South America: systematic review. Trop Med Int Health 21:166–175.

Cvejic D et al (2016) Efficacy of a single dose of milbemycin oxime/praziquantel combination tablets against adult *Echinococcus multilocularis*. Parasitol Res 115:1195–1202.

Eckert J et al (1983) Proliferation and metastases formation of larval *Echinococcus multilocularis*. I. Animal model, macroscopical and histological findings. Parasitol Res 69:737–748.

Eckert J et al (1993) Further evidence of the occurrence of a distinct strain of *Echinococcus granulosus* in European pigs. Parasitol Res 79:42–48.

Eckert J et al (2011) Alveolar echinococcosis. In: Palmer SR et al (eds) Textbook of zoonoses, 2nd edn. Oxford University Press, Oxford.

Grenouillet F et al (2014) *Echinococcus ortleppi* infections in humans and cattle, France. Emerg Inf Dis 20:2100–2112.

Hüttner M, Romig T (2009) *Echinococcus* species in African wildlife. Parasitology 136:1089–1095.

Hüttner M et al (2007) Genetic characterization and phylogenetic position of *Echinococcus felidis*. Int J Parasitol 38:861–868.

Knapp J et al (2015) Taxonomy and molecular epidemiology of *Echinococcus multilocularis*. Vet Parasitol 213:85–91.

Larrieu E et al (2015) Pilot field trial of the EG95 vaccine against ovine echinococcosis. PLoS Negl Dis: e0004134.

Liu CN et al (2015) Discrimination between *E. granulosus sensu stricto, E. multilocularis* and *E. shiquicus* a multiplex PCR assay. PLOS Negl Trop Dis: e0004084.

Mehlhorn H et al (1983) Proliferation and metastasis formation of larval *Echinococcus multilocularis*: II. Ultrastructural investigation. Parasitol Res 69:749–763.

Oksanen A, Lavikainen A (2015) *Echinococcus canadensis* transmission in the North. Vet Parasitol 213:182–186.

Raoul F et al (2015) Trophic ecology, behaviour and host population dynamics of *E. multilocularis*. Vet Parasitol 213:162–171.

Simsek S et al (2015) First report of *Echinococcus equinus* in a donkey in Turkey. Korean J Parasitol 53:731–735.

Soares MCP et al (2013) Anatomo-chemical and molecular description of liver neotropical echinococcosis caused by *E. oligarthrus* in human host. Acta tropica 125:110–114.

Soares MPC et al (2014) Neotropical echinococcosis: *E. vogeli* natural infection and its main definitive host: the bush dog (*Speathos venaticus*). Parasitol Int 63:485–487).

Thompson RC (2008) The taxonomy phylogeny and transmission of *Echinococcus*. Exp Parasitol 119:439–446.

Thompson RCA (2015) Neglected zoonotic helminth: *Hymenolepis nana, Echinococcus Canadensis* and *Ancylostoma ceylanicum*. Clin Microbiol Infect 21:426–432.

Zait H et al (2016) Molecular characterization of *E. granulosus* and *E. Canadensis*.

5.2.2 Tapeworms of Ruminants and Pigs

Ruminants may act as **final hosts** (harbouring the adult worms) as well as **intermediate hosts** (harbouring the larves) (Table 5.4).

5.2.2.1 Adult Worms in Ruminants

1. **Name**: Moniez = French helminthologist (1852–1936); Greek: *thysanos* = fringe. Latin: *ovitellus* = without yolk; *globus* = ball; *punctatus* = with dots; *expandere* = spreading. Stiles = English researcher.
2. **Geographic distribution/epidemiology**: Worldwide, often with prevalence rates of 10–30 % (!).
3. **Biology, morphology**:
   (a) ***Moniezia* species (e.g. *M. expanda*)**:
   Up to 10 m long, scolex with four slit-like suckers, without hooks (Figs. 5.61, 5.62 and 5.64). The proglottids contain two sets of sexual organs, which are located at both lateral sides of each proglottid. The eggs have a diameter of about 60 μm and are polymorphous (Fig. 5.64) containing an **oncosphaera** larva. **Intermediate hosts** are soil mites belonging to the family of Oribatida.
   (b) ***Avitellina* species (*A. centripunctata*)**:
   These worms reach a length of up to 3 m. The strobila gives the impression that it is unsegmented, since even the last proglottids have only a height of about 3 mm (Fig. 5.63b). The genital openings are laying only at one side. The eggs measure 20 × 45 μm. **Intermediate hosts** are soil mites (Oribatida) and dust mites (Psocoptera) and collembols.
   (c) ***Thysaniezia* species (*T. giardi*)**:
   These worms reach a length of up to 2 m; the proglottids have a width of about 1 cm. The genital openings alternate from one side to the other in the next proglottis. The testes have their position laterally outside of the

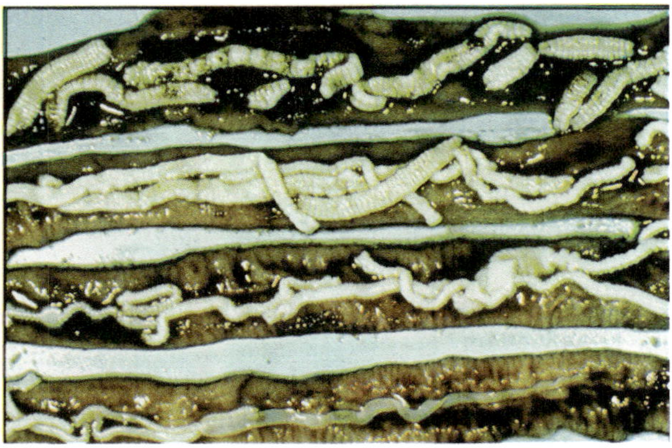

Fig. 5.61 Macrophoto of portions of a dissected intestine of a sheep filled with the strobilae of the tapeworm *Moniezia expansa*

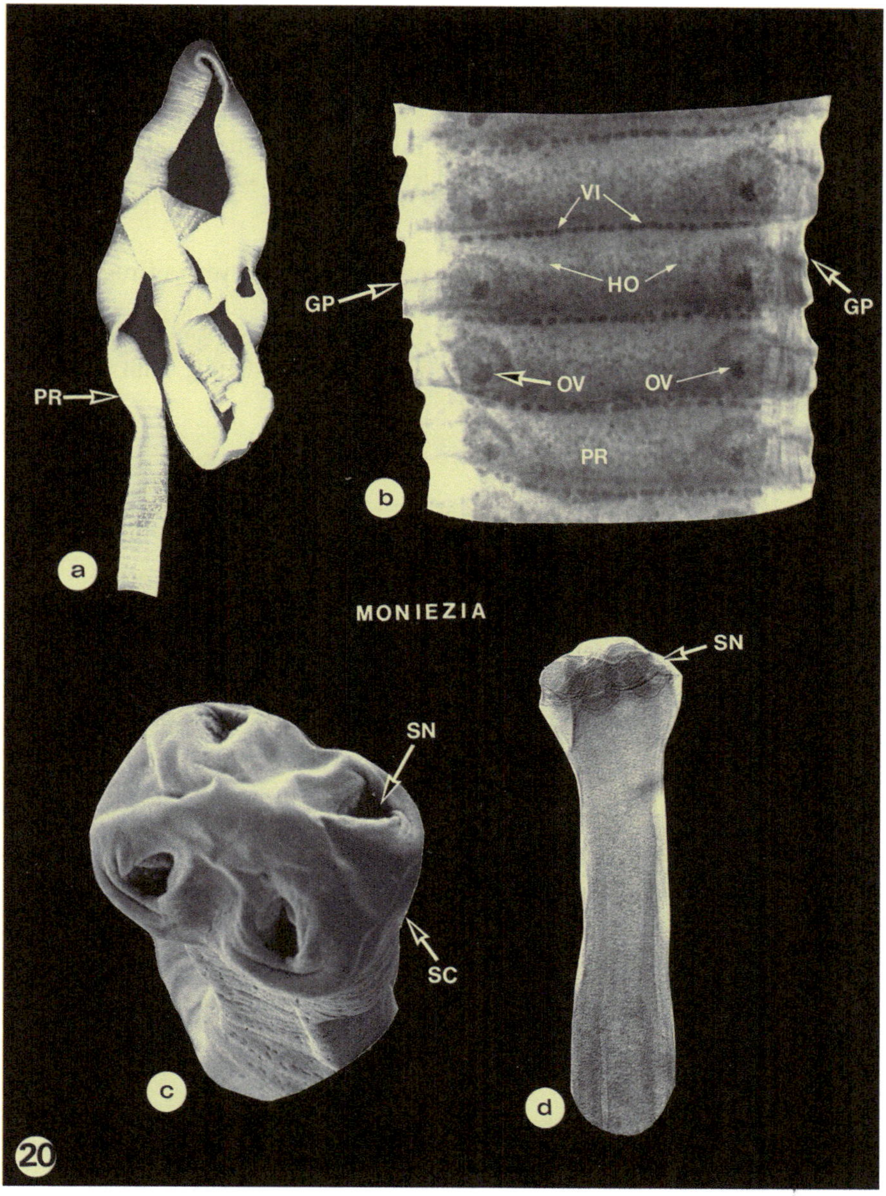

Fig. 5.62 Scanning electron micrograph of different aspects of the tapeworm *Moniezia expansa*. (**a**) Portion of the strobila. (**b**) Proglottids of the middle region of the worm. (**c**, **d**) Scolex and anterior portion of the adult tapeworm. GP = genital porus; HO = testis sphere; OV = ovary region; PR = proglottis; SC = scolex; SN = sucker; VI = vitellarium

longitudinal excretion channels. The uterus contains several so-called paruterine organs (Fig. 5.63a).

(d) ***Stilesia* species (*S. globipunctata, S. hepatica*)**:

The adults reach a length of 60 cm. The posterior proglottids reach a width of 2.5 cm and contain each two characteristic **paruterine organs** (Fig. 5.63c), which are closely filled with 25 μm sized eggs. Intermediate hosts are soil mites (oribatids).

(e) ***Thysanosoma actinoides***:

This species occurs in bile ducts and small intestine of ruminants in North and South America.

4. **Symptoms of disease**: In the cases of *Moniezia* infections only young animals show severe symptoms of disease, which appear starting with weakness,

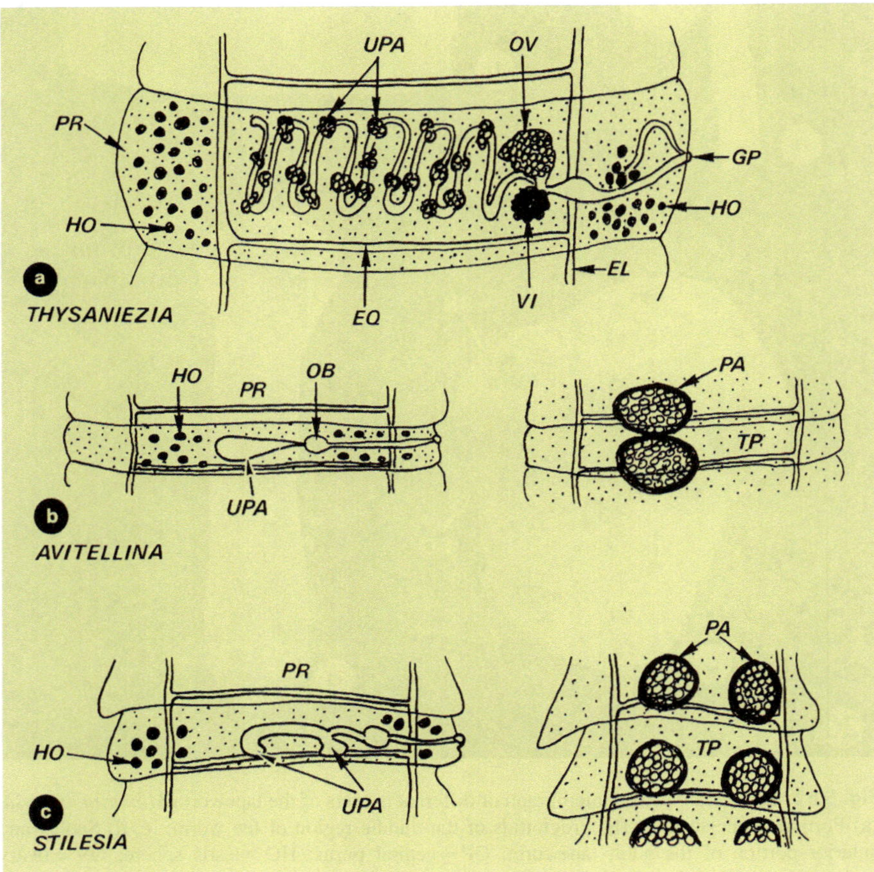

Fig. 5.63 Diagrammatic representation of the proglottids of several tapeworm species of cattle. EL = longitudinal excretion channel; EQ = cross-running excretion channel; HO = testis sphere; OV = ovary region; PA = paruterine organ, containing eggs; PR = proglottis in the midbody region; TP = terminal proglottids; UPA = uterus with anlagen of the paruterine organ; VI = vitellarium

cramps, colic, paresis, anaemia, loss of hair, repeated changes between diarrhoeas and intestinal blocking. Very young animals may even die. The other species lead to mostly low-grade symptoms of disease, so that a definitive clinical diagnosis is difficult.

5. **Diagnosis**: Demonstration of tapeworm proglottids in fresh feces or microscopical approval of eggs within feces after using enrichment methods.
6. **Pathway of infection**: Oral uptake of larva-containing intermediate hosts within the food (Fig. 5.64: 3.1 + 4.1).
7. **Prophylaxis**: Nearly impossible in cases when sheep, etc., stay on the meadow.

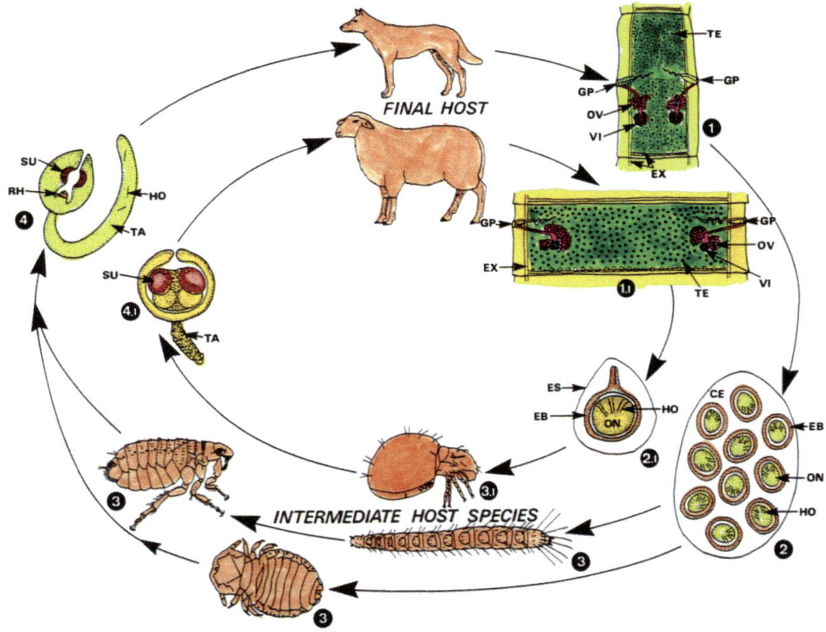

Fig. 5.64 Diagrammatic representation of the life cycles of tapeworms with two sets of sexual organs per proglottid: *Dipylidium caninum* (1–4) of carnivores and **humans** reaching a length of about 50 cm and *Moniezia expansa* (1.1–4.1) of ruminants with a maximum length of 6 cm. **1**, **1.1** Premature proglottids of adult worms parasitizing the intestines of their final hosts. **2**, **2.1** The uteri of fecally excreted cucumber-like proglottids are filled with typical eggs, which in the case of *D. caninum* are always enclosed in capsules (2). **3**, **3.1** *D. caninum* uses larval and adult fleas (*Ctenocephalides* spp.) or chewing lice (*Trichodectes canis*) as intermediate hosts, whereas *M. expansa* develops in oribatid mites (3.1). When the eggs are eaten by such intermediate hosts, the oncosphaera hatches and migrates to the haemocoel. **4** Inside the haemocoel transformation to the second larval type (cysticercoid) occurs. The growth rate is dependent upon the ambient temperature. Infection of the final host is accomplished when infected intermediate hosts are swallowed. The cysticercoids evaginate in the intestine and develop directly into adult tapeworms; this takes 2–3 weeks for *D. caninum* and 4–8 weeks for *M. expansa*. CE = capsule containing eggs; EB = embryophore; ES = eggshell; EX = excretory vessels; GP = genital pore; HO = hooks of oncosphaera; ON = oncosphaera; OV = ovary; RH = rostellar hooks; SU = sucker; TA = tail of cysticercoid; TE = testis; VI = vitelline gland

8. **Incubation period**: Varying: severeness of symptoms depends on the amount of ingested cysticercoid larvae (Fig. 5.64).
9. **Prepatent period**: 4–8 weeks.
10. **Patency**: Depending on the species: 4, 8 or 9 months.
11. **Therapy**: It is economically important to treat lambs in order to avoid severe diseases or death cases. Due to an increase of the number of parasites occurring on often used meadows, also calves should be treated as early as possible. Known cestocides are **niclosamide** (100 mg/kg bodyweight) and **praziquantel** (5 mg/kg bodyweight per sheep, but it is only officially registered for dogs and cats). However, the nematocidal compounds **albendazole** (7.5 mg/kg bodyweight), **fenbendazole** (5–10 mg/kg bodyweight), **mebendazole** (15–20 mg/kg sheep) or **oxfendazole** (5 mg/kg bodyweight) are highly active against tapeworms, too.

Further Reading

Bashtar AR et al (2011) Studies on *Moniezia* of sheep. Parasitol Res 108:177–186.

Dever ML et al (2015) Removal of the tapeworm (*Moniezia* spp.) did not increase growth rates. Vet Parasitol 208:190–194.

Irie T et al (2013) Continuous *Moniezia benedeni* infection in confined cattle. J Vet Med Sci 75:1585–1589.

Moazeni M, Nili M (2004) Mixed infection with tapeworms. Trop biomed 21:23–26.

Mungrube EO et al (2006) The prevalence and economic significance of *Fasciola gigantica* and *Stilesia hepatica*. Trop Anim Health Prod 38:475–483.

Prchal L et al (2015) Biotransformation of anthelmintics and the activity of drug-metabolizing enzymes in the tapeworm *Moniezia expansa*. Parasitology 142:648–659.

Sissay M et al (2008) Prevalence and seasonal incidence of larval and adult cestode infections of sheep and goats. Tro Anim Health Prod 40:387–394.

Strobel H et al (2013) Comparison of three different treatment schedules for praziquantel (Cestocur®, Bayer) in the treatment of tapeworm infections (*Moniezia* sp.). Parasitol Res 112:S139–147.

Venkatesh C, Ramalingam K (2007) Lactic acid, pyruvic acid and lactate pyruvate ratio in the anoplocephalid tapeworm *Stilesia globipunctata* infecting sheep. Vet Parasitol 144:176–179.

5.2.2.2 Larval Tapeworms in Ruminants and Pigs

1. **Name**: Greek: *kystis* = bladder; *kerkos* = tail; *koinos* = together; *ura* = tail. Latin: *saginatus* = well fed; *tenuis* = thin; *collare* = collar; *bos* = cattle; *ovis* = sheep; *collis* = protrusion, little hill; *cerebrum* = brain.
2. **Geographic distribution/epidemiology**: Worldwide.

3. **Biology, morphology**:
   (a) ***Cysticercus bovis***:
   This name describes the up to 10 × 4.5 mm large cysticercus larvae of the cattle-human tapeworm *Taenia saginata* (Figs. 5.65 and 5.66) in the muscles. **Disease**: Cysticercosis.
   (b) ***C. tenuicollis***:
   This is the larva of the tapeworm of dogs: *Taenia hydatigena*, which reaches in cattle a length of 8–15 cm (with a scolex) being situated in the momentum region.
   (c) ***C. ovis***:
   These 10 × 5 mm sized larvae of the tapeworm *Taenia ovis* of dogs are found in the momentum, heart and masseter of sheep.
   (d) ***Coenurus cerebralis***:
   This stage represents the up to 5 cm long larva of the dog tapeworm *Taenia multiceps*. Inside the transparent wall of the larva, many protoscolices can be seen (Fig. 5.47).
   (e) **Hydatid of *Echinococcus cysticus***:
   This is the larva of the tiny dog tapeworm *E. granulosus* (see Fig. 5.56).
   (f) ***Taenia solium***:
   Cysticerci occur in muscles of pigs and are infectious for humans.
   (g) ***Taenia asiatica***:
   Cysticerci occur in muscles of pigs and cause sypmtoms in humans like those due to *Taenia saginata*.
4. **Symptoms of disease**: Organ-specific dysfunctions.
5. **Diagnosis**: finding of these specific looking larvae occurs mostly when the hosts are slaughtered. Serology is possible (but expensive and often not very specific), beginning from day 14 p.i. (ELISA). Infected organs have to be discharged.
6. **Pathway of infection**: Oral uptake of tapeworm eggs being discharged from proglottids during drying. Mostly these eggs are ingested with feces-contaminated food (grass, hay).
7. **Prophylaxis**: Avoiding to feed potential final hosts with raw meat or inner organs of animals.
8. **Incubation period**: Organ specific: 14 days up to months.

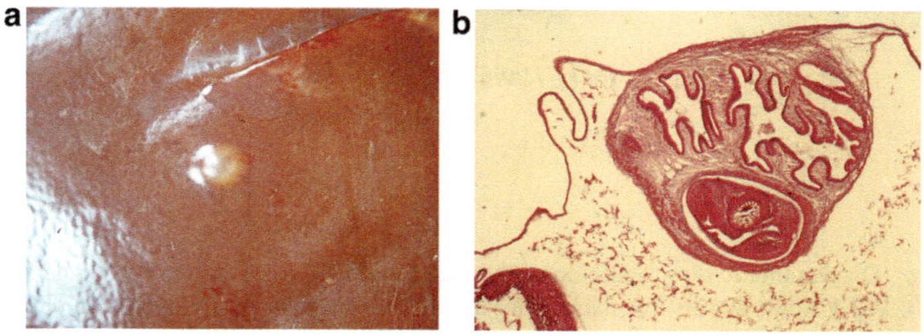

Fig. 5.65 (**a**) Macrophoto of a *Cysticercus bovis* (=larva of the human tapeworm *Taenia saginata* in cattle). (**b**) Section through a *Cysticercus bovis* larva showing the anlage of the scolex

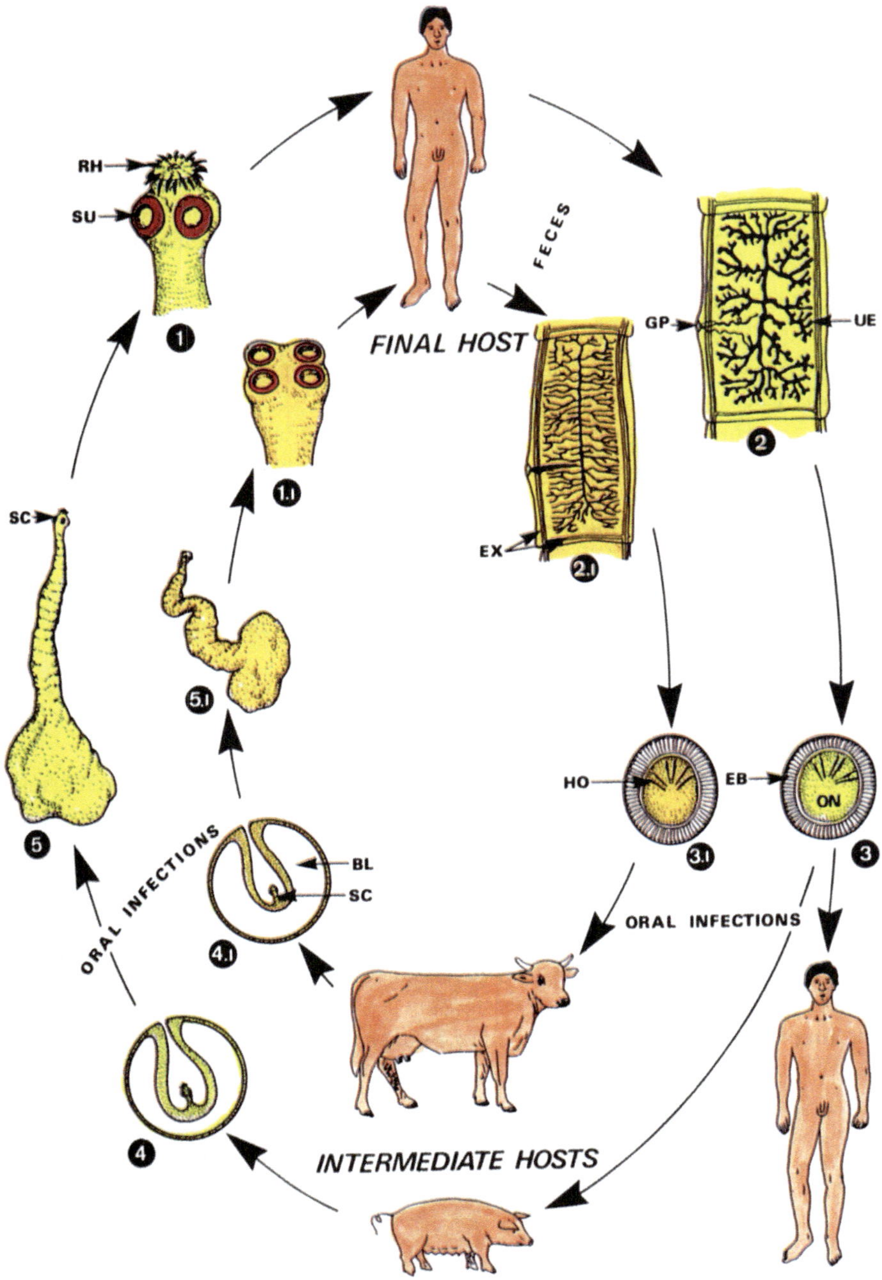

Fig. 5.66 Diagrammatic representation of the life cycles of *Taenia solium* (1–5) and *T. saginata* (1.1–5.1). **1–2.1** Adult worms live exclusively in the intestine of man and reach a length of 4–6 m (*T. solium*) or 6–10 m (*T. saginata*) often with about 2000 proglottids. The scolex of *T. solium* is endowed with an armed rostellum (**1**). The terminal proglottids (10–20 × 5–7 mm) are characterized by a typically branched uterus filled with up to 100,000 eggs. On each day 6–7 of

9. **Prepatent period**: The larvae reach (depending on the species) infectivity starting from 5–6 weeks up to 6–8 months.
10. **Patency**: Larval stages inside intermediate hosts remain infectious for many months up to years.
11. **Therapy**: Chemotherapy of cysticerci of *C. bovis* is possible with **praziquantel** (1 × 100 mg/kg bodyweight subcutaneously). Treatment with **albendazole** and **mebendazole** had varying success. The cysticercus cellulosae was treated successfully with **fenbendazole** (7 × 5 mg/kg bodyweight, oral) as well as the *Coenurus cerebralis* was killed by application of 3 × 25 mg/kg bodyweight orally. Treatment of hydatids is rather difficult, although long-time application of mebendazole and albendazole shows activities on the size decrease of protoscolices and cysts.

Further Reading

See literature on adult tapeworms.

5.2.3 Tapeworms of Horses (Equids)

1. **Name**: Greek: *para* = besides; *anoplos* = unarmed; *kephale* = head. Latin: *foliatus* = surrounded by leafs = the name refers to the leaf (lobe)-like protrusions at the hook-less scolex of these organisms; *mamilla* = nipple; *magna, magnum* = large.
2. **Geographic distribution/epidemiology**: Worldwide.
3. **Biology, morphology**:
   (a) ***Anoplocephala perfoliata***:

   This tapeworm parasitizes in ileum, caecum and colon and reaches a length of up to 25 cm when situated inside the intestine, but shrinks in feces of equids to about 5 cm (Figs. 5.67 and 5.68). The rostellum-less and hook-less scolex is provided with four suckers and four lobe-like protrusions.

Fig. 5.66 (continued) these proglottids detach and may either pass out with the feces or actively migrate out of the anus. **3, 3.1** As an excreted proglottid begins to dry up, a rupture occurs along the midventral and terminal regions and allows eggs to escape. The spherical eggs (40–45 μm; indistinguishable between species) originally have a hyaline outer membrane (eggshell) which is usually lost by the time the eggs are voided with the feces. Thus, the eggs are bordered by a thick, striated embryophore surrounding the oncosphaera (ON). **4, 4.1** When eaten by the intermediate host, the oncosphaera hatches in the duodenum, penetrates the mucosa, enters a venule and is carried throughout the body. A Bladder worm (cysticercus) of about 7–9 × 5 mm is formed, reaching infectivity in about 2 months (*C. cellulosae* in *T. solium*; *C. bovis, C. inermis* in *T. saginata*). When humans ingest eggs of *T. solium* or a terminal proglottid is destroyed inside the intestine, cysticerci may also readily develop in many organs including brain and eyes. These infections lead to severe dysfunctions depending on the parasitized organ (cysticercosis). **5** A person becomes infected when a bladder worm is eaten along with raw or insufficient cooked meat. The evaginating scolex becomes attached to the mucosa of the small intestine and matures in about 5–10 weeks. BL = bladder of cysticercus; EB = embryophore; EX = excretory vessels; GP = genital pore; HO = hooks of oncosphaera; ON = oncosphaera; RH = rostellar hooks; SC = scolex; SU = sucker; UE = uterus filled with eggs

Fig. 5.67 Macrophoto of two contracted horse tapeworms of the genus *Anoplocephala* from the horse feces

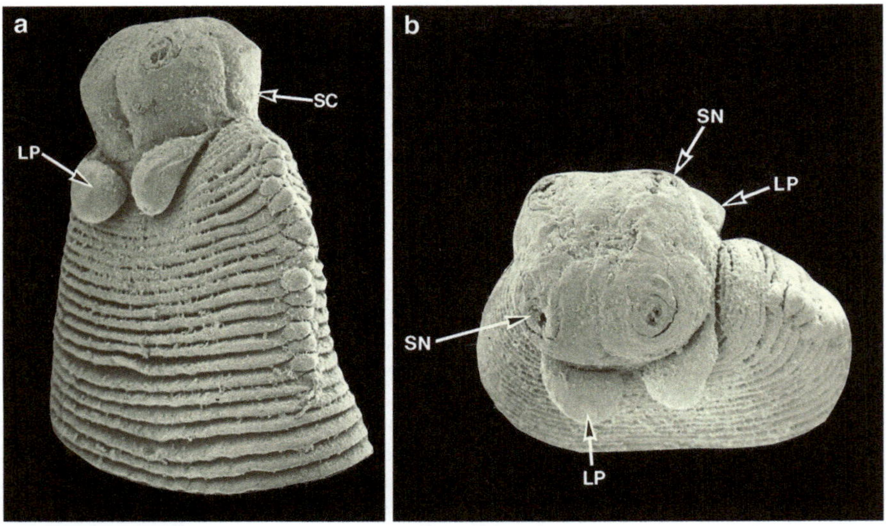

Fig. 5.68 Scanning electron micrographs of the anterior end (**a**) and of the scolex (**b**) of the horse tapeworm *Anoplocephala perfoliata*. LP = lobe-like protrusion; SC = scolex; SN = sucker

Intermediate hosts are soil mites (oribatids). This tapeworm is the most common species in equids reaching prevalence rates of up to 70 %.

(b) ***Anoplocephala magna***:

This tapeworm reaches a length of 80 cm, when being situated in the intestine. Its scolex does not possess protrusions (lobes) of any kind. In feces (outside of the host) this worm is strongly contracted.

(c) ***Paranoplocephala mamillana***:

This rarely occurring tapeworm is very short reaching only a length of 4 cm. Its scolex is provided with lateral, slit-like suckers. The genus *Paranoplocephala* contains several species in voles.

(d) **Related species**:

Anoplocephala gorilla is found in *Gorilla* monkeys, *A. gigantea* occurs in rhinoceros and *A. manubriata* in elephants.

4. **Symptoms of disease**: Clinical symptoms occur rather seldom, but are found in high-grade infections. Then disturbances in digestion, colics, diarrhoeas and loss of weight occur. In some cases of high-grade infections (more than 100 worms per animal) also haemorrhagic enteritis was reported. The parasitaemia persists for about 6 months.
5. **Diagnosis**: Macroscopically visible occurrence of adult worms and/or proglottids in the feces. Demonstration of the 60–80 μm sized eggs with the help of the flotation method. Their shape is polymorphous and they already contain the oncosphaera larva inside (Fig. 5.69). Serologic and coproantigen tests are less safe (Fig. 5.70).
6. **Pathway of infection**: Oral uptake of the intermediate hosts (oribatid mites) together with their grass food.
7. **Prophylaxis**: Regular use of control measurements of the oribatids (keeping grassing grounds rather dry). Treatment of the horses twice (June and autumn) with 1 mg/kg bodyweight praziquantel.

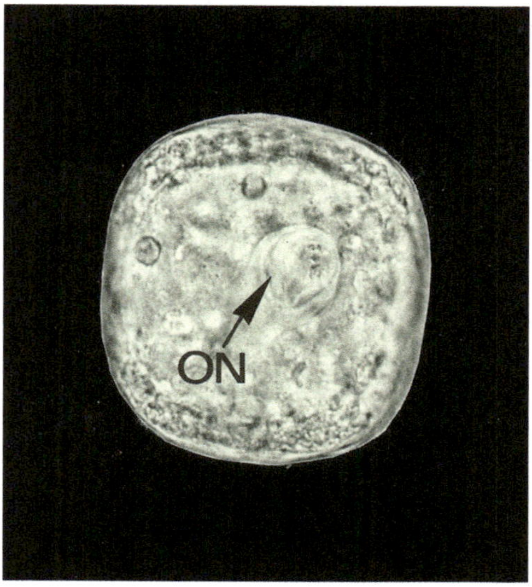

Fig. 5.69 Light micrograph of an egg of the species of the genus *Anoplocephala*. ON = oncosphaera

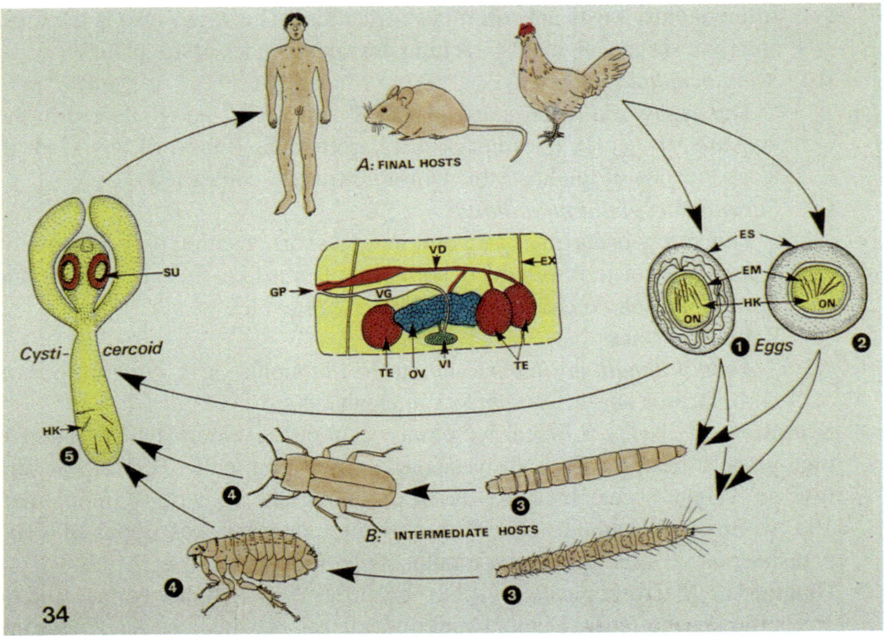

Fig. 5.70 Diagrammatic representation of the life cycle of tapeworms of the family Hymenolepididae. (**A**) Species and final hosts ***Rodentolepis* (*Vampirolepis, Hymenolepis*) *nana*** (*fraterna*) of mice and humans, 4–6 cm long and 1 mm broad, scolex with 24–27 rostellar hooks; *Hymenolepis diminuta* of rats, mice, dogs and humans, up to 6 cm long and 3.5 mm broad, no rostellar hooks; *Echinolepis* (*Hymenolepis*) *carioca* of chickens and birds. Strobila up to 8 cm long and 3–5 mm broad; scolex has no rostellar hooks. The sexual mature proglottids are characterized by three spherical testes (TE); there is no distinct border wall between the proglottids (*dotted lines*). (**B**) Intermediate hosts. **1–4** Eggs containing the oncosphaera larva (**1** *H. nana*, 40–60 × 30–50 μm; **2** *H. diminuta* 60–80 × 70 μm) are infectious to various insects (larvae, adults) as intermediate hosts (**3**, **4**). **5** Inside the body cavity of these hosts a second larva (cysticercoid) is formed, which grows to be a mature tapeworm when the intermediate host is swallowed by the final host. In *H. nana* the intermediate host is optional; when eaten by a man or a rodent, the egg (**1**) hatches in the duodenum, releasing the oncosphaera, which penetrates the mucosa. Here it develops directly into a cysticercoid (**5**). In about 6 days the cysticercoid emerges into the lumen of the small intestine, where it attaches and grows to be a mature worm. EM = embryophore (layer surrounding the oncosphaera); EX = excretion system (longitudinal); ES = eggshell; GP = genital pores; HK = hooks of oncosphaera; ON = oncosphaera; OV = ovary (germarium); SU = sucker; TE = testes; VD = vas deferens; VG = vagina with enlarged seminal vesicle; VI = vitellarium

8. **Incubation period**: In cases of heavy infestations: about 2 weeks.
9. **Prepatent period**: About 4–6 weeks.
10. **Patency**: 6–8 months.
11. **Therapy**: 1 mg/kg bodyweight praziquantel.

Further Reading

Back H et al (2013) The association between *Anoplocephala perfoliata* and colics in Swedish horses. Vet Parasitol 197:580–585.

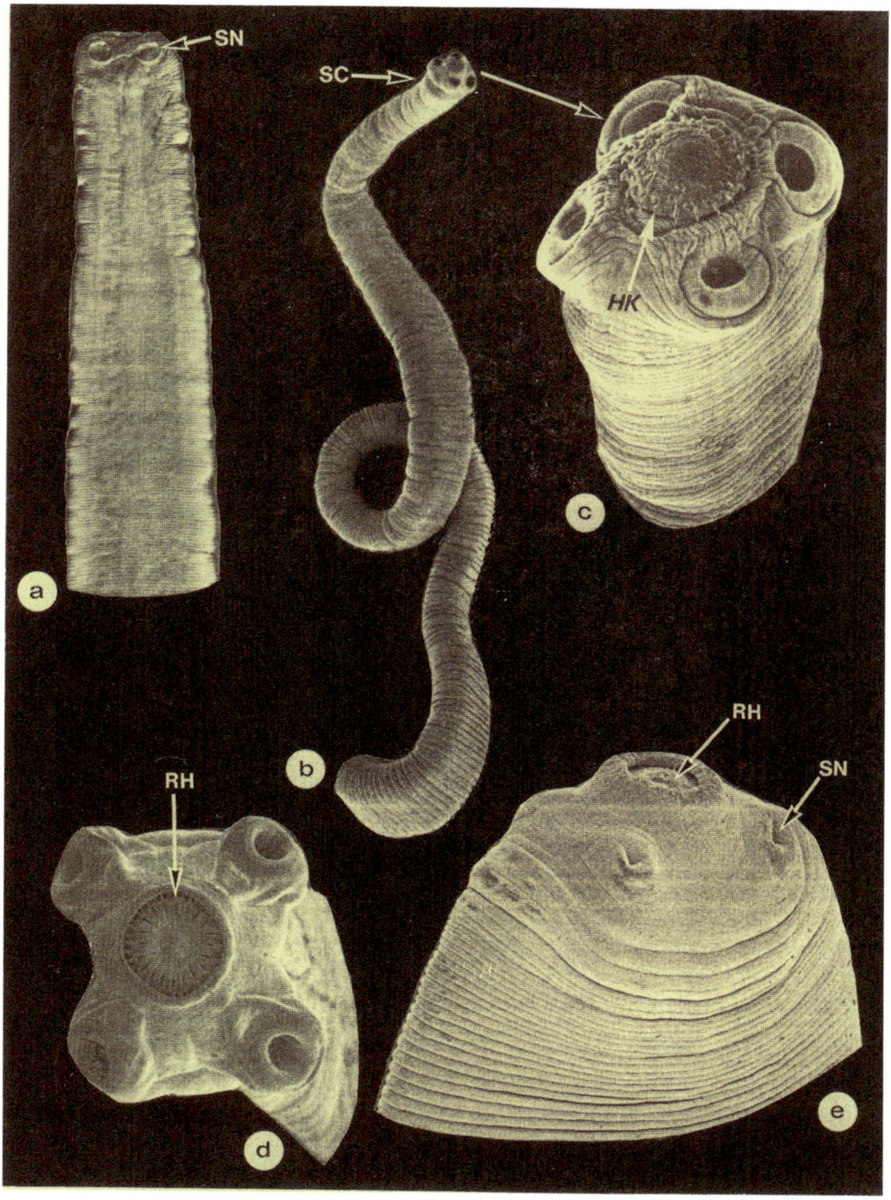

Fig. 5.71 Light and scanning electron micrographs: (**a**) *Hymenolepis diminuta*: without hook crown; (**b**, **c**) *H. fraterna* with rostellum bearing 22–27 hooks; (**d**) *H. microstoma*: aspect of a retracted rostellum; (**e**) *Taenia taeniaeformis*: scolex retracted. HK = crown of hooks; RH = retracted rostellum; SC = scolex; SN = sucker

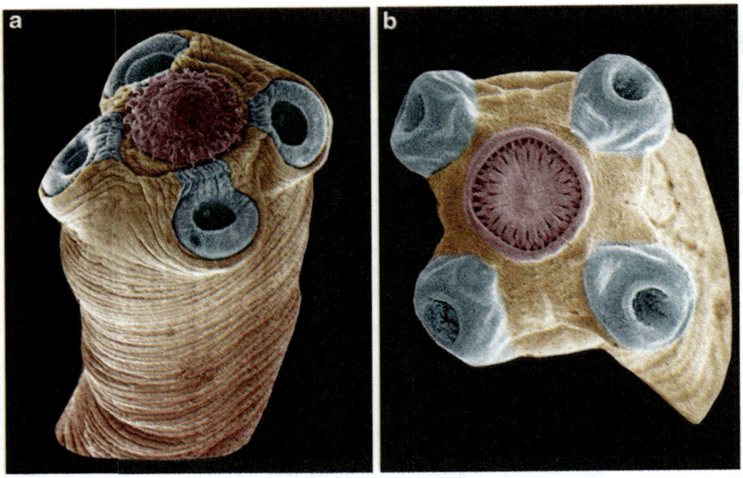

Fig. 5.72 Scanning electron micrographs of scolices: (**a**) *Hymenolepis nana*; (**b**) *H. microstoma*

Bohorquez A et al (2014) Coprologically diagnosing *Anoplocephala* in the presence of *A. magna*. Vet Parasitopl 204:396–401.
Haukisalmi V et al (2014) Phylogenetic relationships and taxonomic revision of *Paranoplocephala*. Zootaxa 3873:371–415.
Pavone S et al (2011) Pathological changes caused by *Anoplocephala perfoliata*. Vet Parasitol 176:43–52.
Rehbein S et al (2011) Evaluation of a double centrifugation technique for the detection of *Anoplocephala* eggs. J Helminthol 85:409–414.
Tomczuk K et al (2015) Seasonal changes of the diagnostic potential in the detection of *Anoplocephala perfoliata*. Parasitol Res 114:767–772.

5.2.4 Tapeworms of Birds

1. **Name**: Davaine (1812–1882), Railliet (1828–1902) = French parasitologists. Greek: *proglottis* = tip of tongue, portions; *tainia* = band, tape; *amoibe* = changing shape; *choanos* = funnel; *hymen* = thin skin; *lepis* = fine scale; *tetra* = four; *gone* = reproduction; *aden* = gland. Latin: *fimbria* = fringe.
2. **Geographic distribution/epidemiology**: Worldwide.
3. **Biology, morphology**:
   (a) ***Davainea proglottina***:

   This species commonly parasitizes in chicken birds. It reaches a length of 0.5 mm–4 mm and possesses in general only 4–9 proglottids with a width of 0.6 mm at the maximum. The scolex is equipped by a protrudable rostrum with two crowns of hooks each containing 30–50 hooks. The proglottids are transparent. These tapeworms are found in the region of the small intestine being firmly anchored in the mucous layer. The genital pores are arranged at the lateral side of the proglottis but regularly alternating from the left to the right border. The eggs belong to the *Hymenolepis* type and are situated in small capsules inside the parenchyma (Fig. 5.73). Intermediate hosts are slugs (e.g. *Limax* species, etc.).

(b) ***Raillietina* species**:

These species occur in the small intestine of many bird species and reach a length of up to 25 cm (e.g. *R. tetragona* in chickens and in turkeys *R. echinobothrida*) and a width of 4 mm. The scolex is provided with a protrudable rostellum which possesses two crowns each with 100–500 single hooks. Their sucker may bear additional hooks. The genital pores are all situated at the same lateral side. The egg capsules contain mostly 6–12 eggs and belong to the so-called *Hymenolepis* type (Fig. 5.73). **Intermediate hosts** are (depending on the species) beetles, snails, flies or ants.

(c) ***Amoebotaenia* species**:

The specimens of these species occur in farmed birds, have a wedge-like shape, reach a length of 4 mm and consist in general of 12–14 proglottids. The rostellum is provided with 10–12 hooks; the genital pores alternate non-regular from one lateral side to the other. The eggs belong to the *Hymenolepis* type (Fig. 5.73) and are collected in a sack-like uterus. **Intermediate hosts** are earthworms (e.g. *Lumbricus* species).

(d) ***Choanotaenia* species**:

They reach a length of up to 25 cm and a width of 3 mm. Their rostellum contains a crown consisting of up to 20 hooks. The genital pores alternate regularly from one side to the other. The eggs are of the *Hymenolepis* type; however, they are equipped with long and relatively broad protrusions. **Intermediate hosts** are beetles but also flies.

Fig. 5.73 Light micrograph of an egg of *Hymenolepis* sp

(e) ***Hymenolepis* species**:

These common species occur in rather large numbers in farmed and in wild living **birds**; however, their impact as agents of disease is mostly rather low. Most important with respect to diseases are

- ***Hymenolepis* (syn. *Drepanidotaenia*) *lanceolatum***:

  Occurring in the small intestine of ducks and geese and reaching a length of up to 15 cm. They have a lancet-like shape since their terminal proglottids are up to 2 cm broader than the anterior ones. Proglottids and eggs are of the *Hymenolepis* type (Fig. 5.73). **Intermediate hosts** are small crustaceans belonging to the genera *Gammarus, Cyclops* and *Diaptomus*.

- ***Hymenolepis* (syn. *Echinolepis*) *carioca*:**

  The adults of this species reach a length of 8 cm, while *E. anatis* specimens are up to 30 cm long. The numerous proglottids are considerably large in width than in length and contain in their centre three large, globular testes. The rostellum is in most of the cases of tapeworms parasitizing in chickens, ducks and geese, not armed by hooks; however, the suckers may contain tiny hooks. The eggs look like those of the above-described genera (Fig. 5.73). **Intermediate hosts** are beetles and flies.

(f) ***Fimbriaria* species**

These species reach a length of up to 40 cm and a width of 5 mm. They parasitize in farm birds and water birds. The pseudoscolex has replaced the strongly reduced scolex. The strobila does not show segmentation. The eggs are excreted unembryonated, appear ovoid, possess a cover (operculum) and measure about 45 μm in length. **Intermediate hosts** are small crustaceans of the genera *Cyclops, Gammarus* and *Diaptomus*.

4. **Symptoms of disease**: In the cases of heavy infections with large numbers of tapeworms, the following symptoms may occur: diarrhoeas, cramps, loss of bodyweight and disturbances of the balance and even death cases are recorded. The same symptoms occur in *Davainea* species.
5. **Diagnosis**: Microscopical demonstration of proglottids and/or single eggs in feces (Fig. 5.73). A species determination just on the shape of eggs is, however, not possible, since the eggs appear too similar. Since treatment is identical, exact species determination is not needed.
6. **Pathway of infection**: Oral by uptake of infected intermediate hosts. Inside their body cavity, the second infectious larva is developed, after the animal has ingested the larva 1 inside the egg.
7. **Prophylaxis**: Important is to keep the stables free from intermediate hosts by regular use of insecticides and application of gazes before windows besides repeated cleaning of the soils.
8. **Incubation period**: Variable (like prepatent period: ~about 2–3 weeks).
9. **Prepatent period**: 2 weeks (*Davainea proglottina, Hymenolepis lanceolatum, Echinolepis carioca*) or 3 weeks (*Choanotaenia* species, *Raillietina tetragona, R. cesticillus*) or 4 weeks (*Amoebotaenia cuneata*).

10. **Patency**: Most only a few weeks.
11. **Therapy**: **Praziquantel** (oral application of 1 × 10 mg/kg bodyweight) acts against adult and immature tapeworms; **niclosamide** (2–6 days: oral 20 mg/kg bodyweight), **fenbendazole** (4–5 days: 100 ppm in food) and **mebendazole** (7 days, each 60 ppm) are recommended in cases of infections with *Raillietina* and *Hymenolepis* species.

Further Reading

Chomicz L, Czubai A (1991) Transmission electron micrograph studies of developmental oncospherical envelopes of *Fimbriaria fasciolaris*. Parasitol Res 77:503–508.

El-Bahy NM, Bazh EK (2015) Anthelmintic activity of ginger, curcumin and praziquantel against *Raillietina cesticillus*. Parasitol Res 114:2427–2434.

Das B et al (2013) Purification and characterization of phosphoenolpyruvate-carboxykinase from *Raillietina echinobothrida*, a cestode parasite of the domestic fowl. Parasitology 140:136–146.

Katoch R et al (2012) Prevalence and impact of gastro-intestinal helminths on body weight gain in backyard chickens. J Parasit Dis 36:49–52.

Kurt M, Acici M (2008) Cross-sectional survey on helminth infections of chickens in the Samsun region, Turkey. Dt Tierärztl Wschr 115:239–242.

Muniz-Pereira LC, Amato SB (1988) *Fimbriaria fasciolaris* and *Cloacotaenia megalops*, cestodes from Brazilian waterfowls. Mem Inst Oswaldo Cruz 96:767–772.

5.2.5 Tapeworms of Rodents and Rabbits

1. **Name**: Greek: *hymen* = thin skin; *lepis* = tiny scale; *nanos* = very small. Latin: *frater* = brother; *denticulatus* = with teeth; *microstoma* = small mouth; *citellus* = order of rodents.
2. **Geographic distribution/epidemiology**: Worldwide.
3. **Biology, morphology**: The species of these tapeworms have different final hosts:

   3.1 **Rabbits and hares**:

   Inside the small intestine, *Cittotaenia* species (e.g. *C. denticulata*, *C. ctenoides*) occur. Their scolex has no crown of hooks and no rostellum. These species possess two sets of genital organs per proglottis (compare see *Moniezia*; Fig. 5.62). The proglottids of the up to 1 m long worms enlarge their width considerably very close behind the scolex; however, they do not reach more than 8 mm in width. The eggs belong to the type of the family Anoplocephalidae (see Fig. 5.69). **Intermediate hosts** are oribatids (soil mites). However, rabbits may also serve as intermediate hosts of tapeworms, which have carnivores as final hosts.

3.2 **Mice and rats**:

(a) Inside the small intestine occurs *Hymenolepis fraterna* (syn. *Vampirolepis nana*). The adult worms reach mostly only a length of 3–5 cm at a width of 1–2 mm. The scolex possesses a rostrum with a single crown of hooks comprising 20–24 hooks (Fig. 5.72b). The eggs (Fig. 5.73) reach a size of 40–60 μm × 30–50 μm and are characterized by typical polar filaments. The development is of the direct type: many oncospheres hatch from the eggs and settle in the same host; others are excreted and infect directly other final hosts, if they are ingested. However, there are also reports that there are included intermediate hosts (e.g. beetles), within which **cysticercoid** larvae are developed and become infectious for final hosts (Fig. 5.74). **Attention**: *Hymenolepis fraterna* parasitizes also in humans.

(b) *Hymenolepis diminuta* parasitizes in the small intestine (Fig. 5.71). The eggs, which have no polar filaments, measure 60–80 μm × 70 μm and thus are considerably larger than those of *H. fraterna (H. nana)*. The scolex has no crown of hooks. The adult worms reach a length of 60 cm and a width of 3–5 mm. The life cycle includes an obligate host change involving beetles as **intermediate hosts**, wherein the infectious larva is developed. **Attention**: also humans may become infected.

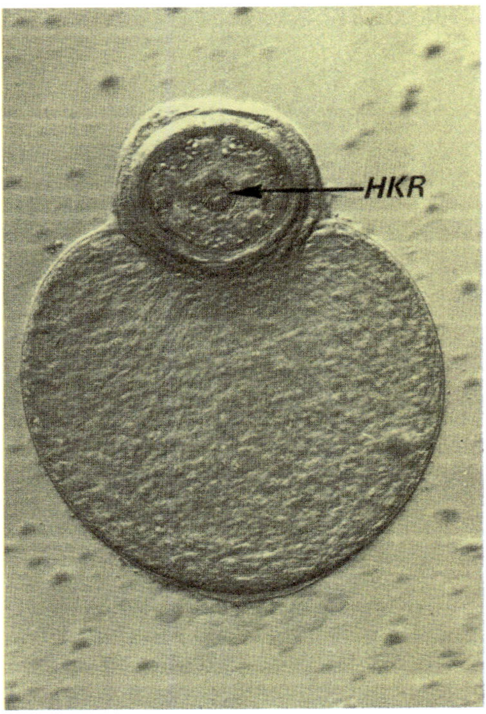

Fig. 5.74 Light micrograph of a typical cysticercoid larva with scolex and anlage of the hooks

(c) A common tapeworm is *Hymenolepis* (syn. *Rodentolepis*) *microstoma*, which may reach a length of up to 20 cm (Figs. 5.71d and 5.72b). It settles inside the small intestine but was also found inside the gallbladder, being attached with the help of their hook-bearing scolex. The eggs do not possess polar filaments but show a concentric cytoplasm around the embryophore.

4. **Symptoms of disease**: In the case of many tapeworms, bloody slimy diarrhoeas or blockage of the intestine may occur. General symptoms are loss of weight and general fitness. From highly infected rabbits also fatal cases are reported.
5. **Diagnosis**: Demonstration of total proglottids or eggs (Fig. 5.73) inside feces after use of enrichment methods.
6. **Pathway of infection**: Oral uptake of infected larvae in intermediate hosts or (in case of *H. fraterna*) by direct uptake of worm eggs in contaminated food.
7. **Prophylaxis**: Practically impossible.
8. **Incubation period**: Mostly 1–2 weeks.
9. **Prepatent period**: 2 weeks; up to 3 weeks in the case of *H. microstoma*.
10. **Patency**: 2 months (up to 1 year in *H. microstoma*).
11. **Therapy**: Praziquantel is effective when using dosages of 5–25 mg/kg bodyweight orally. Also benzimidazoles are effective, e.g. fenbendazole (Panacur®) in the case of *H. fraterna* (5 days × 300 ppm in food), *H. diminuta* (5 days × 30–50 ppm), not registered for rabbits. Waiting times: cattle 7 days.

Further Reading

Allan JC et al (1999) Helminth parasites of the wild rabbit *Oryctolagus cuniculus* near Malham Tarn, Yorkshire, UK. J Helminthol 73:289–294.

Boag B et al (2001) Patterns of parasite aggregation in the wild European rabbit (*Oryctolagus cuniculus*). Int J Parasitol 31:1421–1428.

Foronda P et al (2011) Distribution and genetic variation of hymenolepitid cestodes in murid rodents on the Canary Islands (Spain). Parasites and Vectors 4:185.

Frank R et al (2013) Parasites of wild rabbits (*Oryctolagus cuniculus*) from an urban area in Germany, in relation to worldwide results. Parasitol Res 112:4255–4266.

Mirjalali H et al (2016) Molecular analysis of isolates of the cestode *Rodentolepis nana* from the great gerbil *Rhombomys opimus*. J Helminthol 90:252–255.

Rushworth RL et al (2015) *Rodentolepis straminea* in an urban population of *Apodemus sylvaticus* in the UK. J Helminthol. doi:10.1017/s0022149x15000632.

5.2.6 Tapeworms of Reptiles and Amphibia

In the intestine of reptiles and amphibia, many parasite species are described. They reach—depending on the size of the hosts—lengths of up to 1 m and occur mostly in great numbers (up to 1000 have been found in some specimens). Thus, special literature has to be checked. Members of the following orders are especially common:

- **Order Protocephalidea**:
  They are diagnosed when examining the feces: the typical eggs contain in a central embryophore the six-hooked oncosphaera larva.
- **Order Pseudophyllidea**:
  The eggs are operculated (compare Fig. 5.34) but contain not yet a larva when they are just excreted.
- **Order Cyclophyllidea**
  Proglottids occur in the feces; in case of use of enrichment methods, only eggs are seen, which often consist only the thick-walled embryophore containing the infectious oncosphaera larva (Fig. 5.42b).
  Infection with a few specimens remains mostly symptomless. However, strong infections lead to loss of bodyweight and fitness. For animals kept in homes, it is recommended to check the feces and to use products such as Vermol® produced by Alpha-Biocare GmbH (Düsseldorf) and distributed by Fa. Sera, Heinsberg, Germany.

5.2.7 Tapeworms of Fishes

1. **Name**: Greek: *karyon* = nucleus; *phyllon* = leaf; *pseudo* = wrong, similar looking; *protheus* = in front, advanced; *cephalon* = head; *kyaneos* = bluish; *bothrion* = groove; *laticeps* = wide reaching.
2. **Geographic distribution/epidemiology**: Worldwide.
3. **Biology, morphology**: In the intestine of most fish species, the adult stages of numerous tapeworms may occur (Figs. 5.75 and 5.76), while the larvae of other species have passed the intestinal wall and thus entered the body cavity. The adult tapeworms are systematically described according to their inner and outer organization:
   (a) **Caryophyllidae**:
   The so-called carnation-head tapeworms (e.g. *Caryophyllaeus laticeps*; Fig. 5.75) reach a length of up to 3 cm. The anterior end of these tapeworms, which appear non-segmented, has a scolex which reminds of the carnation flower. Inside exists only a single set of both sexual systems (♀♂). The testes comprise of about 400 single small spheres. The 65 μm × 37 μm sized eggs are set free within the host's feces. Infections occur mainly in spring and autumn. **Intermediate hosts** are oligochaete worms settling on the floor of freshwater ponds (e.g. Tubificidae such as *Tubifex* sp.). **Final hosts** are most carp-like fishes.
   (b) **Pseudophyllidea**
   Most of the species listed hereunder possess only two slit-like suckers at the otherwise unarmed scolex.

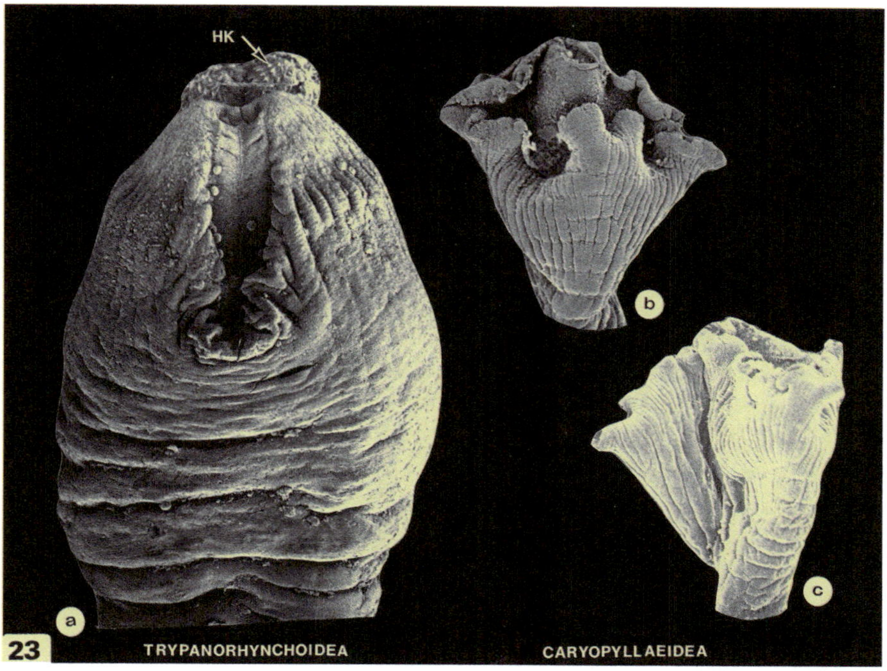

Fig. 5.75 Scanning electron micrograph of armed and unarmed scolices of different tapeworms inside fishes. (**a**) *Hepatoxylon*; (**b**, **c**) *Caryophyllaeus* sp.; HK = hooks

1. In the case of *Triaenophorus* species (Fig. 5.76a, b), there occur in addition four hooks. These species are the most pathogenic tapeworms for fish hosts. *T. lucii* (syn. *Nodulosus*) occurs in pikes and perches and reaches a length of 15 cm and a width of 4 cm, while *T. crassus* reaches in pikes a length of 37 cm. Both species use small crustaceans (genus *Cyclops*) as first intermediate hosts and fishes as second ones, which contain the infectious 6–10 cm long plerocercoids that destroy the liver of the final hosts. The operculated eggs measure 50 μm × 30 μm (Fig. 5.34). The same fish may contain plerocercoids as well as adult worms.
2. ***Eubothrium* species** are also distributed worldwide (Fig. 5.76f, g). *E. crassum* reaches in salmonids a length of up to 80 cm and a width of 6 mm and is characterized by an unarmed scolex with two lateral suckers. The eggs measure 55 μm × 40 μm and contain a **coracidium** larva, which does not have cilia (which occur in other species). First intermediate hosts are small crustaceans (*Cyclops* species); second intermediate hosts are several non-raptor fishes but also perches.
3. ***Cyathocephalus* species** occur in many non-raptor fishes. They reach a length of 5 cm and possess a jug-like, unarmed scolex. The eggs measure 48 × 32 μm. **Intermediate hosts** are gammarid crustaceans.

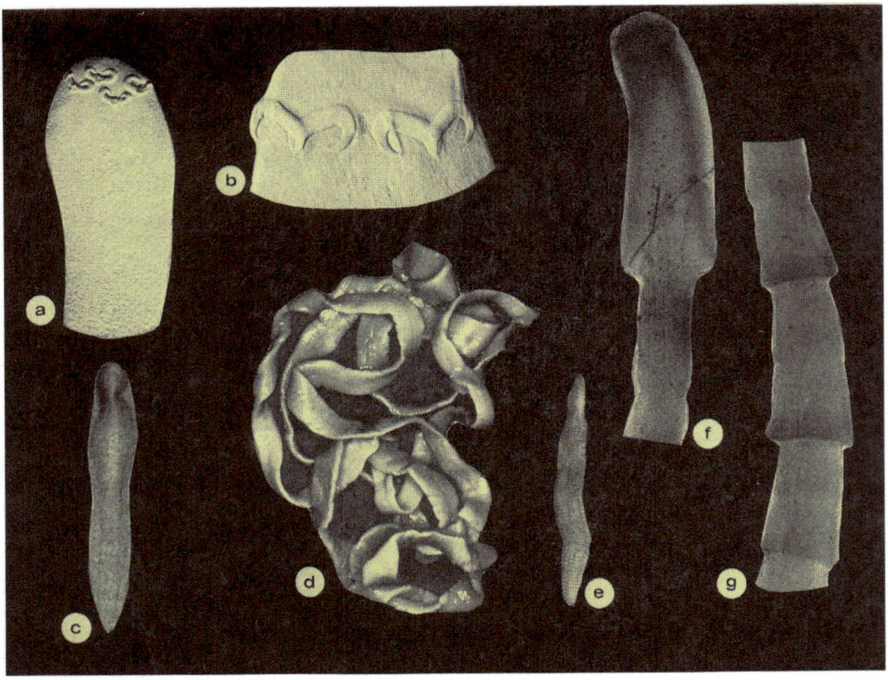

Fig. 5.76 Tapeworms of fishes. (**a**, **b**) Anterior end of pseudophyllidean worms (*Triaenophorus lucii* from pikes); (**c**) Anterior end of the larva of *Diphyllobothrium latum* with lateral suckers. (**d**) *Ligula intestinalis* plerocercoid in the abdominal cavity of carps. (**e**) Plerocercoid of *Schistocephalus solidus* from the body cavity of sticklebacks. (**f**, **g**) Anterior end of strobila of *Eubothrium* sp.

4. ***Bothriocephalus* species** (e.g. *B. scorpii, B. cuspidatus*) parasitize in many fresh and saltwater fishes. They reach a length of up to 1 m and a width of 6 mm. The scolex shows a small apical discus (sucker). The operculated eggs reach a size of 70 μm × 30 μm. First intermediate hosts are copepods, while fish (e.g. those forming hollows in the sand) act as second intermediate hosts which become ingested by the final host.

(c) **Tetraphyllidea**:

The adults of these fish species occur mostly in marine predator fishes. The scolex is provided with four leaf-like protrusions, which bear in some species additional suckers. *Rhinebothrium* species reach in sharks and rays lengths of up to 7 cm. **First intermediate hosts** are always copepods, while several plankton-feeding fish species serve as second intermediate hosts. The eggs are spherical, measure 40–50 μm in diameter and bear in some species polar filaments (as it is the case in some *Hymenolepis* species).

(d) **Proteocephalidae**

The scolex of the adult of these species shows four equally sized lateral suckers and an additional apical one. Important species belong to the genus *Proteocephalus*. The species are mostly small (only 1–10 mm long), but a few other species reach lengths of 20, 40 or even 100 cm. They excrete spherical eggs of about 20–50 μm in diameter depending on the species. **Intermediate hosts** are *Cyclops* species, within which the plerocercoid larva is developed.

4. **Symptoms of disease**: Depending on the species and on the infestation rate, the symptoms of disease may vary considerably reaching from loss of weight, intestinal lesions, blockage of the intestine until death (in mass infections). Since severely infected fishes are slow, they become easily prey of raptor fishes, which thus also become infected. The blood picture shows a considerably increased numbers of leucocytes.
5. **Diagnosis**: Microscopical demonstration of the eggs (Fig. 5.34) within the feces. Furthermore, whole proglottids can be seen in feces and in sections the typical worm strobila can be found.
6. **Pathway of infection**: Oral uptake of infected intermediate hosts.
7. **Prophylaxis**: Difficult up to impossible in fishes living in lakes. In rearing ponds, the empty ponds should be disinfected with chalk (4000 kg/ha). However, high-grade precautionary measurements are obligatory. If ponds are left empty for some weeks, tapeworm eggs lose their infectivity. During the rearing phase, the behaviour of fish should be observed and test catches should be done to check eventual parasites inside the fish.
8. **Incubation period**: Species specific reaching from 2 to 10 weeks.
9. **Prepatent period**: Species specific: 2–10 weeks.
10. **Patency**: Several months, up to lifelong.
11. **Therapy**: Application of R.P. Blue concentrate (Verman®) into ponds. This product consists of chlortetracycline with Di-n-butylcinnoxide plus vitamins. Dosage for carp ponds: 10 g/kg mixed food for 1 day. Repetition after 5 days. Waiting period for eatable fish: 30 days. As addition to food also **praziquantel** can be used (1 × 5 mg/kg in separate basins). In the case of *Eubothrium* worms in salmonids, **fenbendazole** was used (application at 2 days 5–8 mg/kg bodyweight in food). In the case of ornamental fishes in aquaria, the product Tremazol® (produced by Fa. Alpha-Biocare GmbH, Düsseldorf, Germany, sold by Fa. Sera, Heinsberg, Germany) is 100 % effective.

Further Reading

Bosi G et al (2005) Changing in the neuromodulators of the diffuse endocrine system of the alimentary canal of farmed rainbow trouts naturally infected with *Eubothrium crassum*. J Fish Dis 28:703–711.

Brunanska M, Kostic B (2012) Revisiting caryophyllidean type of spermiogenesis in Eucestoda based on *Caryophyllaeus laticeps*. Parasitol Res 110:141–149.

De Chambrier A et al (2014) Taxonomic status of Woodland's enigmatic tapeworms (Proteocephalidea). Syst Parasitol 87:1–19.

Golestaninasab M, Malek M (2015) Two new species of *Rhinebothrium*. J Helminthol. doi:10.1017/s0022149x15000553.

Jirsa E et al (2008) The occurrence of *Caryophyllaeus laticeps*. J Helminthol 82:53–58.

Kralova-Hermadova I et al (2015) Development of microsatellite markers in *Caryophyllaeus laticeps*. Parasitol Res 114:721–726.
Masson G et al (2015) Impact of the cestode *Triaenophorus nodulosus* on the exotic *Lepomis gibbosus* and the autochthonous *Perca fluviatilis*. Parasitology 142:745–755.
Mlocicki D et al (2010) Functional ultrastructure of the hexacanth larvae in the bothriocephalidean cestode *Eubothrium salvelini*. Parasitol Int 59:539–548.
Reyda FB et al (2011) Diversification and species boundaries of *Rhinebothrium*. PLoS One 6(8):e22604.
Schähle Z et al (2016) Infection levels of plerocercoids of the tapeworm *Triaenophorus crassus*. J Helminthol 90:54–61.

5.3 Nematodes (Roundworms)

The nematodes (Table 5.6 and Fig. 5.77) appear—depending on the species—as cylindrical or filament-like organisms. These worms appear as females and males and live in the soil, fresh or saltwater but also as parasites in plants, animals and humans. Their systematic relationships are since long under constant discussion especially since modern molecular methods allow deeper insights.

The recent at many places used classification of the different "groups" is based on the presence or absence of caudal, even in the light microscope hardly visible, sense organs which are called **phasmids** and the ability of these parasites to attach themselves at surfaces including at the host skin with the help of the secretions of caudal glands. In addition, the shape of the oesophagus is used and/or molecular details. In order to get a helpful morphological survey of the nematodes, the following (rather old-fashioned) system is used to present a selection of important nematodes of animals.

Phylum: NEMATHELMINTHES (Selected extract)
Subphylum: Nematoda
Class: Adenophorea (Aphasmidea)
Order: Enoplida
Family: Trichuridae (Trichurinae, Capillariinae)
Family: Trichinellidae
Family: Dioctophymatidae
Order: Mermithida
Family: Mermithidae
Class: Secernentea (Phasmidea)
Order: Rhabditia
Family: Rhabditidae
Family: Strongyloididae
Order: Strongylida
Superfamily: Ancylostomatoidea
Family: Ancylostomatidae

(continued)

Family: Uncinariidae
Superfamily: Trichostrongyloidea
Family: Trichostrongylidae
Family: Dictyocaulidae
Family: Heligmosomatidae
Superfamily: Metastrongyloidea
Family: Metastrongylidae
Family: Angiostrongylidae
Family: Protostrongylidae
Superfamily: Strongyloidea
Family: Strongyloidae
Order: Ascaridida
Superfamily: Ascaridoidea
Family: Ascarididae
Family: Toxocaridae
Family: Anisakidae
Family: Cosmocercidae
Superfamily: Oxyuroidea
Family: Oxyuridae
Superfamily: Heterakoidea
Family: Heterakidae
Family: Ascaridiidae
Order: Spirurida
Superfamily: Spiruroidea
Family: Spiruridae
Family: Spirocercidae
Superfamily: Physalopteroidea
Family: Gnathostomatidae
Family: Physalopteridae
Superfamily: Filarioidea
Family: Filariidae
Family: Onchocercidae
Order: Camallanida
Superfamily: Camallanoidea
Family: Camallanidae
Superfamily: Dracunculoidea
Family: Dracunculidae
Family: Philometridae
Family: Micropleuridae
Family: Anguillicolidae
Order: Diplogasterida
Order: Aphelenchida
Order: Tylenchida
Superfamily: Sphaerularioidea
Family: Sphaerulariidae

Table 5.6 Important parasitic nematodes

Family and species	Length of the adult worm (mm)		Egg resp. larval size (µm)	Final host/location of the adult worms	Intermediate hosts	Prepatency in the final hosts (weeks)
Family Strongyloididae	♀	♂				
Strongyloides papillosus	(a) 4–6 (b) 0.7–1.1	– 0.6	40–60 × 32–40	(a) Ruminants; small intestine (b) outdoors	–	1.5
S. stercoralis	(a) 2 (b) 0.8–1.0	– 0.7	40 × 30 30	(a) Dogs, **humans**; small intestine (b) outdoors	–	2.5–4
Family Ancylostomatidae						
Ancylostoma caninum	14–18	11–14	53–69 × 36–53	Dogs; small intestine	–	2.5
A. duodenale	11–13	8–11	60	**Humans**; small intestine	–	5–6
Necator amerricanus	9–11	7–9	55	**Humans**; small intestine	–	5–6
Family Strongylidae						
Strongylus vulgaris	20–24	10–17	80–93 × 47–54	Horses; colon	–	24
S. equinus	36–48	25–35	72–92 × 41–54	Horses; colon, caecum	–	32–36
S. edendatus	28–40	20–28	72–88 × 90–92	Horses; colon	–	40–44
Syngamus trachea	5–20	6	78–110 × 43–46	Gamefowl; trachea	Earthworms	2.5–3
Family Trichostrongylidae						
Trichostrongylus sp. (=*T. axei*, *T. colubriformis*)	4–6	3–5	75–90 × 40–43	Ruminants, horses, possibly **humans**; stomach	–	3

Ostertagia circimcincta	9–12	7–9	80–100 × 50–50	Sheep; abomasum	–	2
O. ostertagi	8–9	6–8	65–80 × 30–40	Cattle; abomasum	–	3
Haemonchus contortus	18–30	18–21	70–85 × 41–44	Ruminants; abomasum	–	3
Family Metastrongylidae						
Dictyocaulus viviparous	60–80	35–55	L1 (420 µm) in feces	Cattle; bronchial tubes, trachea	–	3–4
D. filaria	30–100	30–80	L1 (550 µm (in feces)	Sheep, goats; lung	–	4–5
Parastrongylus cantonensis	21–25	18	L1 (300 µm) in feces	Rodents, **humans**; lung, brain	Snails, crabs	6–7
Family Oxyuridae						
Enterobius vermicularis	8–13	3	50–60 × 20–30	**Humans**; colon, caecum	–	4–5
Oxyuris equi	40–180	10–20	80–95 × 40–45	Horses; colon, caecum	–	16–20
Fam. Heterakidae						
Heterakis gallinarum	10–15	7–13	65–80 × 35–46	Gamebirds; caecum	–	3–4
Family Ascaridae						
Ascaris lumbricoides	200–410	150–250	50–75 × 40–50	**Humans**, pigs; small intestine	–	6–11
A. suum	200–300	150–250	65–85 × 40–60	Pigs; small intestine	–	6–11
Parascaris equorum	60–380	60–280	90–120 × 60	horses; small intestine	–	4
Toxocara canis	120–180	100–120	90 × 75	Dogs; small intestine		
Toxocara vitulorum	210–270	150–250	69–93 × 62–77	Calves; small intestine	–	3

(continued)

Table 5.6 (continued)

Family and species	Length of the adult worm (mm)		Egg resp. larval size (μm)	Final host/location of the adult worms	Intermediate hosts	Prepatency in the final hosts (weeks)
Family Dracunculidae						
Dracunculus medinensis	500–1200	30	Larvae (600 × 20)	**Humans**, dogs; subcutaneous connective tissue	*Cyclops* (Krebs)	40–56
Family Onchocerdidae						
Onchocerca volvulus	350–700	20–40	Larvae (non-sheathed) (260 × 7)	**Humans**, hypodermal connective tissues	*Simulium* sp.; Simuliids	32–52
O. gutturosa	40–60	40	Larvae (non-sheathed) (260 × 7)	Ruminants; hypodermal connective tissues	*Odagmia* sp.; Simuliids	28
Family Filariidae						
Wuchereria bancrofti	100	40	Larvae (sheathed) (275 × 7)	**Humans**, lymphatic vessels	*Aedes* sp.; *Culex* sp.; mosquitoes	52
Brugia malayi	80–90	30	Larvae (sheathed) (275 × 8)	**Humans**; lymphatic vessels, connective tissues	*Mansonia* sp.; Anopheles sp. mosquitoes	12
Loa loa	70	35	Larvae (sheathed) (290 × 8)	**Humans**, hypodermal connective tissues	*Chrysops* sp.; blackflies	52
Litomosoides carinii	60–120	20–25	Larvae (sheathed) (94 × 7)	Rats; pleural space	*Bdellonyssus* sp. mites	10–11
Dirofilaria immitis	250–300	120–180	Larvae (non-sheathed) (200–300 × 8)	Dogs, cats; heart, pulmonary artery	*Culex* sp.; *Anopheles* sp.; mosquitoes	25
Mansonella perstans	70–80	45	Larvae (non-sheathed) (200 × 4)	**Humans**, dogs; abdominal cavity	*Culicoides* sp. midges	36

Family Trichuridae						
Trichuris trichiura	50–60	50	50	**Humans**; colon	–	4–12
T. ovis	35–70	50	70–80 × 30–42	Ruminants; caecum	–	12
Capillaria annulata	10–50	10–25	60–62 × 24–27	Gamebirds; gullet	Earthworms	3
Family Trichinellidae						
Trichinella spiralis	3–4	1.5	Larvae (100 × 10)	Carnivores, **humans**; intestine	–	1

(a) Parthonegenetic generation (in many species larvae leave the egg already in the intestine and are found in the feces)
(b) Dioecious free-living generation
(c) Prepatency of summer ostertagiosis is about 3 weeks. Prepantency might be extended to 4–5 months, since larvae take a repose period in autumn and winter in the host's intestine

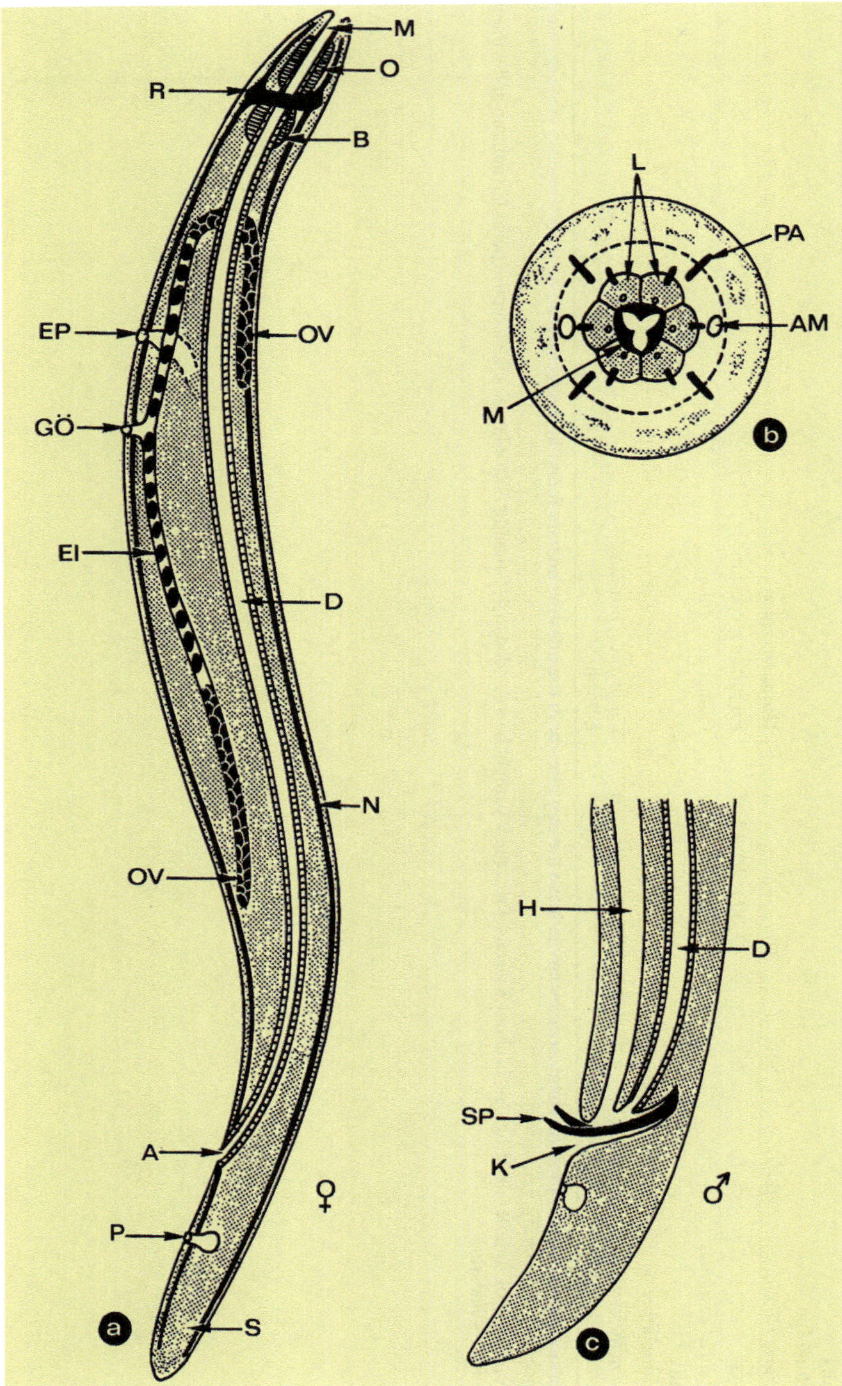

Fig. 5.77 Diagrammatic representation of the morphology of female and male nematodes in longitudinal sections (**a**, **c**) and in cross sections (**b**, **d**). (**a**) Female in longitudinal section;

5.3.1 Morphology of the Nematodes

1. **Name**: Greek: *nema, nematos* = filament, fine thread.
2. **Body characteristics**: Some nematodes are characterized by a series of exclusive criteria concerning their outer appearance:

   2.1 **Lips, teeth, hooks**: These different systems occur often in a trifold arrangement (Figs. 5.77 and 5.92). It is claimed that they are remnants of originally six such structures, which should have been melted. One lip is dorsally located and the two others ventro-laterally. As an example the ventral lips of *Ancylostoma duodenale* and *Necator americanus* are apparently transformed into teeth, hooks and plates (Figs. 5.105 and 5.106). In the case of other species, "teeth and/or hooks are formed inside the mouth (=buccal hollow) (e.g. in *Syngamus* species). In addition, species-specific folds or valves surround the mouth, which thus increase the ability to suck food at the host's surface. Peculiar **headplates** protect the mouth of the specimens of the family Spiruidae and are important for species diagnosis. Their larvae are equipped with peculiar mouth hooks, which also allow species determination.

   2.2 **Stilets**: Several species (often only their larvae) are equipped with stiletto-like bore systems inside their buccal cavity, which are prodrudible and hardened by a thick cuticular layer. Such stiletto-like structures occur e.g. in the mouth of larvae of *Trichinella spiralis* but are absent in their adults.

   2.3 **Cordon, alae**: **Cordons** are apical, bulbus-like protrusions of the cuticle, while **alae** appear as wing-like structures in the mouth region or along the anterior surface (Figs. 5.97 and 5.98). The latter are described as cervical, lateral or caudal alae. Similar functions have cuticular hooks of the *Haemonchus* species.

   2.4 **Bursa copulatrix, spicula**: These cuticular structures, which often contain hook-like components, occur at the posterior end of the males of some species (e.g. *Ancylostoma* species) and help the male to mate with the female by clutching it in the region of its genital opening (Figs. 5.79 and 5.106). The males of other species (e.g. Ascaridae, Strongylidae) have developed two prodrudible spicula being situated at the cloacal wall. They are injected during copulation with the female into its genital opening in order to widen it. Length and size of these spicula are used for species determination (Fig. 5.79).

Fig. 5.77 (continued) (**b**) Cross section through the mouth region; (**c**) Male terminal end in longitudinal section; A = anus; AM = amphids; B = bulbus; D = intestine; EL = oviduct, with eggs; EP = excretion pore; GÖ = genital opening; H = testis; K = cloaca; L = lip; M = mouth; N = longitudinal nerve; Ö = oesophagus; OV = ovarial tube; P = phasmids; PA = sense papillae; R = ring of nerves; S = tail; SP = two spicula

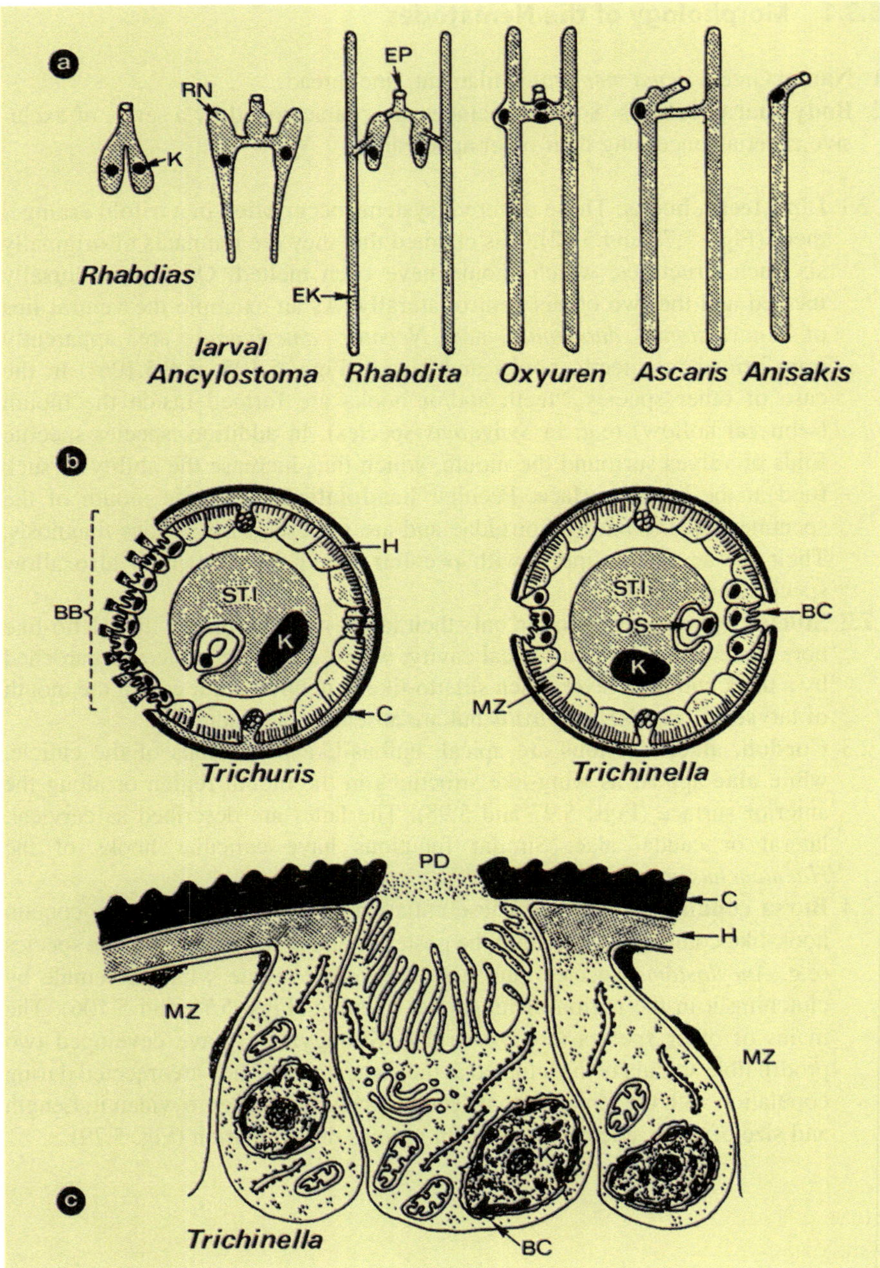

Fig. 5.78 Diagrammatic representation of the excretion systems of nematodes and of the arrangement of the bacillary cells. (**a**) Excretory cell types; (**b**) position and arrangement of the bacillary cells; (**c**) fine structural aspects of a bacillary cell. BB = bacillary band; BC = bacillary cell; C = cuticle; EK = excretion channel; EP = excretion pore; H = hypodermis; K = nucleus; MZ = muscle cell; ÖS = oesophagus; PD = posterior wall of the pore of the BC; RN = renette; STI = stichosome cell

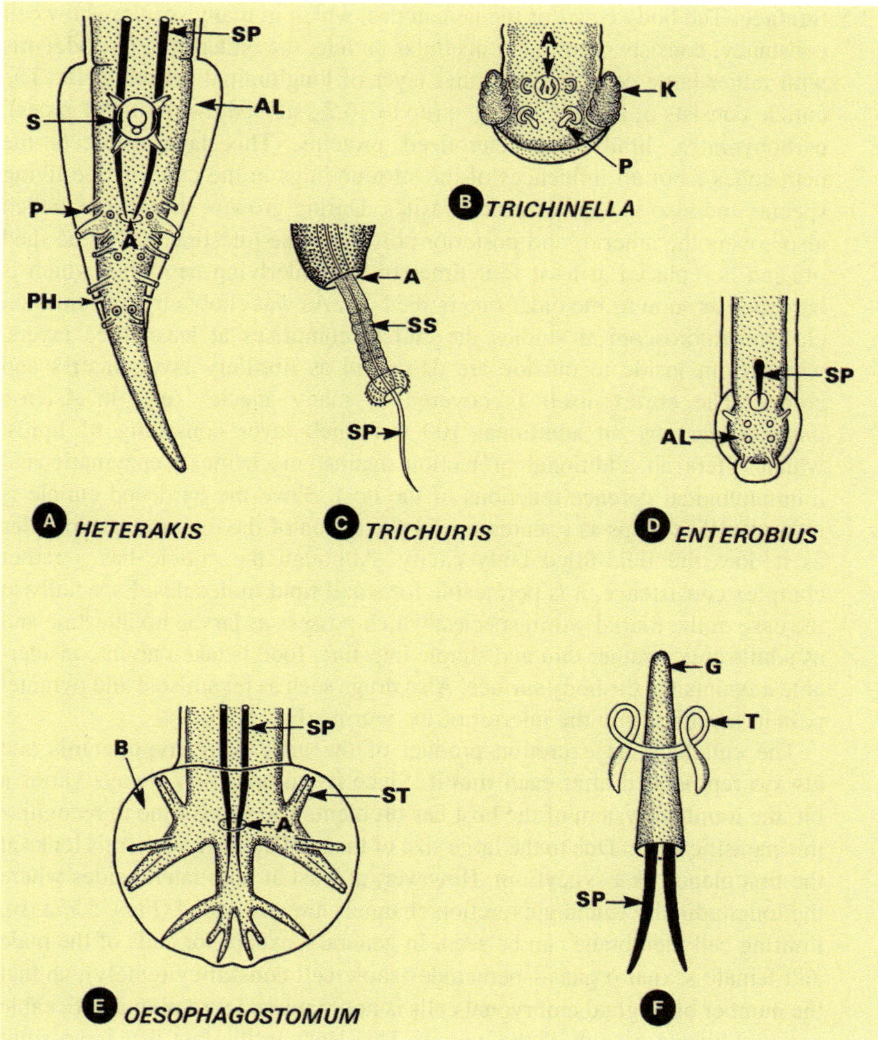

Fig. 5.79 Diagrammatic representation of the terminal ends of males of different genera of nematodes together with their accessory copulation systems. (**a**) *Heterakis* sp.; ventral aspect (alae, 1 sucker, 11 pairs of papillae). (**b**) *Trichinella spiralis*; ventral aspect (copulation bulbus); (**c**) *Trichuris ovis* (the spicula sheath may become protruded); (**d**) *Enterobius vermicularis* (only spiculum hook); (**e**) *Oesophagostomum* sp. (example for strongylids with a stabilized bursa copulatrix); (**f**) Copulation system of a trichostrongylid worm. A = anus; AL = ala, wing-like structure; B = bursa copulatrix; G = gubernaculum; K = copulation system; P = papilla; PH = phasmids; S = sucker; SP = speculum; SS = sheath of spicula; ST = fortifying structures; T = telamon (gland surrounding the spicula sheath)

2.5 **Surface**: The body cover of the nematodes, which in many species show cell constancy, consists of an outer acellular cuticle, the underlying **hypodermis** with rather large cells and the inner layer of longitudinal **muscle cells**. The cuticle consists of keratin, collagen (up to 30 %, steered by up to 150 genes), carbohydrates, lipids and sclerotized proteins. This layer protects the nematodes from all influences of the surroundings in the case of free-living species and also in case of endoparasites. During growth, the cuticle, which also covers the anterior and posterior portion of the intestine, has to be shed off and is replaced at least four times by an underlying new one, which is hardened as soon as the older one is shed off. As was shown by transmission electron microscopical studies the cuticle comprises at least three layers, which from inside to outside are described as fibrillary layer, matrix and cortex. The cortex itself is covered in many species (e.g. in *Ascaris lumbricoides*) by an additional 100 μm thick layer consisting of lipids, which offers an additional protection against mechanical, enzymatic and immunological defence reactions of the host. Since the hardened cuticle is rather "stiff, it helps as counterpart for the action of the longitudinal muscles as it does the fluid-filled body cavity. Although the cuticle has a rather complex consistence, it is permeable for small lipid molecules. Especially in the case of the filarial worm species, which possess as larvae no intestine and as adults only a rather thin and simple intestine, food is taken up in considerable amounts via the body surface. Also drugs such as levamisole and pyrantel pass in juveniles into the interior of the worms (Fig. 5.80).

The cuticle is an excretion product of the underlying **hypodermis** and always reproduced after each **moult**. Since its composition always varies a bit, the immune system of the host has problems and needs time to recognize this parasitic stage. Due to the large size of their cells, the hypodermis looks at the first glance as a syncytium. However, at least at both lateral sides where the longitudinally running excretion channels are embedded (Figs. 5.77a, b), limiting cell membrane can be seen. In general—except for cells of the male and female sexual organs—nematodes show **cell constancy** (eutely), so that the number of original embryonal cells is not increased (or not in a noticeable amount) during growth of the worms. This leads to the fact that large adult worms contain cells reaching often many centimetres in length. The diameter of the hypodermis itself is often rather small compared to the diameter of the cuticle. In some cases, the hypodermis measures only a few μm in diameter and thus is thin compared to the cuticle, which reaches diameters of e.g. 50 μm in the case of *Strongylus equinus* or 80 μm in *Ascaris lumbricoides* or even one-tenth of the worm's diameter in the case of *Ancylostoma duodenale*.

2.6 **Nerve system**:

Nerve strands are located in both of the dorsal and ventral protrusions, which are interconnected in the anterior region of the worm by a ring nerve system (Figs. 5.77a, b and 5.81), which represents a simple central nerve system. Each nerve cell of the dorsal strand is connected a cell protrusion with the ventral strand so that also lateral innervation is guaranteed. The total

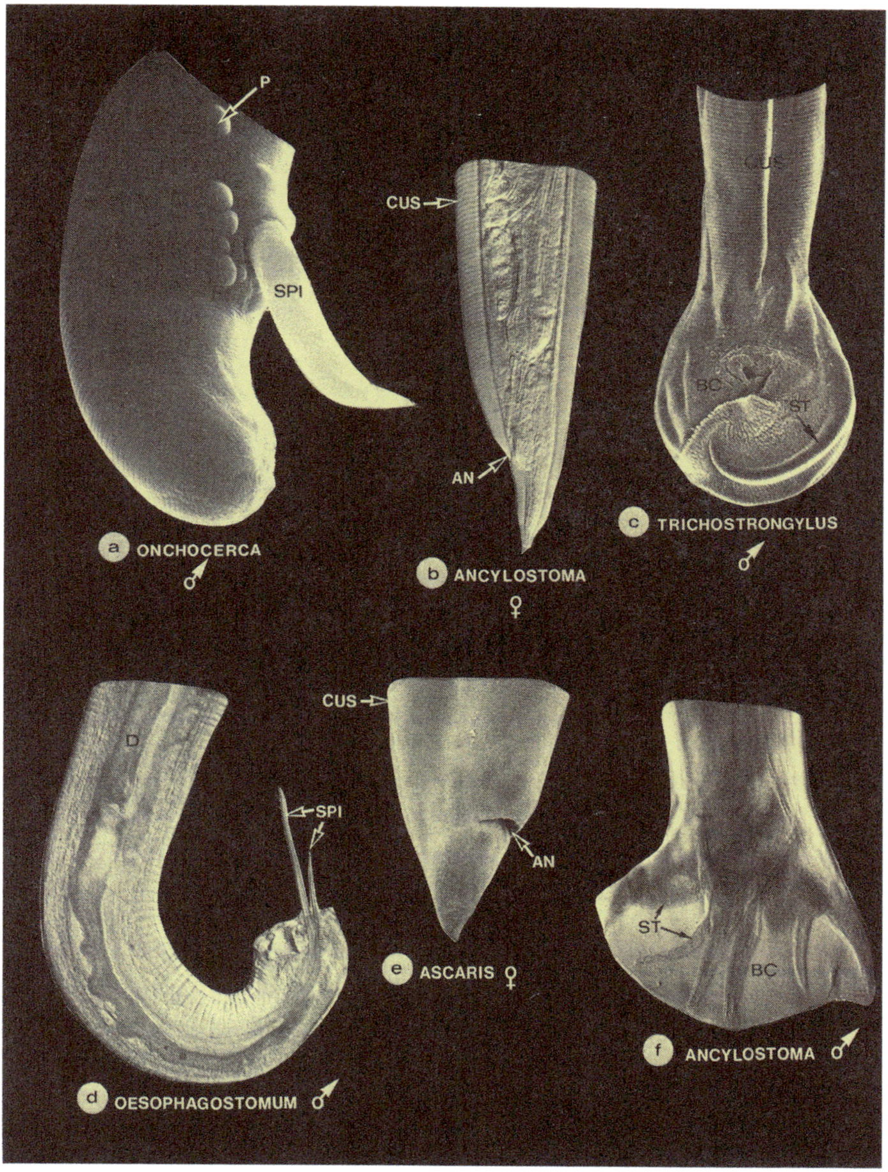

Fig. 5.80 Scanning electron micrograph of the posterior end of nematodes. AN = anus; BC = bursa copulatrix; CUS = cuticular striation; D = intestine; P = sense papilla; SPI = spiculum; ST = fortifications of the BC

number of nerve cells is rather low. For example, the up to 40 cm long worm *Ascaris suum* possesses only 298 nerve cells, while the only 2 mm long species *Caenorhabditis elegans* has 302. In *A. suum* and *Rhabditis* sp., the neurons of the ventral nerve strand are present in 5 groups with always

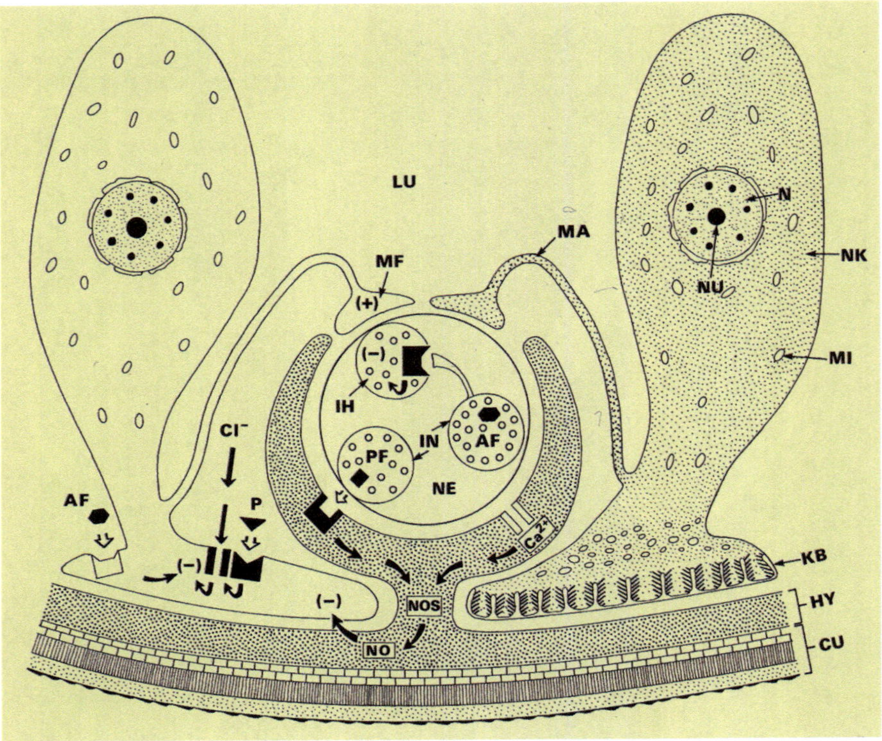

Fig. 5.81 Redrawn diagrammatic representation (according to Maule) of some nerve-muscle relations of the nematode *Ascaris suum*. Inside the ventral nerve strands (NE) which belong to the central nerve system of the nematodes exist interneurons and motoneurons. The inhibitory acting (GABA-erge) motoneuron (IH) becomes blocked by peptides (AF) originating from the interneuron. This leads to the stimulation of the muscle cell (+), which via a muscle projection (muscle finger, MF) and a "muscle arm" (MA) creates a contact between its contractile region and the nerve. Then flows inside the contact zone the stimulating neurotransmitter acetylcholine. However, AF peptides are able to block again the muscle fibre (see left, AF receptor). A further blocking of the muscle cell may occur by actions of another neurobased peptide (PF), which is activated by a Ca++-depending, hypodermal nitrogensynthase (NOS) during NO binding. Again another peptide (P) leads to a Cl-depending inactivity of the muscle cells of the worm. F = FMR = famid-like peptide (isolated from *Ascaris suum*); CU = layers of the cuticle; HY = hypodermis; IH = inhibitory motoneuron; IN = interneuron; KB = contractile region of the muscle cell; LU = lumen of the body cavity of the worm; MA = muscle arm; MF = muscle finger; MI = mitochondrion; N = nucleus; NE = nerve in cross section; NK = non-contractile region of NE; NO = nitrogen; NOS = nitrogenoxidase; NU = nucleolus; P/PF = FMR = famid-like peptide, which was first isolated from the free-living species *Panagrellus* sp.

11 cells. They are able to block as well as stimulate the contacted longitudinal muscle cells (Fig. 5.81). The ventral nerve strand contains also the protrusions of interneurons, which are situated in the apical nerve ring (simple brain). The typical snake-like (winding) movements of the nematodes were induced by

these interneurons of the "brain" and by the motoneurons of the ventral nerve strand.

2.7 The **muscle cells** of the nematodes differ in many aspects from those of other animal phyla. There exist only cross-striated longitudinal ones, which are provided with cytoplasmic protrusions stretching to either the dorsal or ventral nerve strand (while e.g. in vertebrates nerves stretch always to muscle fibres). The muscle cells of nematodes are located within four quadrants, which are separated from each other by the dorsal and ventral nerve strands and by the two lateral protrusions containing the excretion channels (Fig. 5.77b). According to the number of muscle cells being situated within each quadrant, different types of muscle cell arrangements are distinguished.
- **Meromyaria**: Four cells are found in each quadrant at the maximum.
- **Polymyaria**: The members of this group contain more (at least six) muscle cells per quadrant. Larvae of the members of this group may appear at first as meromyarians.

According of the arrangement of the contractile filaments inside the muscle cells (which in general are cross-striated) three basic types can distinguished:
- **Platymyaria**: In this case, the contractile elements are exclusively situated at the side close to the hypodermis.
- **Coelomyaria**: The contractile elements are situated at the lateral sides of the muscle cells so that a central cytoplasmic region without fibres occurs.
- **Circomyaria**: The contractile elements are arranged around the nucleus and thus are located at all cell membranes.

2.8 **Sense organs**: The sense system consists of accumulations of rather simple sensillae, which are found at the anterior end in the mouth region as well as in the region around the anus at the terminal end of the worms. There are two types: (**a**) **Phasmids** (exclusively in the group of the Plasmidea = Secernentia) and (**b**) **amphids**, which are investigated in more detail. Their structures and arrangements are used as criteria for taxonomic determination. **Phasmids** are situated as pairs caudally behind the anus being surrounded by glands. They are considered to be able to detect odours. **Amphids** act apparently as chemoreceptors. They are composed of two laterally situated depressions at the anterior end of the worms (=front side) containing ciliated sense cells. In addition to these structures, 16 papillae can be seen arranged in three rings, which apparently have further sense cell functions (Fig. 5.77).

So-called **bacillary cells** are present in the members of the families Trichuridae and Trichinellidae. These structures are characterized by terminal enlargements which are in contact with the air by an opening, through which gland cell products are released that are apparently used to cover the surface of the worm's body.

2.9 **Intestine**: The intestine appears as a rather long tube, which opens ventrally in some distance before the terminal end of the worm (=subterminally). This leads to the aspect that the males and females possess a species-specific long tail, which in the case of female is more pointed than in males. The females of *Dracunculus medinensis* and *Mermis* sp., however, do not possess an anus at all.

The **intestine** can be subdivided into the following regions:

(a) **Mouth**:

The mouth and its surroundings show species-specific fortifications, which are useful for anchoring in or at tissues and are used for uptake of the targeted food.

(b) **Oesophagus (pharynx)**:

This anterior portion of the intestine, which acts often as a sucker, is lined by a cuticular layer and has been adapted in many species to specific functions allowing the uptake of the species-specific food. In the case of *Trichinella* and *Trichuris* species, so-called **stichosome cells** surround the intestine (Fig. 5.78b).

(c) **Small intestine (central intestine)**:

This intestinal region appears tube-like and possesses a single layer of endothelial cells which are placed on a basal membrane. The size and especially the length of the microvilli is species specific.

(d) **Rectum**:

This upmost posterior portion of the intestine is lined by a cuticle. With the exception of the group of Aphasmidea, the ductules of single-cell glands open into the lumen of the rectum. There occur three cells in the case of females and six in males. Since in males also the single testis opens into the rectum the region is defined as **cloaca** (Fig. 5.79).

2.10 **Excretion systems**:

This system varies considerably in size and function in the group of nematodes (Fig. 5.78). However, protonephridia (like those in the group of Platyhelminthes) do not occur, but two morphologically different organs occur:

(a) a gland system (renette),

(b) a channel system (H-cell).

In the case of the **renette** one or two large gland systems that are situated in the anterior region of the body excrete their products via a ductile that opens ventrally within a porus. This type of excretion mostly occurs in free-living nematodes and in juvenile stages of parasitic species. It is considered as precursor of the so-called **H-cell channel** system (Fig. 5.78a). This H-channel system may appear asymmetric in some species, due to the length reduction of one of the lateral channels, which run in the lateral ledges of the hypodermis (Fig. 5.77b). The **bacillary cells** of the members of the Trichuridae and Trichinellidae act also as excretion organs, since the above-described systems do not occur in these worm groups (Fig. 5.78c).

2.11 **Sexual organs**:

The parasitic nematodes are subdivided into male and female organisms, while some free-living species are protandric hermaphrodites. Both sexes of the parasites can mostly already be distinguished by their exterior aspects. Males are in general smaller than females and are equipped with typical structures that are needed to proceed copulation with a female stage (Fig. 5.79). One or two **spicula** (not present in *Trichinella spiralis*), genital papillae, a **bursa copulatrix**, a **gubernaculum** or a so-called **telamon apparatus** (only present in Strongyloidea) are common helpful systems. The hind end of males is often rolled in and thus name giving in *Onchocerca volvulus*.

(a) **Male sexual system**:

The male testis consists of a single long, often winding tube, which is filled with developmental and fertile sperm, which are released via the vas efferens into the terminally laying cloaca (Fig. 5.77). Originally apparently two testes had been present. However, among the recent nematodes one of these tubes has been reduced. The wedge-like appearing sperm do not possess a flagellum but move like amoebae. In many species, the nuclear membrane of the sperms is no longer present so that they appear "nucleus free". However, the genetic material is present inside the cytoplasm of these sperm and can be used when fusing with a female gamete. The size and shape of the sperms differs somewhat in the various species of the nematodes, but some species-specific features remain persistent. The size of sperm, however, varies considerably. The sperm of ascarids, of *Trichinella spiralis* and of the filariae appear spherical with diameters of about 10 μm, while those of the strongylids and oxyurids are filament like and measure 20 μm in strongylids and up to 150 μm in oxyurids.

(b) **Female sexual system**:

With the exception of the group of the Aphasmidea, most species of the nematodes contain two ovaries (didelphic forms), each of which discharges eggs via oviduct and uterus into the single vagina (Fig. 5.77c). The vagina, which is lined by a cuticle, opens into the female sexual porus, which in the case of many species occurs midventrally or in the first or second third of the female body. Due to this position, couples of nematodes show Y-looking copulation positions. However, in the case of the species of the members of the superfamily Strongylidea (e.g. genera *Strongylus, Syngamus*), both male and female sexual openings are situated at the terminal end of the worms. The different regions (ovary, oviduct, uterus) of the female sexual organs are different with respect to the endothelial layer. The terminal region of this sexual tube is characterized by a central strand, which is termed **rachis** and serves as transportation system to supply the growing egg cells with food. In general the uterus is swollen due to masses of stored eggs and by sperms, which had been injected during repeated copulations. Thus, this region may also be considered as a **receptaculum**

seminis. Eggs are produced in large masses in many species. For example, *Ascaris lumbricoides* females may excrete up to 200,000 eggs per day—however, not all of them are fertilized. Other nematode species produce much lower amounts of eggs, but nearly all are fertilized.

2.12 **Sex determination**:

In many species of the group of Secernentea, the determination of the sex occurs by sex chromosomes. Males have in general one sex chromosome (XO) less than females (XX). But in some groups of nematodes also the XY-XX system occurs. The exact number of chromosomes is still unknown for many nematode species. For example, even for *Ascaris lumbricoides* the number of sex chromosomes is not yet finally fixed. Most realistic seems the number $2n = 48$, whereby the females possess 2×19 **autosomes** and 2×5 X chromosomes. The males with a XO sexual determination possess 2×19 autosomes and 1×5 monovalent X-chromosomes. *Dictyocaulus arnfeldi*, however, possesses only a single set of sex chromosomes (in a XO constellation). Females $= 2n = 12$, males $= 2n - 11$). In the case of Trichuridae, the **XO type** may occur (e.g. *Trichuris trichiura* ♀ $= 2n = 8$) as well as the **XY-type** (*T. ovis* $2\,n = 6$).

Parthenogenic females of *Strongyloides papillosus* occur in the 3n status and contain six chromosomes, while both sexes of the free-living generation possess four chromosomes (in a XY determination). The closely related species *Strongyloides ratti*, however, belongs to the XO type. As further examples the number of chromosomes is depicted below for some other nematode species:

- *Contracaecum spiculicerum* ($2n = 16$; XO);
- *Strongyloides edentatus* ($2n = 12$; XO);
- *Haemonchus caninum* ($2n = 12$; XO);
- *Setaria equina* ($2n = 12$; XO).

Another type of sexual determination apparently occurs in the toad nematode *Rhabdias bufonis*, which lives in the lungs of its host. The parasitic adults are **protandric hermaphrodites**. At first, sperms are produced in the paired sexual organs and become stored in the receptaculum seminis. Afterwards, the same gonads produce the eggs, which become fertilized by the stored own sperms. This gives rise outside of the body to the larvae 1 which grow up to larvae 3 and later to a **hermaphroditic generation** after penetration into another toad. However, some of the L3 larvae develop outside of the body to a free-living generation of males and females. The larvae 3 of their progeny, however, may again enter the toad and grow up to hermaphroditic adults.

The phenomenon of **chromosome (chromatin) diminution** and **elimination** occurs in several families of nematodes. In the case of the horse worm species *Parascaris equorum* (syn. *Ascaris megalocephala*), this process is widely understood. During the first divisions of the fertilized egg cell, a terminal chromosomal fragmentation occurs leading to an elimination of heterochromatin inside somatic cell lines, while the percursors of the sexual cells remain unchanged. This leads as final result to the fact that sexual and somatic cells contain chromosomes in a different quantity and quality. In the cases of *P. equorum* and *P. univalens*, where only a long "collecting chromosome" is present in the haploid stage, this chromosome becomes disrupted during the diminution process of the soma cells in about 60 fragments and some larger portions, which are dissolved within the cytoplasm and thus disappear. The function of this process is (since years) under discussion. This discharge of terminal heterochromatin must not necessarily lead to a reduction of the number of chromosomes as can be seen in *Ascaris suum*, where the number of chromosomes remains stable (♂ = 2n = 38 autosomes plus five sex chromosomes and ♀ = 38 autosomes plus 10 X chromosomes).

5.3.2 Development of the Nematodes

The females of the parasitic nematodes depone eggs and larval developmental stages at different stages of development:

- **Eggs**, which start development, as soon as they have reached the soil (Fig. 5.82). This are **oviparous** species: e.g. belonging to the genera *Ascaris, Trichuris, Ancylostoma*, etc.;

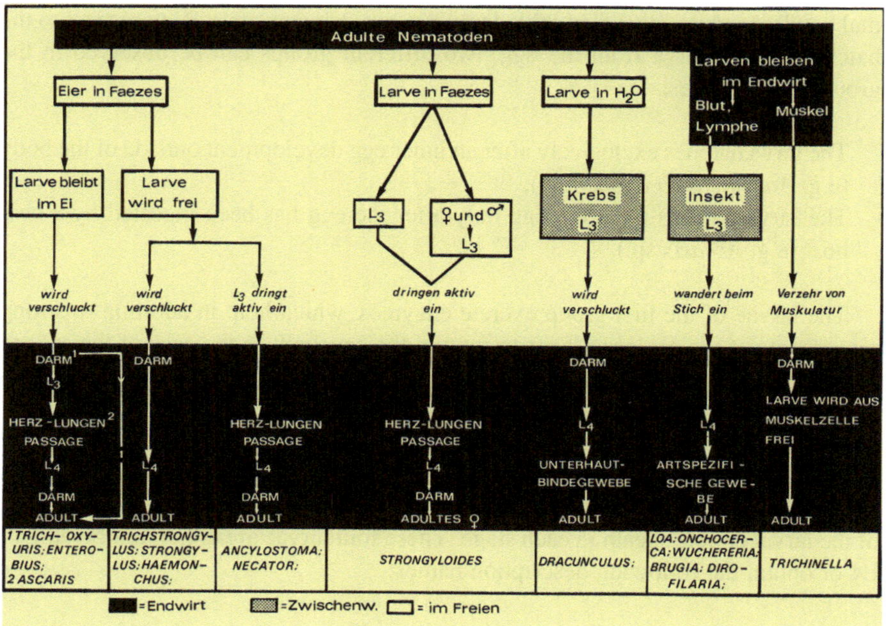

Fig. 5.82 Diagrammatic representation of the developmental stages and hosts of important nematodes

- **Eggs**, from which the larva hatches already within the intestine of the host: e.g. in the case of the parthenogenetic generation of *Strongyloides stercoralis, Dictyocaulus* sp., *Protostrongylus* sp., etc.;
- **Eggs**, from which the larvae hatch during the excretion process (=**ovoviviparous** species): e.g. in the case of several filarial species and in Spirurida. **Larvae** of **viviparous** species: e.g. *Dracunculus medinensis, Trichinella* species, etc. There are clear differences in the diameter of the eggshells. Eggs, which proceed larval development outside of the body, have a very thick and/or stable wall, which offers protection from adverse temperatures or other bad conditions. The larvae of species, which hatch from the egg inside the host's body or just after being excreted, are also firmly protected, since they are covered by an amorphous **cuticle**. Some filarial larvae are even double protected since they do possess a cuticle and are in addition surrounded by the stretched eggshell, which then is called **sheath** (e.g. in *Loa loa, Wuchereria bancrofti*). Again other species have developed similar protection methods: e.g. larvae 3 of *Stephanurus* sp. stay inside the discharged cuticle of their larva 2 and thus also appear "sheathed".

The eggshells of parasites, which are excreted undeveloped or in a low-grade status, are very thick and thus allow an exposition to very high and very low environmental temperatures (e.g. ascarids survive temperatures between 40 °C minus and 40 °C plus, so that sophisticated methods must be used to eliminate them from stables and from other grounds).

Embryonation (=development of the larvae inside the eggshell) affords optimal conditions/humidity, agreeable temperatures, oxygen, etc.). With respect to the hatching of the larva from the egg, two different groups can be observed in the group of nematodes:

- The larva hatches exclusively after an inner egg development outside of the body (e.g. *Ancylostoma duodenale*);
- The larva hatches from the egg only after the egg has been ingested by a new host (e.g. *Ascaris* sp.).

The larvae of the first group excrete enzymes, which help in addition to strong movements of the enclosed larva to rupture the eggshell. In the case of the second group, enzymes of the host in combination with a suitable concentration of CO_2 initiate the destruction of the eggshell and thus the release of the larva.

Many parasitic nematodes undergo a host change, which is often obligatory and direct and without involvement of different generations. They grow up to adults passing four larval stages and a preadult one. Growth is possible due to the release of the larval cuticle sheath in each stage. These four larvae are either numbered L1–L4 or obtain the following description names:

- The first and the second larvae of some species are called **rhabditiform** due to their stab-like oesophagus. They occur e.g. in *Necator americanus, Ancylostoma duodenale* and *Strongyloides stercoralis* (Figs. 5.104 and 5.116).

- The third stage larva of many species is called **filariform** and has its origin in the hatched rhabditiform larva 2. Its oesophagus becomes stretched, and its typical constriction (**isthmus**) disappears (Fig. 5.116).
- The first larva of the group of the filarial worms is called **microfilaria**, which lives in the blood or lymph of their hosts. These larvae are born (=excreted from the female's uterus) either **sheathed** (=surrounded by the stretched eggshell) or **unsheathed** (=eggshell is discharged). The microfilariae of the genera *Wuchereria* and *Loa* belong to the first group and *Onchocerca* larvae to the second one (Fig. 5.137).

The microfilarial sheath does not contain chitin, although it has been present in the eggshell, but contains 54 % amino acids (glutamine, proline), 8 % sugars and 11 % inorganic substances (Na, K, P), but only 0.2 % fats besides 26 % mostly not very well-known other materials in species-specific combinations. This very complex layer protects successfully the microfilariae from the attacks of the host's immune system.

The **third-stage larvae** (L3) of many parasitic nematodes appear also sheathed, since they remain protected from the larval cover (=cuticle) of the larva 3 (e.-g. *Ancylostoma* species). But in any way—sheathed or not—the larvae 3 are able to penetrate into the body of their final hosts, where they are enabled to reach the adult stage after two further moults. However, if these larvae enter other hosts than their special one, they may migrate therein for a while but finally are killed by the host's defence system before reaching maturity.

The individual development via four larval stages may occur within a single host but may also include another one (intermediate host) (Figs. 5.82 and 5.104). Both possibilities exist independent from the fact whether during the life cycle larvae are set free or not.

A further type of the life cycle of nematodes occurs in the genus *Strongyloides*, which includes a facultative alternation between a free-living generation of males and females besides another generation of females which reproduce themselves parthenogenetically, while living as parasites inside their hosts. This type of reproduction is described as **heterogony** (Fig. 5.116). However, this type of life cycle is rather rare among the nematodes, but probably is a relict of the phase, when nematodes started in the evolution to become internal parasites. In this heterogonic life cycle, the intestinal females excrete three types of eggs:

1. **3n eggs**, which become again parthenogenetic females.
2. **2n eggs**, which give rise outside of the host to free-living females.
3. **1n eggs**, which produce outside of the host the free-living males.

In Fig. 5.82, the most common pathways of the individual development of important nematode species are diagrammatically depicted.

A. **Monoxenous developmental life cycle**

In this type of life cycle, which includes only one host, three different basic types can be differentiated:

1. **The first type** is characterized by the fact that the final host is infected by ingesting eggs containing already an infectious larva, which hatches from the egg in the intestine of the host after being taken up (e.g. *Enterobius* species, *Trichuris suis, Ascaris* species). It is possible that the first larva (L1) is transformed to the second larva (L2) inside the egg prior to uptake of the host. The liver–heart–lung–trachea–oesophagus passage of *Ascaris* species may, however, be an indication for the fact that this species formerly belonged to the **second type of development**. The speed of the development of the larvae in the eggs depends on the outer temperatures and thus may take a few days up to several weeks.
2. **The second type** of development occurs in species, where the first larva hatches from the egg outside of the host's body and grows up via two moults to the infectious larva 3. This larva 3 is either ingested by the host within its food (e.g. *Strongylus* species) or it enters actively the skin of the host (e.g. *Ancylostoma* species). In the latter case, it is obligatory for the penetrated larva to start a liver–heart–lung–trachea passage in order to reach the intestine as final site of living. This type of life cycle is mainly restricted to countries with higher temperatures.
3. **The third type**, which occurs in *Trichinella* species, is completely different from those two described above, since in this case neither eggs nor larvae reach a phase outside of a host's body. Final hosts are exclusively carnivores and omnivores (which have to ingest raw meat containing *Trichinella* larvae in order to become infected. In the intestine of the new host, they grow up to male or females. After copulation the females excrete larvae, which enter directly the blood vessels of this host and finally penetrate muscle fibres, where they "wait" until a new final host ingests these infected pieces of meat (Fig. 5.87). In recent times, also horses had been found being infected by *Trichinella* species. This was apparently due to ingestion of beetles or fly maggots (hidden within grass) that had ingested meat of *Trichinella*-infected dead bodies (e.g. of rats, etc.).

B. **Dixenous life cycle of nematodes**

This dixenous (heteroxenous) life cycle includes **obligatorily** an **intermediate host** and may follow one of three different pathways:

1. In the case of **filarial worms**, bloodsucking ectoparasites (insects, ticks, mites) serve as specific intermediate hosts, which ingest during their bloodsucking (or lymph uptake) microfilariae (Fig. 5.82). These larvae 1 enter inside the intermediate hosts the thorax and are transformed to a sausage-like appearing second larva, which develops into the infectious larva 3. This 0.8–1 mm long stage enters the mouthparts of its vector. During the bloodsucking act of the host, the larva 3 passes through the thin connection bands of the mouthparts and enters the **final host** by penetration into the skin, deepening after the vector had redrawn its mouthparts. Important species of filarial nematodes are listed in Table 5.6.

2. In other nematode species, **further types of intermediate hosts** are used. For example, *Dracunculus medinensis* uses small water crustaceans, which finally bear the infectious larva 3. *Parastrongylus cantonensis* uses snails as intermediate hosts and *Porrocaecum* species use earthworms, while *Anisakis* species and *Contracaecum* species even use vertebrates (fishes) as second hosts.
3. **A third type** of a dixenous life cycle occurs in species of the family Protostrongylidae, where the larva 3 is developed in terrestric snails, but leaves this host and becomes attached at food plants, which are ingested by the final hosts.

Further Reading

Deplazes P et al (2013) Lehrbuch der Parasitologie für die Tiermedizin, 3rd edn. Enke, Stuttgart.

Fitzpatrick JL (2013) Global of food security: the impact of veterinary parasites. Vet Parasitol 195:233–248.

Dorny P et al (2009) Emerging food-borne parasites. Vet Parasitol 163:196–206.

Grencis RK et al (2014) Immunity to gastrointestinal nematodes. Immunol Rev 260:183–205.

Hui YH et al (2002) Food-borne disease handbook, 2nd edn. Marcel Becker, New York, Basel.

Kaminsky R et al (2013) Worms—a license to kill. Vet Parasitol 195:286–291.

Mehlhorn H (2017) Encyclopedia of parasitology, 4th edn. Springer, New York (and online version 2016).

5.3.3 Nematodes as Agents of Disease in Animals

5.3.3.1 *Trichuris* Species (Whipworms)

1. **Name**: Greek: *thrix, trichos* = hair; *ura* = tail. Latin: *ovis* = sheep; *sus* = pig; *vulpes* = fox; *mus* = mouse, *lepus* = hare; *discolour* = with two colours; the

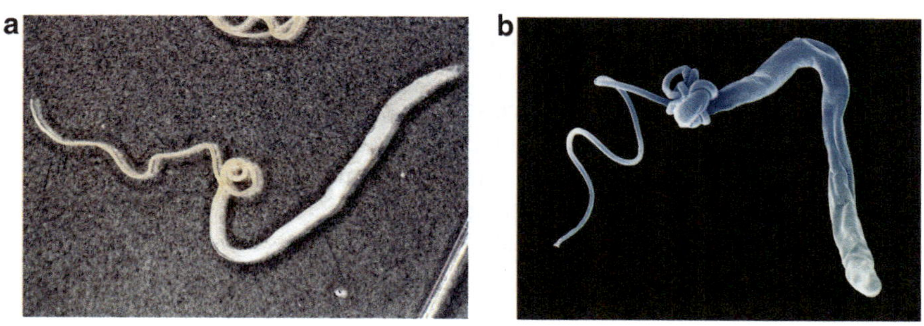

Fig. 5.83 (**a**, **b**) Light (**a**) and scanning electron micrograph (**b**) of adult *Trichuris* worms. Their anterior ends are very thin, so that they obtain a whip-like shape

name *Trichuris* refers to the hair-like anterior end, which led also to the trivial name whipworm (Fig. 5.83).

2. **Geographic distribution/epidemiology**: In the case of cultured farm animals: worldwide; often mass infections occur in cases of overcrowded stables.
3. **Biology, morphology**: In the colon and caeca of many mammals, various species of the genus *Trichuris* are described:

T. discolor (ruminants)	♀, ♂ 4–8 cm
T. leporis (hares)	♀, ♂ up to 2.5 cm
T. suis (pigs)	♀ = 5.5 cm, ♂ 4.5 cm
T. ovis (sheep)	♀, ♂ 4–8 cm
T. muris (rodents)	♀, ♂ up to 2.5 cm
T. trichiura (primates, humans)	♀ = 5–6 cm, ♂ 5 cm
T. vulpis (dogs, foxes)	♀, ♂ 7.5 cm

Some of these species reach extreme high prevalence rates (e.g. 50–70 % in sheep; 25 % in pigs, etc.; dogs/foxes 2–30 %).

T. serrata and *T. campanula* occur in rare cases in cats and *T. skrjabini* and *T. globulosa* in wild and farmed ruminants, while from equids no *Trichuris* species are known.

The filament-like anterior end of these nematodes is anchored in the mucous layer of the intestine of their hosts. The mouth region is as fine, that it is even able to enter into single cells and to suck in cytoplasm, thus destroying these cells. Also fine blood vessels are entered in this way. The hind end, which looks like the grip of a whip, ranges free inside the lumen of the intestine. The females produce large numbers of eggs (e.g. *T. trichiura* ca. 15,000 per day!), which are excreted in an embryonated status within the feces. Under favourite conditions (humidity 20–25 °C), the larva is produced inside the eggshell within 6–12 weeks lately. If these eggs are ingested by the specific host, the 200 μm long larvae hatch from the egg (in the region of the colon and caecum), enter there the mucosa, where they moult after 10–12 days, and thus reach the status of the larva 2. Then they are placed just below the epidermis and introduce later their terminal end into the intestinal lumen. After three further moults, these larva reach maturity and start after fertilization the excretion of eggs, which have a species-specific size and are characterized by two polar plugs (see subheading **Diagnosis**).

4. **Symptoms of disease (Trichuriasis)**: Since each worm ingests besides cell contents also blood (about 5–10 ml per day), the intestinal wall appears spotted with oedemas and the feces may contain traces of blood. In cases of high-grade infections, the following symptoms are common: anaemia, loss of weight, retardation of the growth, weakness and immune deficiency. Low-grade infections, however, are mostly not noted but endanger especially young animals.

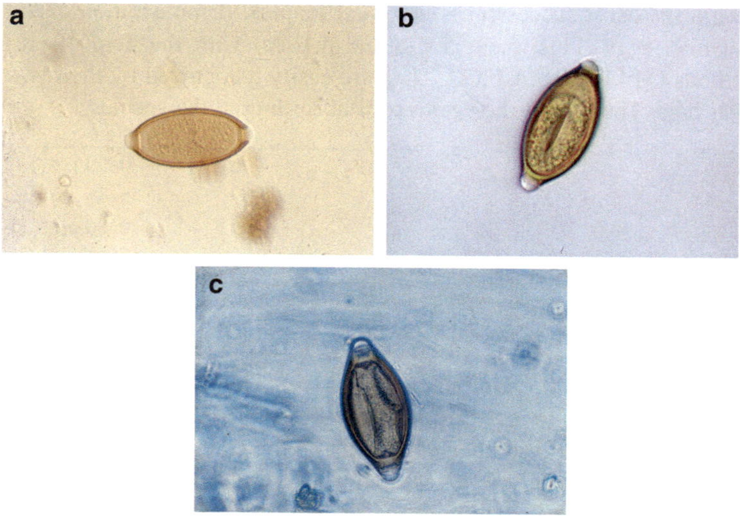

Fig. 5.84 Light micrographs of unembryonated (**a**) and embryonated (**b**, **c**) eggs of the genus *Trichuris*

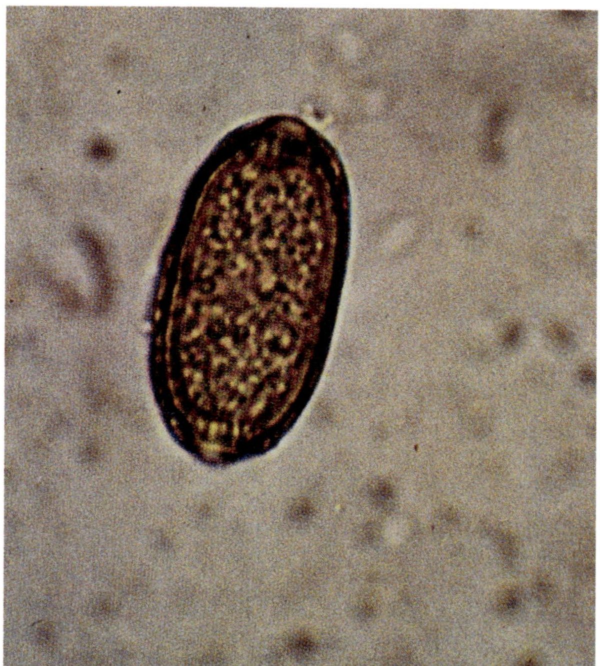

Fig. 5.85 Light micrograph of a *Capillaria* egg, which shows plugs at the two poles which are not much protruding

5. **Diagnosis**: Most successful is the microscopical demonstration of the typical brownish eggs (Figs. 5.84, 5.99b and 5.102c) with the help of enrichment methods (M.I.F.C.; S.A.F.C.). They are easily recognized by their two unique polar lugs. The size of the eggs varies according to the species.

T. vulpis	70–90 µm × 30–40 µm
T. suis	45–75 µm × 27–35 µm
T. ovis	70–80 µm × 25–40 µm
T. muris	65–70 µm × 35–40 µm
T. trichiura	50–55 µm × 21–25 µm
T. vulpis	70–85 µm × 36–40 µm

6. **Pathway of infection**: Oral uptake of larva-containing eggs.
7. **Prophylaxis**: Regular deworming; keep floors of stables free from feces.
8. **Incubation period**: Variable, depending on the grade of infection (1–2 weeks; pigs often: 6 weeks).
9. **Prepatent period**: This time is species dependent on the grade of infection. Examples are:
   - *T. vulpis* 11–15 weeks
   - *T. suis* 7–8 weeks
   - *T. ovis* 7–12 weeks
   - *T. muris* 8–9 weeks

10. **Patency**: Dependent on the grade of infection and strength of the immune system, the patent period may be considerably shortened to 3–4 months. However, low-grade infections may induce longer patent periods:

T. vulpis	1.5 years
T. suis	6–8 months
T. ovis	12 months
T. muris	only 3–4 months

11. **Therapy**: High efficacy rates are reached by application of benzimidazole preparations and macrocyclic lactones. In the case of dogs also emodepside was successfully used.

Further Reading

Callejon R et al (2015) Taxonomy and phylogeny of *Trichuris globulosa*. Inf Genetics Evol 34:61–74.

Kutzer E (1978) Notes on the hare-whipworm *Trichuris leporis*. Parasitol Res 56:69–72.

Liu GH et al (2012) Characterization of the complete mitochondrial genome of *Trichuris ovis* and *T. discolor*. Inf Gen Evol 12:1635–1641.

Petersen HH et al (2014) Parasite population dynamics in pigs infected with *Trichuris suis* and *Oesophagostomum dentatum*. Vet Parasitol 199:73–80.
Petry G et al (2013) Efficacy of emodepside plus toltrazuril oral suspension for dogs (Procox®, Bayer) against *Trichuris vulpis* in naturally infected dogs. Parasitol Res 112:S133–S138.
Skallerup P et al (2015) Detection of a quantitative trait locus associated with resistance to *Trichuris suis* in pigs. Vet Parasitol 210:264–269.
Vega-Sanchez MC et al (2015) Induction of ulcerative colitis in mice influences the course of infection with *Trichuris muris*. J Helminthol 89:593–600.
Wang Y et al (2013) Genetic variability among *Trichuris ovis* isolates. Mitochondrial DNA 2013:50–54.

5.3.3.2 *Capillaria* Species (Hairworms)

1. **Name**: Latin: *capillus* = head hair; *annulatus* = with ring-like proportions; *anatis* = belonging to ducks; *contortus* = turned; *bos* = cattle; *hepar* = liver; *aerophilus* = air loving; *plica* = folding.
2. **Geographic distribution/epidemiology**: Worldwide, leading to intense outbreaks under poor rearing conditions of animals.
3. **Biology, morphology**: The very tiny (=hair-like) adult worms are able to develop themselves directly (=without or indirectly via an intermediate host). After the eggs had been excreted by the females, they develop (outside of the body) the larva within 1–2 weeks (Figs. 5.99a and 5.102a). These eggs have to be ingested by an **intermediate host** (e.g. earthworms in the cases of *C. annulata* and *C. aerophila*) or directly by the **final host** (*C. bovis, C. contorta*). The habitats of the adult worms depend on the species (e.g. *C. annulata* = crop, oesophagus; *C. obsignata, C. bovis* = small intestine; *C. plica* = bladder; *C. hepatica* = liver; see Table 5.7). **Attention**: *C. hepatica, C. aerophila* and *C. philippinensis* may infect humans and thus induce a zoonosis.
4. **Symptoms of disease (Capillariasis)**: Organ-specific symptoms occur. Their intensity depends on the amount of parasitizing worms: catarrhalic and

Table 5.7 Important hair worms

Species	Hosts	Habitat	Length ♀	Length ♂
C. annulata	Gamebirds	Foregut	3 cm	1.5 cm
C. contorta	Chicken/ducks	Foregut	1–2 cm	1 cm
C. obsignata	Many bird species	Small intestine	1.2 cm	0.8 cm
C. anatis	Geese, ducks	Caecum	~1.8 cm	1 cm
C. bovis	Ruminants	Small intestine	~2.2 cm	1.5 cm
C. aerophila	Carnivores/**humans**	Breathing systems	~2 cm	1.7 cm
C. plica	Carnivores	Urinary bladder	Up to 6 cm	Up to 3 cm
C. hepatica	Rodents, e.g. rabbits, **humans**	Liver	Up to 8 cm	Up to 4 cm

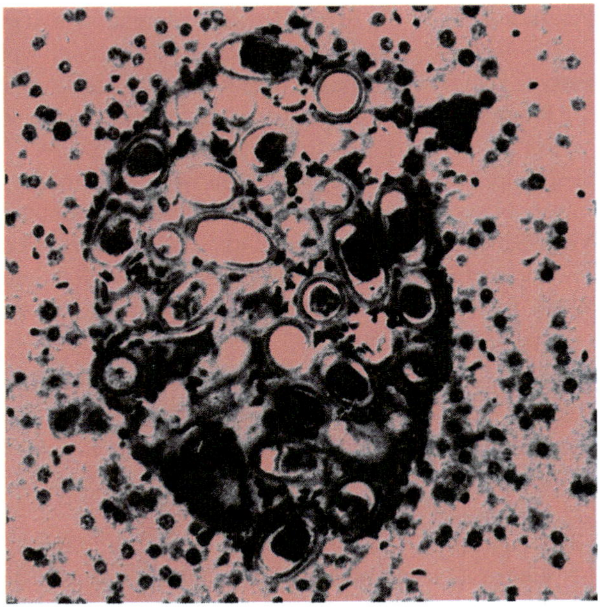

Fig. 5.86 Light micrograph of a section through the liver of a rodent containing a "nest" of eggs of *Capillaria hepatica*

haemorrhagic inflammations, fibrosis of the mucous layers, anaemia, resorption problems, diarrhoeas and loss of weight. In the case of the species *Capillaria hepatica*, massive metabolic problems may occur leading to icterus, anorexia, vomiting, etc.

5. **Diagnosis**: Miscroscopical determination of the eggs inside the body excretions, in liver biopsies or in sections (*C. hepatica*, *C. aerophila*) (Figs. 5.85 and 5.86). The eggs look similar to those of the genus *Trichuris*. They also possess two polar plugs, which, however, are less protruding than those of *Trichuris* species (Fig. 5.85). The egg size ranges between 45 and 60 μm × 30–35 μm. The eggs of *C. hepatica* are not excreted but are located in "nests" in liver tissues (Fig. 5.86).
6. **Pathway of infection**: Infection occurs depending on the species by oral uptake of eggs containing the larva 1 within contaminated food, ingestion of earthworms containing the larva 3 or in the case of *C. hepatica* ingestion of raw meat of herbivorous animals (rodents), containing the eggs of *C. hepatica* in the case of carnivores or humans (see Table 5.7).
7. **Prophylaxis**: Avoiding introduction of eggs by wild animals into stables, regular removal of feces in stables, prophylactic treatment of the animals with nematocidal drugs.
8. **Incubation period**: Depending in the species (examples):

C. aerophila	7–10 days
C. bovis	7–14 days
C. anatis	7–8 days
C. hepatica	1–2 days

9. **Prepatent period**: Examples

C. aerophila	30–40 days
C. anatis	~28 days
C. annulata	18–26 days
C. bovis	~60 days
C. contorta	~50 days
C. hepatica	21–28 days
(syn. *Calodium hepaticum)*	(eggs stick in the liver)
C. obsignata	~22 days
C. plica	~60 days
(syn. *Pearsonema plica*)	

10. **Patency**:

C. aerophila	~1 year
C. annulata	~1 year
C. bovis	~1 year
C. hepatica	Only 4–6 weeks
C. obsignata	~1 year

11. **Therapy**:
   - **Birds**: flubendazole, levamisole;
   - **Mammals**: ivermectin, levamisole, mebendazole, fenbendazole.

Further Reading

Alic A et al (2015) *Pearsonema* (*Capillaria*) *plica* infection and associated urinary bladder pathology in red foxes. Parasitol Res 114:1933–1938.

Galecki R et al (2015) Parasits of wild animals as a potential source of hazard to humans. Annals Parasitol 6:105–108.

Mukaratirwa S, Khumalo MP (2010) Prevalence of helminth parasites in free-range chickens. J S Aft Vet Assoc 81:97–101.

Sharma R et al (2015) *Capillaria hepatica* infection. Annals Parasitol 61:61–64.

Tararu M et al (2015) Morphological and molecular genetic characterization of three *Capillaria* species (*C. anatis, C. pudendotecta, C. madseni*) and *Baruscapillaria obsignata* in avians. Parasitol Res 114:4011–4022.

5.3.3.3 *Trichinella* Species (Trichinosis)

1. **Name**: Greek: *thrix, trichos* = hair; *speira* = winding. Latin: *ella* = diminutive form = tiny hair; *nativus* = present since birth.

2. **Geographic distribution/epidemiology**: Worldwide among carnivores and omnivores.
3. **Biology, morphology**: Recently eight species have been described to belong to the genus ***Trichinella***:
   - ***T. britovi***, occurring (among other hosts) in lynx, foxes, wolves, wild cats, brown bears, wild pigs, horses and **humans**.
   - ***T. murelli***: occurring (among other hosts) in foxes, racoons, horses, dogs, cats, and **humans**.
   - ***T. nativa***: occurring (among other hosts) in walruses, seals, tigers, bears, pigs, wolves, foxes, lynx, dogs, cats and **humans**.
   - ***T. nelsoni***: lions, leopards, hyenas, other game animals and **humans** in Africa.
   - ***T. papuae***: boars, domestic pigs and **humans**.
   - ***T. pseudospiralis***: foxes, lynx, racoons, pigs, rats, raptor birds, cangoroos and **humans**.
   - ***T. spiralis***: Pigs, foxes, bears, horses, cats, dogs and **humans**.
   - ***T. zimbabwensis***: Crocodiles, warans.

   The characteristic muscle larvae are situated inside a muscle cell (Figs. 5.89 and 5.90), which becomes surrounded in some species by an amorphic capsule (e.g. Fig. 5.90b; *T. spiralis, T. britovi, T. nativa, T. murelli* and *T. nelsoni*) but not in *T. pseudospiralis, T. papuae* and *T. zimbabwensis*.

   The adult worms (♀ up to 4 mm long, ♂ up to 1.5 mm long; Fig. 5.88a) live only a few days up to 6 weeks in the intestine of their final hosts (omnivores, carnivores) being anchored with their anterior end in the mucous layer of the small intestine. The females excrete up to 2000 tiny larvae, which reach a size of only 100 μm × 8 μm and enter finally (being transported by the bloodstream) into striated muscle cells (Figs. 5.87, 5.89 and 5.90). Therein they stay until a carnivore or omnivore ingests these pieces of meat. In literature, it is described that they remain infectious for up to 20 years. The whole life cycle is depicted in Fig. 5.87. It can be seen in the above-listed *Trichinella* species that apparently horses may become infected, although they do not feed raw meat that might be infected. Their infections occur by unwilling ingestion of carrion feeding beetles or fly larvae, which have ingested trichine larvae when feeding dead bodies of e.g. rats, which are infected with *Trichinella* larvae.
4. **Symptoms of disease (Trichiniasis, trichinelliasis)**:
   1. **Phase in the intestine**: Diarrhoeas, fever of the typhoid type.
   2. **Phase in the muscles**: Myositis, muscle pain, breathing problems, stiffness of movements, oedemas and high eosinophilia. In severe cases (=extreme high numbers of larvae inside muscle fibres): death due to paralysis of the muscles of the respiratory tract.
5. **Diagnosis**: Demonstration of the larvae in digested muscle probes (as it is done by law in slaughter houses in Germany) or by use of serologic methods (IIFT, ELISA) in suspected infections of farmed animals.

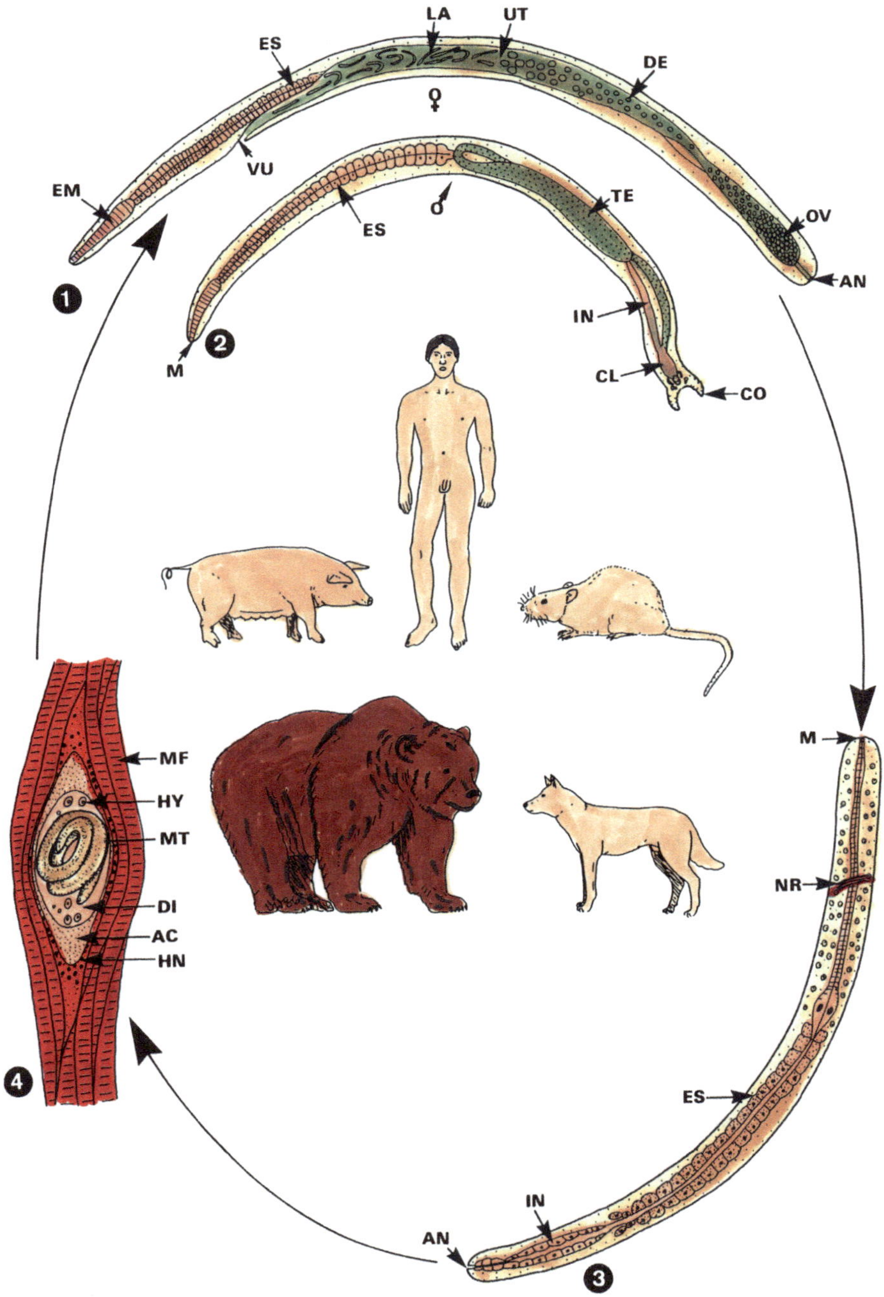

Fig. 5.87 Diagrammatic representation of the life cycle of *Trichinella spiralis*. **1–2** The adults (male 1.5 mm × 40 μm, female 3–4 mm × 60 μm) live for 6 weeks (at the maximum) in the small intestines (being anchored in the mucous layer) of many carnivorous and omnivorous animals including man. Beginning from the 5th day after infection the females release (over a period of 4–16 weeks) in total about 2000 larvae (viviparous), which hatch from their eggshell while still

6. **Pathway of infection**: Oral by uptake of raw or undercooked meat of animals containing muscle larvae (e.g. dead rats or beetles/fly larvae that had ingested such infected meat).
7. **Prophylaxis**: Avoid feeding raw meat to animals. Humans should avoid to eat raw meat in countries without meat control (e.g. in the USA, France).
8. **Incubation period**: 5–7 days.
9. **Prepatent period**: 9–11 days.
10. **Patency**: Years
11. **Therapy**: Treatment of infected pigs is not common, since infected meat is excluded from further use. Worthful animals like horses, lions, dogs, etc., can be treated as reported in the literature:
    - **Tiabendazole**: 5–10 days with 25–50 mg/kg bodyweight,
    - **Mebendazole**: 6 or 10 days with 600 mg per animal
    - **Fenbendazole**: 14 days: 2 × 250 mg
    - **Albendazole**: Dogs, cats: 50 mg/kg bodyweight for 7 days twice a day.

Further Reading

Dubey JP et al (2013) Isolation and characterization of new genetic type of *Toxoplasma gondii* and prevalence of *Trichinella murrelli* from black bear (*Ursus americanus*). Vet Parasitol 196:24–30.

FSA (2008) Trends and sources of zoonotic agents and zoonosis and food-borne outbreaks in the European Union. EFSA J 8(1496):1–368.

Fig. 5.87 (continued) inside the single uterus (UT). **3** The hatched larvae (LA) measure about 100 × 8 μm and are characterized by rounded poles and an extremely long oesophagus (ES); they eventually enter the wall of the intestine and are carried away by the hepatic portal system through the liver, heart and lung and thus are distributed by the arterial system throughout the body of the same host (which thus is the final and intermediate host). **4** When larvae reach skeletal muscles, they penetrate individual fibres and begin to grow, reaching up to 1 mm in length (without moult); up to seven larvae have been seen within a single fibre which is altered due to the parasitism. At first, the region around the worm becomes amorphous (due to disappearance of sarcomeres) and finally a broad dense outer, but still intracellular, zone is produced, apparently by deposition of primarily metabolic (AC) leading to some sort of a capsule. Outside this capsule thickening may be brought about by infiltration of leucocytes and calcification (beginning about 10 months after infection). Such encysted worms are infectious for many years; transmission occurs again when such larvae are ingested by another omnivorous or carnivorous host. In the intestine excystation proceeds; however, the number and location of the following moults are still a matter of controversy. AC = anlage of a capsule (at the inner periphery of an infected muscle fibre); AN = anus; CL = cloaca; CO = copulatory appendages; DE = development of fertilized eggs; DI = disintegrated cytoplasm of the host muscle fibre; EM = oesophagus (muscular region); ES = oesophagus (stichosomal region); HN = host cell nucleus (unchanged); HY = hypertrophied host cell nucleus; IN = intestine; LA = larvae; M = mouth; MF = muscle fibre (uninfected); MT = muscle trichine; NR = neural ring; OV = ovary with eggs; TE = testis; UT = uterus; VU = vulva

a
b
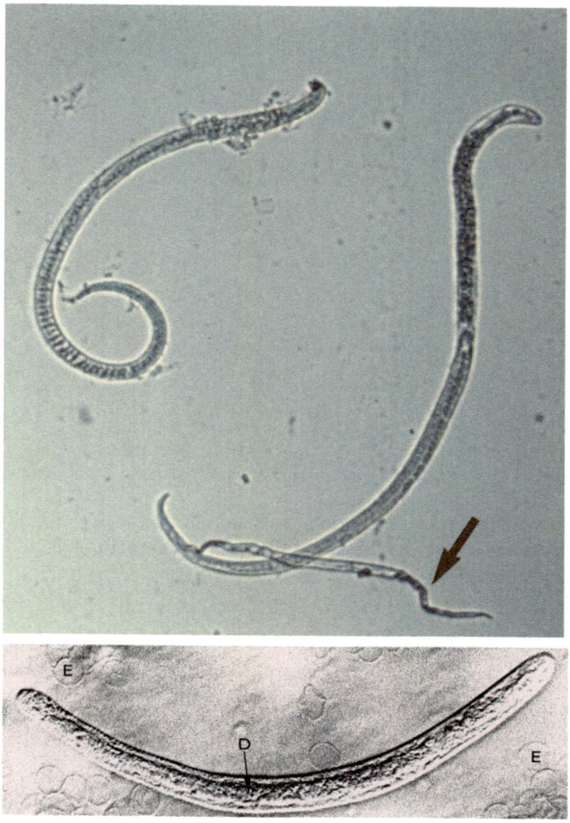

Fig. 5.88 Light micrograph of two adult females and a male of *Trichinella spiralis* (**a**) and a larva 1 (from the host's intestine on its way to the muscle (**b**)

Fig. 5.89 *Trichinella spiralis* light micrographs of a fresh squeeze preparation showing two muscle cells: one contains three larvae, the other only one

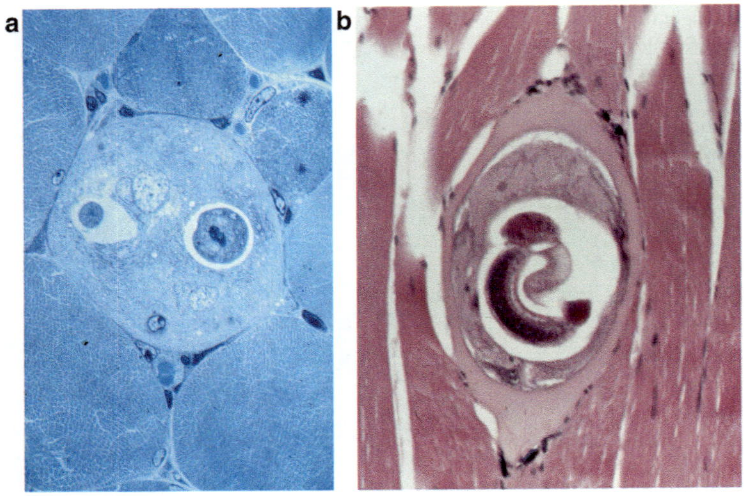

Fig. 5.90 Light micrographs (**a**) Semithin section of a muscle cell showing two cross sections of a *Trichinella spiralis* larva. (**b**) Paraffin section through a larva of *T. spiralis* in an already transformed host muscle cell

Fransen F et al (2015) Genetic evidence of interspecies introgression of mitochondrial genomes between *Trichinella spiralis* and *T. britovi* under natural conditions. Int Gen Evol 36:323–332.

Goltstein B et al (2009) Epidemiology, diagnosis, treatment, and control of trichinellosis. Clin Microbiol Rev 22:127–145.

Onkoba WN et al (2015) Differential immune responses in mice infected with *Trichinella zimbabwensis*. J Helminthol. doi:10.1017/s0022149x15000723.

Pozio E (2016) *Trichinella pseudospiralis*. Vet Parasitol 205:214–222.

Marucci G et al (2009) *Trichinella nelsoni* and *Trichinella T8* mixed infection in a lion. Vet Parasitol 159:225–228.

Mayer-Scholl A et al (2014) PCR for the differentiation of *Trichinella nativa* and *T. britovi*. Vet Parasitol 203:247–249.

Mayer-Scholl A et al (2016) Study on the occurrence of *Trichinella* spp. in racoon dogs. Vet Parasitol 205:205–209

Marucci G et al (2016) Proficiency testing to detect *Trichinella* larvae in meat in the EU. Vet Parasitol 350–359.

Owen IL et al (2014) The probable role of cannibalism in spreading *Trichinella papuae* in a crocodile farm. Vet Parasitol 201:335–338.

5.3.3.4 Ascarids (Roundworms)

1. **Name**: Greek: *askaris* = worm in the intestinal tract; *toxon* = arch; *pro* = before; *kyon* = dog. Latin: *sus* = pig: *para* = besides; *equus* = horse; *leo* = lion.

2. **Geographic distribution/epidemiology**: Worldwide; in farm animals as well as in wild animals.
3. **Biology, morphology**: This group includes many species, which all occur in large numbers of specimens in many hosts. Some examples and their main hosts are:

Ascaris suum (pigs)	♀ up to 35 cm long
Baylisascaris procyonis (dogs, raccoons)	♀ up to 12 cm long
Parascaris equorum (horses, donkeys)	♀ up to 40 cm long
Toxascaris leonina (dogs, cats)	♀ up to 12 cm long
In humans only their larvae occur	

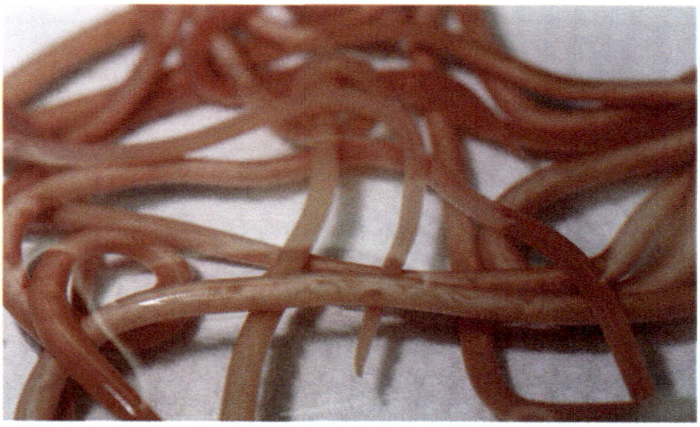

Fig. 5.91 Macrophoto of adult *Ascaris suum* worms

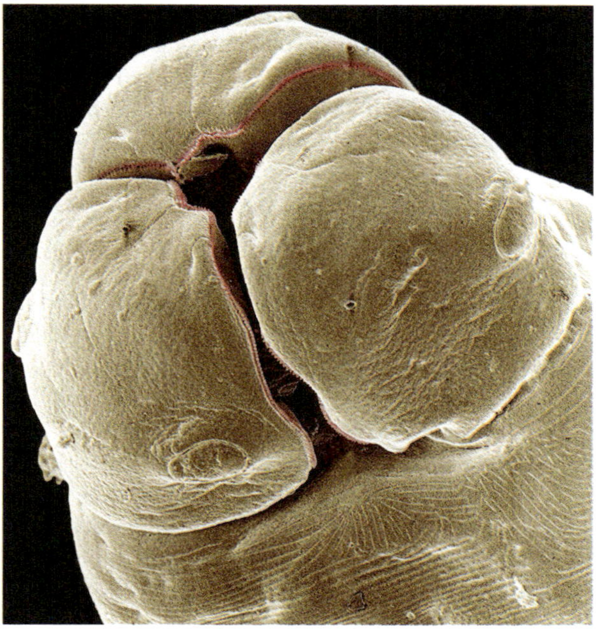

Fig. 5.92 Scanning electron micrograph of the mouth region of *Ascaris* spp. showing the typical three lips

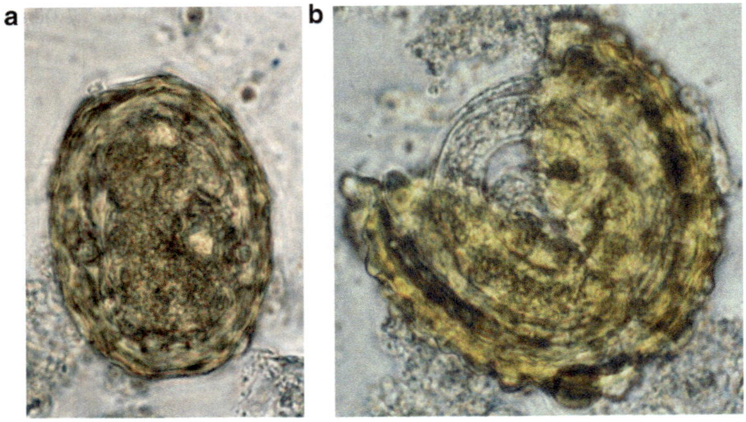

Fig. 5.93 Light micrograph of eggs of *Ascaris suum*. (**a**) Just excreted; (**b**) The eggshell has been split off showing the larva 2

Life cycle:

Ascaris suum: The females (35 cm long) and males (25 cm) live in the lumen of the small intestine of their hosts and appear yellowish (Figs. 5.91 and 5.92). The about 60 × 45 μm sized eggs are covered by a thick, yellowish shell with wrinkles (Fig. 5.93a, b). They are excreted in an unembryonated status but develop outside of the body in a temperature-dependent time (15–40 days) the infectious larva. It is still under discussion whether this is the larva 2 or larva 3, which, however, stays inside the egg until it is ingested by a new host, where it hatches in the intestine. Then the larva leaves the intestinal lumen and reaches within 24 h via blood vessels the liver, where it enters the parenchyma and stays there for about 1 week. Then it passes heart, lung and trachea and reaches the small intestine (mostly on day 8 after infection). There occurs about 25 days p.i. the moult from larva 3 to larva 4 and finally the growth to the adult stage.

The development of the species of other three genera (*Baylisascaris, Parascaris, Toxascaris*) runs similarly.

4. **Symptoms of disease (Ascariasis)**:
   (a) **Phase of body wandering**: Reduced food uptake, pneumonia, coughing, increased breathing, eosinophilia, eventually hepatitis interstitialis and parasitaria multiplex. In the case of pigs, whitish lesions appear on the surface of the liver (Fig. 5.94), and thus, it cannot be used anymore for human consumption.
   (b) **As adults** these roundworms induce loss of bodyweight, diarrhoeas, icterus and occlusion of the intestine due to large amounts of worms (especially in the case of young animals).
5. **Diagnosis**: Microscopical demonstration of typical eggs in the feces after use of concentration methods (M.I.F.C., S.A.F.C.) (Fig. 5.93).

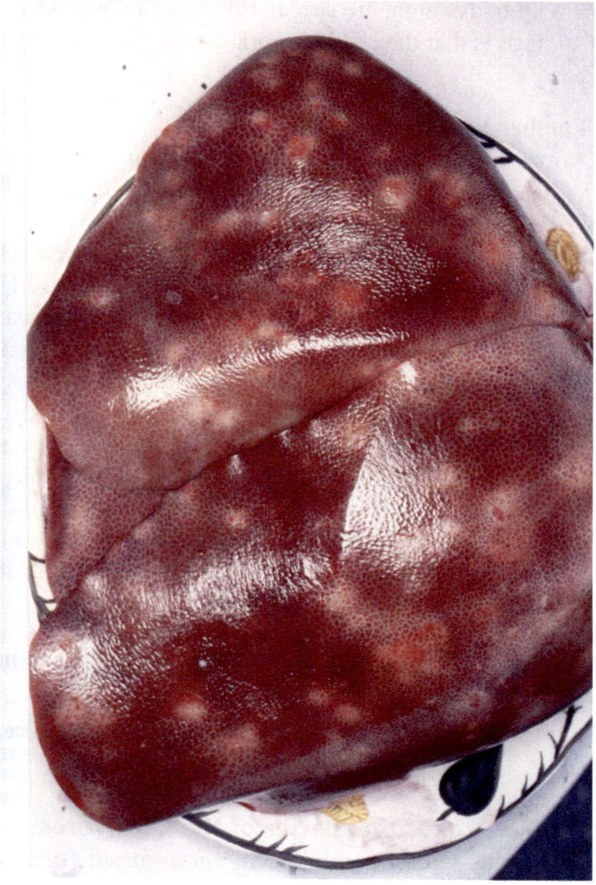

Fig. 5.94 Macrophoto of a pig liver showing the typical "milk spots" indicating altered tissues due to the presence of *Ascaris* larvae

6. **Pathway of infection**: Oral by uptake of larva-containing eggs within contaminated food.
7. **Prophylaxis**: Regularly repeated removal of the feces from the stables; disinfection of the floors of the stables using hot steam (1 × per week). **Chemoprophylaxis** by application of benzimidazole preparation within the food (especially in young animals).
8. **Incubation period**: *Ascaris suum*: 4 days; *P. equorum*: 7 days; *T. leonina* and *B. procyonis* 4–5 days.
9. **Prepatent period**: 8–9 weeks: *A. suum; 10 weeks: T. leonina, B. procyonis;* up to 15 weeks: *T. equorum.*
10. **Patency**: 1–3 years.
11. **Therapy**: Drugs of choice are modern **benzimidazoles** (oral application) and compounds like **ivermectin** and **doramectin** (used as injections); they also eliminate wandering stages. Also **piperazine** salts in dosages of 100–400 mg/

kg bodyweight are successfully used being added to the food. Their disadvantage is that they do not work during the wandering phase of the worms.

Further Reading

Armstrong SK et al (2014) The efficacy of ivermectin, pyrantel and fenbendazole against *Parascaris equorum* infections in foals on farms in Australia. Vet Parasitol 205:575–580.

Fogt-Wyrwas R et al (2013) Intraspecific variation between the ITS sequences of *Toxocara canis, T. cati* and *Toxascaris leonina*. J Helminthol 87:432–442.

Graeff-Teixera C et al (2016) Update of bayliascariasis. Clin Microbiol Rev 29:375–384.

Janssen IJI et al (2013) Genetic variants and increased expression of *Parascaris equorum* P-glycoprotein-11 in populations with decreased ivermectin susceptibility. PLoS One: e61635.

Knaus M et al (2014) Efficacy of a novel topical combination of fipronil, methoprene, eprinomectin and praziquantel against experimental infections of *Toxascaris leonina* in cats. Vet Parasitol 202:40–44.

Leles D et al (2012) Are *Ascaris lumbricoides* and *A. suum* single species? Parasitol Vectors 5:42.

Page LK (2013) Parasites and the conservation of small populations: the case of *Baylisascaris procyonis*. Int J Parasitol: Parasites Wildlife 2:203–210.

Vlaminck J et al (2014) Advances in the diagnosis of *Ascaris suum* infections in pigs and their possible application in humans. Parasitology 141:1904–1911.

5.3.3.5 *Toxocara* Species of the Carnivores

1. **Name**: Greek: *toxon* = arch; *kara* = head; *katta* = cat; *mystax* = beard. Latin: *canis* = dog; *leo* = lion.
2. **Geographic distribution/epidemiology**: Worldwide. Young dogs show infection rates of up to 70 % and older dogs in towns still 10 %. These high infection rates endanger children on playgrounds, where dogs and cats excrete their feces. Transmission of worm eggs by flies is proven.
3. **Biology, morphology**: The carnivorous dogs and cats are hosts of several roundworm species.
   (a) ***Toxocara canis***: This species occurs in dogs and foxes: ♀ = up to 18 cm, ♂ = up to 10–12 cm. The cervical wings are striated and about 2.5 mm long (Figs. 5.96 and 5.97a).
   (b) ***Toxocara mystax* (syn. *cati*)**: This species occurs in felids and is the most common worm of cats; ♀ = up to 12 cm, ♂ = up to 7 cm; the cervical wings appear broad and show deepenings.
   (c) ***Toxascaris leonina***: This species is not very host specific and occurs in many species of felids and canids. The females reach a length of 12 cm (Fig. 5.98), and the males vary between 4 and 7 cm in length.
   (d) ***Baylisascaris procyonis*** (see Sect. 5.3.3.4)

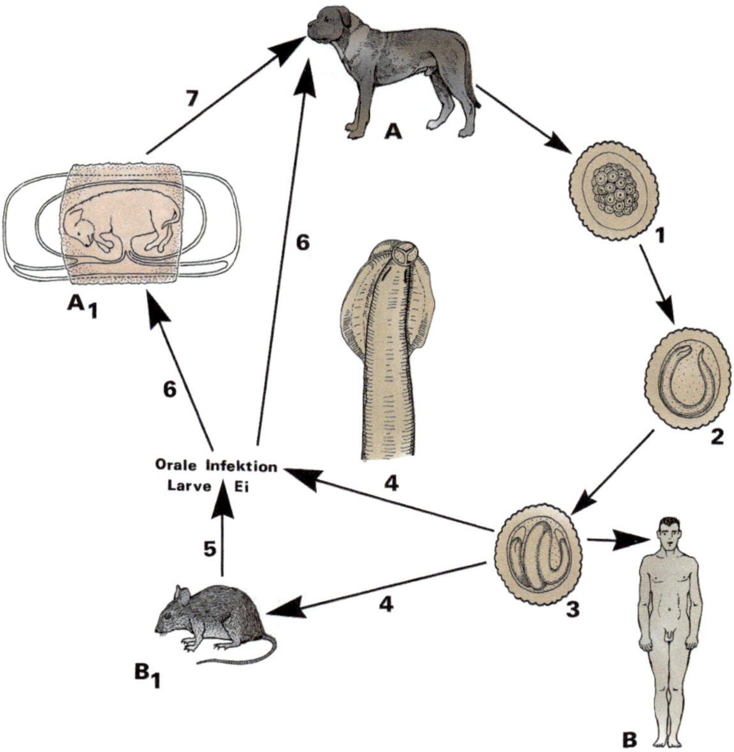

Fig. 5.95 Diagrammatic representation of the life cycle of *Toxocara canis*. **A**: Dog = final host; **B**: = humans = terminal (dead end) hosts without further propagation. **B1** = Mice, other rodents = paratenic hosts: enrich numbers of worm larvae. **1** The eggs are excreted in an unembryonated stage. **2, 3** Outside the body the second stage = infectious larva 2 is developed inside the egg. This stage is then ingested by intermediate hosts (e.g. rodents) (**4**). **4, 5, 6** The adult female dog can be infected by ingesting embryonated eggs (**4, 6**) or by feeding raw mouse meat containing larvae (**5, 6**). If a female dog is pregnant (**A1**) and becomes infected by ingesting eggs (**4, 6**) or infected mice (**5, 6**) the wandering larvae may also enter the fetus, so that the puppies are already infected when born

The typical eggs (Figs. 5.95 and 5.99h) of these species are excreted unembryonated and are found in very large numbers in the feces of their hosts. Under suitable conditions (15–30 °C) and sufficient humidity, the larva 2 or even the larva 3 is developed inside the egg within 3–5 days in the case of *Toxascaris* and 10–15 days in *Toxocara*. The larvae of all species hatch in the intestine. In the case of *Toxascaris leonina*, the larvae enter the intestinal wall but move back into the lumen and reach there maturity, while the larvae 3 of *T. canis* and *T. mystax* (syn. *cati*) start a heart–lung–trachea–oesophagus passage before returning into the intestine and reach maturity there (Fig. 5.95). In the case of non-specific hosts (e.g. mice, humans), the wandering larvae do not reach maturity, but stick

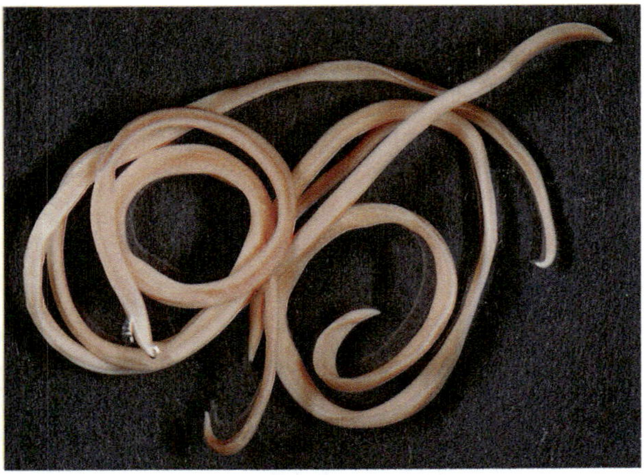

Fig. 5.96 Macrophoto of several adult *Toxocara* worms excreted after deworming

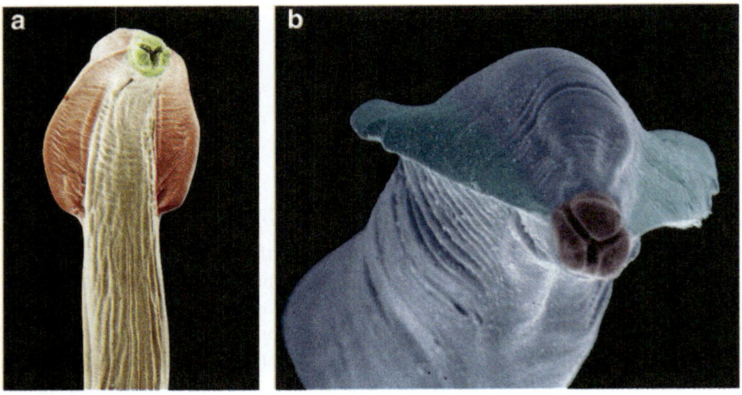

Fig. 5.97 (**a**, **b**) Scanning micrograph of the anterior pole of an adult *Toxocara canis* worm. Note the triangular mouth opening and the two lateral alae (wing-like protrusions)

finally somewhere in different organs, which they had reached within lymph or blood. The host defence system encloses these wandering stages in **granulomas**. If such granulomas are later ingested by final hosts, they go on with their further development inside the intestine. In humans, such larvae are termed **larva migrans visceralis**, which may induce severe lesions (e.g. in brain and eyes). In the case of dog and cat *Toxocara* species, the transmission of the larvae may occur inside the placenta and also within the liver of the mother of the puppies. The galactogenous transmission is especially common in cats.

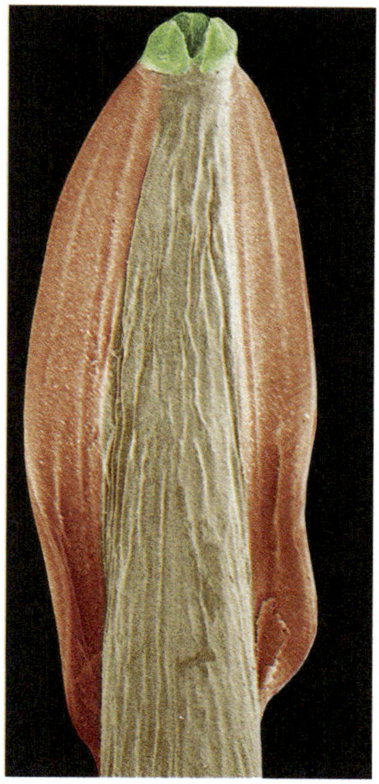

Fig. 5.98 Scanning electron micrograph of the anterior end of the cat roundworm *Toxascaris leonina* showing the mouth and the lateral alae which are longer than those in *T. canis*

4. **Symptoms of disease (Toxocariosis, toxascariosis)**:
   (a) **Puppies of dogs** are harmed by wandering larvae, but may even die when inhaling vomited intestinal fluids, which occur often in cases of high-grade infections during the second and third week after birth (Figs. 5.100 and 5.101). The puppies become sensitive to pressures, excrete slimy feces and show destructions of the intestine. Further symptoms are anaemia, considerable loss of weight and rachitic signs due to lack of vitamin D. Their fur appears more and more shaggy.
   (b) **Older dogs**: The general symptoms are in principle the same as those of the puppies; however, due to progressing immunization, the severity of the symptoms decreases, since most of ingested larvae are killed by immune reactions when they enter and/or pass the intestinal wall.
   (c) **Cats** show similar symptoms like those in dogs. Leading symptoms are mainly the shaggy fur and rachitic symptoms (especially in young animals).
5. **Diagnosis**: Appearance of dead worms in the feces or in vomited fluids; demonstration of the typical thick-walled eggs, which measure 90×75 μm, do not contain a larva and appear yellowish/brownish (Fig. 5.99h).
6. **Pathway of infection**:
   (a) Praenatal in *T. canis* due to infections of the mother.
   (b) Galactogenous in the case of puppies when sucking milk at the mother.

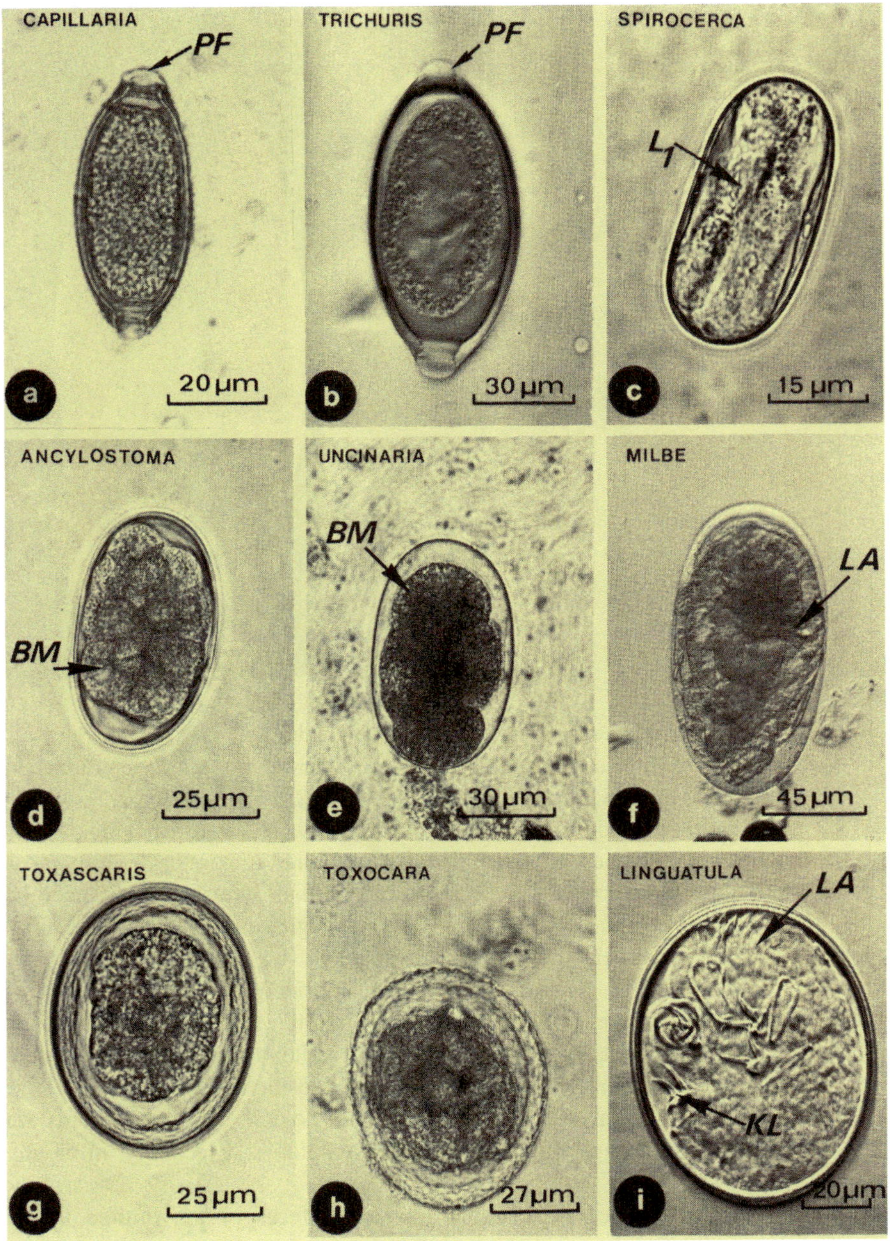

Fig. 5.99 The table shows light micrographs of nematode eggs of carnivores, a mite egg (**f**) and an egg of the pentastomid worm *Linguatula serrata* (**i**). BM = blastomeres; KL = claws of the linguatulatid larva; LA = worm larva; PF = polar plug

   (c) Oral by ingestion of larvae in muscles of mice, etc.
   (d) Oral uptake of eggs containing the larva 2 or 3.
7. **Prophylaxis**: Repeated treatment of adult animals. Puppies of cats and dogs should be treated 2, 4 and 12 weeks after birth by application of relevant medicaments. Pregnant female dogs should be treated 40 and 10 days before birth and 2 and 4 weeks *post-partum*. Disinfection of the resting places of the dogs by hot steam, hot water (>60 °C) and/or by use of chemical disinfection products such as Bergo-Endodes, Schaumann-Endosan, Venno-Endo VI, etc. (all in a 5 % solution for 2–3 h.
8. **Incubation period**:

Puppies	1–2 days
Adult dogs	5–7 days
Cats	3–5 days

9. **Prepatent period**:

First infection of dogs with *T. canis*:	4–5 weeks
Prenatal infection with *T. canis*:	3–4 weeks
Infections with *T. cati*	8 weeks
Infections with *Toxascaris leonina*	10 weeks

10. **Patency**: Variable: weeks to months.
11. **Therapy**: For the treatment of infections with adult worms, several very active compounds are available, while migrating worms/worm larves are much less sensitive.
   (a) **Piperazine salts** eliminate intestinal stages: piperazine base: 75–100 mg/kg bodyweight shows good effects. In the case of highly infected puppies, the dosage should be given divided on 2–3 days.
   (b) **Several broad spectrum anthelmintics** are highly active: **nitroscanate** (1 × 50 mg/kg bodyweight), **flubendazole** (22 mg/kg bodyweight), **levamisole** (5 mg/kg bodyweight) or **pyrantel** (1 × 5 mg/kg bodyweight in the case of dogs and 1 × 20 mg/kg bodyweight in the case of cats).
   (c) **Further broad spectrum anthelmintics** such as **benzimidazoles** do not kill only nematodes but also cestodes: **fenbendazole** (on 3 days each time 50 mg/kg bodyweight) or **mebendazole** (2 × daily for 2 days in case of nematodes). If double infections with tapeworms and nematodes have to be treated, the dosage should be given 2 times daily for 5 days!). The dose depends on the bodyweight and must be checked on the product leaflet.
   (d) The combination of **praziquantel** + **pyrantel** + **febantel** is effective when given once (=1 pill per 10 kg bodyweight) and acts against adult nematodes and cestodes.
   (e) Prenatal as well as galactogenic infections can be treated with **selamectin** (6 mg/kg bodyweight, spot-on) or by application of **moxidectin** and **milbemycin oxime**. It is strongly recommended to substitute Vitamin D deficits.

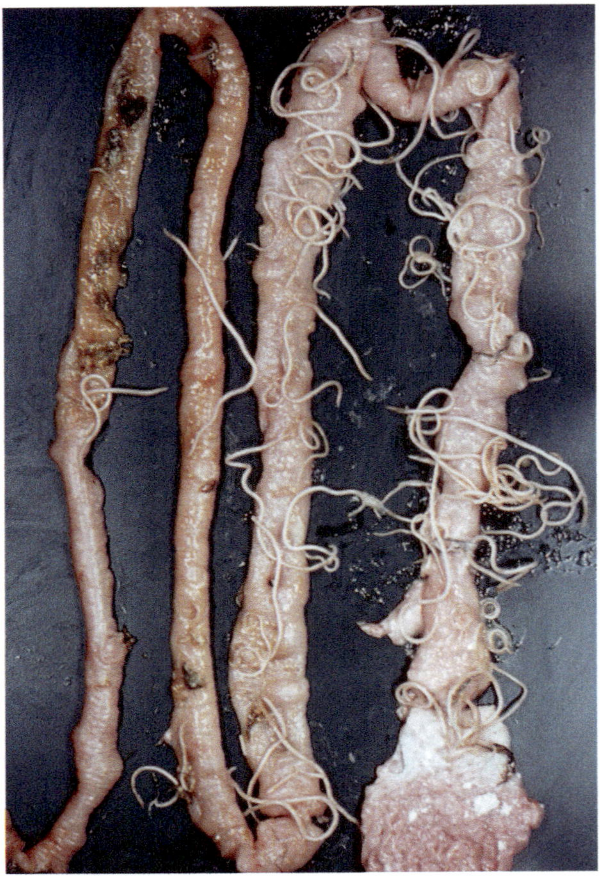

Fig. 5.100 Macrophoto of the opened intestine of a young puppy that died due to the presence of large numbers of toxocarid worms

Further Reading

Azam D et al (2012) Temperature and the development and survival of infective *Toxocara* larvae. Parasitol Res 110:649–656.

Chen J et al (2012) Advances in molecular identification, taxonomy, genetic variation and diagnosis of *Toxocara* sp. Inf Gen Evol 72:1344–1348.

Nijsse R et al (2015) *Toxocara canis* in household dogs. Parasitol Res 114:561–569.

Poulsen CS et al (2015) Differential serodiagnosis of *Toxocara canis* and *T. cati*—is it possible? Parasite Immunol 37:204–207.

Schuster RK (2009) The parasite fauna of stray domestic cats (*Felis catus*). Parasitol Res 105:125–134.

Sommerfelt IE et al (2006) Prevalence of *Toxocara cati* parasites in cat's feces. Vet Parasitol 140:296–301.

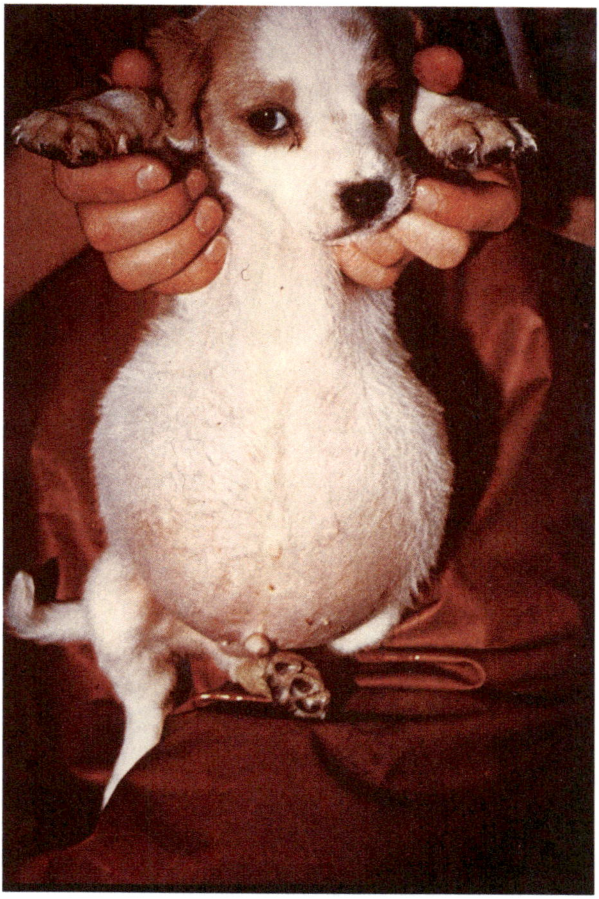

Fig. 5.101 Photo of a dog puppy which suffers from an extreme high *Toxocara* worm load

5.3.3.6 *Toxocara vitulorum* of Ruminants (Large Roundworm)

1. **Name**: Greek: *toxon* = arch; *kara* = head. Latin: *vitula* = little female calf.
2. **Geographic distribution/epidemiology**: Worldwide, but mainly in the tropics and subtropics in cattle, in water buffaloes and rarely in sheep. In Europe mainly in zoo animals.
3. **Biology, morphology**: Inside ruminants only *Toxocara* (syn. *Neoascaris*) *vitulorum* reaches fertility. The females reach a length of 30 cm, while males are smaller (up to 25 cm). The adult worms live in the intestine; their eggs measure 70–90 μm × 60–75 μm and possess a rather thick wall which has numerous depressions (Fig. 5.102E). The development runs mainly like that of *Ascaris suum*; however, also galactogenic and prenatal infections are possible.
4. **Symptoms of disease (Toxocariasis)**: Infections with few worms or the presence of a few larvae in waiting position (hypobiosis) remain symptomless. Infected animals excrete a peculiar smell due to a valerian acid that is found in

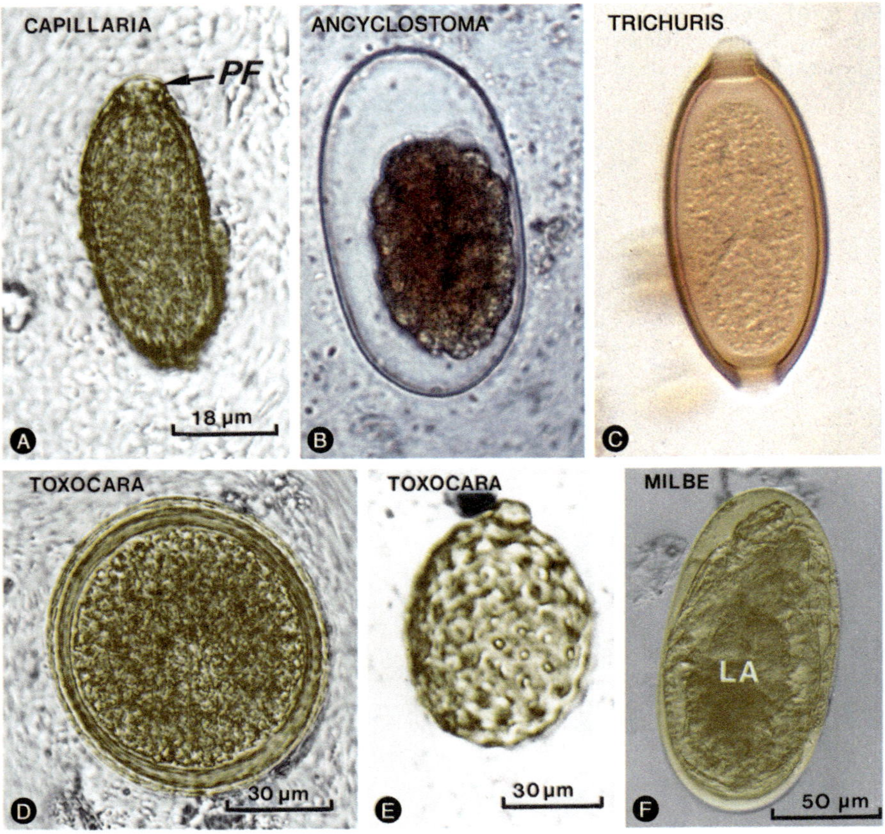

Fig. 5.102 Light micrographs of typical worm eggs of the genera *Capillaria, Ancylostoma, Trichuris, Toxocara* (in through view and surface view) and of an egg of a mite

the urine and in meat resembling the smell of butyric acid. Heavy infections may lead to **pneumonia** (due to migrating larvae), colics, abdominal swellings and alternation of diarrhoea and obstipation. These symptoms are often paired with rejection of food, ileus verminosus and perforations of the intestinal wall, which all together may lead to death.

5. **Diagnosis**: Microscopical demonstration of the typical eggs (Fig. 5.102D, E) after use of enrichment methods (Flotation, S.A.F.C. or M.I.F.C.). If several animals are infected, the stable extrudes a very peculiar smell.
6. **Pathway of infection**: Oral by uptake of eggs, which have developed the infectious larva 2 or 3. Prenatal infections due to wandering larvae 3 are possible.
7. **Prophylaxis**: Regular removal of the feces from stables; **quarantine** of imported animals.
8. **Incubation period**: Lung symptoms may start 4–7 days after infection. Intestinal symptoms do not occur before the 2–4 first weeks have passed.

9. **Prepatent period**: After galactogenic infections, calves show first symptoms starting about 28 days after infection.
10. **Patency**: 3–6 months.
11. **Therapy**: The treatment with the help of benzimidazoles may start 2 weeks after birth. Macrocyclic lactones act also successfully during treatment.

Further Reading

Chelladura JJ et al (2014) *Toxocara vitulorum* infection in a cohort of beef calves in Iowa. Vet Parasitol 214:96–99.

Rast L et al (2014) Why are simple control options for *Toxocara vitulorum* not being implemented by cattle and buffalo smallholder farmers in South-East Asia? Prev Vet Med 113:211–218.

Roberts JA (1993) *Toxocara vitulorum* in ruminants. Helminthol Abstr 62:151–174.

5.3.3.7 Ascarids of Birds (Roundworms)

1. **Name**: Greek: *askaris* = worm in intestinal tissues; *idios* = own. Latin: *gallus* = cock; *columba* = dove; *compar* = similar; *dissimilis* = not similar, different.
2. **Geographic distribution/epidemiology**: Worldwide.
3. **Biology, morphology**:
   (a) ***Ascaridia galli*** occurs in **chicken birds** and **geese**. The females reach a length of up to 11 cm (males up to 7 cm).
   (b) ***Ascaridia columbae*** infects mainly doves and reaches as female only 9 cm in length and 7 cm as male.

   These typical yellowish-whitish roundworms are rather thick reaching diameters of 4 mm and have mouthparts similar to those depicted in Fig. 5.92). The females excrete thick-walled unembryonated eggs, which possess a smooth surface (Fig. 5.103f). They reach a size of 80–-90 µm × 45–50 µm and can only hardly be differentiated at the species level. Outside the body, the larva 1 is developed inside the egg within 10–14 days. While still inside the eggshell, the larva discharges twice their cuticle and thus reaches the infectious stage of the larva 3. These eggs may become ingested by earthworms, wherein they remain infectious. If such eggs in earthworms or others directly from soil were ingested by birds, the larva 3 hatches to the larva 4 inside the intestine of the new host. This stage stays for about 8 days in the lumen of the intestine and starts then to enter the mucous layer of the intestine, where they stay for another 2 weeks. Thereafter, they enter again the lumen of the intestine, where they reach maturity within 6–8 weeks in case of **chickens** and in 5–6 weeks in the case of **doves**. Further species are *Ascaridia compar* (worldwide in **chickens**) and *A. dissimilis* (in Europe, the USA in **turkeys**) or *Ascaridia nymphii* (in **cockatiels**).

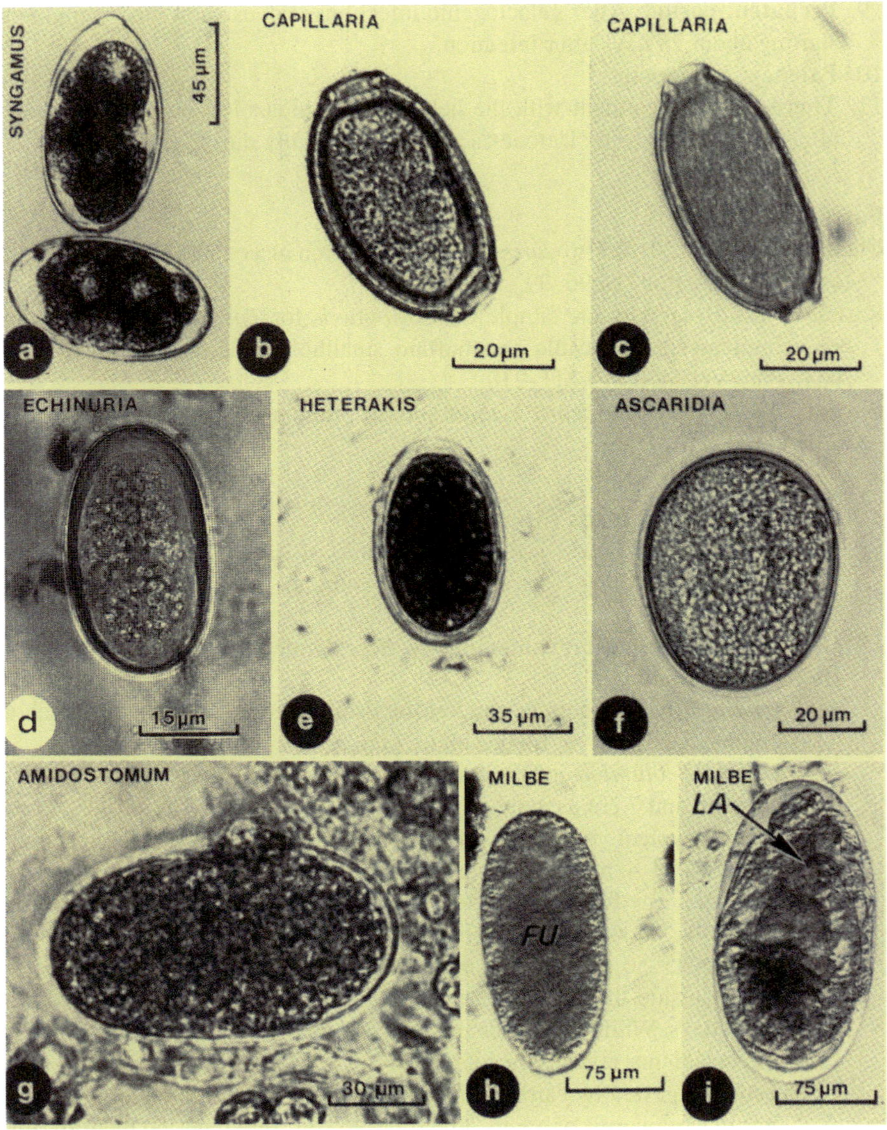

Fig. 5.103 Light micrographs of the eggs of different genera of nematodes of birds (**a–g**) and eggs of mites in their food (**h**, **i**). FU = cleavage stage; LA = larva; Milbe = mite

4. **Symptoms of disease (Ascaridosis)**: Especially young animals may suffer from severe diarrhoeas, loss of feathers and weight, weakness, pale crests (anaemic symptoms), blockage of the intestine (in mass infections), eventually death within 8 days due to mass invasion of larvae into the intestinal wall.
5. **Diagnosis**: Miscroscopical demonstration of the eggs (Fig. 5.103f) in the feces after use of enrichment methods. **Attention**: *Heterakis* eggs (Fig. 5.103e) may look similar depending on the age.

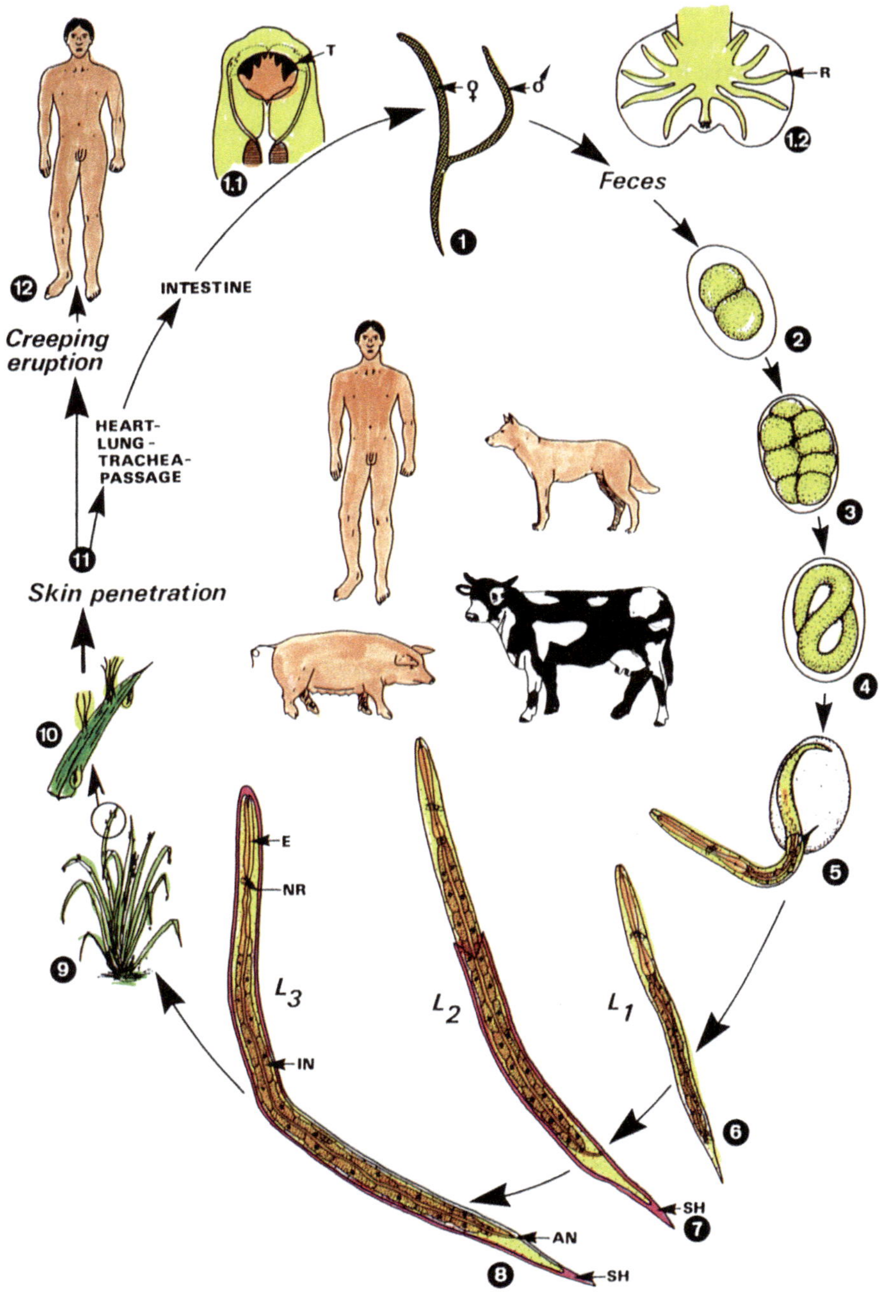

Fig. 5.104 Diagrammatic representation of the life cycle of **hookworms** (e.g. *Ancylostoma duodenale, A. caninum, A. braziliense, Necator americanus, Globocephalus* sp., *Bunostomum* sp.) **1** As shown here for *A. caninum* the adults inhabit (in copula; 1) the small intestine of their hosts, attaching by means of their buccal cavity to the mucous layer and sucking blood using their

6. **Pathway of infection**: Oral by uptake of infectious eggs within food respectively stages within earthworms (e.g. *Lumbricoides* spp.).
7. **Prophylaxis**: Regular removal of feces; cleaning of the food and water vessels; keeping chicken on rusts; separation of freshly hatched birds from adults; regular microscopical check of the feces in order to diagnose worms as early as possible; disinfection of the stables by hot steam or by the use of chemicals such as BERGO-Endodes, Endosan or Venno-Endo VI.
8. **Incubation period**: 3–5 days (=symptoms occur due to larval invasion).
9. **Prepatent period**: 6–8 weeks in *Ascaridia galli* and 5–6 weeks in *A. columbae*.
10. **Patency**: Eventually lifelong.
11. **Therapy**: Treatment can be done by several broad spectrum anthelmintics (e.g. flubendazole, levamisole) being accompanied by application of vitamins A, B, B12 and minerals.

Further Reading

Abe N et al (2015) *Ascaridia nymphii* n. sp. from the alimentary tract of a severely emaciated dead cockatiel (*Nymphicus hollandicus*) Parasitol Res 114:4281–4288.

Bachaya HA et al (2015) Prevalence of *Ascaridia galli* in white leghorn. Trop Biomed 32:11–16.

Das G, Gauly MC (2014) Response to *Ascaridia galli* in growing chicken in relation to their body weight. Parasitol Res 113:1985–1988.

Ferdushy T et al (2012) *Ascaridia galli* in chickens. Parasitol Res 111:2273–2279.

Seaton EM et al (2001) Presence and recovery of *Ascaridia dissimilis* ova on the external shell surface of turkey eggs. Avian Dis 45:500–503.

Fig. 5.104 (continued) species-specific teeth (1.1). With the help of their copulatory bursa (fortified with specific rays; 1.2) the males are attached to the female vulva (location varies according to species), thus giving rise to the typical Y-shaped copulatory aspects. **2–4** Eggs are excreted unembryonated and develop the L1 in the soil. **5–8** The L1 which is called a rhabditiform larva (due to its oesophagus) escapes from the eggshell and feeds on organic material; it then undergoes the first moult, completely shedding its cuticle. After a time spent feeding, the L2 (still rhabditiform) moults to the infectious filariform L3. The second-stage cuticle may be retained (8) as a loose-fitting sheath or it may be lost earlier (7). **9–10** L3 live in the upper few millimetres of soil, migrate to the surface and are often found in groups of thousands on the soil or on plants (moving synchronously). **11** Infection of final hosts occurs when L3 contacts the skin and burrows into it. After a heart–lung–trachea passage, the L3 reach the intestine, moult twice and become sexually mature. (In some species transplacental transmission of L3 or transmission in mother's milk is possible). **12** If a human becomes invaded by L3 of a species or strain that normally matures in animals (e.g. *A. caninum, A. braziliense*), the larvae migrate for months through the cutaneous layers, leading to a disease called creeping eruption. AN = anus; E = oesophagus; IN = intestine; NR = nerve ring; R = rays of bursa copulatrix; SH = sheath (originating from the moulted cuticle of the proceeding larval stage); T = tooth (here each with e peaks)

Tarbiat B et al (2016) The efficacy of flubendazole against different developmental stages of the poultry roundworm *Ascaridia galli* in laying hens. Vet Parasitol 218:66–72.

5.3.3.8 Hookworms of Carnivores

1. **Name**: Greek: *ankylos* = bended; *stoma* = mouth; *kephale* = head; *stenos* = narrow. Latin: *uncinatus* = hook-like; *tubaeformae* = tube-like; *canis*, *caninum* = belonging to dogs; *braziliense* = brasilean; *ceylanicum* = occurring in Ceylon.
2. **Geographic distribution/epidemiology**: Worldwide, especially in warm countries but also in commercial rearing facilities in many countries.
3. **Biology, morphology**: The adult hookworms (Figs. 5.105, 5.106 and 5.107) are characterized by the differentiations of their mouth, which allow them to suck blood at the intestinal wall of their hosts. The males possess a so-called **bursa copulatrix**, which allows it to clutch firmly the females at the sexual opening and thus to enter sperms therein (Fig. 5.106). The eggs of the different species look very similar and except for some differences in size they do not offer criteria that would allow species differentiation (Fig. 5.108). The following three species are most important besides several region ones:

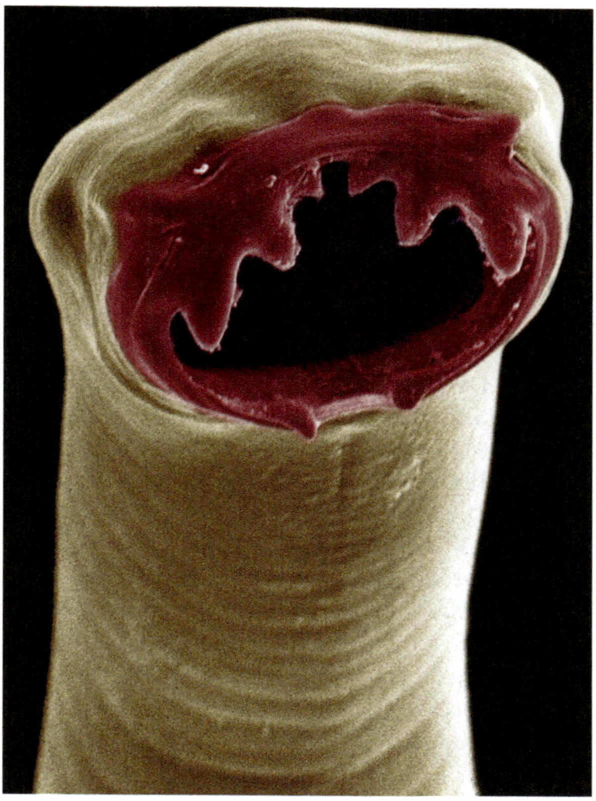

Fig. 5.105 Scanning electron micrograph of the anterior end of the hookworm *Ancylostoma caninum*

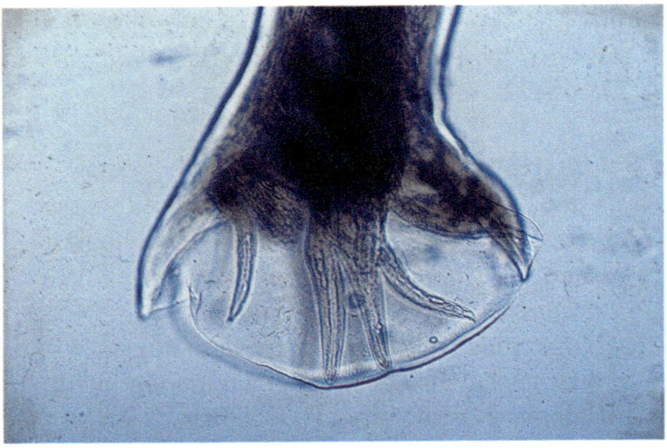

Fig. 5.106 Light micrograph of the bursa copulatrix of a male hookworm

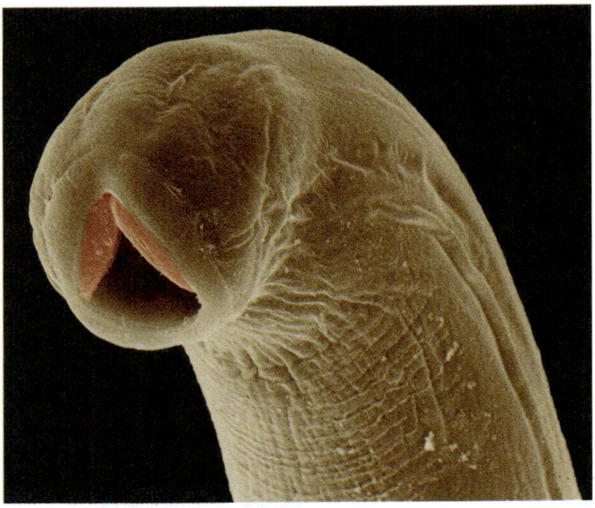

Fig. 5.107 Scanning electron micrograph of the anterior end of the fox and dog hookworm *Uncinaria stenocephala*

(a) ***Ancylostoma caninum***: This species occurs in dogs, wolves, jackals, foxes and rarely in cats. Characteristic is the possession of two large teeth plates in the mouth, which show both three teeth-like protrusions (Fig. 5.105). The females measure 15–21 mm in length and the males 10–14 mm. They parasitize in the small intestine mostly in a copulation position (see Fig. 5.104, 1).

(b) ***Ancylostoma tubaeforme***: This species is mainly found in cats and in a few related species. The outer appearance is similar to that of *A. caninum* except for the length of the oesophagus and the spicula. The males reach a length of 5–12 mm and the females 5–15 mm.

(c) ***Uncinaria stenocephala***: This species occurs mainly in dogs, martens, foxes, wolves and only rarely in cats and other felids. The mouth is

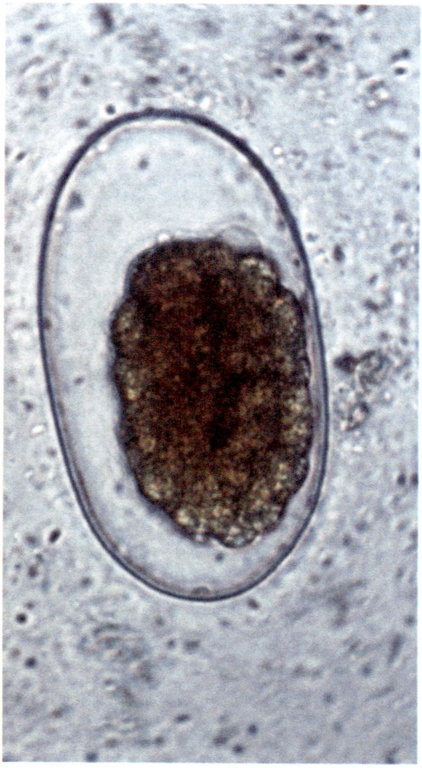

Fig. 5.108 Light micrograph of a typical hookworm egg

weapened by two cutting plates (Fig. 5.107). The females reach a length of 7–16 mm and the males of 5–11 mm.

(d) The species ***Ancylostoma braziliense*** and ***A. ceylanicum*** show similar morphological aspects like *A. caninum*.

(e) Species of the genus ***Uncinaria*** (e.g. *U. sanguinis*) also occur in sea lions, which are apparently infected at common resting places at the shore.

The life cycle of the hookworms is diagrammatically depicted in Fig. 5.108. Their eggs are excreted by the hosts in an early stage of cleavage (6–8 cells). Outside of the body and under favourite temperatures (~20 °C), the larva 1 hatches from the egg and reaches infectivity as larva 3 after two moults within 15 days (Figs. 5.104, 9, 10 and 5.109b). After percutaneous penetration (often via hair follicles but also via oral or galactogenic pathways), the larva 3 starts a heart–lung–trachea–oesophagus passage to reach the small intestine as final site, where maturity is reached after two further moults. The adult worms become anchored mainly at the mucous layer of the jejunum. There they suck blood (about 0.1 ml per day) and change often the sucking place (Fig. 5.109a). If the larvae 3 of these worms enter unspecific hosts, wherein maturity

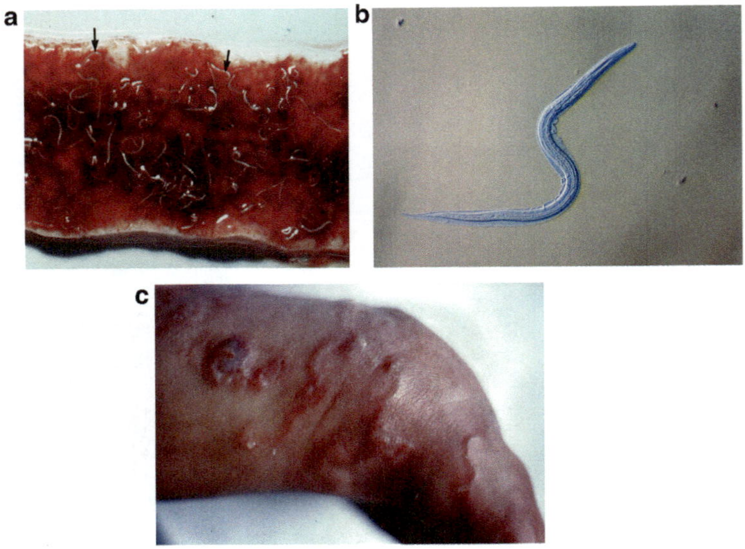

Fig. 5.109 Hookworms. (**a**) Dog intestine with bleedings due to numerous whitish appearing hookworms. (**b**) Hookworm larva 3. (**c**) Arm of a man with traces of numerous wandering hookworm larvae

cannot be reached, they may wander for years inside these bodies as so-called **larva migrans** (Fig. 5.109b, c).

4. **Symptoms of disease (Ancylostomatiasis)**: Young animals may die in cases of heavy infections. Older animals show during infections bloody, fluid feces, considerable loss of weight, rough fur, microcytary, hypochromic anaemia and massive lack of iron. Penetrating larvae may induce local erythema (especially along the belly and between the toes. **Attention**: Larvae of these animal parasites have also a very high zoonotic potential as **larva migrans** in humans! (Fig. 5.109c).
5. **Diagnosis**: Microscopical demonstration of eggs in feces using enrichment methods (e.g. S.A.F.C.). The ovoid eggs of *A. caninum* and *A. tubaeforme* have rounded poles (Fig. 5.108) and curved lateral sides, contain (when laid) 2–8 blastomeres and measure about 60×40 µm. The eggs of *Uncinaria stenocephala* are somewhat larger (73×40 µm). Their poles are not similar and their side walls appear somewhat flattened.
6. **Pathway of infection**:
   (a) **Percutaneous**: The free larva 3 enters (eventually via a hair follicle) into the skin of a host and reaches (in cases of regular hosts) the intestine after a wandering phase inside the body.
   (b) **Oral**: The free larva 3 is ingested within the food and thus reaches without body wandering phase the intestine, where it stays.
   (c) **Galactogenous**: The larva 3, which rests as "sleeping" larva inside the mother's tissue, becomes stimulated by hormones during pregnancy,

wanders into the milk glands and reaches the young animals during milk sucking.

7. **Prophylaxis**: Avoid to keep animals on soil, which cannot be cleaned from feces. Additional cleaning should be done after removing the feces by hot steam/hot water at least once a week. Additionally it is recommended to deworm the animals in a regular time frame. Means for vaccinations are under development.
8. **Incubation period**: Variable, depending on the mode of infection: a few days up to weeks.
9. **Prepatent period**:
   (a) ***A. caninum***: 15–18 days in percutaneous infections and 12–16 days in galactogenous infections.
   (b) ***A. stenocephala***: 13–15 days.
   (c) ***A. tubaeforme***: 18–23 day.
10. **Patency**: Years.
11. **Therapy**:

Nitroscanate	1 × 50 mg/kg bodyweight
Pyrantel	in dogs: 1 × 5 mg/kg bodyweight
	in cats: 1 × 20 mg/kg bodyweight
Fenbendazole	3 days each 50 mg/kg bodyweight against *Uncinaria*
Flubendazole	22 mg/kg bodyweight only against *Uncinaria*
Levamisole	5 mg/kg bodyweight
Mebendazole	2 × daily for 5 days: .... mg/kg bodyweight

New compounds like **moxidectin**, **selamectin** and **emodepside** are also effective if available in the different countries. They are used as spot-on formulations.

Further Reading

Dracz RM et al (2014) Parasitological and hematological aspects of co-infection with *Angiostrongylus vasorum* and *Ancylostoma caninum* in dogs. Vet Parasitol 200:111–116.

Haynes BT et al (2014) Unexpected absence of genetic separation of a highly diverse population of hookworms from geographically isolated hosts. Inf Genetics Evol 28:192–200.

Lee ACY et al (2014) Assessing the speed of kill of hookworms (*Ancylostoma caninum*) by Advantage Multi® for dogs using endoscopic methods. Vet Parasitol 204:402–406.

Marcus AD et al (2015) Ivermectin treatment of free-ranging endangered Australian sea lion pups: effect on hookworm (*Uncinaria sanguinis*) and lice infection status. Parasitol Res 114:2743–2755.

Millan J, Blasco-Costa J (2012) Molecular evidence of shared hookworm *Ancylostoma tubaeforme* haplo types between Iberian lynx and sympatric domestic cats. Vet Parasitol 186:518–522.

Ng-Nguyen D et al (2015) Re-evaluation of the species of hookworms infecting dogs in Central Vietnam. Parasites Vectors 8:401.

Petry G et al (2011) Efficacy of Procox® oral suspension for dogs against *Toxocara cati* and *Ancylostoma tubaeforme* infection of cats. Parasitol Res 109:S37–43.

Postigo I et al (2006) *Uncinaria stenocephala*: assessment of antigens for the immunodiagnosis of canine uncinariosis. Exp Parasitol 114:215–219.

Prullage JB et al (2014) Efficacy of a novel topical combination of fipronil (S)-methoprene, eprinomectin and praziquantel against induced infections of *Ancylostoma* spp. nematodes of cats. Vet Parasitol 202:30–33.

Schimmel A et al (2011) Efficacy of emodepside plus toltrazuril (Procox®) against *Toxocara canis, Uncinaria stenocephala* and *Ancylostoma caninum*. Parasitol Res 109:S1–S8.

5.3.3.9 Hookworms of Ruminants and Pigs

1. **Name**: Greek: *bunos* = little hill, protrusion; *kephale* = head; *stoma* = mouth; *phlebs, phlebos* = vessel, blood vessel. Latin: *trigonus* = triangle; *mucronatus* = provided with a pointed tip; *longus* = long; *globiceps* = round headed.
2. **Geographic distribution/epidemiology**: Worldwide.
3. **Biology, morphology**: The hookworms of both sexes are characterized by their hook-like structures within the oral capsule. While the females have more or less pointed posterior ends, the posterior ends of males are equipped with a holdfast system called **bursa copulatrix** (Fig. 5.106).

   Bunostomum species (e.g. *B. trigonocephalum* of sheep, goats; *B. phlebotomum* of cattle; *Globocephalus urosubulatus* of pigs) are firmly attached at the intestinal wall of the duodenum and ileum with the help of their mouthparts and suck blood in both sexes. Male worms reach only a length of 1.8 cm, while the females measure often 2.5–3 cm. The females excrete thin-walled eggs, which contain 2–4 (maximum 16) blastomeres and measure 85-–104 μm × 45–55 μm. Outside of the body and after a waiting time of 6–8 days, the larva 1 is formed inside the egg, which reaches within another week and after 2 further moults the infectious stage (L3). This larva is either taken up orally (in food) or enters percutaneously into the skin. In the latter case, this larva reaches the intestine after a heart–lung–trachea–oesophagus passage, where it reaches maturity via two further moults within 5–6 weeks p.i.

   The species *Globocephalus urosubulatus* is mainly found in wild pigs. Its adults measure 8 mm in length in the case of females and about 5–6 mm in males. It lives also in the small intestine sucking blood with the help of their teethed mouth capsule. The eggs measure 55 × 35 μm and the infection and growth takes the same pathways as in the above-described species of cattle.
4. **Symptoms of disease (Hookworm disease)**: Infected animals are often recognized by increased licking activity at their feet apparently induced by itching at sites, where the larvae 3 had entered the skin. Anaemia, loss of weight, colics, diarrhoeas alternating with obstipation, darkened feces (due to blood contents), oedemas in the skin and pneumonias during lung passage occur often as consequences of penetrated larvae and their movements inside

the host. In the case of infected pigs also numerous subclinical cases had been described.

5. **Diagnosis**: Microscopical demonstration of the thin-walled eggs (compare Fig. 5.108) after use of enrichment methods of fecal probes.
6. **Pathway of infection**: Percutaneously or orally. The larvae 3 strip off their sheath during penetration into the host's body.
7. **Prophylaxis**: Since the infectious larvae die during dryness within 3 days, it is recommended to keep stables dry and to remove feces within 5–6 days.
8. **Incubation period**: Lung symptoms occur already within the first week after the infection.
9. **Prepatent period**: 30–55 days.
10. **Patency**: At least 1 year.
11. **Therapy**: See chemotherapy and chemoprophylaxis of hookworms of carnivores (Sect. 5.3.3.8).

Further Reading

Ahn KS et al (2015) Identification and prevalence of *Globocephalus samoensis* among wild boars (*Sus scrofa coreanus*) from Korea. Korean J Parasitol 53:611–618.

Gao JF et al (2014) Comparative analysis of the complete mitochondrial genomes of the two ruminant hookworms *Bunostomum trigonocephalus* and *B. phlebotomum*. Gene 541:92–100.

Okoro CK et al (2016) Gastrointestinal helminths of wild hogs. J Helminthol 90:139–143.

Senlik B et al (2011) Helminth infections of wild boars (*Sus scrofa*). J Helminthol 85:404–408.

Wang CR et al (2012) Characterization of *Bunostomum trigonocephalus* and *B. phlebotomum* from sheep and cattle. Res Vet Sci 92:99–102.

Wang CR et al (2013) Sequence variability in four mitochondrial genes among *Bunostomum trigonocephalus* isolates from China. J Helminthol 87:416–421.

5.3.3.10 Heterakidae and Oxyuridae

1. **Name**: Greek: *heteros* = different; *akis* = pointed tip; *oxys* = sharp; *ura* = tail; *passalos* = hook. Latin: *gallus* = chicken; *equus* = horse; *ambiguus* = multifaced; *ovis* = sheep *vermicularis* = belonging to or looking like worms; *spuma* = foam.
2. **Geographic distribution/epidemiology**: Worldwide in huge numbers.
3. **Biology, morphology**: Important species in this group are:

Heterakis spumosa (rodents, hedgehogs)	Colon	♀ 1.3 cm
Heterakis gallinarum (birds)	Caecum	1–2 cm
Oxyuris equi (horses)	Caecum/colon	up to 12 cm
Passalurus ambiguus (hares, rabbits)	Caecum/colon	1 cm
Syphacia **species** (rodents)	Caecum/colon	0.8 cm

(continued)

***Aspiculurus* species** (rodents)	Caecum/colon	0.8 cm
Skrjabinema ovis (sheep, goats)	Colon	0.8 cm
Enterobius vermicularis (monkeys, humans)	Colon	1.2–1.5 cm

Morphological features/development

(a) ***Heterakis* species** (e.g. *H. gallinarum*) (**chickens, ducks, geese**); *H. dispar* (**geese, ducks**); *H. isolonche* (**pheasants, turkeys**) are situated always in large numbers in the caeca of their hosts. The females have mostly a length of 10–15 mm (*H. dispar* up to 23 mm) and are characterized by a very pointed posterior pole (Fig. 5.110a; 5. B-f). After fertilization by the smaller males (5–13 mm, *H. dispar* up to 18 mm), the females excrete per day 800–1000 unembryonated eggs (Fig. 5.111) which appear thick walled, have a smooth surface and

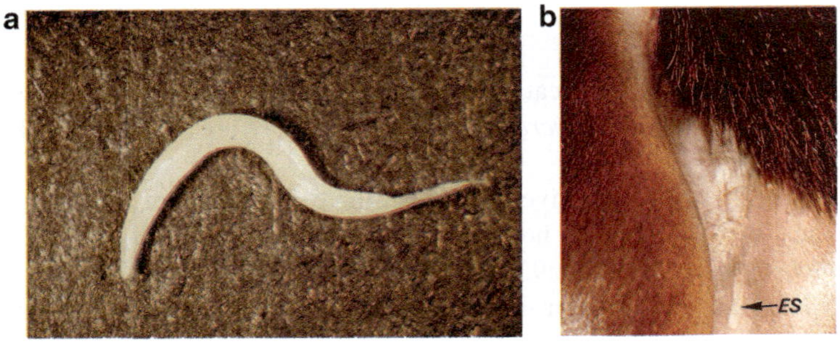

Fig. 5.110 (**a**) Light micrograph of an adult female of the genus *Oxyuris*. (**b**) Macrophoto of the eggs of *O. equi* (glueing together in the region of the anus; ES)

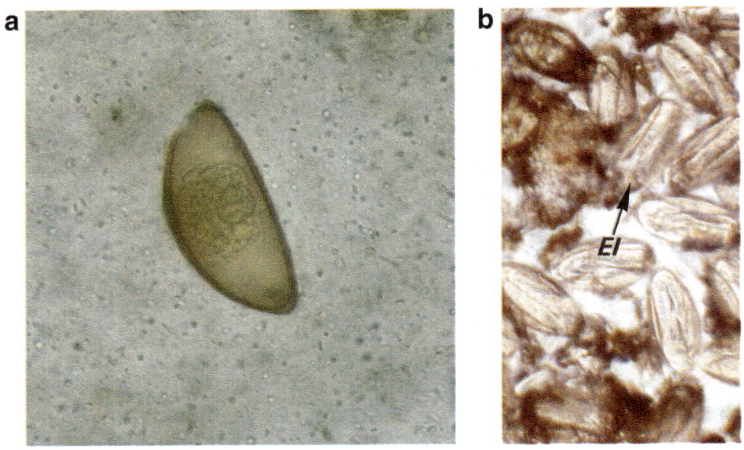

Fig. 5.111 (**a**, **b**) Micrographs of the typical *Oxyuris* eggs (EI) showing the characteristic depression (**a**) and being clutched together around the host's anus (**b**)

measure 60–80 μm × 40 μm. One side of the eggs appears somewhat depressed. There are no significant morphological differences among the different *Heterakis* species. After a temperature-dependent development (14 days at 27 °C, but 2 months at 10–15 °C), the larva 2 or 3 has reached infectivity inside the egg. Intermediate hosts like earthworms (e.g. *Lumbricus* sp.) often ingest these eggs so that they often contain large numbers, which lead to mass infections of hosts even in the case they ingest only a single earthworm. Inside the duodenum of the final hosts the larva 2 or 3 hatches from the egg and wanders into the mucous layer of the caecum, where it stays for 2–5 days. Then it returns into the lumen, where it reaches via further moults the adult status.

(b) ***Passalurus ambiguus*** (in **hares** and **rabbits**) is economically of high importance. Males reach a length of 5 mm, while females are considerably larger (8–11 mm) (Fig. 5.112). The terminal ends of both sexes are pointed. Inside the rectum of their hosts, the females discharge the typical eggs (Fig. 5.113; 5. B-h) which are glued at the fecal strands and measure about 90–105 μm × 45 μm. Larval development inside the egg often starts already in the rectum of the host, so that several excreted eggs already contain the infectious larva 3. Outside of the body, the larva 3 does not

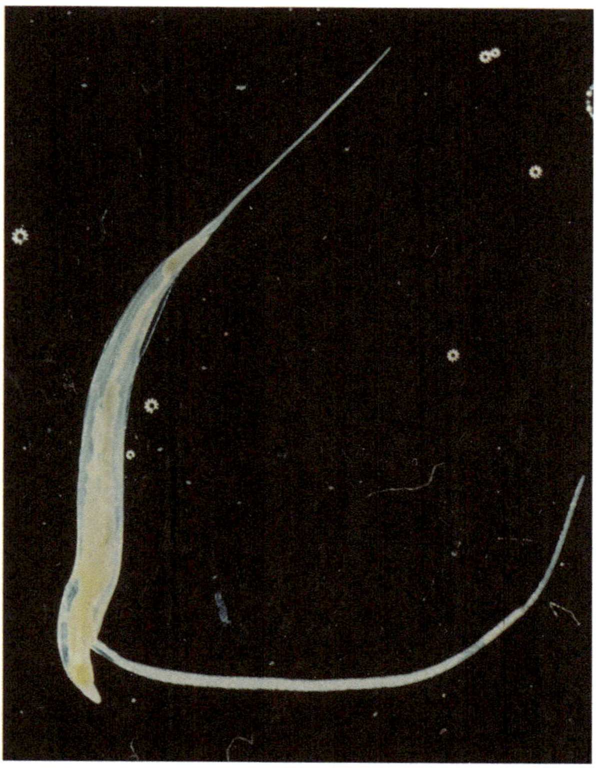

Fig. 5.112 Light micrograph of a paired couple of *Passalurus ambiguus*

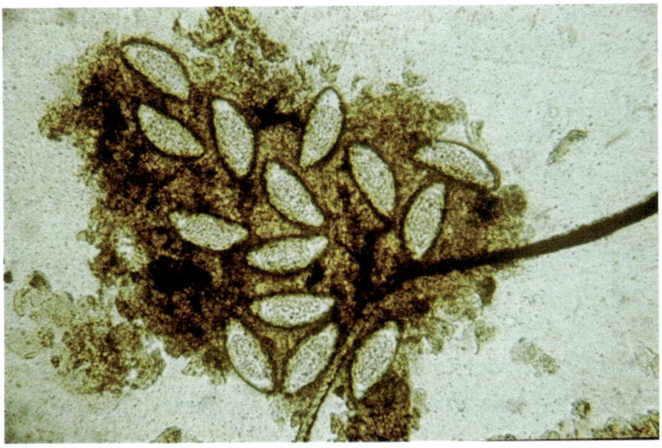

Fig. 5.113 Light micrograph of the typically shaped eggs of *Passalurus ambiguus* in feces of rabbits

leave the eggshell but may be ingested by a new host (but also from the previous one leading to autoinfection). After hatch inside the intestine of the new (or former) host the larva 3 enters at first the mucous layer of the small intestine or of the caeca, where they reach maturity. However, finally they stay in the lumen of the intestine.

(c) **Rats and mice**:
   - ***Aspiculuris tetraptera***: The females reach a length of 3–4 mm, while the males measure only 2–3 mm. Both settle inside the colon, where the females excrete the eggs, which measure 70–100 μm × 30–50 μm and appear symmetric with rather pointed poles (Fig. B 5–1). When discharged by the females, the eggs contain just a morula stage, which develops within a few hours a sausage-like larva 1. The further development occurs within a few hours as described in the species listed under (**a**) and (**b**).
   - ***Syphacia* species**: The adults reach as males a length of only 1.5 mm, while as females they measure up to 4–5 mm in length. They parasitize as larvae inside the caeca and as adults in the colon. The eggs of *S. muris* measure 75 × 30 μm and contain often already a larva when being discharged inside the feces, while the eggs of *S. obvelata* (Fig. B-5 j), which measure 150 μm × 50 μm, are mostly excreted in the morula stage.

4. **Symptoms of disease**:
   (a) **Heterakiasis**: Strong clinical symptoms are mostly rare, but are characterized by disturbances in digestion due to reduced metabolism of the cellulose in the food, reduced egg laying activity and ulcers in the intestine followed by excretion of blood containing feces. Infected animals may also be hit by secondary infection with other agents of disease.
   (b) **Other species** (described in **3 b–d**): Itching occurs around the anus, unrest of infected animals, rabbits “drum” legs often onto soil, diarrhoeas

especially in very young animals, loss of weight due to reduced function of the caeca, reduced food uptake and lethargy.

5. **Diagnosis**: Microscopical demonstration of the typical eggs (Figs. 5.111 and 5.113: B 5–f, h, j) after use of concentration methods (S.A.F.C.). In dead animals, the larvae L3 and L4 can be found in scrapings of the mucosa of the caeca.
6. **Pathway of infection**: Oral by uptake of worm eggs (containing the larva 3) within food, drinking water or dust.
7. **Prophylaxis**: Increased exchange of the bed of straw and disinfection of the floor by hot steam.
8. **Incubation period**: 2 weeks.
9. **Prepatent period**: 8–10 weeks; in the case of *H. spumosa*: 6 weeks.
10. **Patency**: 1–2 months (however, often occurs self-healing by excretion of adult worms).
11. **Therapy**: **Fenbendazole** (25 ppm inside food for 5 days) eliminates immature and mature stages from the intestine. **Attention**: This compound is not officially registered for rabbits which are used for human food production. At least 7 days waiting time is needed as is the case when prescribed for cattle. *Oxyuris* species of rodents can be eliminated by **tiabendazole** and **fenbendazole**. In the case of monkeys and humans, **mebendazole** and **albendazole** are commonly used; however, the treatment must be repeated at intervals of 15–18 days, since repeated autoinfection may occur. In horses as in ruminants, the registered compounds are safe and effective.

Further Reading

Behnke JM et al (2015) Bank voles (*Myodes glareolus*) and house mice (*Mus musculus musculus, M. m. domesticus*) in Europe are each parasitized by their own distinct species of *Aspiculuris*. Parasitology 142:1493–1505.

Das G et al (2014) Egg production dynamics and fecundity of *Heterakis gallinarum* residing in different caecal environments of chickens induced by fibre-rich diets. Vet Parasitol 2005:606–618.

Dewi K, Hasegawa H (2014) Two new species of *Syphacia* in endemic murid rodents from Sulawesi, Indonesia. J Helminthol 88:41–49.

Rimaldi R et al (2007) *Passalurus ambiguus*: new insights into copromicroscopic diagnosis. Parasitol Res 101:557–561.

Snabel V et al (2014) Molecular identification of *Heterakis spumosa* obtained from brown rats (*Rattus norvegicus*) and its infectivity in mice. Parasitol Res 113:3449–3455.

Wolf D et al (2014) *Oxyuris equi*: lack of efficacy in treatment with macrocyclic lactones. Vet Parasitol 201:163–168.

5.3.3.11 *Gnathostoma spinigerum* (Thorny Stomach Worm)

1. **Name**: Greek: *gnathos* = jaw; *stoma* = mouth. Latin: *spinosus* = covered with thorn-like structure.

2. **Geographic distribution/epidemiology**: Asia, America, Africa, imported in single cases in Europe.
3. **Biology, morphology**: This species, which belongs to the suborder Spirurina, occurs in cats, dogs and several wild carnivores, where the worm specimens are found in blood-filled bladders in the stomach (Fig. 5.115b). Up to nine worms had been counted in those cyst-like structures. The adult males reach a length of up to 3 cm, while the females often measure 5 cm. Both sexes are characterized by 11 rows of larger cuticular hooks at the swollen anterior region (bulbus). The remnant surface is covered by smaller, scale-like hooks. The males are equipped with two differently sized spicula. The fertilized eggs measure about 68 μm × 37 μm, appear slightly greenish and possess only at one of their two poles a so-called polar plug (Figs. 5.114 and 5.115a). They are excreted within the feces after they had been squeezed out of the cyst-like intestinal bladder, which is provided with a fine pore. In water, the sheathed larva hatches from the egg within 4 days. This larva has to become ingested by a small crustacean as first intermediate host. Therein the larva grows up to a length of about 0.7–1 mm and moults to become the larva 2. In the case that this host (*Cyclops* sp.) is ingested by a second intermediate host (e.g. a fresh water fish, frogs, reptiles), the larva 3 is developed inside the muscles, where it reaches a size of up to 4.5 mm. This larva 3 is finally infectious for the final hosts, in case that they ingest this second intermediate host raw or in an undercooked status. After a phase of wandering through the liver or other organs, the larva 3 becomes fertile after two more moults. If **humans** ingest these now about 1 cm large larvae, they can be found as **larva interna** in inner tissues or as **larva externa** inside the skin. In pigs, the species *Gnathostoma hispidum*, which also parasitizes inside the stomach, occurs. Gnathostomid worms have also been found in eels, which are used for human consumption. In total, more than 20 *Gnathostoma* species have been described, 6 of which have also been found in humans (*G. binucleatum, G. doloresi, G. hispidum, G. malaysiae, G. nipponicum* and *G. spinigerum*), while others occur exclusively in animals.
4. **Symptoms of disease (Gnathostomiasis)**: Main symptoms are disturbances in digestion, gastritis, eventually also peritonitis, if the larvae break through the stomach wall. Mass infections may lead to death.
5. **Diagnosis**: Microscopical demonstration of the eggs (Fig. 5.115a). **Attention**: Eggs are not excreted every day; thus feces controls have to be done on several consecutive days.
6. **Pathway of infection**: Oral during ingestion of raw, larva-containing fish, amphibia or reptiles.
7. **Prophylaxis**: Avoid to feed raw fish meat to house animals.
8. **Incubation period**: 3–7 days.
9. **Prepatent period**: 21–28 days.
10. **Patency**: Years.
11. **Therapy**: See Ascarids.

Fig. 5.114 Diagrammatic representation of the life cycle of *Gnathostoma spinigerum* using different types of hosts (**A–E**). **1** Adults (male 10–30 mm, female up to 50 mm), which are characterized by a spiny anterior bulbus and many body rows of spines, live in the stomach wall of the final host (cats) leading to tumour-like growths. **2–4** The eggs (65–70 × 40 μm) are unembryonated when passed by feces (2). After 1 week of embryonation the first-stage larva (4) hatches in water and swims actively. It remains covered by a thin sheath representing the inner eggshell. **5** The first intermediate host (**B**) is a cyclopoid copepod, where the L1 penetrates the haemocoel and develops further into a second-stage larva (L2) within 7–10 days. The L2 already has a swollen head bulbus covered with 4 transverse rows of spines. **6** Fish acts as second intermediate hosts when they ingest infected copepods. The L2 penetrates the intestine of its new host and migrates to muscles or connective tissues, where it moults to form the L3 which becomes encapsulated. The L3, measuring about 4 mm, is infective to the final host, where it reaches maturity. However, the encapsulated L3 is also infective to other hosts (**D**) including humans (**E**). **7** If L3 are eaten by other than final hosts (including man), they wander in that host's tissues without further development; in humans and animals, this leads to symptoms of *gnathostomiasis externa* or *interna*, depending on the infected tissues. If humans eat such uncooked paratenic hosts (**D**), they may become infected but no further development occurs, although the ingested L3 remain infective for final hosts within such paratenic hosts (*arrow* not drawn to **A**). In the intestine of humans, worm stages lead to nodule formation (7)

Further Reading

Chai JY et al (2015) Larval *Gnathostoma spinigerum* detected in Asian swamp eels (*Monopterus albus*) purchased from a local market in Yangon, Myamar. Korean J Parasitol 53:619–625.

Cole RA et al (2014) *Gnathostoma* spp. live in Asian swamp eels (*Monopterus*) from food markets and wild populations, USA. Emerg Inf Dis 20:634–640.

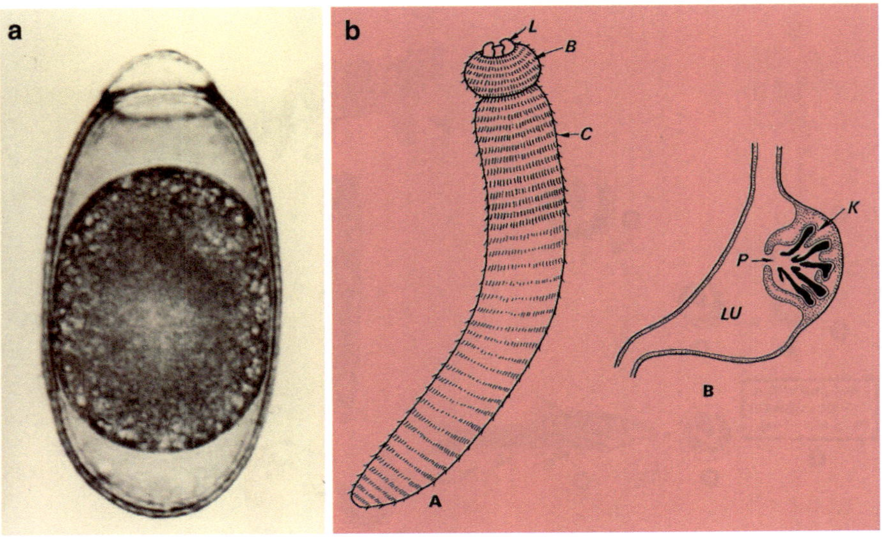

Fig. 5.115 (**a**) Light micrograph of a freshly excreted unembryonated egg, which is characterized by a single polar plug, which is later removed by the hatching larva. (**b**) Diagrammatic representation of an adult worm (A) and a larva-containing cyst (B) inside the intestine of a host. *B* bulbus; *C* cuticular hooks; *K* chamber-like hollow; *L* lip; *LU* lumen of stomach; *P* pore

Saksirisampant W et al (2012) *Gnathostoma spinigerum*: immunodepression in experimental infected mice. Exp Parasitol 132:320–326.

5.3.3.12 *Strongyloides* Species (Dwarf Threadworms)

1. **Name**: Greek: *strongylos* = rounded; *oides* = looking similar. Latin: *papillosus* = covered with slight protrusions; *stercoralis* = within feces; *rattus* = rat; *aves* = birds. Fülleborn and Wester = a German and an English helminthologist.
2. **Geographic distribution/epidemiology**: Worldwide.
3. **Biology, morphology**:

   The following species are common worldwide; however, a much larger number of species exists, which to date are only scarcely considered (Table 5.6):

S. ransomi (syn. *S. suis*; pigs)	4.5–5.5 mm
S. papillosus (ruminants, rabbits)	6–8 mm
S. westeri (horses, donkeys)	8–9 mm
S. ratti (rodents, rats)	1.7–3.4 mm
S. avium (birds)	2–2.5 mm
S. stercoralis (carnivores, monkeys, **humans**)	2–3 mm
S. fuelleborni (monkeys, **humans**)	3–4 mm

Further species are *S. planiceps* (carnivores in Japan, Malaysia), *S. tumefaciens* (cats; USA), and *S. cebus* (New world monkeys). All these thin species with diameters of only 0.1 mm develop two different generations into their life cycle (Fig. 5.116).

(a) A **free-living generation** comprising males (0.7 mm) and females (0.9 mm) with diameters of only 0.05–0.1 mm. They are characterized by a so-called rhabditiform oesophagus.

(b) **Parthenogenetic females**: These stages are considerably larger than the specimens of the free-living generation (see above). They live inside the mucous layer of the small intestine. Their hind end is characterized by a pointed tip, which starts just behind the anus. The vulva opens crossways and is situated at the end of the second third of their body. These parthenogenetic females produce without copulation fully embryonated (=larva-containing) eggs (Fig. 5.117), which are excreted within the feces. Outside of the body the larva 1 hatches from the egg. These larvae grow up (depending on their set of chromosomes) within 48 h into males and females of the free-living generation or within 4 days into the unsheathed, infectious larvae (Fig. 5.116). The L3 larvae may survive—depending on the outside temperatures—for 3 months. In case they get in contact with a host, they penetrate into the skin and reach after a heart–lung–trachea–oesophagus passage the small intestine (Figs. 5.116 and 5.118), where they grow up to females, which start already 5–7 days p.i. with the egg production. In older hosts (apparently after previous infections), the larva 3 does stop its further development and remains in the host's muscles. In the case of lactating mother animals, the larvae may enter the young animals during their milk sucking.

4. **Symptoms of disease**:
   (a) **Skin reactions** such as itching in case of large numbers of penetrating larvae.
   (b) **Lung damages** due to wandering larvae inducing also coughing, pneumonia, weakness and eosinophilia.
   (c) **Catarrhalic enteritis** and the accompanying symptoms such as diarrhoea, hydrothorax, ascites, loss of hair and loss of weight. In the case of heavily infected cattle death rates occurred reaching 15–18 %.
5. **Diagnosis**: Demonstration of the typical, larva-containing eggs, which have a size ranging from 45–65 μm × 25–30 μm (Figs. 5.116 and 5.117). In feces, which had been excreted days before, already larvae can be found when using the Baermann funnel (see Fig. 3.1).
6. **Pathway of infection**: Percutaneously; the larvae 3 enter actively into the skin of the hosts. In addition, galactogenic (via mother milk) and prenatal infections are rather common.
7. **Prophylaxis**: Systematically repeated removal of the feces from stables, which additionally should be kept dry. Disinfection of the stables with hot steam/hot

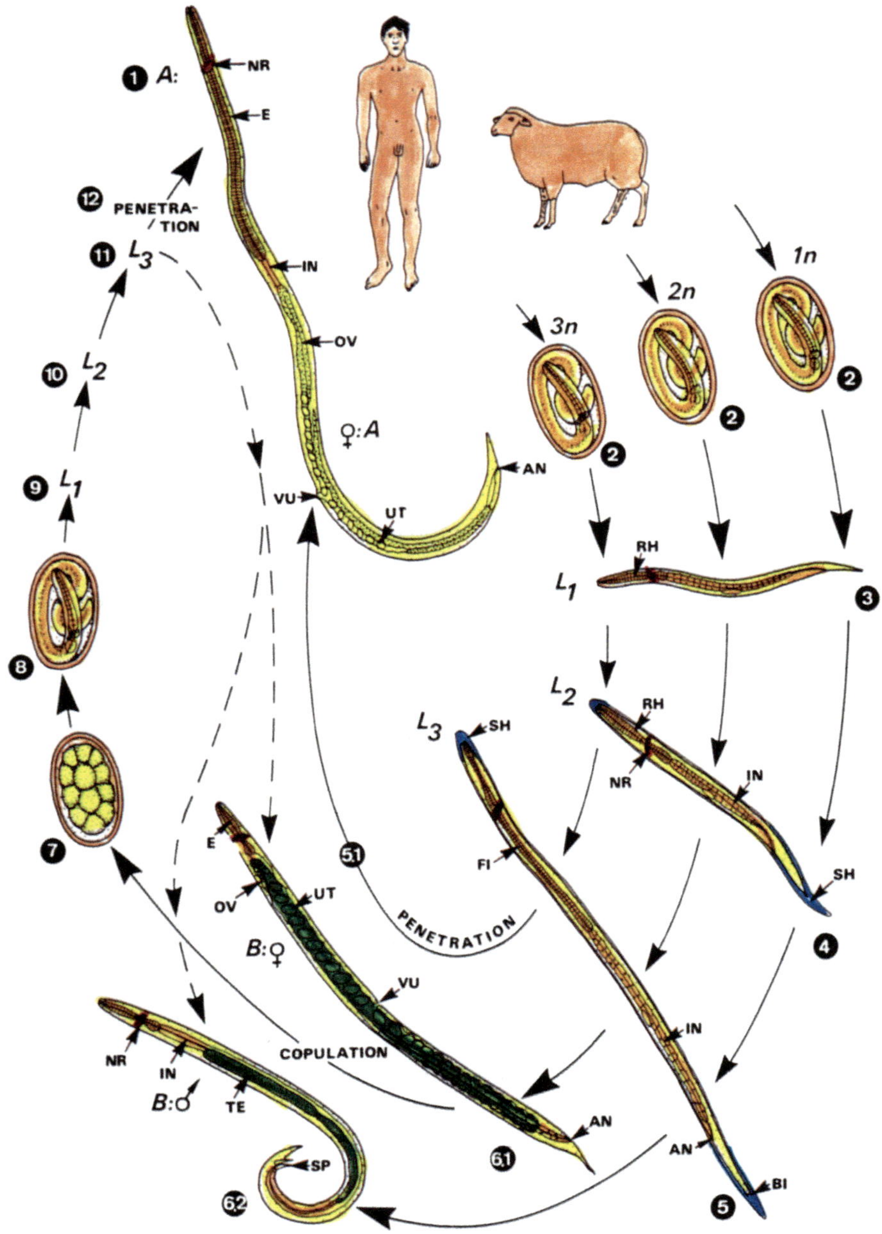

Fig. 5.116 Diagrammatic representation of the life cycles (tentative) of *Strongyloides* spp. (e.g. *S. stercoralis* of man, *S. papillosus* of ruminants). (**A**) Parthenogenetic female-homogonic generation. (**B**) Free-living heterogonic generation. **1–4** Parthenogenetic females live embedded in the mucosa of the small intestine and produce eggs with different numbers of chromosomal sets (n). Larvae may escape from eggshells inside the intestine and then be passed with feces. **3—5.1** The 3n type egg develops directly via L1-L3 into the homogonic female (1). This may occur inside

water or use of chemical disinfectants. **Attention**: Free larvae may survive for months, if humidity is high.

8. **Incubation period**: 1–2 days.
9. **Prepatent period**:

S. papillosus, S. westeri	9–14 days
S. suis	6–9 days
S. ratti, S. venezuelensis	5–7 days
S. stercoralis	9–19 days
S. fuelleborni	4–5 weeks

10. **Patency**: Several months.
11. **Therapy**: A broad spectrum of benzimidazoles, probenzimidazoles and macrocyclic lactones are used; moxidectin in the case of dogs (see Sect. 7).

Further Reading

Anderson J et al (2012) *Trichuris* sp. and *Strongyloides* sp. infections in a free-ranging baboon colony. J Parastiol 98:205–208.

Aranyo JM et al (2012) Control of *Strongyloides westeri*. Rev Bras Parastiol Vet 21:157–160.

Dimitrijevic B et al (2012) Effects of the infection with *Strongyloides papillosus* and albendazole treatment. Vet Parasitol 186:364–375.

Eamudomkarn C et al (2015) Comparative evaluation of *Strongyloides ratti* and *S. stercoralis* larval antigen for diagnosis of strongyloidiasis. Parasitol Res 114:2543–2551.

Kabululu ML et al (2015) Risk factors for prevalence of pig parasitosis in Tanzania. Vet Parasitol 212:460–464.

Labes EM et al (2011) Genetic characterization of *Strongyloides* sp. from captive, semi-captive and wild Bornean orangutans (*Pongo pygmaeus*). Parasitology 138:1417–1422.

Fig. 5.116 (continued) the host's intestine (**autoinfection**) or via free L3 on soil (5.1). **3—6.1** The 2n type eggs produce the heterogonic free-living males (6.1). **3—6.2** The 1n eggs give rise to the free-living males (6.2). **7–12** The progeny of the free-living generation develops via (non-feeding) L3 into parthenogenetic females upon entering the vertebrate host. Some L3, however, may give rise to another free-living generation (apparently endowed with a different chromosomal pattern). After penetration into the vertebrate host, the L3 are carried passively through the bloodstream to the heart and lung and after a moult accidentally break out into the alveolar space. From there, they wander up the respiratory tract to the pharynx and are swallowed. In the intestine the L4 undergoes a final moult and becomes mature, starting (according to some authors) a protandric reproduction. This includes the initial development of a male gonad, followed by the female gonad, and self-fertilization; thus a true parasitic male does not appear. AN = anus; BI = bifurcated posterior pole; E = oesophagus; FI = filariform oesophagus; IN = intestine; L = larval stages; N = number of chromosomal sets; NR = nerve ring; OV = ovary; RH = rhabditiform oesophagus; SH = sheath (cuticle of preceding larval stage); SP = spicula; TE = testis; UT = uterus with eggs

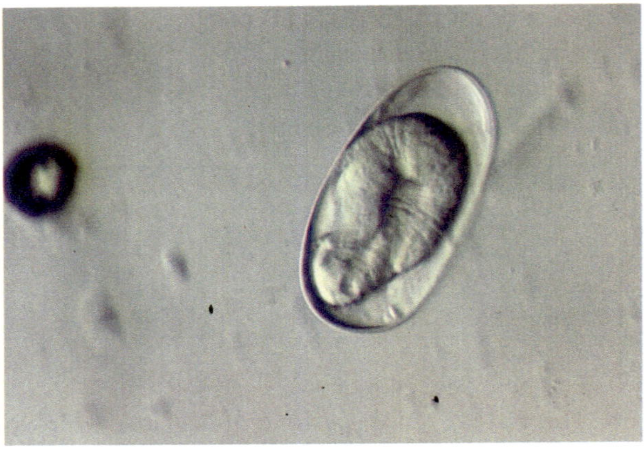

Fig. 5.117 Light micrograph of an egg of *Strongyloides* sp. containing an infectious larva

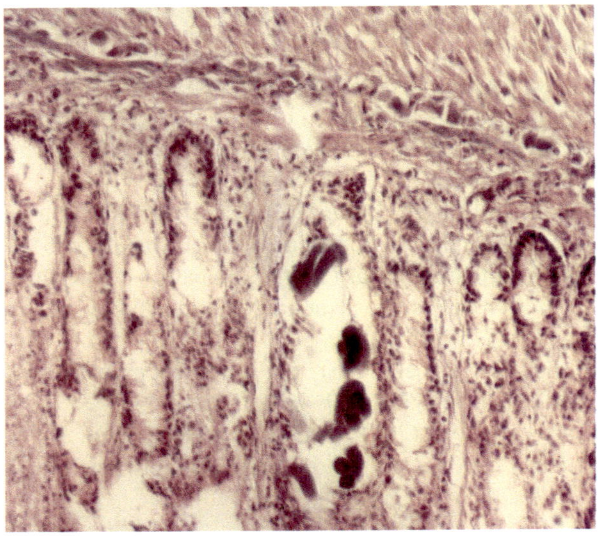

Fig. 5.118 Light micrograph of an intestinal section showing female *Strongyloides* worms being situated between intestinal villi

Lyons ET, Tollivers SC (2014) Prevalence of patent *Strongyloides westeri* infections in thoroughbred foals in 2014. Parasitol Res 113:4163–4164.

Toledo R et al (2015) Strongyloidiasis with emphasis on human infections and its different clinical forms. Adv Parasitol 88:165–241.

Viney M, Kikuchi T (2016) *Strongyloides ratti* and *S. venezuelensis*—rodent models of *Strongyloides* infections. Parasitology. doi:10.1017/s0031182016000020.

Zanetti-Lopez WD et al (2014) Anthelmintic efficacy of ivermectin and abamectin in naturally infected pigs. Res Vet Sci 97:546–549.

5.3.3.13 *Strongylus* Species of Horses

1. **Name**: Greek: *strongylos* = rounded. Latin: *edentatus* = without teeth; *vulgaris* = trivial; *equinus* = belonging to horse.

2. **Geographic distribution/epidemiology**: Worldwide.
3. **Biology, morphology**:

   Important species are:

   (a) ***Strongylus equinus***: Females measure up to 4.7 cm in length and males up to 3.5 cm with a diameter of 2–3 mm. The mouth capsule appears spherical and is provided with 4 tooth-like protrusions at the base of the mouth capsule (Fig. 5.119a). This species parasitizes in the colon and caecum of their hosts (Table 5.6).

   (b) ***S. edentatus***: The females reach a length of 4 cm and the males 3 cm. The mouth capsule appears cone like and does not contain teeth but is surrounded by leaf-like structures (Fig. 5.119b). These parasites live in the colon.

   (c) ***S. vulgaris***: The females reach a length of 2.5 cm and the males up to 1.6 cm. The mouth capsule appears mug like and is equipped with two teeth at its base (Fig. 5.119c). The adults live in the colon.

   These rather stiff nematodes are attached at the mucous layer of the intestine with the help of their mouth region, which acts like a sucker. They lead to bleeding which they suck in. Their eggs are excreted within the feces and outside of the body a larva is formed inside the egg. The larva 1 hatches from the egg and grows up, but remains inside the sheath, which represents the cuticle of the larva 1. Although this larva has already an oesophagus, it does not take up food outside of a host. The further development is specific:

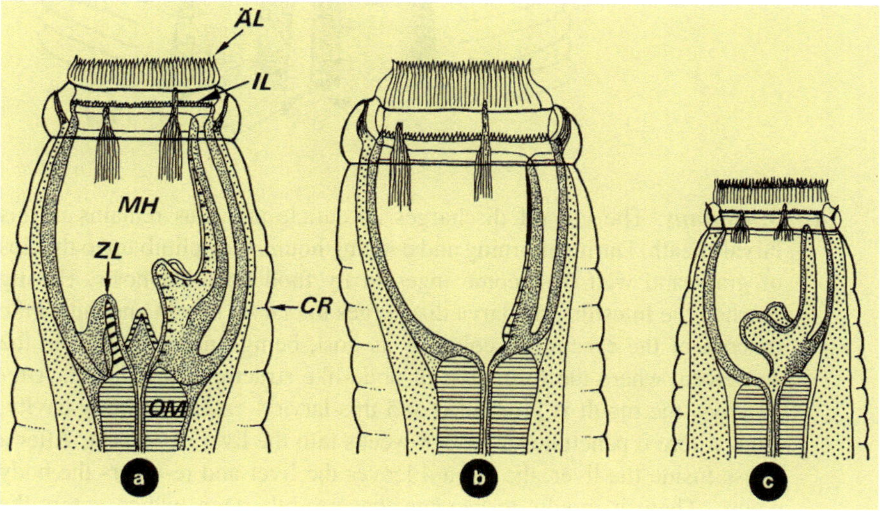

Fig. 5.119 Diagrammatic representation of the anterior ends of three *Strongylus* species. (**a**) *S. equinus*, (**b**) *S. edentatus*; (**c**) *S. vulgaris* (according to Soulsby). AL = outer crown of lamellae; CR = cuticular rings; IL = inner crown of lamellae; MH = buccal hollow (mouth); OM = muscles of oesophagus; ZL = teeth-like cuticular protrusions

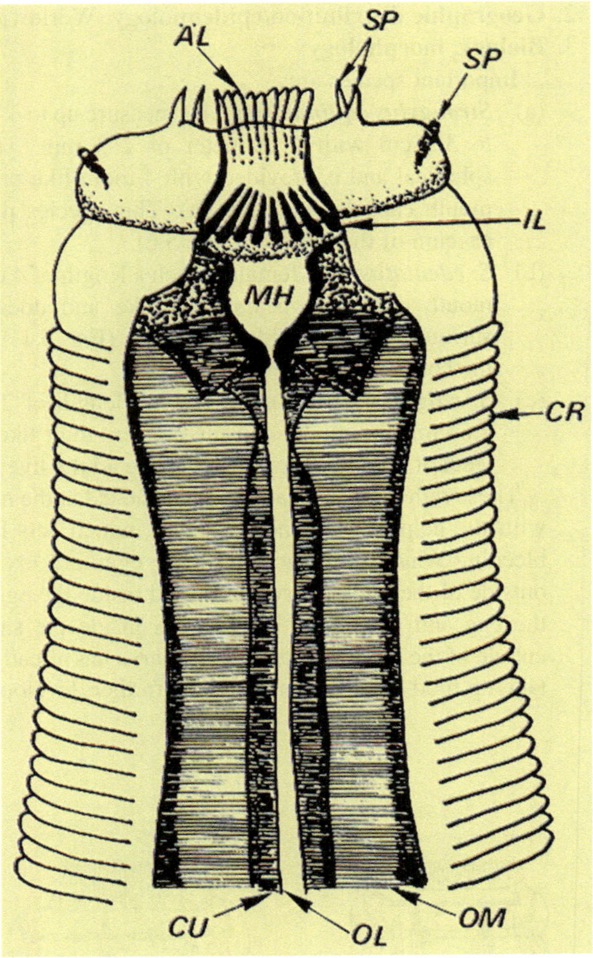

Fig. 5.120 Diagrammatic representation of the anterior end of a small strongylid nematode (*Cylicostephanus* sp.). AL = outer crown of lamellae; CR = cuticle rings; C = cuticle; IL = inner crown of lamellae; MH = mouth cavity; OL = lumen of the oesophagus; OM = muscles of the oesophagus; SP = sense papillae

(a) ***S. equinus***: The larva 1 discharges its cuticle and thus remains in this larval sheath. During morning and evening hours, they climb up to the tips of grass and wait to become ingested by food uptaking hosts. Having reached the intestine, the larva discharges the larval sheath and enters the mucosa of the caecum or colon of its host, being finally situated in the subserosa, where the formation of node-like structures is induced. After 11 days, the moult is proceeded and this larva 4 enters the body cavity, from where it penetrates after 6–8 weeks into the liver of its host. After a phase inside the liver, the larva 4 leaves the liver and re-enters the body cavity. There it moults to become the preadult stage, which enters the intestinal lumen. Therein the worm reaches maturity, starts—depending on its sex—to produce sperm or eggs and copulates with a sexual partner.

(b) ***S. edentatus***: In this species, the larval development until larva 3 proceeds outside of the body. The larva 3 enters the intestinal wall and reaches via

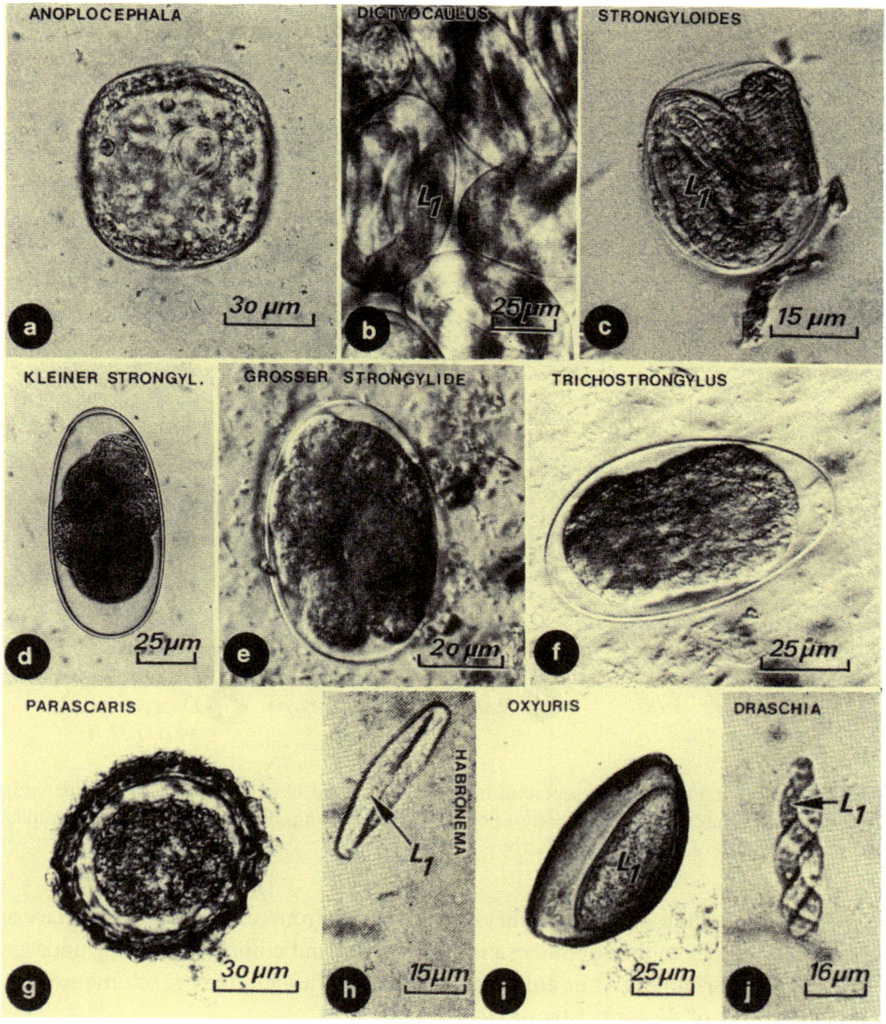

Fig. 5.121 Light micrograph of worm eggs from horses. L1 = larva 1

the portal vein system the liver, where it moults within 11–18 days to become the larva 4. This larva enters the liver tissue, stays there for about 9 weeks and wanders then via the peritoneal wall into the abdominal region, where they induce nod-like structures with diameters of 1–5 cm. Inside these structures, they moult to become preadults, which enter the wall of the caecum and colon and induce there further hooks. Having reached sexual maturity, the male and female worms enter again the intestinal lumen, where copulation and later egg excretion occur.

(c) ***S. vulgaris***: In this species, the larva 3 enters the intestinal wall, where the moult to the larva 4 occurs. These larvae enter the intima of the arterioles of the submucosa of the small intestine, where they induce thrombosis-

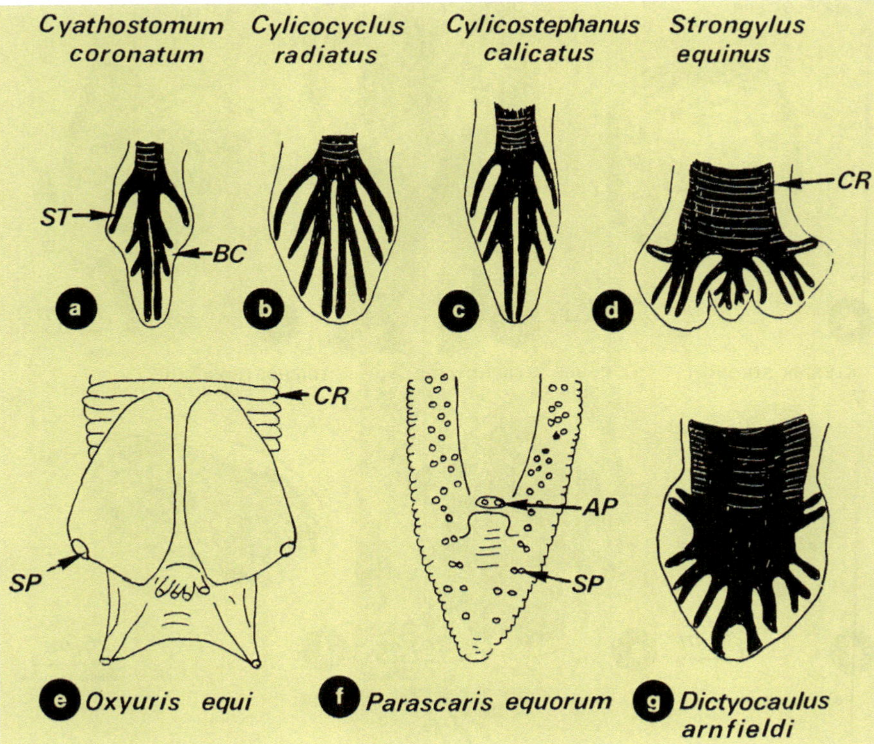

Fig. 5.122 Diagrammatic representation of the terminal end of several nematodes (enlarged). AP = papillae around the anus; BC = bursa copulatrix; CR = rings of cuticle; SP = sense papillae; ST = fortifications of the bursa copulatrix

like symptoms and aneurysms. Starting from day 45 p.i., the larvae 4 wander to the submucosa of the caecum and colon, where they undergo another moult. After entering the intestinal lumen, they reach maturity.

4. **Symptoms of disease (Strongylosis)**:
   (a) **Symptoms due to adult worms**: Anaemia, oedemas, loss of weight, weakness, reduced food uptake and heavy diarrhoeas.
   (b) **Symptoms due to wandering larvae**: Mucosa bleedings in the colon, fever, leucocytosis, aneurysms with embolism, local endarteritis chronica, thrombosis of the arteria of the sexual organs. In the case of *S. vulgaris* infections, the so-called intermittent limping often occurs.
5. **Diagnosis**: Microscopical demonstration of the eggs (in the state of cleavage) inside the feces. The infectious eggs of the large strongyloids have a length of about 90 μm (Fig. 5.121e). The method of choice is flotation. Species diagnosis can be done after larval cultivation.
6. **Pathway of infection**: Oral uptake of larvae within grass food. **Attention**: Larvae survive up to two months and are even able to overwinter inside stables.

7. **Prophylaxis**:
   (a) **Hygienic measurements** in stables: Removal of the feces from stables at short intervals; disinfection of the floors; fresh air inside stables.
   (b) **Pasture**: Alternating of use by cattle and sheep; drying of wet places; young animals should be held on clean pastures; avoidance of overcrowding pastures with too many animals.
   (c) **Chemoprophylaxis**: Repeated treatment of the animals with therapeutic dosages of varying compounds.
8. **Incubation period**: 2–15 days.
9. **Prepatent period**: *S. equinus*: 8–10 months; *S. edentatus*: 10–11 months; *S. vulgaris*: 5–7 months.
10. **Patency**: 1–2 years.
11. **Therapy**: For the treatment of the adult worms, several compounds are registered. However, it is important that treatment also covers the immature stages. Table 5.8 shows active compounds and their efficacy in different hosts.

    According to worldwide experiences, the following recommendations can be given:

    Treatment should be done by alternating use of compounds at intervals of 2 months:
    (a) **Group I** (with effects on nematodes and *Easterophilus* flies): **ivermectin** alone or in **combination** of benzimidazole preparations with **metrifonate.**
    (b) **Group II**: Use of broad spectrum anthelmintics such as the benzimidazoles **cambendazole, fenbendazole, mebendazole, parbendazole, thiabendazole, fenbendazole** as well as **ivermectin** and **oxibendazole**.

Table 5.8 Nematocides for horses

Chemical short term	Dose mg/kg orally	Big strongylids: immatures	Big strongylids: adults	Small strongylids: immatures/adults	Small strongylids: benz. resistant
Pyrantel-H-Pamonat	19.0		+++	+++	+++
Cambendazole	20.0		+++	+++	
Dichlorvos	30–60[a]	?	+++	+++	+++
Ivermectin	0.2	+++	+++	+++	+++
Parbendazole	2 × 2.5		+++	+++	
Fenbendazole	7.5	+++[b]	+++	+++	
Febantel	6.0		+++	+++	
Mebendazole	10.0		+++	+++	
Tiabendazole	50.0	+++[c]	+++	+++	
Oxibendazole	10.0		+++	+++	+++

[a]Depending on age, weight and species (horse, donkey)
[b]According to literature: 75–10 mg/kg p.o. × 5 days
[c]According to literature: 440 mg/kg p.o. × 2 days

(c) **Group III**: **Dichlorvos, ivermectin,** pyrantel or combinations of benzimidazoles with **piperazine** or **oxibendazole**. Macrocyclic lactones can be used, too.

Timing of the treatment: In Europe:
November: application of products of the **group I**;
January: use of products of **group II**;
March use of products of **group I or III**;
May: use of products of **group II**,
July: use of products of **group II** (or **I** or **III**);
September: use of products of **group III**.

Further Reading

Bellaw JL, Nielsen MK (2015) Evaluation of Baerman apparatus sedimentation time on recovery of *Strongylus vulgaris* and *S. edentatus* third stage larvae from equine coprocultures. Vet Parasitol 211:99–101.

Bonneau S et al (2009) Anthelmintic efficacies of a tablet formula of ivermectin-praziquantel on horses experimentally infected with three *Strongylus* species. Parasitol Res 105:817–823.

Duan H et al (2015) Complete mitochondrial genome of an equine intestinal parasite: *Triodontophorus brevicauda* (Strongylidae). Parasitol Int 64:429–434.

Kuzmina TA et al (Fecundity of various species of strongylids (Strongylidae) parasites of domestic horses. Parasitol Res 111:2265–2271.

Nielsen MK et al (2015) Serum *Strongylus vulgaris* specific antibody responses to anthelmintic treatment in naturally infected horses. Parasitol Res 114:445–451.

Studzinska MB et al (2012) The strongylidae belonging to *Strongylus* genus in horses. Parasitol Res 111:1417–1421.

Xu WW et al (2015) The complete mitochondrial genome of *Strongylus equinus*: comparison with other closely related species and phylogenetic analyses. Exp Parasitol 199:95–99.

5.3.3.14 Small Strongylids of Horses

1. **Name**: Greek: *kyaneos* = dark blue; *stoma* = mouth; *kephale* = head.
2. **Geographic distribution/epidemiology**: Worldwide.
3. **Biology, morphology**: The group of the so-called small stronylids of horses comprises several nematodes which belong to two different families. Important and common genera are *Triodontophorus, Craterostomum, Oesophagodontus, Cyathostomum, Cylicocyclus, Cylicostephanus, Cyalocephalus*, etc. (Figs. 5.120, 5.122 and Table 5.6). The males reach a maximum length of 15 mm; the females grow up to 25 mm. Both sexes appear reddish, since they all suck blood. The larvae hatch from the eggs outside of the body and grow up to the larva 3, which is ingested within grass food. Inside the intestine, they penetrate into the intestinal wall—**hypobiosis** is possible. The larva 4 finally leaves the intestinal wall and grows up to maturity inside the lumen of the

intestine. The species diagnosis is based on the appearance of the tiny teeth inside/around the ton-like buccal capsule and thus is rather difficult (Fig. 5.120).

4. **Symptoms of disease**: Since these species mostly occur together with specimens of large strongylids, typical symptoms of pure infections with small strongylids are unknown. In case of mass infections, apparently the same symptoms occur as in infections with large strongylids. Therefore, the disease is also called **strongylosis**.
5. **Diagnosis**: Microscopical demonstration of the eggs, which are excreted in a morula/cleavage stage and measure in general 90–100 μm in length (Fig. 5.121d). Culture of hatched larvae also offer the chance to obtain further diagnostic criteria.
6. **Pathway of infection**: Oral uptake of the sheathed larvae 3 with fecally contaminated food. **Attention**: The larvae 3 survive up to 1 year outside of the body.
7. **Prophylaxis**: See large strongylids.
8. **Incubation period**: 1–5 weeks.
9. **Prepatent period**: 8–20 weeks.
10. **Patency**: Several months.
11. **Therapy**: See Sect. 5.3.3.13.

Further Reading

Cutolo AA et al (2011) Field study on the efficacy of an oral 2% ivermectin formulation in horses. Rev Bras Parasitol Vet 20:171–175.

Cwiklinski K et al (2013) Transcriptome analysis of a parasitic clade V nematode: comparative analysis of potential molecular anthelmintic targets in *Cylicostephanus goldi*. Int J Parasitol 43:917–927.

Dalal S et al (2015) Cross antigenicity of immunodominant polypeptides of somatic antigen of *Oesophagostomum columbianum* with other helminths by western blotting. Vet World 8:1279–1285.

Mitchell MC et al (2016) Development of a recombinant protein-based ELISA for diagnosis of larval cyathostomin infection. Parasitology 143:1055–1066.

Morariu S et al (2016) The prevalence, abundance and distribution of cyathostomins (small stongyles) in horses from Western Romania. Vet Parasitol 223:205–209.

5.3.3.15 *Chabertia* and *Oesophagostomum* Species (Colon Worms, Nodular Worms, Bowel Worms)

1. **Name**: Philip Chabert (1737–1814): French scientist and veterinarian in Lyon. Latin: *ovis* = sheep; *oesophagus* = gullet.
2. **Geographic distribution/epidemiology**: Worldwide.
3. **Biology, morphology**:
   (a) ***Chabertia ovina***: This species is not host specific and parasitizes inside the intestine many ruminants. The adult worms have a cylindrical body, and

their mouth region is thickened due to their large but toothless buccal (mouth) capsule. The adult males have a length of 14 mm, while the females reach 20 mm. The larva 4 and the adults suck blood of the wall of the colon. The females excrete unembryonated eggs (see Addendum Fig. B.4). Outside of the body, the larva 1 hatches from the egg and reaches within 5–6 days (after two moults) the infectious stage (L3; Fig. 5.124). The further development occurs similar to that of the Trichostrongylidae (Sect. 5.3.3.16).

(b) ***Oesophagostomum* species** (e.g. ***O. venulosum, O. radiatum***) (Figs. 5.123b, d, 5.124 and 5.125). These species occur in ruminants and reach as females a size of 2 cm × 0.4 mm and as males about 15–17 mm. They parasitize inside the colon. However, they do not suck blood as do *Chabertia* species. The females excrete unembryonated eggs (see Fig. B.4 in the Addendum). After excretion, the infectious larva 3 is developed within 6–8 days. If these larvae are ingested by their hosts, most of them reach maturity in 6 weeks. Some of them, however, enter the intestinal wall and become enclosed in 2–10 mm sized nodes (Fig. 5.126) and start further development after some time when the other directly grown up worms have left the intestine. *O. dentatum* in pigs is very important. The females reach a length of 14 mm, while males measure 10 mm (Fig. 5.125 and Table 5.6).

4. **Symptoms of disease**: The clinical symptoms occur mostly only in cases of mass infection: diarrhoeas (slimy yellowish), loss of weight, weakness, anaemia, loss of hair and oedemas in the regions of throat and neck as well as in the breast. In the case of infections with *Oesophagostomum* species, peritonitis may occur due to penetrating larvae accompanied by bacterial secondary infections. In case of rectal investigations, the nodes in the intestine are palpable (Fig. 5.126).
5. **Diagnosis**: Microscopical demonstration of the eggs (Fig. B.4 in the Addendum) after use of enrichment methods (M.I.F.C., S.A.F.C.).
6. **Pathway of infection**: Oral by uptake of infectious, sheathed larvae 3 within grass food.
7. **Prophylaxis**: See large strongylids.
8. **Incubation period**: 5–7 days.
9. **Prepatent period**: *Chabertia*: 7 weeks; *Oesophagostomum*: 6 weeks.
10. **Patency**: 1–6 months.
11. **Therapy**: The anthelmintics shown to be active against trichostrongylid nematodes are used also against these worms in the colon.

Further Reading

Arias P et al (1999) Characterization of *Chabertia ovina* by isoenzyme gel electrophoresis: comparative study within *Oesophagostomum venulosum*. Parasitol Res 85:884–886.

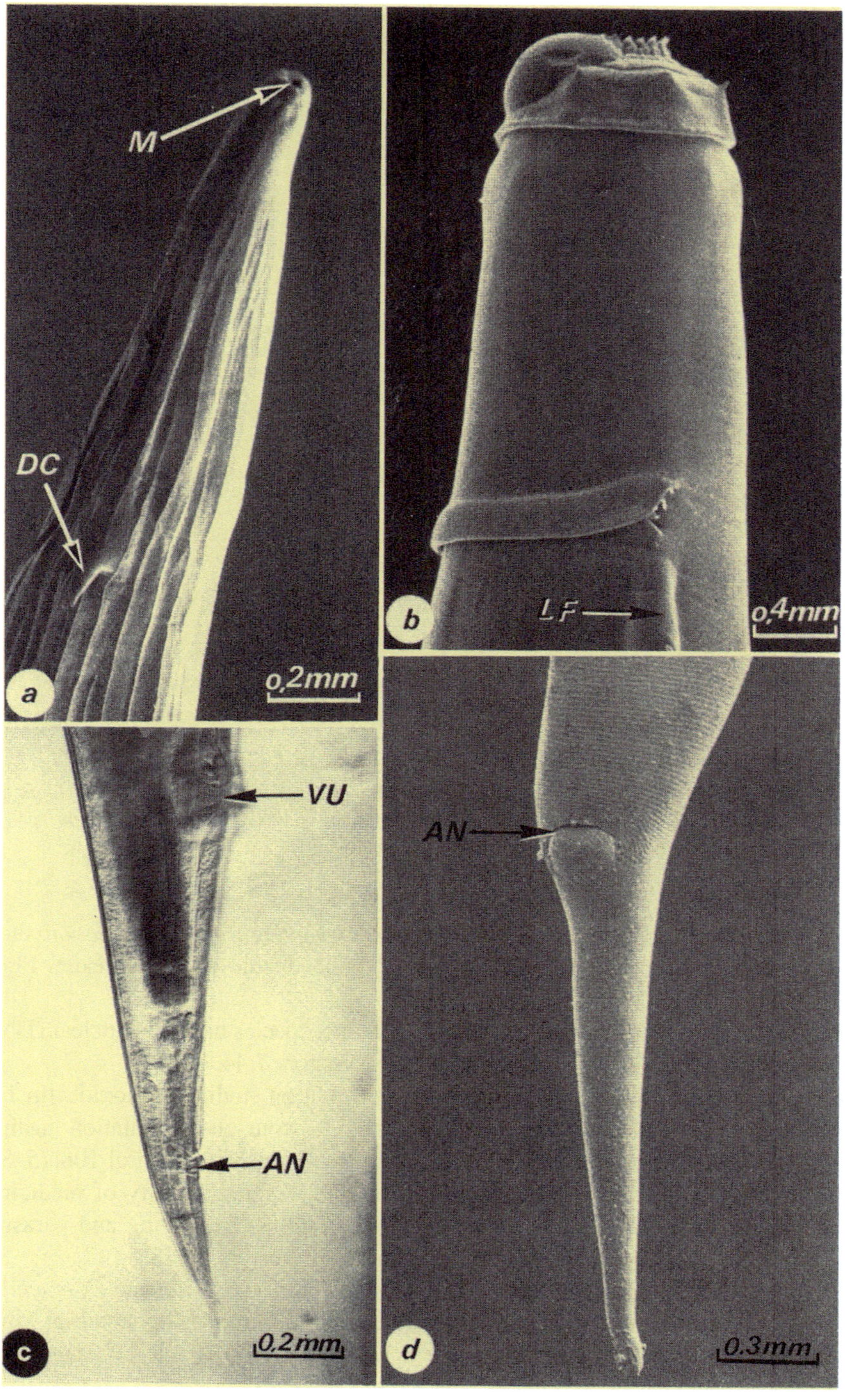

Fig. 5.123 Adult nematodes. (**a**, **b**, **d**) = scanning electron micrographs; (**c**) = light micrograph. (**a**, **c**) Anterior and posterior end of *Haemonchus contortus*; (**b**, **d**) *Oesophagostomum* sp. AN = anus; DC = teeth at the cuticle; LF = lateral wing; M = mouth; VU = vulva

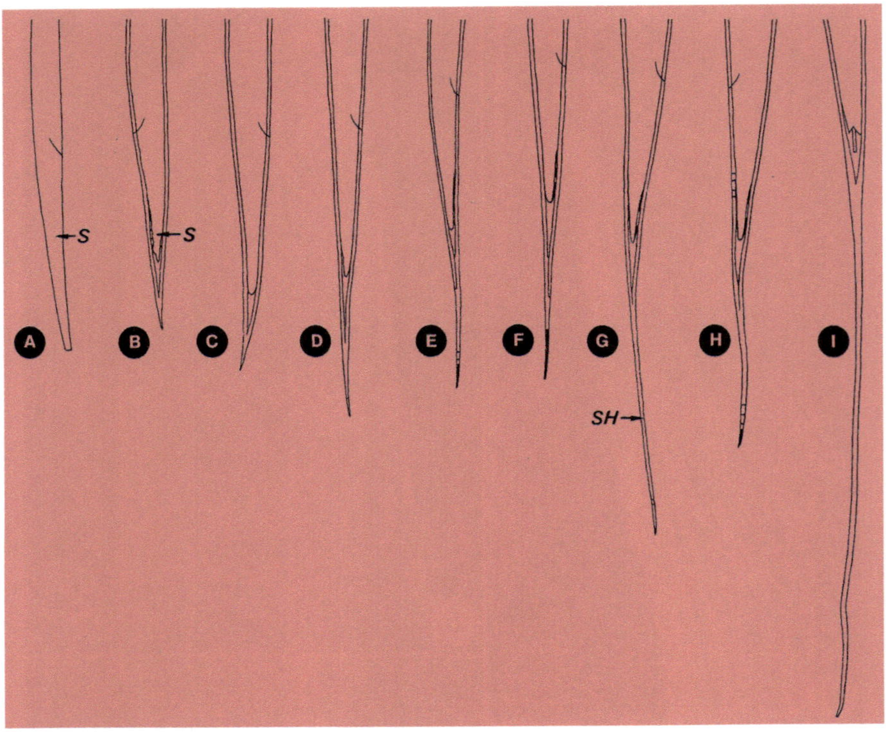

Fig. 5.124 Diagrammatic representation of the hind ends of different infectious larvae (L3) of nematodes of sheep. (**A**) *Strongyloides papillosus*; (**B**) *Trichostrongylus* sp.; (**C**) *Ostertagia* sp.; (**D**) *Cooperia* sp. (**E**) *Haemonchus* sp.; (**F**) *Bunostomum* sp.; (**G**) *Oesophagostomum* sp. (**H**) *Chabertia* sp.; (**I**) *Nematodirus* sp.; S = tail; SH = sheath

Hansen TVA et al (2014) *Trichuris suis* and *Oesophagostomum dentatum* show different sensitivity and accumulation of fenbendazole and levamisole. PLoS Negl Trop Dis: e2752.

Liu GH et al (2014) *Chabertia erschowi* is a distinct species based on nuclear rDNA sequences and mDNA sequences. Parasites Vectors 7:44.

Reinemeyer CR, Cleale RM (2002) Dose confirmation studies of moxidectin 1 % non-aqueous injectable and moxidectin 0.5 % pour-on formulation against *Oesophagostomum radiatum* and *Trichuris discolour*. Vet Parasitol 108:75–81.

Williams AR et al (2014) Assessment of the anthelmintic activity of medicinal plant extracts and purified condensed tannins against free-living and parasitic stages of *Oesophagostomum dentatum*. Parasites Vectors 7:518.

Zhao L et al (2015) Genetic differences between *Chabertia ovina* and *C. erschowi* revealed by sequence analysis of 4 mitochondrial genes. Mitochondrial DNA 26:167–170.

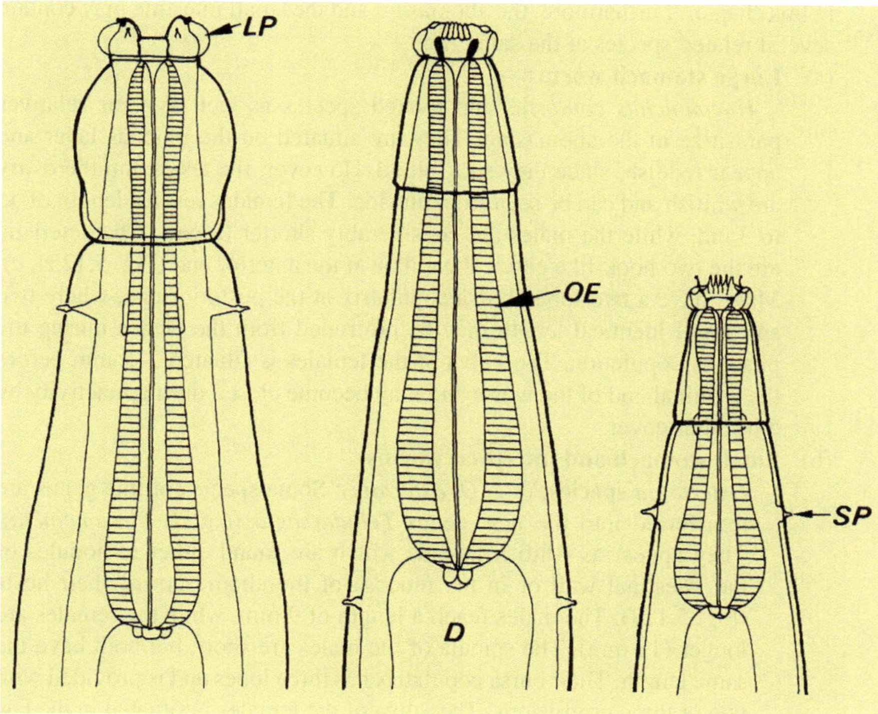

Fig. 5.125 Diagrammatic representation of the anterior end of *Oesophagostomum* species. From *left*: *O. radiatum* (cattle); *O. venosum* (sheep, goat); *O. dentatum* (pigs). D = intestine; LP = lips; OE = oesophagus; SP = sense papilla

5.3.3.16 Trichostrongylids and Relatives

1. **Name**: Greek: *thrix, thrichos* = tiny hair; *strongylos* = rounded; *nema* = thread; *onkos* = hook; *phorein* = to bear. Latin: *contortus* = twisted; *circum* = around; *cinctus* = surrounded by a belt; *tela* = tissue; *curtus* = shortened; *colubriformis* = snake-like; *filicollis* = with a small, fine neck; *helvetianis* = from Swiss; *dorsum* = back; Robert von Ostertag and C.F. Cooper = a German and an English helminthologist.
2. **Geographic distribution/epidemiology**: Worldwide. The selected nematodes parasitize worldwide in plant feeding vertebrates with the exception of fishes. They live—depending on the species—in the stomach or in the small intestine. Massive infections may introduce severe symptoms of disease (Figs. 5.123a, c, 5.124 and Table 5.6).
3. **Biology, morphology**:

   Species parasitizing in ruminants:

   The different species of the genus *Trichostrongylus* live in the abomasum of their hosts. Since the clinical symptoms are not species specific and since their morphology is also not very different, different groups are considered together in this chapter. Furthermore, the abomasum and the small intestine may contain several related species at the same time.

(a) **Large stomach worms**:

Haemonchus contortus and related species as well as their relatives parasitize in the abomasum. They are situated on the mucous layer and appear reddish, since they suck blood. However, the testes and the ovary are whitish and can be seen from outside. The females reach a length of up to 3 cm, while the males are considerably shorter (2 cm). Characteristic are the two hook-like cervical papillae at the anterior end (Fig. 5.123a, c). Males have a two-lobed bursa copulatrix at the posterior end, where two spicula of identical length may be protruded from the cloaca during the phase of copulation. The vulva of the females is situated 3–5 mm before the terminal end of the worm and may become closed during inactivity by a peculiar cover.

(b) **Small stomach and intestinal worms**

- ***Ostertagia* species**: (e.g. *O. ostertagi*): Some species of this genus are transferred into the new genus *Teladorsagia* (e.g. *T. circumcincta*). They appear as whitish threads which are found either in nodules of the intestinal wall or in the mucosa of the abomasum of their hosts (Fig. 5.127). The males reach a length of 9 mm, while the females are longer (12 mm). The spicula of the males are short, but both have the same length. Their bursa copulatrix has three lobes and is provided with two or three protrusions. The vulva of the females is situated in the last fifth of the body. Depending on the species a vulva cover is present or not. The larva 3 is diagrammatically depicted in Fig. 5.124C).
- ***Cooperia* species**: (e.g. *C. oncophora*; (mainly) cattle; *C. curticei* (mainly) sheep) are very tiny, reddish appearing worm being characterized by their spirally rolled-in anterior pole. Their cuticle shows clear ring-like protrusions. The vulva of the females, which have as the males a length of 9 mm, is situated in the last quarter of their body. These worms parasitize in the small intestine and are rarely found in the abomasum. The species-specific tail of the larva 3 is diagrammatically depicted in Fig. 5.124D.
- ***Trichostrongylus* species** (e.g. *T. axei* in the abomasum, *T. colubriformis* in the small intestine) appear as very thin, reddish-brownish, anteriorly pointed worms with a length of 2.5–8 mm. Their cuticle appears with ring-like waves. The different species are not host specific and can be found in many ruminants. The two spicula of the males are mostly different in size. The vulva of the females is situated species specifically in the posterior half of the body. The larva 3 is diagrammatically depicted in Fig. 5.124B).
- ***Nematodirus* species** (e.g. *N. batti* (sheep), *N. filicollis, N. helveticus*) parasitize inside the small intestine. They appear filament like, and their anterior body has a longitudinal protrusion of the cuticle. The males measure 17 mm in length, while the females reach even 25 mm.

The spicula of the males are comparatively rather long reaching 1.5 mm. The vulva is found in the posterior half of the body. The terminal end of the larva 3 is depicted in Fig. 5.124I).

The females of the trichostrongylids excrete eggs in the morula status (Fig. 5.121f; B3 in the annex). Outside of the body, the larva 1 hatches from the eggshell and grows up via two moults to the larva 3, which is enclosed in the detached cuticle (=sheath) of the larva 2. The larva 3 are very sensitive against high temperatures and dryness but survive in moderate temperatures and even in winter. After oral uptake, the larva 3 is developed via two further moults within 3 weeks into the adult male or female. In some species, however, **hypobiosis** occurs, if they are taken up in autumn. In this case, the larva 3 does not proceed its further development but stays there unchanged until spring time. Then the adults are developed and the eggs have better chances to survive outside of the hosts. However, egg production is increased in springtime (**springrise phenomenon**): apparently induced by host hormone activity.

Trichostrongylids in further animals (examples):

(a) **Horses**: *T. axei*: ♀ 8 mm, ♂ 5 mm; stages occur in the stomach and have a direct development.

(b) **Pigs**: *Hyostrongylus rubidus*: ♀ 12 mm, ♂ 7 mm. These worms live in the stomach, suck blood and thus appear reddish; their development is direct.

(c) **Rabbits**: *Graphidium strigosum*: ♀ **20 mm,** ♂ **15 mm**; *Obeliscoides cuniculi*. Both species live in the stomach; *T. retortaeformis* is found in the small intestine.

(d) **Birds**: *Trichostrongylus tenuis*: ♀ 9 mm, ♂ 6 mm; they parasitize in the small intestine and caeca of their hosts and appear reddish. Their development is direct (=without hypobiosis).

4. **Symptoms of disease**: The severity of the symptoms depends on the amount of worm stages inside a host and on the fact whether the specimens of a given species suck blood or not. Severe symptoms occur often in young animals (=during the first period on the meadow in the case of ruminants). Since mostly mixed infections with several species occur, clear and specific symptoms are rare. However, apathia, watery and bad smelling diarrhoeas, loss of weight and hair are common. Oedemas in the throat region, breast and belly are common, too. The latter symptoms are indications of hydraemia, anaemia, and anorexia. Lymph node swellings, increased pulse, breathing problems, fever and intense anaemia (due to bloodsucking species) are symptoms in cases of mass occurrence of worms inside their hosts. Especially young animals may die during such infections. **Attention**: Some of the trichostrongylid species may also infect humans!
5. **Diagnosis**: Microscopical demonstration of the eggs after use of enrichment methods (flotation; M.I.F.C.; S.A.F.C.).
6. **Pathway of infection**: Oral by uptake of sheathed larvae inside food. In the case of the thick-walled eggs of *Nematodirus* species, the larva 3 develops

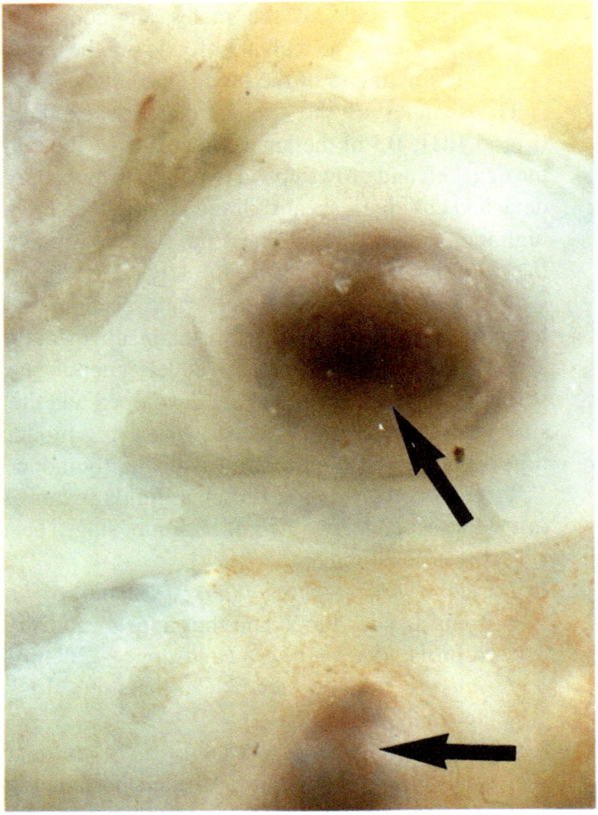

Fig. 5.126 Macrophoto of nodules (*arrows*) in the intestinal mucous layer of a cow, harbouring larvae 3 of *Oesophagostomum* sp

inside the egg and hatch occurs in springtime only after a phase of low temperature.

7. **Prophylaxis**: See therapy: avoid to overcrowd meadows. Recently vaccines have been developed.
8. **Incubation period**: 18–24 days or several months due to overwintering (as hypobiosis).
9. **Prepatent period**: 18–24 days in the case of *Haemonchus*, 18–23 days in the case of *Ostertagia*, 17–21 days in *Trichostrongylus*,14–22 days in *Cooperia*, 15–26 days in *Nematodirus*, 21–32 days in *Marshallagia* and 16–18 days in *Skrjabinagia*.
10. **Patency**: Species specific: reaching from a few days (18–32) in the case of *Nematodirus* species up to months or even more than 1 year in *Trichostrongylus* species.
11. **Therapy**: Due to constant use of anthelmintics, many parasite groups have developed resistances, which are differently intense in the different regions around the world. Thus, it is needed to check the status in each country. Resistances occur especially against **benzimidazoles**. Recently treatment was

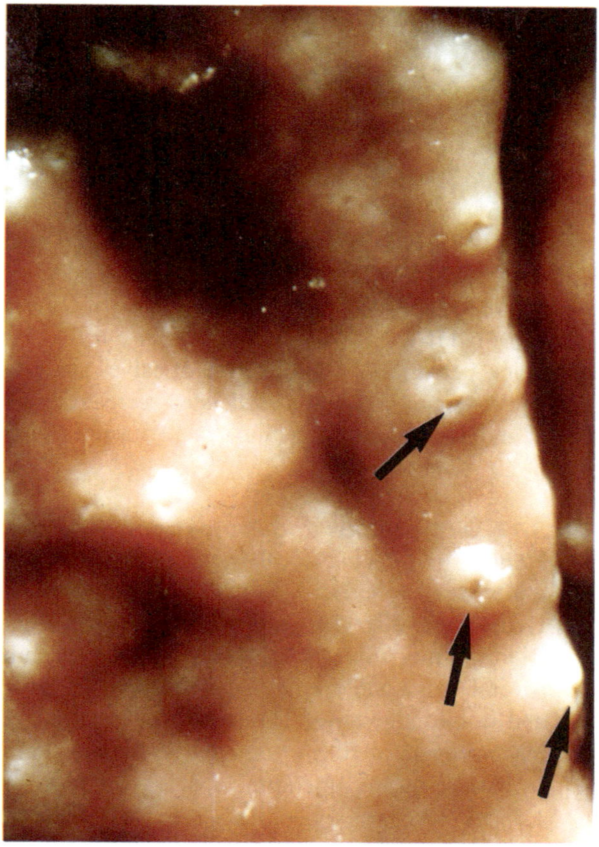

Fig. 5.127 Macrophoto of nodes (*arrows*) due to larvae of *Ostertagia ostertagi* in the intestinal wall of a sheep

established with macrocyclic lactones (as bolus, spot-on, etc.), e.g. ivermectin, doramectin, moxidectin and eprinomectin.

Further Reading

Chaudhry U et al (2015) Genetic evidence for hybridisation between *Haemonchus contortus* and *Haemonchus placei* in natural field populations and its implication for interspecies transmission of anthelmintic resistance. Int J Parasitolo 45:149–154.

Dever ML et al (2015) Persistent challenge with *Trichostrongylus colubriformis* and *Haemonchus contortus* larvae does not affect growth of meat-breed lambs, suppressively treated with anthelmintics when grazing. Vet Parasitol 209:76–83.

Hunter III JS et al (2013) The efficacy of eprinomectin extended-release injection against naturally acquired nematode parasites of cattle, with special regard to 4th stage inhibited *Ostertagia* larvae. Vet Parasitol 192:346–352.

Kaminsky R et al (2008) A new class of anthelmintics effective against drug-resistant nematodes. Nature 452:176–180.

Karanikola SN et al (2015) Development of a multiplex fluorescence immunological assay for the simultaneous detection of antibodies against *Cooperia oncophora, Dictyocaulus viviparus* and *Fasciola hepatica* in cattle. Parasites Vectors 8:335.

Khadija S et al (2013) Soil moisture influences the development of *Haemonchus contortus* and *Trichostrongylus colubriformis* to third stage larvae. Vet Parasitol 196:161–171.

Silva MRI et al (2015) Host-specificity and morphometrics of female *Haemonchus contortus, H. placei* and *H. similis* in cattle and sheep. J Helminthol 89:302–306.

Verschave SH et al (2014) The parasite phase of *Ostertagia ostertagi*. Int J Parasitol 44:1091–1104.

Vlaminck J et al (2015) Vaccination of calves against *Cooperia oncophora* with a double-domain activation-associated secreted protein reduces parasite egg output and pasture contamination. Int J Parasitol 45:206–213.

5.3.3.17 *Dictyocaulus* and *Protostrongylus* Species (Lungworms)

1. **Name**: Greek: *diktyon* = net; *kaulos* = stalk; *protos* = first; *strongylos* = rounded.
2. **Geographic distribution/epidemiology**: Worldwide.
3. **Biology, morphology**: In the lungs of ruminants, horses and rabbits/hares, several nematode species occur, which are roughly differentiated in **large** and **small** lungworms. The large ones (e.g. *Dictyocaulus* species; Fig. 5.128) reach a length of 10 cm, while the small ones vary in size (e.g. *Protostrongylus* species: 2–4 cm; *Muellerius* species: 1–3 cm; *Neostrongylus* species: up to 1.5 cm; *Cystocaulus* species: 3–5 cm). The species determination is very difficult when based on morphological criteria but not needed as basis for a successful treatment (Table 5.6).

   The following features are common to all lungworms:

   – The adult worms have their sites in lung alveoli and in the bronchial system, where they are situated within nodes, or in the trachea (Fig. 5.128).

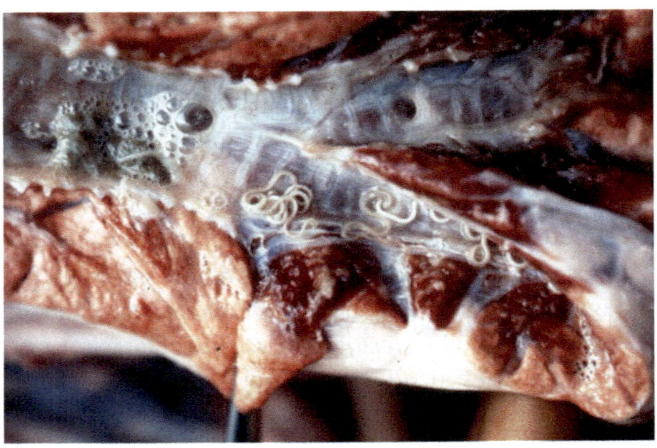

Fig. 5.128 Macrophoto of developmental stages of *Dictyocaulus viviparus* in the opened lung of a calf

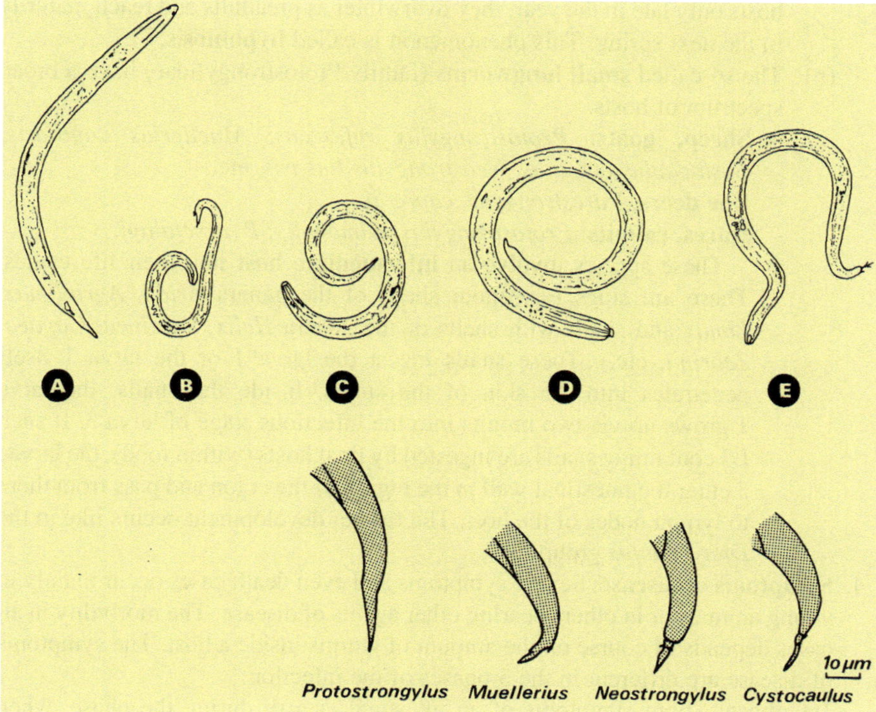

Fig. 5.129 Diagrammatic representation of the larvae (in toto) and their hind ends of different species of lungworms (after Soulsby). (**A**) *Dictyocaulus viviparus*; (**B**) *Muellerius capillaris*; (**C**) *Protostrongylus rufescens*; (**D**) *Dictyocaulus filaria*; (**E**) *Cystocaulus ocreatus*

– The females excrete fully embryonated eggs, from which the larva 1 hatches already inside the lung. Thus, these larvae occur later mainly in the feces and (rarely) in the sputum. According to their morphology, the different species can be diagnosed (Fig. 5.129).

The large and small lungworms have a different life cycle:

(a) **Large lungworms**:

Dictyocaulus species (e.g. *D. viviparus*—cattle; *D. filaria*—sheep, goats; *D. arnfieldi*—horses; *D. eckerti*—roe deers, stags) proceed a life cycle **without intermediate hosts**. The larvae 1 grow up outside the body within 6–10 days and without food uptake into the larva 3 stage, which is sheathed (=sticking side the second larval cover). After oral uptake within food, these larvae 3 reach the small intestine which is left in order to enter via lymph ductules the lymph nodes, where they reach the 4th larval stage. Within lymph, these larvae 4 are transported via the ductus thoracicus and right heart chamber into the lung where they settle inside the bronchioles. There they reach maturity after a further moult. If such larvae reach their hosts only late in the year, they overwinter as preadults and reach maturity in the next spring. This phenomenon is called **hypobiosis.**

(b) The so-called **small lungworms** (family Protostrongylidae) have a broad spectrum of hosts.
   - **Sheep, goats**: *Protostrongylus rufescens; Muellerius capillaris; Cystocaulus ocreatus; Neostrongylus linearis*, etc.
   - **Roe deers**: *Varestrongylus capreoli.*
   - **Hares, rabbits**: *Protostrongylus pulmonalis; P. oryctalagi.*

   These species involve an **intermediate host** into their life cycles. These are slugs (=without shell) of the genera *Arion, Agriolimax, Limax* and snails with shells of the genera *Helix, Succinea, Capaea, Zebrina,* etc.). These snails ingest the larva 1 or the larva 1 itself penetrates into the skin of the snails. Inside the snails, the larva 1 grows up via two moults into the infectious stage of larva 3. If such L3 containing snails are ingested by final hosts (within food), the larvae 3 enter the intestinal wall in the region of the colon and pass from there to lymph nodes of the host. The further development occurs like in the *Dictyocaulus* group.

4. **Symptoms of disease**: Severe symptoms and even death cases occur mainly in young animals or in others bearing other agents of disease. The morbidity in all cases depends of course on the amount of worms inside a host. The symptoms of disease are different in the 3 phases of the infection:

   1st phase: Slight symptoms of an intestinal catarrh during the phase, when larvae penetrate the intestinal wall.

   2nd phase: More significant symptoms occur: coughing with rattling noises, nose fluids, increased speed of heart beats and breathing. High fever (41 °C), lung oedemas, eventually pneumonia, ascites, eventually death.

   3rd phase: Chronical coughing, reduced food uptake, loss of weight, occasional diarrhoeas; eventually reduction of clinical symptoms due to growing immunity.

5. **Diagnosis**: Demonstration of larvae 1 in the feces with the help of the Baermann funnel system or in the mucus of the trachea. Species diagnosis: Fig. 5.129.
6. **Pathway of infection**: Oral.
   (a) *Dictyocaulus* species: Direct ingestion of larvae within grass food.
   (b) Protostrongylidae: Oral uptake within food of infected intermediate hosts (snails), which contain the infectious larva 3.
7. **Prophylaxis**: Such measurements make only sense in the case of the large lungworms, since they have a rather long lifetime.
   (a) Calves should be kept on **separate meadows** (larvae 3 may overwinter!).
   (b) **Pasture rotation.**
   (c) **Pasture hygienic measurements**: e.g. melioration, drainage, etc.
   (d) **Vaccination**: Using lungworm vaccine.

(e) **In stables**: Exchange of straw in short intervals; keeping floors dry.
(f) **Metaphylactic treatment** of animals in stables during winter time.
(g) **Chemoprophylactic treatment**. Using anthelmintics in endangered regions about 6–8 weeks after the animals had been placed onto the meadow and repetition of treatment in July/August.

8. **Incubation period**: Depending on the season. In cases of first infections in springtime: within the first 5–7 days (=invasion phase of larvae). In the case of the first infection in autumn, symptoms occur mostly only after several months, mostly in next springtime due to the phenomenon of **hypobiosis**.
9. **Prepatent period**: In general species specific.
   (a) **Large lungworms**: The excretion of larvae by their hosts starts 24–28 days after infection. In case infections occur in autumn, larval excretion starts in springtime.
   (b) **Small lungworms**: 4–5 weeks in case of *Protostrongylus* species; 1 month in *Muellerius* species and *Cystocaulus ocreatus*; 2 months in *Neostrongylus* species.
10. **Patency**: Species specific:
    (a) **Large lungsworms**: about 2–3 months (rarely up to 6 months).
    (b) **Small lungworms**: about 2 years in *Protostrongylus* species and *Neostrongylus* species, but up to 5–6 years in the case of *Cystocaulus ocreatus* and *Muellerius* species.
11. **Therapy**: Use of macrocyclic lactones (e.g. **ivermectin**: 1 × 0.2 mg/kg bodyweight; **moxidectin**: 1 × 0.4 mg/kg bodyweight; **fenbendazole**: 5 days 1–2 × 10 mg/kg bodyweight).

Further Reading

Bauer BU et al (2010) Comparison of the FLOTAC technique with the McMaster method and the Baermann technique to determine counts of *Dictyocaulus eckerti* L1 and strongylid eggs in faeces of red deer (*Cervus elaphus*). Parasitol Res 107:555–560.

Bloemhoff Y et al (2015) Prevalence and seasonality of bulk milk antibodies against *Dictyocaulus viviparus* and *Ostertagia ostertagi*. Vet Parasitol 209:108–116.

Gethings OJ et al (2015) Asynchrony in host and parasite phenology may decrease disease risk in livestock under climate warming: *Nematodirus battus* in lambs as a case study. Parasitology 142:1306–1317.

Jabbar A et al (2013) The mitochondrial genome of *Protostrongylus rufescens*—implications for population. Parasites Vetors 6:232.

Kuchboev AE, Krücken J et al (2016) Infection levels of protostrongylid nematodes in definite caprine and intermediate gastropod hosts from Uzbekistan. J Helminthol

Lesage C et al (2014) *Protostrongylus pulmonalis* and *P. oryctolagi* parasites of the lungs of the European hare. Parasitol Res 113:2103–2111.

Manga-Gonzales MY, Morrondo-Pelago MP (1994) Larval development of ovine *Neostrongylus linearis* in four experimentally infected mollusc species. J Helminthol 68:207–210.

Ranjan S et al (2010) Evaluation of the persistent activity of moxidectin against *Dictyocaulus viviparus, Haemonchus placei, Trichostrongylus axei* and *Oesophagostomum radiatum* in cattle. Vet Parasitol 167:50–54.

Sevimli FC et al (2011) Efficacy of eprinomectin pour-on treatment in sheep naturally infected with *Dictyocaulus filaria* and *Cystocaulus ocreatus.* J Helmintol 85:472–475.

Strube C et al (2015) Vaccination with recombinant paramyosin against the bovine lung worm *Dictyocaulus viviparus.* Parasites Vectors 8:119.

Vadlejch J et al (2016) Efficacy and persistent activity of moxidectin against natural *Muellerius capillaris* infections in goats and pathological consequences of muelleriosis. Vet Parasitol 218:98–101.

Van Dijk J, Morgan ER (2012) The influence of water and humidity on the hatching of *Nematodirus battus* eggs. J Helminthol 86:287–292.

Veneziano V et al (2011) Efficacy of eprinomectin pour-on against *Dictyocaulus arnfieldi* infection in donkeys (*Equus asinus*). Vet J 190:44–415.

5.3.3.18 *Metastrongylus* Species (Lungworms of Pigs)

1. **Name**: Greek: *meta* = after; *strongylos* = rounded, cylindrical. Latin: *pudendum* = region of sexual organs; *tegere* = covering; *elongatus* = prolongated; *salmo* = salmon; *confusus* = unclear, mixte; *asymmetricus* = not symmetric.
2. **Geographic distribution/epidemiology**: Worldwide.
3. **Biology, morphology**:

   In Europe in pigs five species of the genus *Metastrongylus* (*M. elongatus* (syn. *M. apri*), *M. pudendotectus, M. salmi, M. confusus, M. asymmetricus*) occur. They can be differentiated from each other only by specialists. However, the exact species determination is not needed for treatment. The males reach a length of 1.5–2.5 cm, while the females are larger (2.5–5.5 cm). They are found as clots in small and middle-sized regions of the bronchioles. The bursa copulatrix of the females is rather small and is subdivided into three lobes. The spicula of the males are provided with hooks at their terminal free end. The development is indirect. The eggs measure 55 × 40 μm (Fig. 5.130), are protected by a rather thick wall and contain already the larva 1, when they are excreted. These eggs have to become ingested by **intermediate hosts** such as earthworms (e.g. *Lumbricus terrestris*). In the body of the intermediate host, the larva 1 moults twice to reach the infectious larva 3 stage, which may remain infectious for years (Fig. 5.131). As soon as final hosts ingest such intermediate hosts, the larvae leave the intestine via lymph vessel, pass the right chamber of the heart and enter finally the lung with its bronchial tubes.
4. **Symptoms of disease**: Very common are swellings of the mesenterial nods during the body migration of the larvae 3. Furthermore, bronchitis, pneumonia

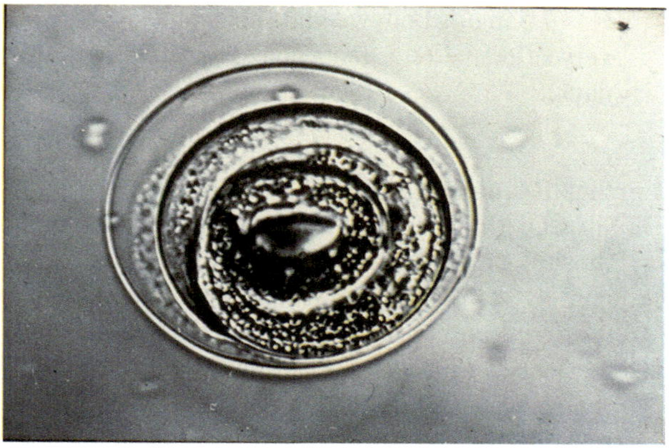

Fig. 5.130 Light micrograph of an egg of a lung worm (*Metastrongylus elongatus*) of pigs

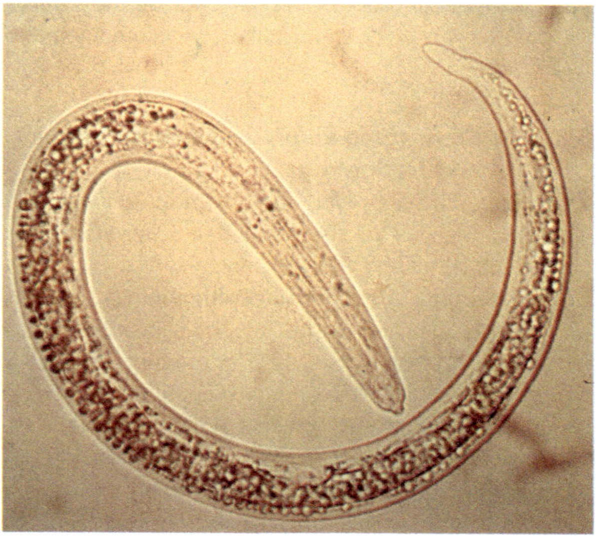

Fig. 5.131 Larva 3 of the nematode *Metastrongylus elongatus* (which stays e.g. in earthworms)

and loss of weight occur. The intensity of the symptoms depends on the amount of worm stages.

5. **Diagnosis**: Microscopical demonstration of the eggs (Fig. 5.130) obtained from feces. Serological tests are also available.
6. **Pathway of infection**: Oral by uptake of infected intermediate hosts.
7. **Prophylaxis**: Keep pigs mainly in regularly cleaned stables.
8. **Incubation period**: About 10–12 days.
9. **Prepatent period**: 28–32 days.
10. **Patency**: Up to 1 year; however, after 6 months, mostly only a few eggs are excreted.
11. **Therapy**: The individual treatment against adult worms can be done with **ivermectin** (1×0.3 mg/kg bodyweight, subcutaneously) or with **levamisole**

(1 × 7.5 mg/kg bodyweight; subcutaneously or orally). Also benzimidazoles are very effective (e.g. fenbendazole: 5 mg/kg bodyweight within food for 5–15 days).

Further Reading

Da Silva D, Müller G (2013) Parasites of the respiratory tract of *Sus scrofa scrofa*. Parasitol Res 112:1353–1356.

Frontera E et al (2005) Evidence and interaction between *Ascaris suum* and *Metastrongylus apri* in experimentally infected pigs. Vet Parasitol 127:295–301.

Gasso D et al (2014) An identification key for the 5 most common species of *Metastrongylus*. Parasitol Res 113:3495–3500.

Marruchella G et al (2012) Fatal bronchopneumonia in a *Metastrongylus elongatus* and porcine circovirus type 2 co-infected pig. Res Vet Sci 93:310–312.

Poglayen G et al (2016) Lung parasites of the genus *Metastrongylus* in wild boar *(Sus scrofa)* in Central-Italy: an eco-epidemiological study. Vet Parasitol 217:45–52.

5.3.3.19 *Crenosoma vulpis* and Other Lungworms of Cats, Canids and Hedgehogs

1. **Name**: Greek: *soma* = body. Latin: *crenatus* = with notches; *vulpes* = fox; *meles* = genus of the Mustelidae; *aerophilus* = liking fresh air; *vas* = little vessel; *abstrusus* = not regular.
2. **Geographic distribution/epidemiology**: *Crenosoma vulpis* is found in Europe, Asia and North America parasitizing in red foxes, polar foxes, marten dogs, wolves and dogs. 40 % of red foxes in some regions of Central Europe are infected. Up to 60 % of dogs coming to vet labs show symptoms of respiratory problems and were infected with *C. vulpis*. Further *Crenosoma* species are *C. goblei* in racoons, *C. melesi* in badges and *C. striatum* in hedgehogs.
3. **Biology, morphology**:

   Selected species in carnivores:

   (a) ***Crenosoma vulpis***: Adults (♀ = 1.5 cm; ♂ = 0.8 cm) live in the bronchioles and trachea; snails are intermediate hosts.

   (b) ***Capillaria aerophila***: Adults (♀ = 3 cm; ♂ = 2–5 cm) live in the trachea, bronchioles and nose cavities. They have a direct life cycle without intermediate hosts.

   (c) ***Aelurostrongylus abstrusus***: Adults (♀ = 1 cm; ♂ = 0.7 cm) live in the alveoli of the lung and in bronchioles. Snails are intermediate hosts.

   (d) ***Angiostrongylus vasorum***: Adult females reach a length of 2–5 cm (males 2.0 cm), appear reddish and parasitize in the arteria pulmonalis and in the right heart ventricle.

   Species in hedgehogs:

   Crenosoma striatum and several related species can be found in large numbers in the bronchioles. The adults reach a length of 1.6 cm as females and 5–7 mm as males. Due to their diameters of only 0.4 mm and their white-reddish bodies, they can be easily found in hedgehogs. Characteristic is furthermore that their cuticle shows a cross striation of the cuticle at the anterior pole. The females depone a large number of eggs, which reach a size of 70 μm × 36 μm, have an ovoid shape and contain already the larva 1 when being excreted. This larva gets mostly already free inside the trachea and reaches within slime the oesophagus and thus is finally excreted within the feces. The larvae obtained by fecal examination have a length of 250–330 μm and are ingested by several species of snails (e.g. *Succinea, Agrolimax*, etc.) Via two moults within 8–10 days, the infectious larva 3 is reached and stored in the intermediate host. If hedgehogs ingest such snails, these larvae 3 penetrate their intestinal wall and reach the lungs via the bloodstream or from the interior of the pleural cavity. Final sites are the bronchioles, where they reach maturity within 8–14 days after arrival.

4. **Symptoms of disease**: Common are chronic coughing, catarrhalic symptoms, discharge of nose fluids, anaemia, pneumonia and a general bad performance. In low-grade infections, symptoms may be completely absent.
5. **Diagnosis**: Larvae can be found in feces with the help of the Baermann funnel system or in smear preparations of the saliva.
6. **Pathway of infection**: Oral by uptake of larva-containing eggs (*Capillaria*) or larvae 3 inside intermediate or transport hosts. In the case of *Filaroides* sp., dog mums may transmit such larvae to the puppies during licking.
7. **Prophylaxis**: Difficult or nearly impossible, since infections occur outdoors.
8. **Incubation period**:
   Crenosoma vulpis: about 11–14 days.
   Capillaria aerophila: about 7–10 days.
   Aelurostrongylus abstrusus: 1–2 days.
   Filaroides sp.: 1–2 days.
   Angiostrongylus vasorum: about 7 days.
9. **Prepatent period**:
   Crenosoma vulpis: about 3 weeks
   Capillaria aerophila: about 6 weeks.
   Aelurostrongylus abstrusus: 4–9 weeks.
   Filaroides sp.: about 5–10 weeks.
   Angiostrongylus vasorum: 33–36 days.
10. **Patency**:
   Crenosoma vulpis: sound data are not available, probably several months.
   Capillaria aerophila: sound data are not available: probably several months.
   Aelurostrongylus abstrusus: 4–24 months.
   Filaroides sp.: several months.
   Angiostrongylus vasorum: 5–6 years

11. **Therapy**: There is no standard cure available. In literature, some treatment schemes had been described as successful:

- **Levamisole**: 2 × daily 7.5 mg/bodyweight subcutaneously for 2 days. Repeated treatment: 2–3 times at intervals of 10 days in the case of *C. aerophila*. A treatment with 7.5 mg/kg bodyweight on 2 successive days was successful in *C. vulpis* and *A. abstrusus*.
- **Albendazole** was effective on *Filaroides* spp. at a dose of 2 × 25–50 mg/kg bodyweight (orally) for 5 days.
- **Fipronil + (S)-methoprene, eprinomectin and praziquantel** (Broadline®) acts against *Capillaria aerophila*.
- **Fenbendazole**: 20 mg/kg bodyweight for 5 days were successful in infections with *Capillaria* spp. ***Filaroides*** spp. and *Aelurostrongylus* spp. The latter species was also eliminated by a dose of 5 mg/kg bodyweight daily for 6 days. Emodepside plus praziquantel (Profender®) killed *Aelurostrongylus abstrusus* in cats.

Further Reading

Albo AM et al (2016) Seroprevalence of circulating *Angiostrongylus vasorum* antigen and parasite specific antibodies in dogs. Parasitol Res. doi:10.1007/s00436-016.

Böhm C et al (2015) Efficacy of emodepside/praziquantel spot-on (Profender®) against adult *Aelurostrongylus abstrusus* nematodes in cats. Parasitol Res 114: S155–164.

Diakou A et al (2015) Occurrence of *Aelurostrongylus abstrusus* and *Troglostrongylus brevior* in domestic cats. Parasites Vectors 8:590.

Houpin E et al (2016) Comparison of three methods for the detection of *Angiostrongylus vasorum* in the final host. Vet Parasitol 220:54–58.

Knaus M et al (2015) Efficacy of Broadline® against *Capillaria aerophila* lungworm infection in cats. Parasitol Res 114:1971–1975.

Latrofa MS et al (2015) *Crenosoma vulpis* in wild and domestic carnivores. Parasitol Res 114:3611–1617.

Magi M et al (2015) Extraintestinal nematodes of the red fox in Italy. J Helminthol 89:506–511.

Tolnai Z et al (2015) Environmental determinants of the spatial distribution of *Angiostrongylus vasorum, Crenosoma vulpis* and *Eucoleus aerophilus* in Hungary. Vet Parasitol 207:355–358.

Traversa D, Di Cesare A (2014) Cardio-pulmonary parasitic nematodes affecting cats in Europe. Frontiers Vet Sci 1:11.

Yao C et al (2011) *Filaroides osleri* (*Oslerus osleri*): two cases and a review of canid infections in North America. Vet Parasitol 179:123–129.

5.3.3.20 *Ollulanus* Species (Stomach Worms)

1. **Name**: Latin: *ollula* = small pot; *tri* = three; *cuspis* = spike; *sus* = pig; Skrjabin = Russian parasitologist.

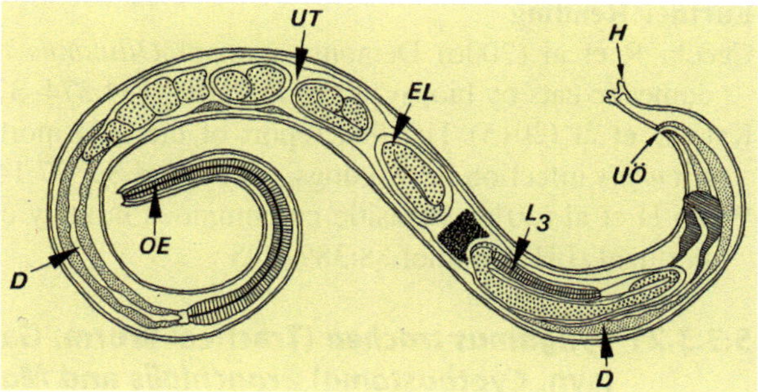

Fig. 5.132 Diagrammatic representation of a female stage of *Ollulanus tricuspis* (after Soulsby). D = intestine; EL = eggs during larva formation; H = hind horn/pike-like structure; L3 = larva 3 in an egg in the uterus; OE = oesophagus; UO = opening of the uterus; UT = uterus

2. **Geographic distribution/epidemiology**: Europe, North and South Europe, focal in many countries worldwide. Main transmission routes: cats to cats and pigs to pigs, but also infections of dogs, foxes and related species have been described.
3. **Biology, morphology**: The adult species of the species *O. tricuspis* (syn. *O. suis, O. skrjabini*) reach as females a size of 0.8–1 mm and as male 0.7–0.8 mm (Fig. 5.132). They parasitize in the mucous layer of the stomach of their hosts. They are characterized by a tiny cup-like mouth region. The females show several horn-like spikes at their terminal end (Fig. 5.132), while the males are provided there with a bursa copulatrix and two spicula which are used during copulation. The females excrete fully embryonated eggs and often already larvae 2 or even 3, which then reach maturity in the same hosts or are vomited and thus set free. If other potential hosts ingest such vomited stages, they become infected. In some animals up to 10,000 worms had been found. Infection rates might be very high. Among free-running cats, up to 40 % had been found infected and foxes up to 7 %.
4. **Symptoms of disease (Ollulanosis)**: Common vomiting, gastritis with increased slime production, formation of ulcers, loss of appetite, weakness and loss of weight.
5. **Diagnosis**: Microscopical investigation of vomited material (vomiting can be induced by application of Rompun®. Investigation of the feces mostly does not show parasitic stages.
6. **Pathway of infection**: Oral uptake of larvae 3 in vomited contents of the stomach. Self re-infections are also very common. The infectious stages may survive up to 3 weeks outside of the body.
7. **Prophylaxis**: Quick removal of vomited material.
8. **Incubation period**: A few days in the case of high-grade infections.
9. **Prepatent period**: About 5 weeks in the case with infections by uptake of larvae 3. Only a few days in cases of oral uptake of adult worms.
10. **Patency**: Years due to repeated self-infections.
11. **Therapy**: **Fenbendazole**: pigs: 5 mg/kg bodyweight orally for 3 days; cats: 50 mg/kg bodyweight on 3 consecutive days; **levamisole** was also used (5 days 5 mg/kg bodyweight subcutaneously).

Further Reading

Cecchi R et al (2006) Demonstration of *Ollulanus tricuspis* in the stomach of domestic cats by biopsy. J Comp Pathol 134:374–377.

Kato D et al (2015) The first report of the ante-mortem diagnosis of *Ollulanus tricuspis* infection in two dogs. J Vet Med Sci 77:1499–1502.

Waap H et al (2014) Parasitic communities in stray cat population from Lisbon, Portugal. J Helminthol 88:389–395.

5.3.3.21 *Syngamus trachea* (Tracheal Worm, Gape Worm), *Syngamus* (syn. *Cyathostoma*) *bronchialis* and *Mammomonogamus* Species

1. **Name**: Greek: *syn* = together; *gamos* = marriage, melting. Latin: *trachea* = air pipe; *bronchus* = branching.
2. **Geographic distribution/epidemiology**: *Syngamus* species: Worldwide in industrial chicken stables and many free-living birds. *Mammomonogamus* species occur in subtropical and tropical regions in many cattle species and buffaloes. The latter species were also described in rare cases in **humans**.
3. **Biology, morphology**:
   (a) ***Syngamus trachea*** (red tracheal worm): The adult stages reach as males a length of 0.8–1 cm and as females up to 2 cm. Both stay permanently in copula being attached with their cup-like mouthparts at the intestinal wall of their host, thus giving the aspect of a fork (Fig. 5.133). There they suck permanently blood and thus appear reddish. The females excrete the 70–-120 μm × 36–55 μm measuring eggs in a morula stage (i.e. containing about 16 cells; Fig. 5.133). The eggshell shows some hyaline thickenings at both poles. The body of the infectious larva 3 is developed within 1–2 weeks inside the egg, which are expectorated (thus reaching the outside) or are even engorged by the same host and excreted via feces. Such eggs or free larva 3 are ingested by several intermediate hosts (annelid worms, insect larvae, snails) and stored there. The infection of the final hosts (birds) occurs by ingestion of eggs, free larvae or larvae in such intermediate hosts. The larvae penetrate the intestinal wall, enter blood vessels and finally reach the lungs where they become mature within 1 week (Fig. 5.134). Egg production (after copulation) starts after another two weeks. The lifespan is short (up to 3 months). These eggs remain infectious for up to 1 year.
   (b) ***Cyathostoma* (syn. *Syngamus*) *bronchialis*.** These specimens parasitize in the trachea and the bronchioles of geese, ducks and many free-living birds. The males reach a length of 6 mm and the females of up to 3 cm. These stages do not live in permanent copulation and suck apparently less amounts of blood than *S. trachea*, since especially the males appear whitish. The developmental cycle runs similar to that of *S. trachea*. Main intermediate hosts are earthworms (e.g. *Lumbricus* stages).
   (c) ***Mammomonogamus laryngeus* (syn. *Syngamus laryngeus*)** occurs in the larynx of cattle and buffaloes in Asia and South America, while

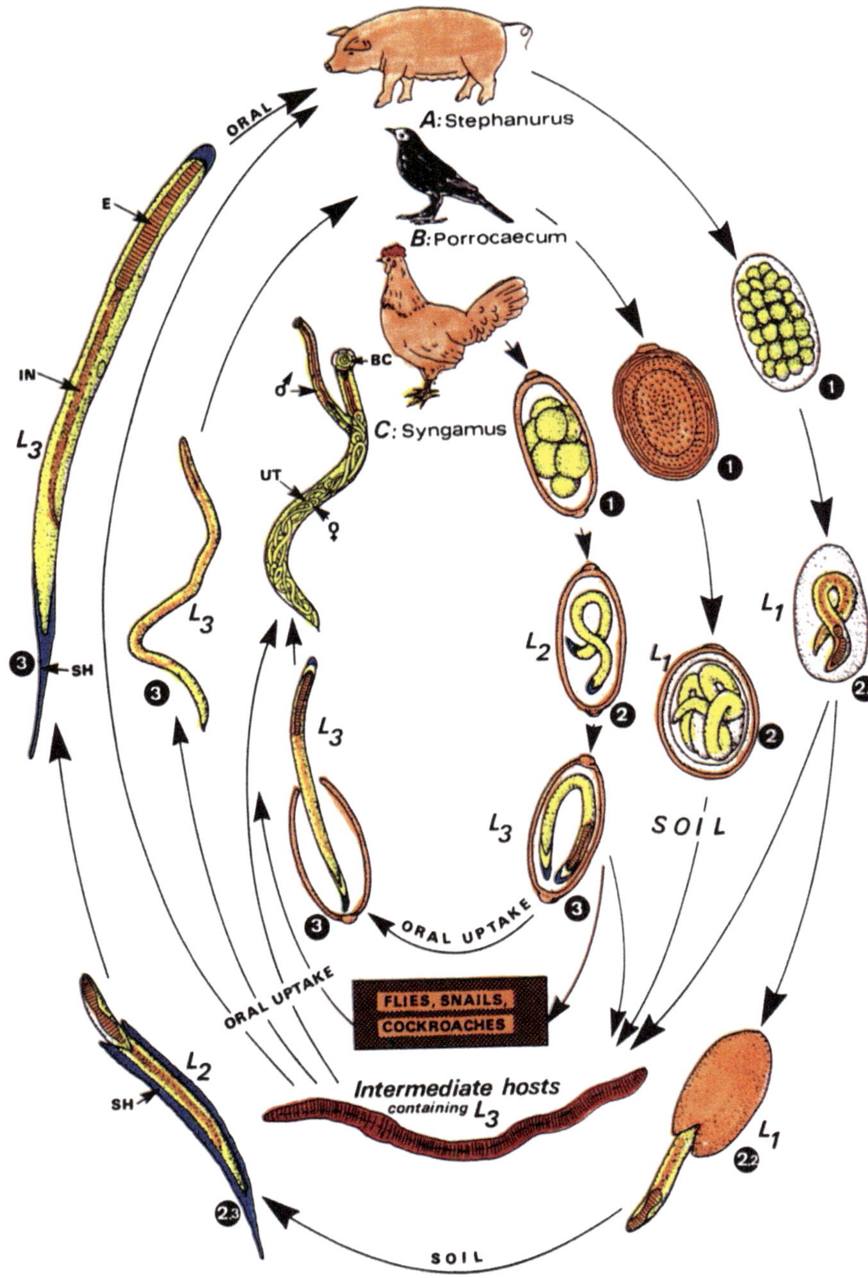

Fig. 5.133 Diagrammatic representation of the life cycles of *Stephanurus* sp. in pigs (**A**), *Porrocaecum ensicaudatum* in blackbirds (**B**) and *Syngamus trachea* in chicken and other birds (**C**). The latter parasites suck blood in both sexes living in permanent copula. **1** Excreted eggs; **2** Larval development; **3** Larva 3 and pathways of infection. BC = mouth cavity; E = oesophagus; IN = intestine; SH = sheath; UT = uterus

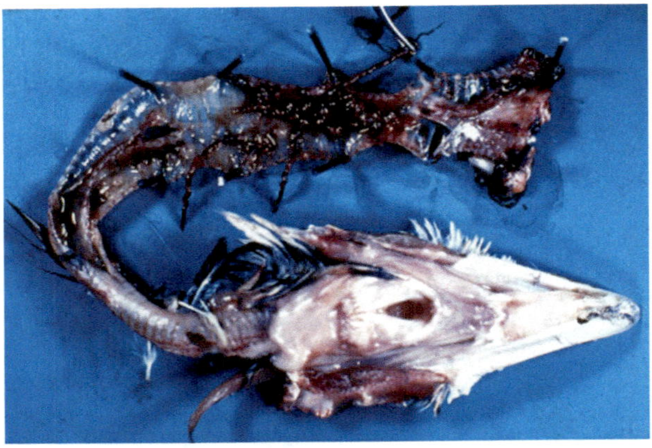

Fig. 5.134 Macrophoto of the opened trachea of a bird containing blood and *Syngamus* worms

M. nasicola (syn. *S. nasicola*) are described from the nose cavity of ruminants in Africa, South America, Philippines and Caribbean regions reaching prevalence rates of up to 40 %. In rare cases also **humans** may become infected.

4. **Symptoms of disease (Syngamidosis)**: Severe symptoms of disease occur mainly in very young animals In general, coughing, breathing problems, apathia, loss of weight, anaemia and eventually death.
5. **Diagnosis**: Demonstration of the eggs in the feces (Fig. 5.133) with the help of enrichment methods. It is characteristic that birds keep the mouth opened. In the trachea of the hosts, the worms may be seen macroscopically. After removal of the feathers, the reddish worms often shine through the skin.
6. **Pathway of infection**: Oral by uptake of larva 3 containing eggs, by ingestion of free larva 3 or by ingestion of larva-containing intermediate hosts (e.g. earthworms).
7. **Prophylaxis**: In stables: repeated, regular removal of feces; outside of stables practically impossible.
8. **Incubation period**: 3–5 days.
9. **Prepatent period**: 16–21 days.
10. **Patency**: Months, especially in old animals may stay low numbers for long.
11. **Therapy**: Fenbendazole (30–60 ppp × 7 days) and levamisole (1 × 20–30 mg/kg bodyweight).

Further Reading

Holand H et al (2013) Temporal and spatial variation in prevalence of the parasite *Syngamus trachea* in a meta-population of house sparrows (*Passer domesticus*). Parasitology 140:1275–1286.

Gethings OJ et al (2015) Spatial distribution of infectious stages of the nematode *Syngamus trachea* with pheasants (*Phasianus colchius*). Vet Parasitol 212:267–274.

Kanarek G (2009) The occurrence of *Cyathostoma microspiculum* in the great cormorant (*Phalacrocorax carbo*). J Helminthol 83:391–398.

Kanarek G et al (2013) *Cyathostoma phenisci* a parasite of the respiratory tract of African penguin (*Spheniscus demersus*). Parasitol Int 62:416–422.

5.3.3.22 *Dioctophyme renale, Hystrichis tricolor* and *Stephanurus dentatus* (Kidney Worms)

1. **Name**: Greek: *diktyon* = net; *phyme* = remnants; *hystrix* = bristle. Latin: *renalis* = belonging to the kidney: *dentatus* = with teeth; *tricolor* = three colours.
2. **Geographic distribution/epidemiology**: Worldwide.
3. **Biology, morphology**:
   (a) ***Dictyophyme renale***: Adult worms (♀ = 100 cm × 1 cm; ♂ = ~ 35 cm × 0.4 cm) appear blood red and live in the kidneys (Fig. 5.135) of dogs, cats, foxes, mustelids, wolves, etc. The typical eggs (Fig. 5.136) measure about 75 × 50 μm and are excreted within the urine in a two-cell stage. Outside of the body (in freshwater, humid soil), the larva develops inside. However, it hatches only when the eggs have been ingested by intermediate hosts (oligochaetes, earthworms, etc.). It takes rather long

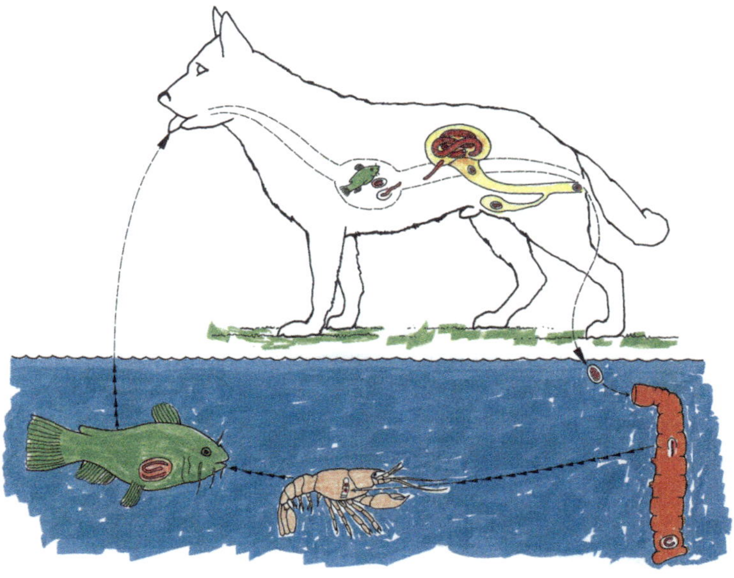

Fig. 5.135 Diagrammatic representation of the life cycle of the kidney worm *Dictyophyme renale*. The blood red appearing adults live in the kidneys and excrete unembryonated eggs. In sweet water the larval development occurs inside the egg. If annelids or crustaceans ingest such eggs, the larva 3 hatches and may also be transported into fish, if they feed infected intermediate hosts. Infection of final hosts occurs as soon as they ingest such intermediate hosts

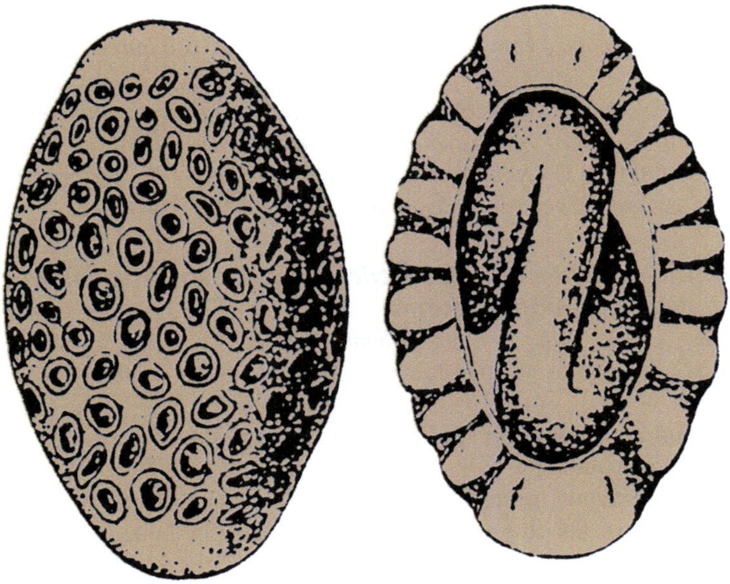

Fig. 5.136 Diagrammatic representation of aspects of the surface and the contents of eggs of *D. renale*

until the larvae have reached infectivity. The **final hosts** mainly become infected by ingestion of infected annelids. However, also paratenic hosts such as fishes or frogs may be involved in case they have ingested infected annelids. If such paratenic hosts are ingested by dogs (or even by humans!), the larva 3 enters the abdominal cavity and penetrates into the kidneys (but also into the pleural cavity), where maturity is reached.

(b) ***Hystrichis tricolor***: These nematodes are found in the gland stomach of ducks, geese and fish feeding birds in Europe.

(c) ***Stephanurus dentatus*** is the kidney worm of pigs in subtropical and tropical regions. The adults live in cysts in the kidneys and in the wall of the ureter and bladder. The infection occurs directly by skin-penetrating larvae 3 or by oral uptake of infected transport (intermediate) hosts such as earthworms. *S. dentatus* may induce severe damages in pigs.

4. **Symptoms of disease**: Haematuria is common, and secondary bacterial infections occur in addition at mechanically produced lesions.
5. **Diagnosis**: Microscopical demonstration of eggs inside urine (after filtering or centrifugation).
6. **Pathway of infection**: Oral during uptake of larva-containing raw fish or larvae in annelids.
7. **Prophylaxis**: Avoidance to feed raw fish.
8. **Incubation period**: 2–4 weeks.
9. **Prepatent period**: 3–6 months.
10. **Patency**: 1–3 years.

11. **Therapy**: Chirurgical removal of the worms. A registered chemotherapy is unknown; however, treatment with **ivermectin** (0.3 mg/kg bodyweight) or **doramectin** (0.3 mg/kg bodyweight) or **fenbendazole** had been shown to be effective.

Further Reading

Sato H et al (2008) Visceral helminths of wild boars (*Sus scrofa leucomystax*) in Japan with special reference of the genus *Morgascaridia*. J Helminthol 82:159–168.

Stewart TB et al (1996) Doramectin efficacy against the kidney worm *Stephanurus dentatus* in sows. Vet Parasitol 66:95–99.

Von Brandt T (1957) Glycogen and lipoids in *Dictyophyme renale*. Z Tropenmed Parasitol 8:21–23.

5.3.3.23 Filariae (Superfamily Filaroidea)

1. **Name**: Latin: *filum* = filament; *oides* = similar; *dirus* = cruel; *seta* = bristle. Greek: *onkos* = hook; *kerkos* = tail; *para* = similar.
2. **Geographic distribution/epidemiology**: See Table 5.9.
3. **Biology, morphology**: This superfamily contains several members of the two families Filariidae and Onchocercidae. All members of these families are transmitted during the bites of bloodsucking insects (Fig. 5.137). This may happen mechanically from wound to wound or due to actions of true (typical) intermediate hosts (bloodsucking ectoparasites), which ingest the larva 1 (microfilaria) during their blood meal and inject the infectious larva 3 during one of the subsequent blood meals. This larva 3 grows finally up and becomes the fertile **macrofilaria**. Inside the hosts, the filariae parasitize in species-specific organs (Table 5.9).
   (a) ***Onchocerca cervicalis*** lives in the subcutaneous tissues of horse and related species and reaches a size of up to 30 cm as females and 20 cm as males. *Onchocerca reticulata* is much longer (♀ = 75 cm, ♂ = 20 cm) and is mainly found in the musculus interosseous and sinews. The non-sheathed microfilariae of *O. cervicalis* reach a length of 200–240 μm, while those of *O. reticulata* measure about 330–370 μm in length (Fig. 5.141B). Vectors are midges of the genus *Culicoides*. Worms of the species *Elaeophora boehmi* occur in the blood vessels of the feet. Their life cycle is, however, unknown.
   (b) ***Onchocerca gutturosa*** parasitizes in cattle and the adults reach as females a length of 60 cm and as males 3 cm. They parasitize in the neck skin region. The microfilaria are sheathed and measure 230–280 × 7 μm. They are finally located in the skin of the neck as well as in that of belly and feet. From there, they were ingested by bloodsucking simuliids of the genus *Odagmia* and by midges of the genus *Culicoides*.
   (c) ***Dirofilaria* species**: ***D. immitis*** parasitizes in canids and cats inside the right heart chamber and in the arteria pulmonalis (Figs. 5.138, 5.139 and

Table 5.9 Important filarial species, their hosts and mode of transmission

Genus/species	Final hosts	Habitat inside final hosts		Microfilariae with (+) or without (−) sheath	Intermediate hosts/vectors
		Adults	Microfilariae		
Onchocerca gutturosa	Cattle/worldwide	Neck bands	SCT	−	Blackflies
O. linealis	Cattle/worldwide	Stomach/ dorsal bands	SCT	−	Blackflies
O. gibsoni	Cattle/Asia, Africa, Australia	Subcutis	SCT	−	Midges
O. ochengi	Cattle/Africa	Subdermal nodes	SCT	−	Blackflies
O. cervicalis	Horses, donkeys/worldwide	Neck bands	SCT	−	Midges
O. reticulata	Horses, donkeys East/South Europe	Tendons	SCT	−	Midges
O. lupi	Canids/USA, Europe	Subcutis, eye	SCT	−	?
Dirofilaria immitis	Canids/South Europe, USA	Heart, Arteria pulmonalis	Blood	−	Mosquitoes (Culicidae)
D. repens	Canids/Asia, Africa, South Europe	Subcutis	SCT, blood	−	Mosquitoes (Culicidae)
Acanthocheilonema reconditum	Canids/Europe, America, Asia, Australia	Subcutis	Blood	−	Fleas
Dipetalonema dracuncoloides	Canids/Europe, Asia, Africa	Peritoneal cavity	Blood	−	Soft tick
Acanthocheilonema viteae	Gerbils/Deserts	SCT	Blood	−	Soft tick
Litomosoides carinii	*Mastomys*, *Sigmodon* rodents/ warm countries	Pleural cavity, eye	Blood	+	*Orthonyssus* mites
Stephanofilaria species	Cattle, horses/Europe, Asia, Africa	SCT, udder	SCT	+	*Musca*, *Stomoxys* flies
Parafilaria species	Cattle, horses/Europe, Asia, Africa	SCT	Wound fluid, lymph	−	*Musca*, *Stomoxys* flies

Setaria equina	Horses/worldwide	Peritoneal cavity	Blood	+	Mosquitoes (Culicidae)
Setaria digitatus, *Setaria labiato-papillosa*	Cattle/worldwide	Peritoneal cavity	Blood	+	Mosquitoes (Culicidae)
Brugia-Arten	Carnivores, monkeys, *humans*	Lymph nodes	Blood	+	Mosquitoes (Culicidae)
Elaeophora böhmi	Horses/worldwide	SCT of metatarsus	SCT	–	Gadflies
Pelectus species	Poultry/worldwide	Connective tissues	SCT	+	Featherlings
Chanolerella species	Poultry/worldwide	Connective tissues	Blood	+	Midges
Cardiofilaria species	Poultry/worldwide	Body cavities	Blood	+	Culicidae

SCT subcutaneous connective tissue, lymph

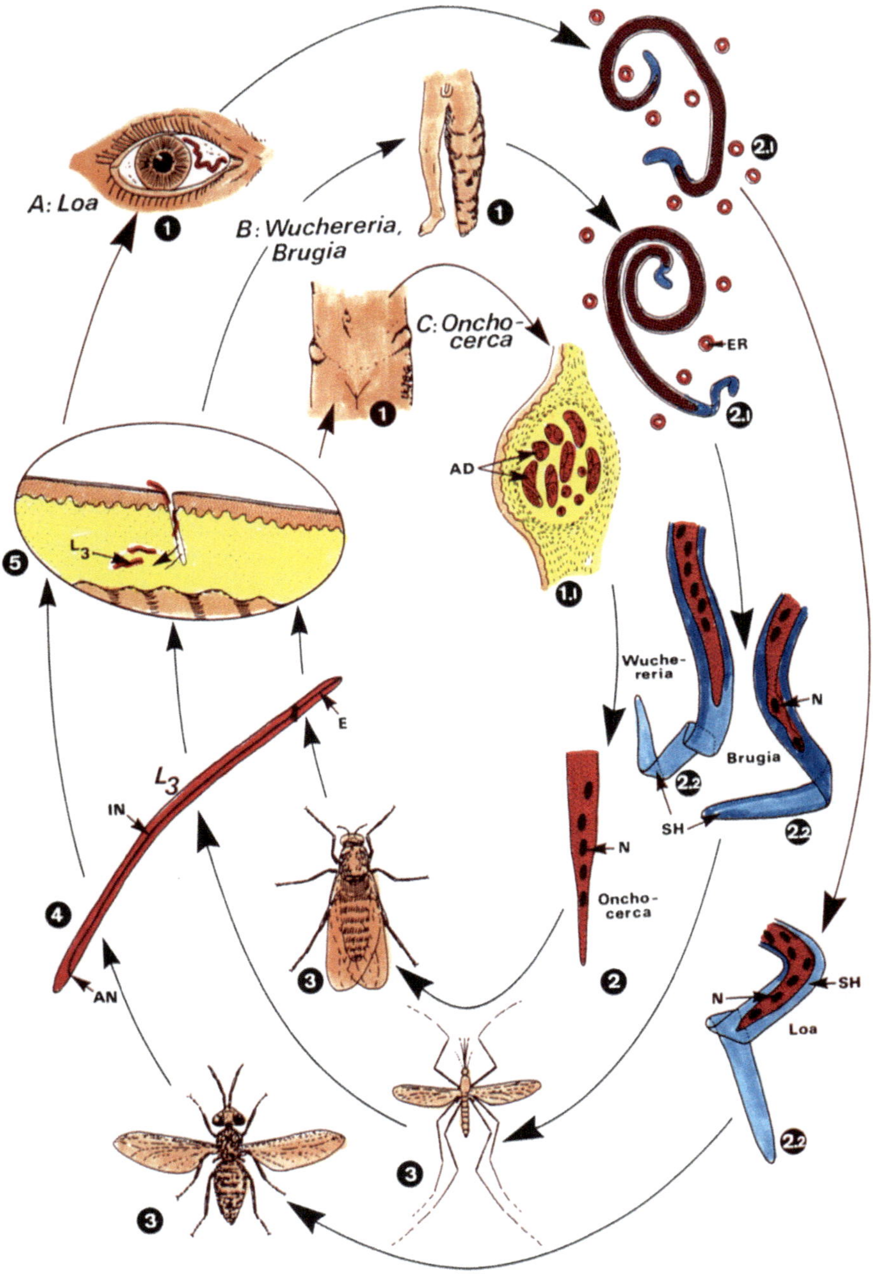

Fig. 5.137 Life cycles of filarial worms. (**A**) *Loa loa* adult worms (=microfilariae: male 3.5 cm, female 7 cm) wander subcutaneously and may pass the anterior chamber of the eye (1). (**B**) *Wuchereria bancrofti* adults (male 4 cm, female 10 cm) and *Brugia malayi* adults (male 3 cm, female 9 cm) live in lymph vessels and lead to a late-stage disease called elephantiasis tropica (1).

5.140). The females reach a length of up to 30 cm (!) and the males up to 18 cm. Their microfilariae are unsheathed, measure 305–320 μm in length and are found permanently in the peripheral blood. Most occur at 6 p.m.; only few are found at 6 a.m. Vectors are mosquitoes of the genera *Aedes, Culex* and *Anopheles*. ***D. repens*** specimens live in the subcutis of canids. Their microfilariae measure 270 × 8 μm and occur in the subcutaneous tissues. Vectors are Culicidae.

(d) ***Stephanofilaria* species**: These worms parasitize in cattle and reach as adult females 8–10 mm and as adult males 2–3 mm. They parasitize in the skin and are especially common in the surface of the udder. The microfilariae reach a length of 150 μm and are often found in the fluids of so-called "**summer wounds**", which occur as follow-up of bites of simuliids and *Stomoxys* specimens. Vectors are in addition *Musca* species.

(e) ***Parafilaria* species** parasitize in horses and cattle. The adults are rather tiny (♀ = 7 cm; ♂ = 3 cm) and live in nods inside the subcutis. The microfilariae reach a size of 300 μm × 8 μm and are often found in the fluids of so-called "**summer wounds**" (Fig. 5.141C).

(f) ***Setaria* species, *Litomosoides carinii* and *Cardiofilaria* species.** The adult worms live in the body cavity of their hosts, while the larvae occur in the blood and are transmitted—depending on the species—by featherlings, mosquitoes or mites (Table 5.9). The *Setaria* species of

Fig. 5.137 (continued) (**C**) *Onchocerca volvulus* adults (male 2–4 cm, female 70 cm) are knotted together in groups in the subcutaneous tissues. Because of host reactions these groups are encapsulated, leading to palpable nodules (1). In sections of these nodules coiled adults are seen (1.1). Microfilariae may induce blindness. **1** Visible signs of diseases. **2** Microfilariae; the long-living females produce (after copulation) thousands of first-stage larvae daily, which measure about 260 × 8 μm. Their shape (2.1), structure (2, 2.2) and diurnal occurrence are species specific: depending on the species they may or may not be sheathed (2.2); their terminal nuclei have a species-typical appearance (2, 2.2); they can be found in blood vessels (*Loa, Brugia, Wuchereria*) or in lymphatic gaps (*Onchocerca*); their occurrence in the peripheral blood can be periodical (*Loa*—during the day; *Wuchereria*—during the night; some subperiodic strains also exist), or they may be permanently present (*Onchocerca*—always present, but in lymph vessels). **3 Intermediate hosts**: Depending on the periodic appearance of microfilariae in host's skin, insects with different biological behaviours are involved as vectors. Daytime feeding vectors (deerflies, *Chrysops* spp., blackflies, *Simulium* spp.) transmit *Loa loa* respectively *Onchocerca volvulus*, whereas night-feeding mosquitoes (*Aedes, Culex, Anopheles*) may be vectors of the nocturnal strains of *Wuchereria* and *Brugia*. When microfilariae are ingested by the intermediate hosts during the blood meal, they penetrate the intestine and enter the abdominal cavity and the thoracic muscles. After a moult the L2 is formed, which has a stumpy shape (sausage stage). Another moult finally leads to filariform infectious L3. **4–5** Larvae 3 reach a length of about 1.5 mm and migrate to the proboscis, from which they escape when the vector is feeding. They enter the skin through the wound channel made by the biting insect (5, *arrow*). Inside the final host (man, animals), the larvae migrate until they reach their favourite site of location, where they mature (after another two moults) within 1 year (prepatent period; Nematodes/Table 5.5). AD = adult worms (in section); AN = anus; E = oesophagus; ER = erythrocyte; IN = intestine; L3 = third larval stage; N = nuclei (their arrangement at the poles of microfilariae is species specific); SH = sheath (eggshell)

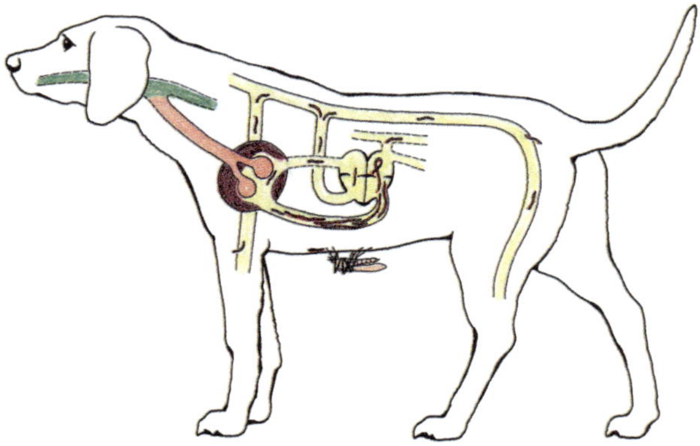

Fig. 5.138 Diagrammatic representation of the life cycle of the filarial *Dirofilaria immitis* in dogs. The larvae 3 are transmitted by *Culex* mosquitoes to dogs. Inside the heart chambers, maturity is reached and microfilariae produced, which appear later in the peripheral blood. These larva 1 can later become ingested by mosquitoes

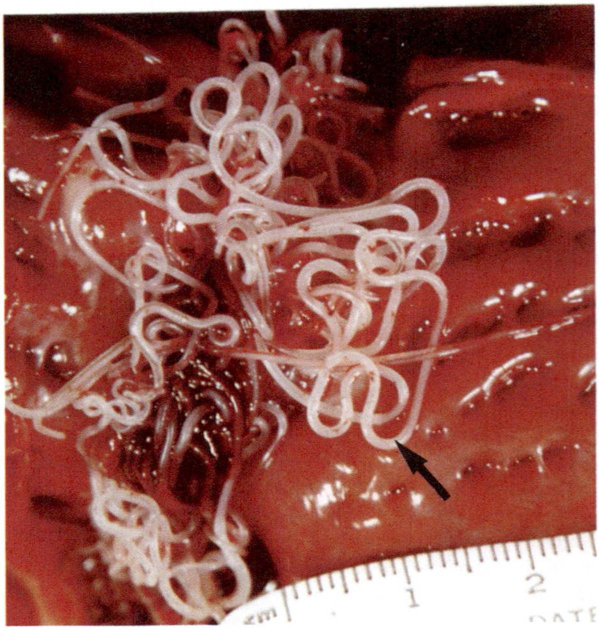

Fig. 5.139 Macrophoto of adult *Dirofilaria* worms in the opened heart of a dog

equids reach a length of up to 15 cm and the male 8 cm. The microfilariae reach a length of 260 μm and show the typical S-shaped tail (Fig. 5.141A).

4. **Symptoms of disease**:
   (a) ***Onchocerca*** **species**: Especially during summer months, the following symptoms may occur: oedemas, fistulae, general weakness, disturbances in movements, reduced sight (due to microfilariae in the eyes) and/or calcifications in skin regions with large amounts of parasites.
   (b) ***Dirofilaria*** **species**: Especially in the case of ***D. immitis***, the following symptoms may occur, which, however, in slight infection are mostly not noticeable: coughing, blood in saliva, dyspnoea, blockage of venules,

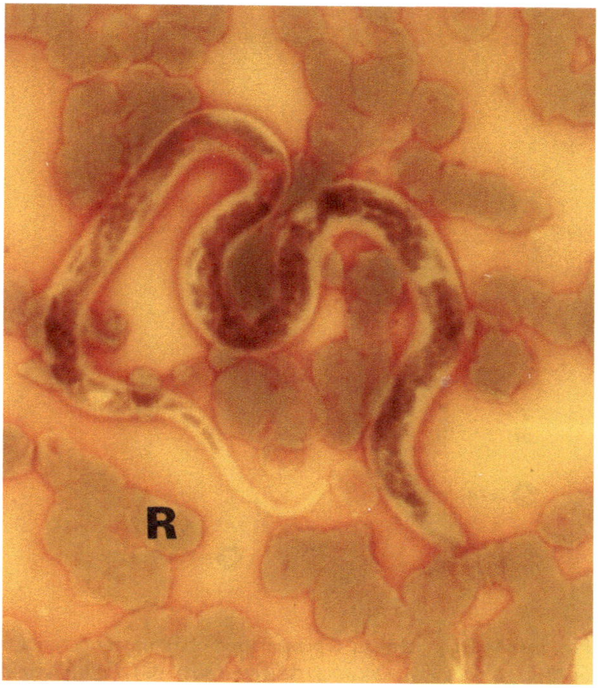

Fig. 5.140 Blood smear preparation showing a microfilaria of *Dirofilaria immitis*. R = erythrocyte

formations of oedemas, liver enlargement, dilation of the right heart chamber, pulmonary high pressure, ascites, eventually death (vena cava syndrome). ***D. repens*** induces in rare cases pruritus and eczemas, but infections are mostly symptomless (being situated in skin nods). **Caution**: *D. repens* may infect also **humans!**

(c) ***Stephanofilaria*** **and** ***Parafilaria*** **species**: These species induce small skin nods. Inflammations, which become open to the surface, may end in so-called "**summer wounds**" (dermatitis granulosa). However, they disappear as soon as colder outside temperatures start to dominate. Infected horses may show strongly reduced fitness.

(d) ***Setaria species***: Infections mostly do not induce symptoms of disease. In the case of the rather rare invasions of *S. digitata* larvae into wrong hosts like horses or sheep, death may occur. *Setaria* species of horses may enter into the scrotum as well as in the eyes introducing infertility and blindness.

5. **Diagnosis**: The microscopical demonstration of microfilariae is possible in smear preparations and after use of larvae-concentration methods.
6. **Pathway of infection**: Depending on the species, larvae are transmitted during bites of bloodsucking insects or by abducting larvae from wounds by feet of flies. **Caution**: some species (e.g. *D. immitis*) have a **zoonotic importance**, although transmitted larvae do not reach maturity in **humans**.
7. **Prophylaxis**: Reduction of biting or licking insects in the surroundings of animals by repellents and/or use of fine gaze before windows of stables.

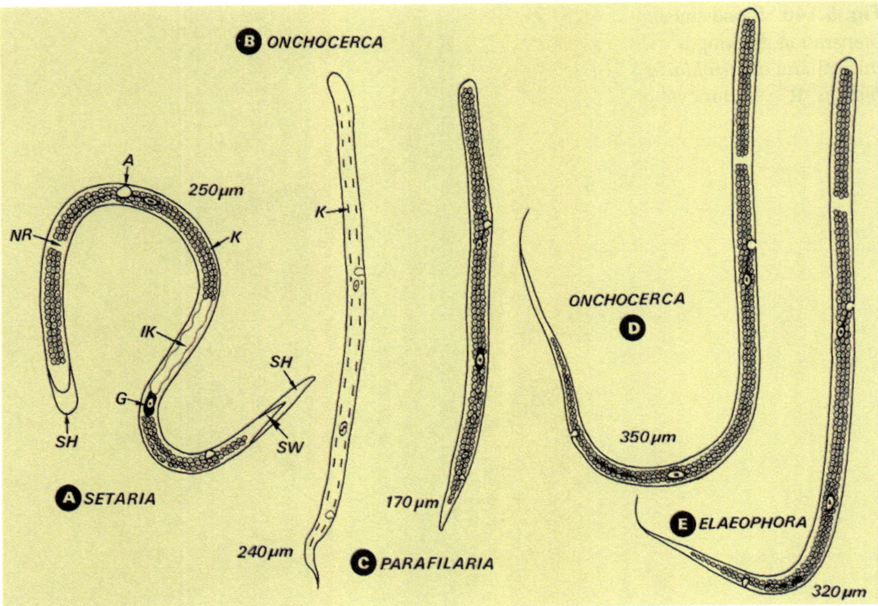

Fig. 5.141 Diagrammatic representation of different microfilariae (larva 1) of filarial worms. (**A**) *Setaria equina*; (**B**) *Onchocerca cervicalis*; (**C**) *Parafilaria multipapillosa*; (**D**) *Onchocerca reticulata*; (**E**) *Elaeophora boehmi*. A = anal pore; EX = excretion pore; G = genital anlage; IK = inner body; K = germ cells; NR = nerve ring; SH = sheath; SW = tail

8. **Incubation period**: Mostly several months.
9. **Prepatent period**: Up to 1.5 years.
10. **Patency**: Years.
11. **Therapy**:
    - Chirurgical removal of adult worms.
    - Use of macrocyclic lactones such as **moxidectin** (0.3 mg/kg bodyweight, orally).
    - In case of *D. immitis*: microfilariae: **melarsamin**; microfilariae: macrocyclic lactones.

Further Reading

Fuehrer HP et al (2016) *Dirofilaria* in humans, dogs, and vectors in Austria. From imported pathogens to the endemicity of *Dirofilaria repens*. PLoS Negl Trop Dis: e0004547.

Galuppi R et al (2013) Evidence of bovine filariosis in Italy (*Parafilaria bovicola*). Vet Parasitol 184:88–91.

Marques SM, Scroferneker ML (2004) *Onchocerca cervicalis* in horses from Southern Brazil. Trop Anim Health Prod 36:633–636.

Mutinda M et al (2012) Putative filariosis outbreak in white and black rhinoceros at Meru National Park in Kenya. Parasites Vectors 5:206.

Senathilake KS et al (2016) Rhizome extracts of *Curcum zedoaria* induce apoptosis in *Setaria digitata*. Exp Parasitol 167:50–60.

Townson S et al (2006) *Onchocerca* parasites and *Wolbachia* endosymbionts. Filaria J 5:4.

Ulesco T et al (2016) Detection of *Dirofilaria repens* and *D. immitis* DNA in mosquitoes from Belarus. Parasitol Res. doi:10.1007/s00436-016-5118-y.

5.3.3.24 Stomach Worms of Equids

1. **Name**: Greek: *habros* = thin, tiny; *megastoma* = large mouth; *nema* = filament; *thrix, trichos* = tiny hair; *strongylos* = rounded, cylindrical. Latin: *axis* = axis.
2. **Geographic distribution/epidemiology**: Worldwide.
3. **Biology, morphology**:
   (a) ***Habronema* species** (e.g. *H. muscae, H. majus* (syn. *H. microstoma*) measure as males 7–16 mm in length and as females they reach 2.5 cm. They live as adults in the mucous layer of the stomach of equids. Larvae of *Musca domestica* ingest the thin-walled eggs being excreted in the feces and adult flies transmit the infectious larva 3 onto the nostril of horses. In the case of *H. majus*, it was shown that stages of *Stomoxys calcitrans* act as vectors.
   (b) ***Draschia megastoma***: The adults appear whitish and reach as females a length of 25 mm. In the feces of horses, the larvae, which are ingested by fly larvae (*Musca domestica*), appear. Ingested larvae enter the intestinal wall.
   (c) ***Trichostrongylus axei***: The females reach a length of 8 mm and the males of 6 mm. They live in the mucous layer of the stomach, where females excrete embryonated eggs (Fig. 5.121h). In contrast to the two above-described species *T. axei* does not need an intermediate host, since free larvae 3 are directly ingested by the hosts.
4. **Symptoms of disease**: The stomach in general shows no or only low-grade symptoms of disease. However, in heavy infections colics, catarrhalic gastritis and loss of weight have been described. In the case of the invasion of skin sores by *Habronema* and *Draschia* larvae, forms of a cutaneous habronematosis are described (so-called **summer sores**), which heal badly and are often superinfected by bacteria (infection via flies).
5. **Diagnosis**: Microscopical demonstration of the thin-walled eggs (see B2 in the terminal annex) in the feces by means of flotation or by demonstration of larvae using of the Baermann funnel system.
6. **Pathway of infection**: Oral by uptake of infectious larvae 3 within food (*Trichostrongylus axei*) or by contact with mouthparts of adult flies, which had been infected as larvae when ingesting larvae 1 in feces. Infections are also possible if horses ingest adult flies.

7. **Prophylaxis**: Regular removal of feces from horse stables and application of fly control measurements.
8. **Incubation period**: 1–4 weeks.
9. **Prepatent period**: 2 months in the cases of *Habronema* sp. and *D. megastoma*, but only about 3 weeks in case of *T. axei*.
10. **Patency**: 4–15 months.
11. **Therapy**:
    (a) *T. axei*: **Ivermectin** (0.2 mg/kg bodyweight), **tiabendazole** (50 mg/kg bodyweight) and **oxfendazole** (10 mg/kg bodyweight; only for cattle/sheep registered).
    (b) The stomach stages of *Habronema* sp. and *Draschia* sp. can be treated with **ivermectin** or **macrocyclic lactones**.
    (c) In the case of cutaneous locations of the larvae (e.g. in summer sores), they should be mechanically removed.

Further Reading

Phosuk I et al (2013) Short report: Molecular evidence of *Trichostrongylus colubriformis* and *T. axei* infections. Am J Trop Med Hyg 89:376–379.

Rakhshandehroo E et al (2014) Molecular and morphological comparison of two different types of *Habronema muscae* in horse. Parasitol Res 113:4439–4445.

Schuster RK, Sivakumar S (2013) A xenodiagnostic method using *Musca domestica* for the diagnosis of gastric habronemosis and examining the anthelmintic efficacy of moxidectin. Vet Parasitol 197:176–181.

5.3.3.25 *Anisakis* Species and Related Species

Anisakis species (e.g. *A. simplex*), *Phocanema* species (e.g. *P. decipiens*) and *Contracaecum* species (e.g. *C. osculatum*) do not harm animals in human environments, since their final hosts are whales, seals or even birds. However, larvae of these worms endanger both humans and house animals, if they eat or are fed with undercooked fish meat containing larval stages. Ingested larvae of this group of nematodes wander in the body of animals and of humans and thus induce a so-called **anisakiasis**.

Further Reading

Abou-Rahma Y et al (2016) First record of *Anisakis* simplex third-stage larvae in European hake (*Merluccius merluccius*) in Egyptian water. J Parasitol Res ID 9609752.

Knoff M et al (2001) Nematodes of elasmobranch fishes from the southern coast of Brazil. Mem Inst Oswaldo Cruz Vol 96:81–87.

Zuloaga J et al (2013) A rat model of intragastric infection with *Anisakis* spp. live larvae. Parasitol Res 112:2409–2411.

5.3.3.26 *Thelazia* Species

1. **Name**: Greek: *thelys* = female. Latin: *gulo* = feeder. Skrjabin = Russian parasitologist. Rhodesia = former name of Zimbabwe.
2. **Geographic distribution/epidemiology**: Worldwide; often with high infection rates (more than 50 %).
3. **Biology, morphology**:
   (a) **Species in ruminants**: Common are *Thelazia gulosa, T. skrjabini* and *T. rhodesi*. The whitish or yellowish appearing females reach a length of 19–20 mm, while the males measure only 8–12 mm in length. The spicula of the males are rather long compared to their body length. The vulva of the females opens close to the anterior end. Some species parasitize in the conjunctival sac of their hosts, others in the channels of the tear glands and again others in the connection channel between nose and eye. Females excrete microfilariae, which are taken up orally by flies of the genus *Musca*. After a development of 9–12 days in the stomach of the flies (intermediate hosts), the larva 3 has reached infectivity and becomes transmitted during the next contact with the eyes of a host. There the worms become mature within 3–4 weeks.
   (b) **Species of horses**: *T. lacrymalis* belongs to the most common species on horses. The females reach a length of 18 mm and the males of 11 mm. They parasitize inside the conjunctival sac, where the females excrete larva-containing eggs. These eggs are ingested by flies (e.g. *Musca autumnalis*) within lacrimal fluids. Inside the flies, it takes about 2 weeks until the larva 3 is ready to become transmitted during the next eye visit of the fly.
   (c) **Species of birds**: The species of the genera *Oxyspirura* are found inside the conjunctival sac.
   (d) **Species of carnivores**: The Asian species *Thelazia callipaeda* is now endemic in Europe, too.

   Caution: All *Thelazia* species have a considerable zoonotic potency and their larvae may be transmitted into the eyes of humans.
4. **Symptoms of disease**: A low-grade infection may remain symptomless; however, in most cases the following symptoms occur: intense tear production, conjunctivitis, iritis, iridocyclitis and oedemas. Degenerating microfilaria lead to clouding of the cornea and to increased light sensitivity.
5. **Diagnosis**: Isolation of the parasites by rinsing eyes and by microscopical inspection.
6. **Pathway of infection**: Transmission via mouthparts of flies.
7. **Prophylaxis**: Fly control in stables.
8. **Incubation period**: 1–2 weeks in cattle; 2–3 days in horses.
9. **Prepatent period**: 3–4 weeks in cattle; 11 weeks in horses.
10. **Patency**: About 12–15 months in cattle, 2–3 months in horses.

11. **Therapy**: Since the mechanical removal from the eyes is very difficult, medical treatment is recommended: **levamisole**: 1–3 days 5 mg/kg bodyweight p.o. or 2 ml injected into the conjunctival sac. Also **ivermectin** (0.5 mg/kg bodyweight as spot-on) are **doramectin** (0.2 mg/kg bodyweight subcutaneously) are highly active.

Further Reading

Gama et al (2016) First report of *Thelazia callipaeda* infection in wild European rabbits (*Oryctolagus cuniculus*). Parasites Vectors 9:230.

Lechat C et al (2015) Comparative evaluation of the prophylactic activity of a slow-release insecticide collar and a moxidectin spot-on formulation against *Thelazia callipaeda* in dogs. Parasites Vectors 8:93.

Naem S (2007) Morphological differentiation among three *Thelazia* species. Parasitol Res 101:145–151.

Otranto D et al (2003) Molecular epidemiological survey on the vectors of *Thelazia gulosa*, *T. rhodesi* and *T. skrjabini*. Parasitology 127:365–373.

Traversa D et al (2005) Molecular characterization of *Thelazia lacrymalis* affecting equids. Mol Cellular Probes 19:245–249.

5.4 Tongue Worms (Pentastomida)

Pentastomida or **Linguatulida** (**tongue worms**) represent a separate group in the animal kingdom showing certain morphological convergences with some other phyla. Therefore, some approaches were done to assign them to the cestodes, nematodes, acanthocephalans, hirudineans, myriapods, crustaceans or arachnids. Recently, the body is divided into the anterior **cephalothorax** and an **abdomen** and therefore placed close to crustaceans. This, however, is not accepted by all authors.

The fine structure of the cuticle indicates a close relationship to the arthropods but differs from that of nematodes due to integration of higher amounts of chitin.

Pentastomids are dioecious and live as adults in the nose, throat and breathing systems of mammals, birds and reptiles (70 % of the final hosts of valid species are snakes). The adults appear annulated externally anchoring with two pairs of oral hooks at the tissues of their hosts (Figs. 5.142, 5.143 and 5.144). They feed species specifically on the endothels/lymph and/or blood of their hosts. The intestine ends either terminally or subterminally. In all developmental stages, an excretion system, blood system and breathing system are missing. In all species, the genital openings are situated in a median-ventral position, in males often in the front of the body and in females the position varies species specifically (ranging from the anterior to the posterior region). After copulation, the female lays eggs, which reach the outside either with the nasal discharge or with the feces after being swallowed. The eggs are taken up orally by intermediate hosts (except for *Righantia*, which shows a direct development). Intermediate hosts are almost exclusively vertebrates (insects in the case of *Raillietiella*). The first larva hatches in the intestine of the intermediate host.

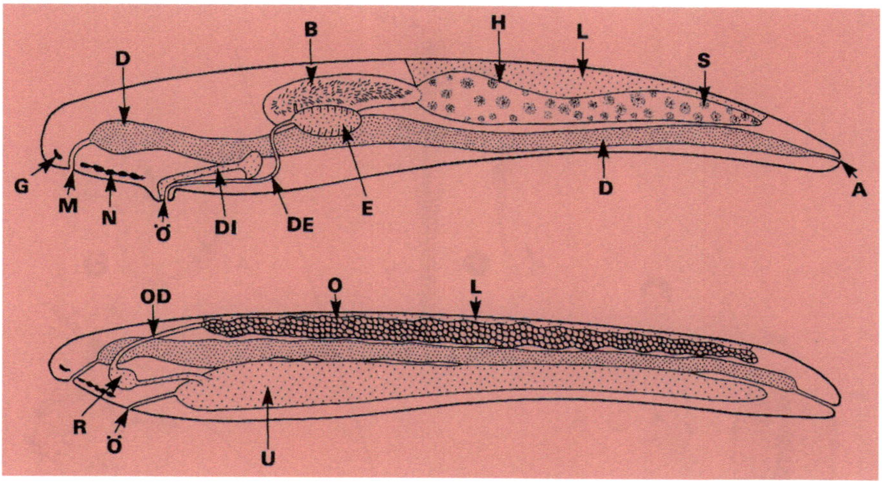

Fig. 5.142 Diagrammatic representation of adults (male: above) of **cephalobaenid** pentastomids (after Riley 1983). A = anus; B = seminal vesicle with filiform sperms; D = intestine; DE = vas deferens; DI = dilatory system; E = ejaculation bulb; G = brain (subpharyngal ganglion); H = testes; L = ligament; M = mouth; N = neural strand; O = ovary; OC = oviduct; Ö = genital opening; R = receptaculum seminis; S = spermatocytes; U = uterus

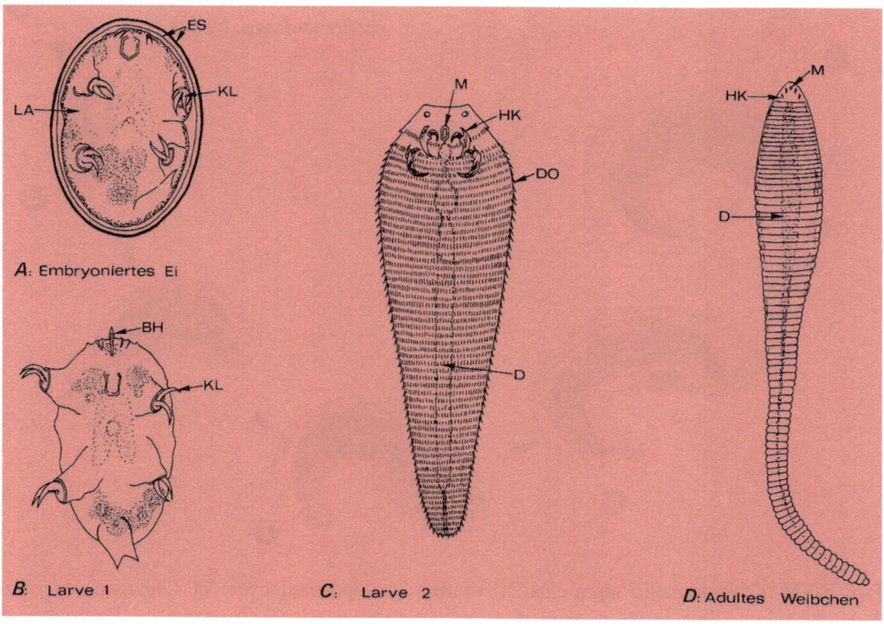

Fig. 5.143 Diagrammatic representation of developmental stages of pentastomids. *A* embryonated egg; *B*, *C* larvae; *D* adult female. BH = drilling hooks; D = intestine; DO = thorn; HK = hook; IES = inner eggshell; KL = claw; LA = larva; M = mouth

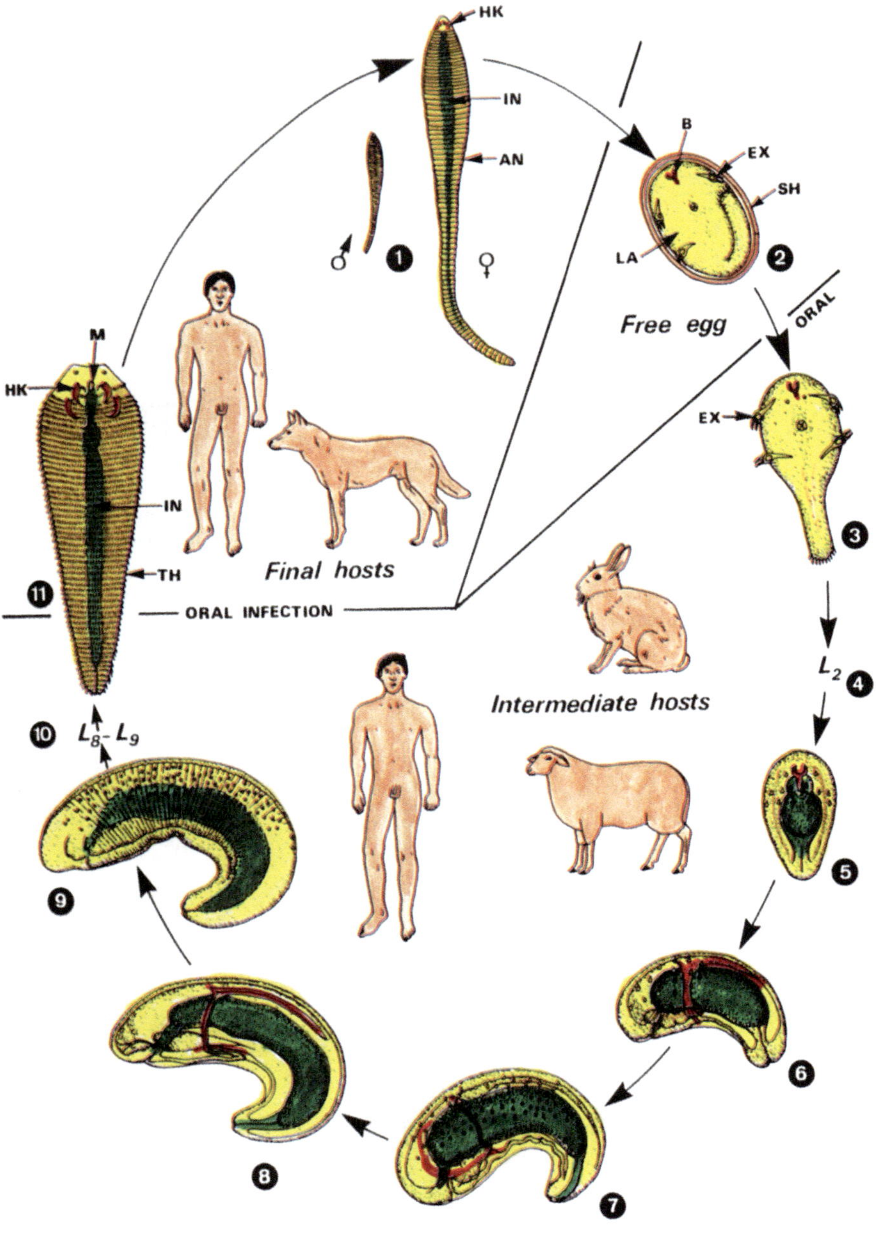

Fig. 5.144 Diagrammatic representation of the developmental cycle of *Linguatula serrata*. **1** Adults (♀, ♂) live in the nose of dogs and (in rare occasions) in humans. **2** Embryonated (free) eggs are released with nasal discharge. **3–11** If intermediate hosts ingest such larva-containing eggs (=**oral infection**), the larva hatches in the intestine and migrates via blood vessels to internal organs, where it grows up via several moults. In case final hosts ingest such larvae (**11**) with raw or undercooked meat, the larvae migrate to the nose and reach sexual maturity. AN = annulus; B = drilling organ; EX = extremity with a claw; HK = mouth hooks; IN = intestine; LA = 1st larva; M = mouth; SH = inner shell; TH = thorn

It penetrates into the intestinal wall and may infest different organs (species specific), where it is encysted and develops into an infectious larva after several moults. The final host (carnivore) is infected by oral uptake of this infectious larva within its tissue. The infection of the throat occurs from the oral cavity already.

System
Phylum: Pentastomida (extract)
Order: Cephalobaenida
Family: Cephalobaenidae
Genus: *Cephalobaena*
Genus: *Raillietella*
Family: Reighardiidae
Genus: *Reighardia*
Order: Porocephalida
Family: Sebekidae
Genus: *Sebekia*
Family: Subtriquetridae
Genus: *Subtriquetra*
Family: Sambonidae
Genus: *Sambonia*
Genus: *Waddycephalus*
Family: Diesingidae
Genus: *Diesingia*
Family: Porocephalidae
Genus: *Porocephalus*
Genus: *Kiricephalus*
Family: Armilliferidae
Genus: *Armillifer*
Genus: *Cubirea*
Family: Linguatulidae
Genus: *Linguatula*

5.4.1 *Linguatula serrata* (Tongue Worms, Pentastomida)

1. **Name**: Greek: *penta* = five; *stoma* = mouth; Latin: *lingua* = tongue; *serratus* = frayed, lacerated.
2. **Geographical distribution/epidemiology**: Worldwide.
3. **Biology/morphology**: The worms are whitish and tongue shaped and appear annulated. Their anterior end is wider than the posterior one and possesses four oral hooks (Figs. 5.143 and 5.144). The adults (females measure up to 13 cm, males up to 2 cm) live in the nasal cavities but also in the respiratory tract. Their either embryonated or larva-containing eggs measure 90 × 70 μm and reach the outside within nasal discharge. If the eggs are ingested by herbivores, the larvae

penetrate the intestinal wall and migrate into the mesenterial lymph nodes, but also into the lung, liver etc., where they reach the infectious nymph stage (4–6 mm) after a development over several months. This stage is called **terminal larva** and is located in nodules filled with liquid, which are called **pentastome nodules** and which are occasionally found during meat inspection. These nodules are left after 1–3 months and the terminal larva reaches either the abdominal cavity or the outside via the oral cavity. If a dog or another final host ingests such larvae orally either within infected innards or as "free" larvae from the outside, the infestation of the nasal cavity occurs. The adult stage is reached after another moult.

4. **Symptoms of disease**: Nose catarrh, fluid slime discharge, sneezing, itching, bacterial secondary infection and reduction/loss of olfaction.
5. **Diagnosis**: Detection of typical eggs (Fig. 5.145c) in nasal discharge or in feces; evidence of adult worms which were spontaneously released when sneezing.

6. **Course of infection**: Orally by uptake of infectious larvae with infected innards or nasally by sniffling at free larvae. **Infection of humans**: Humans can often act as intermediate hosts (=erroneous host). Infection occurs either by:
   - Oral uptake of eggs with drinking water,
   - Oral uptake of eggs with salad, etc.,
   - Accidental uptake of eggs during taxidermy of snakes (e.g. preparation of food in Asia).

7. **Prophylaxis**: No feeding of raw innards.
8. **Incubation period**: Few weeks.
9. **Prepatent period**: 6 months.
10. **Patency**: 2–3 years.
11. **Therapy**: Mechanical removal, eventually provocation of a strong sneezing spell. **Chemotherapy**: Vermol® (Fa. Alpha-Biocare, Neuss).

5.4.2 Other Pentastomids of Reptiles

(a) ***Armillifer annulatus***: This species (♀ 8–12 cm, 25–31 rings/annuli; ♂ 1.5–3 cm, 28–30 rings/annuli) shows a similar developmental cycle like *A. armillatus* (Fig. 5.145) but prefers cobras (*Naja* species) as hosts. The relatively narrow neck of their females is specific.

(b) ***Porocephalus crotali***: The life cycle of this species (♀ 4–7 cm; ♂ 2–3 cm) is similar compared to that of *Armillifer* species, but a number of *Crotalus* species may serve as final hosts. Intermediate hosts are numerous species of rodents. In rare cases infections of humans occur, while monkeys are more often infected.

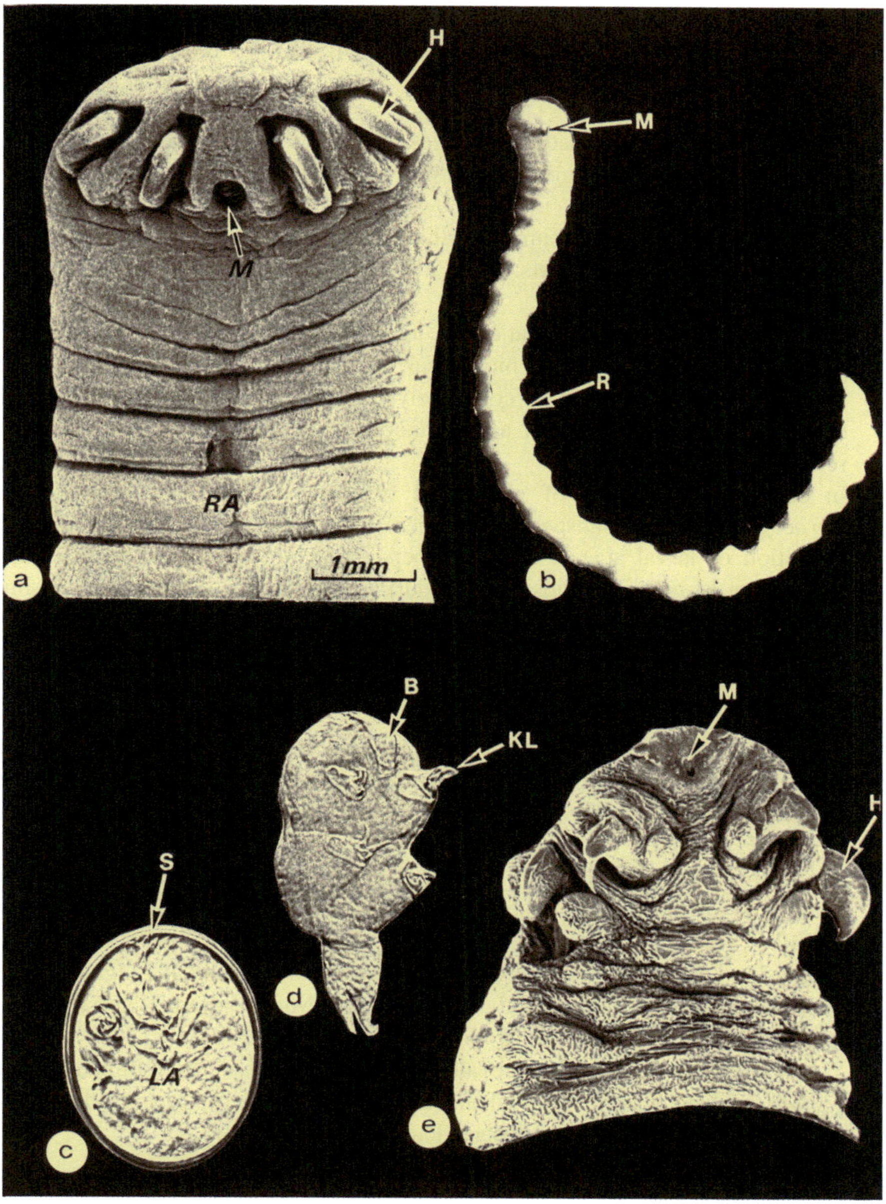

Fig. 5.145 Light and scanning electron micrograph of pentastomids. (**a–d**) *Armillifer* sp. (**a**) Anterior end. (**b**) Total view. (**c**, **d**) Larvae inside an egg (**c**) or free larval drilling stage. e) *Raillietiella* sp., anterior end. B = drilling organ; H = hook; KL = claw; M = mouth; R = annulus; RA = appendage of annulus; S = eggshell

(c) ***Reighardia sternae***: This single species occurs in birds (in the air sacks of seagulls; ♀ 3–4.5 cm; ♂ 0.7 cm) and develops directly without any intermediate host (like some *Raillitiella* species of lizards).

These and many other pentastomid species of reptiles can be treated safely with Vermol® (Alpha-Biocare, Neuss), which was especially developed for treatment of reptiles.

Symptoms of infection of humans with pentastomids depend on their role either as an **intermediate** or **final hosts**:

- In the case of a nasal infestation by adults of *L. serrata,* symptoms of the so-called **Halzoun syndrome** may occur, which means that the airways of the nasal space is partially up to completely blocked. Also numbness and facial edema may occur. In case of a spontaneous sneezing adult worms might be released spontaneously.
- In cases of infestation of abdominal organs by larvae of *Armillifer* or *Porocephalus* species-unspecific disorders may occur due to penetration and migration of larval stages. High numbers of larvae may often cause death. Infestation is often diagnosed only during autopsy or accidentally during X-ray examination (showing calcified, dead stages). **Chemotherapy** for humans is not possible.

Further Reading

Gjerde B (2013) Phylogenetic position of *Linguatula arctica* and *Linguatula serrata* (Pentastomida) as inferred from the nuclear 18S rRNA gene and the mitochondrial cytochrome c oxidase subunit I gene. Parasitol Res 112:3517–3525.

Kelehear C et al (2015) An introduced pentastomid parasite (Raillietiella frenata) infects native cane toads (*Rhinella marina*) in Panama. Parasitology 142:675–679.

Oluwasina OS et al (2014) *Linguatula serrata* (Porocephalida: Linguatulidae) infection among client-owned dogs in Jalingo, North Eastern Nigeria: prevalence and public health implications. J Parasitol Res 2014:916120.

Yabsley MJ et al (2015) High prevalence of *Porocephalus crotali* infection on a barrier island (Cumberland Island) off the coast of Georgia, with identification of novel intermediate hosts. J Parasitol 101:603–607.

5.5 Acanthocephala (Thorny-Headed Worms)

For a long time acanthocephalans were considered as a subphylum of Nemathelminthes. In recent times, electron microscopical investigations showed that there are significant morphological differences of the developmental stages and therefore acanthocephalans have to be considered as an own phylum.

System
PHYLUM: ACANTHOCEPHALA
Class: Archiacanthocephala
Order: Moniliformida
Genus: *Moniliformis*
Order: Oligacanthorhynchida
Genus: *Macracanthorhynchus*
Genus: *Prosthenorchis*
Class: Palaeacanthocephala
Order: Echinorhynchida
Genus: *Acanthocephalus*
Genus: *Echinorhynchus*
Genus: *Pomphorhynchus*
Order: Polymorphida
Genus: *Polymorphus*
Genus: *Filicollis*
Class: Eoacanthocephala
Order: Neoechinorhynchida
Genus: *Neoechinorhynchus*
Genus: *Paratenuisentis*

These classes of acanthocephalans contain exclusively parasitic species, which may reach up to 70 cm in length (Table 5.10) and appear cylindrical and unsegmented externally (Fig. 5.146). They are called **thorny-headed worms** because of their protrudable proboscis, which is equipped with species-specific hooks (Fig. 5.147). With the help of these hooks, the intestine-lacking worms anchor deeper and tighter in the mucosa of vertebrates than tapeworms, which also lack an intestine and feed on the intestinal contents of their hosts. The food uptake is done via the tegument, which, however, has a thicker and more dense structure compared to cestodes and trematodes, but as a syncytial layer it shows certain structural similarities. It is different to the limiting layers of nematodes though.

Acanthocephalans are dioecious and the gonads being located either in 1 or 2 ligament bags fill the body cavity, which is formed as a pseudocoel, or they occur directly in the body cavity without a ligament bag. The **sperm** of acanthocephalans are filiform, measure 20–80 μm in length and do not possess mitochondria nor acrosomes, but show in contrast to nematodes a typical flagellum protruding at the anterior pole from a centriole. This flagellum runs parallel to the widened anterior end and becomes free at the posterior end. Its anterior portion contains besides amorphous DNA, which is not limited by a membrane, also regularly arranged, ovoid, electron-dense intrusions of unknown function. Meiosis of oocysts, which are located in ovarial clusters, occurs after penetration of the sperms. Then the fusion of polar bodies occurs being followed by formation of a fertilization membrane. Some species of the dioecious acanthocephalans show **sex chromosomes** of the type XY in males (heterogametic) and XX (homogametic) in females (e.g. *Macracanthorhynchus hirudinaceus*). In some other species, the males lack

Table 5.10 Important acanthocephalan species

Species	♀ Length (cm)	Final host	Intermediate host
Echinorhynchus truttae	0.7–2.2	Salmonids	*Gammarus* sp. (Amphipoda)
Neoechinorhynchus cylindratus	0.7–1.5	Perches, breams (*Micropterus* sp.)	*Cypria* sp. (Ostracoda)
N. rutili	0.5–1	Trouts and other fishes	Ostracoda
Acanthocephalus anguillae	1–3.5	Whitings and other fishes	Water slater (*Asellus aquaticus*)
Pomphorhynchus laevis	1–3.5	Predatory fishes, eels, whitings	*Gammarus* sp. (Amphipoda)
Paratenuisentis ambiguus	0.8–1.4	Eels	*Gammarus tigrinus*
Macracanthorhynchus hirudinaceus	20–65	Pigs, dogs, **humans**[a]	Beetle larvae
Filicollis anatis	1–2.5	Waterfowls	Water slater
Prosthenorchis elegans	3–8	Monkeys, **humans**?	Cockroaches (*Blattella*)
Moniliformis moniliformis	14–27	Rats, rodents, **humans**	Cockroaches (*Periplaneta*)
Polymorphus minutus	0.4	Waterfowls	Amphipods

[a]Worms do not reach sexual maturity but lead to abscesses or intestinal perforations

the Y chromosome. They appear as X0 and possess one chromosome less than females (e.g. *Moniliformis moniliformis*, *Echinorhynchus truttae*, *Neoechinorhynchus cylindratus*). *Acanthocephalus ranae* shows a XY combination in both sexes though. The number of chromosomes is relatively low. For example, females of *M. hirudinaceus* and *N. cylidratus* possess in the diploid status six chromosomes (including the sex chromosomes), while in *M. moniliformis* and *E. truttae* eight chromosomes are formed. Eventually the males contain one chromosome less.

After fertilization the developing eggs are "sorted" by a special organ called **uterine bell**; it is assumed that eggs, which are not completely developed, are restrained there. The female deposits embryonated eggs, which reach the outside with the feces of the host. From each egg a hooked larva called **acanthor** hatches in the intestine of insects or crustaceans (=intermediate hosts), which differentiates to an **acanthella** larva and finally develops into the infectious **cystacanth** larva (Fig. 5.146). In case a specific vertebrate (=final host) ingests such a stage, this stage develops into the adult. Occasionally also vertebrates (e.g. small fishes) can act as stack host.

Apart from a few accidental infections of humans acanthocephalans are in Europe only for domestic animals of economic importance. For example, in ducks up to 150 acanthocephalans can be found, which can cause severe lesions of the mucosa and serious diarrhoea. In aquacultures, acanthocephalans play an

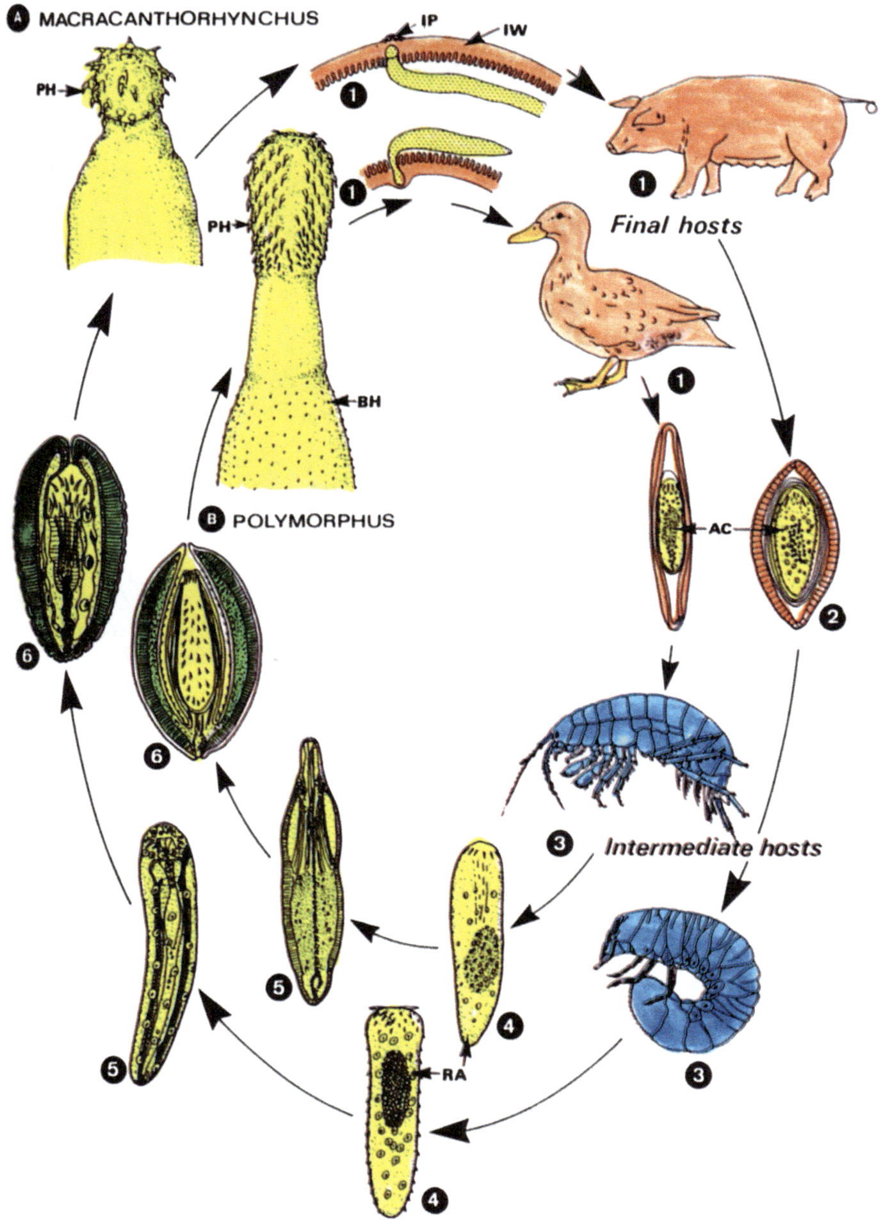

Fig. 5.146 Diagrammatic representation of the life cycle of thorny-headed worms of swine (and humans) and ducks. 1 Infestation of the intestine of final hosts. 2 Deposited worm eggs. 3 Intermediate hosts. 5–6. Development of further larval stages (acanthella, cystacanth), which grow up to adult worms in the final host. AC = acanthor; BH = body hooks; IP = inflamed intestinal wall; IW = intestinal wall; PB = proboscis hooks; RA = appendage of annulus

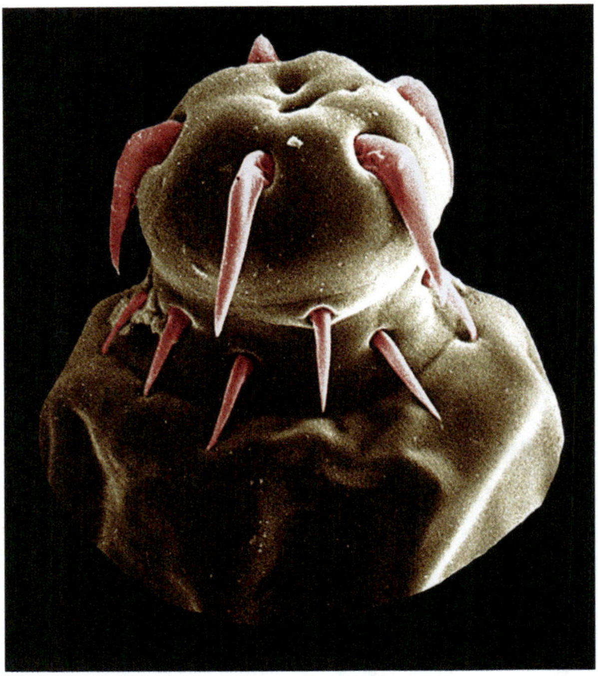

Fig. 5.147 Scanning electron micrograph of the acanthocephalan worm *Neoechinorhynchus* sp. with the typical spines

important role. Young fishes, which were infected via the ingestion of ostracodes with *Neoechinorhynchus rutili*, retard in growth.

5.5.1 *Macracanthorhynchus hirudinaceus* (Giant Thorny-Headed Worm of Swine)

1. **Name**: Greek: *makros* = big, large; *akantha* = hook; *rhynchos* = snout (=long animal with thorns on the anterior end).
2. **Geographical distribution/epidemiology**: Worldwide, recently in Europe almost only in the Balkans. It also occurs in humans and is the main cause for intestinal surgery of children in some Chinese provinces.
3. **Biology/morphology**: The so-called giant thorny-headed worm of swine *Macracanthorhynchus hirudinaceus* is characterized by a protrudable proboscis, which is equipped with many thorns (Fig. 5.146), with the help of which it is anchored in the intestinal mucosa. The females reach a length of up to 60 cm and the males up to 10 cm. Intermediate hosts are larvae of several bug species (grubs), which ingest the eggs (size 70–110 × 40–65 μm) that are deposited completely embryonated, inside the feces. The contained larva 1 (**acanthor**) reaches the infectious stage in the body cavity via development of further larval stages (**acanthella**).

4. **Symptoms of disease**: Unspecific intestinal disorders and inflammations caused by mechanical lesions; occasionally even peritonitis occurs due to perforation of the intestinal wall. In piglets, extreme diarrhoeas, muscular cramps and cachexia also occur. The severity of symptoms, which also can be caused by preadults (in a short incubation period), depends on the number of worms in the intestine of the host.
5. **Diagnosis**: Detection of eggs (Fig. 5.146) in the feces.
6. **Pathway of infection**: Orally by uptake of infected intermediate hosts (bugs).
7. **Prophylaxis**: Not possible on pastures.
8. **Incubation period**: Variable, depending on the infestation grade.
9. **Prepatent period**: 10–12 weeks.
10. **Patency**: Months.
11. **Therapy**: Agent of choice is loperamide hydrochloride (e.g. Imodium®, etc.) (Mehlhorn et al. 1990). An oral dose of 1.5 mg/kg bodyweight (2 × per day on 3 consecutive days) is effective to kill all adult and preadult worms without noticeable side effects. To maintain regular feces deposition, 100 ml castor oil was added to the food. Fenbendazole (orally, 20 mg/kg bodyweight on 5 days) is reported to eliminate thorny-headed worms in monkeys successfully as well.

Further Reading

Foata J et al (2005) Ultrastructure of spermiogenesis and the spermatozoon of *Macracanthorhynchus hirudinaceus* (Pallas, 1781) (Acanthocephala: Archiacanthocephala), a parasite of the wild boar *Sus scrofa*. J Parasitol 91:499–506.

Gassó D et al (2016) Coprological tests underestimate *Macracanthorhynchus hirudinaceus* burden in wild boar. Parasitol Res 115:2103–2105.

Mehlhorn H et al (1990) Loperamid, an efficacious drug against the acanthocephalan *Macracanthorhynchus hirudinaceus* in pigs. Parasitol Res 76:624–626.

5.5.2 Thorny-Headed Worms of Birds

1. **Name**: Greek: *polys* = many; *morphe* = shape; Latin: *filium* = thread; *collare* = collar; *minutus* = very small; *anas* = duck.
2. **Geographical distribution/epidemiology**: Worldwide.
3. **Biology/morphology**: Some acanthocephalans occur in the small intestine (especially in waterfowls), but their economical importance is rather low:
   (a) ***Filicollis anatis***: Females reach a length of up to 25 mm and males reach up to 8 mm. The proboscis of the females is spherical and hooks occur only on the crest in contrast to the males showing 18–122 longitudinal rows with 10–11 hooks each. The eggs containing the acanthor larva measure 65 × 20 μm. Intermediate hosts are aquatic sowbugs, in which the infectious terminal acanthella larva is developed and encysted.

(b) ***Polymorphus minutus* (syn. *P. boschadis*)**: The females reach a length of up to 1 cm and males up to 3 mm. Both sexes have a reddish-orange colour. The proboscis of both sexes is equipped with 16 rows of small hooks. The spindle-shaped, acanthor larva-containing eggs reach a size of 65–70 × 20 μm and appear yellowish. Intermediate hosts are amphipods (Fig. 5.146).

4. **Symptoms of disease**: Bloody feces occur due to mechanical damage of the intestinal wall by the hooks of the proboscises as well as diarrheas, intestinal perforations and secondary bacterial infections leading to death.
5. **Diagnosis**: Detection of the typical eggs in the feces.
6. **Pathway of infection**: Oral uptake of infected intermediate hosts.
7. **Prophylaxis**: Almost impossible due to outdoor infections.
8. **Incubation period**: Several days p.i. due to the drilling activities of the infectious larvae.
9. **Prepatent period**: 1–2 months.
10. **Patency**: 6–12 months.
11. **Therapy**: A special therapy is not known. With respect to the efficacy of treatment against other thorny-headed worm species, fenbendazole (20 mg/kg bodyweight for 5 days orally) or loperamid hydrochloride could be an option, which are agents of choice for treatment of acanthocephalans in swine and fishes.

Further Reading

Amin OM et al (2013) The description of *Mediorhynchus africanus* n. sp. (Acanthocephala: Gigantorhynchidae) from galliform birds in Africa. Parasitol Res 112:2897–2906.

Chen HY et al (2015) Effects of the acanthocephalan *Polymorphus minutus* and the microsporidian *Dictyocoela duebenum* on energy reserves and stress response of cadmium exposed *Gammarus fossarum*. PeerJ 3:e1353.

Komorová P et al (2015) Acanthocephalans of the genus *Centrorhynchus* (Palaeacanthocephala: Centrorhynchidae) of birds of prey (Falconiformes) and owls (Strigiformes) in Slovakia. Parasitol Res 114:2273–2278.

5.5.3 Acanthocephala of Fishes

1. **Name**: Greek: *akantha* = hook, spike; *kephale* = head; *pomphos* = swollen; *rhynchos* = snout; *neos* = new; *echinos* = spike; Latin: *rutilus* = reddish; *lux* = light.
2. **Geographical distribution/epidemiology**: Worldwide.
3. **Biology/morphology**: Fishes are parasitized by a number of acanthocephalan species, which can be distinguished due to the protrudable proboscis and its dentition (Fig. 5.147). The dioecious adults of more than 300 species parasitize in the intestine of fishes, which are the final hosts. In addition, fishes can also harbour larvae of acanthocephalans of other final hosts, and therefore, they may

serve as intermediate (stack) hosts. The intestine-less adults acanthocephalans anchor with the help of their proboscises in the intestinal wall and take up the food via the cuticle of the body surfaces. In all species, the female lays species-specific eggs, which already contain a hooked larva (=**acanthor**). If this larva is ingested by an intermediate host (amphipods for acanthocephalans of fishes), an **acanthella** larva develops from the acanthor larva, which reaches the terminal stage via several moults (non-predatory fishes can serve as stack hosts). When it is taken up by the final host (with the intermediate host), it reaches the adult stage (♀, ♂) in its intestine. Mass infestation, especially in farmed freshwater fishes, may lead to enormous economical losses. Important species are:

(a) ***Pomphorhynchus laevis*** (Fig. 5.148): It occurs in Central Europe in predatory and non-predatory fishes in seawater, brackish water and freshwater. The females reach a length of up to 25 mm. Species of this genus are anchored by a bulb of the neck after perforating them with the help of the hooked proboscis (=18–20 longitudinal rows of 10–12 hooks). Intermediate hosts are amphipods (*Gammarus*); whitings can serve as stack hosts. The eggs measure 66 × 13 μm.
(b) ***Neoechinorhynchus rutili***: This species (♀ 5–13 mm, ♂ 2–6 mm) occurs in freshwater and brackish water fishes in Central and North Europe. The proboscis is spherical with three rings of six hooks each. Intermediate hosts are ostracods. The eggs measure 38 × 20 μm.
(c) ***Acanthocephalus lucii***: This species occurs in many fish species (**perches, eels**) in Central Europe. The females reach a length of up to 17 mm and males 4–8 mm. The proboscis is almost cylindrical with 16, 14 and 12 longitudinal rows of 7–8 hooks each. Intermediate hosts are water slaters. The eggs measure 120 × 16 μm.
(d) ***A. anguillae***: This species occurs in many fish species in Central and North Europe, often in salmonids. The females reach a length of up to 20 mm and males 5–7 mm. The proboscis is clubbed with 10 longitudinal

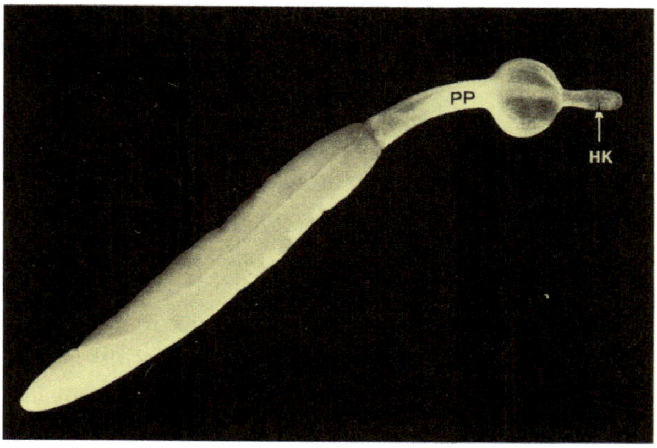

Fig. 5.148 Scanning electron micrograph of the fish acanthocephalan *Pomphorhynchus laevis*. HK = hooks of the proboscis; PP = not enlarged part of the neck

rows of 5–7 hooks each. Intermediate hosts are water slaters. The eggs measure 100 × 12 μm.

(e) ***Echinorhynchus gadi***: This species occurs in codfish and other seawater fishes in Central Europe. The females reach a length of up to 80 mm and males up to 20 mm. The proboscis is equipped with 18–22 longitudinal rows of 10–13 hooks each, of which the biggest are in the basal position. Intermediate hosts are amphipods. The eggs measure 76 × 13 μm.

(f) ***Echinorhynchus truttae***: This species occurs in salmonids in Central Europe. The females reach a length of up to 20 mm and males 8–11 mm. The proboscis measures 1.3 mm with 20 longitudinal rows of 10–13 hooks each. Intermediate hosts are amphipods (gammarids). The eggs measure 110 × 29 μm.

4. **Symptoms of disease**: Mechanical damage of the intestinal wall and obstipation in cases of heavy infestations (300 specimens or more). Dyspepsia and reduced gain of weight. Clinical evidence during section is yellowish papules at the outer surface of the intestine due to the perforating proboscis of the acanthocephalans.
5. **Diagnosis**: Microscopical evidence of eggs in the feces, detection of adults during sections.
6. **Pathway of infection**: Oral uptake of infectious intermediate hosts.
7. **Prophylaxis**: Frequent controls of feces. In ponds: control of intermediate hosts by draining dry and liming.
8. **Incubation period**: Depending on grade of infestation and virulence of the species: days up to weeks.
9. **Prepatent period**: 3–6 weeks (temperature dependent).
10. **Patency**: Often only months.
11. **Therapy**: Agent of choice is loperamid hydrochloride (e.g. Imodium®, which is approved for humans but not for edible fish) which should be applied with food in a dose of 50 mg/kg body weight on 3 consecutive days. In trouts, which did not show any side effects, all worms were killed, even the preadult stages. In vitro assays also niclosamide and levamisole showed an efficacy (Taraschewski et al. 1990). For fishes, R.P. Blaukonzentrat (Verman) is approved for acanthocephalans and cestodes. In ornamental fish, Nematol® of Alpha-Biocare (Neuss) is effective.

Further Reading

Caddigan SC et al (2014) Intra-population variation in behavior modification by the acanthocephalan *Acanthocephalus dirus*: are differences mediated by host condition? Parasitol Res 113:4307–4311.

Dianne L et al (2014) Parasite-induced alteration of plastic response to predation threat: increased refuge use but lower food intake in *Gammarus pulex* infected with the acanothocephalan *Pomphorhynchus laevis*. Int J Parasitol 44:211–216.

Katahira H, Nagasawa K (2014) Helminths from the giant mottled eel *Anguilla marmorata* Quoy; Gaimard in Japan, with a description of *Acanthocephalus longiacanthus* n. sp. (Acanthocephala: Echinorhynchidae). Syst Parasitol 88:91–102.

Katahira H, Nagasawa K (2011) *Echinorhynchus hexagrammi* Baeva, 1965 (Acanthocephala: Echinorhynchidae) from marine fishes off Hokkaido, Japan, with morphological observations and new host records. Syst Parasitol 80:35–40.

Nachev M, Sures B (2016) Seasonal profile of metal accumulation in the acanthocephalan *Pomphorhynchus laevis*: a valuable tool to study infection dynamics and implications for metal monitoring. Parasite Vectors 9(1):300.

Sobecka E et al (2012) Genetic and morphological variation in *Echinorhynchus gadi* Zoega in Müller, 1776 (Acanthocephala: Echinorhynchidae) from Atlantic cod *Gadus morhua* L. J Helminthol 86:16–25.

Taraschewski H et al (1990) Loperamid, an efficacious drug against fish-pathogenic acanthocephalans. Parasitol Res 76:619–623.

5.6 Leeches (Hirudinea, Annelida)

1. **Name**: Latin: *hirudo* = leech; *annulus* = small ring.
2. **Biology/morphology**: The annelids received their name due to their numerous homogeneous segments, which contain a pair of coelom cavities (secondary body cavity) each. Their distinguishing features and developmental cycle differ significantly from those of **flatworms** and **roundworms**. Especially the latter are often mixed up with the annelids due to the round body diameter of many species. For example, worms found in the feces of children and mistaken as earthworms turned out to be the reddish-brown specimens of the genus *Ascaris*, which, however, show cuticular rings which look like the segmentation of earthworms. Most species of the predominantly free-living species are hermaphrodites and live in the soil but also in the ground or freshwater or marine habitats. Only very few species of the Hirudinea–especially the **Hirudinea**–changed to parasitic lifestyle.

System
PHYLUM: ANNELIDA (extract)
Class: Polychaeta (most species free-living)
Class: Myzostomida (parasites of Comatolida)
Class: Clitellata (worms with a clitellum)
Order: Oligochaeta (mostly free-living, e.g. earthworm)
Order: Hirudinea (leeches, many parasitic species)
Family: Rhynchobdellidae ("jawless" leeches)
Family: Pharyngobdellidae (pharynx leeches)
Family: Gnathobdellidae ("jawed" leeches)
Family: Acanthobdellidae (leeches with bristles)

Hirudinea preferably live in freshwater habitats (but also in humid terrestrial biotopes) and have a predatory way of life. Only very few species became bloodsuckers. Due to their specialized way of life, they developed some special features distinguishing them from the other annelids:

- Their body shape is slender to leaf shaped. They almost look like flatworms and have the same ability to change their shape (length) extremely.
- Their 32 (29 in *Acanthobdella*) internal segments are externally not visible (each segment correlates to 2–14 external annuli).
- The coelom is secondary reduced and was replaced by massive muscle fibres and connective tissues.
- At the anterior and posterior ends of the Hirudinea, a strong sucker has been developed (Fig. 5.149; *Acanthobdella* has only a posterior sucker). Both suckers are used for attachment but also for looper-like movements.

The published taxonomy of the Hirudinea is controversial. However, it is accepted in general that this taxon contains four monophyletic groups, of which one is equipped with chetae. The Acanthobdellidae (with chetae) resemble regarding their internal organization most likely oligochaetes showing e.g. a widely differentiated blood system. *Acanthobdella pedellina* reaches a length of up to 3.5 cm and parasitizes in salmonids of freshwater. The cheta-less hirudineans are distinguished with the help of the organization of their mouth and the resulting varying food uptake. They are split into three families:

(a) **Rhynchobdellidae**: These leeches are able to protrude their jawless pharynx and suck blood and particles of the epidermis by causing low pressure. Their blood is—typical for this group—colourless. They do not have eyes and infest their hosts, when they touch the leech, which is lurking at the substrate. Many important fish parasites belong to this group, e.g. members of the genus *Pisciola* (measuring up to 7 cm), which may cause enormous economic losses in fish ponds. These losses are increased if parasitic protozoans of the genera *Haemogregarina*, *Cryptobia* or *Trypanoplasma* occur simultaneously. To this family belong the largest leeches known (e.g. *Haementaria* from South America), which belongs to the so-called nose leeches. While their host is drinking at banks of waters, they creep into the nasal cavity and grow while constantly sucking blood. A group of leeches can cause death of cattle or dogs due to secondary infections and weakening due to constant blood loss. In ducks, the species *Theromyzon tessulatum* occurs, which leaves the nasal cavity after the blood meal. The blood coagulation is inhibited by the protease **hermentin**, which degrades both fibrin and fibrinogen.

(b) **Pharyngobdellidae**: These gullet leeches do not possess neither a jaw, teeth nor a protrudable proboscis but suck blood just using their muscular pharynx. Their blood appears reddish. They possess 3–4 pairs of eyes which enable them to seek their host actively. Many important fish parasites are members of this group and cause damages not only by bloodsucking but also by transmission of parasitic protozoans. The dog leech *Erpobdella octoculata*, reaching a

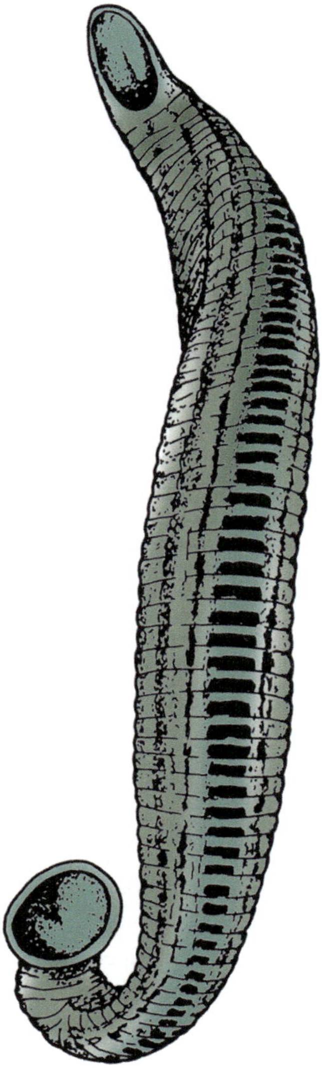

Fig. 5.149 Diagrammatic representation of the medicinal leech *Hirudo medicinalis* with its anterior and posterior sucker

length of up to 6 cm, is rather common in Germany. In contrast to its name it does not suck blood but has a predatory way of life feeding on insect larvae etc.

(c) **Gnathobdellidae**: In general these often very large species (up to 25 cm) possess jaws equipped with teeth (Fig. 5.150b). An exception is the so-called horse leech *Haemopis sanguisuga*, which occurs on wet soil close to horse watering places. It does not suck blood but feeds on earthworms. The "jawed leeches" possess five pairs of eyes and contain reddish blood. In order to find a host, these leeches note movements of the water caused by a potential host. By wave-like swimming motions, they reach their victims, heading finally for them with the help of their olfactory sense organs (chemoreceptors). They

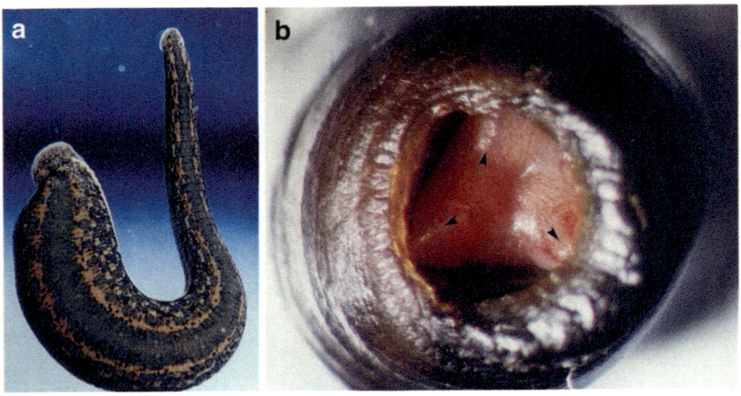

Fig. 5.150 (**a**) Light micrograph of the leech *Hirudo verbana* in total view. (**b**) Light micrograph of the anterior (oral) sucker of *H. medicinalis* showing the three blades, to which surfaces of numerous unicellular salivary glands lead

move on their host by looper-like motions to find the final place for sucking blood. The most famous species is the medicinal leech *Hirudo medicinalis* (Fig. 5.150), which has been used in human medicine for bleeding as a cure for many centuries in medicine. It was assumed that "bad juices in a human body" could be removed with this method. In 1826, only in Germany and France more than 30 millions of leeches were used. Just like in other leeches, the saliva, which is injected into the wound, contains besides an anaesthetic (the bite is not noticed by the host) an anticoagulant, which inhibits coagulation during and after (for some time!) bloodsucking. The saliva, which is secreted by unicellular glands and channelled to the blades of the jaw in fine channels, has besides antithrombotic properties a hypotensive effect on the blood pressure.

Although the species of Gnathobdellidae are able to transmit agents of diseases, but the most important feature is the enormous loss of blood, especially continuing bleeding after the bite. A large leech is able to ingest up to 15 g of blood, which is tenfold of its own body weight. Also the two- to threefold amount of blood is lost due to secondary bleeding. Due to the huge amount of sucked blood, the leech is also able to starve for a long time until a next blood meal is available. Once infested waters or wetlands remain contaminated, since leeches are able to starve for up to 1½ years and accept all kinds of vertebrate hosts.

Copulation between two hermaphroditic specimens occurs with the help of a relatively long, protrudable penis. The fertilized eggs are enveloped in cocoons (=foamy aggregation) containing 1–200 eggs and become attached to the substrate; only very few species carry these along. The development runs directly in general without a **trochophora** larva, which is typical for most annelids. The high reproduction rate causes often a "flooding" of waters by hirudineans. To this group belongs also the jawed land leech *Haemodipsa ceylanica* reaching a length of up to

10 cm, which often attacks humans and animals in huge numbers and then causes massive bacterial secondary infections. It is also capable of transmitting some *Trypanosoma* species.

The **transmission** of agents of diseases by jawed leeches is known for a long time, even though this aspect has been investigated experimentally in rather low numbers. Many reports from the tropics indicate that terrestrial leeches of the genus *Haemodipsa* often attack humans and domestic animals but also birds in masses and cause crippling or even death due to bacterial or viral secondary infections. It was also shown that the medicinal leech *Hirudo medicinalis* is capable of transmitting a number of bacteria and also viruses. As shown by Nehili et al. (1994), bacteriophages (=viruses of bacteria), bacteria and host lymphocytes are able to survive at least 1½ years in the intestine of the leeches. Other parasitic protozoans such as *Toxoplasma gondii*, *Trypanosoma brucei* and *Plasmodium* sp. remain infectious in the intestine of the leeches at least for 1 month after the blood meal. The agents of malaria *Plasmodium* sp. even proliferate resulting in a complete lysis of all erythrocytes of the blood meal. The phenomenon that many agents of diseases and also blood cells survive unchanged for more than 18 months in the intestine of the leeches is caused by the fact that the leeches do not produce any digestion enzymes in the lateral intestinal sacs (diverticula). The lysis of red blood cells is performed by a single endosymbiontic bacteria species (*Pseudomonas hirudinis*), which is transferred to the offspring in low numbers. Protease inhibitors of the leeches even block the lytic processes, which could be induced by ingested leucocytes. The leeches possess digestion enzymes in the intestine, which, however, are difficult to detect but which cause an enormous increase of digestion activity after a blood meal. In experiments was shown that the agents of disease listed above remained in the intestine and that none of them were detected in the unicellular salivary glands by electron microscopic investigations. Therefore, active transmission seems to be possible only by direct blood-to-blood contact, which is given by squeezing a leech attached to the skin or by vomiting of intestinal contents into the wound. This so-called **regurgitation** can be caused e.g. by dropping saline onto the sucking leech.

All this indicates that leeches due to the general uptake of enormous amounts of blood may play a role as potential vectors of agents of diseases for humans and animals. The risk is increasing due to the fact that in endemic (mostly tropical) regions the population numbers as well as the infection rates with agents of diseases rise distinctly. Therefore, it is not surprising that blood derived from leeches caught in the wilderness in Cameroon had been tested serologically positive for HIV1 and HIV2 as well as for hepatitis viruses of types A and B.

Control of leeches is impossible in the outdoors. Masoten® is recommended for ponds in fish farms and disinfection of empty facilities with quicklime. Tremazol® by Alpha-Biocare (Neuss, distributed by Sera, Heinsberg) can be used in aquaria.

Further Reading

Cichocka JM et al (2015) A new leech species (Hirudinida: Erpobdellidae: *Erpobdella*) from a cave in the West Azerbaijan province of Iran. Zootaxa 4013:413–427.

Kang JG et al (2016) Molecular detection of *Bartonella* spp. in terrestrial leeches (*Haemadipsa rjukjuana*) feeding on human and animal blood in Gageo-do, Republic of Korea. Parasite Vectors 9:326.

Müller C et al (2016) More than just one: multiplicity of hirudins and hirudin-like factors in the medicinal leech, *Hirudo medicinalis*. Mol Genet Genomics 291:227–240.

Nehili M et al (1994) Experiments on the possible role of leeches as vectors of animal and human pathogens: a light and electron microscopy study. Parasitol Res 80:277–290.

Nelson MC et al (2015) *Mucinivorans hirudinis* gen. nov., sp. nov., an anaerobic, mucin-degrading bacterium isolated from the digestive tract of the medicinal leech *Hirudo verbana*. Int J Syst Evol Microbiol 65:990–995.

Reese K et al (2015) Infections associated with treatment with *Hirudo medicinalis*—report of two cases and review of the literature. Handchir Mikrochir Plast Chir 47:206–209.

Further Reading (Joint List for This Chapter)

Abe N et al (2015) *Ascaridia nymphii* n. sp. from the alimentary tract of a severely emaciated dead cockatiel (*Nymphicus hollandicus*). Parasitol Res 114:4281–4288

Abou-Rahma Y et al (2016) First record of *Anisakis* simplex third-stage larvae in European hake (*Merluccius merluccius*) in Egyptian water. J Parasitol Res ID 9609752

Ahmed ME et al (2013) First report on circulation of *Echinococcus ortleppi* in humped camel. BMC Vet Res 9:127

Ahn KS et al (2015) Identification and prevalence of *Globocephalus samoensis* among wild boars (*Sus scrofa coreanus*) from Korea. Korean J Parasitol 53:611–618

Albo AM et al (2016) Seroprevalence of circulating *Angiostrongylus vasorum* antigen and parasite specific antibodies in dogs. Parasitol Res. doi:10.1007/s00436-016

Alic A et al (2015) *Pearsonema* (*Capillaria*) *plica* infection and associated urinary bladder pathology in red foxes. Parasitol Res 114:1933–1938

Allan JC et al (1999) Helminth parasites of the wild rabbit *Oryctolagus cuniculus* near Malham Tarn, Yorkshire, UK. J Helminthol 73:289–294

Alves PV et al (2015) A redescription of *Lobatostoma kemostoma* from the Florida pompano fish *Trachinatus carolinus*. J Helminthol 89:335–344

Amin OM et al (2013) The description of *Mediorhynchus africanus* n. sp. (Acanthocephala: Gigantorhynchidae) from galliform birds in Africa. Parasitol Res 112:2897–2906

Anderson J et al (2012) *Trichuris* sp. and *Strongyloides* sp. infections in a free-ranging baboon colony. J Parasitol 98:205–208

Arafa SZ et al (2012) Surface features of the monogenean gill parasites *Pseudodactylus anguillae* and *P. bini* from the European eel *Anguilla Anguilla*. Parasitol Res 110:127–133

Aranyo JM et al (2012) Control of *Strongyloides westeri*. Rev Bras Parasitol Vet 21:157–160

Arias P et al (1999) Characterization of *Chabertia ovina* by isoenzyme gel electrophoresis: comparative study within *Oesophagostomum venulosum*. Parasitol Res 85:884–886

Armstrong SK et al (2014) The efficacy of ivermectin, pyrantel and fenbendazole against *Parascaris equorum* infections in foals on farms in Australia. Vet Parasitol 205:575–580

Ash A et al (2015) Controlling *Taenia solium*: impact of a triple dose albendazole-regime. Acta Trop. doi:10.1016/j.actatropica.2015.05.018
Ashour DS et al (2014) Insights into regulatory molecules of intestinal epithelial cell turnover during experimental infection by *Heterophyes heterophyes*. Exp Parasitol 143:48–54
Ashrafi K et al (2014) Fascioliasis: a worldwide parasitic disease of importance in travel medicine. Travel Med Inf Dis 12:636–649
Azam D et al (2012) Temperature and the development and survival of infective *Toxocara* larvae. Parasitol Res 110:649–656
Azizi H et al (2015) Experimental life cycle of *Hypoderaeum conoideum* (Echinostomatidae). Iran J Parasitol 10:102–104
Bachaya HA et al (2015) Prevalence of *Ascaridia galli* in white leghorn. Trop Biomed 32:11–16
Back H et al (2013) The association between *Anoplocephala perfoliata* and colics in Swedish horses. Vet Parasitol 197:580–585
Badets M et al (2013) Alternative development in *Polystoma gallieni* (Monogenea) and life cycle evolution. Exp Parasitol 135:283–286
Bashtar AR et al (2011) Studies on *Moniezia* of sheep. Parasitol Res 108:177–186
Bauer BU et al (2010) Comparison of the FLOTAC technique with the McMaster method and the Baermann technique to determine counts of *Dictyocaulus eckerti* L1 and strongylid eggs in faeces of red deer (*Cervus elaphus*). Parasitol Res 107:555–560
Beck M et al (2015) Comparative recruitment, morphology and reproduction of a generalist trematode, *Dicrocoelium dendriticum*, in three species of hosts. Parasitology 142:1297–1305
Becker B, Mehlhorn H (1980) Light and electron microscopic studies on the effects of praziquantel on *Schistosoma mansoni, Dicrocoelium dendriticum* and *Fasciola hepatica*. Parasitol Res 63:113–128
Behnke JM et al (2015) Bank voles (*Myodes glareolus*) and house mice (*Mus musculus musculus, M. m. domesticus*) in Europe are each parasitized by their own distinct species of *Aspiculuris*. Parasitology 142:1493–1505
Bellaw JL, Nielsen MK (2015) Evaluation of Baerman apparatus sedimentation time on recovery of *Strongylus vulgaris* and *S. edentatus* third stage larvae from equine coprocultures. Vet Parasitol 211:99–101
Bernal D et al (2014) Surface analysis of *Dicrocoelium dendriticum*. The molecular characterization of exosomes reveals the presence of miRNAs. J Proteomics 105:232–241
Beugnet F et al (2013) Preventive efficacy of Frontline® Combo and Certifect® against *Dipylidium caninum* infestation of cats and dogs using a natural flea infestation model. Parasite 20:7
Beugnet F et al (2014) Occurrence of *Dipylidium caninum* from client-owned cats and dogs in Europe using a new PCR detection assay. Vet Parasitol 205:300–306
Bian CR et al (2015) Serological prevalence of *Schistosoma japonicum* in motile populations in previously endemic and now non-endemic regions in China. PLoS One. doi:10.1371/journal.pone.0128896
Bloemhoff Y et al (2015) Prevalence and seasonality of bulk milk antibodies against *Dictyocaulus viviparus* and *Ostertagia ostertagi*. Vet Parasitol 209:108–116
Boag B et al (2001) Patterns of parasite aggregation in the wild European rabbit (*Oryctolagus cuniculus*). Int J Parasitol 31:1421–1428
Böhm C et al (2015) Efficacy of emodepside/praziquantel spot-on (Profender®) against adult *Aelurostrongylus abstrusus* nematodes in cats. Parasitol Res 114:S155–S164
Bohorquez A et al (2014) Coprologically diagnosing *Anoplocephala* in the presence of *A. magna*. Vet Parasitol 204:396–401
Bombarova M et al (2015) Cytogenetics of *Aspidogaster limacoides*: karyotype, spermatocyte division and genome size. Parasitol Res 114:1473–1483
Bonfana B et al (2015) *Echinococcus equinus* and *E. granulosus* from the UK. Int J Parasitol 95:161–166
Bonneau S et al (2009) Anthelmintic efficacies of a tablet formula of ivermectin-praziquantel on horses experimentally infected with three *Strongylus* species. Parasitol Res 105:817–823

Bosi G et al (2005) Changing in the neuromodulators of the diffuse endocrine system of the alimentary canal of farmed rainbow trouts naturally infected with *Eubothrium crassum*. J Fish Dis 28:703–711
Boulana B et al (2013) First report of *Echinococcus shiquicus* in dogs (China). Acta Trop 127:21–24
Boyce K et al (2012) Identification of a new species of digenean *Notocotylus malhamensis* from the bank vole *Myodes glareolus*. Parasitology 139:1630–1639
Brunanska M, Kostic B (2012) Revisiting caryophyllidean type of spermiogenesis in Eucestoda based on *Caryophyllaeus laticeps*. Parasitol Res 110:141–149
Caddigan SC et al (2014) Intra-population variation in behavior modification by the acanthocephalan *Acanthocephalus dirus*: are differences mediated by host condition? Parasitol Res 113:4307–4311
Callejon R et al (2015) Taxonomy and phylogeny of *Trichuris globulosa*. Inf Genetics Evol 34:61–74
Carmena D, Cardona GA (2014) *Echinococcus* in wild carnivorous species. Vet Parastiol 202:69–94
Castellano PP et al (2015) *Dicrocoelium dendriticum* induces autophagic vacuoles accumulation in human hepatocarcinoma cells. Vet Parasitol 212:175–180
Cecchi R et al (2006) Demonstration of *Ollulanus tricuspis* in the stomach of domestic cats by biopsy. J Comp Pathol 134:374–377
Chai JY et al (2015) Larval *Gnathostoma spinigerum* detected in Asian swamp eels (*Monopterus albus*) purchased from a local market in Yangon, Myamar. Korean J Parasitol 53:619–625
Chaignat V et al (2015) The brown hare as novel intermediate host for *E. multilocularis* in Europe. Parasitol Res 114:3167–3169
Chaligannis I et al (2015) *Echinococcus granulosus* infection dynamics in livestock in Greece. Parasitol Res 114:3167–3169
Chaudhry U et al (2015) Genetic evidence for hybridisation between *Haemonchus contortus* and *Haemonchus placei* in natural field populations and its implication for interspecies transmission of anthelmintic resistance. Int J Parasitol 45:149–154
Chelladura JJ et al (2014) *Toxocara vitulorum* infection in a cohort of beef calves in Iowa. Vet Parasitol 214:96–99
Chen J et al (2012) Advances in molecular identification, taxonomy, genetic variation and diagnosis of *Toxocara* sp. Inf Gen Evol 72:1344–1348
Chen HY et al (2015) Effects of the acanthocephalan *Polymorphus minutus* and the microsporidian *Dictyocoela duebenum* on energy reserves and stress response of cadmium exposed *Gammarus fossarum*. Peer J 3:e1353
Chomicz L, Czubai A (1991) Transmission electron micrograph studies of developmental oncospherical envelopes of *Fimbriaria fasciolaris*. Parasitol Res 77:503–508
Choudhary V et al (2015) Transcriptome analysis of the adult rumen fluke *Paramphistomum cervi* following next generation sequencing. Gene 570:64–70
Cichocka JM et al (2015) A new leech species (Hirudinida: Erpobdellidae: *Erpobdella*) from a cave in the West Azerbaijan province of Iran. Zootaxa 4013:413–427
Cole RA et al (2014) *Gnathostoma* spp. live in Asian swamp eels (*Monopterus*) from food markets and wild populations, USA. Emerg Inf Dis 20:634–640
Coucher MA et al (2016) Cystic echinococcosis in South America: systematic review. Trop Med Int Health 21:166–175
Craig P et al (2015) *Echinococcus granulosus* epidemiology and stat-of-art of diagnostics in animals. Vet Parasitol 213:132–148
Cutolo AA et al (2011) Field study on the efficacy of an oral 2% ivermectin formulation in horses. Rev Bras Parasitol Vet 20:171–175
Cvejic D et al (2016) Efficacy of a single dose of milbemycin oxime/praziquantel combination tablets against adult *Echinococcus multilocularis*. Parasitol Res 115:1195–1202

Cwiklinski K et al (2013) Transcriptome analysis of a parasitic clade V nematode: comparative analysis of potential molecular anthelmintic targets in *Cylicostephanus goldi*. Int J Parasitol 43:917–927

Da Silva D, Müller G (2013) Parasites of the respiratory tract of *Sus scrofa scrofa*. Parasitol Res 112:1353–1356

Dalal S et al (2015) Cross antigenicity of immunodominant polypeptides of somatic antigen of *Oesophagostomum columbianum* with other helminths by western blotting. Vet World 8:1279–1285

Dao TM et al (2014) *Opisthorchis viverrini*-like liver fluke in birds from Vietnam. J Helminthol 88:441–446

Das G, Gauly MC (2014) Response to *Ascaridia galli* in growing chicken in relation to their body weight. Parasitol Res 113:1985–1988

Das B et al (2013) Purification and characterization of phosphoenolpyruvate-carboxykinase from *Raillietina echinobothrida*, a cestode parasite of the domestic fowl. Parasitology 140:136–146

Das G et al (2014) Egg production dynamics and fecundity of *Heterakis gallinarum* residing in different caecal environments of chickens induced by fibre-rich diets. Vet Parasitol 2005:606–618

De Chambrier A et al (2014) Taxonomic status of Woodland's enigmatic tapeworms (Proteocephalidea). Syst Parasitol 87:1–19

Deplazes P et al (2013) Lehrbuch der Parasitologie für die Tiermedizin, 3rd edn. Enke, Stuttgart.

Dever ML et al (2015a) Removal of the tapeworm (*Moniezia* spp.) did not increase growth rates. Vet Parasitol 208:190–194

Dever ML et al (2015b) Persistent challenge with *Trichostrongylus colubriformis* and *Haemonchus contortus* larvae does not affect growth of meat-breed lambs, suppressively treated with anthelmintics when grazing. Vet Parasitol 209:76–83

Dewi K, Hasegawa H (2014) Two new species of *Syphacia* in endemic murid rodents from Sulawesi, Indonesia. J Helminthol 88:41–49

Diakou A et al (2015) Occurrence of *Aelurostrongylus abstrusus* and *Troglostrongylus brevior* in domestic cats. Parasites Vectors 8:590

Dianne L et al (2014) Parasite-induced alteration of plastic response to predation threat: increased refuge use but lower food intake in *Gammarus pulex* infected with the acanothocephalan *Pomphorhynchus laevis*. Int J Parasitol 44:211–216

Dimitrijevic B et al (2012) Effects of the infection with *Strongyloides papillosus* and albendazole treatment. Vet Parasitol 186:364–375

Dorny P et al (2009) Emerging food-borne parasites. Vet Parasitol 163:196–206

Dracz RM et al (2014) Parasitological and hematological aspects of co-infection with *Angiostrongylus vasorum* and *Ancylostoma caninum* in dogs. Vet Parasitol 200:111–116

Duan H et al (2015) Complete mitochondrial genome of an equine intestinal parasite: *Triodontophorus brevicauda* (Strongylidae). Parasitol Int 64:429–434

Dubey JP et al (2013) Isolation and characterization of new genetic type of *Toxoplasma gondii* and prevalence of *Trichinella murrelli* from black bear (*Ursus americanus*). Vet Parasitol 196:24–30

Dupuy C et al (2014) Prevalence of *Taenia saginata* cysticercosis in French cattle in 2010. Vet Parasitol 203:65–72

Eamudomkarn C et al (2015) Comparative evaluation of *Strongyloides ratti* and *S. stercoralis* larval antigen for diagnosis of strongyloidiasis. Parasitol Res 114:2543–2551

Eckert J et al (1983) Proliferation and metastases formation of larval *Echinococcus multilocularis*. I. Animal model, macroscopical and histological findings. Parasitol Res 69:737–748

Eckert J et al (1993) Further evidence of the occurrence of a distinct strain of *Echinococcus granulosus* in European pigs. Parasitol Res 79:42–48

Eckert J et al (2011) Alveolar echinococcosis. In: Palmer SR et al (eds) Textbook of zoonoses, 2nd edn. Oxford University Press, Oxford

EFSA (2008) Trends and sources of zoonotic agents and zoonosis and food-borne outbreaks in the European Union. EFSA J 8(1496):1–368
Eichenberger RM et al (2013) Multi-test analysis and model-based estimation of the prevalence of *Taenia saginata* in naturally infected dairy cows. Int J Parasitol 43:853–859
El-Bahy NM, Bazh EK (2015) Anthelmintic activity of ginger, curcumin and praziquantel against *Raillietina cesticillus*. Parasitol Res 114:2427–2434
El-Seifi MA et al (2012) ELISA and some biochemical tests of Heterophyidae infection in laboratory animals. Parasitol Res 110:679–687
Elshazly AT et al (2008) Comparison of three immunodiagnostic tests for experimental *Heterophyes heterophyes* infection in dogs. Vet Parasitol 151:196–202
Evans NA (1977) The site preferences of two digeneans (*Asymphylodora kubanicum* and *Sphaerostoma bramae* in the intestine of the roach. J Helminthol 51:197–204
Ferdushy T et al (2012) *Ascaridia galli* in chickens. Parasitol Res 111:2273–2279
Fitzpatrick JL (2013) Global of food security: the impact of veterinary parasites. Vet Parasitol 195:233–248
Foata J et al (2005) Ultrastructure of spermiogenesis and the spermatozoon of *Macracanthorhynchus hirudinaceus* (Pallas, 1781) (Acanthocephala: Archiacanthocephala), a parasite of the wild boar *Sus scrofa*. J Parasitol 91:499–506
Fogaca RC et al (2014) Phage-displayed peptides as capture antigens in an innovative assay for *Taenia saginata* infected cattle. Appl Microbiol Biotechnol 98:8887–8894
Fogt-Wyrwas R et al (2013) Intraspecific variation between the ITS sequences of *Toxocara canis*, T. cati and Toxascaris leonina. J Helminthol 87:432–442
Foronda P et al (2011) Distribution and genetic variation of hymenolepitid cestodes in murid rodents on the Canary Islands (Spain). Parasites Vectors 4:185
Fourie JJ et al (2013) Prophylactic treatment of flea-infested dogs with an imidacloprid/flumethrin collar (Seresto®, Bayer) to preempt infection with *Dipylidium caninum*. Parasitol Res 112: S36–S46
Frank R et al (2013) Parasites of wild rabbits (*Oryctolagus cuniculus*) from an urban area in Germany, in relation to worldwide results. Parasitol Res 112:4255–4266
Fransen F et al (2015) Genetic evidence of interspecies introgression of mitochondrial genomes between *Trichinella spiralis* and *T. britovi* under natural conditions. Int Gen Evol 36:323–332
Frontera E et al (2005) Evidence and interaction between *Ascaris suum* and *Metastrongylus apri* in experimentally infected pigs. Vet Parasitol 127:295–301
Fuehrer HP et al (2016) *Dirofilaria* in humans, dogs, and vectors in Austria. From imported pathogens to the endemicity of *Dirofilaria repens*. PLoS Negl Trop Dis: e0004547
Galecki R et al (2015) Parasits of wild animals as a potential source of hazard to humans. Ann Parasitol 6:105–108
Galuppi R et al (2013) Evidence of bovine filariosis in Italy (*Parafilaria bovicola*). Vet Parasitol 184:88–91
Gama et al (2016) First report of *Thelazia callipaeda* infection in wild European rabbits (*Oryctolagus cuniculus*). Parasites Vectors 9:230
Gao JF et al (2014) Comparative analysis of the complete mitochondrial genomes of the two ruminant hookworms *Bunostomum trigonocephalus* and *B. phlebotomum*. Gene 541:92–100
Gasso D et al (2014) An identification key for the 5 most common species of *Metastrongylus*. Parasitol Res 113:3495–3500
Gassó D et al (2016) Coprological tests underestimate *Macracanthorhynchus hirudinaceus* burden in wild boar. Parasitol Res 115:2103–2105
Georgieva S et al (2014) *Echinostoma revolutum*. Parasites Vectors 7:520
Gethings OJ et al (2015a) Asynchrony in host and parasite phenology may decrease disease risk in livestock under climate warming: *Nematodirus battus* in lambs as a case study. Parasitology 142:1306–1317
Gethings OJ et al (2015b) Spatial distribution of infectious stages of the nematode *Syngamus trachea* with pheasants (*Phasianus colchius*). Vet Parasitol 212:267–274

Giese EG et al (2015) *Rohdella amazonica* n. sp. (Aspidogastrea) from the Amazonian banded puffer fish *Colomesus psittacus*. J Helminthol 89:288–293
Gjerde B (2013) Phylogenetic position of *Linguatula arctica* and *Linguatula serrata* (Pentastomida) as inferred from the nuclear 18S rRNA gene and the mitochondrial cytochrome c oxidase subunit I gene. Parasitol Res 112:3517–3525
Golestaninasab M, Malek M (2015) Two new species of *Rhinebothrium*. J Helminthol. doi:10.1017/s0022149x15000553
Goltstein B et al (2009) Epidemiology, diagnosis, treatment, and control of trichinellosis. Clin Microbiol Rev 22:127–145
Graeff-Teixera C et al (2016) Update of bayliascariasis. Clin Microbiol Rev 29:375–384
Grencis RK et al (2014) Immunity to gastrointestinal nematodes. Immunol Rev 260:183–205
Grenouillet F et al (2014) *Echinococcus ortleppi* infections in humans and cattle, France. Emerg Inf Dis 20:2100–2112
Hanna REB (2015) *Fasciola hepatica*: histology of the reproductive organs and differential effects of triclabendazole on drug-sensitive and drug-resistant fluke isolates. Pathogens 4:431–456
Hanna REB et al (2016) *Fasciola hepatica*: a light and electron microscope study of the ovary and of the development of oocytes within eggs in the uterus provides an insight into reproductive strategy. Vet Parasitol 221:93–103
Hansen TVA et al (2014) *Trichuris suis* and *Oesophagostomum dentatum* show different sensitivity and accumulation of fenbendazole and levamisole. PLoS Negl Trop Dis: e2752
Haukisalmi V et al (2014) Phylogenetic relationships and taxonomic revision of *Paranoplocephala*. Zootaxa 3873:371–415
Haynes BT et al (2014) Unexpected absence of genetic separation of a highly diverse population of hookworms from geographically isolated hosts. Inf Genet Evol 28:192–200
Heneberg P et al. Integrative taxonomy of Central European parasitic flatworms of the family Prosthogonimidae. Lühe, 1909 (Tremadoda: Plagiorchiidae). Parasitol Int 64:264–273
Higon M et al (2011) Screening trematodes for novel intervention targets: a proteomic and immunological comparison of *Schistosoma haematobium*, *S. bovis* and *E. caproni*. Parasitology 138:1607–1619
Holand H et al (2013) Temporal and spatial variation in prevalence of the parasite *Syngamus trachea* in a meta-population of house sparrows (*Passer domesticus*). Parasitology 140:1275–1286
Houpin E et al (2016) Comparison of three methods for the detection of *Angiostrongylus vasorum* in the final host. Vet Parasitol 220:54–58
Hui YH et al (2002) Food-borne disease handbook, 2nd edn. Marcel Becker, New York
Hunter JS III et al (2013) The efficacy of eprinomectin extended-release injection against naturally acquired nematode parasites of cattle, with special regard to 4th stage inhibited *Ostertagia* larvae. Vet Parasitol 192:346–352
Hüttner M, Romig T (2009) *Echinococcus* species in African wildlife. Parasitology 136:1089–1095
Hüttner M et al (2007) Genetic characterization and phylogenetic position of *Echinococcus felidis*. Int J Parasitol 38:861–868
Irie T et al (2013) Continuous *Moniezia benedeni* infection in confined cattle. J Vet Med Sci 75:1585–1589
Jabbar A et al (2013) The mitochondrial genome of *Protostrongylus rufescens* – implications for population. Parasites Vectors 6:232
Janssen IJI et al (2013) Genetic variants and increased expression of *Parascaris equorum* P-glycoprotein-11 in populations with decreased ivermectin susceptibility. PLoS One: e61635
Jirsa E et al (2008) The occurrence of *Caryophyllaeus laticeps*. J Helminthol 82:53–58
Kabululu ML et al (2015) Risk factors for prevalence of pig parasitosis in Tanzania. Vet Parasitol 212:460–464
Kaminsky R et al (2008) A new class of anthelmintics effective against drug-resistant nematodes. Nature 452:176–180

Kaminsky R et al (2013) Worms – a license to kill. Vet Parasitol 195:286–291

Kanarek G (2009) The occurrence of *Cyathostoma microspiculum* in the great cormorant (*Phalacrocorax carbo*). J Helminthol 83:391–398

Kanarek G et al (2013) *Cyathostoma phenisci* a parasite of the respiratory tract of African penguin (*Spheniscus demersus*). Parasitol Int 62:416–422

Kang JG et al (2016) Molecular detection of *Bartonella* spp. in terrestrial leeches (*Haemadipsa rjukjuana*) feeding on human and animal blood in Gageo-do, Republic of Korea. Parasite Vectors 9:326

Karanikola SN et al (2015) Development of a multiplex fluorescence immunological assay for the simultaneous detection of antibodies against *Cooperia oncophora, Dictyocaulus viviparus* and *Fasciola hepatica* in cattle. Parasites Vectors 8:335

Katahira H, Nagasawa K (2011) *Echinorhynchus hexagrammi* Baeva, 1965 (Acanthocephala: Echinorhynchidae) from marine fishes off Hokkaido, Japan, with morphological observations and new host records. Syst Parasitol 80:35–40

Katahira H, Nagasawa K (2014) Helminths from the giant mottled eel *Anguilla marmorata* Quoy; Gaimard in Japan, with a description of *Acanthocephalus longiacanthus* n. sp. (Acanthocephala: Echinorhynchidae). Syst Parasitol 88:91–102

Kato D et al (2015) The first report of the ante-mortem diagnosis of *Ollulanus tricuspis* infection in two dogs. J Vet Med Sci 77:1499–1502

Katoch R et al (2012) Prevalence and impact of gastro-intestinal helminths on body weight gain in backyard chickens. J Parasit Dis 36:49–52

Kavana NJ et al (2014) The life cycle of *Spirometra* species from Peninsular Malaysia. Trop Biomed 31:487–495

Keiser J, Utzinger J (2009) Food-borne trematodiases. Clin Microbiol Rev 22:466–483

Kelehear C et al (2015) An introduced pentastomid parasite (Raillietiella frenata) infects native cane toads (*Rhinella marina*) in Panama. Parasitology 142:675–679

Kelley JM et al (2016) Current thread of triclabendazole resistance in *Fasciola hepatica*. Trends Parasitol. doi:10.1016/j.pt.2016.03.002

Khadija S et al (2013) Soil moisture influences the development of *Haemonchus contortus* and *Trichostrongylus colubriformis* to third stage larvae. Vet Parasitol 196:161–171

Khedri J et al (2015) Prevalence and intensity of *Paramphistostomum* spp. Iran J Parasitol 10:268–272

Knapp J et al (2015) Taxonomy and molecular epidemiology of *Echinococcus multilocularis*. Vet Parasitol 213:85–91

Knaus M et al (2014) Efficacy of a novel topical combination of fipronil, methoprene, eprinomectin and praziquantel against experimental infections of *Toxascaris leonina* in cats. Vet Parasitol 202:40–44

Knaus M et al (2015) Efficacy of Broadline® against *Capillaria aerophila* lungworm infection in cats. Parasitol Res 114:1971–1975

Knoff M et al (2001) Nematodes of elasmobranch fishes from the southern coast of Brazil. Mem Inst Oswaldo Cruz 96:81–87

Komorová P et al (2015) Acanthocephalans of the genus *Centrorhynchus* (Palaeacanthocephala: Centrorhynchidae) of birds of prey (Falconiformes) and owls (Strigiformes) in Slovakia. Parasitol Res 114:2273–2278

Köse M et al (2015) Infections of larval stages of *Dicrocoelium dendriticum* and *Brachylaima* sp. in brown garden snail (*Helix aspersa*) in turkey. Korean J Parasitol 53:647–651

Koski P et al (2016) Killing *Cyrodactylus salaris* by heat and chemical disinfection. Acta Vet Scand 58:21

Kralova-Hermadova I et al (2015) Development of microsatellite markers in *Caryophyllaeus laticeps*. Parasitol Res 114:721–726

Kravlova-Hromadova I et al (2011) Multiple origins of European populations of the giant liver fluke *Fascioloides magna*. Int J Parasitol 41:373–383

Krecek RC et al (2008) Prevalence of *Taenia solium* cysticercosis. Vet Parasitol 154:38–47

Kuchboev AE, Krücken J et al. (2016) Infection levels of protostrongylid nematodes in definite caprine and intermediate gastropod hosts from Uzbekistan. J Helminthol
Kuhn JA et al (2015) Parasite communities of two three-spined stickleback populations. Parasitol Res 114:1327–1339
Kurt M, Acici M (2008) Cross-sectional survey on helminth infections of chickens in the Samsun region, Turkey. Dt Tierärztl Wschr 115:239–242
Kutzer E (1978) Notes on the hare-whipworm *Trichuris leporis*. Parasitol Res 56:69–72
Kuzmina TA et al (2012) Fecundity of various species of strongylids (Strongylidae) parasites of domestic horses. Parasitol Res 111:2265–2271
Labes EM et al (2011) Genetic characterization of *Strongyloides* sp. from captive, semi-captive and wild Bornean orangutans (*Pongo pygmaeus*). Parasitology 138:1417–1422
Larrieu E et al (2015) Pilot field trial of the EG95 vaccine against ovine echinococcosis. PLoS Negl Dis: e0004134
Latchumikanthan A et al (2014) Occurrence of *Schistosoma nasale* infection in bullocks of Puducherry. J Parasitol Dis 38:238–240
Latrofa MS et al (2015) *Crenosoma vulpis* in wild and domestic carnivores. Parasitol Res 114:3611–3617
Lechat C et al (2015) Comparative evaluation of the prophylactic activity of a slow-release insecticide collar and a moxidectin spot-on formulation against *Thelazia callipaeda* in dogs. Parasites Vectors 8:93
Lee ACY et al (2014) Assessing the speed of kill of hookworms (*Ancylostoma caninum*) by advantage Multi® for dogs using endoscopic methods. Vet Parasitol 204:402–406
Leles D et al (2012) Are *Ascaris lumbricoides* and *A. suum* single species? Parasitol Vectors 5:42
Leok CS et al (2002) Morphology of the oviduct fluke *Prosthogonimus ovatus* from Indonesian native chickens. J Vet Med Sci 64:1129–1131
Lesage C et al (2014) *Protostrongylus pulmonalis* and *P. oryctolagi* parasites of the lungs of the European hare. Parasitol Res 113:2103–2111
Li YS et al (2014) The *Schistosoma japonicum* self-cure phenomenon in water buffaloes. Int J Parasitol 44:167–171
Lightowlers MW (2010) Eradication of *Taenia solium* cysticercosis. Int J Parasitol 40:1183–1192
Ling F et al (2015) Anthelmintic efficacy of cinnamaldehyde and cinnamic acid from cortex cinnamon essential oil against *Dactylogyrus intermedius*. Parasitology 142:1744–1750
Liu GH et al (2012) Characterization of the complete mitochondrial genome of *Trichuris ovis* and *T. discolor*. Inf Gen Evol 12:1635–1641
Liu GH et al (2014) *Chabertia erschowi* is a distinct species based on nuclear rDNA sequences and mDNA sequences. Parasites Vectors 7:44
Liu CN et al (2015) Discrimination between *E. granulosus sensu stricto, E. multilocularis* and *E. shiquicus* a multiplex PCR assay. PLoS Negl Trop Dis: e0004084
Lopez WBZ et al (2014) Historic of therapeutic efficacy of albendazole sulphoxide administered in different routes, dosages and treatment schemes against *Taenia saginata*. Exp Parasitol 137:14–20
Lyons ET, Tollivers SC (2014) Prevalence of patent *Strongyloides westeri* infections in thoroughbred foals in 2014. Parasitol Res 113:4163–4164
Magi M et al (2015) Extraintestinal nematodes of the red fox in Italy. J Helminthol 89:506–511
Mahdavi SA et al (2013) Survey of migratory birds (Anatidae: *Anas* sp.) for schistosome parasites. Iran J Parasitol 8:333–336
Manga-Gonzales MY, Morrondo-Pelago MP (1994) Larval development of ovine *Neostrongylus linearis* in four experimentally infected mollusc species. J Helminthol 68:207–210
Marcus AD et al (2015) Ivermectin treatment of free-ranging endangered Australian sea lion pups: effect on hookworm (*Uncinaria sanguinis*) and lice infection status. Parasitol Res 114:2743–2755
Marques SM, Scroferneker ML (2004) *Onchocerca cervicalis* in horses from Southern Brazil. Trop Anim Health Prod 36:633–636

Marruchella G et al (2012) Fatal bronchopneumonia in a *Metastrongylus elongatus* and porcine circovirus type 2 co-infected pig. Res Vet Sci 93:310–312

Marucci G et al (2009) *Trichinella nelsoni* and *Trichinella T8* mixed infection in a lion. Vet Parasitol 159:225–228

Marucci G et al (2016) Proficiency testing to detect *Trichinella* larvae in meat in the EU. Vet Parasitol 350–359

Masson G et al (2015) Impact of the cestode *Triaenophorus nodulosus* on the exotic *Lepomis gibbosus* and the autochthonous *Perca fluviatilis*. Parasitology 142:745–755

Mayer-Scholl A et al (2014) PCR for the differentiation of *Trichinella nativa* and *T. britovi*. Vet Parasitol 203:247–249

Mayer-Scholl A et al (2016) Study on the occurrence of *Trichinella* spp. in racoon dogs. Vet Parasitol 205:205–209

McFadden AMJ et al (2011) Investigation of an outbreak of *Taenia saginata*. Vet Parasitol 176:177–184

Mehlhorn H (2017) Encyclopedia of parasitology, 4th edn. Springer, Berlin (and online version 2016)

Mehlhorn H et al (1983) Proliferation and metastasis formation of larval *Echinococcus multilocularis*: II. Ultrastructural investigation. Parasitol Res 69:749–763

Mehlhorn H et al (1990) Loperamid, an efficacious drug against the acanthocephalan *Macracanthorhynchus hirudinaceus* in pigs. Parasitol Res 76:624–626

Mendoza-Palmero CA et al (2015) Molecular phylogeny of neotropical monogeneans from cat fishes (Siluriformes). Parasites Vectors 8:164

Millan J, Blasco-Costa J (2012) Molecular evidence of shared hookworm *Ancylostoma tubaeforme* haplo types between Iberian lynx and sympatric domestic cats. Vet Parasitol 186:518–522

Mirjalali H et al (2016) Molecular analysis of isolates of the cestode *Rodentolepis nana* from the great gerbil *Rhombomys opimus*. J Helminthol 90:252–255

Mitchell MC et al (2016) Development of a recombinant protein-based ELISA for diagnosis of larval cyathostomin infection. Parasitology 143:1055–1066

Mlocicki D et al (2010) Functional ultrastructure of the hexacanth larvae in the bothriocephalidean cestode *Eubothrium salvelini*. Parasitol Int 59:539–548

Moazeni M, Nili M (2004) Mixed infection with tapeworms. Trop Biomed 21:23–26

Moné H et al (2015) Introgressive hybridization of *Schistosoma haematobium* by *S. bovis* at the origin of the first case report of schistosomiasis in Corsica, France. Parasitol Res 114:4127–4133

Monteiro CM et al (2011) Helminth parasitism in the Neotropical cormorant (*Phalacrocorax brasilianus*) in Southern Brazil: effect of host size, weight, sex and maturity state. Parasitol Res 109:849–855

Morariu S et al (2016) The prevalence, abundance and distribution of cyathostomins (small stongyles) in horses from Western Romania. Vet Parasitol 223:205–209

Morley NJ et al (2010a) The effects of host size and temperature on the emergence of *Echinoparyphium recurvatum* cercariae from *Lymnaea peregra* under natural light conditions. J Helminthol 84:307–316

Morley NJ et al (2010b) The effects of host size and temperature on the emergence of *Echinoparyphium recurvatum* cercariae from *Lymnaea peregra* under natural light conditions. J Helminthol 84:317–326

Mukaratirwa S, Khumalo MP (2010) Prevalence of helminth parasites in free-range chickens. J S Afr Vet Assoc 81:97–101

Müller C et al (2016) More than just one: multiplicity of hirudins and hirudin-like factors in the medicinal leech, *Hirudo medicinalis*. Mol Genet Genomics 291:227–240

Mungrube EO et al (2006) The prevalence and economic significance of *Fasciola gigantica* and *Stilesia hepatica*. Trop Anim Health Prod 38:475–483

Muniz-Pereira LC, Amato SB (1988) *Fimbriaria fasciolaris* and *Cloacotaenia megalops*, cestodes from Brazilian waterfowls. Mem Inst Oswaldo Cruz 96:767–772

Murthy GSS et al (2013) Evaluation of a polyclonal antibody based on sandwich ELISA for the detection of faecal antigens in *Schistosoma spindale* infections in bovines. J Parasitol Dis 37:47–51
Mutinda M et al (2012) Putative filariosis outbreak in white and black rhinoceros at Meru National Park in Kenya. Parasites Vectors 5:206
Nachev M, Sures B (2016) Seasonal profile of metal accumulation in the acanthocephalan *Pomphorhynchus laevis*: a valuable tool to study infection dynamics and implications for metal monitoring. Parasite Vectors 9(1):300
Naem S (2007) Morphological differentiation among three *Thelazia* species. Parasitol Res 101:145–151
Nagataki M et al (2015) Mitochondrial DNA sequences of 37 collar-spined echinostomes. Inf Genetics Evol 35:56–62
Nehili M et al (1994) Experiments on the possible role of leeches as vectors of animal and human pathogens: a light and electron microscopy study. Parasitol Res 80:277–290
Nelson MC et al (2015) *Mucinivorans hirudinis* gen. nov., sp. nov., an anaerobic, mucin-degrading bacterium isolated from the digestive tract of the medicinal leech *Hirudo verbana*. Int J Syst Evol Microbiol 65:990–995
Ng-Nguyen D et al (2015) Re-evaluation of the species of hookworms infecting dogs in Central Vietnam. Parasites Vectors 8:401
Nielsen MK et al (2015) Serum *Strongylus vulgaris* specific antibody responses to anthelmintic treatment in naturally infected horses. Parasitol Res 114:445–451
Nijsse R et al (2015) *Toxocara canis* in household dogs. Parasitol Res 114:561–569
Okello AL et al (2016) Assessing the impact of a joint human-porcine intervention package for *Taenia solium* control. Acta Trop 159:185–191
Okoro CK et al (2016) Gastrointestinal helminths of wild hogs. J Helminthol 90:139–143
Oksanen A, Lavikainen A (2015) *Echinococcus canadensis* transmission in the North. Vet Parasitol 213:182–186
Oluwasina OS et al (2014) *Linguatula serrata* (Porocephalida: Linguatulidae) infection among client-owned dogs in Jalingo, North Eastern Nigeria: prevalence and public health implications. J Parasitol Res: 916120
Onkoba WN et al (2015) Differential immune responses in mice infected with *Trichinella zimbabwensis*. J Helminthol. doi:10.1017/s0022149x15000723
Otranto D et al (2003) Molecular epidemiological survey on the vectors of *Thelazia gulosa, T. rhodesi* and *T. skrjabini*. Parasitology 127:365–373
Otranto D et al (2007) Morphological and molecular differentiation between *Dicrocoelium dendriticum* and *D. chinensis*. Acta Trop 104:91–98
Otronto D, Traversa D (2002) A review of dicrocoeliosis of ruminants including recent advances in the diagnosis and treatment. Vet Parasitol 107:317–335
Owen IL et al (2014) The probable role of cannibalism in spreading *Trichinella papuae* in a crocodile farm. Vet Parasitol 201:335–338
Page LK (2013) Parasites and the conservation of small populations: the case of *Bayliascaris procyonis*. Int J Parasitol 2:203–210
Pakharukova MY et al (2015) The first comprehensive study of praziquantel effects in vivo and in vitro on the European liver fluke *Opisthorchis felineus*. Int J Antimicrobiol Agents 46:94–100
Pavone S et al (2011) Pathological changes caused by *Anoplocephala perfoliata*. Vet Parasitol 176:43–52
Petersen HH et al (2014) Parasite population dynamics in pigs infected with *Trichuris suis* and *Oesophagostomum dentatum*. Vet Parasitol 199:73–80
Petney TN et al (2013) The zoonotic fish borne liver flukes *Clonorchis sinensis, Opisthorchis felineus* and *O. viverrini*. Int J Parasitol 43:1031–1046
Petry G et al (2011) Efficacy of Procox® oral suspension for dogs against *Toxocara cati* and *Ancylostoma tubaeforme* infection of cats. Parasitol Res 109:S37–S43

Petry G et al (2013) Efficacy of emodepside plus toltrazuril oral suspension for dogs (Procox®, Bayer) against *Trichuris vulpis* in naturally infected dogs. Parasitol Res 112:S133–S138

Phosuk I et al (2013) Short report: molecular evidence of *Trichostrongylus colubriformis* and *T. axei* infections. Am J Trop Med Hyg 89:376–379

Poddubnaya LG et al (2012) Ultrastructure of the vitellarium in the digeneans *Phyllodistomum angulatum* and *Azygia lucii*. Acta Parasitol 57:235–246

Poglayen G et al (2016) Lung parasites of the genus *Metastrongylus* in wild boar *(Sus scrofa)* in Central-Italy: an eco-epidemiological study. Vet Parasitol 217:45–52

Pomaznoy M et al (2013) Adult *Opisthorchis felineus* major protein fractions: comparison with liver flukes *O. viverrini* and *Clonorchis sinensis*. Exp Parasitol 135:297–306

Postigo I et al (2006) *Uncinaria stenocephala*: assessment of antigens for the immunodiagnosis of canine uncinariosis. Exp Parasitol 114:215–219

Poulsen CS et al (2015) Differential serodiagnosis of *Toxocara canis* and *T. cati* – is it possible? Parasite Immunol 37:204–207

Pozio E (2016) Trichinella pseudospiralis. Vet Parasitol 205:214–222

Prchal L et al (2015) Biotransformation of anthelmintics and the activity of drug-metabolizing enzymes in the tapeworm *Moniezia expansa*. Parasitology 142:648–659

Prullage JB et al (2014) Efficacy of a novel topical combination of fipronil (S)-methoprene, eprinomectin and praziquantel against induced infections of *Ancylostoma* spp. nematodes of cats. Vet Parasitol 202:30–33

Quilichini Y et al (2007) Ultrastructural study of spermiogenesis and the spermatozoon of *Crepidostomum metoecus*: a parasite of *Salmo trutta*. J Parasitol 93:458–468

Rakhshandehroo E et al (2014) Molecular and morphological comparison of two different types of *Habronema muscae* in horse. Parasitol Res 113:4439–4445

Ranjan S et al (2010) Evaluation of the persistent activity of moxidectin against *Dictyocaulus viviparus, Haemonchus placei, Trichostrongylus axei* and *Oesophagostomum radiatum* in cattle. Vet Parasitol 167:50–54

Raoul F et al (2015) Trophic ecologie, behaviour and host population dynamics of *E. multilocularis*. Vet Parasitol 213:162–171

Rast L et al (2014) Why are simple control options for *Toxocara vitulorum* not being implemented by cattle and buffalo smallholder farmers in South-East Asia? Prev Vet Med 113:211–218

Reese K et al (2015) Infections associated with treatment with *Hirudo medicinalis* – report of two cases and review of the literature. Handchir Mikrochir Plast Chir 47:206–209

Rehbein S et al (2011) Evaluation of a double centrifugation technique for the detection of *Anoplocephala* eggs. J Helminthol 85:409–414

Reinemeyer CR, Cleale RM (2002) Dose confirmation studies of moxidectin 1% non-aqueous injectable and moxidectin 0.5% pour-on formulation against *Oesophagostomum radiatum* and *Trichuris discolour*. Vet Parasitol 108:75–81

Reyda FB et al (2011) Diversification and species boundaries of *Rhinebothrium*. PLoS One 6(8): e22604

Rimaldi R et al (2007) *Passalurus ambiguus*: new insights into copromicroscopic diagnosis. Parasitol Res 101:557–561

Roberts JA (1993) *Toxocara vitulorum* in ruminants. Helminthol Abstr 62:151–174

Rondeland D et al (2015) Larval trematode infections ion *Lymnaea glabra* populations. Parasite 22:38

Rushworth RL et al (2015) *Rodentolepis straminea* in an urban population of *Apodemus sylvaticus* in the UK. J Helminthol. doi:10.1017/s0022149x15000632

Saijuntha W et al (2011) Genetic characterization of *Echinostoma revolutum* and *Echinoparyphium recurvatum*. Parasitol Res 108:751–755

Saksirisampant W et al (2012) *Gnathostoma spinigerum*: immunodepression in experimental infected mice. Exp Parasitol 132:320–326

Santoro M et al (2011) Helminth community structure of the Mediterranean gull (*Ichthyaetus melanocephalus*) in Southern Italy. J Protozool 97:364–366

Saowakon N et al (2013) *Paramphistomum cervi*: the in-vitro effect of plumbagin on motility, survival and tegument structure. Exp Parasitol 133:179–186
Sato Y et al (2008a) Mitochondrial DNA sequence and gene order of the Sri Lankan *Schistosoma nasale* is affiliated to the African/Indian group. Parasitol Int 57:460–464
Sato H et al (2008b) Visceral helminths of wild boars (*Sus scrofa leucomystax*) in Japan with special reference of the genus *Morgascaridia*. J Helminthol 82:159–168
Schähle Z et al (2016) Infection levels of plerocercoids of the tapeworm *Triaenophorus crassus*. J Helminthol 90:54–61
Schimmel A et al (2011) Efficacy of emodepside plus toltrazuril (Procox®) against *Toxocara canis, Uncinaria stenocephala* and *Ancylostoma caninum*. Parasitol Res 109:S1–S8
Schuster RK (2009) The parasite fauna of stray domestic cats (*Felis catus*). Parasitol Res 105:125–134
Schuster RK, Sivakumar S (2013) A xenodiagnostic method using *Musca domestica* for the diagnosis of gastric habronemosis and examining the anthelmintic efficacy of moxidectin. Vet Parasitol 197:176–181
Schuster RK et al (2007) Liver flukes in dogs and treatment with praziquantel. Vet Parasitol 150:362–365
Scioscia NP et al (2014) The Pampas fox (*Lycalopex gymnocercus*) as a new definitive host for *Spirometra erinacei*. Acta Trop 133:78–82
Seaton EM et al (2001) Presence and recovery of *Ascaridia dissimilis* ova on the external shell surface of turkey eggs. Avian Dis 45:500–503
Senathilake KS et al (2016) Rhizome extracts of *Curcum zedoaria* induce apoptosis in *Setaria digitata*. Exp Parasitol 167:50–60
Senlik B et al (2011) Helminth infections of wild boars (*Sus scrofa*). J Helminthol 85:404–408
Sereno-Uribe AL et al (2015) Morphological and molecular analyses of larval and adult stages of *Echinoparyphium recurvatum*. J Helminthol 84:458–464
Sevimli FC et al (2011) Efficacy of eprinomectin pour-on treatment in sheep naturally infected with *Dictyocaulus filaria* and *Cystocaulus ocreatus*. J Helmintol 85:472–475
Shah HB et al (2013) Seasonal changes in infra-populations of *Diplozoon kashmirensis*. Parasitol Res 112:3347–3356
Sharma R et al (2015) *Capillaria hepatica* infection. Ann Parasitol 61:61–64
Silva MRI et al (2015) Host-specificity and morphometrics of female *Haemonchus contortus, H. placei* and *H. similis* in cattle and sheep. J Helminthol 89:302–306
Simsek S et al (2015) First report of *Echinococcus equinus* in a donkey in Turkey. Korean J Parasitol 53:731–735
Sissay M et al (2008) Prevalence and seasonal incidence of larval and adult cestode infections of sheep and goats. Tro Anim Health Prod 40:387–394
Skala V et al (2014) Insights into the development of *Notocotylus attenuatus*. Parasitol Int 63:94–99
Skallerup P et al (2015) Detection of a quantitative trait locus associated with resistance to *Trichuris suis* in pigs. Vet Parasitol 210:264–269
Skirnisson K et al (2016) Occurrence of *Mesocestoides canislagopodis* in mammals. Parasitol Res. doi:10.1007/s00436-016-5006-5
Snabel V et al (2014) Molecular identification of *Heterakis spumosa* obtained from brown rats (*Rattus norvegicus*) and its infectivity in mice. Parasitol Res 113:3449–3455
Soares MCP et al (2013) Anatomo-chemical and molecular description of liver neotropical echinococcosis caused by *E. oligarthrus* in human host. Acta Trop 125:110–114
Soares MPC et al (2014) Neotropical echinococcosis: *E. vogeli* natural infection and its main definitive host: the bush dog (*Speathos venaticus*). Parasitol Int 63:485–487
Sobecka E et al (2012) Genetic and morphological variation in *Echinorhynchus gadi* Zoega in Müller, 1776 (Acanthocephala: Echinorhynchidae) from Atlantic cod *Gadus morhua* L. J Helminthol 86:16–25
Sommerfelt IE et al (2006) Prevalence of *Toxocara cati* parasites in cat's feces. Vet Parasitol 140:296–301

Southgate VR et al (1995) Mating behaviours in mixed infections of *Schistosoma haematobium* and *S. mattheei*. Parasitol Res 81:651–656

Stewart TB et al (1996) Doramectin efficacy against the kidney worm *Stephanurus dentatus* in sows. Vet Parasitol 66:95–99

Strobel H et al (2013) Comparison of three different treatment schedules for praziquantel (Cestocur®, Bayer) in the treatment of tapeworm infections (*Moniezia* sp.). Parasitol Res 112:S139–S147

Strube C et al (2015) Vaccination with recombinant paramyosin against the bovine lung worm *Dictyocaulus viviparus*. Parasites Vectors 8:119

Studzinska MB et al (2012) The strongylidae belonging to *Strongylus* genus in horses. Parasitol Res 111:1417–1421

Swiderski Z et al (2011) Egg formation and the early embryonic development of *Aspidogaster limacoides* with comments on their phylogenetic significance. Parasitol Int 60:371–380

Szell Z et al (2015) Environmental determinants of the spatial distribution of *Mesocestoides* sp. Vet Parasitol 212:427–430

Tararu M et al (2015) Morphological and molecular genetic characterization of three *Capillaria* species (*C. anatis, C. pudendotecta, C. madseni*) and *Baruscapillaria obsignata* in avians. Parasitol Res 114:4011–4022

Taraschewski H et al (1990) Loperamid, an efficacious drug against fish-pathogenic acanthocephalans. Parasitol Res 76:619–623

Tarbiat B et al (2016) The efficacy of flubendazole against different developmental stages of the poultry roundworm *Ascaridia galli* in laying hens. Vet Parasitol 218:66–72

Thaenkhan U et al (2012) Families Opisthorchiidae and Heterophyidae: are they distinct? Parasitol Int 61:90–93

Thompson RC (2008) The taxonomy phylogeny and transmission of *Echinococcus*. Exp Parasitol 119:439–446

Thompson RCA (2015) Neglected zoonotic helminth: *Hymenolepis nana, Echinococcus Canadensis* and *Ancylostoma ceylanicum*. Clin Microbiol Infect 21:426–432

Titi A et al (2014) Ruminal paramphistomosis in cattle from Northeastern Algeria: prevalence, parasite burdens and species identification. Parasite 21:50

Toet H et al (2014) Liver fluke vaccines in ruminants. Int J Parasitol 44:915–927

Toledo R et al (2015) Strongyloidiasis with emphasis on human infections and its different clinical forms. Adv Parasitol 88:165–241

Tolnai Z et al (2015) Environmental determinants of the spatial distribution of *Angiostrongylus vasorum, Crenosoma vulpis* and *Eucoleus aerophilus* in Hungary. Vet Parasitol 207:355–358

Tolstenkov OO et al (2012) The neuro-muscular system in fresh water furcocercaria from Belurus: I. Schistosomatidae. Parasitol Res 110:185–193

Tomczuk K et al (2015) Seasonal changes of the diagnostic potential in the detection of *Anoplocephala perfoliata*. Parasitol Res 114:767–772

Townson S et al (2006) *Onchocerca* parasites and *Wolbachia* endosymbionts. Filaria J 5:4

Trambow SR et al (2015) Prevalence of gastrointestinal helminth infections in ovine populations of Kashmir valley. Vet World, EISSN: 2231-0916

Traversa D, Di Cesare A (2014) Cardio-pulmonary parasitic nematodes affecting cats in Europe. Front Vet Sci 1:11

Traversa D et al (2005) Molecular characterization of *Thelazia lacrymalis* affecting equids. Mol Cellular Probes 19:245–249

Traversa A et al (2012) A large outbreak of *Opisthorchis felineus* in Italy. Clin Microbiol Inf Dis 31:1089–1093

Trevisan C et al (2016) Severe seizures in pigs naturally infected with *Taenia solium*. Vet Parasitol 220:67–71

Ulesco T et al (2016) Detection of *Dirofilaria repens* and *D. immitis* DNA in mosquitoes from Belarus. Parasitol Res. doi:10.1007/s00436-016-5118-y

Usmanova NM, Kazakou VI (2010) The DL1 repeats in the genome of *Diphyllobothrium latum*. Parasitol Res 107:449–452

Vadlejch J et al (2016) Efficacy and persistent activity of moxidectin against natural *Muellerius capillaris* infections in goats and pathological consequences of muelleriosis. Vet Parasitol 218:98–101

Van Dijk J, Morgan ER (2012) The influence of water and humidity on the hatching of *Nematodirus battus* eggs. J Helminthol 86:287–292

Vega-Sanchez MC et al (2015) Induction of ulcerative colitis in mice influences the course of infection with *Trichuris muris*. J Helminthol 89:593–600

Veneziano V et al (2011) Efficacy of eprinomectin pour-on against *Dictyocaulus arnfieldi* infection in donkeys (*Equus asinus*). Vet J 190:44–415

Venkatesh C, Ramalingam K (2007) Lactic acid, pyruvic acid and lactate pyruvate ratio in the anoplocephalid tapeworm *Stilesia globipunctata* infecting sheep. Vet Parasitol 144:176–179

Verschave SH et al (2014) The parasite phase of *Ostertagia ostertagi*. Int J Parasitol 44:1091–1104

Vicente B et al (2014) A *Fasciola hepatica* derived fatty acid binding protein induces protection against schistosomiasis caused by *Schistosoma bovis*. Exp Parasitol 145:145–151

Vimalraj PG, Latchumikanthan A (2015) *Schistosoma spindale* infection in a captive jackal (*Canis aureus*). J Parasitol Dis 39:120–121

Viney M, Kikuchi T (2016) *Strongyloides ratti* and *S. venezuelensis* – rodent models of *Strongyloides* infections. Parasitology. doi:10.1017/s0031182016000020

Vlaminck J et al (2014) Advances in the diagnosis of *Ascaris suum* infections in pigs and their possible application in humans. Parasitology 141:1904–1911

Vlaminck J et al (2015) Vaccination of calves against *Cooperia oncophora* with a double-domain activation-associated secreted protein reduces parasite egg output and pasture contamination. Int J Parasitol 45:206–213

Von Brandt T (1957) Glycogen and lipoids in *Dictyophyme renale*. Z Tropenmed Parasitol 8:21–23

Waap H et al (2014) Parasitic communities in stray cat population from Lisbon, Portugal. J Helminthol 88:389–395

Wang CR et al (2012) Characterization of *Bunostomum trigonocephalus* and *B. phlebotomum* from sheep and cattle. Res Vet Sci 92:99–102

Wang Y et al (2013a) Genetic variability among *Trichuris ovis* isolates. Mitochondrial DNA 2013:50–54

Wang CR et al (2013b) Sequence variability in four mitochondrial genes among *Bunostomum trigonocephalus* isolates from China. J Helminthol 87:416–421

Wang F et al (2014) *Spirometra* severely infecting wild-caught snakes from food market in Guangzhou and Shenzhen. Sci World J ID874014

Weyher A et al (2010) Molecular identification of *Schistosoma mattheei* from feces of Kinda (*Papio cynocephalus kindae*) and Grayfoot baboon (*Papio ursinus griseipes*) in Zambia. J Parasitol 96:184–190

Wicht B et al (2010) Inter- and intraspecific characterization of tapeworms of the genus *Diphyllobothrium* from Switzerland using nuclear and mitochondrial DNA targets. Parasitol Int 59:35–39

Williams AR et al (2014) Assessment of the anthelmintic activity of medicinal plant extracts and purified condensed tannins against free-living and parasitic stages of *Oesophagostomum dentatum*. Parasites Vectors 7:518

Wolf D et al (2014) *Oxyuris equi*: lack of efficacy in treatment with macrocyclic lactones. Vet Parasitol 201:163–168

Xu WW et al (2015) The complete mitochondrial genome of *Strongylus equinus*: comparison with other closely related species and phylogenetic analyses. Exp Parasitol 199:95–99

Yabsley MJ et al (2015) High prevalence of *Porocephalus crotali* infection on a barrier island (Cumberland Island) off the coast of Georgia, with identification of novel intermediate hosts. J Parasitol 101:603–607

Yakhchali M et al (2013) Detection of infection with larval stages of *Ornithobilharzia turkestanicum* using PCR. Iran J Parasitol 8:627–633

Yank X et al (2015) Mitrochondrial genome of *Hypoderaeum conoideum*-comparison with selected trematodes. Parasites Vectors 8:97

Yao C et al (2011) *Filaroides osleri* (*Oslerus osleri*): two cases and a review of canid infections in North America. Vet Parasitol 179:123–129

Yoneva A et al (2014) First study of vitellogenesis of the broad fish tapeworm *Diphyllobothrium latum*. Parasitol Int 63:747–753

Zait H et al (2016) Molecular characterization of *E. granulosus* and *E. Canadensis*

Zalesny G, Hildebrand J (2012) Molecular identification of *Mesocestoides* spp. Parasitology 110:1055–1061

Zanetti-Lopez WD et al (2014) Anthelmintic efficacy of ivermectin and abamectin in naturally infected pigs. Res Vet Sci 97:546–549

Zhao L et al (2015) Genetic differences between *Chabertia ovina* and *C. erschowi* revealed by sequence analysis of 4 mitochondrial genes. Mitochondrial DNA 26:167–170

Zintl A et al (2014) Bovine paramphistomes in Ireland. Vet Parasitol 204:199–208

Zuloaga J et al (2013) A rat model of intragastric infection with *Anisakis* spp. live larvae. Parasitol Res 112:2409–2411

Ectoparasites (Arthropods: Acari, Insecta, Crustacea) 6

Name: greek: *arthron* = joint; *pus*, *podos* = foot; *hex* = six; *pus*, *podos* = foot (=hexapods); *acari* = mite; *zoon* = animal. Latin: *crusta* = crust, bark, shell.

Animals with a heterogeneous segmentation are termed **arthropods**. Their more or less rigid exoskeleton composed of chitin and other elements has to be sloughed regularly during periods of growth. The heteronomous body and leg segments of arthropods are connected through membranous elements and thus movable. Depending on the structure of their **mouthparts**, arthropods are subdivided into amandibulates and mandibulates. The amandibulates basically comprise the Chelicerata, whereas the **mandibulates** include Crustaceans and Insecta (=insects). Some authors additionally rank the Pentastomida among the amandibulates, which are herein referred to as a distinct phylum.

From the parasitological point of view, arthropods mainly include ectoparasites. They are of particular importance because of their high numbers and their frequent function as carriers (from host to host) of viruses, rickettsiae (=obligate intracellular bacteria), other bacteria and different stages of endoparasites. Besides this direct function in infections of various hosts, winged arthropods (e.g. gnats, flies) or fast-moving species (e.g. bugs, cockroaches) play a substantial role during the geographical propagation of pathogens. Without harming themselves, they become **vectors** during epidemics. In addition to using such living vectors, many stages of parasites (e.g. cysts of amoeba, worm eggs and larvae), however, utilize non-living vector systems such as dust, droplets, water, wind, etc., as means of propagation.

6.1 Arachnids, Chelicerata

The parasitic species of **arachnids** (ticks, mites) are taxonomically closely related to free-living representatives like spiders and scorpions (Figs. 6.1 and 6.2), which are predators and only threaten humans or their pets and livestock by "unintended" poison bites or stings. They are characterized by a direct development, meaning that

© Springer International Publishing Switzerland 2016

H. Mehlhorn, *Animal Parasites*, DOI 10.1007/978-3-319-46403-9_6

Fig. 6.1 Typical "Garden cross spider" inside its catching net hanging anterior end downwards. The anterior end bearing the chelicerae is directed to the ground

a stage, which resembles the adults, hatches from the egg and reaches the final size and sexual maturity through numerous moults.

In contrast, the parasitic species of **arachnids** (ticks, mites) pass differentially shaped developmental stages (**larvae**, **nymphs** and **adults**). Larvae have six legs, whereas adults possess eight legs. Each of which is subdivided into seven segments (coxa, trochanter, femur, patella, tibia, tarsus, pretarsus). Both the free-living and the parasitic arachnids are characterized by two pairs of mouthparts: the more or

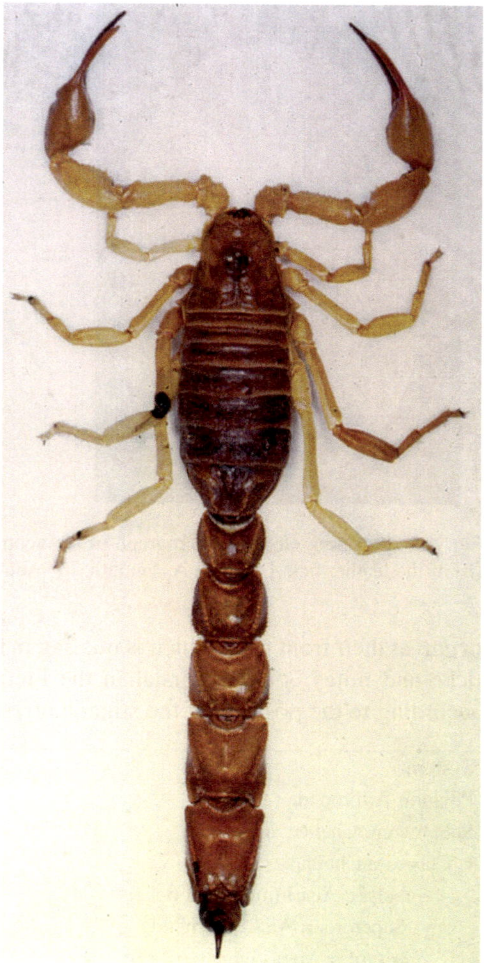

Fig. 6.2 Typical scorpion showing the injection pike of the poison gland at the end of the tail, which in aggression position is placed between the two anterior claws

less scissor-like **chelicerae** and the segmented palps (**pedipalps**). While most spiders (Araneae) and scorpions are **predators**, most Acari (Acarina) are ectoparasites with the exception of some mites living as burrowing endoparasites. The Acari differ from the other arachnids in that their body segments are fused and their body does not seem to be subdivided into prosoma and opisthosoma. Only the front part of the body, containing the mouthparts, is well defined and called **gnathosoma** (Fig. 6.3). The chelicerae and the pedipalps are inserted at the **basis capituli**, both of which are differentially modified between species. The subclass Acari, which can roughly be further divided into **ticks** and **mites**, is characterized by the location and the arrangement of paired openings of the tracheae called **stigmata** (respiratory system). In ticks, these openings are always positioned behind the coxa ("hips") of the third or fourth pair of legs as **metastigmata**, while the position is variable in mites. In addition, ticks possess a so-called **Haller's**

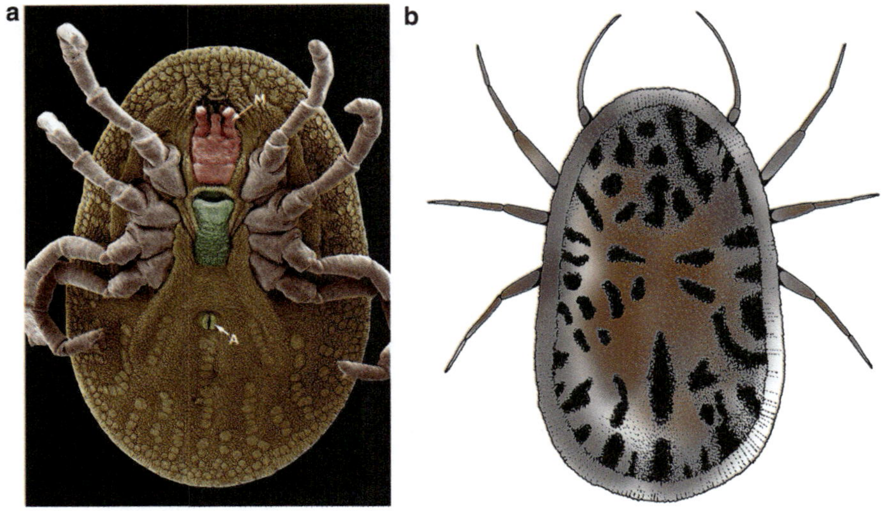

Fig. 6.3 Scanning electron micrograph of the ventral side (**a**) and diagrammatic representation (**b**) of the leather tick *Argas* sp. A = mouth; M = mouth

organ at their front tarsi, which is missing in mites. The Zoological classification of ticks and mites is controversial in the literature. Many modern authors classify according to the position of the stigmata (respiratory openings) at the ventral side:

System
Phylum: Arthropoda (abstract)
Subphylum: Chelicerata
Class: Arachnida
Subclass: Acari (mites and ticks)
Superorder: Anactinotrichida
Order: Metastigmata (ticks)
Family: Argasidae (soft ticks)
Family: Ixodidae (hard ticks)
Order: Mesostigmata
Family: Dermanyssidae (bird mites)
Family: Liponyssidae (rodent mites)
Family: Varroidae (bee mites)
Superorder: Actinotrichida
Order: Prostigmata
Family: Demodecidae (hair follicle mites)
Family: Trombiculidae (harvest mites)
Order: Astigmata (=Cryptostigmata)
Family: Acaridae
Family: Glycyphagidae (dust mites)
Family: Sarcoptidae (itch mites)
Family: Psoroptidae (mange mites)

For that reason, there will only be a rough distinction between ticks and mites in this book. Yet there will be a subdivision into families, because between them the fewest differences are to find.

6.1.1 Ticks

Ticks can grow up to 3 cm when soaked and are of special importance in veterinary and human medicine because of their function as transmitters of major diseases even in temperate zones. The transmission of pathogens like viruses, extracellular bacteria, rickettsiae, anaplasmas and protozoans as well as nematodes occurs during the blood meal. Thereby ticks do not sting small veins like other bloodsuckers. Instead, they use their **hypostome** equipped with barbed hooks to dig a pit, which becomes filled with blood several times during the sucking process, which may take several minutes up to days. Thus, ticks are "**pool feeders**". Their flexible cuticle allows the ticks to swell up to their multiple body size when having an extremely filled gut. Therefore, numerous ticks, of which all stages of both gender suck with only a few exceptions, can cause heavy blood loss (anaemia) even in big animals. As a consequence, these animals are more prone to infectious diseases. Some ticks are anchored with their chelicerae and their barbed hooks within the hypostome so deeply in the host that they can hardly be pulled out. In doing so, to avoid the release of the pathogens, they should not be stunned or squeezed. The injected saliva of the ticks contains, besides vasodilatory, narcotic and anticoagulant substances, also neurotoxic components, which can provoke **ascending tick paralysis** in some species. Thereby the musculature of the back limbs of hosts (including humans) is paralysed first, followed by paralysis of the front limbs. As soon as the paralysis reaches the breast muscles, this leads to death. However, as long as the ticks are pulled out early enough, the paralyses are reversible during 12–14 h. With a single blood meal, which has to be taken before a moulting or oviposition, ticks are able to survive for a surprisingly long time (several months, some hard ticks survived in laboratories for up to 10 years).

According to their morphology and biology, ticks are separated into two families:

1. **Soft ticks** (Argasidae, Table 6.1; Figs. 6.3 and 6.4)
2. **Hard ticks** (Ixodidae, Table 6.1; Figs. 6.5 and 6.6)

In both families, the sexually mature adults evolve from the eight-legged nymphs via the six-legged larvae, which hatched from the eggs. Each stage is completed with a moult being essential for further growth. To find their host, both tick families use a set of chemoreceptors located along their forelegs, in particular in an organized group of them called the **Haller's organ** (Fig. 6.7). The major differences between both families are summarized in Table 6.1.

Table 6.1 Important differences between soft ticks (Argasidae) and hard ticks (Ixodidae)

	Soft ticks (Argasidae)	Hard ticks (Ixodidae)
Cuticle	The cuticle appears leather-like	The cuticle is relatively hard
Scutum	A shield (**scutum**) does not exist, therefore sexual dimorphism does not exist	A shield (**scutum**) of rather rigid cuticle covers the entire back of males, but only partially the back of females, nymphs and larvae. The scutum, however, is not stretched during blood feeding
Captiulum	The capitulum is visible from dorsal only in larvae and thus invisible in the other stages	The capitulum with mouthparts is protruding the anterior end of the tick
Stigmata	Stigmata are located between the coxae of the third pair of legs	Stigmata are located behind the coxae of the fourth pair of legs
Eyes	Most species do not have eyes	Often each an eye is formed at both edges of the scutum being composed of a cuticular lense and sensory cells (except for e.g. *Ixodes*, *Haemaphysalis*)
Blood feeding	Nymphs and adults suck blood several times (e.g. every 4–6 weeks for 30–60 min); larvae suck for several days	Larvae, nymphs and adults suck blood only once (for several days)
Nymphs	Usually two, in some species up to eight nymph stages may occur (*Argas*: often 4)	There occurs only one nymph stage in the life cycle
Mating	Multiple mating. Oviposition: a few hundreds after each copulation after various blood meals (*Argas:* 4–6 times: 100–300 eggs each)	The males die after several copulations, which occur during the blood meal of the female. The females die after the one-time oviposition (Egg numbers: 3000 for *Ixodes*; 6000 for *Dermacentor*; 15,000 for *Amblyomma*)
Way of living	The species mainly live in grooves, cracks, etc., of stables, nests and attack their hosts while sleeping	Species mainly live outdoors; during their developmental cycle they attack one up to three hosts

Circulatory system. Ticks exhibit an open circulatory system, into which the dorsally located heart is incorporated. The transported fluid called haemolymph passes all internal organs and (unfortunately) is an ideal carrier of pathogens. It is composed of plasma and at least five other cell types.

Respiratory system. Nymphs and adults possess a tracheal system which is missing in most larvae. Two so-called stigmata positioned at the hind legs (name = metastigmata) are the external connection.

Nervous system. It is composed of a central nervous system (CNS) of 0.5 mm size. It arises from the fusion of the suboesophageal and the supraoesophageal ganglion and is therefore called **synganglion**. A definite ventral nerve cord is missing in ticks, yet single retrograde dispersed nerve fibres are present.

Sensory organs. Ticks have developed several sensory organs. Most frequently are **thermoreceptors** and **mechanoreceptors** without pores called **setae** (=bristles) spread throughout the whole body. A unique accumulation of setae with pores is located at the Haller's organ. These setae serve as chemoreceptors to detect the host's perspirations. These systems are of particular importance when no eyes are present as it is the case in the genera *Ixodes* and *Argas*. If formed, the eyes of Ixodidae are located laterally and those of Argasidae ventrally and are present in all three developmental stages. These eyes are composed of a large cuticular, transparent lens under which several bundles of unipolar photoreceptor cells are clustered. However, sharp pictures cannot be generated with this system.

Sexual systems. The female sexual system is composed of a vagina, which opens up ventrally to the outside and of a pair of ovarial tubes. They often join up at their ends (creating a U) or are even reduced to one single tube. The oocytes are located along the outside of the tube, where they mature and are released to the inside. Soon after the fertilization a quite thick eggshell evolves, which is covered with a wax film by the so-called Gené's organ during egg deposition. The oocytes of the Ixodidae do not start to grow before the tick has finished the sucking process at the surface of the host. They still contain a thinned spot (so-called micropyle) through which sperms could invade. However, as the ovary is already filled with sperms, it is thought that the sperms enter the oocytes via the so-called **funicles** (=short stalks) before the thick eggshell is formed. This ovary's basic structure explains why young and thin-walled oocytes are predominantly attacked by pathogens from the haemolymph. Among accessory glands, higher developed species like the mestastriate Ixodidae (e.g. *Dermacentor Rhipicephalus, Amblyomma*) additionally have a sperm storage, in which the so-called endospermatophores are kept. This storage is missing in prostriate genera (e.g. *Ixodes*).

The **male sexual system** is composed of paired seminiferous tubules, which are U-shaped, and accessory glands. The sperm formation starts with the formation of **prosperms** of which several are enclosed together into one spermatophore. The process of subsequent maturation is dependent on the genera. The sperm formation in male *Ixodes* can already be completed during the nymph stage. Consequently, the mating can take place directly after the moult (still on the soil). However, most male *Ixodes* mate several times. The up to 0.5 mm long prosperms, which are embedded into the spermatophores, do not develop further until they are transferred to a female. This finally 24 h lasting maturation process is called **capacitation** or **spermateleosis**. The mature sperm is about 1 mm (!) long and gets motile by myofibrils located peripherally. During fertilization only the nucleus of the sperm invades the oocyte (karyogamy). The sex of ticks is determined by sex chromosomes. Hard ticks usually are type XX-X0 (possibly with different X chromosomes) and soft ticks are type XX-XY. Most ticks are diploid, although some parthenogenetic species (e.g. member of the *Haemaphysalis longicornis*) are triploid or even polyploid. The diploid numbers of chromosomes differ in females depending on the species between 12 and 36. Yet, there occur high variations within one genus (e.g. *Ornithodoros gurneyi*—12 and *O. alactiagalis*—

34). Males of these ticks always harbour one chromosome less than the females because they possess only one sex chromosome (e.g. *Haemaphysalis sulcata* 21 instead of 22). The genus *Argas* has a diploid genotype of 24 autosomes and 2 sex chromosomes. In general, the X chromosome is much bigger (3–4 times) compared to the autosomes in all ticks.

The **host shift** of hard ticks (Ixodidae) usually processes in a species-specific manner. They are classified as three-, two- or one-host ticks (see Tables 6.2 and 6.3). Soft ticks (Argasidae) are characterized by falling of their host after each quite short blood meal (**repletion**). One-host *Ixodes* (e.g. all species of the genus *Boophilus*) attack their host as soil hatched larvae and stay on this host as nymphs and adults. Solely females leave the host to oviposit on the soil. The development is temperature dependent and thus takes about 8–12 weeks. Two-host ticks (e.g. *Rhipicephalus bursa*, *R. evertsi*) attack their hosts as larvae and leave it only as blood-filled nymphs. After their moult on the soil, the adults search for the second host, on which the next sucking action and the mating takes place. To oviposit, the females fall on the ground afterwards. In three-host ticks (e.g. *Ixodes ricinus*, *Dermacentor marginatus*, *Haemaphysalis* species), every blood meal is followed by a moult on the soil. Afterwards, each stage searches for a new host, which is often bigger than the previous host (e.g. mouse → hare → cow). Depending on the climate, the entire development may last from 3 months up to 3 years. After a blood meal, ixodid ticks can possibly survive several years (in laboratories up to 10 years have been reported) without sucking at another host.

Table 6.2 Argasid and ixodid ticks: trivial and scientific names (examples)

American dog tick	*Dermacentor variablilis*
Brown dog tick	*Rhipicephalus sanguineus*
Brown ear tick	*Rhipicephalus appendiculatus*
Castor bean tick	*Ixodes ricinus*
Cayenne tick	*Amblyomma cajennense*
European sheep tick	*Ixodes ricinus*
Eyed tampan	*Ornithodorus moubata*
Karoo tick	*Ixodes rubicundus*
Kennel tick	*Rhipicephalus sanguineus*
Lone star tick	*Amblyomma americana*
Moose tick	*Dermacentor albipictus*
Pigeon tick	*Argas reflexus*
Red legged tick	*Rhipicephalus evertsi*
Rocky mountain wood tick	*Dermacentor albertsoni*
South African bont tick	*Amblyomma hebraeum*
Spinose ear tick	*Otobius megnini*
Taiga tick	*Ixodes persulcatus*
Tropical bont tick	*Amblyomma variegatum*
Tropical horse tick	*Anocentor nitens*
Winter tick	*Dermacentor albipictus*
Yellow dog tick	*Haemaphysalis leachi*

Table 6.3 Important tick species and transmitted pathogens

Species	Length of unfed adults (mm)[a]	Hosts	Main host[b]	Disease/agents of disease	Type of agent of disease[c]
Argasidae—Soft ticks					
Ornithodorus moubata	♀ 10 ♂ 8	Many	**Humans**, many animals	Recurrent fever (*Borrelia duttoni*)	S
Argas persicus	♀ 5.5–11 ♂ 5.5–8	Many	Poultry, many animals	Spirochaetosis	
A. reflexus	5–8	Many	Pigeons, many animals, **humans**	Spirochaetosis (*Borrelia anserina*)	S
Otobius megnini *O. lagophilus*	4–8	Many	Dogs, ruminants, horses, pigs, **humans**	Secondary bacterial infection	–
Ixodidae—Hard ticks					
Ixodes ricinus	♀ 2.8–3.4 (7–8) ♂ 2.8–4	3	Dogs, cats, cattle, **humans**	Borreliosis, TBE, Redwater (*Babesia divergens*, *B. microti*)	B V P
I. dammini[d] *I. pacificus* *I. scapularis*	♀ 3–4 (12) ♂ 3–4.2	3 3 3	Deer, cattle, **humans**	Borreliosis, Encephalitis; Babesiosis	B V P
Dermacentor marginatus	5 (16)	3	Many mammals	Tularemia (*Francisella tularensis*) Rocky Mountain spotted fever (*Rickettsia rickettsii*)	B R
D. reticulatus	5 (10)	3	Many mammals	Anaplasmosis, Piroplasmosis (*Babesia canis*, *Theileria equi*)	A P
D. andersoni	5	3	Many mammals, humans	Anaplasmosis Piroplasmosis (*Babesia canis*, *Theileria equi*)	A P
Boophilus annulatus	♀ 2–2.5 (6–8) ♂ 2	1	Cattle	Texas fever (*Babesia bigemina*), bovine piroplasmosis (*B. bovis*)	P
B. microplus	♀ 2–2.5 (6–8) ♂ 2	1	Cattle, horses	Q fever (Coxiella *burneti* = *R. burneti*) Anaplasmosis (*Anaplasma marginale*)	R A

(continued)

Table 6.3 (continued)

Species	Length of unfed adults (mm)[a]	Hosts	Main host[b]	Disease/agents of disease	Type of agent of disease[c]
Amblyomma species *A. variegatum* *A. hebraeum*	♀ 6–7 (-20) ♂ 5–6	3	Many mammals, **humans**	Tularemia (*Francisella tularensis*) Rocky Mountain spotted fever (*Rickettsia rickettsii*) Theileriosis	B R
Hyalomma species *H. anatolicum* *H. marginatum*	4–6 (10–14)	2–3	Ruminants	Mediterranean theileriosis (*Theileria annulata*)	P
Rhipicephalus appendiculatus	♀ 2–4 (8–10) ♂ 4–5	3	Cattle, goats, horses, dogs	East coast fever (*Theileria parva*)	P
R. bursa	4 (9–11)	2	Cattle, goats, horses, dogs	Piroplasmosis (*Babesia ovis*, *B. canis*, *Theileria ovis*)	P
R. evertsi	4 (9–11)	2	Many mammals	East coast fever (*Theileria parva*) Biliary fever (*Theileria equi*) Q fever (*R. conori*) Spirochaetosis (*Borrelia theileri*)	P P R S
R. sanguineus	♀ 2–3 (6–7) ♂ 2	3	Dogs, **humans**	Boutonneuse fever (*Rickettsia conori*) Piroplasmosis	P
Haemaphysalis punctata	♀ 2.8–3.5 (8–9) ♂ 2.5–3.1	3	Ruminants, **humans**	Encephalitis Piroplasmosis Anaplasmosis	V P A
H. leachi leachi	♀ 2.8–3.5 (8–9) ♂ 2.5–3.1	3	Carnivores, small mammals	Canine Piroplasmosis Tick bite fever (*Rickettsia conori*) Q fever (*R. burnetii;* syn. *Coxiella*)	P R

[a]Size of fed females in brackets
[b]Hosts were chosen because of transmission of agents of disease; others are possible
[c]These agents of disease do not occur in all hosts and may be transmitted by other tick species as well
[d]Some authors consider *I. dammini* as *I. scapularis*
A Anaplasma, *B* Bacteria, *P* Protozoans, *R* Rickettsiae, *S* Spirochaetes, *V* Virus, *TBE* Tick-borne encephalitis

This explains why agents of disease (transmitted by ticks) reported to have disappeared in a region reappear after several years.

Nutrition. The search of hosts and the species-specific procedure of the blood meal of the ticks led to the development of specific behavioural patterns. First of all, ticks have to find a host. In all ticks, this happens through sensing of body odours, which can prompt e.g. Argasidae to migrate several metres (from the attic to the flat). **The host specificity** is generally low as any endothermic individual is accepted. That is why ticks are often imported by rodents into gardens where they then attack people or their pets. The blood meals of ticks (Argasidae: max. 1 h, Ixodidae 5–7 days) usually last quite a long time compared to the short sucking action of gnats or mosquitoes. Therefore, the injection of locally narcotic and anticoagulant substances is indispensable. According to recent studies, the tick saliva also contains **prostaglandin** (=unsaturated hydroxyl acid). Likewise, arachidonic acid is often supplemented. It modulates the immune response of the host and thus attenuates the inflammatory reaction, which is of particular importance with regard to the long-lasting sucking action. Secretions originate from the botryoidal structured acini of the salivary glands (Fig. 6.4), which contain both granular and agranular cell types. While hard ticks possess a high variety of those cells, soft ticks only have one type of granular and agranular acini, respectively. The sucking action of hard ticks occurs in the following phases:

1. Preparation of the sucking tube (without sucking, max. up to 24 h),
2. Slow sucking (2–4 days),
3. Fast sucking (12–36 h),
4. Fast release and potentially falling off the host.

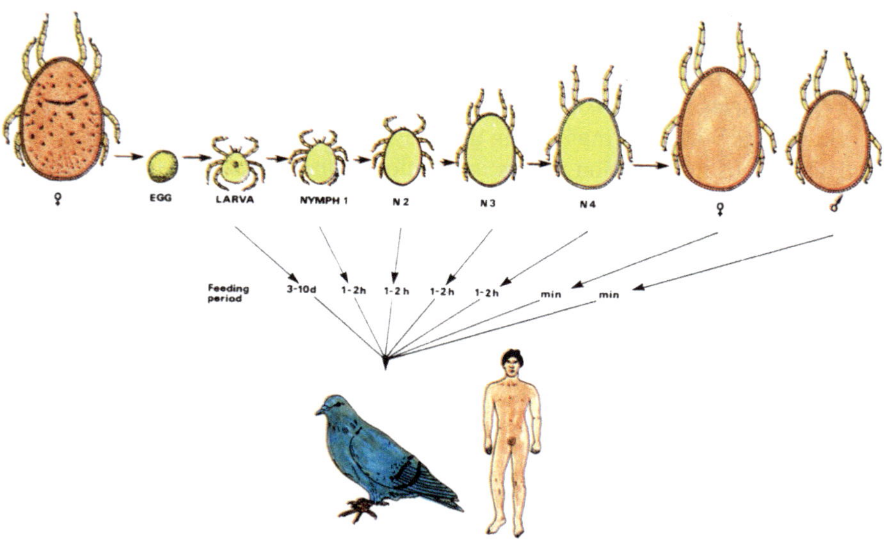

Fig. 6.4 Diagrammatic representation of the developmental stages of *Argas* sp. h = hours of sucking time; N1–N4 = nymph stages

The ingested blood is phagocytosed by enterocytes within the gut. As a consequence, the cells swell up and possibly displace the whole intestinal lumen. Bursted enterocytes get replaced by growing surrogate cells. Pathogens are also taken up with the blood and reach blood digesting phagolysosomes. Only if the pathogens are adapted to the respective enzymes, they are able to survive and to pass into the haemolymph or into other organs of the tick. Numerous pathogens are successful in doing so, but thereby received a quite high level of host specificity. Therefore, some pathogens are only transmitted by one species or, in some cases, by only one race within a species. Ticks are of great importance as vectors as it is shown for some pathogens (out of more than 50 existing ones) in Tables 6.2 and 6.3. Since they suck blood of potentially infected animals, a variety of pathogens are detected in ticks. This does not necessarily mean that they in turn transmit these pathogens. Following **types of pathogens** have been observed in ticks:

Viruses:		Intra- and extracellular
Bacteria:	*Borrelia*:	Intra- and extracellular
	Rickettsiales:	Intracellular and intranuclear
	Anaplasma:	Intracellular
Protozoa:	Piroplasms:	Intracellular within cytoplasm
		Free within the haemolymph
		Within the saliva
Roundworms:		Intra- and extracellular

There are four possible **transmission routes** of pathogens into the inner of a tick:

1. **Transstadial transmission**: The pathogens enter the salivary gland cells (Fig. 6.4) and are later released within the saliva. This mode has been proven for viruses, bacteria, piroplasms and larvae of roundworms.
2. **Transovarial transmission**: The pathogens reach the undifferentiated oocytes via the tick's haemolymph and in that way infect the following tick generation. This way of transmission occurs in addition to the transstadial transmission in the case of the transmission of the FSME viruses, *Borrelia* stages, *Babesia* stages and probably also in the case of *Rickettsia* bacteria.
3. **Transmission by regurgitation**: During an ongoing meal, parts of the intestinal contents are pumped into the host. *Borrelia* stages and trypanosomes are claimed to be transmitted via this route.
4. **Transmission by eating ticks**: *Hepatozoon* species or larvae of roundworms are transmitted in this way when their reptile host "nibbles" its own or foreign ticks from the skin. Bacteria are also supposed to be transmitted like this.

It is remarkable that some pathogens can be transmitted by several tick species mostly belonging to different genera. This is probably due to the unspecific host choice and the geographical distribution of pathways of ticks. Tick larvae and nymphs are often transported by birds, leading to a fast spread of pathogens along the flight paths of visitant or migratory birds. However, the pathogens can develop only in a few vertebrates as they cannot overcome immunological barriers that easily. For people in Europe two pathogens being transmitted by *Ixodes ricinus* are of particular importance and their propagation is ongoing: the FSME virus and the spirochaetes of the *Borrelia burgdorferi* complex.

Further Reading

Bowman AS, Nuttal PA (eds) (2004) Ticks: biology, disease and control. Cambridge University Press, Cambridge.
Estrada-Pena A et al (2004) Ticks of domestic animals. Houten, Atlantia.
Jongejan F et al (eds) (2003) Ticks and tick-borne pathogens. Academic, Dordrecht.
Mehlhorn B, Mehlhorn H (2009) Ticks on progress. Düsseldorf University Press, Düsseldorf.
Sonenshine DE, Roe RM (eds) (2014) Biology of ticks. 2 Volumes, Oxford University Press, Oxford.

6.1.1.1 Argasidae (Soft Ticks)

1. **Name**: greek: *argos* = shiny; *ornithes* = birds; *doros* = leather pouch
2. **Geographical distribution/epidemiology**: worldwide
3. **Biology/morphology**:

   ***Argas* species** (including *A. reflexus, A. polonicus, A. persicus*) are characterized by their ovoid, dorsoventral body (Fig. 6.3). Males grow up to 8 mm and females up to 1.1 cm and appear in a grey-brown colour. The mouthparts (except for the six-legged larvae) are not visible dorsally. Nymphs and adults suck blood from their hosts for about half an hour at night. Thereby, adults take up about 0.3 ml of blood (=vast blood loss during mass infestations). Larvae, however, stay feeding up to 10 days on their host. The developmental period is temperature dependent and requires a suitable host. It can therefore last from 3 months up to 3 years. *Argas* species are not host specific and can thus attack several bird species (**pigeons, poultry and free-living birds but also many mammals including humans**) (Fig. 6.4).

   In arid or semi-arid areas, *Ornithodoros* species (including *O. moubata* = *eyeless tampan*; *O. savignyi* = *eyed tampan, sand tampan*) are found, which hide away in barns or shacks during the day.

   In India, Africa and the USA, the species ***Otobius megnini*** occurs. Females of this species do no longer suck blood, while larvae and nymphs attack the host (cattle, horses) and feed within the ears (Tables 6.2 and 6.3).
4. **Symptoms of disease**: Due to the extensive blood loss, e.g. pigeons suffer from severe exhaustion or even an inability to fly. In young animals, this can lead to death. Other animals suffer from skin inflammation.
5. **Diagnosis**: Detection of larvae in the plumage, search for hiding places (cracks, chinks) of adult stages in the barn.
6. **Path of infection**: Softs ticks actively seek their hosts at night.

7. **Prophylaxis**: Hygienic measures (cleaning of the barn, sealing cracks and chinks) and disinfection with contact insecticides; add acaricide to sand baths. *Caution*: Humans can be infested as well.
8. **Therapy/Control**: Burning old litter; contact insecticides (aerosol treatment, besprinkle, sprays) to be used preferably in the morning.

Further Reading

Boxler B et al (2016) Host finding of the pigeon tick *Argas reflexus*. Med Vet Entomol 30:193–199.

Díaz-Martín V et al (2015) New salivary anti-haemostatics containing protective epitopes from *Ornithodoros moubata* ticks: assessment of their individual and combined vaccine efficacy. Vet Parasitol 212:336–349.

Díaz-Martín V et al (2015) Development of vaccines against *Ornithodoros* soft ticks: An update. Ticks Tick Borne Dis 6:211–220.

Montasser AA et al (2011) Efficacy of abamectin against the fowl tick, *Argas* (*Persicargas*) *persicus* (Oken, 1818) (Ixodoidea: Argasidae). Parasitol Res 109:1113–1123.

Niebuhr CN et al (2014) Efficacy of chemical repellents against *Otobius megnini* (Acari: Argasidae) and three species of ixodid ticks. Exp Appl Acarol 64:99–107.

Niebuhr CN et al (2013) Off-host collection methods of the *Otobius megnini* (Acari: Argasidae). J Med Entomol 50:994–998.

Pourseyed SH et al (2010) *Metarhizium anisopliae* (Ascomycota: Hypocreales): an effective alternative to chemical acaricides against different developmental stages of fowl tick *Argas persicus* (Acari: Argasidae). Vet Parasitol 172:305–310.

Uspensky I (2014) Tick pests and vectors (Acari: Ixodoidea) in European towns: Introduction, persistence and management. Ticks Tick Borne Dis 5:41–47.

6.1.1.2 Ixodidae (Hard Ticks)

1. **Name**: greek: *ixodes* = sticky; *hyalinos* = transparent; *omma* = eye; *rhipidion* = fan; *kephale* = head; *amblys* = weak; *bos* = cattle: *philein* = loving, adoring; *derma* = skin; *kentetes* = spiny; *haima* = blood; *physa* = bladder.
2. **Geographical distribution/epidemiology**: Worldwide. In Europe, there exist mainly species belonging to four different genera (Fig. 6.5). Sometimes masses of hard ticks are found on a broad spectrum of animals.
3. **Biology/morphology**: The following hard ticks are most common on **dogs** and **cats** (Table 6.2):
   (a) ***Ixodes ricinus*** (Wood tick): Males grow up to 4 mm, not engorged females up to 5 mm, while engorged ones may reach up to 1.5 cm (Figs. 6.5–6.13), eyes are missing, anal cleft is situated in front of the anus, palps are long and slim.
   (b) ***Rhipicephalus sanguineus*** (Brown dog tick, recently entering in Germany, but only viable in barns, flats, etc., while outside of buildings in warm

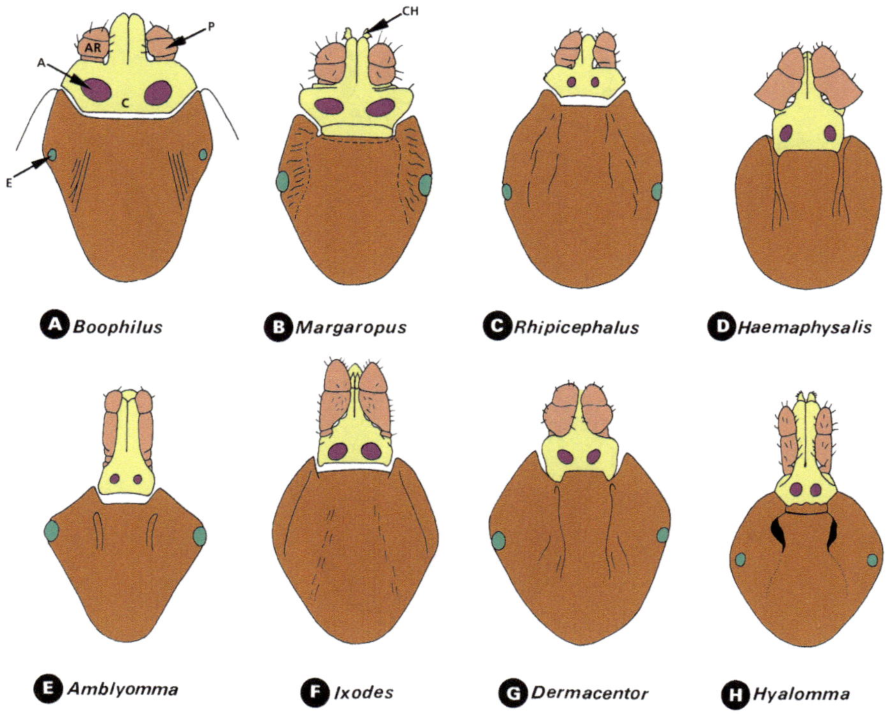

Fig. 6.5 Diagrammatic representation of the anterior portion of the body and of the shield (=scutum) of hard tick genera. A = area porosae; AR = segment of the pedipalps; C = capitulum; CH = chelicerae; E = eye; P = pedipalps

countries). Males grow up to 3.5 mm, not engorged females up to 3 mm, engorged ones reach up to 1.2 cm of high. Eyes exist, anal cleft behind the anus, palps short and wide, basis capituli is hexagonal (Figs. 6.14, 6.15 and 6.16).

(c) Particularly on dogs ***I. hexagonus*** (Hedgehog tick), ***I. canisuga*** (Fox tick) (Table 6.4), ***Haemaphysalis concinna*** (Game ticks), ***Dermacentor marginatus*** (Sheep tick), ***Hyalomma marginatum*** (Fig. 6.19) and ***Dermacentor reticulatus*** (Ornate cow tick (Figs. 6.17, 6.18 and Table 6.2).

All these species are three-host ticks, meaning that the larvae (three pairs of legs), the nymphs (four pairs of legs) and the adults each parasitize another host. The developmental period depends (since it takes place outdoor) on the temperature, the humidity and the success in finding of a host (all are non-host specific). These ticks mainly appear in larger numbers during spring and autumn. This is due to the fact that they need up to 3 years for their entire development in temperate zones with distinct winters. Therefore, after dormancy, several developmental stages (**larvae**, **nymphs** and **adults**) of different tick generations may infest one host at the same

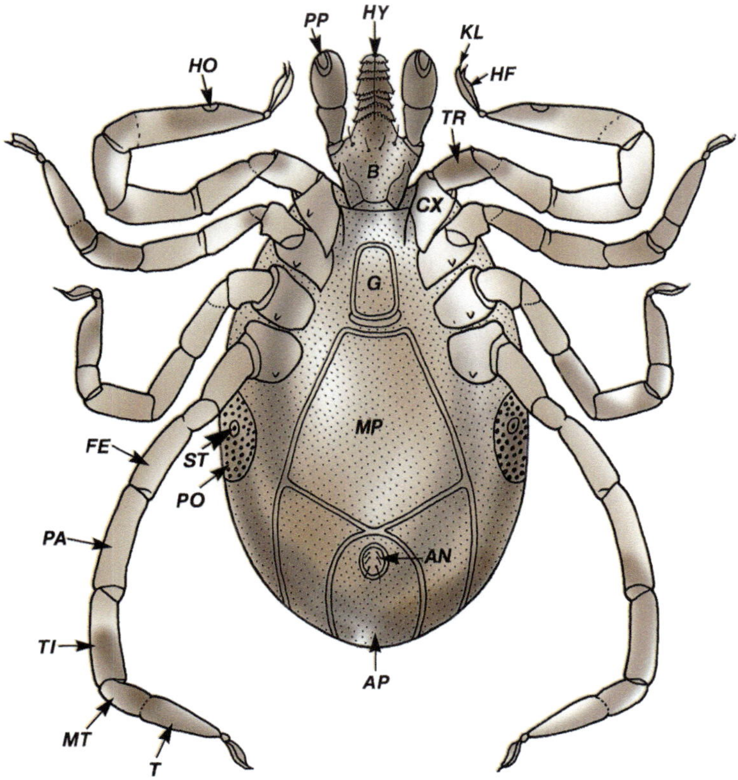

Fig. 6.6 Diagrammatic representation of the ventral side of a hard tick (Ixodidae). AN = anus; AF = anus plate; B = basis of the capitulum; CX = coxa; G = genital plate; FE = femur; HF = holdfast papilla at the foot; HO = Haller's organ (sense); HY = hypostome; KL = claw; MP = median plate; MT = metatarsus; PA = patella; PO = pore plate; PP = pedipalps; ST = stigma (breathing opening); T = tarsus; TI = tibia; TR = trochanter

time. After the moult on the ground during late summer/spring, they seek a new host (depending on the opportunity) often resulting in mass infestations. *I. ricinus* larvae suck for 4–5 days, nymphs for 3–5 days and females for 5–14 days, while males suck several times but only shortly and copulate thereafter with a female during its sucking action. Females ingest up to 200 times blood of their own weight (up to 400 mg). This also holds true for other hard tick species.

Ruminants are mostly attacked by hard ticks (Ixodidae), while soft ticks (Fig. 6.3) are only rarely found thereon. This is surely due to fact that local species prefer avian hosts and only suck blood for a short time at night before hiding again. Hard ticks on ruminants may be diagnosed looking at the arrangement and structure of their mouthparts. With a few exceptions (including the two-host species *Rhipicephalus bursa*), they are three-host ticks, meaning that the three developmental stages (larvae, nymphs, adults)

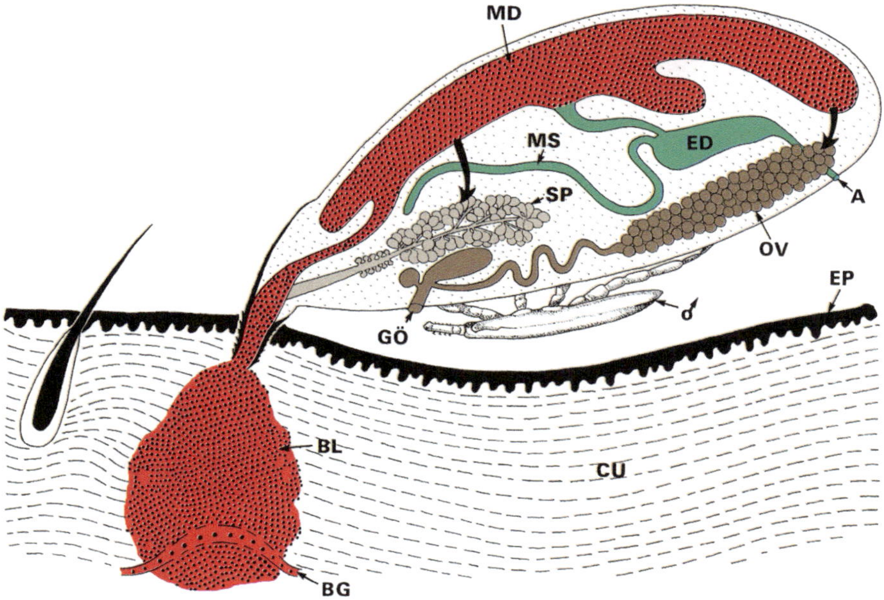

Fig. 6.7 Diagrammatic representation of a female hard tick during bloodsucking, while the male is in copulation position. The arrows point onto the pathways of the agents inside the tick. A = anus; BG = blood vessel; BL = blood lacune; CU = cutis; ED = hind gut; EP = epidermis; GO = genital opening; MD = blood filled midgut; MS = Malpighi ductules; OV = ovarial tube with oocytes; SP = salivary glands

suck blood each on a different host, respectively, and drop off after the blood meal, which lasts for several days.

Species determination

(1a) Pedipalps are long. Anal cleft in front of the anus. Without eyes (Fig. 6.5F). (♂ up to 3 mm, ♀ engorged up to 14 mm; eggs up to 0.5 mm; three-host ticks)... ***Ixodes ricinus***

(1b) Pedipalps are short, anal cleft behind the anus...........................2

(2a) Without eyes, pedipalps are short cone-shaped (Fig. 6.5D)........... ....Genus ***Haemaphysalis*** (including *H. punctata*—♂ up to 3 mm, ♀ engorged up to 14 mm; *H. leachi* yellow dog tick; three-host ticks)

(2b) With eyes (Fig. 6.5)... 3

(3a) With festoons (festoon like wrinkles at the dorsum).................. 4

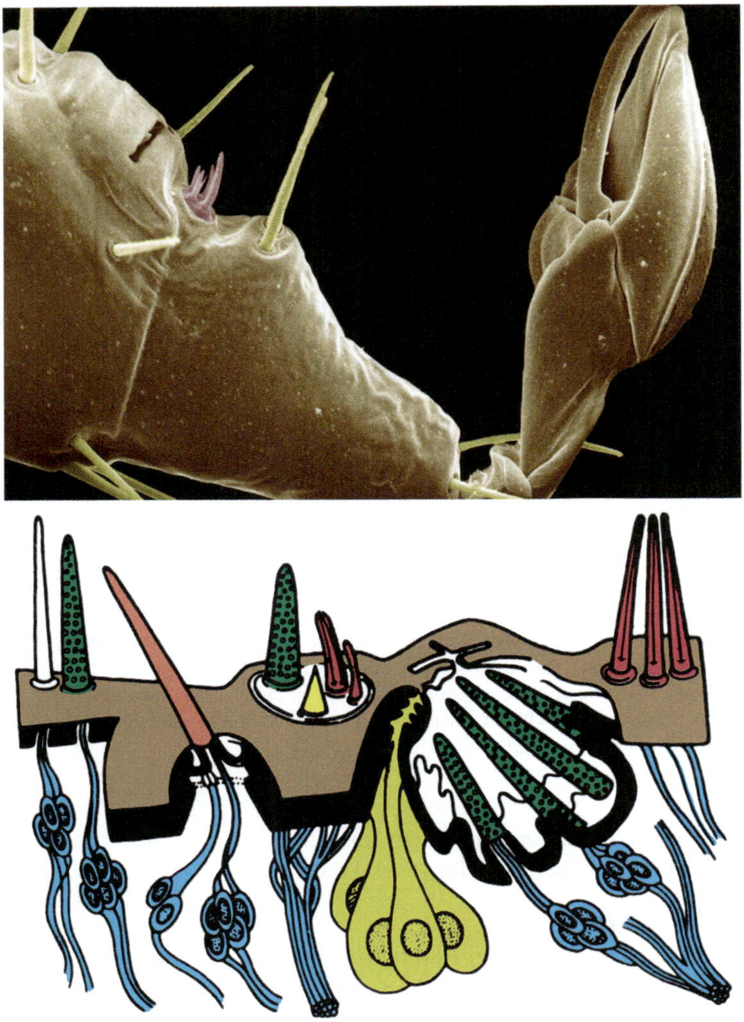

Fig. 6.8 Scanning electron micrograph (**a**) and diagrammatic representation (**b**) of the so-called Haller's organ at the tarsus of the forelegs of *Ixodes riciunus*. From the invagination protrude the tips of sensory organs

(3b) Without festoons....................................Genus ***Boophilus***
(including *B. annulatus* in the USA; without anal cleft, short three-part pedipalps (Fig. 6.5A); one-host ticks)
(4a) Constantly with 11 festoons... 5
(4b) Irregular festoons... 6

a

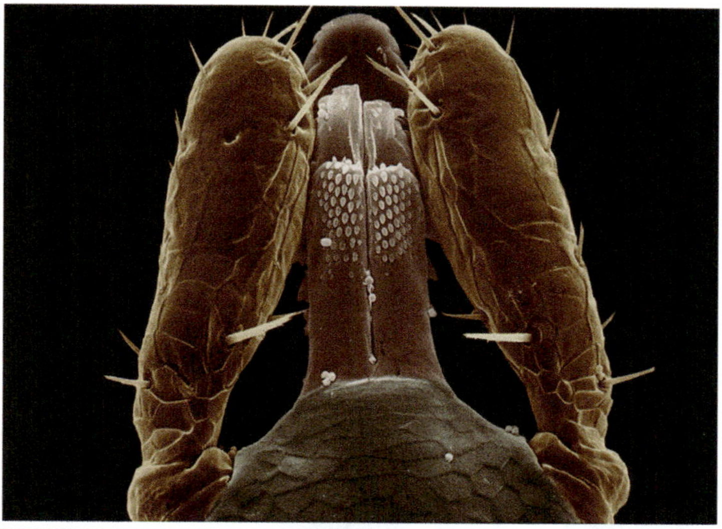

b

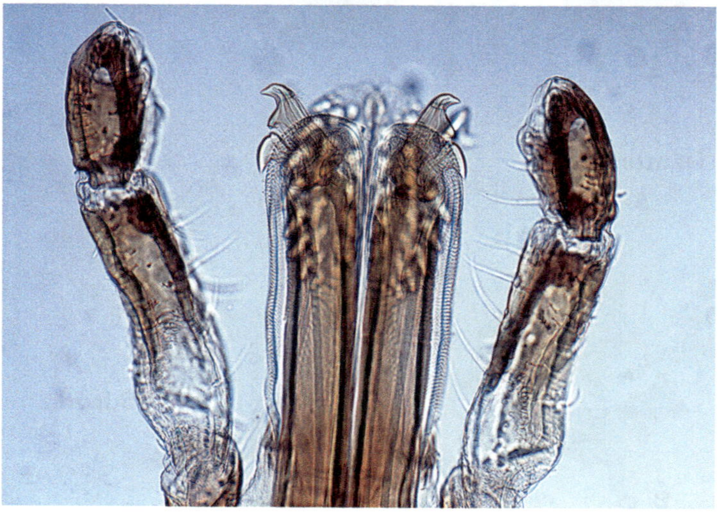

Fig. 6.9 Scanning electron micrograph (**a**) and light micrograph (**b**) of the anterior end of an ixodid tick (*I. ricinus*) showing the sucking channel, chelicerae and pedipalps

(5a) Long pedipalps.. Genus ***Amblyomma*** (including *A. hebraeum, A. variegatum* in Africa (Fig. 6.20); three-host ticks)

(5b) Short pedipalps, square basis capituli (Fig. 6.17)...............Genus ***Dermacentor***

(6a) Short pedipalps, hexagonal basis capituli (Fig. 6.14–6.16)........Genus ***Rhipicephalus*** (including *R. sanguineus*, brown dog tick, all three-host

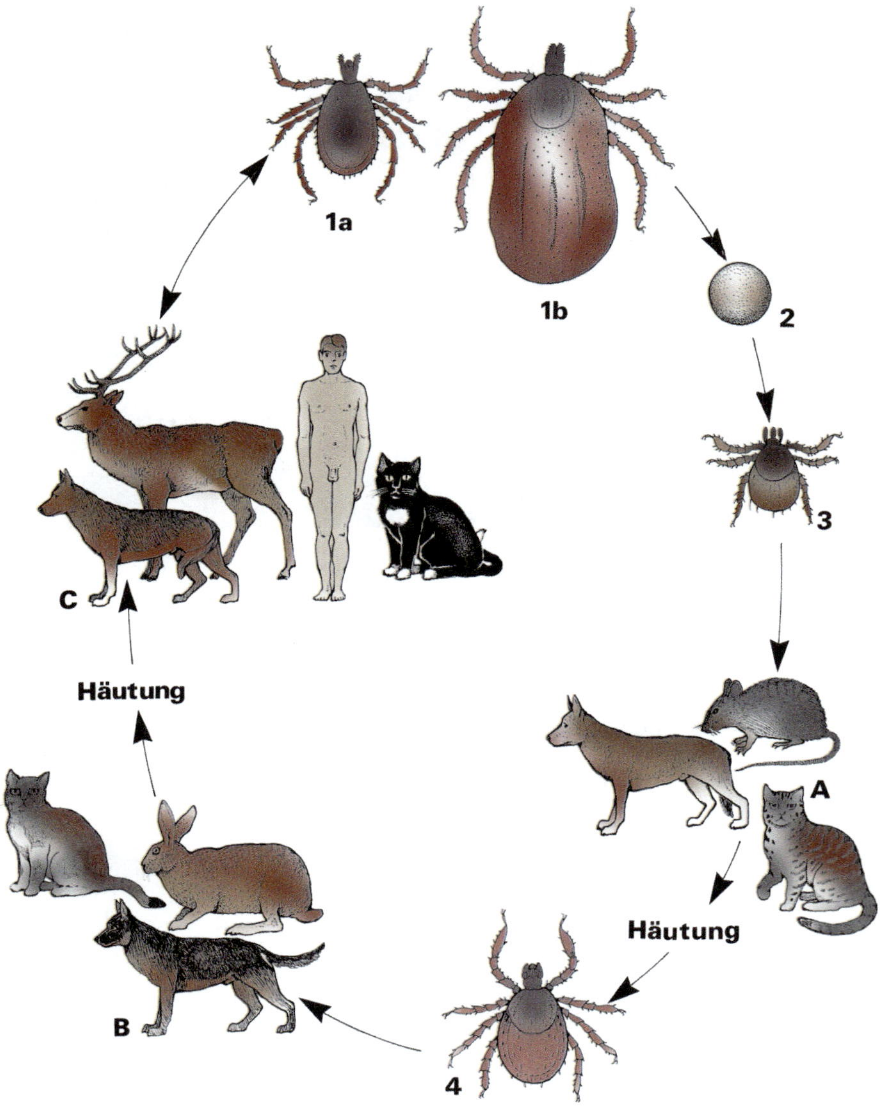

Fig. 6.10 Diagrammatic representation of the developmental cycle of *Ixodes ricinus*. **1a** Male; **1b** Engorged female; **2** Egg; **3** Larva; **4** Nymph; Häutung: engl. molt

ticks; *R. bursa*, brown tick, two-host ticks, ♂ up to 3 mm, ♀ engorged up to 1.5 cm (Fig. 6.14))

(6b) Long pedipalps, comma-shaped spiracle (Fig. 6.19)...........Genus ***Hyalomma*** (including *H. anatolicum* excavatum in the Mediterranean area; mostly two-host but also three-host ticks, ♀ engorged up to 25 mm)

Fig. 6.11 Photos of *Ixodes ricinus* stages. Above: larva; middle: left: nymph, right: male; below: unsucked, lurking female

Fig. 6.12 Macrophoto of a couple of *Ixodes ricinus* in copulation. The female is already fully sucked

Fig. 6.13 Macrophoto of two females of *Ixodes ricinus* excreting eggs on leafs

4. **Symptoms of a disease**:

   In **dogs/cats**: Local inflammations in the case of juvenile tick stages. Intoxications (rarely with paresis), restlessness, strong itching after mass infestations; intense inflammations.

   In **ruminants/horses**: Restlessness, itching, hyperkeratosis, skin inflammations with ulcerations (especially after pulling of the tick body),

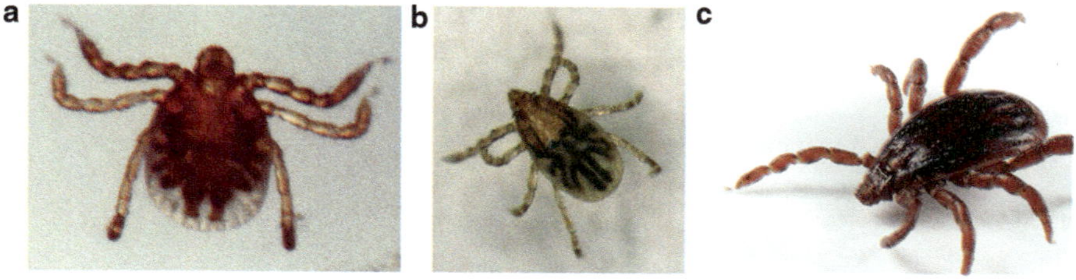

Fig. 6.14 Microscopical (**a**, **b**) and macrophoto of stages of *Rhipicephalus sanguineus*. (**a**) Larva; (**b**) Nymph; (**c**) Adult male

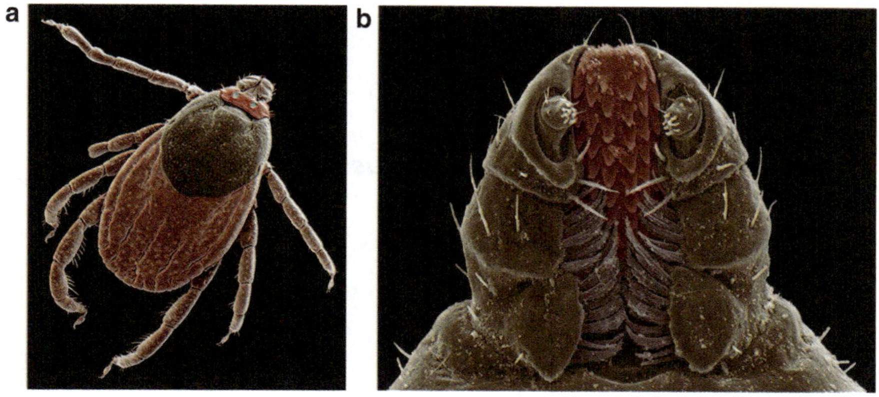

Fig. 6.15 Scanning electron micrographs of *Rhipicephalus sanguineus*. (**a**) Unsucked female; (**b**) Ventral aspect of the anterior pole

developmental disorders, emaciation and anaemia in case of massive infestation. Ticks serve as reservoirs for animal and human pathogenic **rickettsiales** (including *Coxiella burnetii*, cause of Q-fever), **viruses** (including pathogens of meningoencephalitis, Crimean-Congo fever, African tick bite fever), **bacteria** (including ***Listeria***, ***Borrelia***) and **protozoans** (*Ixodes* transmits, e.g. *Babesia bovis* and *B. divergens*, *Haemaphysalis* transmits *B. major*, babesiosis or theileriosis of horses). **Tick paralysis**: the saliva of some species contains a neurotoxin, which causes high fever and subsequently leads to paralysis of the back limbs, followed by an ascending paralysis of the front musculature. Paralyses of the muscles of the respiration tract are fatal. If the ticks are removed early enough, the animal recovers within 24 h.

5. **Diagnosis**: Microscopical inspection of ticks (adults, nymphs or larvae) after their mechanical removal from the skin with pointed forceps without anaesthesia of the tick to prevent the release of potential pathogens.
6. **Pathway of infection**: Ticks climb onto plants (larvae up to 30 cm, nymphs up to 1 m, adults up to 1.5 m) and attach at passing animals (directed by their Haller's organ, which is situated at the tarsus of the first pair of legs) (Fig. 6.8).

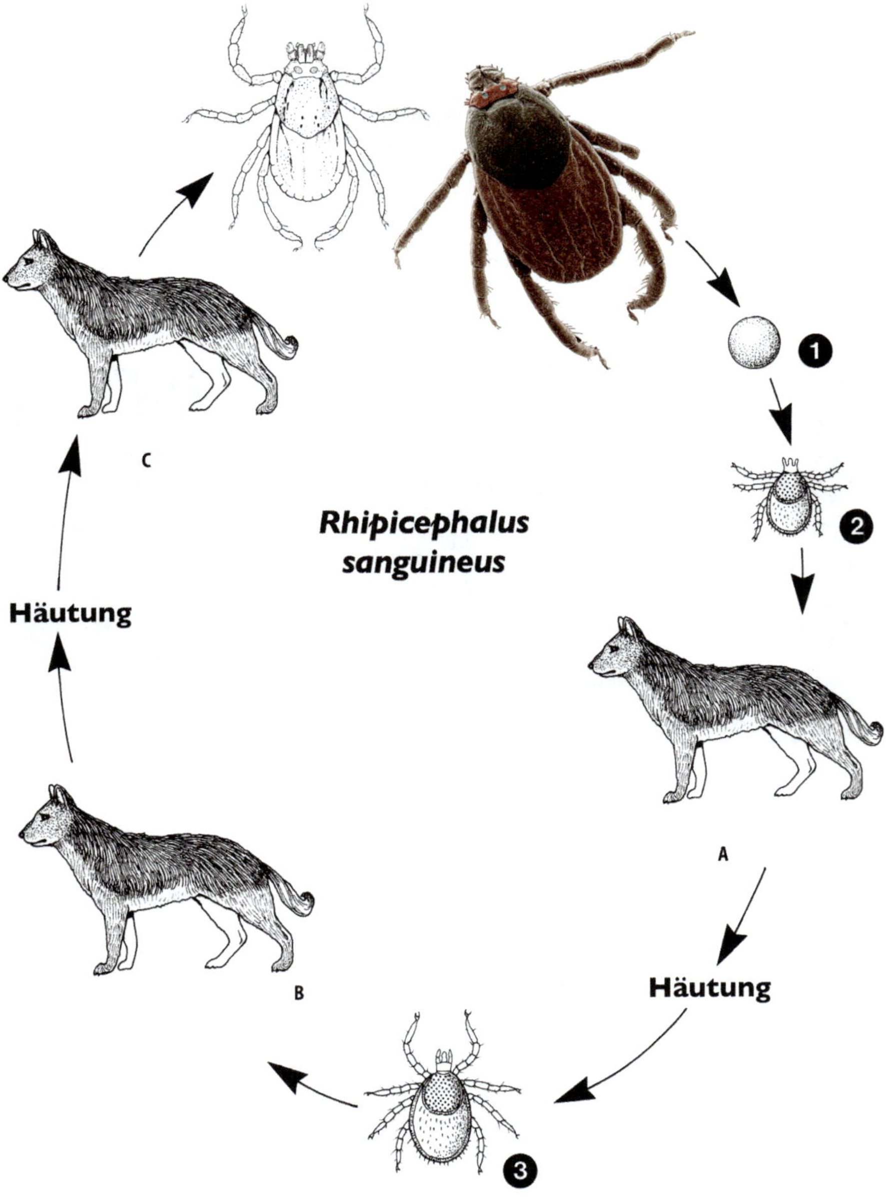

Fig. 6.16 Diagrammatic representation of the life cycle of *Rhipicephalus sanguineus*. **1**. Egg; **2**. Larva; **3**. Nymph; (**A**, **B**, **C**) = hosts (besides dogs also cats, sheep, cattle, pigs, hares, rabbits and humans are attacked) Häutung = molt

7. **Prophylaxis**: Inspection of the coats and treatment of all animals on meadows with acaricides (e.g. **cyfluthrin**, **deltamethrin**) on a regular basis. For dogs and cats, application of repellents in pour-on mode or as collars (see Table 6.5). *Caution*: Ticks can survive for years without sustenance in nature.

Table 6.4 *Ixodes* species of animals in Europe

Species	Size (♀ unfed)	Hosts
Ixodes ricinus	3–4.5 mm (11 mm fed)	More than 250 species
I. hexagonus	3–3.5 mm	Hedgehogs, canids
I. canisuga	3–4 mm	Fox, badger, also dogs
I. trianguliceps	2.8–3.3 mm	Bank vole, hedgehog, mole
I. persulcatus	3 mm	Mammals, birds
I. apronophorus	3 mm	Waterrat

8. **Incubation period**: Hard ticks require a few hours to become finally attached and several days to suck blood in larger quantities. Therefore, skin reactions, itches and restlessness only appear a few days later. Besides, each particular pathogen, which is transmitted by blood contact or airborne within the tick feces (e.g. *Rickettsia* stages), can cause diseases. Two- and three-host ticks parasitizing ruminants stay on their host for several days. The one-host ticks, on the other hand, stay their entire developmental period on one host, which can lead to mass infestations.
9. **Therapy/control**: Avoidance of infestation of cattle and other ruminants by cattle ticks is necessary everywhere in the countries with high rates of cattle rearing. The treatment is chosen according to the number of hosts with respect to the tick species, the properties of the active compound (with or without residual effects), the formulation of the agent and according to the diagnosis (composition of tick population and tick density). Careful attention should be paid to the selection and dosage of the preparation (tolerance, formation of resistances of the ticks, residues in meat and milk) especially in the case of repetitive treatments (usually after 2 weeks). Application of the compounds is realized by spraying, dipping, infusions or similar procedures. Compounds which degrade relatively fast are less harmful to surrounding nature. Pyrethroids are, at least under laboratory conditions, toxic e.g. for fishes! Approved substances for fighting ticks are e.g. Cyfluthrin and Deltamethrin (Table 6.5).

Further Reading

Boxler B et al (2016) Host finding of the pigeon tick *Argas reflexus*. Med Vet Entomol 30:193–199.

Khater HF et al (2013) The acaricidal efficacy of peracetic acid and deltamethrin against the fowl tick, *Argas persicus*, infesting laying hens. Parasitol Res 112:259–269.

Montasser AA (2010) The fowl tick, *Argas* (*Persicargas*) *persicus* (Ixodoidea: Argasidae): description of the egg and redescription of the larva by scanning electron microscopy. Exp Appl Acarol 52:343–361.

Montasser AA et al (2011) Efficacy of abamectin against the fowl tick, *Argas* (*Persicargas*) *persicus* (Oken, 1818) (Ixodoidea: Argasidae). Parasitol Res 109:1113–1123.

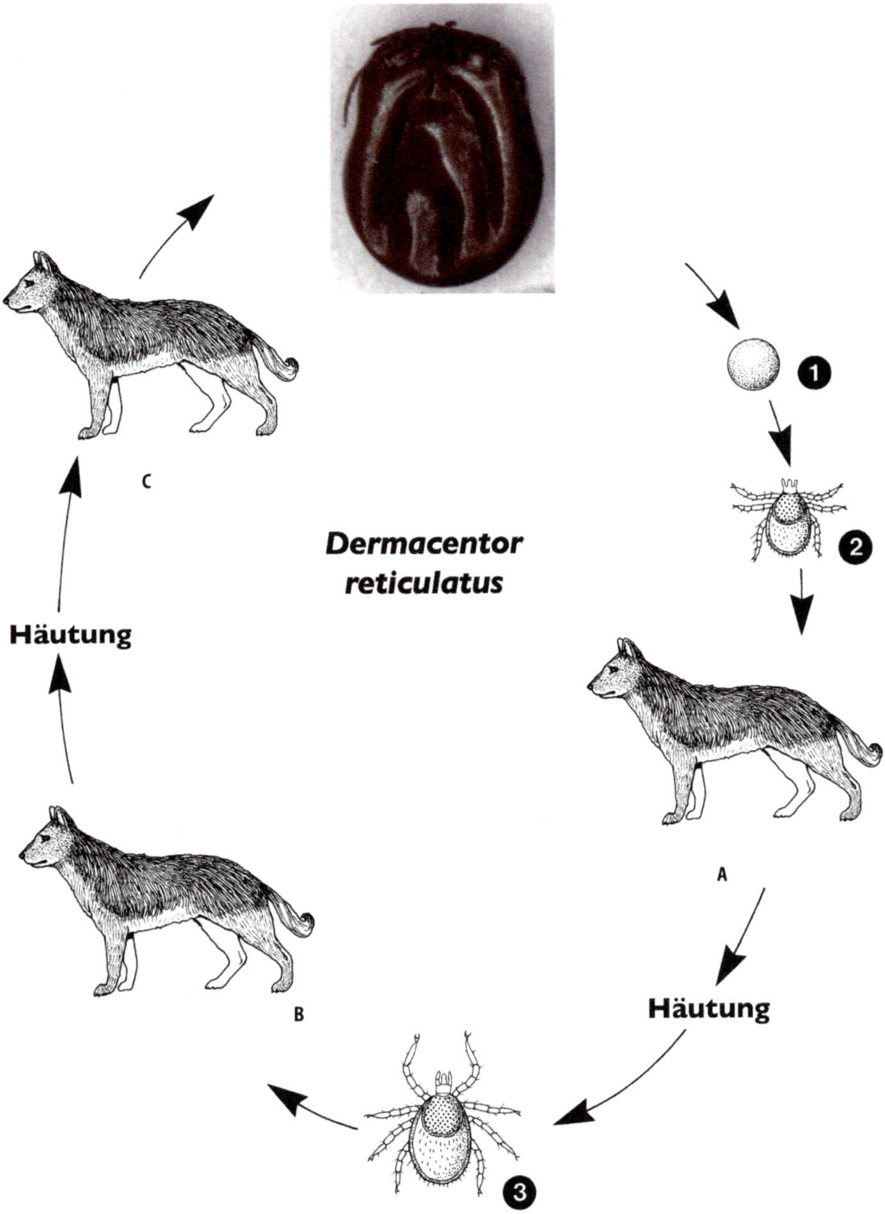

Fig. 6.17 Diagrammatic representation of the life cycle of *Dermacentor reticulatus*. **1**. Egg; **2**. Larva; **3**. Nymph; (**A**, **B**, **C**) = hosts (**A**, **B** = mostly small mammals; **C** = dogs, foxes, ruminants, humans) Häutung = molt

Fig. 6.18 Microscopic photo of an adult unsucked female of *Dermacentor reticulatus* lurking for hosts

Fig. 6.19 Diagrammatic representation of the engorged female and the male of *Hyalomma marginatum*, the vector of the Crimean-Congo fever

Santos HA et al (2010) The influence of the fasting period on the number of nymphal instars and on the sex ratio of *Argas* (*Persicargas*) *miniatus* (Acari: Argasidae). Rev Bras Parasitol Vet 19:164–168.

6.1.2 Mites

With a length of 0.2–1 mm, **mites** are rather small animals, which are, partly due to their protective colouring, hard to recognize with the naked eye. Depending on the way of life, their body shape and the characteristics of their limbs, they have

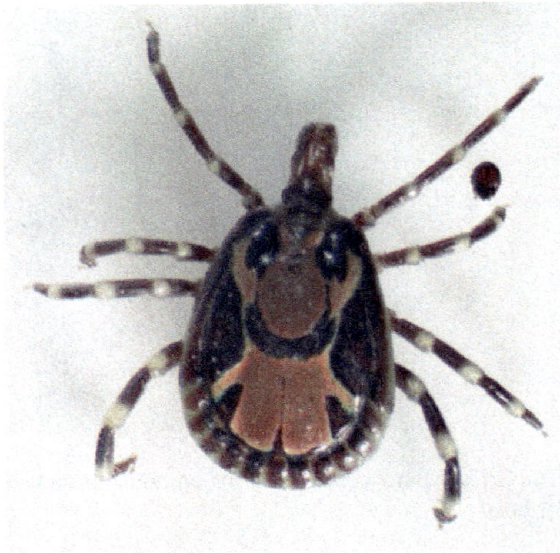

Fig. 6.20 Light micrograph of an unsucked female of *Amblyomma hebraeum*

Table 6.5 Compounds and products against hard ticks on the European market (selection)

Active compound/Substance group	Product	Target group
Flumethrin (**P**) Deltamethrin (**C**)	**Necklaces** Sultix® Scalibor®	Dogs Dogs
Permethrin (**P**) Pyriprol (**V**) Imidacloprid (**V**) Metaflumizon (**V**) Amitraz (**F**) Fibronil (**V**)	**Spot-on** Exspot® Practic® Advantix® ProMerisDuo® Frontline®	Dogs Dogs Dogs Dogs and cats Dogs and cats
Permethrin (**P**)+ Pyriproxifen (**MI**)	**Spray** Duowin®	Dogs
Amitraz (**V**)	**Spray + Bath**	Cattle, sheep, goats
Cyhalothrin (**P**) Flumethrin (**P**) Permethrin (**P**) Phoxim (**O**)	**Pour-on** **Pour-on** **Earmark** **Spray + Bath**	Cattle, sheep Cattle Cattle Sheep

C carbamate, *F* foramidine, *MI* moulting inhibitor, *O* organophosphate, *P* pyrethroid, *V* different other active groups

reached highly diverse variations. Systematically they are classified as **Meso-**, **Pro-** and **Astigmata** (**Cryptostigmata**) according to the position of their breathing openings. Particularly remarkable in mites is their frequently long "hair coat" in relation to their body size, which comprises the mechano- and chemoreceptors (trichobothria, solenidia). Like ticks, mites have to slough their chitinous exoskeleton several times during growth and thus usually undergo three stages:

(a) **Larvae** (only three pairs of legs), (b) **nymphs** and (c) **adults** (four pairs of legs each). In fact, apart from the egg, six stages exist: prelarva, proto-, deuto-, tritonymph, male and female adult. However, at least one stage is missing in most mites nowadays: the deutonymph. This is often a dormant/resting stage, which is omitted under favourable conditions. Most mites (e.g. the house dust mite *Dermatophagoides pteronyssinus*; Fig. 6.21) feed on organic material in the soil, on plants or on food remnants in the house. Thereby, they often obtain economic importance as pests of food in stockpiling, especially in the case of mass appearance.

Some of these species, e.g. the so-called dust mites, *Tyrophagus putrescentiae*, *Glycyphagus domesticus* and *Acarus siro*, inhabit flour and other stored materials and can cause allergic reactions due to persistent irritations by their characteristic long bristles and sloughed skins. (The human itch mite was formerly named *Acarus* mite (today *Sarcoptes*!). Today the genus *Acarus* comprises species, which had been formerly listed as flour mites in the genus *Tyroglyphus*). The symptoms of disease occur in different variations as so-called **pseudoscabies** which is approved as occupational disease (e.g. baker's dermatitis or itch; see Table 6.6). Other mites similar in appearance to dust mites (e.g. *Dermatophagoides pteronyssinus*; Fig. 6.21) may cause severe allergic reactions in the pharynx and the respiratory system particularly in humans but also in birds. This leads to asthmatic symptoms in infested persons or animals.

Fig. 6.21 Light micrograph of house dust mites (*Dermatophagoides pteronyssinus*) on detritus being used as food

Table 6.6 Important mites

Species	Size (mm)	Hosts	Disease
Glycyphagus domesticus	♀ 0.4–0.75 ♂ 0.3–0.5	**Humans**[a], many animals	Grocer's itch
Tyrophagus putrescentiae	♀ 0.4 ♂ 0.4	**Humans**[a], many animals	Copra itch
Acarus siro (=Tyroglyphus)	♀ 0.4–0.6 ♂ 0.4	**Humans**[a], many animals	Baker scabies
Dermatophagoides pteronyssimus	♀ 0.4 ♂ 0.4	**Humans**[a], many animals	Dermatosis, allergic asthma
Dermanyssus gallinae	♀ 0.7 ♂ 0.6	**Humans**[a], many animals	St. Louis encephalitis **V**; Chicken anaemia
Trombicula akamushi	Larva 0.25–0.5	Larva: on **humans**	Tsutsugamushi fever **R**
Neotrombicula autumnalis	Larva 0.25–0.5	Larva: on cattle, pigs, **humans**, dogs, cats	Dermatosis, so-called scrub itch
Sarcoptes scabiei	♀ 0.3–0.45 ♂ 0.2–0.3	**Humans**	Scabies
S. bovis	♀ 0.3–0.5 ♂ 0.2–0.3	Cattle	Mange[b]
S. suis	♀ 0.4–0.5 ♂ 0.25	Pigs	Mange[b]
Notoëdres cati	♀ 0.2–0.3 ♂ 0.15–0.18	Cats	Mange[b]
Otodectes cynotis	♀ 0.4–0.5 ♂ 0.3–0.4	Dogs	Mange[b]
Knemidocoptes mutans	♀ 0.4–0.5 ♂ 0.2–0.25	Poultry	Knemidokoptic mange
Demodex folliculorum	♀ 0.4 ♂ 0.3	**Humans**	Poss. acne, rosacea
Demodex canis	♀ 0.3 ♂ 0.25	Dogs	Eczema, pustules
Psoroptes sp.	♀ 0.6–0.8 ♂ 0.5–0.65	Ruminants	Mange[b]
Chorioptes sp.	♀ 0.4–0.6 ♂ 0.3–0.45	Ruminants	Mange[b]
Varroa jacobsoni	♀ 1.2–1.7 ♂ 0.8	Bees	Varroaosis

[a]These dust mites feed on food residues
[b]Mange of animals can be caused also by other species of mites
V Viruses, *R Rickettsia* (=intracellular bacteria)

The formation of the **allergen** is thought to proceed in the following steps: The sleeping human individual loses dandruff in bed (=0.5 g/day), on which after a while millions of fungi grow. They are ingested by mites, whose natural habitat are mattresses, blankets, underlays, etc. In the intestinal tract of mites strong allergens are generated, which are released with fecal blobs and reach the air during "making the bed" or beating the mattresses. These particles are then inhaled by humans and cause the above-mentioned symptoms especially in sensitive persons. More than 25 % of the German population are listed as so-called **atopic** (allergic) **patients** being genetically predisposed, which is thought to be caused by a gene on the 11th chromosome. A consistent humidity in the bed turned out to be the most important prerequisite for mite growth (humans transpire about 1 l water per night!). It is therefore recommended to hoover beds and the surroundings on a regular basis, to change bed sheets and pillows frequently and to display mattresses in the sun in order to dry out the mites or to "freeze" them in winter as mites do not tolerate temperatures below zero. In extreme cases, severe contaminated beds and carpets have to be devastated. A **house-dust allergy** can be diagnosed with the help of a skin test. By injecting increasing amounts of house dust extract persons can become desensitized. At temperatures below 17 °C, the development of house dust mites stops, so that low temperatures in the bedroom offer a good protection against the propagation of mites. Lowering the humidity likewise suppresses the mite growth. Indeed, about 75 % of humidity is optimal for mites and fungi (e.g. *Aspergillus*). Animals may also suffer from the symptoms in the presence of large amounts of such "dust mites". Thus, sleeping places of pet animals should be kept clean and placed at dry spots avoiding humidity, which would allow growth of fungi.

As vectors of human and pet diseases the usually non-host-specific mites are of little importance (Table 6.6). The impairment of the host mostly happens by mite infestation of rooms. According to their feeding behaviour, whereby their mouthparts (chelicerae and pedipalps) underwent specific adaptations, they are classified into surface pests, non-burrowing and burrowing mites.

Surface mites feed on skin scales of their hosts and thereby may cause dermatitis and loss of hair. Species of the genus *Chorioptes* are abundant in cattle, horses and sheep. *Otodectes cynotis* (Fig. 6.31) is abundant in the external auditory meatus of dogs and cats. This species is characterized by an extremely shortened pair of legs.

In **non-burrowing mites**, the mouthparts form a proboscis with the help of which both sexes suck blood or lymph from their host. Of particular importance are *Dermanyssus gallinae* (Figs. 6.43 and 6.44), the red fowl mite, and some species of the genus *Ornithonyssus* (tropical chicken mites or rat mites) (Fig. 6.45). *Dermanyssus gallinae* leaves the host after the blood meal (at night!), while many specimens of other species (e.g. the Northern fowl mite *Ornithonyssus sylviarum*) stay throughout their entire development on the host. During bloodsucking, viruses (e.g. pathogens of the St Louis encephalitis) and various species of the genus *Rickettsia* can be transmitted. In humans and animals, stings are quite painful and, depending on the individual reaction, up to 2 cm sized bladder-like swellings may appear. Among the non-burrowing mites, the species of the family *Trombiculidae* are of particular importance because as adults they are free-living, but their larvae and eventually also the nymphs suck lymph from vertebrates with the help of their mouthparts. Severe infestations by the so-called harvest mite (*Neotrombicula autumnalis*; Figs. 6.39, 6.40 and 6.41) cause strong allergic reactions (the saliva of the mites acts as an allergen) and induce the so-called scrub itch. Larvae of *Neotrombicula akamushi* (Japanese: *okamushi* = red tiny animals) can transmit the pathogen of the so-called tsutsugamushi fever, *Rickettsia tsutsugamushi*, from rodents to humans.

Mange mites are of particular importance as human parasites. They build burrow as deep as into the *stratum germinativum* of the epidermis and thereby cause, as a consequence of inflammatory reactions, the so-called **mange** in animals or **scabies** in humans (worldwide 300 million people are infested). Those ovoid, only 0.3–0.5 mm sized mites merely possess stubby legs, of which the front ones are protruded. Besides, the legs are characterized by inarticulate, petiolate sucker-like structures which, like the dorsal chitinous crochets and the long, mostly at the hind legs attached bristles, guarantee a safe anchorage in the burrows during their meal. Only female nymphs and males reach the skin surface, where they mate at the latest after 3–4 days. After moulting the adult female deposits eggs (2–4 times a day for about 2 months) in the burrows, where later the larvae hatch. Males require about 12 days and females about 15 days until sexual maturity.

The most important species in human is *Sarcoptes scabiei* (Fig. 6.28), in cattle *S. bovis*, in dogs *S. canis*, in pigs *S. suis*, in horses *S. equi* and in hedgehog *S. sp.*, all of which penetrate into shorthaired parts of the body but have the potency to propagate. According to molecular biological investigations, the above-cited *Sarcoptes* species are breeds or variations (subspecies) of the species *S. scabei* and are known to be able to infest humans. *Notoedres cati* (Fig. 6.29), the pathogen inducing a scaly face in felids, is distinguished from the above-mentioned species by its dorsal opening of the anus but also triggers the typical symptoms: drastic itch (**pruritus**), hair loss (**alopecia**) and severe epidermal cornification especially at the ears together with saniopurulent exudation caused by bacterial secondary infections (**pyodermia**). In gamefowls the species *Knemidocoptes mutans* causes characteristic skin alterations: the so-called **scaly leg**. This disease may substantially impair the laying performance of hens. An intermediate status between surface mites and burrowing mites represent the members of the genus *Demodex*, which like e.g. *D. folliculorum* and *D. canis* live in their hosts' hair follicles and feed on sebum. In case of extremely high numbers of mites, hair loss and bacterial secondary infections occur in the infested skin areas.

The mite species *Varroa destructor* (syn. *V. jacobsoni*) (Fig. 6.52) has recently been introduced from Asia to Europe and currently spreads out in indigenous honeybees (*Apis mellifica*, syn. *A. mellifera*). The lateral-ovoid female mites (1.2 × 1.7 mm) primarily harm the breed of bees. Infestations are, however, initially not recognized, but the infested bee colonies collapse due to lacking offspring in autumn. Cause of death is thereby the so-called **bee dysentery**, which is triggered by the microsporidian protozoan *Nosema apis* and can be spread severely due to the impairment caused by *Varroa* mites. Males of *Varroa* have a circular body shape, are notably smaller (0.8 mm) than the females and die immediately after copulation. A satisfying **control** of the microsporidians causing bee dysentery (nosemosis, nosematosis) is not yet available. *Varroa*-infested colonies were exterminated in the past. Nowadays, an incense stick with the active substance isopropyl-4,4-dibromobenzilate (=Folbex®) or sprinkling with Perizin® as systemic treatments are recommended. Besides, a number of preparations containing pyrethroids or amitraz, which are not yet approved in Germany, are available. The *Varroa*-mite disease is a notifiable epizootic, because a spread within 3 months within a radius of 6–11 km has been observed.

Further Reading

Arian LG, Morgan MS (2015) Reproductive biology of *Euroglyphus macynei* with comparisons to *Dermatophagoides farinae* and *D. pteronyssinus*. Exp Appl Acarol 66:1–9.

Hervas D et al (2013) IgE levels due to *Dermatophagoides pteronyssinus*. Allergy Immunol 160:383–386.

Vidal-QuistJC et al (2015) Allergen expression in the European house dust mite. Med Vet Entomol 29:137–146.

6.1.2.1 *Demodex* Species

1. **Name**: Greek: *demos* = sebum; *dex* = woodworm larva; *katta* = cat. Lat.: *canis* = dog; *felis* = cat; *bos* = cattle; *ovis* = sheep; *capra* = goat; *mus* = mouse; *rattus* = rat.
2. **Geographical distribution/epidemiology**: Worldwide.
3. **Biology/morphology**:

   The ***Demodex*** species of animals are named after their host. They are supposed to be very host specific. However, molecular biological investigations indicate that there are several strains. The fact that some authors claimed to have detected different "species" on the same host argues for that.

   (a) **Species in canids/felids**: ***Demodex canis*** (frequent) and ***D. felis*** (syn. *D. cati*) (rare) are whitish, elongated mites (females measure 250 μm, males about 150 μm in length). All stages are located in hair follicles (sebaceous glands) as well as occasionally in lymph nodes and arterioles of the subcutis of their host; certain skin areas are infested preferentially, too (Figs. 6.22, 6.23, 6.24, 6.25 and 6.26). *D. canis* is also found in small numbers in healthy skin of almost every dog. T-cell defects (**primary immunosuppression**) and iatrogenic treatments (cortison or cytostatic therapy) as well as organ damages (hepatitis, adrenocortical hyperplasia, tumours) as triggers of **secondary immunosuppression** lead often to clinical manifestations (Gothe 1989). Most common infested areas in **dogs** are head region, dorsal surfaces of the front limbs (rarely on the inner legs), lateral belly and the chest region (Fig. 6.23). All adult predator mites (Fig. 6.26) are characterized by their eight stubby legs and their rather long abdomen (with cuticle curls). The whole development takes place in the hair follicles. The copulation is carried out on the surface of the host (males die 3–7 days thereafter). Females invade the openings of the hair follicles, and after 1 day, they start to lay the spindle-shaped eggs (about 70–90 × 25 μm), within which during 2 days the first larva is formed and which hatches soon after. The development of adults proceeds via this first larva stage and two nymphs (proto- and deutonymph) within

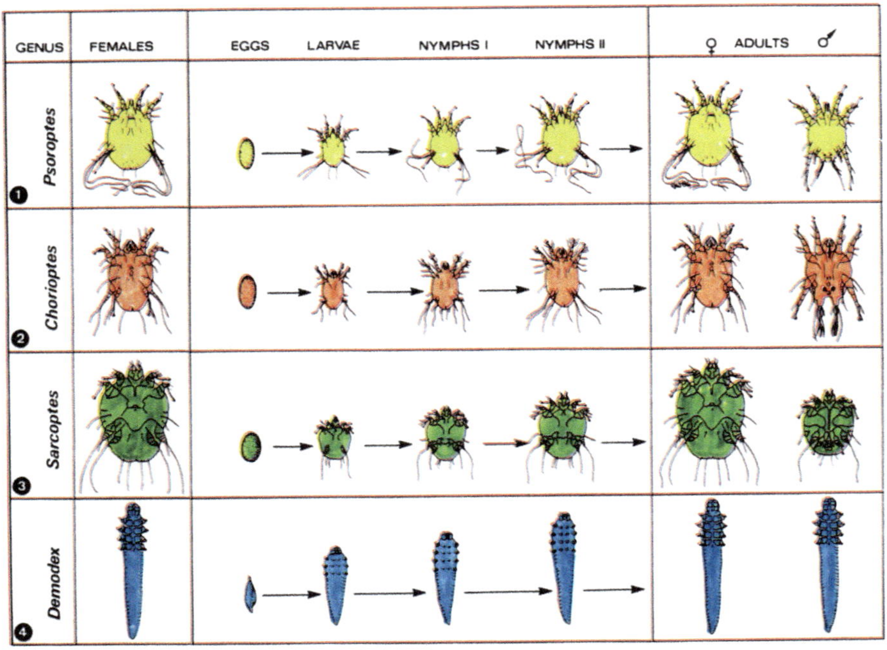

Fig. 6.22 Diagrammatic representation of the developmental stages of important mite genera

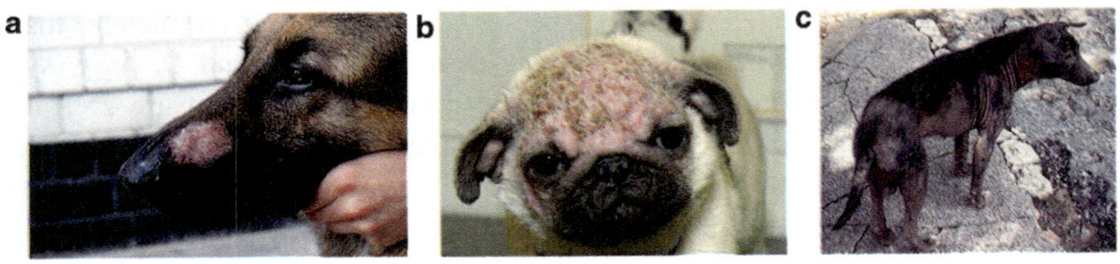

Fig. 6.23 Aspects of demodicosis. (**a**) local form; (**b**, **c**) generalized stage

9–21 days. The larva as well as the protonymph possesses only three pairs of legs. The deutonymph and the female have four pairs of legs and they destroy the hair matrix of their hosts. Initially, they are located in the upper third of the hair follicle; then they migrate deeper, whereby the follicle is stretched and the outer and inner sheaths get separated. This leads finally to hair loss by rupture of the hair follicles and destruction of the hair bulb.

(b) The ***Demodex*** species in **pigs, ruminants** and **horses** are relatively rare in Germany, but under certain circumstances (e.g. immunosuppression) they cause lump formations, which contain up to 5000 mites.

4. **Symptoms of disease**: Infections of dogs with *Demodex canis* are characterized by a high infestation density (up to 8000 mites per cm^3 skin). In general, two main forms of mange are defined:

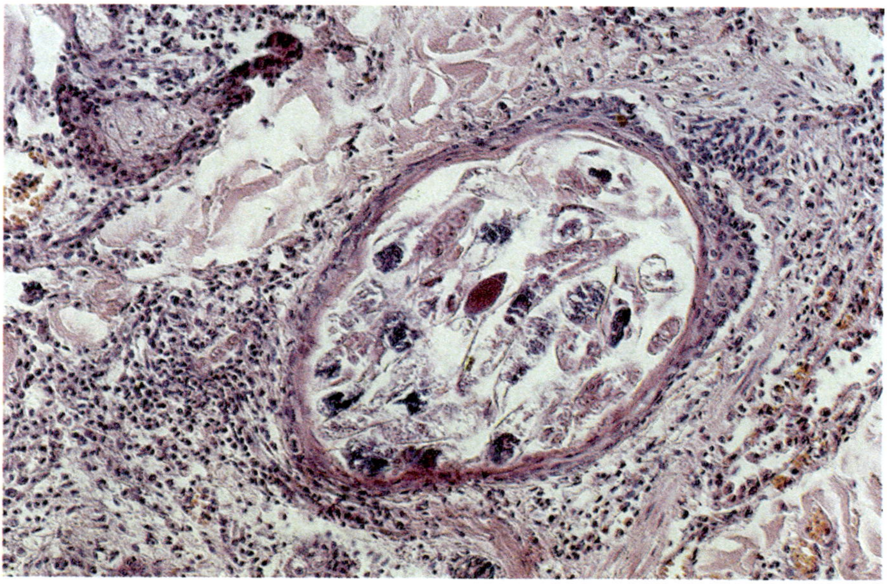

Fig. 6.24 Microscopical aspect of a skin section of a dog with many *Demodex* mites

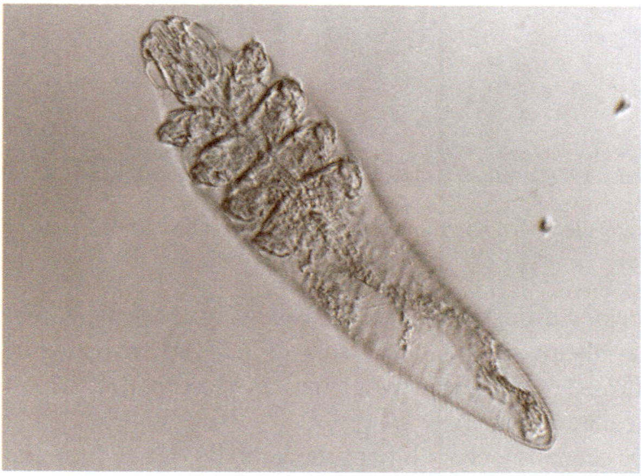

Fig. 6.25 *Demodex* mite after maceration of the surrounding skin tissues

(a) **Localized squamous demodicosis**: This demodicosis, which has a good prognosis, is characterized by several distinct localized loci of an erythematous, scaly alopecia (periocular, at lips and at forelegs) and occurs mostly in 3–6 months old dogs. Up to 90 % of the cases undergo a spontaneous cure, the rest evolves the generalized form despite treatment.

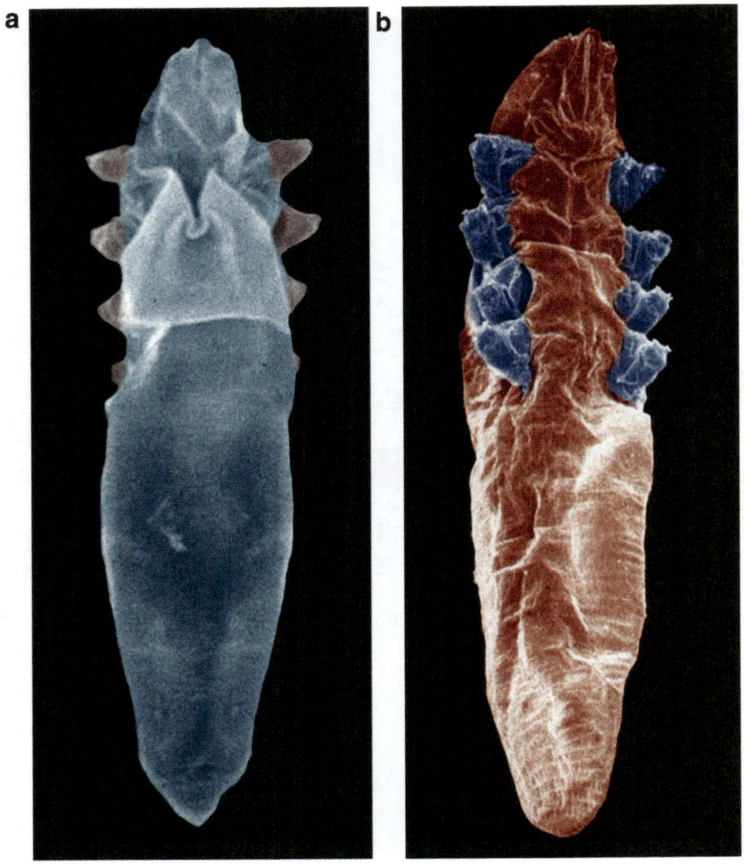

Fig. 6.26 Scanning electron micrographs of the dorsal (**a**) and ventral (**b**) side of a *Demodex* mite

(b) **Generalized pustular demodicosis** being characterized by initial alopecia, erythema and flaking. This disease has a bad prognosis. It is a manifested seborrhoea, pyoderma and pruritus (large parts of the body surface are affected).

Occasionally occurs bronchopneumonia due to bacterial contributions (*Aerobacter* spp., *Proteus* spp.).

After secondary bacterial infections of the skin, interdigital pyodermia occurs (50 % of dogs) with generalized demodicosis, which leads particularly in addition to *Pseudomonas* infections to septicaemia with abscess formation. In horses, ruminants and pigs, the quality of leather can be impaired by nodule formation in the skin, especially in the case of infestation along the dorsum.

5. **Diagnosis**: Microscopical detection of mites in skin scrapings or skin biopsies.

6. **Pathway of infection**: By body contact, mainly in puppies older than 3 months!

   Adult mites may pass over into the fur from one partner to the other during copulation.
7. **Prophylaxis**: Animals with generalized demodicosis and especially bitches, whose puppies have demodicosis, should be excluded from breeding.
8. **Incubation period**: Variable, usually not relevant.
9. **Prepatent period**: 1–3 weeks, since the life cycle of *Demodex* species takes about 3 weeks.
10. **Patency**: Single individuals only live for a short time but due to permanent reproduction the parasites may be present lifelong in a host.
11. **Therapy**: The following information concerns mainly **dogs**. Chemotherapy altogether is problematic because a spontaneous healing occurs in the case of the localized (about 90 %) as well as in the case of the generalized infection (about 50 %) during the first year of life, even in the absence of a specific therapy. During this phase shearing hair, purging, scab-solving washings and skin greasing provide relief. As a specific therapy baths with the amidine product, **Amitraz** (Derasect®, Demodectic Mange Wash, Extodex®) proved to be very effective (high cure rates after 3–6 times of topical full-body treatment (at an interval of 14 days) with 0.05 % of active agent. The treatment should be continued until no living mites can be detected in skin scrapings anymore. In case of the pustular form, it is recommended to use immune system strengthening measures (e.g. administration of vitamins, proteins, deworming products). For treatment of pigs, baths and sprinkles with Amitraz are recommended as well as therapies with **ivermectin** and **doramectin**. Treatment with **imidacloprid** and **moxidectin** (10 mg/kg body weight + 2.5 mg/kg body weight, repetition after 4 weeks) also proved to be successful, as well as spot-on use of **selamectin** 6 mg/kg body weight (likewise repetition after 6 weeks).

Further Reading

Abu-Samra MT, Shnaib YA (2014) A study of the nature of association between *Demodex* mites and bacteria in lesions of demodectic mange in cattle. Vet Med Int Article 413719.

Ravera I et al (2015) Serum detection of IgG antibodies against *Demodex canis* in healthy dogs and dogs with juvenile demodicosis. Res Vet Sci 101: 161–164.

Rojas M et al (2012) Molecular study of three morphotypes of *Demodex* mites from dogs. Parasitol Res 111:2165–2172.

Silbermayr K et al (2015) Phylogenetic relationships and new genetic tools for the detection and discrimination of the three feline *Demodex* mites. Parasitol Res 114:747–752.

Zhao YE et al (2014) Molecular identification on phylogenetic study of *Demodex caprae*. Parasitol Res 113:3601–3608.

6.1.2.2 Mange Mites in Carnivores

1. **Name**: Greek: *sarx* = flesh; *koptei* = to wound; *noton* = dorsum; *hedraios* = fixed position; *otos* = ear; *dectikos* = snappy; *kynos* = dog; *kata* = cat. Latin: *canis* = dog.
2. **Geographical distribution/epidemiology**: Worldwide with high prevalence rates.
3. **Biology/morphology**:

   Species:

   (a) ***Sarcoptes canis*** stages (females: 300–400 μm long (Fig. 6.29); males: 200–250 μm) are found in the skin of the whole body. They are particularly frequent in the surface of the nose, ear tips and eye area. Today it is assumed that all *Sarcoptes* species of animals are strains of the same species (*Sarcoptes scabiei var. canis*).

   (b) ***Notoedres cati*** stages (females: 230–300 μm long (Fig. 6.29), males: 150–180 μm) are localized in the **skin** of **cats**.

   (c) ***Otodectes cynotis*** stages (females : 350–500 μm long (Figs. 6.29 and 6.31), males : 150–180 μm) are predominantly found on the **skin** of the external ear of **dogs** and **cats**.

   Unlike *Otodectes* mites, *Sarcoptes* and *Notoedres* mites live in burrows of the epidermis. The development to the adult stage always proceeds via a larval stage (which has three pairs of legs) and two nymph stages (proto-, telonymph; each with four pairs of legs). This takes in males 2 weeks and 3 weeks in females. Male and female telonymphs leave the burrows to copulate. After the copulation, the males die and the females burrow back into the skin.
4. **Symptoms of disease**: As a consequence of secondary infections, ulcerations, scab formation, scaling and hair loss arise in the skin regions (Fig. 6.27). Sepsis is common. In the case of ear mange, there is a risk of eardrum affection by potential perforation.
5. **Diagnosis**: Detection of the mites in burrows after extraction with a disinfected needle or in skin scrapings (using KOH maceration).
6. **Pathway of infection**: Males and telonymphs are transmitted by body contacts.
7. **Prophylaxis**: Isolation of infested animals. ***Caution***: There is a risk of infection for humans (in the case of *Sarcoptes* infections).
8. **Incubation period**: 1–2 weeks; timing depends on the infestation density.
9. **Prepatent period**: 3 weeks.
10. **Patency**: Years (due to repeated growth of several generations of mites).
11. **Therapy**: In the case of *Sarcoptes* mange and *Notoedres* infestations of dogs, Amitraz baths (Ectodex®) are as successful as treatment with **selamectin** (6 mg/kg body weight) or application of the combination of **imidacloprid** and **moxidectin** (Advocate®: 10 mg/kg body weight + 2.5 mg/kg body weight). *Otodectes* infestation (Fig. 6.31) in dogs and cats was likewise controlled with these two agents.

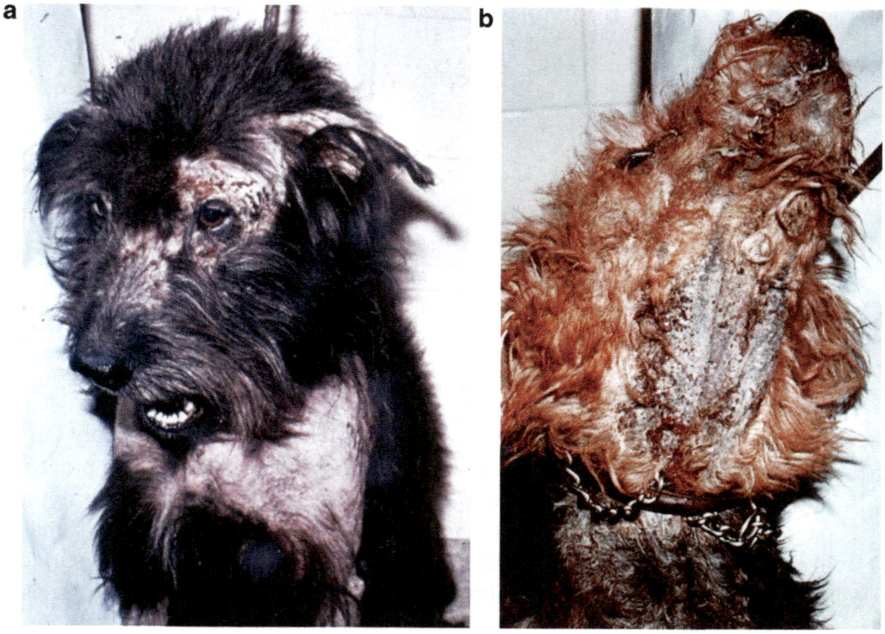

Fig. 6.27 Stray dogs with large spots of scabies (**a**, **b**)

Further Reading

Arther RG et al (2015) Clinical evaluation of the safety and efficacy of 10 % imidacloprid +2.5 % moxidectin topical solution for the treatment of the ear mite (*Otodectes cynotis*) in dogs. Vet Parasitol 210:64–68.

Gothe R (1989) Die Demodikose des Hundes – eine Faktorenkrankheit? Berl Münch Tierärztl Wschr 102:293–297.

Hellmann K et al (2013) Treatment of naturally *Notodectes cati* infested cats with a combination of imidacloprid 10 %/moxidectin 1 % spot on (Advocate®/Advantage® Multi, Bayer). Parasitol Res 112:S57–S66.

Matsuyama R et al (2015) Coexistence of two different genotypes of *Sarcoptes scabiei* derived from companion dogs and wild racoon dogs. Vet Parasitol 212:356–360.

Six RH et al (2016) Efficacy of solaner, a novel oral isoxazoline against two common mite infestations in dogs: *Demodex* spp. and *Otodectes cynotis*. Vet Parasitol 222:62–66.

Zhao YE et al (2014) Population identification of *Sarcoptes hominis* and *S. canis* in China using DNA sequences. Parasitol Res 114:1001–1010.

6.1.2.3 *Sarcoptes* Mange in Pigs

1. **Name**: Greek: *sarx* = meat; *koptei* = initiating a wound. Latin: *sus* = pig.
2. **Geographical distribution/epidemiology**: Worldwide.

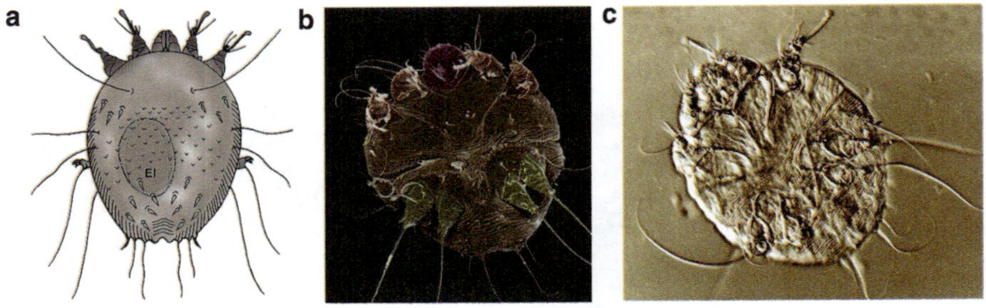

Fig. 6.28 Diagrammatic representation (**a**), scanning electron micrograph (**b**) and light micrograph (**c**) of a *Sarcoptes scabiei* mite

3. **Biology/morphology**: *S. suis* shows the typical habitus of *Sarcoptes* stages (Fig. 6.28). With a maximum size of 0.5 mm, females are considerably larger than males (0.25 mm). As in other *Sarcoptes* species, the development proceeds from the egg via one larval stage (with three pairs of legs) and two nymph stages to the adult stage (all with four pairs of legs; females need 21 days, males 14 days). Males copulate with the terminal female nymphs on the skin surface. After moult, the fertilized females burrow hollows into the skin, which reach until the *stratum germinativum* of the epidermis. There they lay eggs, from which larvae hatch and dig channels to the surface.
4. **Symptoms of disease (Mange)**: Papules, scab formation, secondary bacterial infections, subsequent extreme itching (restlessness of animals (Fig. 6.30).
5. **Diagnosis**: Detection of the mites in skin scrapings.
6. **Pathway of infection**: Transmission from animal to animal by body contact. Free-living mites die after approximately 2 weeks.
7. **Prophylaxis**: Regular cleaning of the stables and equipment by hot steam or disinfection with contact insecticides (e.g. carbamates or pyrethroids with long-term effects, e.g. CBM 8® or Ins 15®, Permethrin 25®). ***Caution***: Only use for disinfection of stables! Avoidance of predisposing factors (inadequate accommodation, bad feeding, worm infestation). Prophylactic treatment of all animals with contact insecticides by spray coating method (see Therapy).
8. **Incubation period**: Variable, often very long, at least 1 week.
9. **Prepatent period**: 14–21 days.
10. **Patency**: Years as a result of repeated formation of several generations.
11. **Therapy**: Whole-animal treatment per bath or spraying with amitraz (Taktic®) or subcutaneous administration of **ivermectin** or **doramectin** (0.2 mg/kg body weight).

Further Reading

Haas C et al (2015) First detection of sarcoptic mange in free-ranging wild boar (*Sus scrofa*) in Switzerland. Schweiz Arch Tierheilkd 157:269–275.

Laha R (2015) Sarcoptic mange infestation in pigs: an overview. J Parasit Dis 39:596–603.

Swe PM et al (2014) Scabies mites alter the skin microbiome and promote growth of opportunistic pathogens in a porcine model. PLoS Negl Trop Dis 8(5):e2897.

6.1.2.4 Mange in Ruminants

1. **Name**: Greek: *sarx* = flesh; *koptein* = to induce wounds; *psora* = mange; *chorion* = leather; Latin: *bos* = cattle; *ovis* = sheep; *capra* = goat; *rupicapra* = chamois; *cuniculus* = rabbit.
2. **Geographical distribution/epidemiology**: Worldwide.
3. **Biology/morphology**:
   (a) ***Sarcoptes*** **species** (*S. bovis*, cattle; *S. ovis*, **sheep**, **cattle**; *S. rupicaprae*, **chamois** and **goat**). The females are 0.6 mm long at the maximum (males up to 0.3 mm). They possess eight short legs, of which only the two front pairs are dorsally visible (Figs. 6.28 and 6.29). The tulip-shaped adhesive pads of the legs are situated at the tips of stems.
   (b) ***Psoroptes*** **species** (*P. ovis*, **cattle**, **sheep**; *P. cuniculi*, **rabbits**). Females reach a length of 0.8 mm, males 0.65 mm. Characteristics (Fig. 6.32) are the sucker-like structures (in males on the 1st–3rd pair of legs, in females only on the first two pairs of legs), which are located terminally at segmented stems. All eight legs are visible from the dorsal side of the mites.
   (c) ***Chorioptes*** **species** (including *C. bovis*, **cattle**, **sheep**, **goat**). Females reach a length of 0.65 mm, males 0.5 mm (Fig. 6.34). The sucker-like structures are bell shaped (they do not occur on the third pair of legs in females) and are located on short, unsegmented pedicles. The species live on the skin and feed on scales. The development period (egg—adult) takes 11 days.

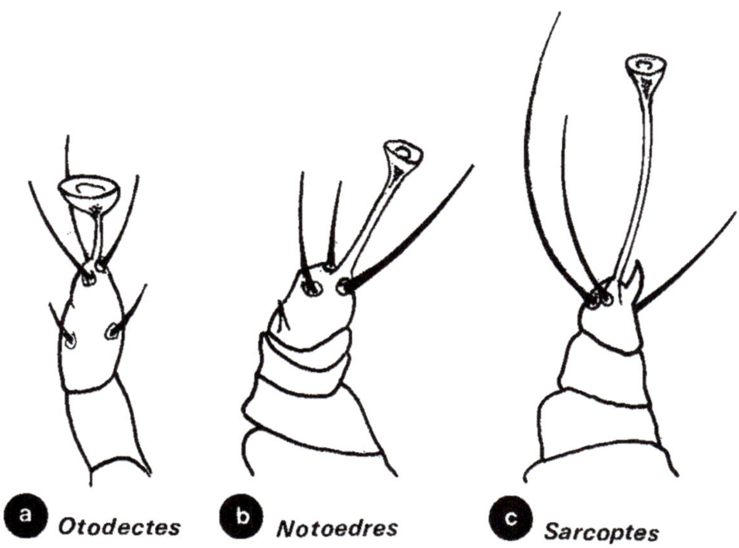

Fig. 6.29 Diagrammatic representation of the stalk-like protrusions at the feet of mite genera

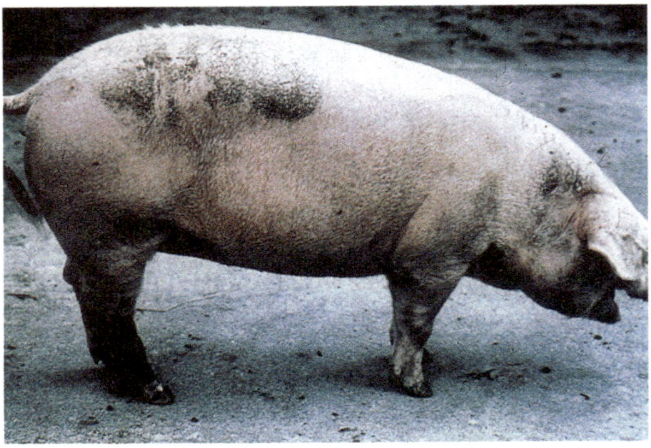

Fig. 6.30 Macrophoto of a pig with large *Sarcoptes*-infected skin regions

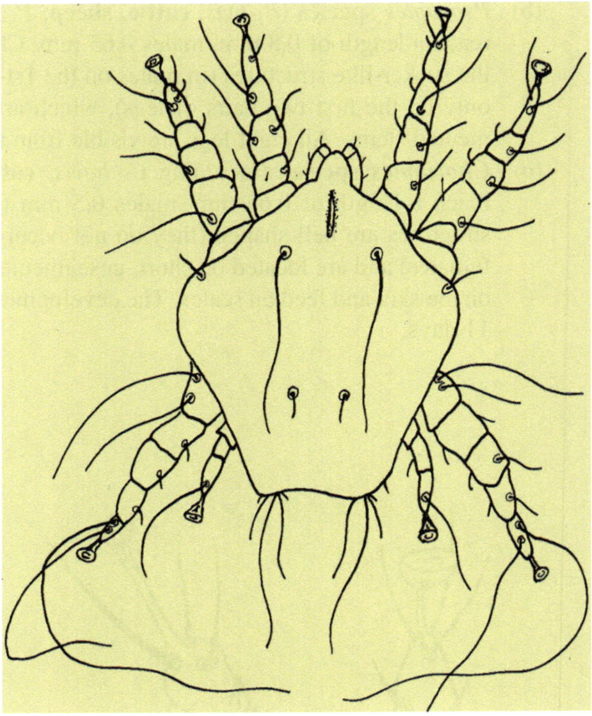

Fig. 6.31 Diagrammatic representation of an *Otodectes cynotis* mite

4. **Symptoms of disease (Mange)**: Extreme damages in the skin are caused especially by *Psoroptes* mange (Fig. 6.33). **Leading symptom** in *Sarcoptes* and *Psoroptes* mange is an extremely severe itching (=severe rubbing reactions), while in *Chorioptes* mange apparently only occurs as a moderate itching. Skin lesions due to intense rubbing, hair loss, thick scabs as well as a purulent exudate are typical indicators of persisting secondary infections.

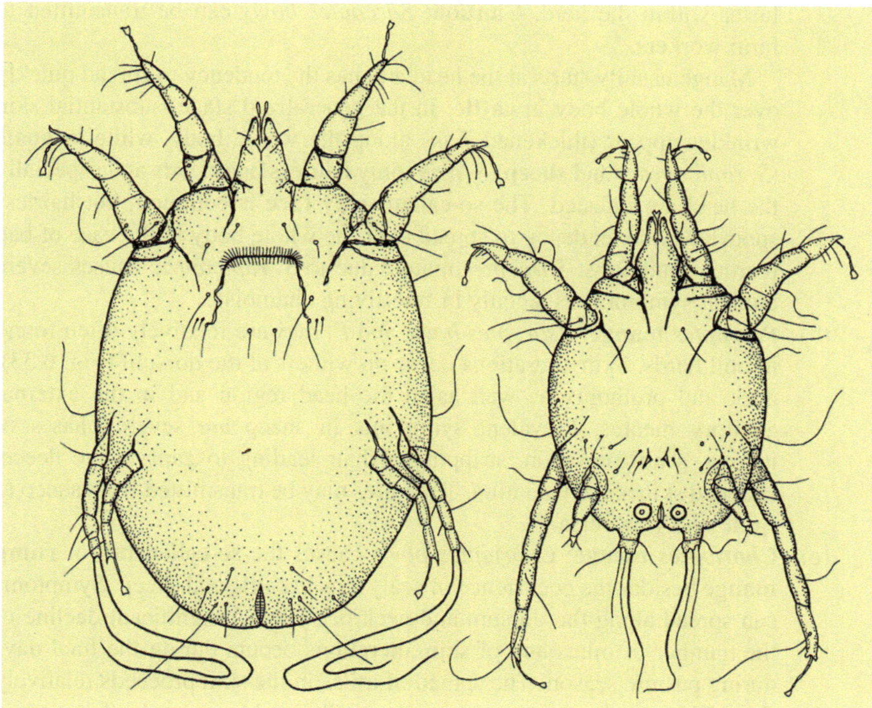

Fig. 6.32 Diagrammatic representation of the female (*left*) and the male stage of *Psoroptes ovis*

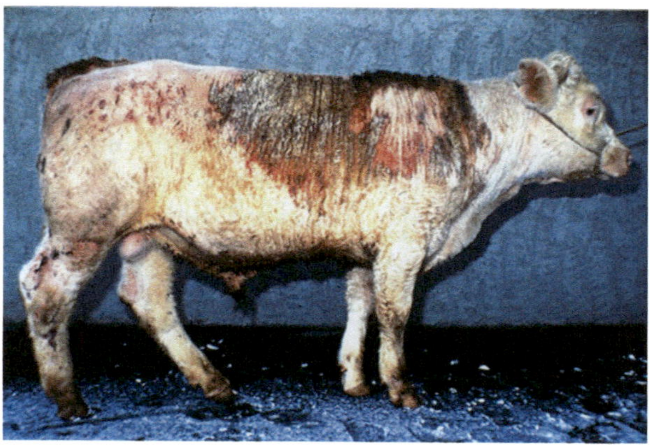

Fig. 6.33 Photo of a young bull suffering from *Psoroptes* mange

(a) ***Sarcoptes* mange**: Its progress is acute and leads to severe clinical symptoms such as massive impairment of the general condition, emaciation, faintness, anaemia and sepsis. *Sarcoptes* mange is observed less frequent than *Psoroptes* mange but progresses more acute and spreads faster within the herd. **Caution**: *Sarcoptes bovis* can be transmitted to farm workers.

Mange usually starts at the head and has the tendency to spread quickly over the whole body in **cattle**. In the generalized stage, substantial skin wrinkles appear (thickened skin) along the whole body, while in **goats** (*S. rupicaprae*) and **sheep** (*S. ovis*) only hairless body parts and especially the head are infested. The so-called **scaly face** is common, but hairless spots have the tendency to spread over the whole body in the case of bad rearing conditions. *Sarcoptes* mange due to *S. rupicaprae* causes severe clinical symptoms especially in free-living chamois.

(b) ***Psoroptes* mange**: *Psoroptes bovis* and *P. ovis* are relatively often found in bull herds. Skin alterations occur on withers at the dorsum (Fig. 6.33), neck and prothorax as well as in the head region and in the external auditory meatus. Prevalent symptoms in sheep are severe phases of itching, scab formation, stripping of hair leading to gaps in the fleece, chapped skin and emaciation. The mites may be transmitted from sheep to cattle and vice versa.

(c) ***Chorioptes* mange**: *Chorioptes bovis* causes the so-called **tail** or **rump mange** besides the occurrence of scaly legs in cattle and sheep. Symptoms can spread along the dorsum and neck of sheep. A significant decline of the number of mites and of skin alterations occurs during the final days during pasture season. The spread of mites in the skin proceeds relatively slow. This peculiar mange occurs especially in older animals of cattle and in sheep (**attention**: **foot rot** has to be excluded in differential diagnosis).

5. **Diagnosis**: Microscopic detection of mite stages with the help of skin scrapings. Samples have to be taken from several skin areas and (in case of herd infestations) from several animals. Especially probes from the border area of altered skin regions have to be analysed. At the onset of an infection, the detection of mites is often difficult because of the low numbers of parasitic stages. **Caution**: *Psoroptes ovis* mange in sheep is **no longer notifiable** in Germany but probably still in several countries.
6. **Pathway of infection**: By body contact and contact with contaminated tools (e.g. tether strings, etc.).
7. **Prophylaxis**: A 3-week quarantine is recommended before entering new animals into a herd. Preventive application of **ivermectin** (Ivomec®) or **doramectin**. Dosages and withholding period: see **Therapy**.
8. **Incubation period**: Several weeks.
9. **Prepatent period**: *Sarcoptes* mites need about 3 weeks until maturity; *Psoroptes* and *Chorioptes* mites: 9–11 days.
10. **Patency**: Several weeks.
11. **Therapy**: The treatment of mange in Germany (and in many other countries) is restricted by the Health authorities to registered pharmaceuticals. **The way of use** and **the physiology of the animal** species determine the application rate, the amount of active agent used and the **withdrawing periods**. Many compounds are used for external application acting as contact acaricides

targeting adult and larval mites but are ineffective against their eggs. Therefore, repetition of treatment is necessary. Some active agents (**Phoxim**, **γ-HCH**) additionally generate a gaseous phase (advantageous in case of hidden mites). Apparently, a sufficient effect against mange mites depends on a long exposure time to the active agent and has the disadvantage of long withdrawing periods. Active agents with relative short withdrawing periods are thus preferable to apply on lactating cows.

The following **rules** should be considered in case of external applications:

(a) Intense cleaning of the stables before treatment.

(b) Cleaning of the animals, removing adhesive feces and scabs, so that the acaricide gets into contact with the mites.

(c) All animals of the herd have to be treated (also clinically inapparent animals).

(d) Appropriate treatment is carried out by sprayers (backpack sprayers, about 5.8 bar). Adult animals are treated with 3–4 l of the prepared formulation.

Treatment starts at the head, follows along the dorsum up to the tail, to the extremities and limbs (do not forget the areas below the belly and the tail or inside the ears!)

(e) The use of the exact concentration of the compounds is obligatory.

(f) For disinfection of tether devices, of stable tools and of the floors and walls of the stable, considerable higher concentrations must be used than during animal treatment. Stables, which are kept empty for at least 4 weeks, become mite free even without disinfection.

(g) Repetition of the treatment (at least 1 repetition, better 2) with contact acaricides should be done at fixed dates. The veterinarian should check whether mites survived and whether the treatment was done correctly according to the indications of the compound. He/she informs the animal owners on the withdrawing periods. **Caution**: Compounds which are not allowed for lactating animals are **γ-HCH** compounds and Sebacil® solutions = **Phoxim**.

(h) Modern endectocides (parenteral/pour-on application) have clearly simplified the mange therapy (e.g. use of **ivermectin** and **doramectin** is recommended).

Further Reading

Bochkov AV et al (2014) Integrated Bayesian species elimination and morphological diagnostics of chorioptic mange mites. Parasitol Res 113:2603–2607.

Botelho-Vieira MI et al (2014) Re-emergence of *Chorioptes bovis*. Braz J Vet Parasitol 23:530–533.

Hamel D et al (2015) Treatment and control of bovine sarcoptic and psoroptic mange infestation with ivermectin (IVOMEC® Gold). Parasitol Res 114:535–542.

Sarre C et al (2015) Comparative immune responses against *Psoroptes ovis* in two cattle breeds. Vet Res 46:131.

Stoeckli MR et al (2013) The effect of *Psoroptes ovis* infection on ovine epidermal barrier function. Vet Res 44:11.
Visser M et al (2013) The treatment of bovine sarcoptic mange (*Sarcoptes scabiei* var. *bovis*) using eprinomectin extended-release injection. Vet Parasitol 192:359–364.

6.1.2.5 Mange in Horses

1. **Name**: Greek: *sarx* = flesh; *koptein* = to induce wounds; *psora* = mange; *chorion* = leather; *demos* = sebum; *dex* = woodworm larva. Latin: *bos* = cattle; *equus* = horse; *cuniculus* = rabbit.
2. **Geographical distribution/epidemiology**: Worldwide.
3. **Biology/morphology**:
   (a) ***Sarcoptes equi*** (Fig. 6.28c): Females reach a length of up to 0.45 mm and males up to 0.25 mm. Only the two anterior ones of the eight stubby legs are visible in the dorsal aspect. Bowl-shaped sucker-like structures exist on the unsegmented pedicles (in males at the 1st, 2nd and 4th pairs of legs and in females at the 1st and 2nd pairs). The mites parasitize mainly on the head and along the backline. This species is rather **rare**.
   (b) ***Psoroptes* species** (*P. equi*, *P. cuniculi*): Females are up to 0.8 mm long and males up to 0.5 mm. They are longer than wide; all legs are dorsally visible. Legs are equipped by a tulip-shaped sucker on a long trisegmented pedicle (Fig. 6.29). In females, the third pair of legs is provided with two long bristles. These mites especially infest protected skin areas (including the base of the tail, the inner surface of the legs, neck, lower abdomen, etc.). The infestation rates are relatively **low**.
   (c) ***Chorioptes bovis***: Femals reach a length of 0.6 mm and males measure up to 0.45 mm; all legs are visible from above. They possess a bell-shaped sucker-like structure on a short unsegmented pedicle (Fig. 6.34) which in females is absent on the third pair of legs. These mites occur in the skin of the feet and ankles (scaly leg). *Chorioptes* infestations are relatively **common**, but lead only rarely to severe symptoms.
   (d) ***Demodex equi***: They are found in low numbers in hair follicles of horses. Adult mites are elongated (up to 400 μm), appear cylindrical and possess stubby legs. Eggs (60–80 × 40 μm) are significantly smaller than those of *D. caballi*, which reach 100 μm in length. *D. caballi* is predominantly found in the eyelids.
4. **Symptoms of disease (Mange)**: Severe itching occurs as well as skin nodules, pustules, dandruff and barks, thickening and wrinkling of the skin, hairless areas.
5. **Diagnosis**: Detection of mite stages in skin scrapings.
6. **Pathway of infection**: Body contact, transmission also via contaminated harness.

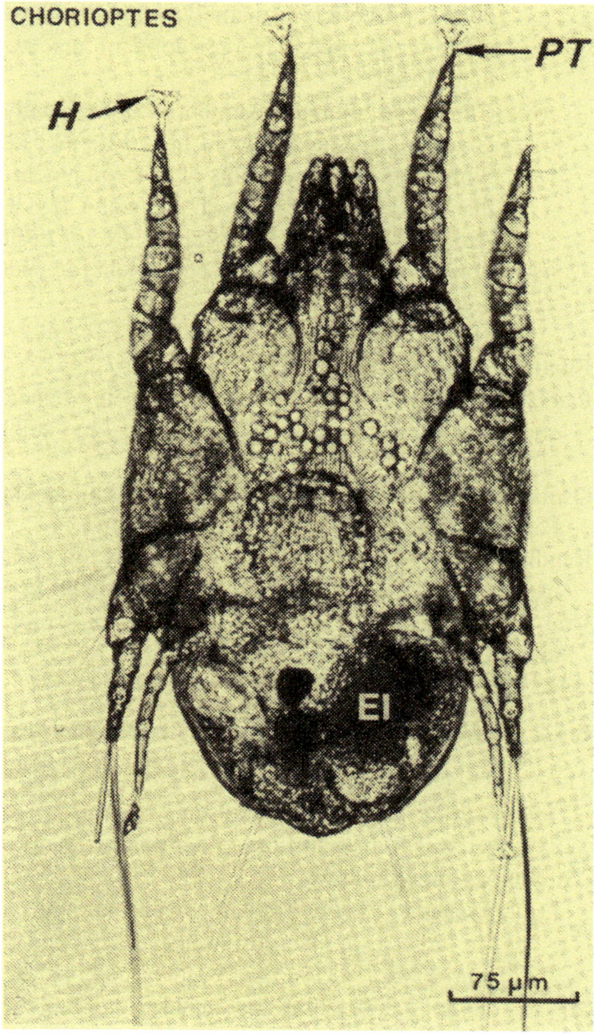

Fig. 6.34 Microscopical aspect of a *Chorioptes* mite. EI = egg; H = holdfast system; PT = pretarsus

7. **Prophylaxis**: Regular disinfection of the stables and equipment using contact insecticides (e.g. CBM 8®, Permethrin 25®, Ardap®, INS 15®). **Caution**: Do not use these products on animals!
8. **Incubation period**: Variable, depends on the general condition of the host.
9. **Prepatent period**: About 9–10 days are needed for the development of infectious stages.
10. **Patency**: The adult females live at the maximum 5 weeks. However, repeated infections may follow each other and increase the number of mites.
11. **Therapy**:
    (a) In cases of ***Sarcoptes*** and ***Psoroptes*** **mange** (burrowing or non-burrowing mites): Isolation of suspicious animals and immediate treatment with

contact insecticides, namely whole-body treatment of the animals by washing or spraying application of macrocyclic lactones (ivermectin, doramectin).

(b) ***Chorioptes* mange** treatment is done locally (distal from the carpal or tarsal joint). The treatment has to be repeated at least 1× after 8 days. Simultaneous disinfection of the equipment and stables by compounds that have a long-term effect, e.g. with CBM 8®, Permethrin 25, INS 15 or Ardap®). **Caution**: The products are for exclusive use in stable disinfection. Not using stables and equipment for a period of 4 weeks also leads to decontamination.

Further Reading

Paterson S, Coumbe K (2009) An open study to evaluate topical treatment of equine chorioptic mange with shampooing and lime sulphur solution. Vet Dermatol 20:623–629.

Osman SA et al (2006) Clinical and therapeutic studies on mange in horses. Vet Parasitol 141:191–195.

Rendle DI et al (2007) Comparative study of doramectin and fipronil in the treatment of equine chorioptic mange. Vet Rec 161:335–338.

Ural K et al (2008) Eprinomectin treatment of psoroptic mange in hunter/jumper and dressage horses. Vet Parasitol 156:353–357.

6.1.2.6 Mange in Hares, Rabbits and Laboratory Rodents

1. **Name**: Greek: *psora* = mange; *notos* = dorsum; *hedraios* = stuck. Latin: *cuniculus* = rabbit; *mus* = mouse.
2. **Geographical distribution/epidemiology**: Worldwide.
3. **Biology/morphology**:
   (a) ***Psoroptes cuniculi*** (agent of **ear mange**) causes especially in **rabbits** (in large-scale plants) significant economic damages, if early and persistent control measures are not accomplished. Size of the adults: males 0.45 × 0.3–0.4 mm and females 0.5–0.7 × 0.4–0.5 mm. Characteristics are the trumpet-like adhesive pads on segmented pedicles of the first three (males) or 1st, 2nd and 4th pairs of legs (females) (Fig. 6.32). The development proceeds via a six-legged larval stage and 2 eight-legged nymphs to the adult mite and takes 2–3 weeks.
   (b) ***Notoedres cuniculi*** (syn. *cati*), agent of the so-called **scaly face**. Adult females burrow in the area of the lips, nasal bridge, forehead, ears and bags below the eyes. However, a spreading over the whole body is also possible. Like in *Sarcoptes cuniculi*, a **scaly face** may occur. This species is very rare and thus without larger economic importance.
   (c) ***Notoedres muris*** (agent of **mouse ear mange**): It resembles *N. cati*. Lesions occur on ears, nose and tail as well as on external genitals.
4. **Symptoms of disease**: Ear mange is characterized by severe itching. Often occur scales, nodules and scab formations. Exudations are frequently

complicated by secondary bacterial infections. Emaciation and hair loss occur along the infested areas. In the case of an infestation inside the central and inner ear, rabbits often tilt and shake their heads. As a consequence of central nervous disorders, movement disorders (stagger) and apathy appear. In the case of severe infestations, death due to sepsis or meningitis may occur.

5. **Diagnosis**: Detection of mite stages in skin scrapings.
6. **Pathway of infection**: Juvenile stages of burrowing mites are transmitted by body contact.
7. **Prophylaxis**: Isolation of infested animals. Regular cleaning and disinfection of the stables together with leaving them empty for at least 3–4 weeks so that migrating nymphs and males die.
8. **Incubation period**: In the case of severe infestations, it takes only a few weeks because the succession of the generations is short (about 2–3 weeks).
9. **Prepatent period**: In case of an initial infestation, the next generation appears after 3–4 weeks in high numbers of individuals.
10. **Patency**: Months up to a lifetime due to the quick succession of generations.
11. **Therapy**: Administration of macrocyclic lactones (e.g. ivermectin).

Further Reading

Anholt H et al (2014) Ear mange mites (*Notoedres muris*) in black and Norway rats. J Wildl Dis 50:104–108.

Galdhar CN et al (2012) Clinico-biochemical and therapeutic studies on notoedric mange in pet rabbits. J Parasit Dis 39:113–116.

Hallal-Calleros C et al (2013) Hormonal and behavioral changes induced by acute and chronic infestation with *Psoroptes cuniculi*. Parasites Vectors 6:361.

Isingla LD et al (1996) Therapeutic trial of ivermectin against *Notoedres cati* var. *cuniculi* infection in rabbits. Parasite 3:87–89.

Izdebska J, Ralbieki L (2013) Sarcoptic mites parasitizing the brown rat (*Rattus norvegicus*). Annals Parasitol 59:125–128.

Wen H et al (2010) The effect of self-licking behaviour on pharmacokinetics of eprinomectin and clinical efficacy against *Psoroptes cuniculi* in topically administered rabbits. Parasitol Res 106:607–613.

6.1.2.7 Mange in Birds

1. **Name**: Greek: *sarx* = flesh; *koptein* = to wound; *psora* = mange; *chorion* = leather; *demos* = sebum; *dex* = woodworm larva. Latin: *mutare* = to mutate; *pila* = ball; *laevis* = soft.
2. **Geographical distribution/epidemiology**: Worldwide.
3. **Biology/morphology**: Specimens of the genus *Knemidocoptes* are relatively small mites (males up to 0.25 mm and females up to 0.5 mm long), which almost appear circular from the top view. They are characterized by stubby legs (Fig. 6.35), which are equipped with claws in females, but in males there occur long, unsegmented pedicles with sucker-like structures. Females mostly deposit the six-legged larvae **viviparously** into the skin burrows of their hosts.

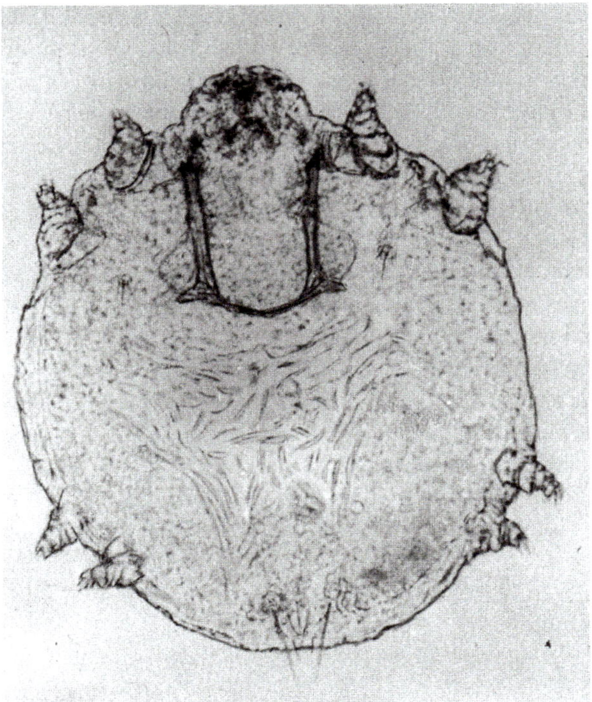

Fig. 6.35 Light micrograph of *Knemidocoptes pilae* from a budgerigar

(a) ***Knemidocoptes mutans*** parasitizes in gamefowl and pigeons in the area of the featherless legs, the toe basis and the dorsal site of the metatarsus (the stands) and causes scaly legs. The development via 1 larva and 2 nymphs to the adult mite takes 30 days in males and 26 days in females.

(b) ***K. pilae*** colonizes in canaries and parakeets as well as in other parrots the skin of the face and of the eyelids, but also is found along the cloaca and on certain areas of the legs. The entire development takes about 3 weeks. Females lay embryonated eggs. The clinical picture is called "**beak mange**".

(c) ***K. laevis*** infests the feathered regions (especially at the rump and the dorsum) in gamefowl and pigeons and thereby causes **body mange**. Females deposit the larvae viviparously.

4. **Symptoms of a disease**:

(a) **Scaly leg disease**: Infestation with *K. mutans* leads to severe itchiness (pattering, pecking at the legs, tucking up the legs under the body), incrustations, scab formation, movement disorders. Especially in older chickens typical hyperkeratosis with severe impairment of the general condition and arteritis in different organs occur. However, in today's animal farming the scaly leg disease has no longer major economic importance because of its rare appearance, though in individual aviaries all animals can be infested at the same time!

(b) **Beak mange**: Infections with *K. pilae* induce grey-yellowish skin proliferations, which are covered with boreholes. As a consequence of bacterial secondary infections, beak and possibly claws anomalies appear. Beak mange is common among birds in rearing facilities.

(c) **Body mange**: This disease due to infestations with *Knemidocoptes laevis* is characterized by feather loss, by calcareous scabs as well as by inflamed skin pustules in the infested areas. This disease is relatively common.

5. **Diagnosis**: Detection of adult mites (Fig. 6.35) and of embryonated eggs in skin scrapings, when exposed to 10 % caustic potash.
6. **Pathway of infection**: By body contact, probably already at the age of fledglings or nestlings.
7. **Prophylaxis**: Regular disinfection of the stables or stands (pigeons) with the help of contact insecticides. Regular inspection of the animals.
8. **Incubation period**: Variable, depending on the incidence. In the case of *K. pilae* infestations, it can take years until symptoms appear.
9. **Prepatent period**: The succession of generations lasts about 4 weeks. Males are already sexually mature after 3 weeks.
10. **Patent period**: Practically lifelong due to the dense succession of generations.
11. **Therapy** of the **scaly leg** or **body mange**: In advanced cases, only an individual animal treatment can lead to success. Before the treatment, the scabs (up to 1 cm thick) have to be macerated and removed with glycerine or soap solution. At the same time, washing or bathing the legs (up to the beginning of feathers) with contact insecticides or e.g. dapping the legs with Odylen®. Even healthy appearing animals have to be subjected to treatment. Repetition of treatment after 10 days. For disinfection measures, see **Prophylaxis**. Treatment of beak mange has to be done accordingly. Covering the burrows with Vaseline® or paraffin oil leads to suffocation of the mites. Application of contact insecticides with cotton buds on altered beak areas after previous removal of the scabs is as effective.

Further Reading

Dabert J et al (2011) The first report of *Knemidocoptes intermedius* in naturally infected European birds. Parasitol Res 109:237–240.

Dabert J et al (2013) Multidisciplinary analysis of *Knemidocoptes jamaicensis* parasitizing the common chaffinch (*Fringilla coelebs*). Parasitol Res 112:2372–2380.

Mete A et al (2014) Knemidocoptic mange in wild Golden eagles, California, USA. Emerg Inf Dis 20:1716–1722.

6.1.2.8 *Cheyletiella* Species

1. **Name**: Greek: *cheilos* = lip; *parasitos* = parasite, assessor Lat.: *vorare* = to devour, to engorge. Jasgur, Blake = Russian and English scientists.

2. **Geographical distribution/epidemiology**: Global distribution, severe infestation in weakened animals (infection rates up to 70 % in rabbits, 25 % in cats and dogs).
3. **Biology/morphology**: The species are not very host specific and can infest many hosts (including humans):
   (a) ***C. yasguri*** (dog)
   (b) ***C. blakei*** (cat, fox)
   (c) ***C. parasitivorax*** (rabbits)

   All stages of the life cycle (**larva, nymphs 1 and 2, adults**) feed on skin and skin derivatives or on lymph extravasating from wounds of the host. Adult mites become up to 600 μm (females) or 400 μm (males) long and are characterized by the massive claws of the pedipalps (Figs. 6.36 and 6.37). The eggs (230 × 100 μm) are stuck on the hair with the help of filamentous structures (Fig. 6.38).
4. **Symptoms of disease**: In case of massive infestation, mange-like skin alterations as well as severe itchiness can occur.
5. **Diagnosis**: Detection of the mites: they are brushed down onto a dark ground and determined with a loupe (microscope) or detected by impression method with the help of transparent plastic adhesive tape. If the animal nibbles at its coat, mite eggs may be found also in feces as passers-by.
6. **Pathway of infection**: Body contact with infested animals (*Cheyletiella*). *Neotrombicula* larvae let themselves drop from plants on passing endotherms (unspecific choice) during late summer or autumn. **Caution**: Also **humans** are in danger of infestation!
7. **Prophylaxis**: Application of vermin collars, treatment with insecticide containing product or use of repellents sprayed onto dogs or cats.
8. **Incubation period**: 1 week
9. **Treatment**: Application of contact insecticides (see ticks).

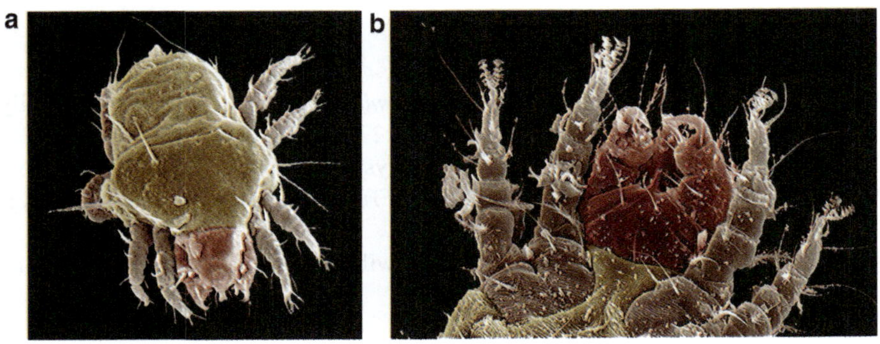

Fig. 6.36 Scanning electron micrographs of *Cheyletiella yasguri* mites. (**a**) Total aspect. (**b**) Anterior end

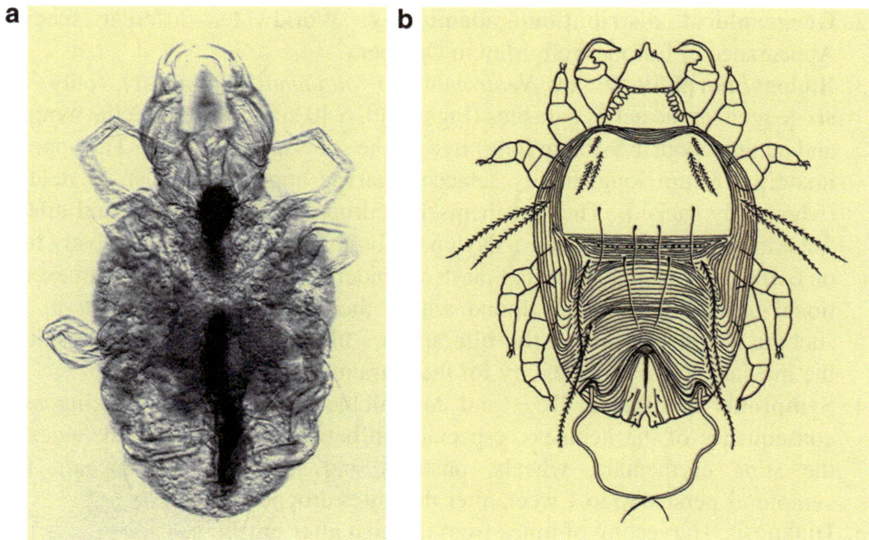

Fig. 6.37 Light micrograph (**a**) and diagrammatic representation (**b**) of an adult stage of *Cheyletiella yasguri*

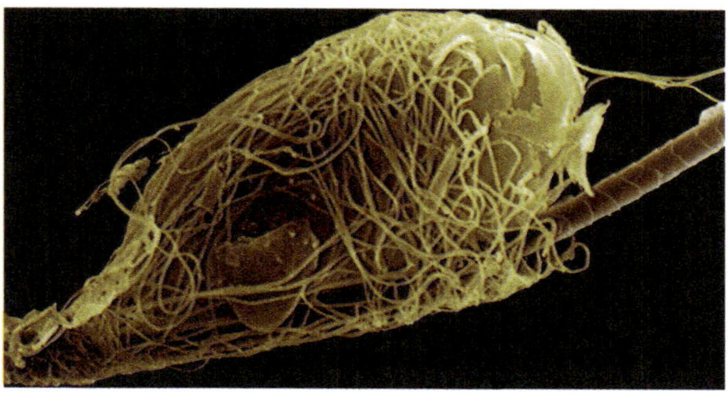

Fig. 6.38 Scanning electron micrograph of an egg of *C. yasguri*, which was attached by the female by help of filament at hair

Further Reading

Ken KM et al (2014) Dermatosis associated with mites other than *Sarcoptes*. Semin Cutan Med Surg 33:110–115.

Salant H et al (2014) Ectoparasites in urban stray cats. Med Vet Entomol 28:314–318.

6.1.2.9 Lymph Sucking Mites in Vertebrates

1. **Name**: Greek: *neo* = new; *thrombos* = clump. Lat.: *autumnalis* = belonging to the autumn.

2. **Geographical distribution/epidemiology**: Worldwide, different species. Appearance in Europe from May to October.
3. **Biology/morphology**: In *Neotrombicula autumnalis* (chigger), only the six-legged larvae act as parasites (Figs. 6.39, 6.40 and 6.41), while the nymphs and adults (about 1.5–2 mm long) live on the soil saprophytically. The approximately 300 μm long, highly setaceous larvae appear yellowish to reddish (when fully sucked). They let themselves drop down onto the ground after a "sucking activity" lasting for a day up to about a week. The larvae mostly feed on extraintestinally predigested mash of epidermal cells, so that only in exceptional cases blood can be found within their intestine. Subsequent to the sucking action the **side of the bite** appears **bloody** (Fig. 6.42) as a result of the injected enzymes necessary for the extraintestinal digestion.
4. **Symptoms of disease**: Dogs and cats nibble at the sites of sucking as a consequence of the itchiness, especially in between the toes. At the sides of the sting erythemata, wheals, pustules and haemorrhages appear. The symptoms persist up to 1 week after the mites dropped off.
5. **Diagnosis**: Harvesting of mites from the coat after brushing.
6. **Pathway of infection**: The mites climb onto the legs from the ground.
7. **Prophylaxis**: Insecticides (see against ticks) and (biological) dipping in neem extract (MiteStop® Co. Alpha-Biocare, Düsseldorf).
8. **Incubation period**: 1–2 days.
9. **Prepatent period**: Mites are only noted when occurring in higher numbers.
10. **Patency**: The sucking action lasts in general 1–3 days.
11. **Therapy**: Application of contact insecticides or acaricides (see ticks), bath in MiteStop® Co. Alpha-Biocare, Düsseldorf).

Further Reading

Grannoulopoulos GD et al (2012) First report on *Neotrombicula autumnalis* infestation in a cat and a dog from Corfu. Vector Borne Zoonotic Dis 12:1065–1067.

Parcell BJ et al (2013) Conjuctivitis induced by a red bodied mite, *Neotrombicula autumnalis*. Parasite 20:25.

Stekolnikov AA, Klimov PB (2010) A revision of chiggers of the minuta species group (Acari, Trombiculidae: *Neotrombicula*). Syst Parasitol 77:55–69.

Stekolnikov AA et al (2014) *Neotrombicula inopinata* : a possible causative of trombiculiasis in Europe. Parasites Vectors 7 :90.

6.1.2.10 Bloodsucking Mites

1. **Name**: Greek: *derma* = skin; *nyssein* = to sting; *ornis* = bird; *lipos* = fat. Latin: *sanguis* = blood.
2. **Geographical distribution/epidemiology**: Worldwide, these mites occur in henhouses and other stables often in huge numbers.
3. **Biology/morphology**:
   (a) ***Dermanyssus gallinae*** **(Red chicken mite)**: Soaked females are up to 3 mm long. They appear plain grey whitish and blood red when sucked

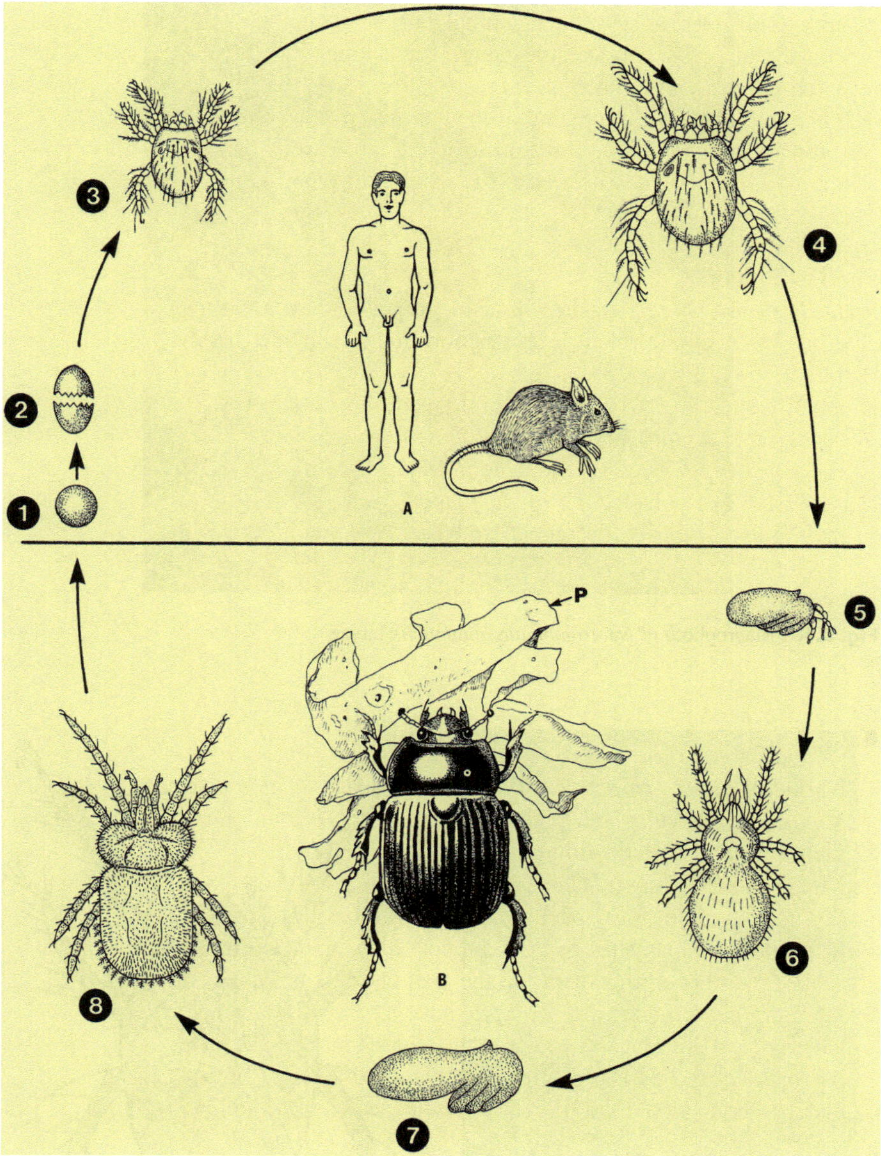

Fig. 6.39 Diagrammatic representation of the life cycle of *Neotrombicula autumnalis*. **1–4**. The six-legged larva hatches from the egg and starts to suck **lymph** at vertebrates. **5–6**. After sucking, the larva drops down and enters the soil and becomes (after a moult) the inactive protonymph. After moulting, she becomes the deutonymph, which is carnivorous. **7–8**. After another inactive phase as tritonymph (**7**), they become sexually active males and females, which feed on remnants of plants and produce eggs after copulation

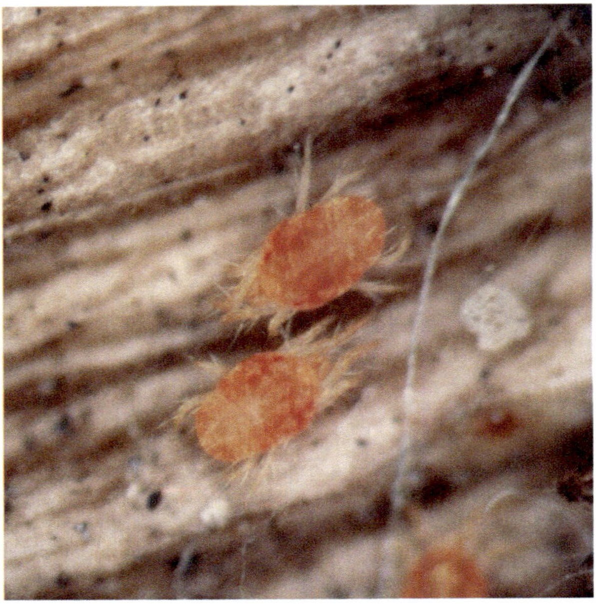

Fig. 6.40 Macrophoto of *Neotrombicula autumnalis* larvae

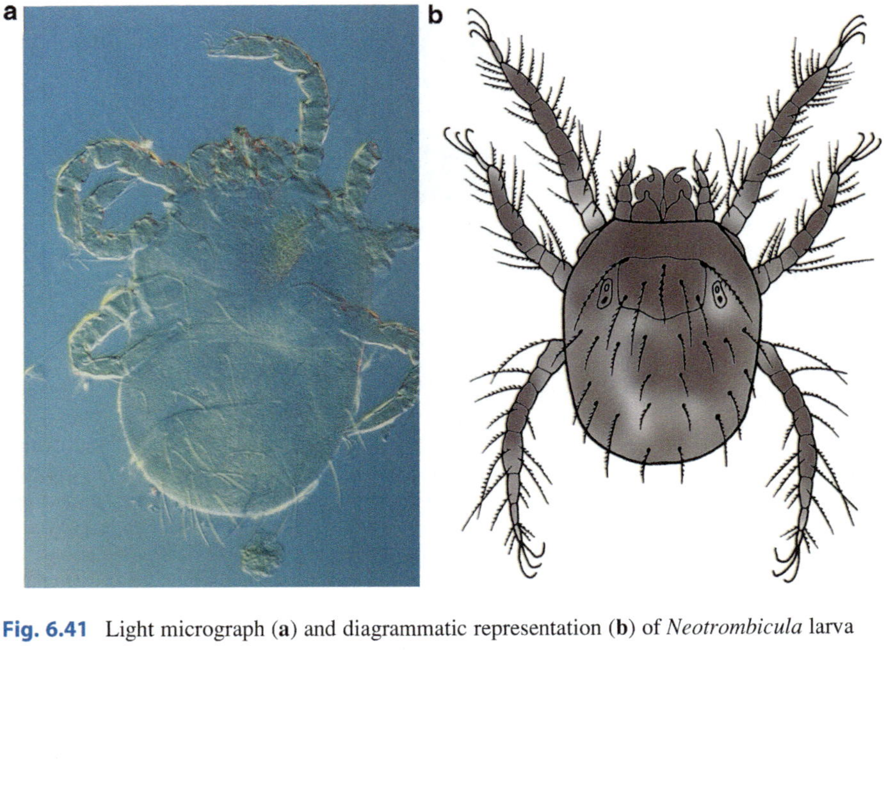

Fig. 6.41 Light micrograph (**a**) and diagrammatic representation (**b**) of *Neotrombicula* larva

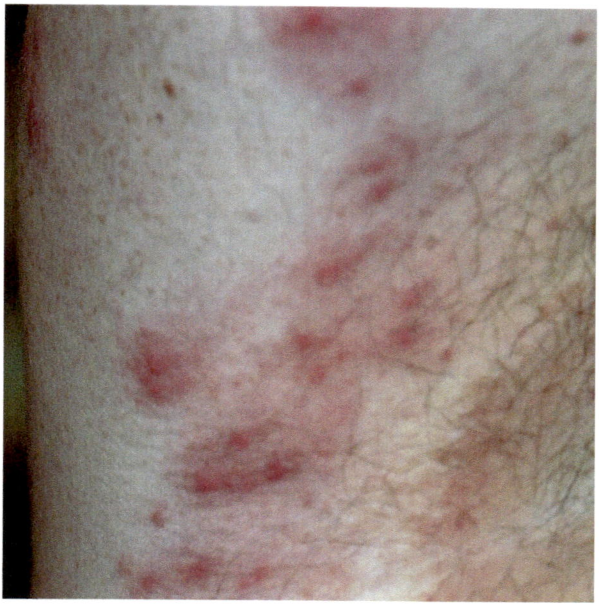

Fig. 6.42 Skin reaction after sucking of *Neotrombicula* larvae

Fig. 6.43 Macrophoto of a fully engorged *Dermanyssus gallinae*

(Figs. 6.43 and 6.44). The development starts from eggs (400 × 200 μm) via six-legged larvae and eight-legged proto- and tritonymphs to the adult males and females. This development takes only 4–12 days. The adult

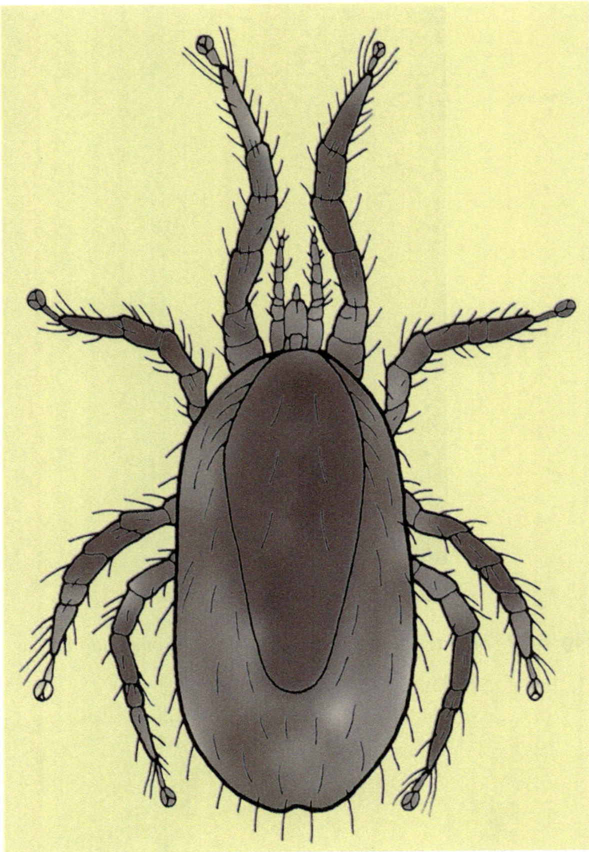

Fig. 6.44 Diagrammatic representation of a red bird mite (*Dermanyssus gallinae*)

mites can survive for 5–9 months without food intake. *Dermanyssus gallinae* infests virtually all **warm-blooded animals** including **humans**. The sucking action happens mostly at night and lasts about 30–60 min. Only after a blood meal the females are able to lay eggs. The larvae do not feed and slough in lairs. Therefore, stables have to be treated several times in any case.

Related species but strictly host specific are *D. bovis* (cattle), *D. ovis* (sheep), *D. caprae* (goat), *D. phylloides* (pigs), *D. cati* (cats), *D. canis*, *D. phylloides*, *D. gatoi* (dogs), *D. caviae* (guinea pigs), *D. musculi* (mice), *D. ratti* (rats), *D. folliculorum* and *D. brevis* (humans).

(b) ***Ornithonyssus sylviarum***: These so-called Northern fowl mites reach as females a length of up to 0.8 mm (Fig. 6.45). These mites remain on the once infested host (from larva to the adult stage), so that large populations occur on a single host.

(c) ***Liponyssoides sanguineus***: This so-called house mouse mite also attacks humans.

Fig. 6.45 Light microscopic photo of an adult Northern fowl mite (*Ornithonyssus sylviarum*)

4. **Symptoms of disease**: Itchiness, pain, restlessness, urticaria, skin irritations and secondary infections; in the case of massive infestation: anaemia and severe impairment of the laying performance.
5. **Diagnosis**: Microscopic inspection of the stages, which have been taken from lairs or from the plumage.
6. **Pathway of infection**: The mites climb onto their host from lairs.
7. **Prophylaxis**: Regular disinfestation of the stables with contact acaricides or with the biological product MiteStop®. **Caution**: If the organophosphate phoxim is used (e.g. in ByeMite®), drinking water, food and eggs have to be removed for 24 h.
8. **Incubation period**: Several days.
9. **Prepatent period**: Days.
10. **Patency**: Lifelong.
11. **Therapy**: Spraying with the biological agent MiteStop®.

Further Reading

Abdel-Ghaffar F et al (2009) In vitro efficacy of ByeMite® and Mite-Stop® on developmental stages of the red chicken mite *Dermanyssus gallinae*. Parasitol Res 105:1469–1471.

Barimani A et al (2016) Traps containing carvacrol, a biological approach for the control of *Dermanyssus gallinae*. Parasitol Res. doi:10.1007/s00436-016-5113-3.

Bartley K et al (2015) Identification and evaluation of vaccine candidate antigens from the poultry red mite *Dermanyssus gallinae*. Int J Parasitol 45:819–830.

De La Riva DG et al (2015) Temperature governs on-host distribution of the Northern fowl mite, *Ornithonyssus sylviarum* (Acari: Macronyssidae). J Parasitol 101:18–23.

George DR et al (2015) Should the poultry red mite *Dermanyssus gallinae* be of wider concern for the veterinary and medical science? Parasites Vectors 8:178

Immedieto D et al (2015) Laboratory evaluation of a native strain of *Beauveria bassiana* for controlling *Dermanyssus gallinae*. Vet Parasitol 212:478–482.

Locher N et al (2010a) In vitro and field studies on the contact and fumigant toxicity of a neem-product (Mite-Stop®) against the developmental stages of the poultry red mite *Dermanyssus gallinae*. Parasitol Res 107:417–423.

Locher N et al (2010b) Light and scanning electron microscopic investigations on MiteStop®-treated poultry red mites. Parasitol Res 107:433–437.

Makert GR et al (2016) A method to identify protein antigens of *Dermanyssus gallinae* for the protection of birds from poultry mites. Parasitol Res. doi:10.1007/s00436-016-5017-2.

Mueller RS (2011) Evidence based therapy of demodicosis of dogs. Tierärztl Praxis (K)39:419–424.

Tabari MA et al (2015) Carvacol as a potent natural acaricidal against *Dermanyssus gallinae*. Parasitol Res 114: 2801–3806.

6.1.2.11 Air Sac Mites or Lung Mites

1. **Name**: Greek: *kytos* = cell; *itis* = suffix; *stoma* = mouth. Latin: *nudus* = naked; *sternum* = breastbone; *tracheus* = rough, stiff; *trachea* = breathing tube.
2. **Geographical distribution/epidemiology**: Worldwide.
3. **Biology/morphology**:
   (a) ***Cytodites* species** (including *C. nudus*) occur in many pigeons, gamefowl and wild birds and are found in the entire breathing system but can also parasitize in the abdominal cavity, in the liver and inside the kidney. Free-living species are unknown.

   Females of *C. nudus* measure 0.6 × 0.5 mm, males 0.42–0.55 × 0.4 mm. These mites are soft-skinned, dorsoventrally flattened and almost bristle-less (Figs. 6.45 and 6.46). Their gnathosoma is wide; chelicerae are missing. On all extremities unsegmented pedicles with sucker-like structures are located. The vulva lies between the two back pairs of the large extremities. The larvae hatch from the egg during or just after leaving the female vulva; via two nymph stages the adult stage is reached.

   (b) ***Sternostoma tracheacolum*** mites parasitize as bloodsuckers especially in the upper air system of **parakeets**, **canaries**, **estrildid finches** and **agapornids** but also of many **wild bird species**. As in *Cytodites nudus*, spreading into the lung, the air sacs and the abdominal cavity have been observed. *S. tracheacolum* reaches a size of about 0.7 × 0.5 mm and is characterized by the transformation of the claws of the first pair of legs to a scissor (Fig. 6.47). The eggs are deponed fully embryonated. The emerging larvae develop via two nymph stages to adults.

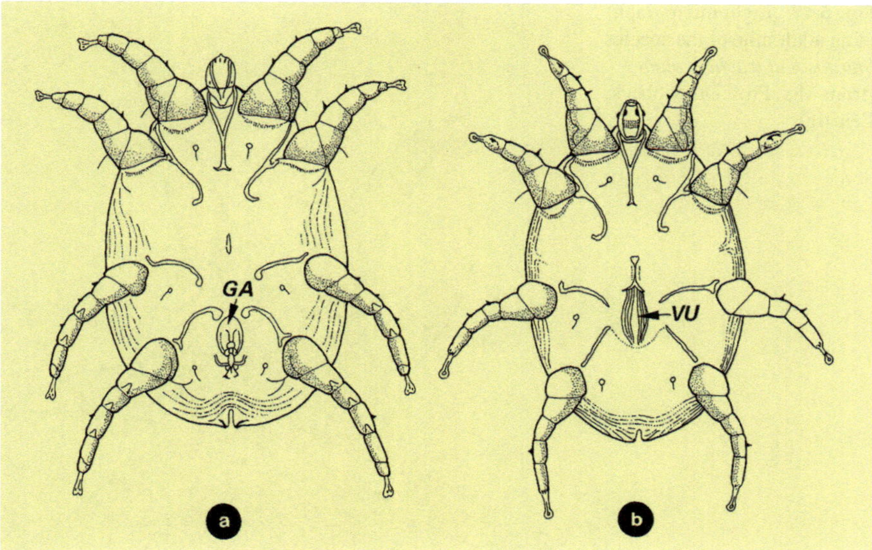

Fig. 6.46 Diagrammatic representation of a male (**a**) and a female (**b**) of *Cytodites nudus* seen from ventral (after Vain and Brohmer). GA = male sexual apparatus; VU = vulva

(c) **Lung mites**. In canids and felids, the mite *Pneumonyssus caninum* occurs in the area of the nasal cavities and in the airways. Their presence leads to nose secretion and coughing as well as to epistaxis. Transmission occurs by contact with infested animals.

4. **Symptoms of disease**: Slight infestation rates remain asymptomatic. In the case of massive infestation—often as a result of lung diseases—aphony, coughing, breathing difficulties (wheezing, opened beak), gagging, pneumonia, bronchitis and peritonitis occur. Death cases are then relatively common; air sacs appear flour strewn when dissected.
5. **Diagnosis**: During butchering, the mites can be detected microscopically in swabs of the infested airways (Figs. 6.46 and 6.47). The large sized *Cytodites* specimens appear as white dots and those of *S. tracheacolum* as dark dots.
6. **Pathway of infection**: The exact infestation is unknown; however, it occurs presumably by oral uptake of different developmental stages (during billing, etc.), active immigration of the mites into the throat or eventually during uptake of drinking water.
7. **Prophylaxis**: Unknown.
8. **Incubation period**: Variable, depends on the infestation density. Oral or nasal infection experiments showed inconsistent results.

9./10. **Prepatent period/patency**: The succession of generations takes about 3 weeks. An infestation can persist lifelong because the propagation within the host is obviously not impeded (cf. *Demodex* species).

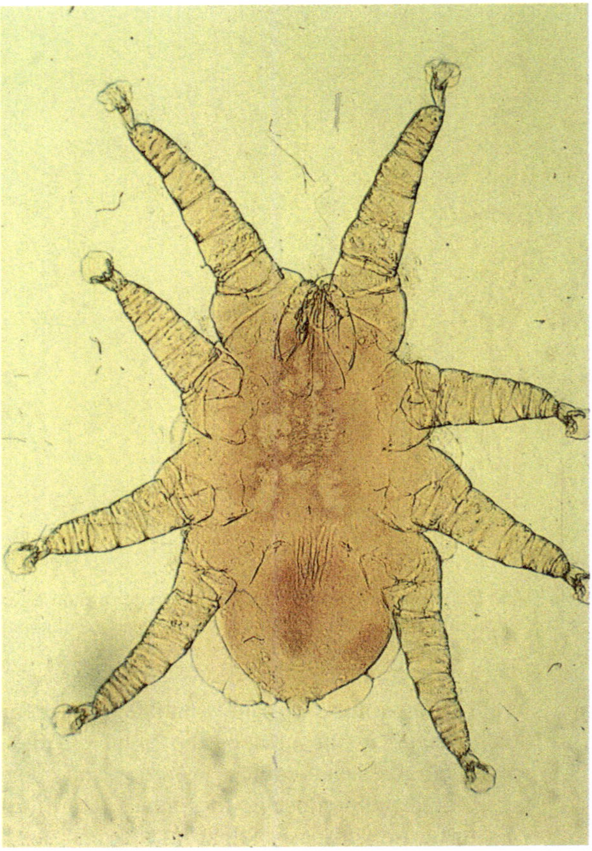

Fig. 6.47 Light micrograph of an adult mite of the species *Sternosoma tracheacolum* (from Mrs. Prof. Dr. Ribbeck, Leipzig)

11. **Therapy: Tentatively**: External application of acaricides (contact insecticides). However, the prognosis is poor. The eradication of *Cytodites nudus* only succeeds with the abolition of sick and infectious birds. Therapeutic tests using macrocyclic lactones had been successful. Moxidectin (Advocate® spot-on 1× per week).

Further Reading

Ayround M, Dies KHC (1992) *Cytodites nudus*-induced granulomatous pneumonia in chickens. Can Vet 33:754–755.

De Rojas M et al (2002) Phylogenetic relationships in rhinonyssid mites based on rDNA sequences. Parasitol Res 88:675–681.

McOrist S (1983) *Cytodites nudus* infestation of chickens. Avian Pathol 12:151–155.

Mascarenhas CS et al (2011) Nasal mites of *Paroaria coronata* (Passeriformes). Neotrop Entomol 40:507–508.

Mendes MM et al (2014) Nasal mites of Tyrannidae (aves) in Brazil. Braz J Biol 74:480–482.
Pence DB, Casto S (1975) Two new species and new records of nasal mites of the genus *Sternostoma* (Acarina: Rhinonyssinae) from birds in Texas. J Parasitol 61:360–368.
Tidemann SC et al (1992) Parasitism of wild Gouldian finches (*Erythrura gouldiae*) by the air-sac mite *Sternostoma tracheacolum*. J Wildl Dis 28:80–84.

6.1.2.12 Feather Quill Mites and Feather Follicle Mites

1. **Name**: Greek: *syrinx* = tube; *philein* = to love; *rhynchos* = beak, snout; *harpyr* = pointed.
2. **Geographical distribution/epidemiology**: Worldwide. The importance of these mites as feather parasites in poultry, pet birds, wild bird species and also pigeons is rather low.
3. **Biology/morphology**: The diagnosis of the mites species living in feather quills in the follicles is reserved to specialists (Figs. 6.48 and 6.49).
4. **Symptoms of disease**:
   (a) **Feather quill mites (Family Syringophilidae)**: The feather quills of the big wing and tail feathers appear opaque (not opalescent, but like flour-covered); occasionally feathers get lost.
   (b) **Feather follicle mites (Family Harpyrhynchidae)**: Nodes and cyst-like protrusions occur on the feather follicles as well as local feather loss. Also skin nodes and dermatitis (non-itching mange form) may occur.
5. **Diagnosis**:
   (a) **Feather quill mites**: Detection of the developmental stages and adults in feather quills (Fig. 6.48). Before the moult, mites are also seen outside of the quill region.
   (b) **Feather follicle mites**: Detection of the mites and their eggs in feather follicles and in surface layers of the skin with the help of microscopical analysis of skin scrapings (Fig. 6.49).
6. **Pathway of infection**: By body contact, feather quill mites leave the feather before moult and invade other growing feathers via the feather poles.
7. **Prophylaxis**: Regular intense disinfection of the stables with the help of suitable contact insecticides or MiteStop®.
8. **Therapy**: Administration of contact insecticides in powder or liquid form (aerosol treatment). Dipping baths have been proven to be successful especially in pigeon stocks. For pet birds, pyrethrum or carbamate derivatives should be chosen because of their good tolerance in warm-blooded animals. Odylen®: local application after plucking of the altered body regions. It is recommended to treat all animals of a stock and to disinfect the stables with contact insecticides or MiteStop®.

Fig. 6.48 Diagrammatic representation of a feather quill mite (*Syringophilus bipectinatus*) (according to Schaefer 2010). P = praepodosoma; PP = propodosomal shield

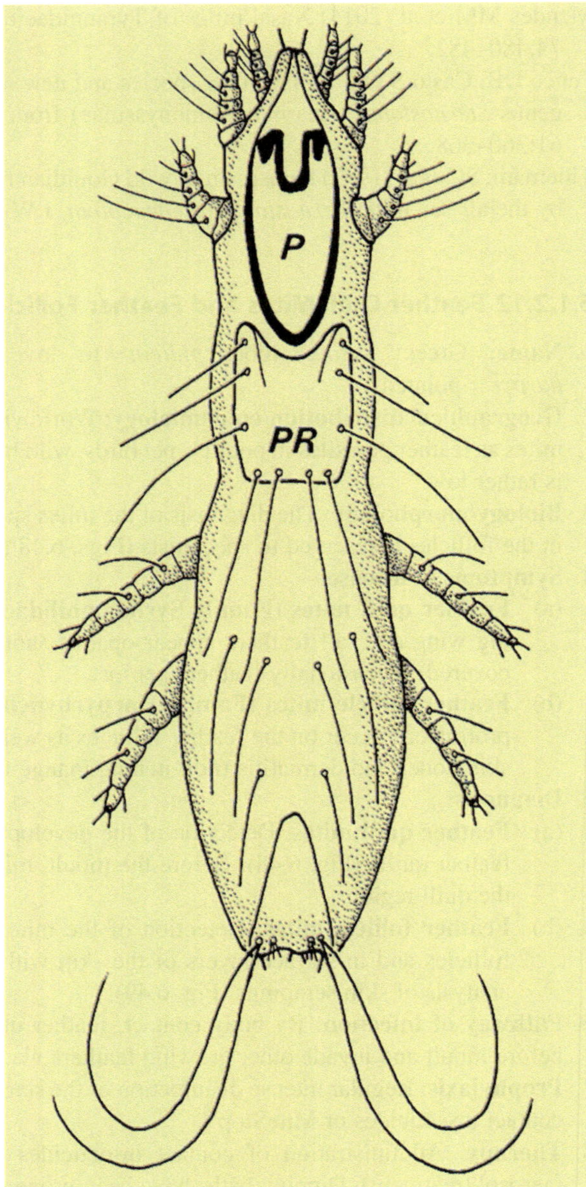

Further Reading

Fain A (1972) Notes on acarina of the family Cheyletidae and Hapyrhynchidae producing mange in birds and mammals. Acta Zool Pathol Antverp 56:37–60.

Moss WW et al (1968) Karyotypes and developmental stages of *Harpyrhynchus* sp. n., a parasite of Northern American birds. J Parasitol 54:377–392.

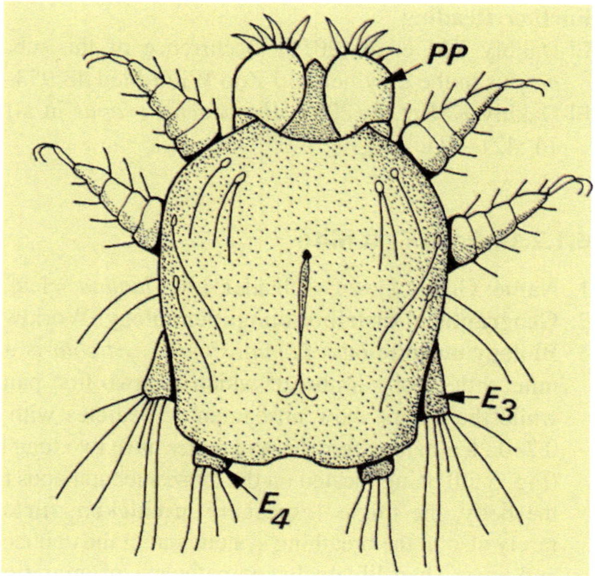

Fig. 6.49 Diagrammatic representation of a male of the feather follicle mite: *Harpyrhynchus* (syn. *Sarcopterinus*) *nidulans*. E3,4 = Third and fourth bristle pairs at the reduced legs; PP = pedipalps

Schaefer M (ed) (2010) Brohmer – Fauna von Deutschland, 23rd edn. Quelle & Meyer, Wiebelsheim.

Shoracki M et al (2014a) Syringophylid mites associated with the rails. Syst Parasitol 88:227–232.

Shoracki M et al (2014b) The new quill mite species of the family Syringophilidae parasitizing the house sparrow. Zootaxa 3765:194–200.

6.1.2.13 Nest Mites

1. **Name**: Greek: *hypo* = hypo; *dex* = woodworm larva.
2. **Geographical distribution/epidemiology**: Worldwide.
3. **Biology/morphology**: *Hypodectes* species live as adults in the nests of **house** and **wild birds**, frequently also of **pigeons** (*H. propus*), and feed on detritus. However, their nymph 2 enter as a so-called **hypopus stage** the subcutis of many birds (frequently as dormant stage).
4. **Symptoms of disease**: Restlessness of the animals, scratchiness, feather loss, replacement by abnormal feathers, flightlessness in case of severe destruction of the plumage.
5. **Diagnosis**: Detection of the adults in the nest and of the nymph 2 by biopsy in "featherless" areas.
6. **Pathway of infection**: Body contact and spreading in nests.
7. **Prophylaxis/therapy**: Treatment of nests with acaricides or spraying with MiteStop® (Fa. Alpha-Biocare, Neuss, Germany).

Further Reading

El-Dakhly KM et al (2012) Occurrence of the subcutaneous mite *Hypodectes propus* in the grey heron. J Zoo Wildl Med 43:953–955.

El-Dakhly KM et al (2013) *Hypodectes propus* in a rufous turtle dove. J Vet Ci 14 :421–424.

6.1.2.14 Fowl Cyst Mite

1. **Name**: Greek: *kystos* = bladder. Lat.: *lamina* = leaf.
2. **Geographical distribution/epidemiology**: Worldwide.
3. **Biology/morphology**: *Laminosioptes cysticola* is a cylindrical, nearly hairless mite with stubby limbs, of which the two first pairs are equipped with claws, while the others show unsegmented pedicles with sucker-like structures. The 0.2–0.26 mm long females and males have two long bristles at their posterior end (Fig. 6.50) being located on the loose subcutaneous tissue of the neck, the breast, the belly, the thighs (especially in **chicken**, **turkey**, **pigeon and pheasant**), rarely also in the breathing system and in the connective tissue of spleen, kidney and liver. Their life cycle and pathways of migration are mostly unknown.

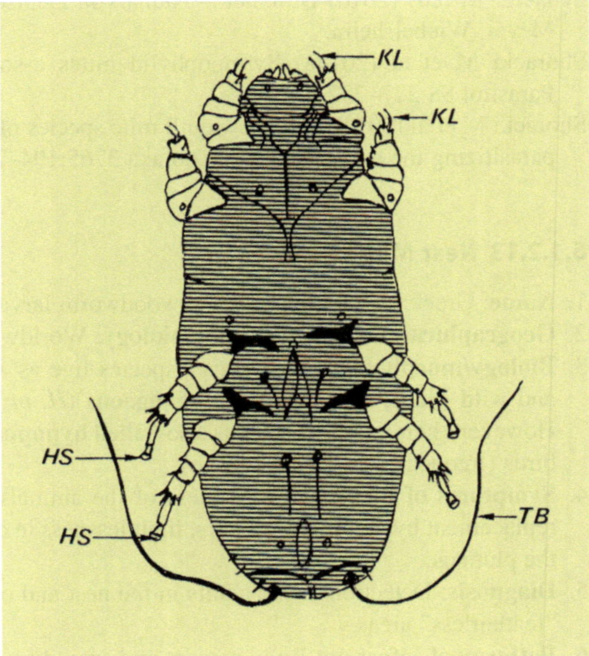

Fig. 6.50 Diagrammatic representation of a female of the nodule mite *Laminosioptes cysticola* (according to Fain). *HS* Holdfast system; *KL* claws; *TB* terminal bristles

Further Reading

Amure J, Stuart JC (1977) A case of *Laminosioptes cysticola* infection. Vet Rec 101:387–389.

Kaliner G (1970) Occurrence of *Laminosioptes cysticola* in the abdominal organs of turkeys. Pathol Vet 7:321–328.

Radfar MH et al (2012) Biodiversity and prevalence of parasites of domestic pigeons. Trop Anim Health Prod 44:225–229.

6.1.2.15 Tracheal Mites

1. **Name**: Greek: *akari* = mite. Lat.: *apis* = bee.
2. **Geographical distribution/epidemiology**: Worldwide.
3. **Biology/morphology**: Among the species of the genus ***Acarapis***, only *Acarapis* (syn. *Tarsonemus*) *woodi* became important while the others (*A. vagans, A. externus*) are considered as harmless. *A. woodi* females measure up to 180 × 90 μm and males 110 × 85 μm. These mites are externally characterized by a distinct segmentation (Fig. 6.51). They live and feed in the major tracheal truncus. Gradually, the breathing system is clogged by the accumulation of mites, leading to breathing impairment, flightlessness and finally death by asphyxiation.

a b

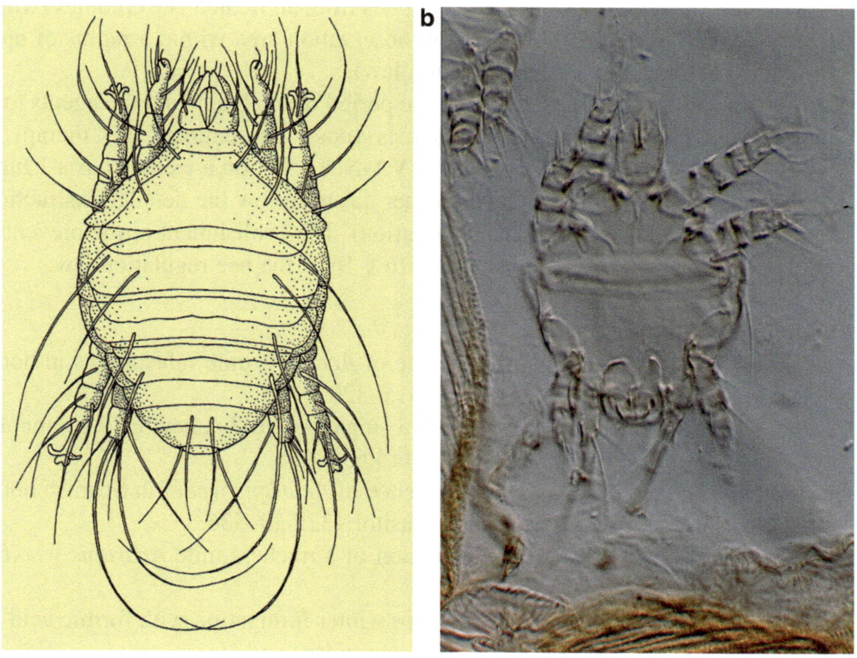

Fig. 6.51 (**a**, **b**) Diagrammatic representation (**a**) and light micrograph (**b**) of the tracheal mite *Acarapis woodi* parasitizing in bees

4. **Symptoms of disease (Acariasis, mite flu)**: Often high numbers of immobile bees are found in/at the hive. Sick bees show a swaying departure from the hive and soon afterwards fall down onto the ground without the ability to fly again.
5. **Diagnosis**: Microscopic detection of the mites (Fig. 6.51b) in the tracheae, possibly after maceration with KOH. Dried bees are not suitable for the examination. A negative result does not definitely exclude an infestation. **Caution**: Acariasis is a notifiable plague according to § 10 (1) of the animal protection act and the regulation of animal epidemics.
6. **Pathway of infection**: By body contacts. Transfer of fertilized females from one bee to others.
7. **Prophylaxis**: Preventive precautions are regulated by § 14 and § 15 of the bee regulation law (in Europe).
8. **Incubation period**: Variable, depends on the incidence.
9. **Prepatency**: The entire development takes about 10–16 days.
10. **Patency**: Lifelong.
11. **Therapy**:
    (a) If **acariasis** is diagnosed officially, according to the vet's instruction, the owner firstly has to treat all bee colonies of a hive and secondly has to eliminate innocuously all dead bees. The governmental authority is allowed to prohibit the removal or movement of bee colonies in the hive in order to prevent the spreading of acariasis. Moreover, the vet authorities can order the killing of contaminated and the remaining bee colonies as well as request the analysis of probes from all treated bee colonies (winter deaths) or define the area as an observation area within a radius of up to 2 km (§ 15 of the bee regulation law).
    (b) The authority also determines the period of treatment and the agents to be used for the control of acariasis (poss. compounds: see therapy of varroosis, including e.g. Folbex-VA-Neu®: 8× on a weekly basis 1 fumigation strip per colony. The owner has to follow the detailed instructions of the Vet (§ 14 of the bee regulation). The abolishment of the preventive measures is regulated according to § 16 of the bee regulation law.

Further Reading

Ahn AJ et al (2015) Molecular prevalence of *Acarapis* mite infestations in honey bees in Korea. Korean J Parasitol 53:315–320.

Cepero A et al (2015) Is *Acarapis woodi* a single species? A new PCR protocol to evaluate its prevalence. Parasitol Res 114:651–658.

Garrido-Bailón E et al (2012) The prevalence of *Acarapis woodi* in Spanish honey bee (*Apis mellifera*) colonies. Exp Parasitol 132:530–536.

Kojima Y et al (2011) PCR-based detection of a tracheal mite *Acarapis woodi*. J Invert Pathol 108:135–137.

Underwood RM, Currie RW (2009) Indoor winter fumigation with formic acid for control of *Acarapis woodi* (Acari: Tarsonemidae) and *Nosema* disease, *Nosema* sp. J Econ Entomol 102:1729–1736.

6.1.2.16 Brood Mites

1. **Name**: Lat.: *Varro* = the name refers to a Roman scientist, who wrote a book on bees 100 years before Christ; *destructor* = destroyer.
2. **Geographical distribution/epidemiology**: In Europe, these mites come from the east, almost already spread throughout Germany. They are distributed worldwide except for Australia and some African countries close to the equator.
3. **Biology/Morphology: *Varroa destructor* (syn. *V. jacobsoni*)** is an ectoparasite; males reach a size of about 0.8 × 0.8 mm and females up to 1.5 × 2 mm (Fig. 6.52). Males and juvenile stages exclusively sit on the bee brood (larvae and pupae); females parasitize on adult bees and especially on the drone brood. The development from the egg to the adult mite and the mating takes place during the bee brood phase and lasts about 6–8 days. Only females overwinter on adult bees.
4. **Symptoms of disease (Varroaosis)**: After an asymptomatic latency of 2–3 years during which the mite infestation within the colony constantly increases, increased numbers up to masses of stunted worker bees (with malformed wings, shortened abdomen) appear. Marked decline of the brood occurs, because the mite larvae and nymphs suck on the bee larvae. The lifetime of adult bees is shortened, because *Varroa* mites may suck about 0.1 mg haemolymph within 2 h. An infestation with these mites can finally lead to the complete loss of a whole bee colony.
5. **Diagnosis**: Detection of the macroscopically visible mites on adult bees (Fig. 6.52). For the detection of mite nymphs, bee pupae are taken from brood cells and analysed with the help of a stereomicroscope. A reliable procedure is the elution of >100 adult bees in petrol, removal of the bees (with grosgrain gauze) and the analysis of the filtrate for *Varroa* stages. Detection can likewise be performed with a so-called diagnostic treatment—it should be done not during honey production—according to the species

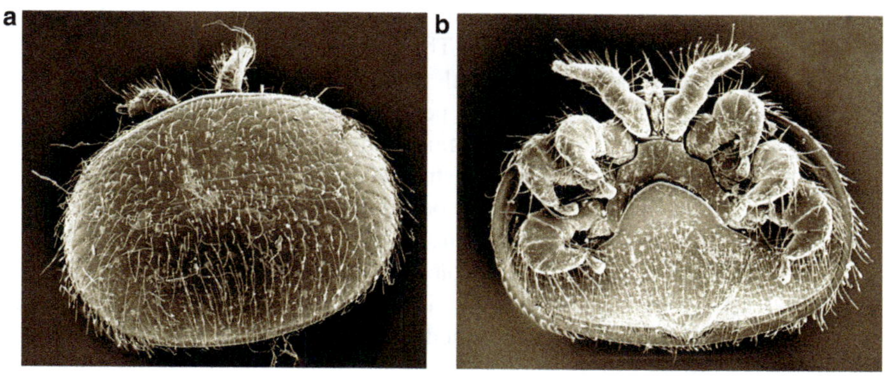

Fig. 6.52 Scanning electron micrograph of a *Varroa destructor* from its dorsal (**a**) and ventral (**b**) side

specifications mentioned in therapy. The varroaosis was a notifiable plague according to § 10 (1) of the animal epidemic law in the past. In **differential diagnosis,** one has to pay attention to **malicious foulbrood**, a likewise **notifiable plague** in Germany. Here, only the brood gets infested. Bee larvae assimilate spores during feeding. A typical characteristic is the ropy nature of the breeding cell contents in case of a recent infection.

6. **Pathway of infection**: Spreading occurs by bee workers and possibly by the drones during the nuptial flight.
7. **Prophylaxis**: Regular examination of dead bees in the winter debris. Treatment of the bees with mite killing agents. Pay attention to falling down of dead bees.
8. **Incubation period**: Severe killing rates within a colony occur usually only after 2–3 years.
9. **Prepatency**: The entire individual development cycle of the mites takes about 6–8 days.
10. **Patency**: An adult female can live up to 7–8 months. Without treatment, the mite infestation in one bee colony persists permanently, if the colony doesn't die out.
11. **Therapy**:
    (a) **Preventive measures** against **varroaosis** are regulated by § 16a of the bee regulation law. Accordingly, if an apiary is infested with *Varroa* mites, the owner has to treat all bee colonies of the apiary annually. The appropriate authority is allowed to prescribe the kind and the period of treatment in a definite area.
    (b) **Chemotherapy**:
    1. **Systemic agents**: Perizon® (**coumaphos**), Cekafix® (coumaphos derivate: substitution of 3-chlorine against 3-bromine) or Apitol® (thiazoline derivate) is administered after the preparation of the ready-to-use agent suspension (solution) along the honeycomb lane of the bees and/or with the feed (e.g. sugar syrup with Apitol®). The substances reach the haemolymph of the bees via the food cycle. If the haemolymph is ingested by mites, a selective toxic effect occurs in the mites. Flumethrin and formic acid are also used. A complete control measure does not exist yet. The control agents are only allowed to be used after the honey harvest!

       For advices regarding the period of treatment (unless determined by the authorities) and the resulting withholding period after the treatment, see manuals of producers/distributors.
    2. **Contact agents**: Treatment occurs by fumigation strips: Folbex-VA-Neu® (isopropyl-4,4-dibromobenzilate) 4× at intervals of 4 days (1 fumigation strip per colony with about 7 beleaguered normal honeycombs) preferentially in autumn (=colonies do not contain brood); for withholding period, see user manual.
    3. **Treatment** by help of **evaporation plates**: Illertissen mite plate®—IMP (1 plate contains: 14.2 g water-free formic acid). The acid escapes rapidly from the plate in free air. The insertion of the plate is carried out

on the honeycomb frame so that the vapour fills all honeycomb lanes. Depending on the size of the beehive, 1–2 plates (covering at least ¾ of the frame surface) over a minimum of 12 h are inserted into the hive. (Success of the treatment is only given at an outside temperature above 12 °C.) In case of severe infested colonies, treatment is repeated 4× at intervals of 4 days. Do not use before the last honey harvest. For more advices, see instruction leaflet.

4. **Additional treatments**: In many countries not yet approved pharmaceuticals against *Varroa* mites are Apistan® (tau-**fluvalinate**, a pyrethroid, Bayvarol® strips (**flumethrin**) or anti-*Varroa* HVG® = liquid formulation of 0.25 % amitraz, which is either used as an aerosol or as "long-acting slow release amitraz" = Miticur® as drug-impregnated polymer strips (ethylene vinyl acetate) hung in between the honeycombs (outside the yield period). Miticur® contains 100 g/kg amitraz and 900 g/kg ethylene vinyl acetate. **Amitraz** is also suitable for the control of acariasis.

(c) **Biological control measures** are not sufficiently effective up to now, e.g. removal of the drone brood and installation of a protected honeycomb (queen is blocked for 1 week) and subsequent extermination of the sealed brood (procedure is restricted to one- to two-time usage).

The application of so-called anti-juvenile hormones (e.g. **Precocene** II = 6,7-dimethoxy-2,2-dimethyl-e-chromium, Sigma) may disturb the reproduction readiness, which is influenced by the juvenile hormone in such a way that the reproduction rate is severely reduced.

Further Reading

Anderson DL, Morgan MJ (2007) Genetic and morphologic variation of bee parasitic *Tropilaelaps* mites. Exp Acarol 43:1–24.

DeGrandi-Hoffmann et al (2016) Population growth of *Varroa destructor*. Exp Appl Acarol 69:21–34.

Kang Y et al (2016) Disease dynamics of honeybees with *Varroa destructor*. Math Biosci 275:71–92.

Mutinelli F, Rademacher (2002) European legislation governing the use of drugs in bee colonies to control varroaosis. Reg Affairs 13:401–406.

Piou V et al (2016) Impact of the Phoretic Phase on Reproduction and Damage Caused by *Varroa destructor* (Anderson and Trueman) to Its Host, the European Honey Bee (*Apis mellifera* L.). PLoS One 11(4):e0153482.

Rosenkranz et al (2010) Biology and control of *Varroa destructor*. J Invertebr Pathol 103(Suppl 1):S96–S119.

6.1.2.17 Mites and Other Agents of Disease in the Skin of Reptiles and Amphibians

1. **Species**: On the body surface, especially of captive animals in the case of poor cleanliness, **mites** often appear as long-term parasitizing forms in a great variety

of forms (Fig. 6.53). In addition to adults also larvae may occur sometimes; *Trombicula* larvae (including *Hannemania* species on amphibians) are seen in wrinkles. Different **tick species** may suck blood for several days, while **biting insects** only enter reptiles or amphibians for a few minutes for their sucking action. Striking **diseases** (e.g. anaemia) only appear in cases of massive infestation with mites. *Ophionyssus natricis* occurs especially on snakes below the scales, in the chin, eye and cloaca regions. Sometimes occur **fly maggots** (*Sarcophaga*, *Calliphora*; see Fig. 6.104), which dig burrows in wounds of the skin and may cause **myiasis** in toads and in turtles.

Also **bot flies** (Fig. 6.114) and related species lay their eggs on toads, on other amphibians and on reptiles: Larvae immigrate into the nasal openings, destroy the mucosa and burrow from there on into the eyeholes. Sometimes they reach the brain and destroy it. In aquatic species, further species parasitize, which are mainly known from fish.

2. **Prophylaxis**: To prevent the infection of other animals by ectoparasites, snakes should be regularly treated against mites before they are introduced into a new animal group. This basically also applies to other reptiles or amphibians, which originate from fresh catches. The hygiene measures that prevent or reduce mite infestation include the removal of sloughs, of bark pieces and of similar furnishings in the terrarium, which offer hiding places for mites.
3. **Therapy**: Removal of the maggots by cutting them out of the wound. A solution of 3 % H_2O_2 is used for disinfection. In the case of mite infestation inside cages, spraying of animals should be done with the agents Acarol® or Reptile spray® of the company Alpha-Biocare sold by Fa. Sera, Heinsberg.

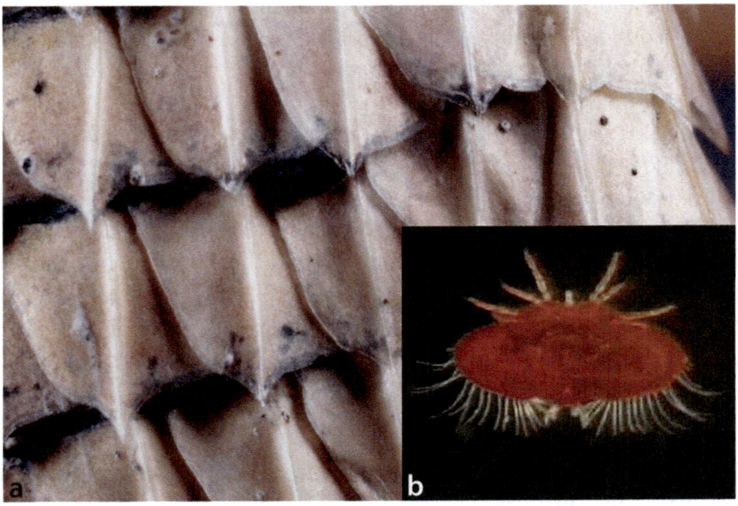

Fig. 6.53 (**a**, **b**) Macrophotos of the skin of reptiles showing sucking mites (enlarged in an inset)

Further Reading

Attademo AM et al (2012) Trombiculid mites (*Hannemania* sp.) in *Leptodactylus chaquensis* (Amphibia: Anura) inhabiting selected soybean and rice agroecosystems of Argentina. J Zoo Wildl Med 43:579–584.

Mariana A et al (2012) Acariasis on pet Burmese python, *Python molurus bivittatus* in Malaysia. Asian Pac J Trop Med 4:227–228.

Schilliger LH et al (2013) *Cheyletus eruditus* (*taurrus*): an effective candidate for the biological control of the snake mite (*Ophionyssus natricis*). J Zoo Wildl Med 44:654–659.

Westfall MC et al (2008) Patterns of trombiculid mite (*Hannemania dunni*) parasitism among plethodontid salamanders in the western Piedmont of North Carolina. J Parasitol 94:631–634.

6.2 Insects

The class Insecta is the most widespread group of the animal kingdom in terms of the number of species and density of individuals. The system of Insecta comprises primarily unwinged lower forms, the so-called **Apterygota**, and primarily winged higher insects, the **Pterygota**. The latter include all parasitic species, which predominantly live as ectoparasites (temporarily or stationary) on their hosts.

System
Phylum: Arthropoda (extract)
Subphylum: Tracheata
Class: Insecta (=Hexapoda; Insects)
Subclass: Apterygota (wingless insects)
Subclass: Pterygota (primarily winged insects)
Order: Phthiraptera (lice)
Suborder: Anoplura (sucking lice)
Suborder: Mallophaga (biting lice)
Order: Rhynchota (true bugs)
Family: Reduviidae (raptor bugs)
Family: Cimicidae (bed bugs)
Order: Diptera (two-winged insects)
Suborder: Nematocera (with large antennae)
Family: Phlebotomidae (sand flies)
Family: Culicidae (mosquitoes)
Family: Ceratopogonidae (biting midges)
Family: Simuliidae (black flies)
Suborder: Brachycera (with short antennae)
Family: Tabanidae (horse flies)
Suborder: Cyclorrhapha (pupae with preformed circular opening depression)
Family: Muscidae (house flies or stable flies)
Family: Glossinidae (tsetse flies)

(continued)

Family: Calliphoridae (blow flies)
Family: Sarcophagidae (flesh flies)
Family: Gasterophilidae (horse bot flies)
Family: Oestridae (bot flies)
Family: Hippoboscidae (louse flies)
Order: Siphonaptera (fleas)

Insects may

- Serve as **intermediate hosts** of important human or livestock parasites (e.g. for protozoans, cestodes, nematodes and others),
- As vectors spreading stages of the groups of rickettsia, bacteria and viruses **and**
- Contribute to the mechanical transmission of protozoans and bacteria (e.g. flies disseminate amoebae cysts, salmonellae and *Shigella* stages, etc.).

Endoparasitic forms may appear in some higher insect groups (=Pterygota), however, less often (see myiasis, sand flea) than ectoparasites.

The body of adult Pterygota is (especially in the case of the parasitic forms) highly modified but always shows the following basics (Fig. 6.54):

1. The segmented body is composed of a head (**caput**), a chest (**thorax**) and a torso (**abdomen**).
2. The chitinous exoskeleton of the dioecious insects is moulted characteristically often during growth. Males and females mostly show a sexual **dimorphism**.
3. The head, whose single segments are fused to a head capsule, carries dorsally 1 pair of segmented antennae (Fig. 6.80) and ventrally three pairs of mouthparts: **Mandibles**, **maxilla 1** and **2**. The head normally shows 1 pair of large compound eyes and in some species in addition also several single lensed eyes (Fig. 6.99).
4. The three thoracic segments (pro-, meso-, metathorax) carry each 1 pair of feet, which are composed of **coxa**, **trochanter**, **femur**, **tibia** and **tarsus**. The latter is equipped with the characteristic supporting structures such as claws, pulvillae, etc.
5. Meso- and metathorax primarily develop by cutaneous folds two pairs of wings, which can be moved by help of strong muscles. However, in some groups they are often completely or partly degenerated. For example, fleas and bed bugs have no wings, while dipterans possess a pair of functional front wings and a pair of reduced hind wings (halteres). The structure of the wings is species specific and thus is used for taxonomy.
6. The segments of the abdomen carry no limbs but possess **functionally different extensions**, which amongst other functions are used for copulation or brood care. The sexual organs, the copulatory systems and the often rather complicated supportive organs (accessory glands, etc.) as well as the Malpighian tubules (excretions system) are all placed in the abdomen.

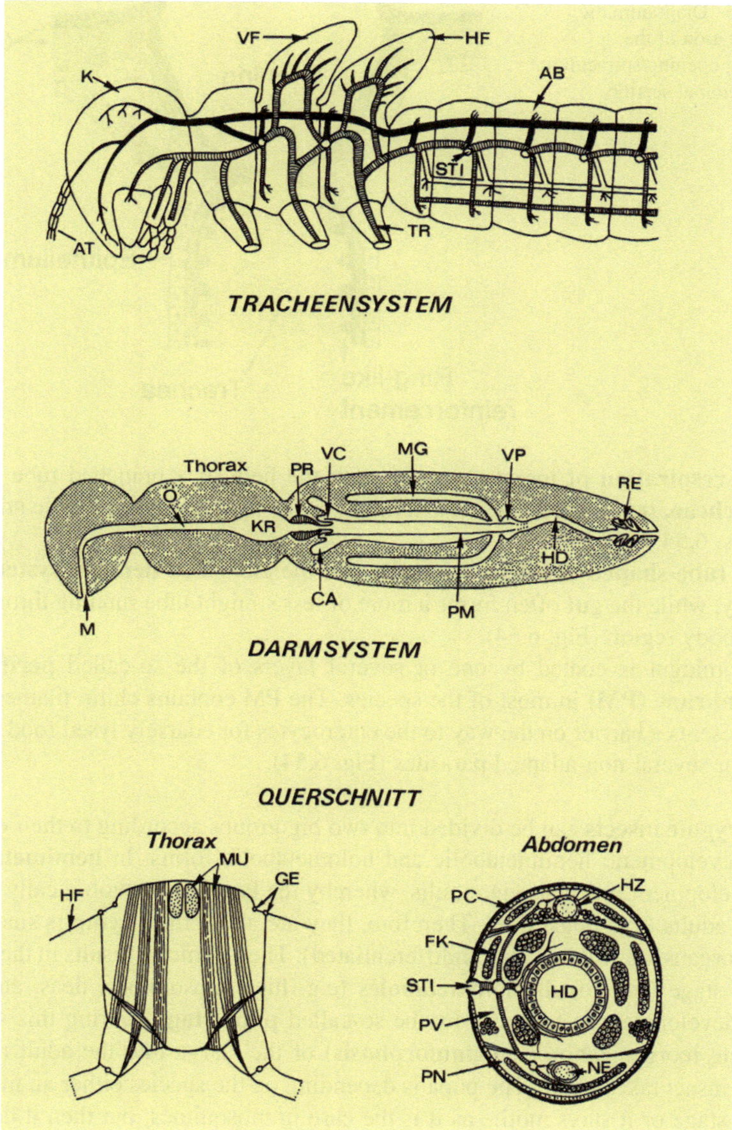

Fig. 6.54 Diagrammatic representation of the body organization of insects: Tracheensystem = tracheal system; Darmsystem = intestinal system; Querschnitt = cross section. AB = abdomen; AT = antennae; CA = caeca; FK = fat body; GE = joints; HD = terminal intestine, hind gut; HF = hind wings; HZ = heart; K = head; KR = crop; M = mouth; MG = Malpighian tubes; MU = wing muscles; NE = nerves; Ö = oesophagus; PC = pericardial sinus; PM = peritrophic membrane; PN = perineural sinus; PR = proventriculus; PV = perivisveral sinus; RE = rectum; STI = stigma (opening of the tracheal system); TR = trachea/tracheal tube; VC = valvula pylorica

Fig. 6.55 Diagrammatic representation of the breathing opening/spiraculum in longitudinal section

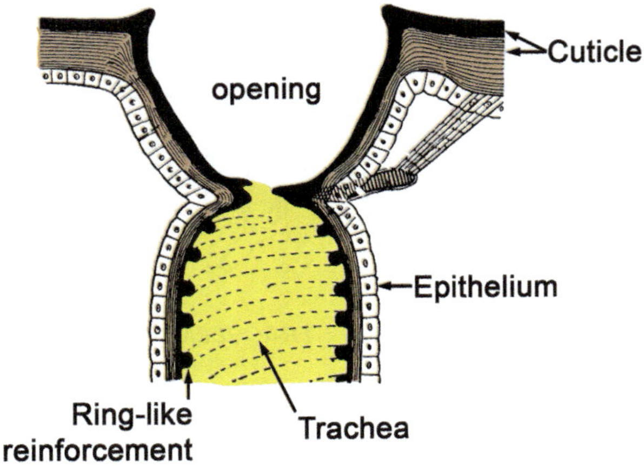

7. The **respiration** of insects happens with the help of a branched tube system (**tracheae, tracheoles**), which externally opens up with characteristic **stigmata** (Figs. 6.54 and 6.55).
8. The **tube-shaped heart** lays dorsally and the elongated **nervous system** ventrally, while the **gut** often forms a more or less straight tube running through the midbody region (Fig. 6.54).
9. The midgut is coated by one or several layers of the so-called **peritrophic membrane (PM)** in most of the species. The PM contains chitin filaments and represents a barrier on the way to the enterocytes for coarsely lysed food as well as for several non-adapted parasites (Fig. 6.54).

Pterygote insects can be divided into two big groups according to their ontogenetic development: hemimetabolic and holometabolic forms. In **hemimetaboles**, the development proceeds via moults, whereby the **larvae** morphologically resemble the adults (e.g. bugs, lice). Therefore, they are also called **nymphs** since their sexual organs are still lacking (undifferentiated). The last moult results in the sexual mature stage (**imago**). In **holometaboles** (e.g. flies, mosquitoes, fleas, etc.), the larval development is followed by the so-called **pupa stage**. During this stage, a complete reorganization (=**metamorphosis**) of the "larva into the adult male or female insect takes place. The pupa is depending on the species either an immotile resting stage or it stays motile as it is the case in mosquitoes, but then it does not take up food.

6.2.1 Lice (Phthiraptera)

1. **Name**: Greek: *phtheir* = louse; Latin: *apterus* = without wings; *pes, pedis* = foot; *pediculus* = small foot; *humanus* = human; *corpus* = body; *pubes* = region of sexual organs; *malleus* = hammer (here used to describe chewing activity).

2. **Some groups** of lice are of special importance as ectoparasites in vertebrates: **the bloodsucking Anoplura** (mammalian lice) and the **biting lice** (Mallophaga: chewing lice, feather-chewing lice).

In **humans**, the following species of Anoplura occur:

1. *Pediculus humanus capitis* (head louse, length 2–3.5 mm) (Fig. 6.56).
2. *Pediculus humanus corporis* (body louse, 3–4.5 mm).
3. *Pediculus pubis* (pubic louse, length 1–1.2 mm).

Related species are *P. schaeffi* of chimpanzees and *P. gorilla* of gorilla monkey. Body and head lice are considered as spatial and microclimatic separated subspecies, which can be crossbred experimentally. Besides these specific lice, the following bloodsucking **animal lice** can temporarily infest humans:

1. *Haematopinus suis* (pig louse, length 4–6 mm) (Figs. 6.57 and 6.58),
2. *H. eurysternus* (cattle louse, length 2.5–3 mm) (Fig. 6.59).

Fig. 6.56 Light micrograph of a female of *Pediculus humanus capitis* showing a protruded egg in the abdominal region

Fig. 6.57 Light micrograph of two adult pig lice (*Haematopinus suis*)

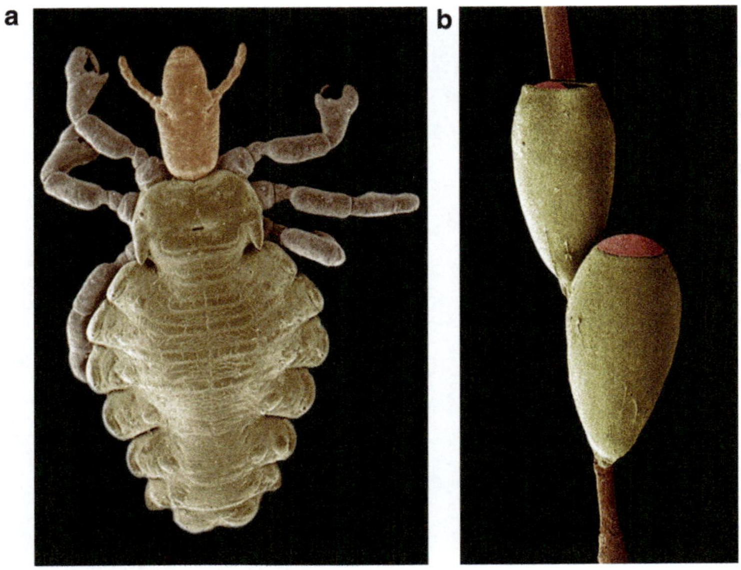

Fig. 6.58 Scanning electron micrographs. (**a**) Dorsal aspect of a pig louse (*Haematopinus suis*). (**b**) Two eggs of the pig louse glueing at hair. The cover (operculum) of the upper stage has already been lifted off

The Anoplura develop wingless larvae and adults; their head is smaller in width than the three fused thoracic segments. They anchor themselves with the help of typical claws at their feet at the hair of their hosts (Fig. 6.56). All lice miss compound eyes; they possess only two simple lense eyes (e.g. ocelli in human lice) or they are even blind (animal lice, Fig. 6.59). All stages have stinging-sucking

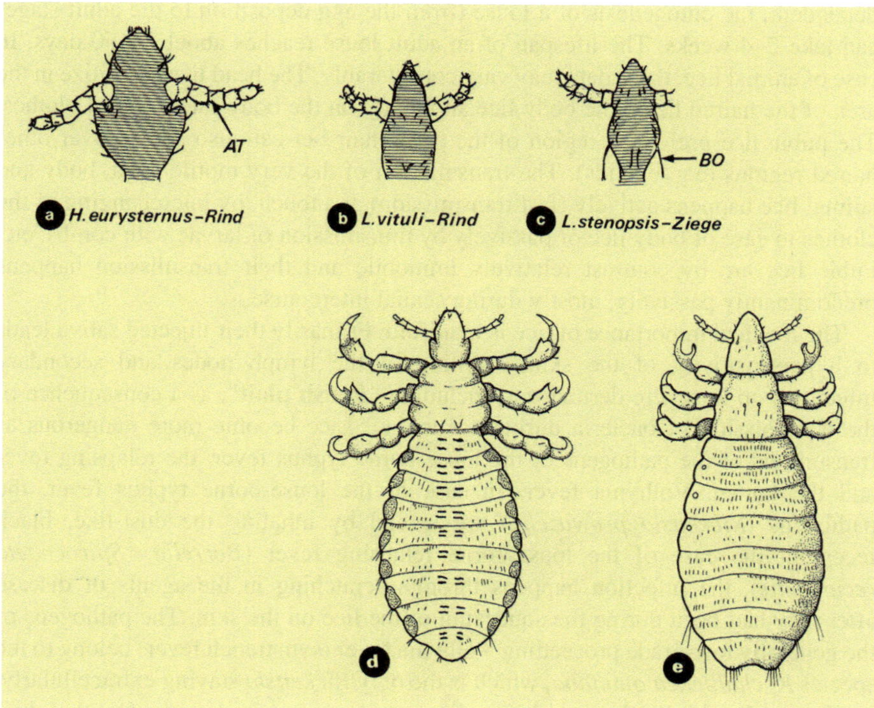

Fig. 6.59 Diagrammatic representation of different life of ruminants. (**a**) Head of a female from ventral: (**b**, **c**) Head of females from dorsal; (**d**, **e**) Dorsal aspects of *Haematopinus eurysternus* (**d**) and *Linognathus vituli* (**e**)

mouthparts, which sit seclusively in the blunt appearing proboscis. These bloodsuckers mostly cannot survive prolonged periods without food but have to suck blood to some extent frequently (at least 2–3× within 24 h) from their host. The head louse survives only less than 24 h without a blood meal, while the body louse can survive (especially at low temperatures) for up to 10 days on the floor. The metabolic processes of these lice depend on symbionts. These bacteria and fungi are located in the appending organ of the gut (so-called **mycetome**) and of the gonads and are transmitted constantly to the offspring.

The female lice lay daily 3–4 eggs after copulation and attach them with a water-insoluble substance as so-called **nits** (Fig. 6.58b) to the hairs. The sucking activity of these lice causes eczemas, so that in combination with the nits entire hair of the host may clutch together. In contrast, the body louse sticks their eggs also often at fibres of clothes. Females of human lice lay at least 90 eggs in total. They are about 0.8–1 mm long and 0.3 mm wide and possess a characteristic cover (operculum). The likewise bloodsucking juvenile stage (**nymph**) hatches after 8–14 days (temperature-dependent) and moults 3× until the adult stage is reached (hemimetabolic insects!). Because the entire development is temperature dependent, the ontogenesis of a louse (from the egg deposition to the adult stage) can take 2–4 weeks. The lifespan of an adult louse reaches about 30–50 days. In case of animal lice, these data may vary considerably. The head lice parasitize in the area of the haired head, the body lice sit mostly on the body-faced site of clothes. The pubic lice prefer the region of the pubic hair but can also spread over other haired regions (e.g. eyelids). The transmission of the very motile head, body and animal lice happens actively (fast transmission) at a touch, by interchanging of the clothes in case of body lice or passively by transmission of larvae with combs, etc. Pubic lice are by contrast relatively immobile and their transmission happens predominantly passively, mostly during sexual intercourse.

The medical importance of lice is manifold. Primarily their injected saliva leads to hypersensitivity of the skin, swelling of the lymph nodes and secondary infections up to severe dermatoses, including "**Polish plait**", as a consequence of the transmission of bacteria during scratching. Lice become more dangerous as transmitters of the pathogens of the louse-borne typhus fever, the relapsing fever and the human Volhynia fever. In case of the louse-borne typhus fever, the pathogens (*Rickettsia prowazekii*) are spread by inhaling the dust-like, black feces. In the case of the louse-borne relapsing fever (*Borrelia* = *Spirochaeta recurrentis*), the infection happens through scratching in the agents of disease after they had been during the squeezing of the lice on the skin. The pathogens of the generally low-grade proceeding Volhynia fever (syn. trench fever) belong to the species *Rochalimaea quintana*, which is the only *Rickettsia* staying extracellularly.

The smaller **Mallophaga** (=**biting lice**) can be recognized by the fact that their head is wider than the thorax and that they have chewing-biting mouthparts. They mainly appear on animals, sometimes in high numbers: *Bovicola bovis* (1.2–1.6 mm) in cattle, *Trichodectes canis* (1.5 mm) in canids and different species in gamefowl (e.g. *Menopon gallinae*). With the help of the mouthparts, they ingest portions of the epidermis. Mass occurrences of lice lead to heavy economic losses in livestock farming. **Chewing lice** (in mammals) or **feather-chewing lice** (in poultry) cause severe itchiness, partially also anaemia and moult-like skin reactions (Figs. 6.60, 6.61, 6.62, 6.63 and 6.64). Both chewing lice and feather-chewing lice can occasionally be transmitted to humans.

Further Reading

Damriyasa IM et al (2004) Prevalence, risk factors and economic importance of infestations with *Sarcoptes scabiei* and *Haematopinus suis* in sows of pig breeding farms in Germany. Med Vet Entomol 18:361–367.

Hornok S et al (2010) Survey on blood sucking lice of ruminants and pigs with molecular detection of *Anaplasma* and *Rickettsia* spp. Vet Parasitol 174:355–358.

Song SD et al (2014) Variation in mitochondrial minichromosome composition between blood sucking lice of the genus *Haematopinus* that infest horses and pigs. Parasite Vectors 7:144.

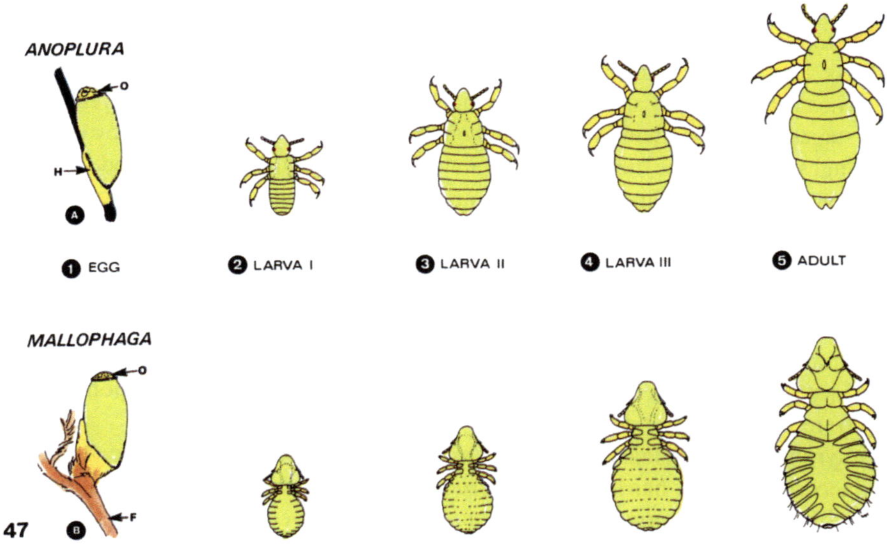

Fig. 6.60 Diagrammatic representation of the developmental stages of sucking lice (Anoplura) and biting lice (Mallophaga). F = feather; H = hair; O = operculum (egg cover)

6.2.1.1 Bloodsucking Lice (Anoplura)

1. **Name**: Greek: *anaplos* = unarmed; *ura* = tail.
   Haematopinus: Greek: *haima* = blood; *pinein* = to drink.
   Haemodipsus: Greek: *haima* = blood; *dipsa* = thirst.
   Linognathus: Greek: *linon* = thread; *gnathos* = jaw.
   Solenopotes: Greek: *solen* = tube; *potes* = drinker.
   Polyplax: Greek: *polys* = many; *plax* = plate.

2. **Geographical distribution/epidemiology**: Worldwide; in mass farming, intensive infestation occurs. Up to 60 % of the animals are infected often in combination with chewing lice.
3. **Biology/morphology**:
   Species:

Haematopinus eurysternus (Fig 6.59)	Cattle	3.5 mm
Linognathus vituli (Fig 6.59)	Cattle	3 mm
Solenopotes capillatus	Cattle	2 mm
Haematopinus suis	Pig	4–6 mm
Haematopinus asini microcephalus	Horse	3.5 mm
H. asini asini	Donkey	3.5 mm
Linognathus setosus	Dog/fox	2.5 mm
Linognathus stenopsis	Goat/sheep	3 mm
Haemodipsus ventricosus	Rabbit	1.5 mm
Polyplax serrata	Mice	1.3 mm
Polyplax spinulosa	Rats	1.5 mm

Females are mostly a bit bigger. All stages (larva or nymphs 1, 2 and 3) and both females and males are always wingless and equipped with terminal chelae at the six legs, which allow clinging to hairs. The development is **hemimetabolic** (i.e. without pupal stage), and the adult stage is reached after 3–4 weeks. All stages suck blood with the help of their retractable mouthparts, whereby blood is ingested with one tube and saliva with a second tube (with blood-thinning components) is simultaneously pumped into the sting wound. The yellow-whitish, fertilized eggs (**nits**), which are provided with a lid (=**operculum**; Fig. 6.58b), get glued to the hair basis with a water-insoluble substance. With the growing hair, they slowly reach the surface of the animal. Then they are already empty because the larvae burst the lid after about 6–9 days, hatch and immediately start to suck blood. The lifespan is relatively short (1–2 months). Lice suck blood every 2–4 h and can mostly survive only short starvation periods (rarely more than 2 days).

4. **Symptoms of disease**: Restlessness occurs induced due to severe itchiness accompanied by hairless, blank scrubbed areas. Thereby secondary infections with exudate formation and incrustation (=weeping dermatosis) may develop. General condition is impaired as a consequence of constant disturbance and blood loss. Partly severe anaemia occurs (especially in young animals). **Caution**: Especially pig lice can be transmitted to humans temporary.
5. **Diagnosis**: Detection of the adults or the species-specific nits glued at hair (Fig. 6.58b). They can be discovered with the naked eye or with the help of a loupe (differential diagnosis: exclude mange).
6. **Pathway of infection**: By body contact, especially in the stable.
7. **Prophylaxis**: Frequent control of the coat, regular spraying with the bio-agents MiteStop® or Anticks®—both produced by Fa. Alpha-Biocare.
8. **Therapy**: **Macrocyclic lactones** have been proven to be successful as pour-on solution or as spray. Besides, the biological agent MiteStop® has turned out to show a good effect (the extract is diluted with water 1:30) without the need of withdrawing periods.

Further Reading

Colwell DD (2014) Life history parameters of the cattle long-nosed sucking louse, *Linognathus vituli*. Med Vet Entomol 28:432–437.

Dong WG et al (2014) Fragmented mitochondrial genomes of the rat lice, *Polyplax asiatica* and *Polyplax spinulosa*: intra-genus variation in fragmentation pattern and a possible link between the extent of fragmentation and the length of life cycle. BMC Genomics 15:44.

Frank R et al (2013) Parasites of wild rabbits (*Oryctolagus cuniculus*) from an urban area in Germany, in relation to worldwide results. Parasitol Res 112:4255–4266.

Hornok S et al (2010) Survey on blood-sucking lice (Phthiraptera: Anoplura) of ruminants and pigs with molecular detection of *Anaplasma* and *Rickettsia* spp. Vet Parasitol 174:355–358.

Mehlhorn H et al (2012) Biting and bloodsucking lice of dogs – treatment by means of a neem seed extract (MiteStop®, Wash Away Dog). Parasitol Res 110:769–773.

Song SD et al (2014) Variation in mitochondrial minichromosome composition between blood-sucking lice of the genus *Haematopinus* that infest horses and pigs. Parasit Vectors 7:144.

Veneziano V et al (2013) Field efficacy of eprinomectin against the sucking louse *Haematopinus asini* on naturally infested donkeys. Vet J 197:512–514.

6.2.1.2 Biting Lice (Mallophaga: Chewing Lice and Feather-Chewing Lice)

1. **Name**: Greek: *mallos* = wool; *phagein* = to consume.
   Bovicola: Latin: *bos, bovis* = cattle; *collum* = neck.
   Werneckiella: Wernecke = German scientist.
   Trichodectes: Greek: *thrix* = hair; *dex* = wood worm larva.
   Felicola: Greek: *koleps* = envelope. Latin: *felis* = cat.
   Menopon: Greek: *men* = months; Latin: *pons* = bridge.
   Lipeurus: Greek: *leipein* = to leave: *ura* = tail.
   Anaticola: Greek: *koleps* = envelope. Latin: *anas* = duck.

2. **Geographical distribution/epidemiology**: Worldwide, in case of dense animal farming frequently mass appearance. In mammals often co-infection with bloodsucking lice.
3. **Biology/morphology**:

Species (extract)			**Suborder**
Bovicola bovis	Cattle	1.5–2 mm	(Ischnocera)
B. ovis	Sheep	1.5–2 mm	(Ischnocera)
B. caprae	Goat	1.5 mm	(Ischnocera)
Werneckiella equi equi	Horse	2.0 mm	(Ischnocera)
W. equi asini	Donkey	1.8 mm	(Ischnocera)
Trichodectes canis	Dog/canids	2 mm	(Ischnocera)
Felicola subrostratus	Cats	1.35 mm	(Ischnocera)
Anaticola anseris	Goose	4 mm	(Ischnocera)
Eomenacanthus stramineus	Turkey/fowls	3.2 mm	(Amblycera)
Menopon gallinae	Chicken	1.8 mm	(Amblycera)
Lipeurus caponis	Chicken	2.3 mm	(Ischnocera)
Columbicola columbae	Pigeons	2.5 mm	(Amblycera)

The chewing lice or feather-chewing lice (Figs. 6.60, 6.61, 6.62, 6.63 and 6.64) feed on epidermis or on the surface of the feathers or hair. The development proceeds hemimetabolic (=without pupa) via three larval stages and takes about 3 weeks until the sexually mature adult has grown up. The lid-provided eggs (=nits) get glued at hairs or at the feather base. The larvae hatch from the eggs

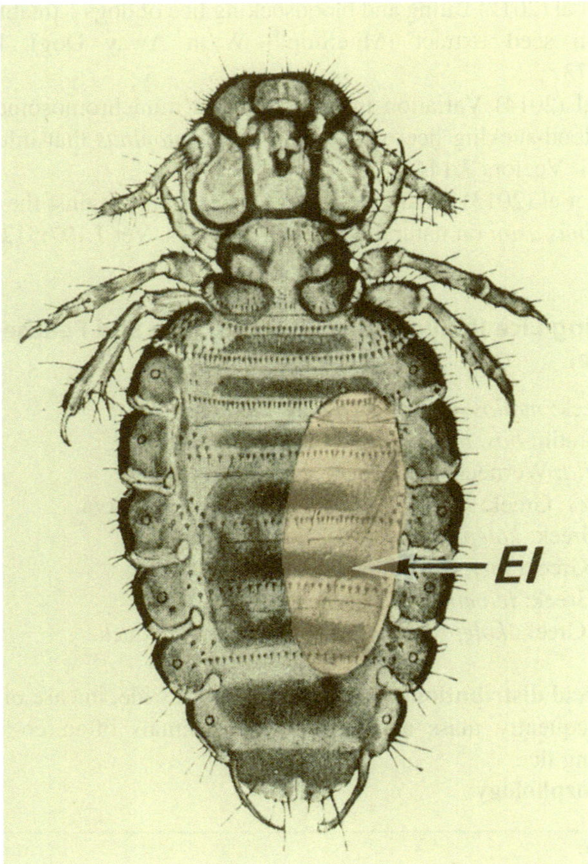

Fig. 6.61 Diagrammatic representation of the horse mallophage *Werneckiella equi* with a transparent egg (EI)

a

b

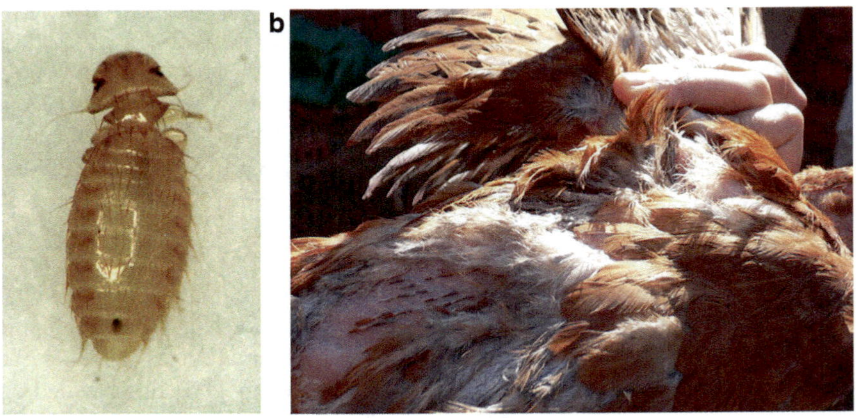

Fig. 6.62 Macrophotos of a feather mallophage (**a**) and chicken skin with featherlings (**b**)

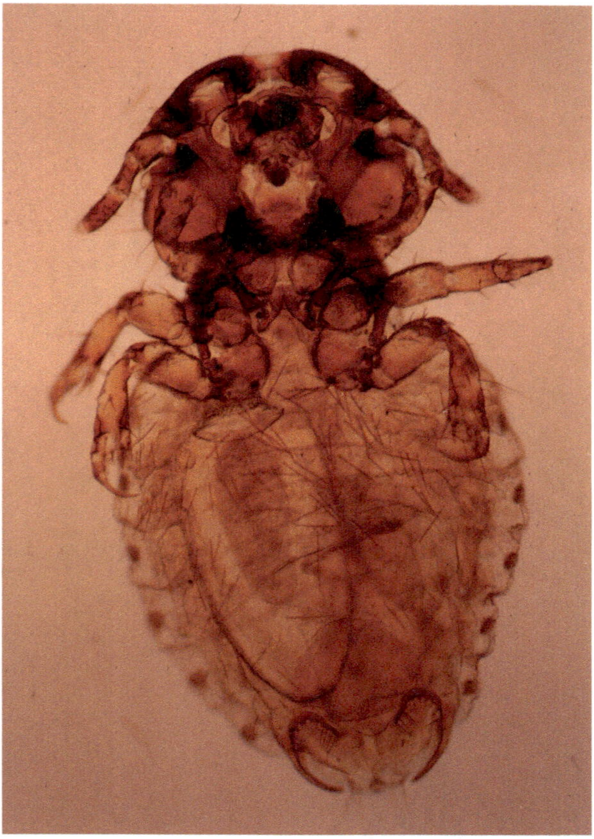

Fig. 6.63 Light micrograph of a dog mallophage (*Trichodectes* sp.)

species specifically after 4–12 days. There exist two big groups within the Mallophaga:

(a) **Ischnocera** (Greek: *ischnos* = thin; *keras* = horn, protrusion).

Antennae, which are clearly visible, possess 3–5 segments.

(b) **Amblycera** (Greek: *amblys* = blunt; *keras* = horn, protrusion).

The two antennae possess four segments and are often not visible, since they may lie in a fold.

Chewing lice possess only one claw at the legs and feather-chewing lice 2. Amblycera can also nibble at blood vessels of the skin.

4. **Symptoms of disease**: Low-grade infestations mostly stay inapparent; severe infestation leads to restlessness, itchiness and loss of hairs or feathers. Severe scaling is often common. Scrubbing at walls or equipment as a consequence of itchiness leads to wool loss in sheep or leather damages. Exudates at severe infested regions are potential breeding grounds for bacteria. In Amblycera, also massive blood loss can cause severe inefficiency.

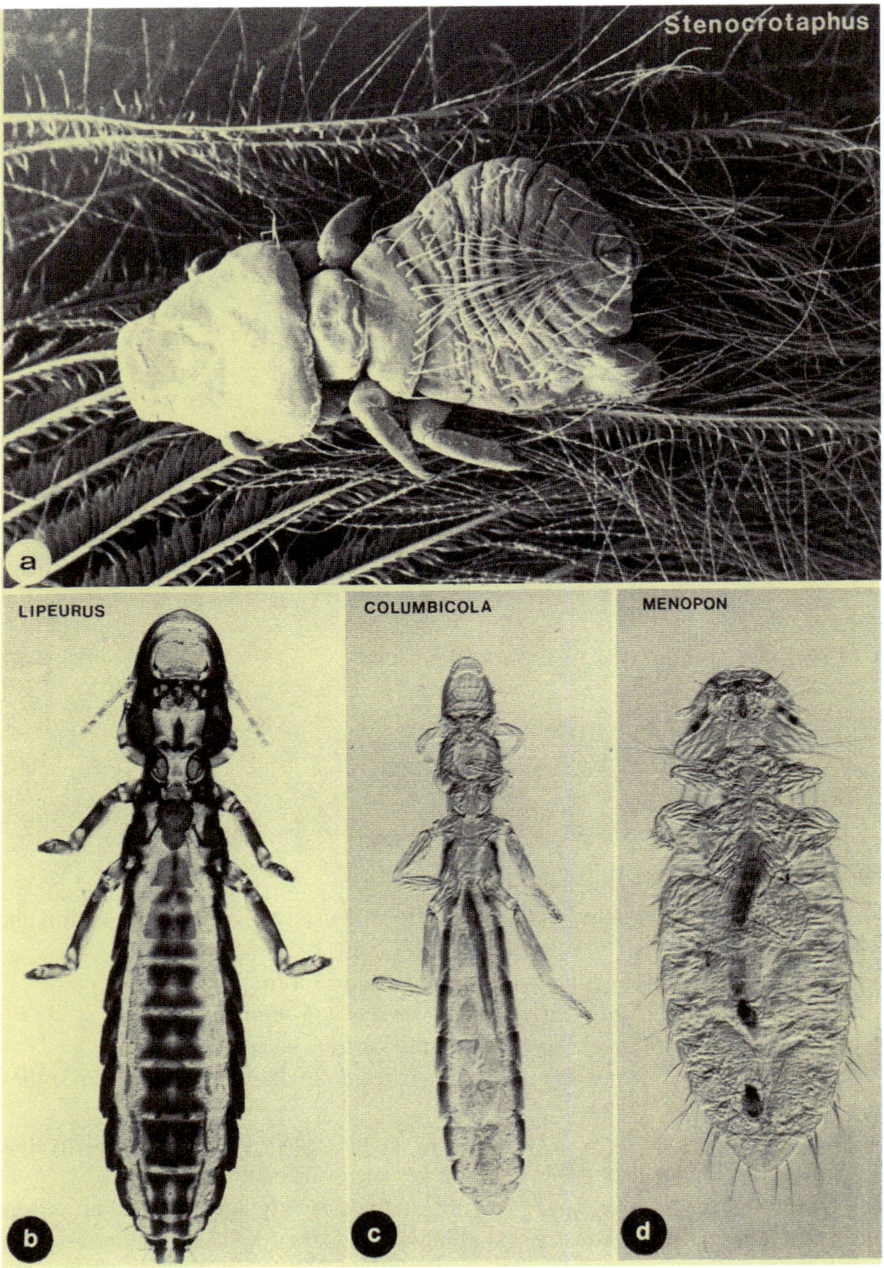

Fig. 6.64 (**a–d**) Featherlings of birds. (**a**) Scanning electron micrograph of *Stenocrotaphus gigas* (large chicken louse). (**b**) *Lipeurus caponis* (wing louse of chickens). (**c**) *Columbicola columbae* (wing louse of doves). (**d**) *Menopon gallinae* (feather shaft louse of birds)

In birds, mainly restlessness, nest escape, large skin and feather lesions, emaciation and loss of efficiency appear. Death of young animals may occur in the case of mass infestation due to anaemia (blood loss). In sick short-beaked pigeons, often a strong proliferation of the feather-chewing lice population occurs.

5. **Diagnosis**: Detection of adults (Fig. 6.60) and eggs (nits) with the naked eye. Ensuring by microscopical examination of hairs or feathers. A species diagnosis is not necessary because all Mallophaga are responsive to contact insecticides due to their similar mode of life.
6. **Pathway of infection**: Transmission occurs from animal to animal, also during dust bath of birds or by phoresy (e.g. attachment to mosquitos or louse flies).
7. **Prophylaxis**: Addition of contact insecticides to the dust bath of birds (in case of extensive avicultures) or regular spraying with the biological agent MiteStop® (Co. Alpha-Biocare, Düsseldorf).
8. **Therapy**: **Macrocyclic lactones** have been proven to be successful in chewing lice. Both in chewing lice and feather-chewing lice, the biological agent MiteStop® (Co. Alpha-Biocare, Düsseldorf) has turned out to show a good effect as washing solution diluted with water 1:40.

Further Reading

Asala SA, Mbajiorgu FE (1996) Epigenetic variation in the Nigerian skull: sutural pattern at the pterion. East Afr Med J 73:484–486.

Bartlow AW et al (2016) Walk or ride? Phoretic behaviour of amblyceran and ischnoceran lice. Int J Parasitol 46:221–226.

Dos Santos AC et al (2011) Phthiraptera (Arthropoda, Insecta) in *Gallus gallus* from isolated and mixed backyard rearing systems. Rev Bras Parasitol Vet 20:17–21.

Gustafsson DR et al (2015) The chewing lice (Insecta: Phthiraptera: Ischnocera: Amblycera) of Japanese pigeons and doves (Columbiformes), with descriptions of three new species. J Parasitol 101:304–313.

Kagira JM et al (2013) Relationship between the prevalence of ectoparasites and associated risk factors in free-range pigs in Kenya. SRN Vet Sci 2013:650890.

Sands B et al (2016) Residual and ovicidal efficacy of essential oil-based formulations in vitro against the donkey chewing louse *Bovicola ocellatus*. Med Vet Entomol 30:78–84.

Smith WA et al (2013) Phylogenetic analysis of symbionts in feather-feeding lice of the genus *Columbicola*: evidence for repeated symbiont replacements. BMC Evol Biol 13:109.

Valim MP, Cicchino AC (2015) A remarkable new genus and a new species of chewing louse (Phthiraptera, Ischnocera, Philopteridae) from Brazil. Zookeys 541:57–70.

Yoon KS et al (2015) In vitro and in vivo evaluation of infestation deterrents against lice. J Med Entomol 52:970–978.

6.2.1.3 Bee Louse (*Braula coeca*)

1. **Name**: Greek: *braula* = louse. Latin: *coecus* = blind.
2. **Geographical distribution/epidemiology**: Worldwide.
3. **Biology/morphology**: Contrary to its trivial name, *Braula coeca* belongs to the **flies**. Bee lice reach a maximum length of 1.5 mm and thus are relatively small. They are wingless in all stages (Fig. 6.65). One single ommatidium is developed on each side of the head. With the help of two dentated crests on the last

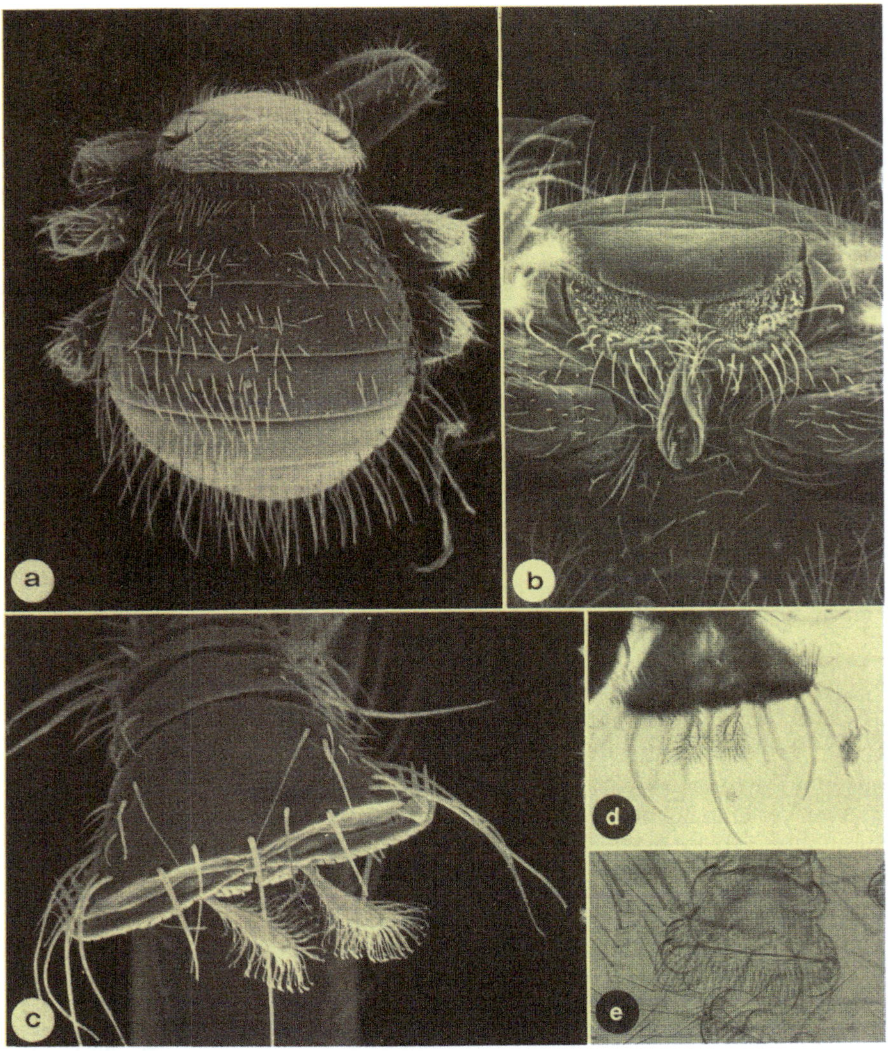

Fig. 6.65 Scanning electron micrographs of the "bee louse" *Braula coeca*. (**a**) From dorsal; (**b**) Head: front aspect; (**c–e**) Attachment systems at the feet of adults and larvae

tarsal segments, often up to 20 specimens of the bee lice may anchor at the "hair coat" of the queen bee and extract food from her proboscis. The eggs (0.8×0.5 mm) are laid on the top of the breeding cell or the honey cell lid. The larvae burrow tubes into the wall and thereby damage the honeycombs.

4. **Symptoms of disease**: Reduced egg laying of the infested queen bee occurs as a consequence of the harassment and withdrawal of the food. The degree of the impairment depends on the amount of parasites.
5. **Diagnosis**: Detection of the wingless flies on the queen.
6. **Pathway of infection**: Spreading from hive to hive by flying bees.
7. **Prophylaxis**: Regular control of the queen and mechanical removal of the bee lice.
8. **Incubation period**: An impairment of the colony is only noticed late.
9. **Prepatent period**: The ontogeny of a fly takes about 2–3 weeks.
10. **Patency**: A bee colony can be continuously infested.
11. **Therapy**:

(a) First **individual treatment of the queen**:

Collecting the bee lice with forceps or a honey-wetted stick; this is, however, not easy because the bee lice move fast. Since the bee queen is the "host of choice" for the succeeding generation, the collection has to be repeated several times. If this removal is not successful, the queen can be blown on with tobacco smoke every 2–3 weeks (2–3×). Effective agent is nicotine, which is more toxic to flies than to bees. Caution and close monitoring of the queen is required because of a possible overdosing of nicotine!

(b) In the case of an obvious **brood decline** due to severe impairment of the laying performance of the queen, **evaporation agents** can be applied, whereby the sensitivity of the bees to the agents and a possible contamination of the honey have to be considered. For example, naphthalene or phenothiazine can be installed in the hive in the form of a soaked cloth (**nappy**). Para-dichlorobenzene (Imkerglobol®) is likewise suitable. According to recent reports, Folbey-VA-Neu® and Perizin® (see *Varroa*) shall also be effective.

Further Reading

Dobson JR (1999) Bee-louse *Braula schmitzi* Örösi-Pál (Diptera: Braulidae) new to the British Isles, and the status of *Braula* spp. in England and Wales. Brit J Entomol Nat Hist 11: 139–148.

Gouin F (1951) The mouthparts of *Braula caeca*. C R Hebd Seances Acad Sci 232:1774–1775.

Markosian AA et al (1971) Biological characteristics of *Braula* and its control. Veterinaria 49:68–69.

6.2.2 Fleas (Order Aphaniptera, Siphonaptera)

Fleas are flattened laterally, secondary wingless, mostly brownish coloured insects, whose adults are able to perform an enormous jump due to their especially strong formed three pairs of legs (Fig. 6.66). Males like females feed on the blood from their hosts; 94 % of the known species suck on mammals and only 6 % on birds. Most of the fleas are **not host specific** but instead can—especially after long starving periods (in case of *Pulex irritans* more than ½ year)—accept blood from other hosts; however, this can reduce the individual fertility. The mouthparts of the adult fleas are formed to 2-canal-stinging organs (Fig. 6.66g). Through the bigger canal, blood is sucked, while through the second, much thinner canal saliva is pumped into the stinging canal simultaneously. This saliva prevents blood coagulation but is responsible for the partially severe skin reactions of the hosts (pustule-like swellings, itching). Besides, flea bites sensitize the skin so that older stings start to itch again. This is even more unpleasant as fleas get easily disturbed during their meal so that they sting again and a number of bites occur next to each other. The blood meals mostly take place on a daily basis and can last 20–150 min, whereby large amounts of the sucked blood (especially the serum!) get immediately excreted anally during the meal.

Apart from their remarkable saltatorial legs, fleas are characterized by their clear segmentation with typical squamous plate formation, by short, in a gutter at the head retractable antennae as well as by a dorsal, from the 10. abdominal segment formed plate (=**pygidial plate** or **sensilium**) with specific numbers of sensory hairs (Fig. 6.67). The receptors (**trichobothria**), which are particularly accumulated in this area, recognize air movements and soil vibrations, which enable the flea to enter the host. Because fleas cannot focus with their ocelli (compound eyes are non-existent), these sensory plates represent one of the most important help to find a host.

Fleas live a maximum of about 1½ years. The copulation often takes place on a host, whereupon, for about 3–6 weeks, the female daily lays 10–25 eggs that drop to the ground. The cat flea deposits about 800–1000 eggs like this, while the human flea deposits only 450 eggs. After approx. 5 days (temperature dependent), each eyeless larva hatches from the egg, which is also called "**wireworm**" because of its bristled appearance (Figs. 6.66a, 6.68f and 6.70) and mainly feeds on detritus. However, it also needs proteins for their development. It obtains them by ingesting portions of dead adults or by ingesting fecal drops, which contain remnants of blood and are regularly excreted by the adults during feeding.

After about 2–3 weeks and two moults, the larvae spin themselves a silky cocoon with the help of their salivary glands. Within about 3 more days, the larva develops into a pupa and stays—depending on the microclimate—in this immobile condition for about 1–2 weeks. The hatching from the pupal cocoon is mostly triggered by an external stimulus, e.g. vibrations indicating the arrival of a host. If this stimulus is omitted—because no host visits the flea populated lair—the adult flea can stay for months in this pupal cocoon, apparently with a reduced metabolism. The first host then provokes a parallel mass exodus in all present pupae, e.g. in case of a

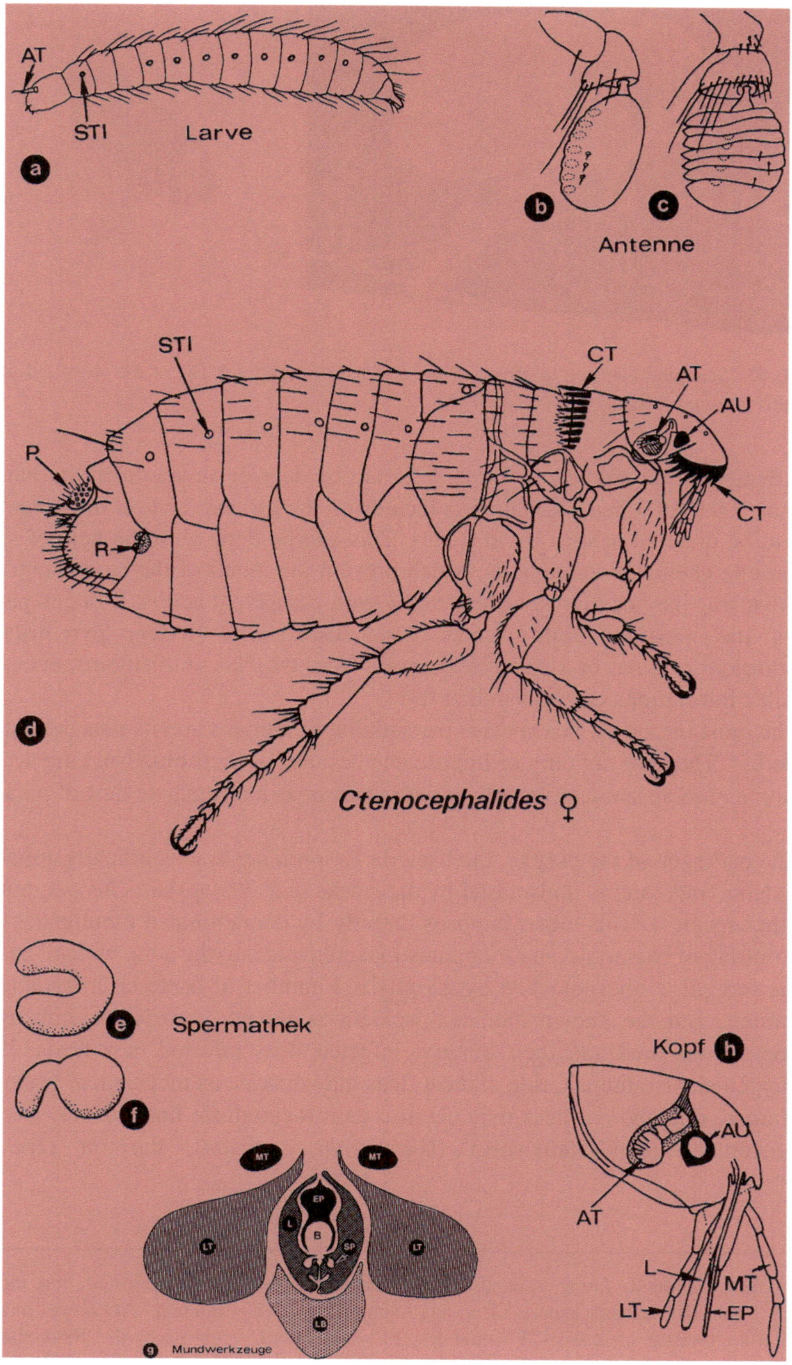

Fig. 6.66 Diagrammatic representation of the morphology of fleas. (**a**) Larva ("wireworm"). (**b**) Antenna of *Tunga penetrans*. (**c**) Antenna of *Pulex irritans*. (**d**) Female of *Ctenocephalides felis*. (**e–f**) Receptacles of *Xenopsyllla cheopis* (**e**) and *Pulex irritans*. (**g**) Cross section through the

a

b

Fig. 6.67 Scanning electron micrograph (**a**) and light micrograph (**b**) of the so-called cat flea (*Ctenocephalides felis*)

resettlement of an old nest by birds, a kennel by dogs or an apartment by humans. Because 99 % of a flea population is located on the ground, control measures have also to be directed against these stages. This can be done with the help of sprays containing chemical compounds, which prevent the moult of the larval stages and like that stop the occurrence of new replication-competent adults. If one in parallel fights the existing adult insects with insecticides (today: pyrethrum or pyrethroid = original or synthesized agents from the chrysanthemum), even massive flea infestations can be handled fast.

The most important flea species parasitizing in humans and its pets are listed in Table 6.7. They are not only of importance because of their bloodsucking activity, but some of them serve as vectors of pathogens or as intermediate host of parasites:

1. The pathogen of the **plague**, the bacteria *Yersinia pestis*, endemically appears in rodents (rats) and is transmitted by their fleas (e.g. *Xenopsylla cheopis*, but also other species). This either happens directly by contaminated mouthparts or by vomiting of bacteria-containing intestinal contents into the sting wound, because the foregut is often blocked by an outsized number of bacteria. Infestations of humans and the known medieval epidemics always arose when infected rat populations died and their hungry, infected fleas infested new hosts. If this happened, also human, dog and cat fleas may operate as mechanical vectors.
2. Human, dog and cat fleas may act also as **intermediate hosts** of the so-called "cucumber seed" tapeworm (*Dipylidium caninum*), the rat tapeworm

Fig. 6.66 (continued) biting apparatus (epipharynx and laciniae are injected into the skin during bite). (**h**) Lateral aspect of a flea (simplified). AT = antenna; AU = eye (ocellus); B = blood = feeding channel; CT = ctenidia; EP = epipharynx (represents the food channel); L = laciniae = maxillar stinging bristles each containing a salivary ductule; LB = labium; LT = labial palpus; M = maxillar palpus; P = pygidial plate = sensillum; R receptaculum seminis; SP = salivary ductule; STI = stigma

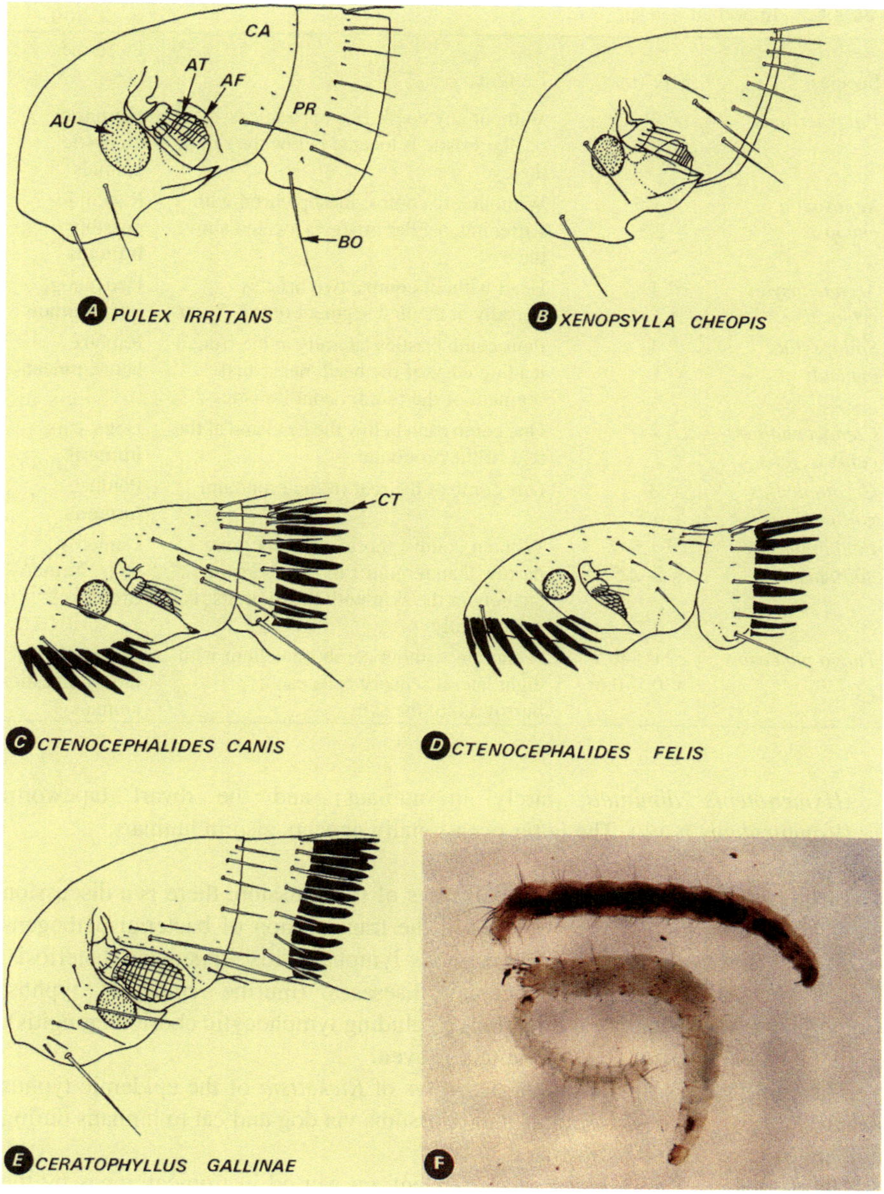

Fig. 6.68 Diagrammatic representation of the anterior ends of fleas (**A–F**) and a light microscopical aspect of the differently aged larvae of the cat flea *Ctenocephalides felis*. AF = groove of the antenna; AT = antenna; AU = eye = ommatidium; BO = bristles; CA = caput = head; CT = ctenidial combs; PR = pronotum

Table 6.7 Important flea species

Species	Size (mm)	Features	Preferred hosts
Pulex irritans	♂ 2–2.5 ♀ –4	Without any comb-like appendices; ocellar bristle is located below the rim of the eye	**Humans**, domestic animals
Xenospylla cheopsis	♂ 1.5 ♀ 2.5	Without any combs; mesopleuron with stiffening, ocellar bristle is located above the eye	Rats, mice, possibly **humans**
Archaeopsylla erinacei	♂ 4 ♀ 3	Head without combs, two bristles dorsally at the first segment of the thorax	Hedgehog, rats, **humans**
Spilopsyllus cuniculi	♂ 3 ♀ 3	Four comb bristles laterally at the frontal leading edge of the head, neck: first segment of the thorax: comb bristles	Rabbits, hares, rodents
Ctenocephalides canis C. felis	♂ 2 ♀ 3	One comb each below the head and at the rear of the pronotum	Dogs, cats, **humans**
Ceratophyllus gallinae	♂ 3 ♀ 3.5	One comb at the rear of the pronotum	Poultry, **humans**
Echidnophaga gallinacea	♂ 1.5–2 ♀ 2–2.5	Without combs; thorax dorsally more narrow than tergum 1 of the abdomen; ♀ anchors in the skin with the mouthparts permanently	Gamefowl, dogs, **humans** (tropics)
Tunga penetrans	♂ 0.5–0.7 ♀ 0.5–0.6	Pronotum without comb; sensillum with eight lateral sensory cells each; ♀ burrows into the skin	**Humans**, large domestic animals

(*Hymenolepis diminuta*; rarely in humans) and the dwarf tapeworm (*Vampirolepis nana*). The latter occasionally appears also in humans.

Besides these definitely proven pathways of transmission, there is a discussion about a possible involvement of fleas in the transmission of bacterial pathogens (including those of the tularaemia, caseous lymphadenitis, erysipelas, listeriosis, brucellosis, salmonellosis), rickettsial disesases (murine epidemic typhus, Boutonneuse fever) and viral infections (including lymphocytic choriomeningitis). In many cases, this is already definitely proven.

Thus, in the USA the infection pathways of *Rickettsia* of the epidemic typhus group (*R. typhi*, *R. felis*) from rats and opossums via dog and cat to humans during epidemics have been demonstrated.

Harmful skin ulcerations to a large extent are caused in tropical zones by the so-called **sand fleas** (especially *Tunga penetrans*; see Fig. 6.74). These fleas do not only temporarily parasitize as ectoparasites like other flea species do, but their females burrow into the skin of humans and pets (predominantly into toes) and grow within 8–10 days to a ball-like shape with a diameter of 4–6 mm, whereby the posterior portion with the oviduct and the stigma protrudes from the skin. After the copulation by the males which roam on the skin of the hosts, the females lay several thousand eggs, which fall on the ground and from which under favourable conditions within 3 weeks via larvae and pupae again adults develop, which seek a new host.

Further Reading

Cucchi-Stefanoni K et al (2008) Fatal anemia and dermatitis in captive agoutis (*Dasyprocta mexicana*) infested with *Echidnophaga* fleas. Vet Parasitol 155:336–339.

Dryden MW et al (2016) Evaluation and comparison of a flumethrin-imidacloprid collar and repeated monthly treatments of fipronil/(s)-methoprene to control flea, *Ctenocephalides f. felis*, infestations on cats for eight months. Parasite Vectors 9:287.

Feldmeier H et al (2004) Investigations on the biology, epidemiology, pathology and control of *Tunga penetrans* in Brazil: IV. Clinical and histopathology. Parasitol Res 94:275–282.

Gamerschlag S et al (2008) Repetitive sequences in the ITS1 region of the ribosomal DNA of *Tunga penetrans* and other flea species (Insecta, Siphonaptera). Parasitol Res 102:193–199.

Gilles J et al (2009) Polymerase chain reaction detection of *Rickettsia felis*-like organism in *Archaeopsylla erinacei* (Siphonaptera: Pulicidae) from Bavaria, Germany. J Med Entomol 46:703–707.

Klimpel S et al (2005) Field trial of the efficacy of a combination of imidacloprid and permethrin against *Tunga penetrans* (sand flea, jigger flea) in dogs in Brazil. Parasitol Res 97:S113–119.

Kreppel KS et al (2016) Effect of temperature and relative humidity on the development times and survival of *Synopsyllus fonquerniei* and *Xenopsylla cheopis*, the flea vectors of plague in Madagascar. Parasite Vectors 9:82.

Lawrence AL et al (2015) Integrated morphological and molecular identification of cat fleas (*Ctenocephalides felis*) and dog fleas (*Ctenocephalides canis*) vectoring *Rickettsia felis* in central Europe. Vet Parasitol 210:215–223.

Márquez FJ (2015) Detection of *Bartonella alsatica* in European wild rabbit and their fleas (*Spilopsyllus cuniculi* and *Xenopsylla cunicularis*) in Spain. Parasite Vectors 8:56.

Nagy N et al (2007) Investigations on the life cycle and morphology of Tunga penetrans in Brazil. Parasitol Res 101(Suppl 2):S233–42.

Oteo JA et al (2014) '*Candidatus Rickettsia asemboensis*' and *Wolbachia* spp. in *Ctenocephalides felis* and *Pulex irritans* fleas removed from dogs in Ecuador. Parasite Vectors 7:455

Sekeyová Z et al (2012) Identification of *Rickettsia africae* and *Wolbachia* sp. in *Ceratophyllus garei* fleas from Passerine birds migrated from Africa. Vector Borne Zoonotic Dis 12:539–543.

6.2.2.1 Fleas of Carnivores (Canids, Felids)

1. **Name**: Greek: *ktenos* = comb; *kephale* = head. Latin: *felis* = cat; *canis* = dog; *pulex* = flea; *irritare* = to irritate.

2. **Geographical distribution/epidemiology**: Worldwide.
3. **Biology/morphology**:
   (a) ***Ctenocephalides canis*** (Fig. 6.68c)

   Females up to 3.5 mm long and males up to 2.5 mm. Dorsum of the tibia with 8 indentations.

   (b) ***Ctenocephalides felis*** (Figs. 6.67 and 6.68d, f)

   Females up to 3 mm long and males up to 2.5 mm. Dorsum of the tibia with six indentations.

   (c) ***Pulex irritans*** (rare in Germany now)

   This flea has approximately the same size as *C. canis* and possesses no combs (see Fig. 6.73; see Table 6.7).

   All 3 species are not very host specific. Their development is **holometabolic** (=complete: egg, three larvae, pupae, adults). Only the adults suck blood (both genders) and sting frequently, if possible on a daily basis. Afterwards, they eventually leave the host again and in the meantime live in its lair. During the 20–100 min lasting bloodsucking time, the fleas take up 20 times of their weight but defecate large amounts undigested, which are ingested by their larvae. About 20–30 of the whitish, about 0.5 mm long eggs get deposited in batches of 4–8 by the females daily, so that a mass infestation can appear quickly! After 4–12 days, each of the eyeless, bristled larva (so-called "wireworms") hatches from its egg shell and starts feeding on flea feces and organic materials from the animal lair. Eventually proglottids of the cucumber seed tapeworm are taken up so that they become intermediate hosts. Via 2 moults, the "wireworm" grows up to the third larva, which spins a cocoon that gets strengthened with detritus, and pupates. After a pupal period of at least 1 week, the fertile fleas hatch, which are viable for about 1½ year. The developmental period for *C. felis* is temperature dependent at least 11 days (*C. canis* 18 days), but it can also last for months. The hatching is triggered by an external stimulus (e.g. vibration), which announces the arrival of a host often inducing a mass exodus of adult fleas from their cocoon in the cases of resettlements of lairs or when entering of long-time empty flats.
4. **Symptoms of disease**: Severe itchiness, eczema and skin alterations as a consequence of secondary infections, allergic reactions (flea allergic dermatitis; see Figs. 6.75 and 6.76) and occurrence of the cucumber seed tapeworm.
5. **Diagnosis**: Detection of adults (combing out, showering in a light bathtub) and detection of reddish flea feces and whitish eggs in the hair or larvae in the animal lair.
6. **Pathway of infection**: Contact with infested animals or their lairs.
7. **Prophylaxis**: Application of a pet flea collar or of repellents (*pour-on*) or use of insecticides (see ticks).
8. **Therapy/control**: See prophylaxis. The spraying of the floors with compounds, which prevent the moult of the flea larva, is also highly recommended. The rustic method according to Wilhelm Busch (see Fig. 6.77) of dipping the dog in a nicotine fluid is not applied anymore nowadays.

Further Reading

Beugnet F et al (2015) Sustained speed of kill and repellency of a novel combination of fipronil and permethrin against *Ctenocephalides canis* flea infestations in dogs. Parasite Vectors 8:52.

Dumont P et al (2014) Curative and preventive efficacy of orally administered afoxolaner against *Ctenocephalides canis* infestation in dogs. Vet Parasitol 201:212–215.

Halos L et al (2016) Knock-down and speed of kill of a combination of fipronil and permethrin for the prevention of *Ctenocephalides felis* flea infestation in dogs. Parasite Vectors 9:57.

Mencke N et al (2009) Transmission of feline calicivirus via the cat flea (*Ctenocephalides felis*). Parasitol Res 105:185–189.

Vobis M et al (2005) Experimental quantification of the feline leukaemia virus in the cat flea (*Ctenocephalides felis*) and its faeces. Parasitol Res 97:S102–S106.

Vobis M et al (2004) Molecular phylogeny of isolates of *Ctenocephalides felis* and related species based on analysis of ITS1, ITS2 and mitochondrial 16S rDNA sequences and random binding primers. Parasitol Res 94:219–226.

Yiguan W et al (2016) Influence of bloodmeal host on blood feeding, egg production, and offspring sex ratio of *Ctenocephalides felis felis* (Siphonaptera: Pulicidae). J Med Entomol 53: 888–893.

6.2.2.2 Fleas of Hares, Rabbits, Rodents

1. **Name**: Greek: *spilos* = spot; *psyllos* = flea. Latin: *cuniculus* = rabbit.
   Greek: *xenos* = foreign; *psyllos* = flea; *cheops* = Egyptian Pharaoh, builder of a giant pyramide in Giza/Cairo;
   Greek: *leptos* = narrow; *psyllos* = flea. Latin: *segnis* = lazy.
   Greek: *megas* = big; *bothrion* = pit. Latin: *turba* = restlessness.
2. **Geographical distribution/epidemiology**: Worldwide.
3. **Biology/morphology**: The fleas temporally suck blood and let themselves fall into the lair of the host after the blood meal.
   (a) *Spilopsyllus cuniculi* (in **hares, rabbits**). Males up to 1.6 mm long and females up to 2 mm. Pronotal ctenidia (Fig. 6.71) with 14–15 thorns; genal ctenidia with 4–6 thorns (=cheek teeth). Transmitter of the myxoma virus and other pathogens (including tularaemia: *Francisella tularensis*).
   (b) *Xenopsylla cheopis* (**Tropical rat flea**). 1.5–2.5 mm long (Fig. 6.72a); ctenidia are missing. Transmitter of the plague (*Yersina pestis*).
   (c) *Nosopsyllus fasciatus* (**European rat flea**). Females and males up to 2 mm long. Pronotal ctenidia with 20 thorns.
   (d) *Leptopsylla segnis* (Fig. 6.72c) (**House mouse flea**) and *Ctenophthalmus assimilis* (**Field mouse flea**) (cf. Fig. 6.72d, e).
   (e) *Megabothris turbidus* (Fig. 6.72b) appears in many *Microtus* species.
   The development cycle of the fleas comprises three free-living larval stages (Figs. 6.68f and 6.70), which feed on detritus and blood remains

(=feces of the adults), as well as one pupal stage. **Caution**: Possible mass hatching of the adults from the pupae may occur after a vibration trigger when touching at an abandoned nest or lair!

4. **Symptoms of disease**: Itchiness and local skin reactions due to bites. Mass infestation leads to eczematous skin alterations, anaemia and emaciation.
5. **Diagnosis**: Inspection of the hair coat and the skin. Detection of the sucked females and males.
6. **Pathway of infection**: Physical contact. **Caution**: Fleas can jump relatively far and high!
7. **Prophylaxis**: Regular and intense cleaning as well as disinfection (contact insecticides see ticks) of the animal lair or changing of the litter. **Caution**: Pupae can persist in a resting stage for a long time (up to 3 years). Adults hatch (after a vibration trigger) mostly after 3 weeks.
8. **Therapy**: Spraying or dusting of the animals with a suitable contact insecticide (preparation see ticks). **Caution**: For rabbits reared for human consumption, approved insecticides do not exist: withholding periods at least as long as in cattle/sheep.

Further Reading

Frank R et al (2013) Parasites of wild rabbits (*Oryctolagus cuniculus*) from an urban area in Germany, in relation to worldwide results. Parasitol Res 112:4255–4266.

Kreppel KS et al (2016) Effect of temperature and relative humidity on the development times and survival of *Synopsyllus fonquerniei* and *Xenopsylla cheopis*, the flea vectors of plague in Madagascar. Parasite Vectors 9:82.

Márquez FJ (2015) Detection of *Bartonella alsatica* in European wild rabbit and their fleas (*Spilopsyllus cuniculi* and *Xenopsylla cunicularis*) in Spain. Parasite Vectors 8:56.

Rzotkiewicz S et al (2015) Novel evidence suggests that a '*Rickettsia felis*-like' organism is an endosymbiont of the desert flea, *Xenopsylla ramesis*. Mol Ecol 24:1364–1373.

6.2.2.3 Fleas of Birds

1. **Name**: Greek: *keras* = horn; *pyhllon* = leaf; *echidna* = hedgehog. Latin: *gallus* = cock; *columba* = pigeon; *gallinaceus* = belonging to the hen.
2. **Geographical distribution/epidemiology**: Worldwide.
3. **Biology/morphology**: *Ceratophyllus* species (*C. gallinae*, *C. columbae*; Fig. 6.69) are not host specific and thus can be found on a number of birds as temporary ectoparasites to suck blood. They mainly are distinguished by the shape and the arrangement of the genital organs. Thus, the approximately 3.5 mm long females of *C. gallinae* have a seminal receptacle which is built of two parts of different size, while in the about 3 mm long females of *C. columbae*, it consists of 2–3 parts of approximately the same size. Both species possess eyes and a pronotal ctenidium (Fig. 6.72c), while a genal ctenidium is missing. In case

Fig. 6.69 Light micrograph of a male flea of *Ceratophyllus* sp.

Fig. 6.70 Scanning electron micrograph of stages of the cat flea on the floor. *Above*: adult flea; *left below*: egg with a hatching larva; *right below*: larva 2 (="wireworm")

of the tropical hen flea (*Echidnophaga gallinacea*), which attaches itself at the skin, many ctenidia are missing. The ontogeny processes via one larval stage (with three moults) and 1 pupa takes about 17–30 days (temperature-dependent). The pupae often overwinter in the nests of their hosts, so that in springtime during a resettlement the parasite infestation occurs immediately. Also humans (e.g. during cleaning of old nests) can get infested and may serve temporarily as

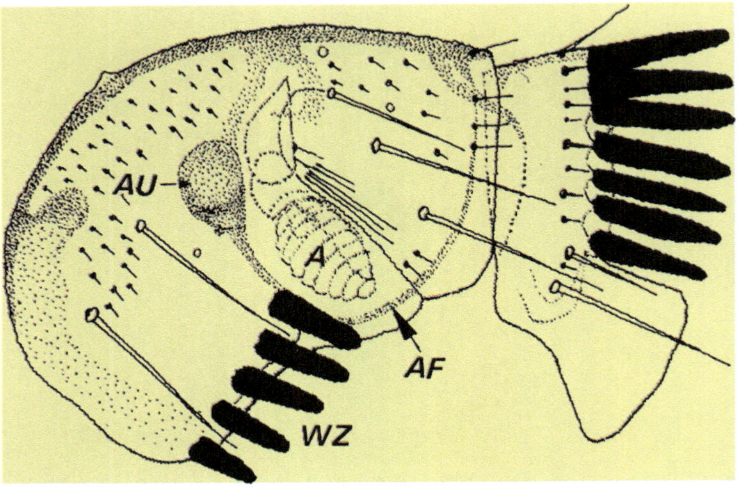

Fig. 6.71 Diagrammatic representation of the head of a rabbit flea *Spilopsyllus cuniculi*. A = groove of the antenna; AN = antenna; AU = ommatidium; WZ = genal ctenidium

a host. Besides, bird fleas are also to find on hedgehogs, from which they may jump over to humans in winter.

4. **Symptoms of disease**: Restlessness due to itchiness of the numerous bites and retarded development in young birds. In case of severe infestations, anaemia appears quite often. Hens or pigeons do not visit their nests, which leads to relocation of the eggs and decrease of the laying performance.
5. **Diagnosis**: Detection of flea larvae and adult fleas (Fig. 6.69) in the nests or discovery of fleas on the birds in case of severe infestations.
6. **Pathway of infection**: Physical contact or contact with the nests or nesting boxes.
7. **Prophylaxis**: Regular inspection of the nests or nesting boxes (**caution**: fleas also suck blood on humans) and eventually spraying of contact insecticides (powder form, preparations: see therapy). Also human housings can fall a victim to mass infestation, if hygienic measures are ignored in case of keeping the birds in the rooms.
8. **Therapy**: In-depth cleaning of nests and nesting boxes (exchange of the litter after a disinfection with contact insecticides). For the treatment of the animals, several preparations are suitable. Repeated treatment should be done after 3–4 weeks. Spraying the nests with MiteStop® (Co. Alpha-Biocare; see mites) is likewise effective and doesn't require a withholding period.

Further Reading

Appelgren A et al (2016) Relative fitness of a generalist parasite on two alternative hosts: a cross-infestation experiment to test host specialization of the hen flea *Ceratophyllus gallinae* (Schrank). Evol Biol 29:1091–1101.

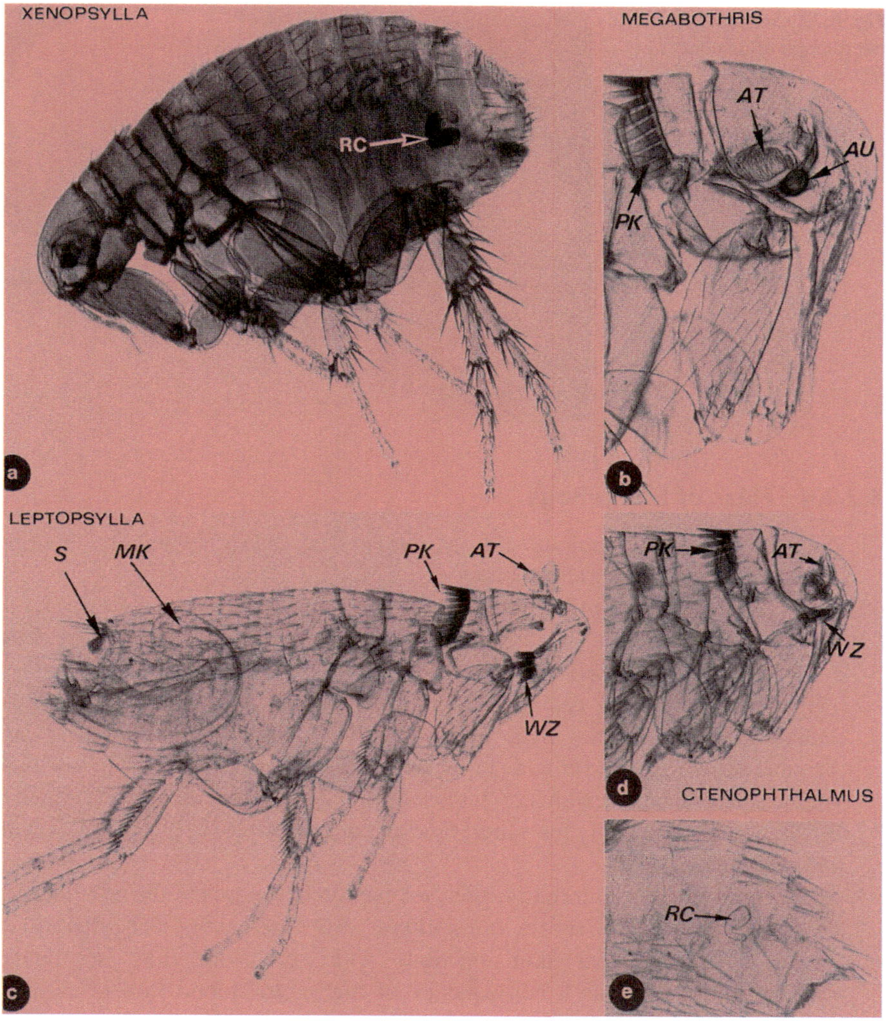

Fig. 6.72 Light micrographs of fleas. (**a**) *Xenopsylla cheopis*. (**b**) *Megabothris turbidus*. (**c**) *Leptopsylla segnis*. (**d**, **e**) *Ctenocephthalmus agyrtes*. AT = antenna; AU = eye = ommatidium; MK = male copulation organs; PK = pronotal comb = ctenidium; RC = receptaculum seminis; S = sensillum = bristle; WZ = genal ctenidium

Gyimesi ZS et al (2007) Sticktight flea (*Echidnophaga gallinacea*) infestation in a Victoria crowned pigeon (*Goura victoria*). J Zoo Wildl Med 38:594–596.

Kapoor R, Elston DM (2012) What's eating you? The sticktight flea (*Echidnophaga gallinacea*). Cutis 89:157–158.

Lipatova I et al (2015) Fleas (Siphonaptera) in the nests of dormice (Gliridae: Rodentia) in Lithuania. J Med Entomol 52:469–474.

Fig. 6.73 Scanning electron micrograph of a human flea (*Pulex irritans*)

6.2.2.4 Fleas of Hedgehogs

1. **Name**: Greek: *archaios* = old; *psyllos* = flea; *hystrix* = prickly. Latin: *erinaceus* = hedgehog; *talpa* = mole.
2. **Geographical distribution/epidemiology**: Europe.
3. **Biology/morphology**: In hedgehogs, besides specific hedgehog fleas (*Archaeopsylla erinacei*, *Hystrichopsylla talpae*), also dog, cat and bird fleas (Figs. 6.67 and 6.72) can appear in huge numbers (the latter also infests humans!). The hedgehog fleas with a length of 4 mm (sucked females) can become quite big, appear dark brown and are host specific (they do not bite humans, but dogs!). Features of *A. erinacei* are 2–3 conical teeth at the cheeks. All fleas suck blood only for a short time and afterwards leave the animal (=fall into the lair).
4. **Symptoms of disease**: Initially, allergic reaction in the area of the bite. In case of frequent bites and severe infestations, also eczematous skin alterations occur. Later confluent skin reactions with scab formation (secondary infections), anaemia as a consequence of high-grade blood loss in case of mass infestation as well as severe restlessness due to itchiness and sometimes even apathy in case of heavy disturbances occur.
5. **Diagnosis**: Detection of the adult fleas in the bathwater, which is supplemented with a HCH preparation, e.g. chlorohexol concentrate: 1 ml to be added to 1 l water (repeated watering of the hedgehog with the supplemented, lukewarm water).
6. **Pathway of infection**: Fleas actively seek their host.
7. **Prophylaxis**: In case of temporary keeping, first use of insecticides e.g. with Alugan®-Spray or pyrethrum spray, e.g. Axis®Natur, or spraying with INS 15 or CBM 8®. Repetition of the treatment every 10–12 days and regular cleaning of the sleeping place (carton with a lateral opening and a heat keeping bottom consisting of wood or thick cardboard).

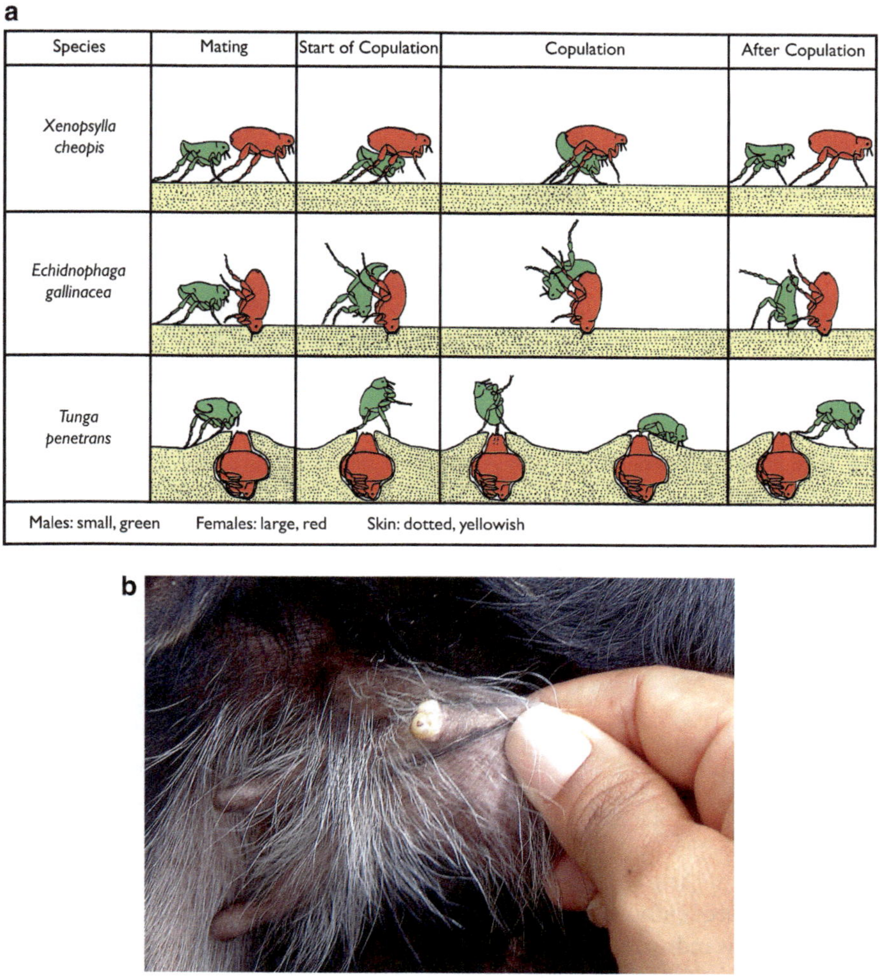

Fig. 6.74 Diagrammatic representation of copula positions of different fleas, which occur on the skin of the host. In the case of two species, the females (*red*) have entered the skin partly or in total

8. **Incubation period**: 1 day after the sucking action; the bites induce severe allergic reaction.
9. **Prepatency**: Fleas infest their host daily.
10. **Patency**: Fleas mostly stay for a short time on their hosts to suck blood (several minutes to hours).
11. **Therapy**:
    (a) Watering of the hedgehog with insecticide-containing water (e.g. 0.7 g Alugan® concentrate per 1 l lukewarm water) or treatment with Alugan® spray or **pyrethrum** spray, e.g. Axis®Natur, Petvital®. (**Caution**: Protect the hedgehog's eyes).

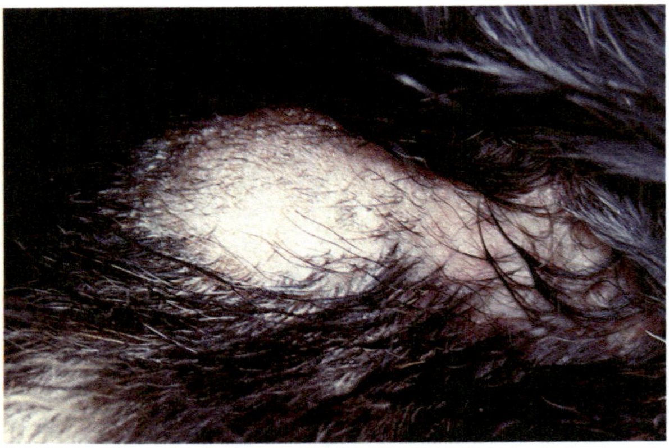

Fig. 6.75 Skin reaction due to flea bite allergy

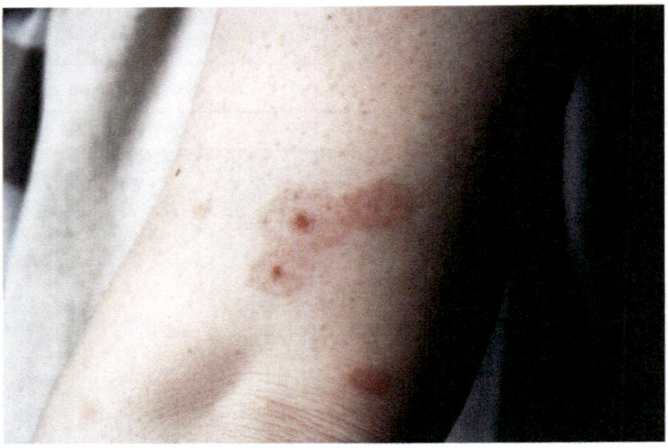

Fig. 6.76 Flea bite reactions of humans. Fleas bite often in rows after being disturbed

(b) Subsequently (after approx. 20 min), the hedgehog gets thoroughly showered with lukewarm water. Thereby, the damaged fleas (partly only inactivated) get washed out of the coat (care for immediate release of the used water and substantial rinsing). **Caution**: Allow the hedgehog to dry *lege artis* in a warm room!

(c) Later, powdering of the hedgehog with chlorohexol powder or CBM 8® (for further preparations, see therapy of mange mites).

Further Reading

Gilles J et al (2009) Polymerase chain reaction detection of *Rickettsia felis*-like organism in *Archaeopsylla erinacei* (Siphonaptera: Pulicidae) from Bavaria, Germany. J Med Entomol 46:703–707.

Hajipour N et al (2014) Hedgehogs (*Erinaceus europaeus*) as a source of ectoparasites in urban-suburban areas of Northwest of Iran. J Arthropod Borne Dis 9:98–103.

Hornok S et al (2014) Vector-borne agents detected in fleas of the northern white-breasted hedgehog. Vector Borne Zoonotic Dis 14:74–76.

Fig. 6.77 Diagrammatic representation showing the rustic flea removement by dipping a dog into a nicotine fluid [drawing by Dr. Walldorf after a design of the famous German poet Wilhem Busch (1832–1908)]

6.2.3 Bloodsucking Bugs (Order: Rhynchota, Heteroptera)

Out of the numerous bug species, two groups of bloodsucking forms have of medical or hygienic importance:

1. **Bed bugs** (Cimicidae)
2. **Raptor bugs** (Reduviidae)

All bugs possess **stinging-sucking mouthparts**, which are situated in a labium-formed sheath (transformed 2. maxillae). This stiletto-like sheath (**proboscis**) is retracted against the ventral side of the head and of the thorax segments and is only erected in case of a bite. However, only the stiletto-like mandibles and the 1. maxillae are injected into the skin of the host. The latter form two hollow tubes: one for the saliva injection and the other for food uptake.

The two front wings of many bugs comprise in Reduviidae of a rough and a skinny region (=typical feature of the **Heteroptera**). In bed bugs, wings are missing completely. Only small, chitinous wing rudiments at the mesothorax bear witness to their relation to winged forms.

The development of these bugs occurs as hemimetaboly starting from the egg and running via five nymph stages to the adult. The **raptor bugs** (Reduviidae) of the genera *Triatoma*, *Rhodnius*, *Dipetalogaster* and *Panstrongylus* reach a length of about 3 cm and can transmit (among others) the pathogens of the Chagas disease (*Trypanosoma cruzi*) throughout their whole lifespan of about 400 days. Transmission does not happen during the bite but by rubbing fecal drops (containing the metacyclic trypanosomes) into the bite channels. In many laboratories, "clean uninfected" raptor bugs are cultured and used for **xenodiagnosis**. Thereby, such bugs are applied to persons with suspected Chagas disease and tested for the parasites in their intestine after about 3 weeks.

The **bed bug** species *Cimex lectularius* as well as the species *C. hemipterus*, *C. columbarius*, *Oeciacus hirundinis* and *Leptocimex boueti* may appear in human housings, nests and stables. Before blood ingestion, they appear dorsoventrally flattened, are about 4–5 mm long and stay hidden during the daytime and start to suck blood from a broad spectrum of animals and humans and their pets at night. The bites often lead to big, severely itching skin inflammations and eventually to allergic skin reactions. Adult bugs live for about 1 year and may survive without feeding up to ½ year, which complicates their eradication. Each female lays about 200–500 eggs, which reach a length of about 1 mm and appear white. Since bed bugs mostly appear as "**nuisance insects**" and transmit pathogens (e.g. bacteria) mechanically by their mouthparts; only in rare occasions, they have a low-grade medical importance. However, their **scent glands** (in adults ventrally situated in the metathorax; in nymphs dorsally in the abdomen) as well as their fluid feces induce a characteristic rather bad smell inside bug-infested rooms. These volatile substances keep the bug population together and accordingly, owing to the given copulation opportunities, secure the persistence of the population.

6.2.3.1 *Cimex* Species

1. **Name**: Latin: *cimex* = true bug; *lectularius* = belonging to the bed; *hirundo* = swallow; *pipistrellus* = bat; *columba* = pigeon.
2. **Geographical distribution/epidemiology**: Worldwide.
3. **Biology/morphology**: ***Cimex* species** (*C. lectularius* = bed bug; *C. columbarius* = pigeon bug; *Oeciacus hirundinis* = swallow bug; *Cimex pipistrellus*, *Leptocimex boueti* = bat bug) only appear in not cleaned **dovecotes**

and **henhouses**. As temporary ectoparasites, they only suck blood at night. The dorsoventrally flattened parasites reach, depending on the species, a length of about 4–6 mm and 3 mm in width (Fig. 6.78). Characteristic is the scent gland, which generates a pungent and typical smell. From the eggs, which are deposited in lairs (cracks, nests and nesting boxes), the larvae hatch. Via five larval stages, the sexual maturity is reached within 6–7 weeks after oviposition (temperature dependent). The hindwings are completely degenerated, while the forewings are preserved as small, squamous rudiments.

4. **Symptoms of disease**: After bites skin wellings are formed, which cause severe itching. Retarded development occurs in young host animals because of restlessness and blood loss. Infested nests or nesting boxes are not visited any longer by birds (breeding activity is thus interrupted).
5. **Diagnosis**: Detection of the bugs in the liars of the stable, in nests or in nesting boxes. The typical, rancid, unpleasant smell facilitates the discovery of the bugs.
6. **Pathway of infection**: Bloodsucking bugs seek their hosts actively at night.
7. **Prophylaxis**: Elimination of liars in the stable (sealing of cracks and gaps!). Spraying the nests and nesting sites with suitable contact insecticides. **Caution**: *Cimex* species can survive for a long time (several months) without feeding!
8. **Control**: In-depth cleaning and disinfection of nests and nesting boxes. For the treatment of animals, several wall-sprayable insecticides (e.g. different pyrethroids) or the biological, non-toxic MiteStop® (Co. Alpha-Biocare, Düsseldorf) are suitable. As precaution, the treatment has to be repeated after 4–6 weeks (thus killing the adults of the next generation).

Further Reading

Bartonička T, Růžičková L (2013) Recolonization of bat roost by bat bugs *(Cimex pipistrelli*): could parasite load be a cause of bat roost switching? Parasitol Res 112:1615–1622.

Brown CR et al (2015) Predation by ants controls swallow bug (Hemiptera: Cimicidae: *Oeciacus vicarius*) infestations. J Vector Ecol 40:152–157.

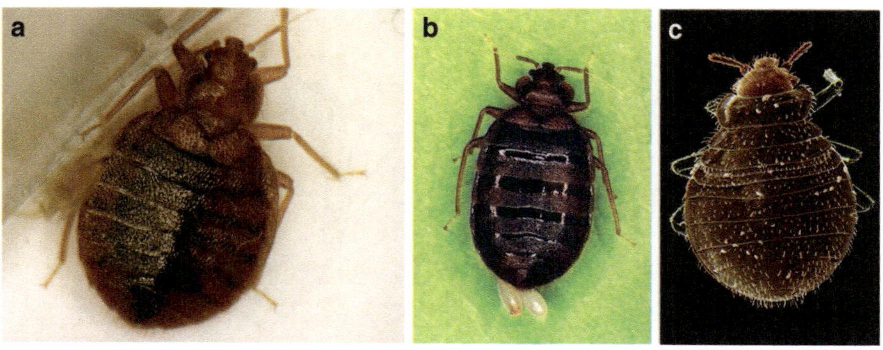

Fig. 6.78 (**a**) Light micrograph of an engorged bed bug (*Cimex lectularius*). (**b**) Light micrograph of a *Cimex* female laying eggs. (**c**) Scanning electron micrograph of a *Cimex* larva

Kamimura Y et al (2014) Duplicated female receptacle organs for traumatic insemination in the tropical bed bug *Cimex hemipterus*: adaptive variation or malformation? PLoS One 9:e89265.

Sheele JM, Ridge GE (2016) Toxicity and potential utility of ivermectin and moxidectin as xenointoxicants against the common bed bug, *Cimex lectularius* L. Parasitol Res. doi:10.1007/s00436-016-5062-x.

6.2.3.2 Raptor Bugs (Reduviidae)

1. **Name**: Latin: *reduvia* = remnant; *personatus* = with 1 larva; *infestare* = to infest; Greek: *tri* = three; *tomos* = to cut (here: the body is segmented into 3 parts).
2. **Biology/morphology**: Raptor bugs are important due to their transmission of the agents of **Chagas disease** (*Trypanosoma cruzi*) in South and Central America, especially important in Venezuela, Brazil and Argentina. Several of the numerous bug species in Latin America may lead to natural infection with *Trypanosoma cruzi,* since they suck blood at numerous farm animals as well as at wild animals in the surroundings of farms.

   For the transmission of the **pathogens** to animals and humans, the following species are of importance: *Triatoma infestans*, *T. dimidiata*, *T. maculata*, *Rhodnius prolixus* and *Panstrongylus megistus.* These Reduviidae are large and strong animals (Figs. 4.26 and 6.79). In contrast to bed bugs, they are able to fly as adults and therefore able to enter new biotopes. Remarkable is the long, nose-like forehead (rostrum), on the lower side of which the proboscis lays,

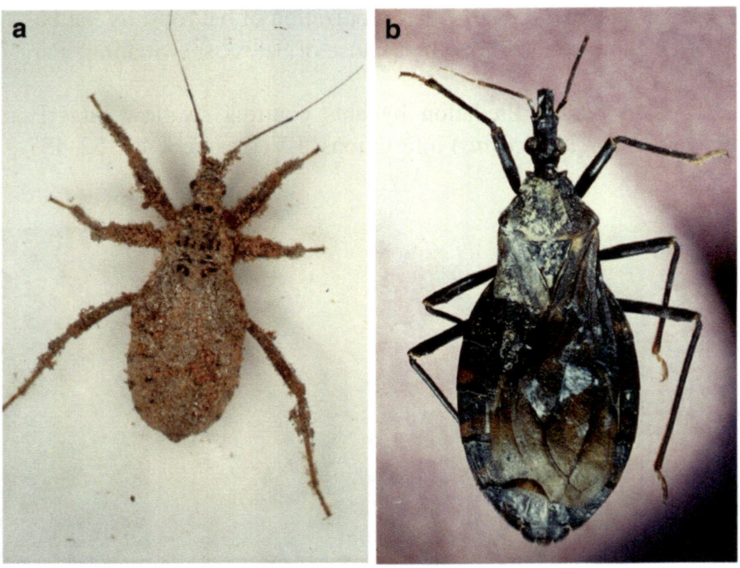

Fig. 6.79 Light micrographs of a dust-covered so-called feces bug (*Reduvius personatus*) (**a**) and so-called Chagas bug (genus *Triatoma*) (**b**)

while at the sides, shortly before the eyes, the segmented antennae are anchored. Only *Rhodnius* species do not possess this long rostrum. The forewings are so-called hemi-elytra without a cuneus. The wing membrane shows 2–3 big fields. The members of the family Triatomae live in similar lairs as the bed bugs. Each female lays about 2000 eggs. The development cycle is species and temperature dependent and takes up to 1 year and more. The masked hunter (*Reduvius personatus*; Fig. 6.79b) likewise belongs to this spectrum. It also appears in Central Europe, lives predatorily and bites humans and animals rather accidentally (but very painful).

3. **Impact of bites**: The Triatomae bite sleeping humans and animals very often in the face region (thus are called kissing bugs) but also suck at hands and feet. The bite itself is barely perceptible. The skin reactions are very diverse.
4. **Transmission of pathogens**: *Trypanosoma cruzi* undergoes morphogenesis within the bug (Fig. 4.26) and finally reaches infectivity in the hindgut. The specimens of Triatomae empty the terminal portion of their gut before or after the blood meal. Thereby, the infectious trypanosomes can enter the skin via the stinging canal, thus reaching the bloodstream of the host. Thus, the infection does not happen via the proboscis. Since hungry bugs also suck blood at other bugs, Chagas pathogens can spread easily within a bug population.

Further Reading

Alzogaray RA (2016) Behavioral and toxicological responses of *Rhodnius prolixus* (Hemiptera: Reduviidae) to the insect repellents DEET and IR3535. J Med Entomol 53:387–393.

Carrasco HJ et al (2014) *Panstrongylus geniculatus* and four other species of triatomine bug involved in the *Trypanosoma cruzi* enzootic cycle: high risk factors for Chagas' disease transmission in the Metropolitan District of Caracas, Venezuela. Parasit Vectors 7:602.

De Souza Rde C et al (2015) Does *Triatoma brasiliensis* occupy the same environmental niche space as *Triatoma melanica*? Parasite Vectors 8:361.

Peterson JK et al (2015) *Rhodnius prolixus* life history outcomes differ when infected with different *Trypanosoma cruzi* strains. Am J Trop Med Hyg 93:564–572.

Rodríguez-Planes LI et al (2016) Selective insecticide applications directed against *Triatoma infestans* (Hemiptera: Reduviidae) affected a nontarget secondary vector of Chagas disease, *Triatoma garciabesi*. J Med Entomol 53:144–151.

6.2.4 Order Diptera

1. **Name**: Greek: *dis* = two; *pteron* = wing.
2. **Biology/morphology**: The adults of the holometabolic dipterans are characterized by the possession of two big, membranous forewings and two so-called **halteres** (=reduced hindwings) (Fig. 6.80b). Exceptions are made by the secondarily wingless louse flies: the Hippoboscidae (Figs. 6.111 and 6.112).

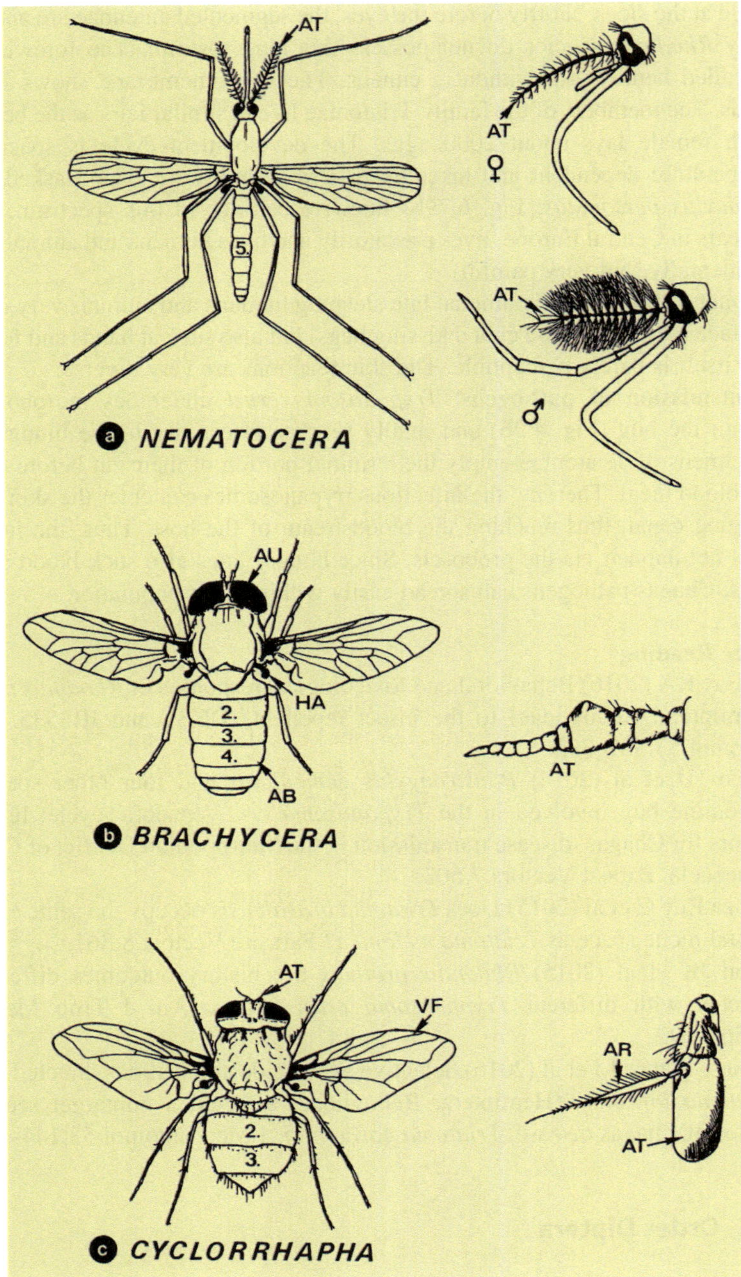

Fig. 6.80 Diagrammatic representation of the adults of three groups of Diptera. AB = abdomen; AR = arista (bristle); AT = antenna; AU = eye; HA = haltere; VF = anterior wing

The forewings contain tracheae ("veins"), the arrangement of which is genus and species specific, so that they can be used for taxonomy. The composition and characteristic of the antennae as well as the mode of the hatching process of the adults from the puparium are used for the systematic classification into three suborders (Figs. 6.80 and 6.81):

(a) **Nematocera** (with 6 and more antenna segments),
(b) **Brachycera** (with 3 antenna segments),
(c) **Cyclorrhapha** (three-segmented antenna with a typical arista (lateral element) and a circular cut in the puparium = lid).

The two first suborders are often also summarized as Orthorrhapha. In all of these three dipteran suborders bloodsucking and non-bloodsucking species appear.

6.2.4.1 Suborder Nematocera

1. **Name**: Greek: *nema* = thread; *keras* = horn.
2. **Biology/morphology**: The Nematocera (**mosquitoes** in close sense) have thread-like antennae comprising at least 6 (mostly more than 10) segments, which are bushy haired in males, while in females only few bristles occur (Fig. 6.82). Their larvae possess a hard head capsule and strong mandibles. Pupae belong to the so-called "free type", meaning that the extremities clearly shine through the wall of the puparium. The adults hatch from the pupa by cutting a dorsal, longitudinal slit in the thoracic region (**orthorrhaph**).

Four families are of huge medical importance as vectors or intermediate hosts (Table 6.8): *Culicidae*, *Simuliidae*, *Phlebotomidae* and *Ceratopogonidae*.

(a) **Family Culicidae (Mosquitoes)**

1. **Name**: Greek: *anopheles* = harmful; *aedes* = house, room. Latin: *culex* = mosquito; Sir Patrick Manson (1844–1922) = English scientist.
2. **Biology/morphology**: The **females** (antennae always with 15 segments) of the genera *Anopheles*, *Aedes*, *Culex*, *Mansonia* and *Culiseta* (mosquitoes) suck **blood** from their host with the help of their proboscis (Figs. 6.83c, 6.85 and 6.87) every 3–4 days, mostly at night (at species specific times). They need blood for the development of the eggs and thus can starve for only 8–10 days. The **males** (14 antenna segments), by contrast, feed on plant juices (Fig. 6.82). The females lay species-specific 40–400 eggs singly (*Aedes* spp., *Anopheles* spp.) or glued together, thus appearing as little "boats" (*Culex* spp.) mostly into moist biotopes or directly into the waters (Fig. 6.84A). There, the eyeless larvae hatch after about 12 h to 2 days (temperature-dependent) (Fig. 6.84B). With the help of their respiratory opening at the posterior end (often provided with a siphon), air is absorbed from the water surface into the tracheal system. Thereby, larval species of the genera *Culex* and *Aedes* take the typical inclined position to

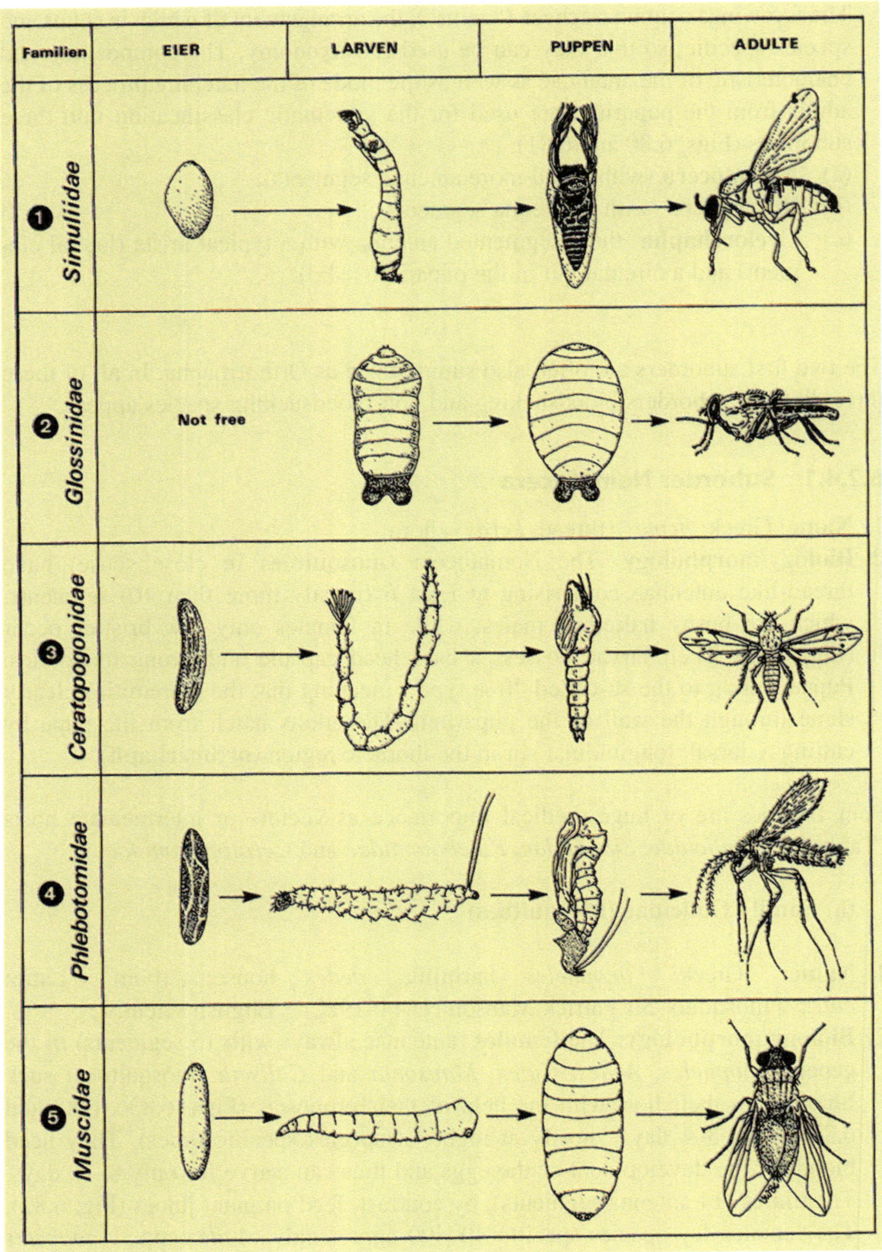

Fig. 6.81 Diagrammatic representation of the life cycles of important Diptera

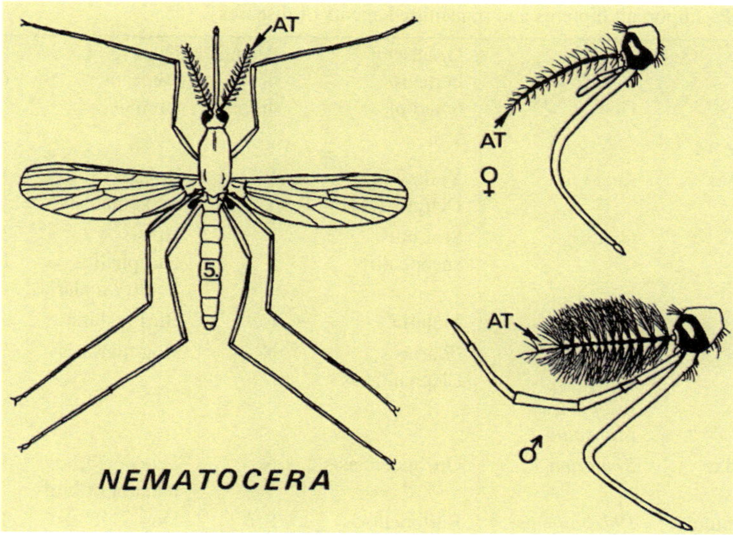

Fig. 6.82 Diagrammatic representation of the antennae (AT) of nematoceran mosquitoes

the water surface, while the *Anopheles* larva (without siphon!) lies in parallel (Fig. 6.84B). The mosquito larvae mostly feed on herbal food in particle size. Four larval stages are completed within about 10–14 days, each with one moult. From the fourth larva, the free pupa arises, which is motile and therefore able to take up atmospheric oxygen at the water surface, but it doesn't feed anymore (Fig. 6.84C). After about 3 days, the imago (Fig. 6.84D) hatches from the puparium, so that the whole, temperature-dependent development takes often more than 2–3 weeks. In case of favourable climate conditions, this time can be reduced considerably. Within the genera ***Aedes*** (Fig. 6.87) and ***Culex***, which can species dependend transmit in certain countries the pathogens of the yellow fever, the tropical elephantiasis, the different filariasis and encephalitis forms, the West Nile fever as well as the dengue fever. However, only a few species of a genus are able to transmit (e.g. *Aedes aegypti*, *Culex quinquefasciatus* (Fig. 6.85), *C. fatigans* and others). By contrast, the pathogens of malaria can be transmitted by more than 60 *Anopheles* species (Figs. 6.86, 6.88 and 6.89). In the transmission cycles of parasites (=protozoans, worms), the pathogens are restricted to the female mosquito and do not pass over to the insect's offspring (Table 6.8). However, during the last 10 years, evidence was repeatedly provided for a transovarian spread of viruses.

During the **sucking action**, the female mosquitoes specifically inject their mouthparts into little blood vessels (**vessel feeder**). In response to the bite **allergic reactions** often arise.

Table 6.8 Important dipteres and transmitted agents of diseases

Family	Genus	Disease of humans (examples)	Agents of disease	Disease of domestic animals	Agents of disease
Nematocera					
Culicidae	*Aedes*	Yellow fever Dengue fever	V V	Rabbit myxomatosis	V
	Culex	St. Louis encephalitis	V	Horse encephalitis Poultry malaria	V P
	Anopheles	Malaria	P	Bird malaria	
	Aedes *Culex* *Anopheles* *Mansonia*	Filariasis, Elephantiasis	N	Dog filariasis	N
Simuliidae	*Simulium*	Onchocerciasis	N	*Leucocytozoon* malaria of birds	P
Phlebotomidae	*Phlebotomus*	Bartonellosis Papataci fever Leishmaniasis	R/B V P	Dog leishmaniasis	P
Brachycera					
Tabanidae	*Chrysops*	Tularaemia Loiasis	B N	Surra	P
Cyclorrhapha					
Muscidae	*Musca*	Poliomyelitis Bacteriosis (Salmonellosis, Cholera) Trachoma Amebiasis Myiasis caused by larvae	V B V P	Virosis Bacteriosis	V P
	Stomoxys	Poliomyelitis Bacteriosis Sleeping sickness	V B P	Poulty spirochaetosis	B
Glossinidae	*Glossina*	Sleeping sickness	P	Nagana Surra	P P
Sarcophagidae	*Sarcophaga*, *Wohlfahrtia*	Myiasis caused by larvae	F	Myiasis caused by larvae	F
Calliphoridae	*Callitroga*	Myiasis caused by larvae	F	Myiasis caused by larvae	F
Gasterophilidae	*Gasterophilus*	Myiasis caused by larvae	F	Myiasis caused by larvae	F
Oestridae	*Oestrus*, *Hypoderma*, *Dermatobia*	Myiasis caused by larvae	F	Myiasis caused by larvae	F
Hippoboscidae	*Melophagus*, *Lipoptena*	Skin irritation	LF	Cachexia	LF

B bacteria, *F* flies, *LF* louse flies, *N* nematodes, *P* protozoans, *R* *Rickettsia* (=intracellular bacteria), *V* viruses

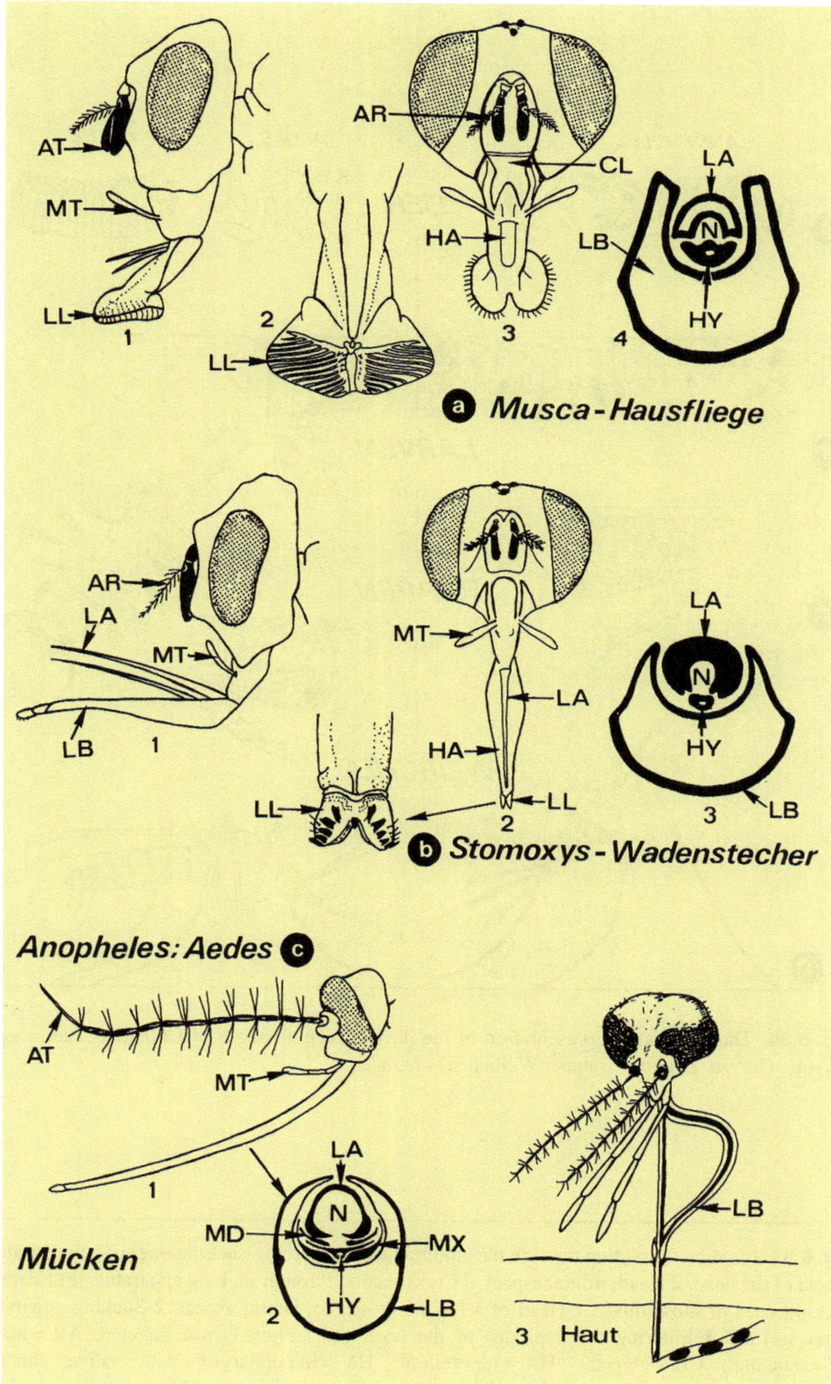

Fig. 6.83 Diagrammatic representation of the heads and mouthparts of Diptera: (**a**) Licking-sucking ones of a fly. **1** Lateral aspect of the head. **2** Labellum. **3** Frontal aspect of the head showing the large compound eyes and at the top of the head three single ocellar lenses. **4** Cross

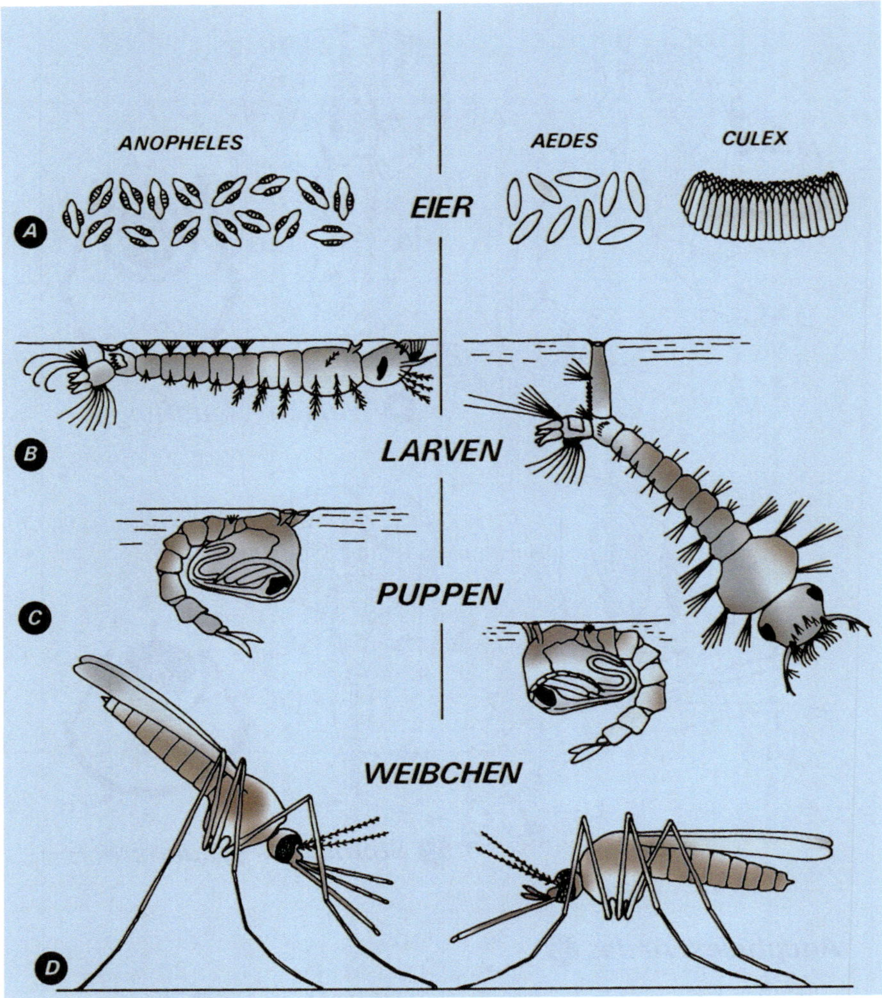

Fig. 6.84 Diagrammatic representation of the developmental stages of Culicidae. Eier = eggs; Larven = larvae; Puppen = pupae; Weibchen = females

---

Fig. 6.83 (continued) section through the mouthparts. (**b**) Piercing-sucking ones of a fly: **1** Lateral aspect of the head. **2** Head, frontal aspect. **3** Cross section through sucking apparatus. (**c**) Piercing-sucking ones of mosquitoes: **1** Head of a female mosquito, lateral aspect. **2** Sucking apparatus: cross section. **3** Entering the capillary of the host; the labium is not injected. AR = arista; AT = antenna; CL = clypeus; HA = haustellum; HY = hypopharynx with saliva channel; LA = labrum; LB = labium; LL = labellum; MD = mandible; MT = maxillar palpus; MX = maxilla; N = food channel

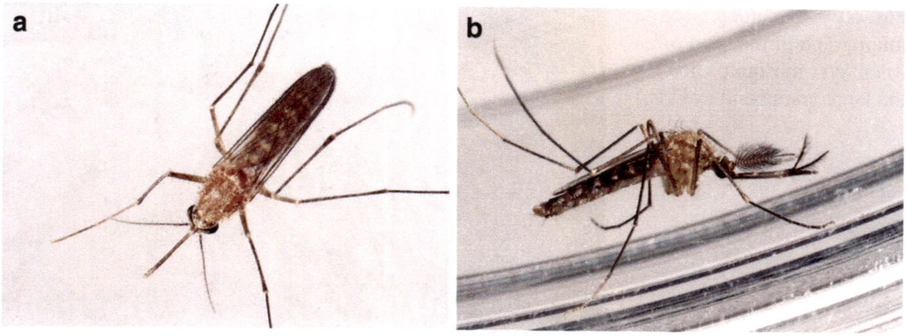

Fig. 6.85 Macrophotos of a female (**a**) and male (**b**) of the mosquito species *Culex quinquefasciatus*

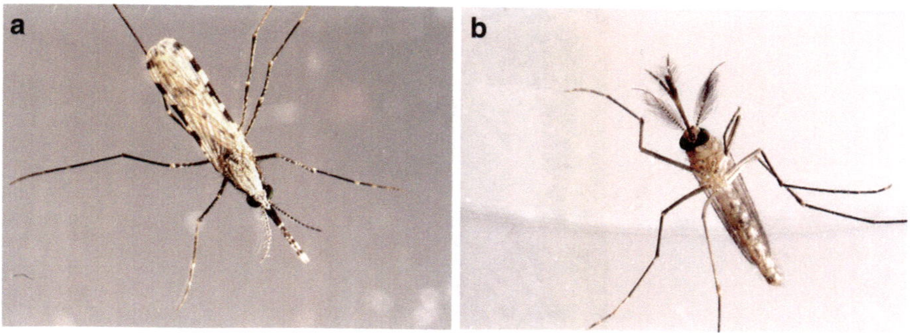

Fig. 6.86 Macrophotos of a female (**a**) and male (**b**) of *Anopheles stephensi*

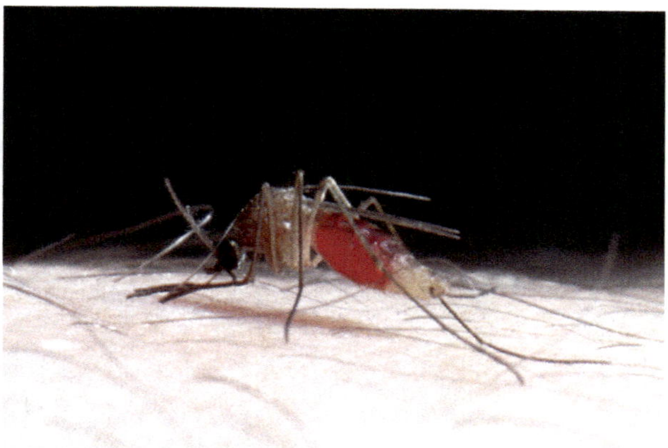

Fig. 6.87 *Culex* mosquito during blood sucking

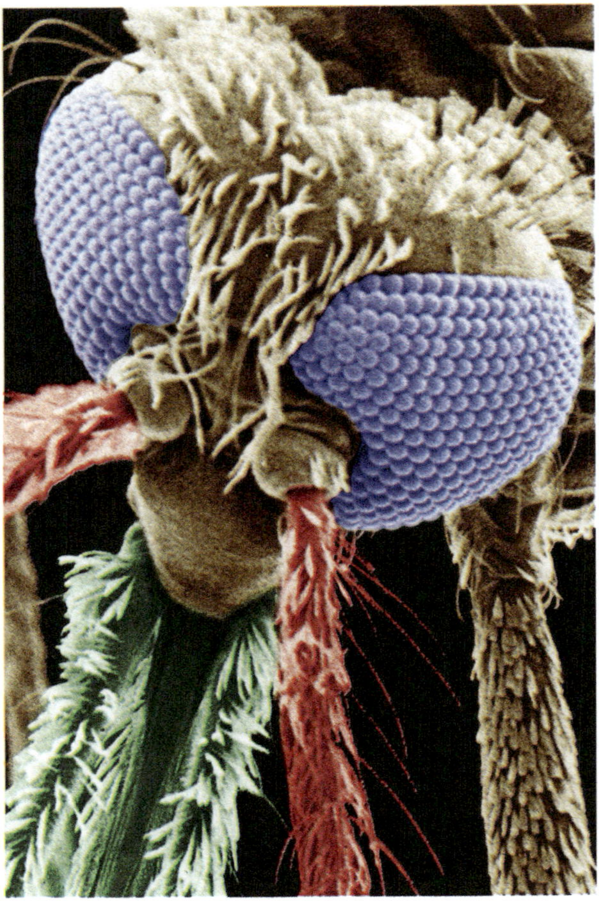

Fig. 6.88 Scanning electron micrograph of the head of an *Anopheles* mosquito showing the large compound eye

Fig. 6.89 Scanning electron micrograph of a female *Anopheles* mosquito showing its long and piercing mouthparts and palps

A **biological "preparation"** made from the *Bacterium thuringiensis*, serotype H 14, especially affects mosquito larvae. It is composed of **spores** and **parasporal crystals**. After the oral uptake, the crystals get dissolved in the gut of the larva. The released toxic by-products cause a feeding stop, whereby, among other reactions, the intestinal wall gets destroyed; germinating bacteria may penetrate into the abdominal cavity and cause a deadly sepsis. It is remarkable that there is no epidemic spread of this bacterium and no harmful effects occur in humans or other vertebrates. Therefore, the successful application of this biological control measure is ensured, which is of big importance for the control of **malaria**, of **onchocerciasis** and of the different forms of **tropical filariasis**.

Further Reading

Derua YA et al (2015) Sibling species of the *Anopheles funestus* group, and their infection with malaria and lymphatic filarial parasites, in archived and newly collected specimens from northeastern Tanzania. Malar J 14:104.

Koch LK et al (2016) Modeling the habitat suitability for the arbovirus vector *Aedes albopictus* (Diptera: Culicidae) in Germany. Parasitol Res 115:957–964.

Lühken R et al (2015) Physico-chemical characteristics of *Culex pipiens* sensu lato and *Culex torrentium* (Diptera: Culicidae) breeding sites in Germany. J Med Entomol 52:932–936.

Molaei G et al (2015) Vector-Host interactions of *Culiseta melanura* in a focus of Eastern equine encephalitis virus activity in Southeastern Virginia. PLoS One 10:e0136743.

Newman CM et al (2016) Decreased flight activity in *Culex pipiens* (Diptera: Culicidae) naturally infected with *Culex* flavivirus. J Med Entomol 53:233–236.

Okorie PN et al (2015) Insecticide susceptibility of *Anopheles coluzzii* and *Anopheles gambiae* mosquitoes in Ibadan, Southwest Nigeria. Med Vet Entomol 29:44–50.

Ughasi J et al (2012) *Mansonia africana* and *Mansonia uniformis* are vectors in the transmission of *Wuchereria bancrofti* lymphatic filariasis in Ghana. Parasite Vectors 5:89.

Weitzel T et al (2015) *Culex pipiens* s.l. and *Culex torrentium* (Culicidae) in Wrocław area (Poland): occurrence and breeding site preferences of mosquito vectors. Parasitol Res 114:289–295.

Werblow A et al. (2015) Hyperparasitism of mosquitoes by water mite larvae. Parasitol Res 114:2757–2765.

Werner D, Kampen H (2015) *Aedes albopictus* breeding in southern Germany, 2014. Parasitol Res 114:831–834.

Werner D et al (2016) First record of *Aedes koreicus* (Diptera: Culicidae) in Germany. Parasitol Res 115:1331–1334.

(b) **Family Simuliidae (Black flies)**

1. **Name**: Lat.: *simulare* = to imitate.
2. **Biology/morphology**: The mostly black coloured adults of the genus ***Simulium*** reach a length of about 2–5 mm and are significantly smaller than the adults of

the Culicidae (Fig. 6.90 and Table 6.8). **Only the females suck blood** but always at daytime (Figs. 6.91 and 6.92a). The males, whose compound eyes differ from those of the females in that they show a separation into small and big lenses, feed on nectar (Fig. 6.92b). The bite of the females (Fig. 6.91) is very painful because of their relatively rough, serrated mandibles and 1. maxillae. In contrast to the Culicidae, which enter their mouthparts into little blood vessels, simuliids are **pool feeders**, meaning that they scratch the skin with their serrated mouthparts and suck blood from the resulting "pool". Mass infestations of single hosts (up to 20,000), which are not uncommon after weather-dependent mass hatching, may lead to the death of livestock animals (including cattle) as a consequence of anaphylactic shocks.

Immediately after the hatching from the puparium, the copulation of males and females takes place. The females (Figs. 6.90 a, b) then need a quick blood meal for oogenesis. 4–5 days later, about 250 eggs are laid on plants or stones of rapid-flowing waters. Already after 4 h, the larvae hatch (Figs. 6.90c, d), which (attached at plants) live as "filter feeder" and after five moults (temperature dependent, in total taking about 5 days) transform into pupae with a typical cover (Figs. 6.90e–g). After about 4 more days of dormancy, the pupae leave the cover and reach—carried by air bubbles—the water surface, where the imago (adult stage) leaves the puparium. The overall duration of this development, which happens in the tropics throughout the whole year, takes about 9 days. In European latitudes, this development period can prolong significantly and stops during winter time. Altogether, the females live for 3 weeks. Some *Simulium* species can transmit the nematode *Onchocerca volvulus*, the pathogen of the human **river blindness**, during the sucking action (4–6 min). For unknown reasons, not all but only a few species (e.g. *S. damnosum, S. neavei* in Africa; *S. ochraceum, S. metallicum, S. callidum* in Central America) are competent vectors.

Black flies in Germany/Northern Europe

1. **Biology/morphology**: Small, darkened, fly-like black flies with a maximum length of 4 mm, with a characteristic head (Figs. 6.91 and 6.92) and sabre-like mouthparts (only in the bloodsucking females). Mass attacks of black fly species of the genera *Simulium*, *Odagmia*, *Wilhelmia* and *Boophthora* disturb the grazing cattle after mass hatching in the months from April to June (thus stormy flight reactions of cattle may occur because of the painful bites). Larvae and pupae of the Simuliidae are attached to plants in rapid-flowing waters. Mostly 2–3 generations of Simuliidae occur within a summer (in Central Europe).
2. **Symptoms of disease**: Slight to severe symptoms (mostly from April till July in Europe) occur: e.g. numerous typical, pinprick-like, partially confluent bleedings at thin-skinned body parts like the udder (Fig. 6.93), the scrotum, the foreskin, the thigh inside, abdominal area, the mouth, nose and around the anus. Subcutaneous oedemata, cardiac cycle failures occur in the case of the so-called **cutaneous-mucocutaneous syndrome** while breathing difficulties due

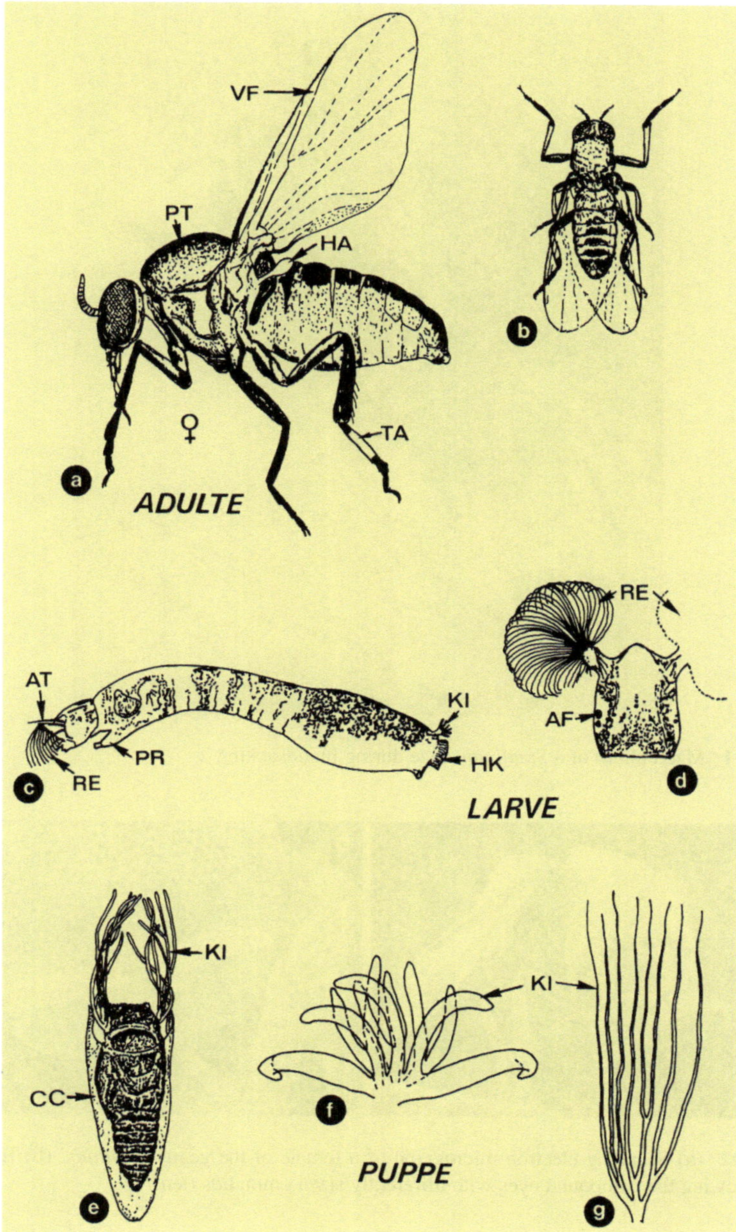

Fig. 6.90 Diagrammatic representation of the developmental stages of the simuliids. (**a**) Adult female of *Simulium damnosum*: lateral. (**b**) Same: dorsal aspect. (**c**) Lateral aspect of a larva. (**d**) Enlargement of the apical food catching fan of the larva. (**e**) Pupa in a cocoon. (**f**) Gills of the pupa of *S. damnosum*. (**g**) Gills of the pupa of *S. neavei*. Adulte = adult stage; Larve = larva; Puppe = pupa. AF = eye spot; AT = antenna; CC = cocoon; HA = haltere; HX = hooks; KI = gills; PR = anterior legs (proleg); PT = prothorax; RE = food catching fan; TA = tarsus; VF = anterior wing

Fig. 6.91 Macrophoto of a *Simulium* stage during bloodsucking

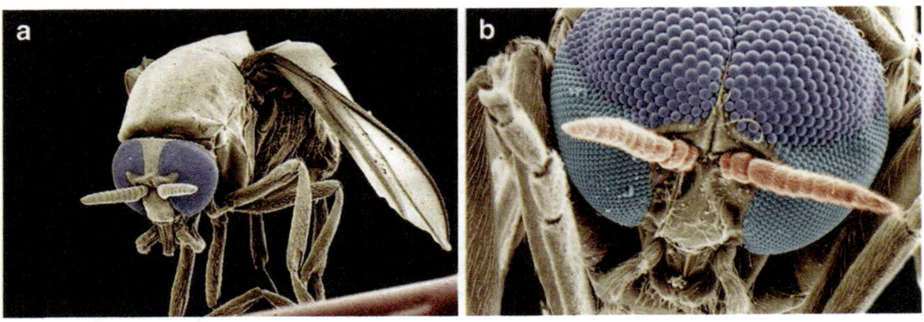

Fig. 6.92 (**a**) Scanning electron micrograph of a female of the genus *Simulium*. (**b**) Head of a male showing the compound eyes with differently sized ommatids (lenses)

to oedemata in the respiratory tract (acute laryngitis, pharyngitis) lead to the so-called **respiratory syndrome** accompanied by circulatory disorders and pareses. Shock reactions due to salvia toxins (in case of the **simuliotoxicosis**) can lead to death within 2–4 h (paralysis of the respiratory centre).

3. **Diagnosis**: Typical bite wounds (Fig. 6.93), detection of females in the nose, the gullet, in between the thighs.

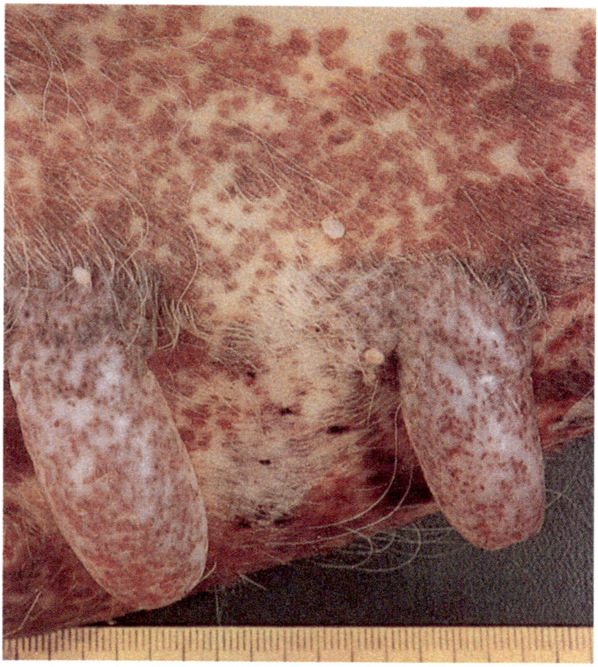

Fig. 6.93 Macrophoto of the udder of a cow with numerous biting sites of simuliids

4. **Pathway of infection**: Attacks of females (often in swarms).
5. **Prophylaxis**:
   (a) Keeping the animals in stables during critical days (humid-warm);
   (b) Control of the black flies' brood in creeks crossing the fields (by use of *Bacterium thuringiensis* formulations or by removing of weeds from the waters).
   (c) Chemoprophylaxis with contact insecticides (Pour-on/spot-on) or repellents.
6. **Incubation period**: Variable, depending on the infestation density; 1 h up to days.
7. **Duration of infestation**: Black flies only suck for a short time.
8. **Therapy**: Extermination of existing insects by spraying contact insecticides. Antiallergenic treatment (locally use of corticosteroids in case of severe cases), support of the circulation, in peracute cases: tracheotomy.

Further Reading

Bernotiene R, Bartkeviciene G (2013) The relationship between water temperature and the development cycle beginning and duration in three black fly species. J Insect Sci 13:1.

Cherairia M et al (2014) Biodiversity and bionomics of the black flies (Diptera: simuliidae) of northeastern Algeria. Zootaxa 3796:166–174.

Ivković M et al (2016) Blackflies (Diptera: Simuliidae) in Croatia: species richness, distribution and relationship to surrounding countries. Zootaxa 4109:16–30.
Volfova V et al (2008) Hyaluronidase of bloodsucking insects and its enhancing effect on leishmania infection in mice. PLoS Negl Trop Dis 2(9):e294.
Zarroug IM et al (2016) Seasonal variation in biting rates of *Simulium damnosum* sensu lato, vector of *Onchocerca volvulus*, in two Sudanese foci. PLoS One 11 (3):e0150309.

(c) **Family Phlebotomidae (Sand flies)**

1. **Name**: Greek: *phleps* = blood vessel; *tomos* = sharp. Latin: *perniciosus* = dangerous, fatal.
2. **Biology/morphology**: The species of the genus ***Phlebotomus*** are mostly only 2.5 mm large and characterized by **dense hair cover at the body** and **wings** (Figs. 6.81 and 6.94). Both genders feed on plant juices. **Females** additionally suck **blood** from different hosts at night. After the copulation and about 1–2 days after a blood meal, the females of most species lay several times 30–50 eggs into humid but loose soil. Approximately 6–12 days later, the larva hatches, moults 4× within 4–6 weeks, feeds on detritus and differentiates finally into the "free pupa". After 6–14 days of the pupal period, the adults hatch during humid nights. They only live for a rather short time (about 14 days) when considering the overall development of about 7 weeks. A number of species transmit the pathogens summarized in Table 6.8. Vectors of special importance are *Phlebotomus perniciosus* (South Europe, Northern France), *P. papatasi* (Asia; Fig. 6.94) and *Lutzomyia* spp. (America). *P. mascittii* occurs in Germany at particularly protected places (e.g. railway tunnels).

Fig. 6.94 Macrophoto of a sand fly of the species *Phlebotomus papatasi*

Further Reading

Battisti JM et al (2015) Colonization of *Lutzomyia verrucarum* and *Lutzomyia longipalpis* sand flies (Diptera: Psychodidae) by *Bartonella bacilliformis*, the etiologic agent of Carrión's disease. PLoS Negl Trop Dis 9(10):e0004128.

Denlinger DS et al (2016) Diagnostic doses and times for *Phlebotomus papatasi* and *Lutzomyia longipalpis* sand flies (Diptera: Psychodidae: Phlebotominae) using the CDC bottle bioassay to assess insecticide resistance. Parasite Vectors 9(1):212.

Gebresilassie A et al (2015) The influence of moonlight and lunar periodicity on the efficacy of CDC light trap in sampling *Phlebotomus* (*Larroussius*) *orientalis* Parrot, 1936 and other *Phlebotomus* sandflies (Diptera: Psychodidae) in Ethiopia. Parasite Vectors 8:106.

Jahanifard E et al (2014) Diversity of sand flies (Diptera, Psychodidae) in southwest Iran with emphasis on synanthropy of *Phlebotomus papatasi* and *Phlebotomus alexandri*. Acta Trop 140:173–180.

Kasap OE et al (2013) The distribution of the *Phlebotomus major* complex (Diptera: Psychodidae) in Turkey. Acta Trop 127:204–211.

Kasap OE et al (2015) Phylogeography of the subgenus *Transphlebotomus* Artemiev with description of two new species, *Phlebotomus anatolicus* n. sp. and *Phlebotomus killicki* n. sp. Infect Genet Evol 34:467–479.

Vieira VR et al (2015) Ecological aspects of phlebotomine sand flies (Diptera, Psychodidae, Phlebotominae) in areas of American cutaneous leishmaniasis, in the municipality of Paraty, Rio de Janeiro, Brazil. I-Index of Abundance by Location and Type of Capture. J Med Entomol 52:886–895.

(d) **Family Ceratopogonidae (Biting midges)**

1. **Name**: Greek: *keras* = horn; *pogan* = beard. Obviously, this name is referred to the haired wings.
2. **Biology/morphology**: These small midges with a length of 1–4 mm have been identified as transmitters of viruses (arthropod borne = **ARBO**) in humans and pets (e.g. it is the pathogen of the African horse sickness and of Bluetongue disease in sheep and cattle, e.g. of the outbreak in Europe 2006–2009, etc.). However, the **filariae species**, which are transmitted to humans by the genus *Culicoides*, are of little medical importance. Adults of the biting midges appear worldwide (especially in wetlands and in the tundra) in high numbers. They also occur in huge quantities and may become an unbearable plague. Only **females** suck **blood** (*Culicoides* species during evening and night hours), whereby they prefer the margins of the clothes in humans and the rims of the eyes, the abdominal walls and the region above the toes in animals. The bites cause an unpleasant burning. The larvae live—depending on the species—in humid soils, at the edge of waters (also brackish and salt water!) or in leaf axils of tropic plants. For their development, the biting midges need several weeks in temperate zones, while 1 week is sufficient in tropical regions (Figs. 6.81 and 6.95). In the last years, the biting midges emerged as vectors of the pathogens of the so-called

Fig. 6.95 Light micrograph of a midge of the species *Culicoides obsoletus*—one of the vectors of the bluetongue virus, which induced in Europe the epidemic during the years 2006–2009

bluetongue disease of farm ruminants like sheep and cattle (while wild animals, e.g. antelopes do not become sick) (Fig. 6.96). Also the so-called Schmallenberg disease (=birth defects in fetus of sheep and cattle) in Europe (Figs. 6.97a–c) had severely hit the farmers in Europe.

Further Reading

Bauer B et al (2009) An appraisal of current and new techniques intended to protect bulls against *Culicoides* and other haematophagous nematocera: the case of Schmergow, Brandenburg, Germany. Parasitol Res 105:359–365.

Brugger K et al (2016) Outdoor and indoor monitoring of livestock-associated *Culicoides* spp. to assess vector-free periods and disease risks. BMC Vet Res 12:88.

Clausen PH et al (2009) Seasonal dynamics of biting midges (Diptera: Ceratopogonidae, Culicoides spp.) on dairy farms of Central Germany during the 2007/2008 epidemic of bluetongue. Parasitol Res 105:381–386.

Mehlhorn H et al (2009) Entomological survey on vectors of Bluetongue virus in Northrhine-Westfalia (Germany) during 2007 and 2008. Parasitol Res 105:321–329.

Mehlhorn H et al (2009) Bluetongue disease in Germany (2007–2008): monitoring of entomological aspects. Parasitol Res 105:313–319.

Steinke S et al (2016) Emergence of *Culicoides obsoletus* group species from farm-associated habitats in Germany. Med Vet Entomol 30:174–184.

6.2.4.2 Horse flies (Suborder: Brachycera; Tabanidae)

1. **Name**: Greek: *brachys* = short; *keras* = horn; *chrysos* = golden; *ops* = appearance; *haima* = blood. Latin: *tabanus* = gadfly; *potare* = to drink.
2. **Biology/morphology**: The members of the family ***Tabanidae*** (horse flies) are insects with a strong body and with a length of up to 30 mm. However, *Chrysops*

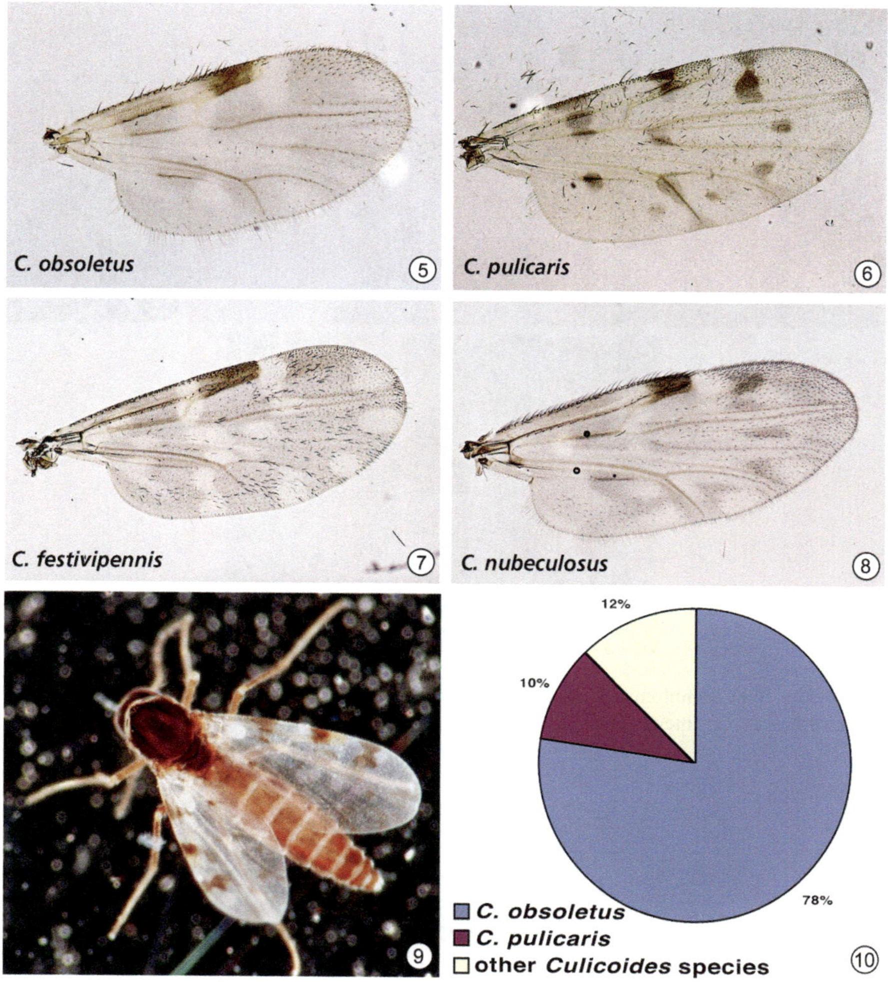

Fig. 6.96 *Culicoides* species are vectors of the bluetongue virus in Europe (2006–2009). **5–8** Wings of European midges. **9** *C. obsoletus*, dorsal aspect. **10** Species distribution caught during epidemics

species, which show a sexual dimorphism of their wings, reach only 1 cm. Only **the females suck blood**. After copulation they deposit about 100–1000 adhesive eggs to stalks of riparian plants. The larvae hatch after 5–7 days, retire into the mud and feed on detritus (*Chrysops*) or live as predators (*Tabanus*). Mostly after nine moults (in a temperature-dependent time, possibly with dormancy), the last larva moves to dry areas of its biotope and pupates within 2–3 weeks before the imago hatches (Figs. 6.98, 6.99, 6.100 and 6.101). The adults detect their hosts visually (they also notice movements from a distance).

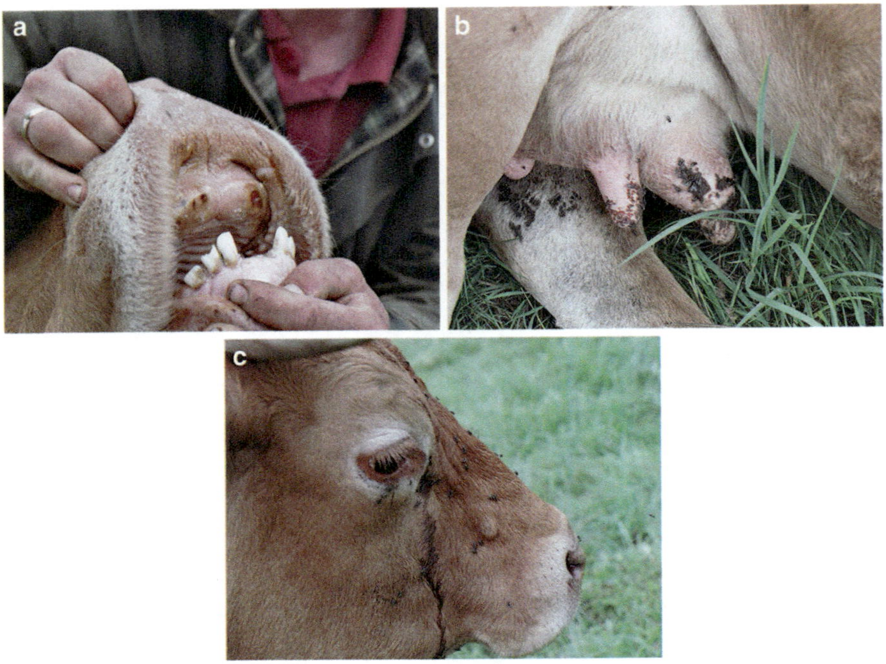

Fig. 6.97 (**a–c**) Symptoms seen in a cow infected by the bluetongue virus (serotype 8) (**a**, **b**) Haemorrhagies; (**c**) Intense tear production

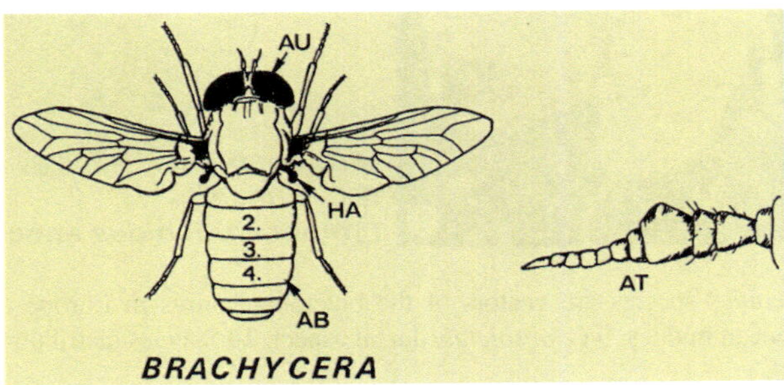

Fig. 6.98 Diagrammatic representation of an adult horse fly and its antenna. AB = abdomen; AT = antenna; AU = eye; HA = haltere

Besides their importance as **nuisance insects**, which in some cases can suck large amounts of blood, their bites can lead to severe skin irritations; *Tabanidae* (especially members of the genus *Chrysops*) are also of importance as vectors (Fig. 6.99). In fact, *Tabanidae* can transmit several pathogens mechanically after interrupted blood meals; however, only four species are known as intermediate hosts of the nematodes *Loa loa*, which parasitize at humans: *Chrysops dimidiatus*, *C. centurionis*, *C. langi and C. silaceus*. Therefore, the disease called **loiasis** is

Fig. 6.99 (**a**) Macrophoto of the head and the mouthparts of the horse fly genus *Chrysops*. (**b**) Dorsal aspect of the same individual

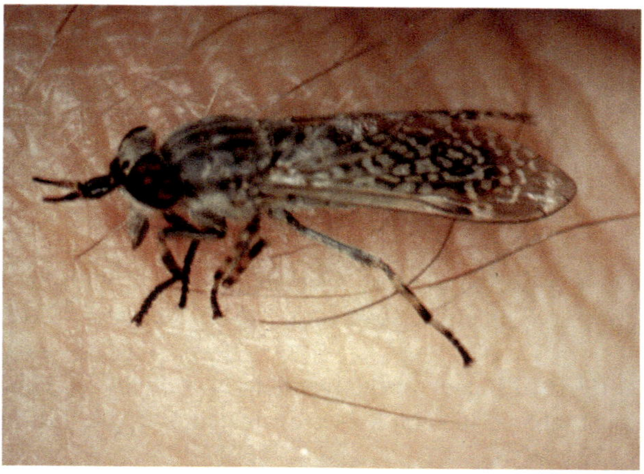

Fig. 6.100 Macrophoto of the so-called rainfly *Haematopota pluvialis*

for unknown reason limited to the habitat of these *Chrysops* species in West and Central Africa and is not found in other, similar climatic zones. *Chrysops caecutiens* and *Haematopota* species, the so-called **clegs**, are widely spread in Europe und become very annoying for humans and cattle especially close to water borders (e.g. the Notch-horned Cleg *H. pluvialis*; Fig. 6.100).

Further Reading

Coelho WM, Bresciani KD (2013) Molecular and parasitological detection of *Leishmania* spp. in a dipteran of the species *Tabanus importunus*. Rev Bras Parasitol Vet 22:605–607.

Iboh CI et al (2012) Occurrence and distribution of *Chrysops* species in Akamkpa community of Cross River State, Nigeria. Pak J Biol Sci 15:1139–1143.

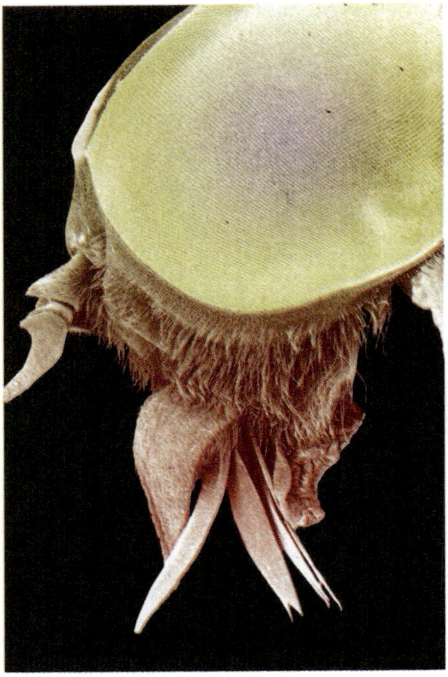

Fig. 6.101 Micrograph of the head of the large horse fly *Tabanus* sp. showing one of its big lateral eyes, the stiff, sword-like mouthparts and the horn-like antenna

Joy JE, Stephens CR (2016) Sensory trichites associated with the food canal of *Chrysops callidus* (Diptera: Tabanidae). J Med Entomol. 2016 Apr 20. pii: tjw015.

Herholz C et al (2016) Efficacy of the repellent N,N-diethyl-3-methyl-benzamide (DEET) against tabanid flies on horses evaluated in a field test in Switzerland. Vet Parasitol 221:64–67.

Ribeiro JM et al (2015) An insight into the sialome of the horse fly, *Tabanus bromius*. Insect Biochem Mol Biol 65:83–90.

Zeegers T, Müller GC (2014) A review of the *Tabanus semiargenteus*-subgroup as part of the *Tabanus bovinus* species-group (Diptera: Tabanidae) with the description of two new species for science. Acta Trop 137:152–160.

6.2.4.3 Flies (Suborder: Circular-Seamed Flies: Cyclorrhapha)

1. **Name**: Greek: *kyklos* = ring, circle; *rhaphis* = needle, crack. The name refers to the circular cut with the help of the rupture of which the adult gets free from the puparium.
2. **Biology/morphology**: This group of flies (Fig. 6.102) includes several families, the members of which are of importance as beneficial destruents, but act also as nuisance insects, as vectors of agents of disease and as important parasites (Table 6.8).

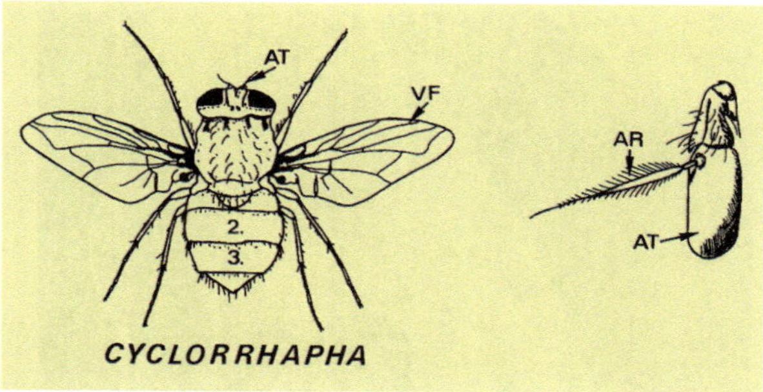

Fig. 6.102 Diagrammatic representation of a typical fly and their antenna. AT = antenna; AR = arista; VF = anterior wing

(a) **Families Muscidae (true flies), Sarcophagidae (flesh flies), Calliphoridae (blow flies)**

1. **Name**: Latin: *musca* = fly. Greek: *kallos* = beautiful; *phorein* = to carry; *sarx* = flesh; *phagein* = to feed.
2. **Examples**: The family of **Muscidae** (true flies) worldwide includes 3900 species, whose adults have licking mouthparts (typical **labellae**; Figs. 6.103, 6.104 and 6.105) or stinging mouthparts, which help to feed on detritus or blood. Both groups became especially important as mechanical transmitters of viruses, bacteria, protozoans or worm eggs (Table 6.8).

The domestic fly ***Musca domestica*** (Figs. 6.81 and 6.103) deposits approx. 1000 eggs onto feces of animals (horses, sheep, cattle) or of humans. From the eggs, the typical maggot-like larvae hatch, which feed on dung but may temporarily live gastrointestinally in different hosts causing the disease called **myiasis**. From a barrel-shaped pupa, the imago hatches through a circular slit in the puparium (**cyclorraph** type). The overall duration of the development ranges (temperature-dependent) between 8 and 50 days.

Actual studies listed more than 100 different microorganisms (e.g. EHEC, *Staphylococcus*) and parasites in or on *Musca* flies, so that there is a high risk of infection in case they visit human or animal food.

The overall developmental period of ***Stomoxys calcitrans*** (horn fly, stable fly; Fig. 6.106) takes about 27–37 days after about 60–100 eggs have been deposited into straw-containing dung by the females, which live approx. for 70 days. Both **males** and **females suck blood** (Fig. 6.107). Due to their frequent change of hosts during repeated blood meals, the adults of *Stomoxys* species are (besides their annoyance and painful bites) of particular importance as mechanical transmitters of pathogens (Table 6.8).

The family **Sarcophagidae** (flesh flies, e.g. *Sarcophaga carnaria*—common flesh fly) comprises about 2000 species worldwide. The females deposit eggs, from which often already during the deposition act the larvae hatch, meaning that they are viviparous. If they are deposited on animals or humans, they are directly able to enter eyes or body openings (e.g. nose, mouth, ears, etc.) or penetrate into

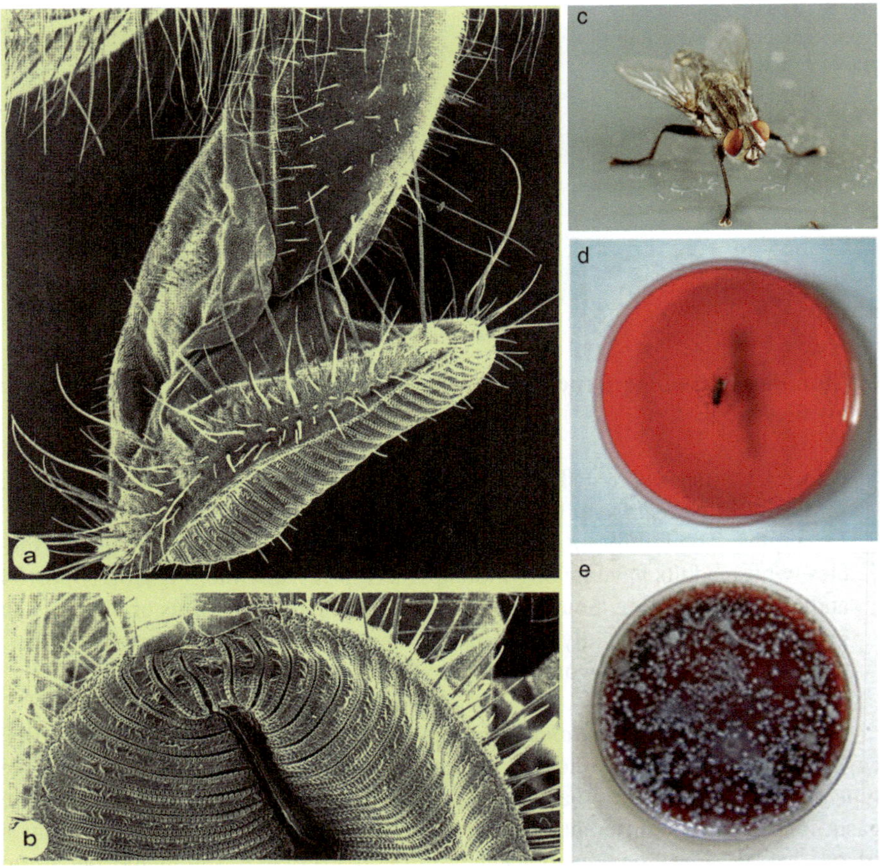

Fig. 6.103 (**a**, **b**) Scanning electron micrographs of the mouthparts of a fly showing the typical labellae used for licking. (**c**) Adult stage of *Musca domestica*. (**d**) *Musca* fly on a fresh agar plate. (**e**) Same agar plate several days after the fly left it, large amounts of apparently transmitted bacteria have grown up

the skin via small wounds. Even if they do not cause a myiasis, they can mechanically transmit a broad spectrum of pathogens from host to host (including humans). The species *S. carnaria*—their females may reach a length of 2 cm, are very common in Europe. Their abdomen shows a chequered black-white alternating pattern.

The family **Calliphoridae** (bottle flies, e.g. *Calliphora vomitoria*—blue bottle fly) comprises more than 1000 species worldwide. *Calliphora vicina* (syn. *erythrocephala)* females reach a length of 8–12 mm and are characterized by their blue-metallic appearing body (Fig. 6.104). They deposit their eggs into carcasses or wounds of animals or humans. The species *C. vomitoria* behaves similarly and reaches a length of 14 mm.

Fig. 6.104 Macrophoto of an adult *Calliphora* fly (=blue flesh fly) and a pupa of this species

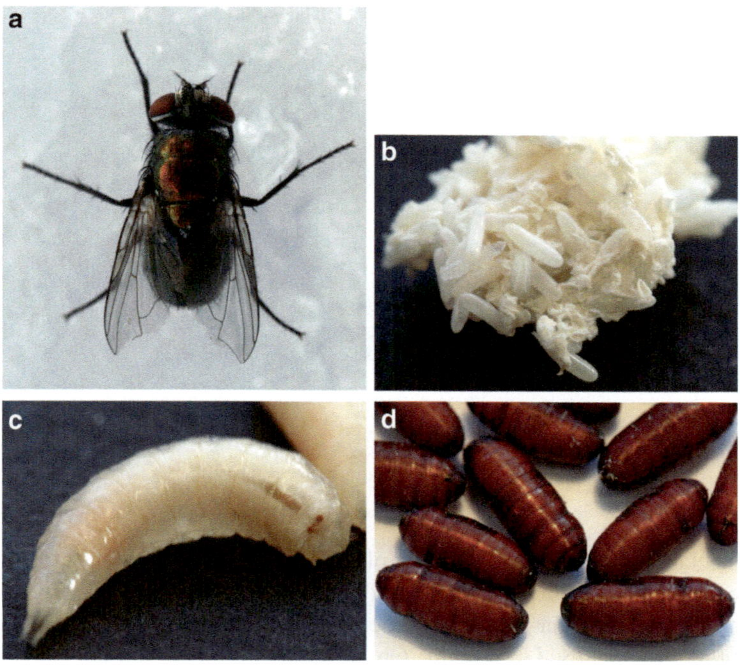

Fig. 6.105 *Lucilia sericata*. Macrophotos of the adult stage (**a**), eggs (**b**), larva (**c**) and pupae (**d**)

The Calliphoridae also comprise the green bottle fly (*Lucilia* spp.). Larvae of the *Lucilia* species (e.g. *L. sericata*, *L. caesar*) are often encountered on feces or putrescent substances but occur also in wounds of animals and humans. The up to 1 cm long females directly lay the larvae onto such substrates, where the further development takes place (Figs. 6.105a–d). The females often suck additionally

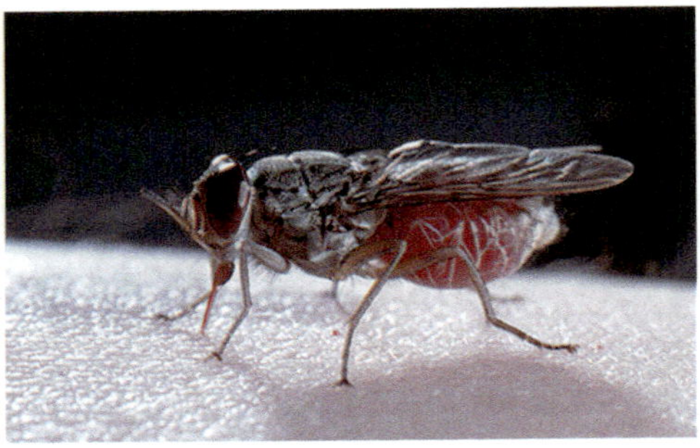

Fig. 6.106 Macrophoto of an adult stage of *Stomoxys calcitrans*

Fig. 6.107 Macrophoto of an adult stage of *Stomoxys calcitrans* from below of a glass plate

flower nectar. Today, *Lucilia* larvae are placed on non-healing wounds to enable the debridement of the wound. The same effect is achieved with the agent Larveel® (developed by Co. Alpha-Biocare, Düsseldorf), which is made of larval extracts of the maggots of the fly *L. sericata*. On wounds is a layer formed that binds at bacteria, thus removing them during change of wound bandage (Figs. 6.108 and 6.109).

(b) **Family: Glossinidae (Tsetse flies)**

1. **Name**: Greek: *glossa* = tongue; *tachinos* = fast. Afrikaans: *tsetse* = flying sound of the flies. Latin: *morsitans* = biting; *palpare* = to palpate; *fuscus* = dark.

Fig. 6.108 Macrophoto of a tsetse fly (genus *Glossina*) sucking blood at a skin surface

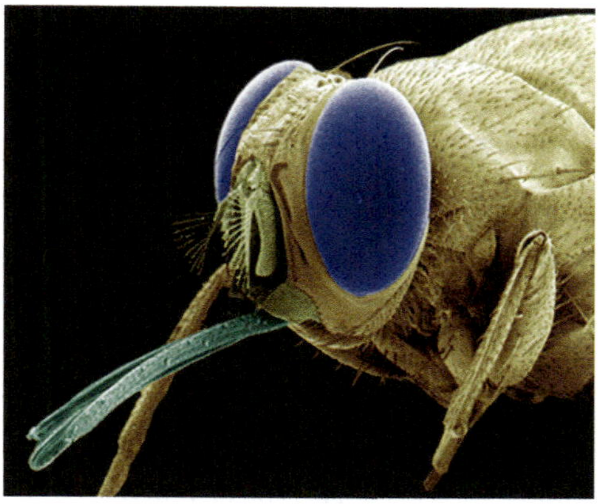

Fig. 6.109 Scanning electron micrograph of the anterior body of a tsetse fly (*Glossina* sp.)

2. **Biology/morphology**: The ***Glossina*** specimens (Fig. 6.109) possess a typical proboscis, which protrudes forward horizontally from the head like a tongue. The smaller species (like *G. tachinoides*) reach a length of about 6–8 mm; the larger species (*G. palpalis*, *G. morsitans*, *G. fusca*) measure about 9–14 mm length. In resting position, *Glossina* specimens are recognized by their wings, which completely overlay (thus appearing **linguiform**, too) and are not held parallel to the body as in the case of other flies (e.g. *Chrysops*). **Both genders suck blood** with the help of their partially rasp-like dentated mouthparts. Like the Simuliidae and the ticks, *Glossina* species are **pool feeders**. By cutting off capillaries of their hosts they cause small haematoma in the skin, which they keep fluid with the help of their saliva containing anticoagulants. As a biological exception the tsetse flies are characterized by an extremely sophisticated brood care. With the help of "mammary glands" (filled by symbionts) within the uterus, **a single larva** is fed and developed until the third larva stage is reached. It is deposited after 8–15 (in average 10–12) days onto a protected place (species-specific biotope!). This larva develops into a pupa within 5–15 h,

from which after 20–35 days the imago hatches. Therefore, the overall holometabolic development of each generation lasts 40–60 days. During their about 90 days lasting adult life, the females lay—instead of hundreds of eggs (like other flies)—only about 8–12 times a differentiated larva (larva 3), which has relative high chances to survive until pupation (Fig. 6.81).

A total of about 19 *Glossina* species are able to transmit trypanosomes to humans and animals. In nine species (including *G. morsitans*, *G. palpalis*, *G. pallidipes*), the pathogens of the human sleeping sickness (*T. brucei rhodesiense*, *T.b. gambiense*) propagate and are developed into infectious (metacyclic) trypomastigote forms. Because of the medical significance for humans and the nowadays again rising importance of the sleeping sickness, different approaches to control *Glossina* flies are taken. However, they all show only temporary a success (due to the peculiar brood care of these species).

(c) **Family: Hippoboscidae (Louse flies)**

A. **Louse flies of cattle and horses**

1. **Name**: Greek: *hippos* = horse, mare; *bos* = cattle.
2. **Biology/morphology**: The **louse flies** received their name since they live in the coat of their hosts just like lice. They have a thick exoskeleton and possess chunky legs with big claws for clinging at hair (Fig. 6.110). The wings are reduced in some species. The horse louse fly *Hippobosca equina* (8 mm long) is winged. Males of *Lipoptena cervi* (deer louse fly, 3–5 mm long) stay winged, while the females skip off the wings after they have reached the host (Fig. 6.111). The misleadingly termed "sheep tick" or "sheep louse" (*Melophagus ovinus*, 5 mm long; Fig. 6.112) has no wings and stays on the host (passage from host to host occurs during body contact with other infested animals). All species can also infest humans. The bite of the louse fly (**males and**

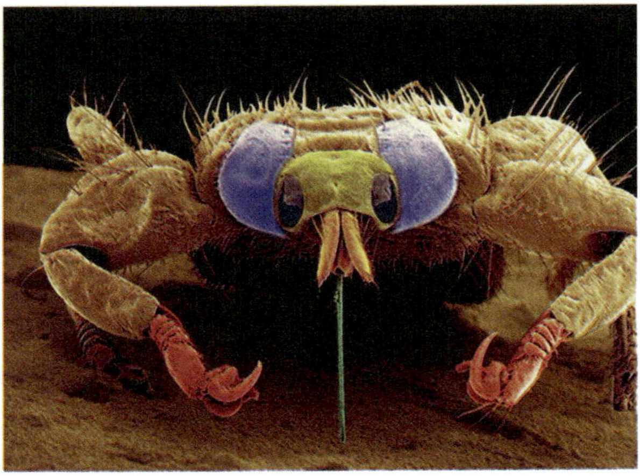

Fig. 6.110 Scanning electron micrograph of the louse fly *Lipoptena cervi*

Fig. 6.111 Macrophoto of the louse fly *Lipoptena* sp., which has thrown off its wing after arriving on a host

Fig. 6.112 Macrophoto of the louse fly of sheep (*Melophagus ovinus*)

Fig. 6.113 Macrophoto of the typical pupa of a louse fly

females suck blood) is painful and leads to severe itchiness, scratching and bacterial secondary infections (inducing weight and wool loss). During their 4–7 months lasting lifetime, the females lay 10–15 larvae 3 individually, which (as in the tsetse fly) pupate within 10 h. The pupae appear barrel shaped, are 3 mm long and are attached at the hairs of the hosts with the help of a glueing surface secret (Fig. 6.113). After the pupal period of about 20–23 days, the adults hatch. They copulate after 3–4 days, if they have had the chance to suck blood.

B. **Louse flies of birds**

1. **Name**: Latin: *lynkeus* = sharp-edged; *pseudo* = wrong.
2. **Geographical distribution/epidemiology**: Worldwide.
3. **Biology/morphology**: *Pseudolynchia* species (including *P. canariensis* = syn. *Lynchia maura*) can be found on many birds. Their stiff, cuneiform body of 9 mm length is finely haired. They possess two 5–7 mm long wings and relatively small compound eyes. The strong, double claws at the powerful legs are characteristic. Louse fly females deposit larvae, which are ready to pupate and normally fall into the nests of their hosts. After the pupal period (max. 1 month), the adults hatch. Both genders feed on blood (the sucking action lasts 5–10 min), and they have a lifespan of about ½ year.
4. **Symptoms of disease**: Restlessness, itchiness, exudates, incrustations with secondary infections, anaemia in case of mass infestation (occasionally death cases).
5. **Diagnosis**: Detection of the adults in the plumage.
6. **Pathway of infection**: Descent of the adult forms or crawling from nests.
7. **Prophylaxis**: Addition of insecticides into dust baths.
8. **Therapy**: Spraying the plumage with contact insecticides. For further indications, see therapy of bird lice.

(d) ***Gasterophilus* species (Horse bot flies)**

1. **Name**: Greek: *gaster* = stomach; *philein* = to adore. Latin: *intestinalis* = belonging to the guts; *equus* = horse.
2. **Geographical distribution/epidemiology**: Worldwide.
3. **Biology/morphology**: Adult flies of the genus ***Gasterophilus*** (most common is *G. intestinalis*; syn. *G. equi*) become up to 1.8 cm long, appear reddish yellow,

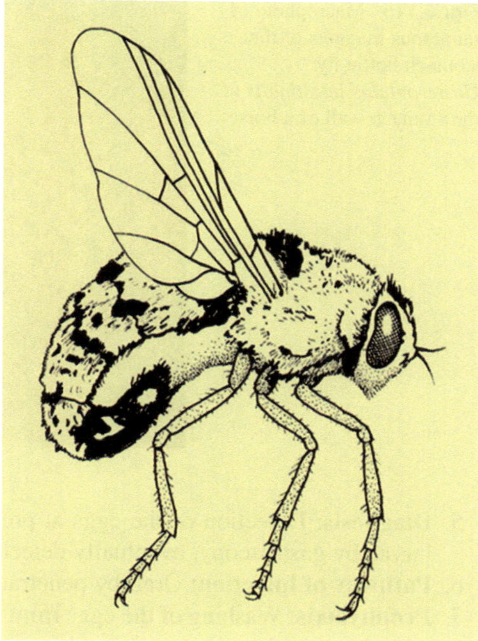

Fig. 6.114 Diagrammatic representation of the female of the stomach bottle fly of horses (*Gasterophilus intestinalis*)

brown black or maroon and do not take up food anymore as adults. The females (Fig. 6.114) directly lay their glueing, capped eggs, which in case of *G. intestinalis* appear light yellow and become about 2 mm long, on different body parts (species specific: mouth, nostrils, legs, shoulder, flanks, etc.) of horses or on specific food plants. The larvae hatch within 5–10 days and penetrate into the mucosa of the tongue of their hosts. Therefore, they either actively migrate into the mouth cavity or get there by licking at the itching coat caused by the larval movements. After a migration within the tongue (21–28 days), the larvae moult and reach the second stage. After being swallowed by the host, they attach at a species-specific region of the intestinal system (stomach and/or duodenum) where they anchor at the mucosa with the help of their two mouth clamps and feed on it. They become up to 2 cm long, appear reddish or buff and feature 11 rings, which are covered with species-specific clamps (Fig. 6.115). After 8–12 months, the larva 3 gets excreted within the host's feces and pupates onto the ground in a very short time. After 3–8 weeks, the adult fly hatches. Main flight time of *G. intestinalis*: July and August (in moderate climates).

4. **Symptoms of disease**: Inflammations of the oral mucosa; in case of mass infestation: gastritis, digestive disorders (in autumn), emaciation, anaemia, faintness, colics. In the case of severe infestations (up to 1000 larvae), also cases of death may occur.

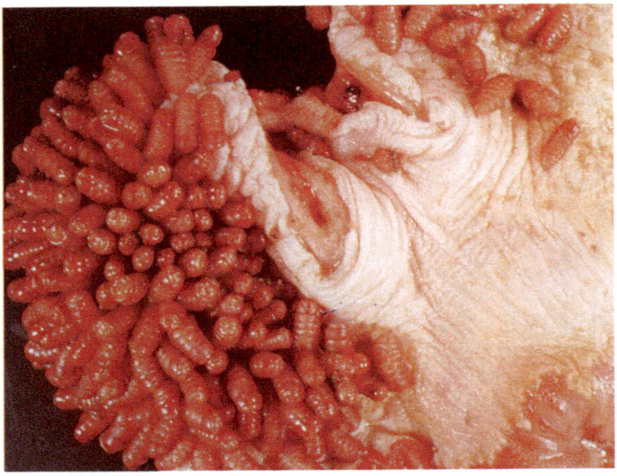

Fig. 6.115 Macrophoto of numerous maggots of the stomach bottle fly *Gasterophilus intestinalis* at the stomach wall of a horse

5. **Diagnosis**: Detection of the eggs at predilection sites, direct detection of the larvae by gastroscopy, eventually detection of dead larvae in the feces.
6. **Pathway of infection**: Oral, by penetration of the larvae 1.
7. **Prophylaxis**: Washing of the eggs from the fur with warm water supplemented with insecticides. Consequent chemotherapy of infested animals (1 up to 2× per year): about 4 weeks after the first frost (when adults do not longer fly) and close to the end of winter to keep the next generation of flies low.

   Control of approaching bot flies with repellents is only successful for hours. Insecticides last longer. Application of MiteStop® keeps insects away for 4–7 days.
8. **Incubation period**: Symptoms in the mouth area occur after 3–4 days; in the intestinal tract it takes 3–4 weeks.
9. **Prepatency**: Larva excretion after about 8 and more months p.i.
10. **Patency**: Infestation lasts 8–12 months.
11. **Therapy**: Treatment of the larvae 2 and 3 within the intestinal tract during autumn with **Metrifonate** (Neguvon®: 30–35 mg/kg body weight, orally), with **Metrifonate** combinations with e.g. **Febantel** = Rintal Plus® or **Mebendazol** = Telmin® Plus, **Dichlorvos** = Equigard® or with Ivermectin (0.2 mg/kg body weight, all orally).

(e) **Bot flies of equids**

1. **Name**: Greek: *rhis*, *rhinois* = nose; *oistros* = biting fly; *hypo* = below; *derma* = skin. Latin: *purpureus* = reddish blue.
2. **Geographical distribution/epidemiology**: Worldwide.
3. **Biology/morphology**:
   - ***Rhinoestrus purpureus***: Nasal bot fly. The larvae 3 appear yellowish and possess thorn fields at the anterior end of the segments; the mouth clamps are very strong and settle in the nasopharynx region and rarely in the eyes.

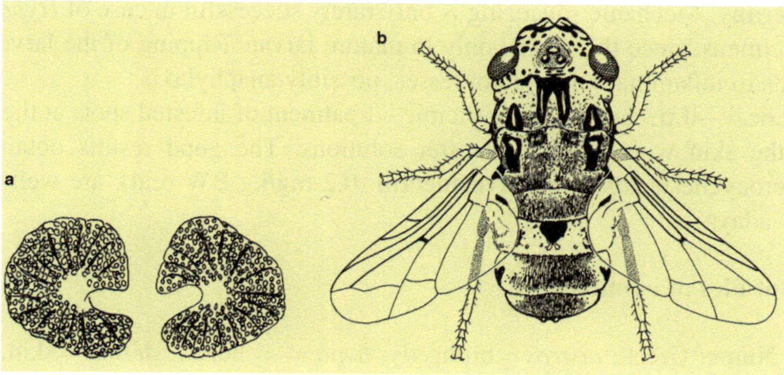

Fig. 6.116 Diagrammatic representation of the adult nose bottle fly *Rhinoestrus purpureus* (**b**) and of the specific abdominal stigma (breathing) plates (**a**)

Up to 700 larvae 1 of this nasal bot fly are dropped "bomb-like" by the female flies at the choanae of the hosts (also at eyelids). After growing up (by feeding at respective skin areas), the L3 is expectorated, thus leaving the host. After a temperature-dependent pupal period (2–5 weeks), the adults hatch (Fig. 6.116).

- ***Hypoderma* species**: Their upper side is equipped with thorns (except for the 10th segment!). They possess kidney-shaped stigma plates, with a bell-mouthed channel. However, larvae do not get ready to pupate.

4. **Symptoms of disease**:
   - *Rhinoestrus*: In case of eye infestation: conjunctivitis; coughing, sneezing spell, inflammations, swellings of the lymph nodes, emaciation, morbidity by sepsis in case of severe infestation of the nasopharyngeal zone.
   - *Hypoderma*: Purulent swellings, inflammations.

5. **Diagnosis**:
   - *Rhinoestrus*: Detection of the larvae in the rinsing fluids of the nose or eyes.
   - *Hypoderma*: Extraction of the larvae from bumps in the skin.

6. **Pathway of infection**: The female deposits larvae (*Rhinoestrus*) or eggs (*Hypoderma*) on the eyes or onto the fur/coat.
7. **Prophylaxis**: Application of repellents (e.g. Repellan® 1× daily: Wellcare-Emulsion every 10–14 days), MiteStop® every 4–7 days.
8. **Incubation period**: 4–7 days.
9. **Prepatency**: Already after 1–2 weeks, small skin bumps may be observed in the case of *Hypoderma* infestation.
10. **Patency**:
    - *Rhinoestrus*: August/September till April/June
    - *Hypoderma*: 8–10 weeks after the invasion. Deposition of the eggs from May till September.

11. **Therapy**: Mechanic squeezing is only rarely successful in case of *Hypoderma* specimens, since this works only in mature larvae. Ripping of the larval body leads to inflammations and abscesses, possibly anaphylaxis.

    Local—if necessary also systemic—treatment of infested spots at the eye or in the skin with phosphoric ester solutions. The good results obtained by macrocyclic lactones, e.g. **Ivermectin** (0.2 mg/kg BW oral), are well-known nowadays.

(f) **Bot flies of ruminants**

1. **Name**: Greek: *oistros* = biting fly; *hypo* = = below; *derma* = skin. Latin: *lineatus* = brindled; *bos* = cattle; *ovis* = sheep.
2. **Geographical distribution/epidemiology**: Worldwide.
3. **Biology/morphology**: Only the larvae of the bot flies are of importance:
   - **Nose bot flies**: *Oestrus ovis* (sheep bot fly). Adult females (Fig. 6.117), about 1 cm long, throw their 500–600 larvae 1 towards the nose (eyes) of sheep (but also of humans!). After attaching to the skin, the L1 enters into the superior nasal meatus, where they overwinter and feed on mucosa cells (with the help of their typical mouth clamps). In spring, it moults to the L2 and L3. Finally, the larva reaches the ground dropping down within nasal mucus and pupates there. This development requires about one month, so that mostly only one fly generation occurs per summer.
   - **Warble flies**: *Hypoderma bovis* (common cattle grub, up to 15 mm), *H. lineatum* (northern cattle grub, up to 13 mm) (Figs. 6.117, 6.118, 6.119 and 6.120).

     After deposition of the eggs by the females, the larvae 1 develop and hatch within 4–6 days from the egg. They burrow into the skin and carry out an extensive migration within the body (also passing internal

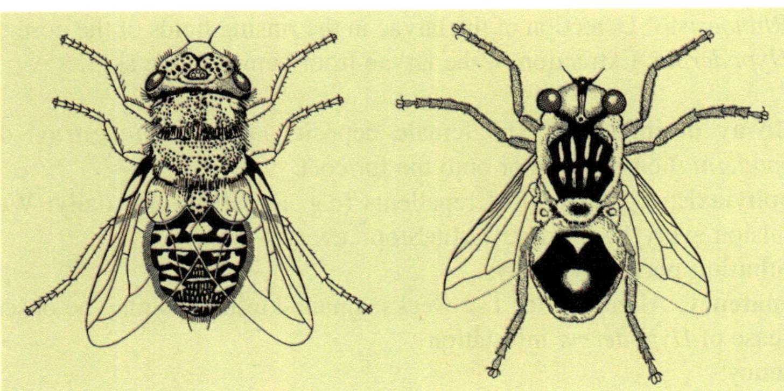

Fig. 6.117 Diagrammatic representation of adult flies, the larvae of which induce myiasis. *Left*: *Oestrus ovis*; *right*: *Hypoderma lineatum*

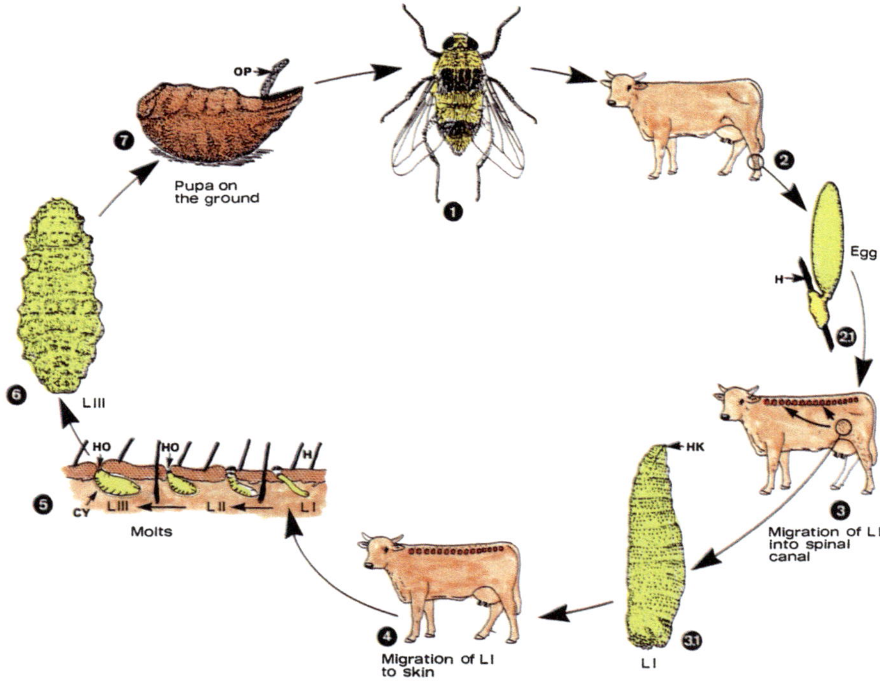

Fig. 6.118 Life cycle of *Hypoderma bovis* (warble fly). **1**, **2** Adult females lay several hundred eggs which are singly deposited and become attached to the host's hair (**2.1**) by an attachment organ. The first-stage larva hatches from the egg in about 4 days, crawls down the hair and penetrates the skin. **3** The exact route taken by migrating larvae in the host is not known, but after several months they reach the final site, the epidural fat of the spinal canal (**3.1**). **4–6** At the beginning of the year, the L_1 leaves the spinal canal and moves to the final site on the back; this is an area of 25 cm on either side of the midline from shoulder to tail and is where the cysts (=warbles) are formed from March to July. The L_1 (**3.1**) measures about 10 mm in length just after arrival; it moults into the second-stage larva soon after reaching the skin and cuts a hole (HO) in it through which it respires by means of paired terminal spiracles. After moulting the third larval instar appears and then develops into the 30-mm-long prepupa (**6**). After several weeks, the yellowish-brown prepupa forces its way through the skin's opening, drops to the ground and moves actively seeking shelter. **7** The pupal stage needs 3–10 weeks depending upon external conditions (*H. lineatum* needs 4 weeks or less). Adults of *Hypoderma bovis* appear from June to mid-September, are unable to feed and live for only 3–5 days; they emerge early in the day and mate within 1 h. CY = cyst or warble; H = hair; HK = larval mouth hooks; HO = hole in the skin; OP = opening (for emergence) in the pupal cocoon

organs), until they finally as L2 and L3 (skin larvae) cause the so-called warbles (Figs. 6.119 and 6.120) (a host reaction).

4. **Symptoms of disease (Myiasis)**:
   - **Nose bot flies**: Sneezing, coughing, rhinorrhoea, lacrimation (in case of eye infestation). Itchiness causes vigorous spinning movements of the head ("gid"); central nervous disorders caused by secondary infections

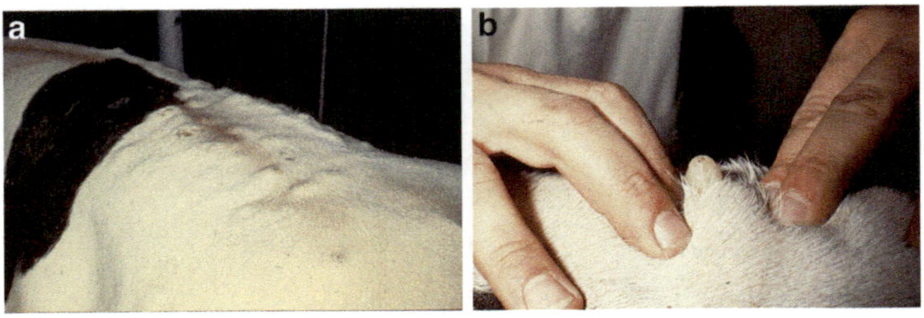

Fig. 6.119 (**a**) Protrusions at the skin of a cow containing *Hypoderma bovis* larvae. (**b**) Squeezing off such a *H. bovis* larva

Fig. 6.120 Larva 3 of a warble fly showing the dark appearing mouth hooks

lead to impaired movement coordination ("turning sickness"). Severe infestation causes weight and wool loss; secondary infections may lead to mortality.

– **Warble flies** induce the so-called **hypodermosis**: Landing of the female flies on the host's skin leads to restlessness of the hosts and eventually to escape reflexes inducing injury risk at fences, etc.

Damages caused by larvae:

Larvae 1: During their migration from the skin to the vertebral canal, they cause damages, bleedings, oedematous alterations and eosinophilic infiltrations. The burrows can be filled with purulent exudates by secondary infections. **Caution**: Larvae may rest in the spinal cord region. If they are killed there by treatment, neurological disorders and eventually paralysis may arise.

Larvae 2, 3: Larvae, which have returned into the skin, cause encapsulated bumps reaching the size of a golf ball (so-called warbles, Fig. 6.119). They show at the outside a so-called spiracle, within which also bacterial pathogens (despite the antibacterial excretions of the larvae) can infiltrate and cause abscesses. Severe infestation eventually causes sepsis

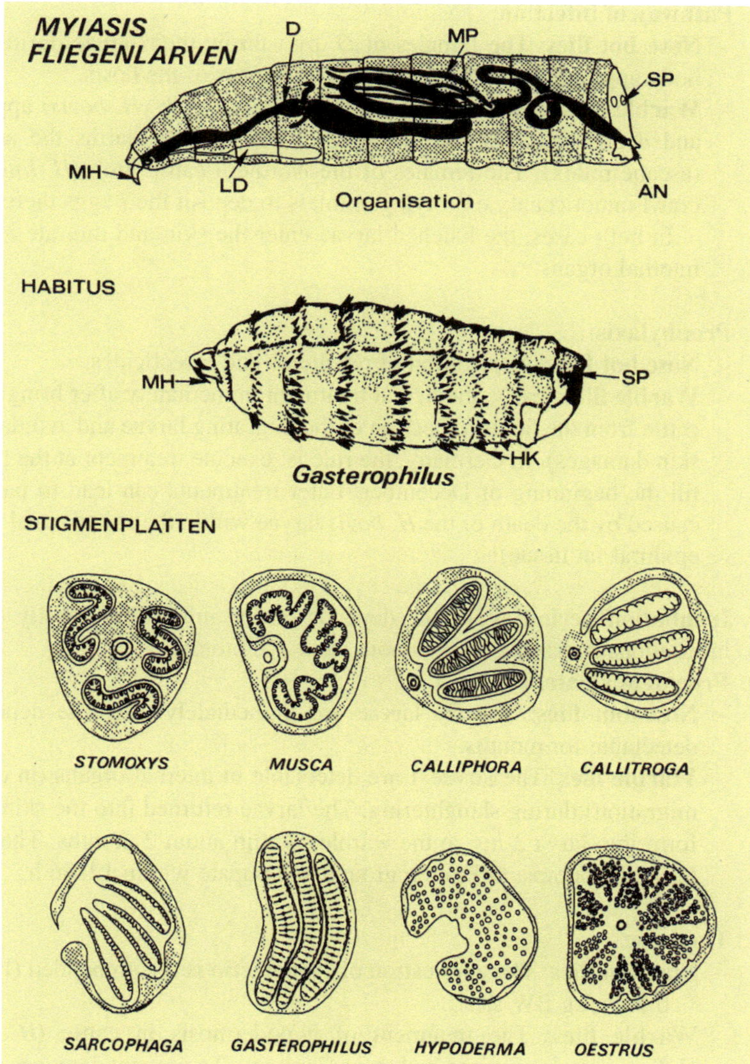

Fig. 6.121 Diagrammatic representation of the characteristics of various fly larvae. Two of the stigma plates (spiracles) shown below the fully represented larva occur always at the posterior end of the larva. AN = anus; D = intestine; HK = hooks; LD = labial glands; MH = mouth hooks; MP = Malpighi ductules; ST = stigma plate

with lethal consequences and/or may induce the total loss of the coat for leather production.

5. **Diagnosis**: The larvae taken from the nose or skin (**bumps**) can be identified based on the structure of the stigma (Fig. 6.121). The paired stigma plates lay at the posterior end of the larvae.

6. **Pathway of infection**:
   - **Nose bot flies**: The females of *O. ovis* throw their larvae during their body approach onto the nose and eye regions of the hosts.
   - **Warble flies**: Females of the common cattle grub (*H. bovis*) approach and deposit the eggs in the hairs. Their descent alarms the animals (escape reflex). The females of the Northern cattle grub (*H. lineatum*) crawl unnoticeably onto lying animals to deposit their eggs there.

     In both cases, the hatched larvae enter the skin and migrate into the internal organs.

7. **Prophylaxis**:
   - **Nose bot flies**: Application of repellents and insecticides.
   - **Warble flies**: Early, medicinal treatment immediately after bringing the cattle from the fields (detection of the migrating larvae and avoidance of skin damages). In Germany, the rule is: execute treatment at the longest till the beginning of December. Later treatments can lead to paralysis caused by the death of the *H. bovis* larvae within the spinal canal (in the epidural fat tissue)!

8. **Incubation period**: Variable, depends on the infestation density (minor infestations possibly occur without clear symptoms).

9./10. **Prepatency/patency**:
   - **Nose bot flies**: The fly larvae are immediately after the deposition detectable for months.
   - **Warble flies**: The larvae 1 are detectable in internal organs (in case of migration) during slaughtering. The larvae returned into the skin transform into larva 3 inside the warbles within about 2 months. Then they leave the bumps, fall to the ground and pupate within 12–36 h.

11. **Therapy**:
   - **Nose bot flies**: The application of **Ivermectin** is recommended (Ivomec ®, 0.2 mg/kg BW s.c.).
   - **Warble flies**: The treatment of hypodermosis in cattle (*H. bovis*; *H. lineatum*) is required by law in Germany. To prevent damages caused by the larvae, the treatment should be carried out during autumn (till the end of November: see prophylaxis). In case of treatment in spring, financial losses have to be taken into account. **Caution**: No treatment from the beginning of December till the end of March (migrating larvae in the spinal canal can cause paralysis or death).

     Treatment in autumn, e.g., with **ivermectin** or **doramectin**. Do not use in the case of lactating animals (withholding period: see eatable tissues). Simultaneous effects against stomach, gut and lung worms (adults and 4. larval stages are killed).

     Treatment in spring only from the end of March/beginning of April immediately before driving up of the cattle. Treatment e.g. with

Fenthion should be done along the spine of the animal (pour-on method). This agent is also suitable to be used after the swarm period of the warble flies (early treatment) for treatment against migrating larvae. **Caution**: Pay attention to obligatory withholding period.

Further Reading

Dominghetti TF et al (2015) *Stomoxys calcitrans* (Diptera: Muscidae) outbreaks: current situation and future outlook with emphasis on Brazil. Rev Bras Parasitol Vet 24:387–395.

Fu Y et al (2016) Genetic diversity and population genetics of the warble flies *Hypoderma bovis* and *H. sinense* in Qinghai Province, China. Parasite Vectors 9:145

Hilali MA et al (2014) Monthly variations of *Rhinoestrus* spp. (Diptera: Oestridae) larvae infesting donkeys in Egypt: Morphological and molecular identification of third stage larvae. J Adv Res 6:1015–1021.

Hoppenheit A et al (2014) Analysis of *Glossina palpalis gambiensis* and *Glossina tachinoides* from two distant locations in Burkina Faso using MALDI TOF MS. Parasitol Res 113:723–726.

Jafari Shoorijeh S et al (2007) Comparative efficacy of pour-on and subcutaneous injection of ivermectin on *Melophagus ovinus* (L.) in Darab ecotype goats of Southern Iran. Vet Parasitol 148:179–183.

Klong-Klaew T et al (2012) Observations on morphology of immature *Lucilia porphyrina* (Diptera: Calliphoridae), a fly species of forensic importance. Parasitol Res 111:1965–1975.

Krcmar S et al (2006) Response of Tabanidae (Diptera) to different natural attractants. J Vector Ecol 31:262–265.

Liu SH et al (2016) The incidence and species composition of *Gasterophilus* (Diptera, Gasterophilidae) causing equine myiasis in northern Xinjiang, China. Vet Parasitol 217:36–38.

Müller GC et al (2012) An annotated checklist of the horse flies (Diptera: Tabanidae) of Lebanon with remarks on ecology and zoogeography: Pangoniinae and Chrysopsinae. J Vector Ecol 37:216–220.

Mysterud A et al (2016) Phenology of deer ked (*Lipoptena cervi*) host-seeking flight activity and its relationship with prevailing autumn weather. Parasite Vectors 9:95

Niederegger S et al (2016) Muscle attachment site (MAS) patterns for species determination in five species of *Sarcophaga* (Diptera: Sarcophagidae). Parasitol Res 115:241–247.

Phasuk J et al (2016) A comparison of attractants for sampling *Stomoxys calcitrans* (Diptera: Muscidae) on dairy farms in Saraburi Province, Thailand. J Econ Entomol 109:942–946.

Ramesh A et al (2016) The impact of climate on the abundance of *Musca sorbens*, the vector of trachoma. Parasite Vectors 9:48.

Rehbein S et al (2013) The efficacy of eprinomectin extended-release injection against *Hypoderma* spp. (Diptera: Oestridae) in cattle. Vet Parasitol 192:353–358.

Rudolf I et al (2016) Molecular survey of arthropod-borne pathogens in sheep keds (*Melophagus ovinus*), Central Europe. Parasitol Res. 2016 Jun 20. [Epub ahead of print]

Sokół R, Michalski MM (2015) Occurrence of *Hippobosca equina* in Polish primitive horses during the grazing season. Ann Parasitol 61:119–124.

Suh SJ et al (2012) Louse flies (Diptera: Hippoboscidae: Ornithomyinae) of the Republic of Korea: an updated checklist, including two new records of bird louse flies. J Med Entomol 49:231–236.

Sukontason KL et al (2014) *Sarcophaga* (*Liosarcophaga*) *dux* (Diptera: Sarcophagidae): A flesh fly species of medical importance. Biol Res 47:14–17.

Szpila K et al (2015) Flesh flies (Diptera: Sarcophagidae) colonising large carcasses in Central Europe. Parasitol Res114:2341–2348.

Wamwiri FN, Changasi RE (2016) Tsetse flies (*Glossina*) as vectors of human African trypanosomiasis: A Review. Biomed Res Int 2016:6201350.

Zeegers T, Müller GC (2014) A review of the *Tabanus semiargenteus*-subgroup as part of the *Tabanus bovinus* species-group (Diptera: Tabanidae) with the description of two new species for science. Acta Trop 137:152–160.

6.3 Crustaceans (Crustacea)

The Crustacea (=**shellfishes**) are named after their cuticle serving as exoskeleton, which—in contrast to those of insects—besides chitin and pigments also contains chalk in higher amounts. Therefore, growth can—like in insects—only take place immediately after one of the several hormone-steered (ecdysone) moults, as long as the new, epidermis-built cuticle is still soft and thus elastic. The different groups of the heteronomously segmented crustaceans are quite diverse in their morphology. However, the lower (**Entomostraca**) and the higher (**Malacostraca**) crustaceans can be distinguished from each other by the outer appearace of the adults as well as when looking at their larvae (**nauplius** or **zoea**). Most of these species live aquatically (in freshwater, saltwater) breathing via **gills**. They possess in contrast to the insects **2 pairs of antennae**. As in insects, the mouthparts are composed of one pair of mandibles and two pairs of maxillae. In most crustaceans, each of the heteronomous segments shows one pair of extremities, which are developed at the head as **maxillipeds**, at the thorax as walking limbs (**pereiopods**; in decapods five pairs, often with scissors) and at the abdomen as swimmerets or copulating organs (**pleopods**). In the case of the segmented lever extremities, it can mostly be differentiated into an **endopodite** and an **exopodite**, whereby during evolution the exopodite developed into a cheliped and the exopodite into an oar-like leg. As in insects, the ontogenesis of crustaceans starts with a fertilized, centrolecithal egg and develops via a superficial sulcus to the larvae (nauplius, zoea), which in the course of several moults more and more resemble the dioecious adults. In almost all groups of the crustaceans parasites occur, whose partially significant morphologic specialization provokes zoological interest. Other forms serve as intermediate hosts for endoparasites of humans. Hence, e.g. the copepods may carry the larvae of the fish tapeworm *Diphyllobothrium latum* or those of the so-called Guinea worm *Dracunculus medinensis*. Owing to their mode of life in the water, such crustaceans mostly parasitize in or on fishes or in other crustaceans (but also in mammals!) and therefore they can especially in fish farms cause severe losses. Below, a few forms are selected that are of special importance for humans.

System
Phylum: Arthropoda (extract)
Subphylum: Branchiata
Class: Crustacea
Subclass: Ostracoda (seed shrimps; crustacean with bivalve shell, e.g. gill parasites of fishes)
Subclass: Copepoda (intermediate hosts for endoparasites; some parasitizing forms as ectoparasites in fishes)
Subclass: Branchiura (fish lice)
Order: Arguloida (carp lice)
Subclass: Cirripedia (barnacles; many endoparasites of sea crabs)
Subclass: Malacostraca (higher crustacean)
Order: Amphipoda (freshwater shrimps; free-living)
Order: Isopoda (wood lice; some ectoparasites in fishes)
Order: Decapoda (ten-footed crabs; typical crustaceans, predators)

***Salmincola* species**. These parasites belonging to the Copepoda anchor already as freshly hatched nauplius larvae at the surface of salmons (salmonids, including trouts) with the help of their antennae and mouthparts, which are transformed into strong hooks and claws. They stay lifelong on the same fish, lose any segmentation during the moults and appear as bulky sacks after the final anchoring. The 3–4 mm sized females stay anchored tightly and feed on the epidermis (similar to Fig. 6.123), while males sized about 1 mm migrate along the surface of the fishes on the search for females. In the case of mass infestation, the fishes often die.

Lernaea* species**. They likewise belong to the Copepoda. As a consequence of their parasitism, they change their shape even more intensively and appear as tumour-like proliferations in infested animals. They mostly need a fish as intermediate host and enter in the definitive host (via the respiratory system) several larger blood vessels. Thus, ***Pennella balaenopterae is able to reach inside such large vessels a length of up to 30 cm in whales. Related *Lernaeocera* or *Phrixocephalus* species cause severe losses in fishing farms (up to 30 % of the offspring may die). *Lernaea cyprinacea* (the so-called anchor worm) may reach a length of about 25 mm (Fig. 6.122). After anchoring, they dedifferentiate totally and appear baggy and finally anchor deeply in the musculature or in the gills of different fishes. Only their two egg bags remind of the relationship to Copepoda. As a consequence of the deep invasion into the skin, often severe secondary infections occur—also in

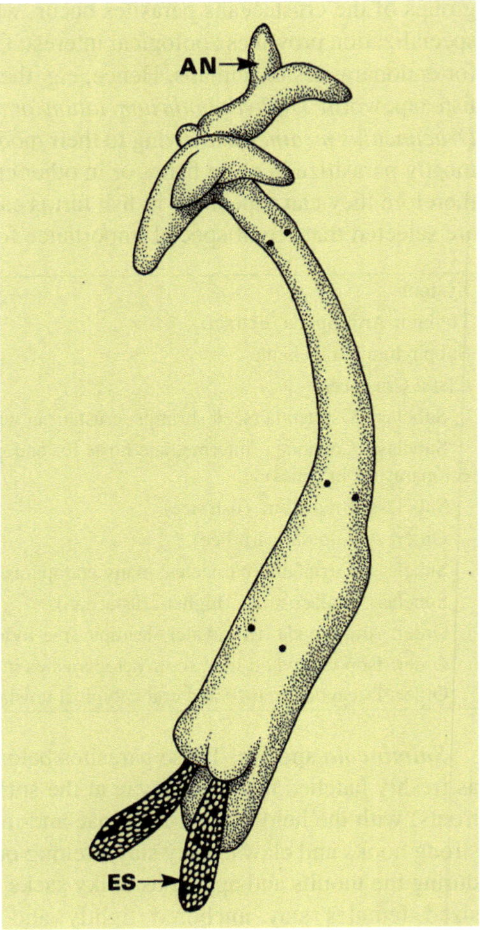

Fig. 6.122 Diagrammatic representation of a so-called "anchor worm" (*Lernaea* sp.). AN = anchor organ; EI = egg sack

internal organs—so that this species often is the cause of mass killing of fishes (especially in the tropics).

***Ergasilus* species**. These at the maximum 3 mm long parasites of freshwater and saltwater fishes have kept the typical Copepoda appearance (Fig. 6.123); only the two antennae are transformed into strong holding clamps. Hereby they stay anchored at the gills of their host for their lifetime and feed on epithelia. The males live free in the water, where also the copulation takes place. Thereafter, the males die and the females change into the parasitic lifestyle. From the deposited eggs, the **nauplius** larva hatches after 10–12 days. Via four further larval stages (so-called **copepodit** stages) and respective moults, they become mature within 10–12 days. In case of mass infestations of the gills, mucous formation, emaciation and often death occur.

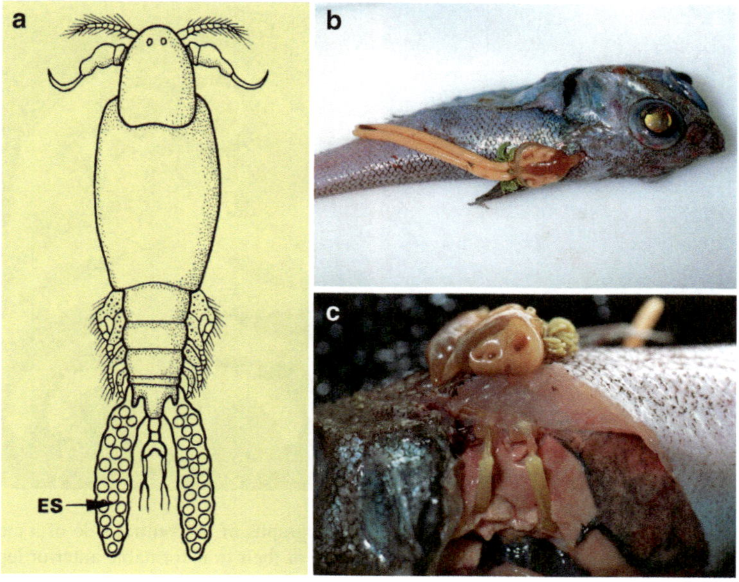

Fig. 6.123 (**a**) Diagrammatic representation of an *Ergasilus* copepod; (**b**, **c**) Parasitic copepod outside of a fish (**b**) and inside (**c**) showing that it is entered deeply into the body. ES = egg sack

Carp lice. These parasites belong to the subclass **Branchiura** (=fish lice) and comprise relatively few species but are of high deleterious importance for fish farming.

The species of the genus *Argulus* are spread worldwide (in freshwater and saltwater), reach a maximal length of 6–22 mm (Fig. 6.124) and live temporarily as ectoparasites (*A. coregoni* on salmons; *A. japonicus* on carps; *A. foliaceus* on carps, pikes, perches, etc.).

The females are able to swim and deposit their eggs onto the ground. They do not give rise to a nauplius larva but directly to a subadult stage, which already resembles the adult. Typical features of carp lice are their antennae, which are transformed into ventral hooks, and their maxillae, which are transformed into enormous suckers (Fig. 6.124). With the help of both systems, they may become attached to the host. Moreover, the extreme flattening of the carapace allows a low flow resistance during their sucking on the swimming fish. Their **ingestion** of food is supported by a stiletto located at front of the mouth (Fig. 6.124b). Extraintestinally digested epidermis is ingested preferably but often also blood. Many fish species can be infested, so that fish farms are highly endangered. Thus, up to 4000 *A. foliaceus* had been counted on a 28 cm long tench. The deleterious damage is not directly caused by the blood loss but by nidation of bacteria or fungi within the wounds. *Argulus* species also transmit the pathogens of the dangerous ascites of fishes (*Pseudomonas punctata*). In addition, *Argulus* also transmits *Rhabdovirus carpio*, which is the pathogen of the spring viremia of carp. This

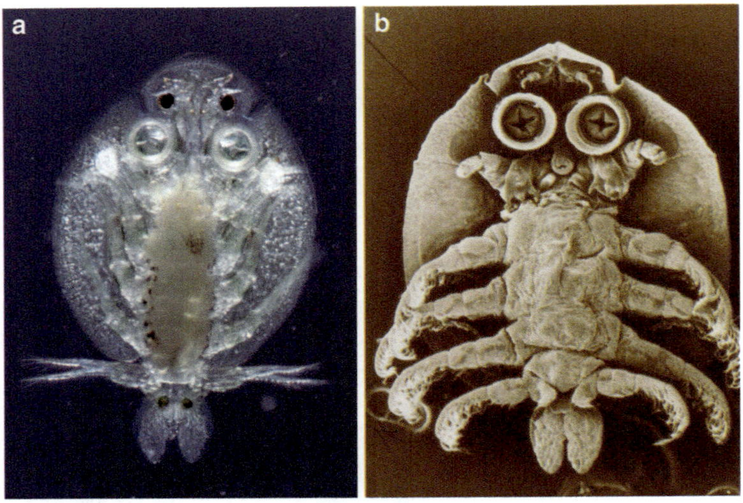

Fig. 6.124 Light (**a**) and scanning electron (**b**) micrographs of the ventral side of a carp louse, which becomes firmly anchored at a fish with the help of their transformable anterior legs

occurs in case of contamination of the mouth stiletto—especially by frequent changes of the host. The bacterium *Aeromonas salmonicida*, which causes carp erythrodermatitis, is also mechanically transmitted.

The **salmon** or **sea lice** (*Lepeophtheirus salmonis*) do not belong to the genus *Argulus* but to the subclass *Branchiura* and are Copepoda just like the directly related *Caligus* species. Both genera have in common that not only females but also males live as parasites and can only hardly be distinguished from females by size and shape. From the deposited eggs a **nauplius** larva hatches, which moults once (Fig. 6.125). The **second nauplius** stage moults into the first **copepodit**, which either has to find a host or dies. After anchoring, it grows into a so-called **chalimus** stage. Three further chalimus and two **preadult stages** follow. While all **chalimus stages** are still attached with their so-called front filament, the preadults and also the adults are able to move freely on the fish surface. In some cases, they also leave the fishes (often during treatment) and return later. This behaviour makes a control of the salmon lice very difficult in the case of rearing ponds in the fjords of Norway or along the Faroe Islands.

***Sacculina* species**. These species belong to the Cirripedia and transform their body into a rhizoid mass, which proliferates within the body of the crab and thus may lead to death.

Isopods. Many species of the isopods suck as larvae and/or adults blood at the gills of fishes. The blood loss and the injuries cause severe losses in fish farms. These low-grade host-specific "**wood lice**" (Fig. 6.126) may reach a similar size as the host fish. The infestation of the oral cavity is common in these parasites. This blockage of the fish's feeding (**lockjaw**), which is accompanied by the blood loss

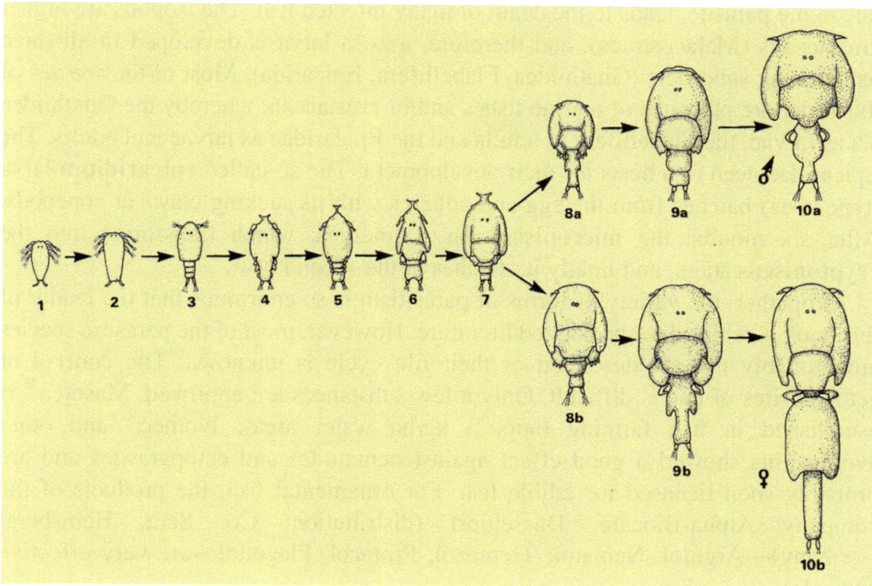

Fig. 6.125 Diagrammatic representation of the life cycle stages of the so-called salmon louse (*Lepeophtheirus salmonis*) (order Copepoda). **1**, **2** Nauplius stages I, II (length 0.54–0.58 mm), free swimming; **3** Copepodit stage (0.7 mm): invasion stage. **4–7** *Chalimus* stages I–V (1.2–2.8 mm long)—attached stages. **8a**, **9b** Preadult males motile on the skin (2.9 or 4.2 mm). **10a**, **b** Adult males (**10a**; 5 mm) and adult female (**10b**; 10 mm). The timing of the development of the eggs and the different stages depends of the water temperature and the water salinity. A generation needs about 6 weeks at a water temperature of 9–12 °C. Salmons are mostly infected for years with rather few amounts of these parasites. However, damages and loss of weight may be considerable

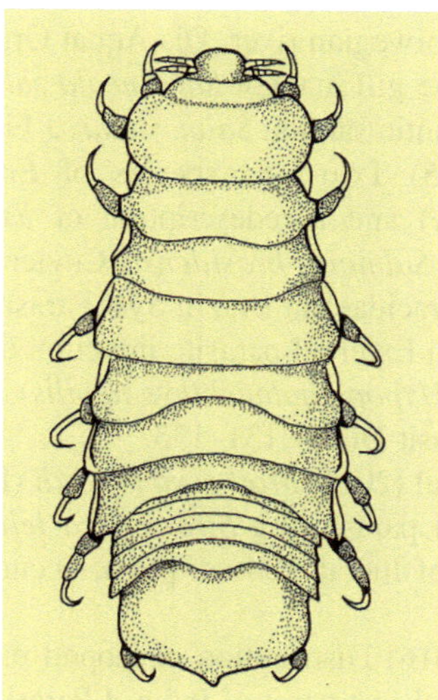

Fig. 6.126 Diagrammatic representation of a wood louse (*Livoneca* sp.) from dorsal

due to the parasite, leads to the death of many infected fish. The isopods are higher crustaceans (Malacostraca), and therefore, a zoea larva is developed in all three parasitizing suborders (Gnathiidea, Flabellifera, Epicarida). Most of the species of this order are parasites of marine fishes and/or crustacean, whereby the Gnathiidea act as larvae, the Flabellifera as adults and the Epicaridae as larvae and adults. The Epicaridae need two hosts for their development. The so-called **epicaridium** larva (type zoea) hatches from the egg and adheres with its sucking claws at copepods. After six moults, the **microniscus** larva emerges, which transforms into the **cryptoniscus** stage, and finally it attaches at the second host.

Altogether, the variety of forms of parasitism is so enormous that the reader of this book is referred to specialized literature. However, most of the parasitic species are probably not yet described or their life cycle is unknown. The **control** of ectoparasites of fish is difficult. Only a few substances are approved. Masoten® is established in fish farming (app. 3 kg/ha water area). Ivomec® and other avermectins showed a good effect against nematodes and ectoparasites and are probably soon licensed for edible fish. For ornamental fish, the products of the company Alpha-Biocare, Düsseldorf (distribution: Co. Sera, Heinsberg, Germany)—Argulol, Nematol, Tremazol, Protazol, Flagellol—are very effective and safe.

Further Reading

Elumalai V et al (2014) Infestation of parasitic barnacle *Sacculina* spp. in commercial marine crabs. J Parasit Dis 38:337–339.

Hanson SK et al (2011) Evaluation of emamectin benzoate for the control of experimentally induced infestations of *Argulus* sp. in goldfish and koi carp. J Aquat Anim Health 23:30–34.

Kar B et al (2016) Variation in susceptibility pattern of fish to *Argulus siamensis*: Do immune responses of host play a role? Vet Parasitol 221:76–83.

Kusterle S et al (2012) Population dynamics of *Salmincola salmoneus* on Atlantic salmon in a northern Norwegian river. Dis Aquat Organ 100:59–70.

Kusterle S et al (2013) The gill maggot *Salmincola salmoneus* as an indicator of repeat spawning in Atlantic salmon *Salmo salar*. J Fish Biol 82:1068–1073.

Marques TM et al (2015) Two new species of *Ergasilus* Nordmann, 1832 (Copepoda: Ergasilidae) and a redescription of *Ergasilus salmini* Thatcher Brazil-Sato, 2008 from *Salminus brasiliensis* Cuvier and *S. franciscanus* Lima Britsky (Teleostei: Characidae) in Brazil. Syst Parasitol 90:81–89.

Nematollahi A et al (2013) External parasite infection of common carp (*Cyprinus carpio*) and big head (*Hypophthalmichthys nobilis*) in fish farms of Mashhad, northeast of Iran. J Parasit Dis 37:131–133.

Panakkool-Thamban A et al (2016) *Mothocya renardi* (Bleeker, 1857) (Crustacea: Isopoda: Cymothoidae) parasitising *Strongylura leiura* (Bleeker) (Belonidae) off the Malabar coast of India: Redescription, occurrence and life-cycle. Syst Parasitol 93:583–599.

Rameshkumar G et al (2016) Distribution of isopod parasites in Carangid fishes from Parangipettai, Southeast coast of India. J Parasit Dis 40:124–128.

Taborda NL et al (2016) A new species of *Ergasilus* (Copepoda: Ergasilidae) from *Geophagus altifrons* and *G. argyrostictus* (Perciformes: Cichlidae) in the Brazilian Amazon. Acta Parasitol 61:549–555.

Further Reading (Joint List for This Chapter)

Abdel-Ghaffar F et al (2009) In vitro efficacy of ByeMite® and Mite-Stop® on developmental stages of the red chicken mite *Dermanyssus gallinae*. Parasitol Res 105:1469–1471

Abu-Samra MT, Shnaib YA (2014) A study of the nature of association between *Demodex* mites and bacteria in lesions of demodectic mange in cattle. Vet Med Int, Article 413719

Ahn AJ et al (2015) Molecular prevalence of *Acarapis* mite infestations in honey bees in Korea. Korean J Parasitol 53:315–320

Alzogaray RA (2016) Behavioral and toxicological responses of *Rhodnius prolixus* (Hemiptera: Reduviidae) to the insect repellents DEET and IR3535. J Med Entomol 53:387–393

Amure J, Stuart JC (1977) A case of *Laminosioptes cysticola* infection. Vet Rec 101:387–389

Anderson DL, Morgan MJ (2007) Genetic and morphologic variation of bee parasitic *Tropilaelaps* mites. Exp Acarol 43:1–24

Anholt H et al (2014) Ear mange mites (*Notoedres muris*) in black and Norway rats. J Wildl Dis 50:104–108

Appelgren A et al (2016) Relative fitness of a generalist parasite on two alternative hosts: a cross-infestation experiment to test host specialization of the hen flea *Ceratophyllus gallinae* (Schrank). Evol Biol 29:1091–1101

Arian LG, Morgan MS (2015) Reproductive biology of *Euroglyphus macynei* with comparisons to *Dermatophagoides farinae* and *D. pteronyssinus*. Exp Appl Acarol 66:1–9

Arther RG et al (2015) Clinical evaluation of the safety and efficacy of 10% imidacloprid +2.5% moxidectin topical solution for the treatment of the ear mite (*Otodectes cynotis*) in dogs. Vet Parasitol 210:64–68

Asala SA, Mbajiorgu FE (1996) Epigenetic variation in the Nigerian skull: sutural pattern at the pterion. East Afr Med J 73:484–486

Attademo AM et al (2012) Trombiculid mites (*Hannemania* sp.) in *Leptodactylus chaquensis* (Amphibia: Anura) inhabiting selected soybean and rice agroecosystems of Argentina. J Zoo Wildl Med 43:579–584

Ayround M, Dies KHC (1992) *Cytodites nudus*-induced granulomatous pneumonia in chickens. Can Vet 33:754–755

Barimani A et al (2016) Traps containing carvacrol, a biological approach for the control of *Dermanyssus gallinae*. Parasitol Res. doi:10.1007/s00436-016-5113-3

Bartley K et al (2015) Identification and evaluation of vaccine candidate antigens from the poultry red mite *Dermanyssus gallinae*. Int J Parasitol 45:819–830

Bartlow AW et al (2016) Walk or ride? Phoretic behaviour of amblyceran and ischnoceran lice. Int J Parasitol 46:221–226

Bartonička T, Růžičková L (2013) Recolonization of bat roost by bat bugs (*Cimex pipistrelli*): could parasite load be a cause of bat roost switching? Parasitol Res 112:1615–1622

Battisti JM et al (2015) Colonization of *Lutzomyia verrucarum* and *Lutzomyia longipalpis* sand flies (Diptera: Psychodidae) by *Bartonella bacilliformis*, the etiologic agent of Carrión's disease. PLoS Negl Trop Dis 9(10):e0004128

Bauer B et al (2009) An appraisal of current and new techniques intended to protect bulls against *Culicoides* and other haematophagous nematocera: the case of Schmergow, Brandenburg, Germany. Parasitol Res 105:359–365

Bernotiene R, Bartkeviciene G (2013) The relationship between water temperature and the development cycle beginning and duration in three black fly species. J Insect Sci 13:1

Beugnet F et al (2015) Sustained speed of kill and repellency of a novel combination of fipronil and permethrin against *Ctenocephalides canis* flea infestations in dogs. Parasite Vectors 8:52

Bochkov AV et al (2014) Integrated Bayesian species elimination and morphological diagnostics of chorioptic mange mites. Parasitol Res 113:2603–2607

Botelho-Vieira MI et al (2014) Re-emergence of *Chorioptes bovis*. Braz J Vet Parasitol 23:530–533

Bowman AS, Nuttal PA (eds) (2004) Ticks: biology, disease and control. Cambridge University Press, Cambridge

Boxler B et al (2016) Host finding of the pigeon tick *Argas reflexus*. Med Vet Entomol 30:193–199

Brown CR et al (2015) Predation by ants controls swallow bug (Hemiptera: Cimicidae: *Oeciacus vicarius*) infestations. J Vector Ecol 40:152–157

Brugger K et al (2016) Outdoor and indoor monitoring of livestock-associated *Culicoides* spp. to assess vector-free periods and disease risks. BMC Vet Res 12:88

Carrasco HJ et al (2014) *Panstrongylus geniculatus* and four other species of triatomine bug involved in the *Trypanosoma cruzi* enzootic cycle: high risk factors for Chagas' disease transmission in the Metropolitan District of Caracas, Venezuela. Parasite Vectors 7:602

Cepero A et al (2015) Is *Acarapis woodi* a single species? A new PCR protocol to evaluate its prevalence. Parasitol Res 114:651–658

Cherairia M et al (2014) Biodiversity and bionomics of the black flies (Diptera: simuliidae) of northeastern Algeria. Zootaxa 3796:166–174

Clausen PH et al (2009) Seasonal dynamics of biting midges (Diptera: Ceratopogonidae, Culicoides spp.) on dairy farms of Central Germany during the 2007/2008 epidemic of bluetongue. Parasitol Res 105:381–386

Coelho WM, Bresciani KD (2013) Molecular and parasitological detection of *Leishmania* spp. in a dipteran of the species *Tabanus importunus*. Rev Bras Parasitol Vet 22:605–607

Colwell DD (2014) Life history parameters of the cattle long-nosed sucking louse, *Linognathus vituli*. Med Vet Entomol 28:432–437

Cucchi-Stefanoni K et al (2008) Fatal anemia and dermatitis in captive agoutis (*Dasyprocta mexicana*) infested with *Echidnophaga* fleas. Vet Parasitol 155:336–339

Dabert J et al (2011) The first report of *Knemidocoptes intermedius* in naturally infected European birds. Parasitol Res 109:237–240

Dabert J et al (2013) Multidisciplinary analysis of *Knemidocoptes jamaicensis* parasitizing the common chaffinch (*Fringilla coelebs*). Parasitol Res 112:2372–2380

Damriyasa IM et al (2004) Prevalence, risk factors and economic importance of infestations with *Sarcoptes scabiei* and *Haematopinus suis* in sows of pig breeding farms in Germany. Med Vet Entomol 18:361–367

De La Riva DG et al (2015) Temperature governs on-host distribution of the Northern fowl mite, *Ornithonyssus sylviarum* (Acari: Macronyssidae). J Parasitol 101:18–23

De Rojas M et al (2002) Phylogenetic relationships in rhinonyssid mites based on rDNA sequences. Parasitol Res 88:675–681

De Souza Rde C et al (2015) Does *Triatoma brasiliensis* occupy the same environmental niche space as *Triatoma melanica*? Parasite Vectors 8:361

DeGrandi-Hoffmann et al (2016) Population growth of *Varroa destructor*. Exp Appl Acarol 69:21–34

Denlinger DS et al (2016) Diagnostic doses and times for *Phlebotomus papatasi* and *Lutzomyia longipalpis* sand flies (Diptera: Psychodidae: Phlebotominae) using the CDC bottle bioassay to assess insecticide resistance. Parasite Vectors 9(1):212

Derua YA et al (2015) Sibling species of the *Anopheles funestus* group, and their infection with malaria and lymphatic filarial parasites, in archived and newly collected specimens from northeastern Tanzania. Malar J 14:104

Díaz-Martín V et al (2015a) New salivary anti-haemostatics containing protective epitopes from *Ornithodoros moubata* ticks: assessment of their individual and combined vaccine efficacy. Vet Parasitol 212:336–349

Díaz-Martín V et al (2015b) Development of vaccines against *Ornithodoros* soft ticks: an update. Ticks Tick Borne Dis 6:211–220

Dobson JR (1999) Bee-louse *Braula schmitzi* Örösi-Pál (Diptera: Braulidae) new to the British Isles, and the status of *Braula* spp. in England and Wales. Brit J Entomol Nat Hist 11:139–148

Dominghetti TF et al (2015) *Stomoxys calcitrans* (Diptera: Muscidae) outbreaks: current situation and future outlook with emphasis on Brazil. Rev Bras Parasitol Vet 24:387–395

Dong WG et al (2014) Fragmented mitochondrial genomes of the rat lice, *Polyplax asiatica* and *Polyplax spinulosa*: intra-genus variation in fragmentation pattern and a possible link between the extent of fragmentation and the length of life cycle. BMC Genomics 15:44

Dos Santos AC et al (2011) Phthiraptera (Arthropoda, Insecta) in *Gallus gallus* from isolated and mixed backyard rearing systems. Rev Bras Parasitol Vet 20:17–21

Dryden MW et al (2016) Evaluation and comparison of a flumethrin-imidacloprid collar and repeated monthly treatments of fipronil/(s)-methoprene to control flea, *Ctenocephalides f. felis*, infestations on cats for eight months. Parasite Vectors 9:287

Dumont P et al (2014) Curative and preventive efficacy of orally administered afoxolaner against *Ctenocephalides canis* infestation in dogs. Vet Parasitol 201:212–215

El-Dakhly KM et al (2012) Occurrence of the subcutaneous mite *Hypodectes propus* in the grey heron. J Zoo Wildl Med 43:953–955

El-Dakhly KM et al (2013) *Hypodectes propus* in a rufous turtle dove. J Vet Ci 14:421–424

Elumalai V et al (2014) Infestation of parasitic barnacle *Sacculina* spp. in commercial marine crabs. J Parasite Dis 38:337–339

Estrada-Pena A et al (2004) Ticks of domestic animals. Houten, Atlantia

Fain A (1972) Notes on acarina of the family Cheyletidae and Hapyrhynchidae producing mange in birds and mammals. Acta Zool Pathol Antverp 56:37–60

Feldmeier H et al (2004) Investigations on the biology, epidemiology, pathology and control of *Tunga penetrans* in Brazil: IV. Clinical and histopathology. Parasitol Res 94:275–282

Frank R et al (2013) Parasites of wild rabbits (*Oryctolagus cuniculus*) from an urban area in Germany, in relation to worldwide results. Parasitol Res 112:4255–4266

Fu Y et al (2016) Genetic diversity and population genetics of the warble flies *Hypoderma bovis* and *H. sinense* in Qinghai Province, China. Parasite Vectors 9:145

Galdhar CN et al (2012) Clinico-biochemical and therapeutic studies on notoedric mange in pet rabbits. J Parasite Dis 39:113–116

Gamerschlag S et al (2008) Repetitive sequences in the ITS1 region of the ribosomal DNA of *Tunga penetrans* and other flea species (Insecta, Siphonaptera). Parasitol Res 102:193–199

Garrido-Bailón E et al (2012) The prevalence of *Acarapis woodi* in Spanish honey bee (*Apis mellifera*) colonies. Exp Parasitol 132:530–536

Gebresilassie A et al (2015) The influence of moonlight and lunar periodicity on the efficacy of CDC light trap in sampling *Phlebotomus* (*Larroussius*) *orientalis* Parrot, 1936 and other *Phlebotomus* sandflies (Diptera: Psychodidae) in Ethiopia. Parasite Vectors 8:106

George DR et al (2015) Should the poultry red mite *Dermanyssus gallinae* be of wider concern for the veterinary and medical science? Parasites Vectors 8:178

Gilles J et al (2009) Polymerase chain reaction detection of *Rickettsia felis*-like organism in *Archaeopsylla erinacei* (Siphonaptera: Pulicidae) from Bavaria, Germany. J Med Entomol 46:703–707

Gothe R (1989) Die Demodikose des Hundes – eine Faktorenkrankheit? Berl Münch Tierärztl Wschr 102:293–297

Gouin F (1951) The mouthparts of *Braula caeca*. C R Hebd Seances Acad Sci 232:1774–1775

Grannoulopoulos GD et al (2012) First report on *Neotrombicula autumnalis* infestation in a cat and a dog from Corfu. Vector Borne Zoonotic Dis 12:1065–1067

Gustafsson DR et al (2015) The chewing lice (Insecta: Phthiraptera: Ischnocera: Amblycera) of Japanese pigeons and doves (Columbiformes), with descriptions of three new species. J Parasitol 101:304–313

Gyimesi ZS et al (2007) Sticktight flea (*Echidnophaga gallinacea*) infestation in a Victoria crowned pigeon (*Goura victoria*). J Zoo Wildl Med 38:594–596

Haas C et al (2015) First detection of sarcoptic mange in free-ranging wild boar (*Sus scrofa*) in Switzerland. Schweiz Arch Tierheilkd 157:269–275

Hajipour N et al (2014) Hedgehogs (*Erinaceus europaeus*) as a source of ectoparasites in urban-suburban areas of Northwest of Iran. J Arthropod Borne Dis 9:98–103

Hallal-Calleros C et al (2013) Hormonal and behavioral changes induced by acute and chronic infestation with *Psoroptes cuniculi*. Parasites Vectors 6:361

Halos L et al (2016) Knock-down and speed of kill of a combination of fipronil and permethrin for the prevention of *Ctenocephalides felis* flea infestation in dogs. Parasite Vectors 9:57

Hamel D et al (2015) Treatment and control of bovine sarcoptic and psoroptic mange infestation with ivermectin (IVOMEC® Gold). Parasitol Res 114:535–542

Hanson SK et al (2011) Evaluation of emamectin benzoate for the control of experimentally induced infestations of *Argulus* sp. in goldfish and koi carp. J Aquat Anim Health 23:30–34

Hellmann K et al (2013) Treatment of naturally *Notodectes cati* infested cats with a combination of imidacloprid 10%/moxidectin 1% spot on (Advocate®/Advantage® Multi, Bayer). Parasitol Res 112:S57–S66

Herholz C et al (2016) Efficacy of the repellent N, N-diethyl-3-methyl-benzamide (DEET) against tabanid flies on horses evaluated in a field test in Switzerland. Vet Parasitol 221:64–67

Hervas D et al (2013) IgE levels due to *Dermatophagoides pteronyssinus*. Allergy Immunol 160:383–386

Hilali MA et al (2014) Monthly variations of *Rhinoestrus* spp. (Diptera: Oestridae) larvae infesting donkeys in Egypt: morphological and molecular identification of third stage larvae. J Adv Res 6:1015–1021

Hoppenheit A et al (2014) Analysis of *Glossina palpalis gambiensis* and *Glossina tachinoides* from two distant locations in Burkina Faso using MALDI TOF MS. Parasitol Res 113:723–726

Hornok S et al (2010) Survey on blood sucking lice of ruminants and pigs with molecular detection of *Anaplasma* and *Rickettsia* spp. Vet Parasitol 174:355–358

Hornok S et al (2014) Vector-borne agents detected in fleas of the northern white-breasted hedgehog. Vector Borne Zoonotic Dis 14:74–76

Iboh CI et al (2012) Occurrence and distribution of *Chrysops* species in Akamkpa community of Cross River State, Nigeria. Pak J Biol Sci 15:1139–1143

Immedieto D et al (2015) Laboratory evaluation of a native strain of *Beauveria bassiana* for controlling *Dermanyssus gallinae*. Vet Parasitol 212:478–482

Isingla LD et al (1996) Therapeutic trial of ivermectin against *Notoedres cati* var. *cuniculi* infection in rabbits. Parasite 3:87–89

Ivković M et al (2016) Blackflies (Diptera: Simuliidae) in Croatia: species richness, distribution and relationship to surrounding countries. Zootaxa 4109:16–30

Izdebska J, Ralbieki L (2013) Sarcoptic mites parasitizing the brown rat (*Rattus norvegicus*). Ann Parasitol 59:125–128

Jafari Shoorijeh S et al (2007) Comparative efficacy of pour-on and subcutaneous injection of ivermectin on *Melophagus ovinus* (L.) in Darab ecotype goats of Southern Iran. Vet Parasitol 148:179–183

Jahanifard E et al (2014) Diversity of sand flies (Diptera, Psychodidae) in southwest Iran with emphasis on synanthropy of *Phlebotomus papatasi* and *Phlebotomus alexandri*. Acta Trop 140:173–180

Jongejan F et al (eds) (2003) Ticks and tick-borne pathogens. Academic, Dordrecht

Joy JE, Stephens CR (2016) Sensory trichites associated with the food canal of *Chrysops callidus* (Diptera: Tabanidae). J Med Entomol. 2016 Apr 20. pii: tjw015

Kagira JM et al (2013) Relationship between the prevalence of ectoparasites and associated risk factors in free-range pigs in Kenya. SRN Vet Sci 2013:650890

Kaliner G (1970) Occurrence of *Laminosioptes cysticola* in the abdominal organs of turkeys. Pathol Vet 7:321–328

Kamimura Y et al (2014) Duplicated female receptacle organs for traumatic insemination in the tropical bed bug *Cimex hemipterus*: adaptive variation or malformation? PLoS One 9:e89265
Kang Y et al (2016) Disease dynamics of honeybees with *Varroa destructor*. Math Biosci 275:71–92
Kapoor R, Elston DM (2012) What's eating you? The sticktight flea (*Echidnophaga gallinacea*). Cutis 89:157–158
Kar B et al (2016) Variation in susceptibility pattern of fish to *Argulus siamensis*: do immune responses of host play a role? Vet Parasitol 221:76–83
Kasap OE et al (2013) The distribution of the *Phlebotomus major* complex (Diptera: Psychodidae) in Turkey. Acta Trop 127:204–211
Kasap OE et al (2015) Phylogeography of the subgenus *Transphlebotomus* Artemiev with description of two new species, *Phlebotomus anatolicus* n. sp. and *Phlebotomus killicki* n. sp. Infect Genet Evol 34:467–479
Ken KM et al (2014) Dermatosis associated with mites other than *Sarcoptes*. Semin Cutan Med Surg 33:110–115
Khater HF et al (2013) The acaricidal efficacy of peracetic acid and deltamethrin against the fowl tick, *Argas persicus*, infesting laying hens. Parasitol Res 112:259–269
Klimpel S et al (2005) Field trial of the efficacy of a combination of imidacloprid and permethrin against *Tunga penetrans* (sand flea, jigger flea) in dogs in Brazil. Parasitol Res 97:S113–S119
Klong-Klaew T et al (2012) Observations on morphology of immature *Lucilia porphyrina* (Diptera: Calliphoridae), a fly species of forensic importance. Parasitol Res 111:1965–1975
Koch LK et al (2016) Modeling the habitat suitability for the arbovirus vector *Aedes albopictus* (Diptera: Culicidae) in Germany. Parasitol Res 115:957–964
Kojima Y et al (2011) PCR-based detection of a tracheal mite *Acarapis woodi*. J Invert Pathol 108:135–137
Krcmar S et al (2006) Response of Tabanidae (Diptera) to different natural attractants. J Vector Ecol 31:262–265
Kreppel KS et al (2016) Effect of temperature and relative humidity on the development times and survival of *Synopsyllus fonquerniei* and *Xenopsylla cheopis*, the flea vectors of plague in Madagascar. Parasite Vectors 9:82
Kusterle S et al (2012) Population dynamics of *Salmincola salmoneus* on Atlantic salmon in a northern Norwegian river. Dis Aquat Organ 100:59–70
Kusterle S et al (2013) The gill maggot *Salmincola salmoneus* as an indicator of repeat spawning in Atlantic salmon *Salmo salar*. J Fish Biol 82:1068–1073
Laha R (2015) Sarcoptic mange infestation in pigs: an overview. J Parasite Dis 39:596–603
Lawrence AL et al (2015) Integrated morphological and molecular identification of cat fleas (*Ctenocephalides felis*) and dog fleas (*Ctenocephalides canis*) vectoring *Rickettsia felis* in central Europe. Vet Parasitol 210:215–223
Lipatova I et al (2015) Fleas (Siphonaptera) in the nests of dormice (Gliridae: Rodentia) in Lithuania. J Med Entomol 52:469–474
Liu SH et al (2016) The incidence and species composition of *Gasterophilus* (Diptera, Gasterophilidae) causing equine myiasis in northern Xinjiang, China. Vet Parasitol 217:36–38
Locher N et al (2010a) In vitro and field studies on the contact and fumigant toxicity of a neem-product (Mite-Stop®) against the developmental stages of the poultry red mite *Dermanyssus gallinae*. Parasitol Res 107:417–423
Locher N et al (2010b) Light and scanning electron microscopic investigations on MiteStop®-treated poultry red mites. Parasitol Res 107:433–437
Lühken R et al (2015) Physico-chemical characteristics of *Culex pipiens* sensu lato and *Culex torrentium* (Diptera: Culicidae) breeding sites in Germany. J Med Entomol 52:932–936
Makert GR et al (2016) A method to identify protein antigens of *Dermanyssus gallinae* for the protection of birds from poultry mites. Parasitol Res. doi:10.1007/s00436-016-5017-2
Mariana A et al (2012) Acariasis on pet Burmese python, *Python molurus bivittatus* in Malaysia. Asian Pac J Trop Med 4:227–228

Markosian AA et al (1971) Biological characteristics of *Braula* and its control. Veterinaria 49:68–69

Marques TM et al (2015) Two new species of *Ergasilus* Nordmann, 1832 (Copepoda: Ergasilidae) and a redescription of *Ergasilus salmini* Thatcher Brazil-Sato, 2008 from *Salminus brasiliensis* Cuvier and *S. franciscanus* Lima Britsky (Teleostei: Characidae) in Brazil. Syst Parasitol 90:81–89

Márquez FJ (2015) Detection of *Bartonella alsatica* in European wild rabbit and their fleas (*Spilopsyllus cuniculi* and *Xenopsylla cunicularis*) in Spain. Parasite Vectors 8:56

Mascarenhas CS et al (2011) Nasal mites of *Paroaria coronata* (Passeriformes). Neotrop Entomol 40:507–508

Matsuyama R et al (2015) Coexistence of two different genotypes of *Sarcoptes scabiei* derived from companion dogs and wild racoon dogs. Vet Parasitol 212:356–360

McOrist S (1983) *Cytodites nudus* infestation of chickens. Avian Pathol 12:151–155

Mehlhorn B, Mehlhorn H (2009) Ticks on progress. Düsseldorf University Press, Düsseldorf

Mehlhorn H et al (2009a) Entomological survey on vectors of Bluetongue virus in Northrhine-Westfalia (Germany) during 2007 and 2008. Parasitol Res 105:321–329

Mehlhorn H et al (2009b) Bluetongue disease in Germany (2007–2008): monitoring of entomological aspects. Parasitol Res 105:313–319

Mehlhorn H et al (2012) Biting and bloodsucking lice of dogs – treatment by means of a neem seed extract (MiteStop®, Wash Away Dog). Parasitol Res 110:769–773

Mencke N et al (2009) Transmission of feline calicivirus via the cat flea (*Ctenocephalides felis*). Parasitol Res 105:185–189

Mendes MM et al (2014) Nasal mites of Tyrannidae (aves) in Brazil. Braz J Biol 74:480–482

Mete A et al (2014) Knemidocoptic mange in wild Golden eagles, California, USA. Emerg Inf Dis 20:1716–1722

Molaei G et al (2015) Vector-Host interactions of *Culiseta melanura* in a focus of Eastern equine encephalitis virus activity in Southeastern Virginia. PLoS One 10:e0136743

Montasser AA (2010) The fowl tick, *Argas* (*Persicargas*) *persicus* (Ixodoidea: Argasidae): description of the egg and redescription of the larva by scanning electron microscopy. Exp Appl Acarol 52:343–361

Montasser AA et al (2011) Efficacy of abamectin against the fowl tick, *Argas* (*Persicargas*) *persicus* (Oken, 1818) (Ixodoidea: Argasidae). Parasitol Res 109:1113–1123

Moss WW et al (1968) Karyotypes and developmental stages of *Harpyrhynchus* sp. n., a parasite of Northern American birds. J Parasitol 54:377–392

Mueller RS (2011) Evidence based therapy of demodicosis of dogs. Tierärztl Praxis (K) 39:419–424

Müller GC et al (2012) An annotated checklist of the horse flies (Diptera: Tabanidae) of Lebanon with remarks on ecology and zoogeography: Pangoniinae and Chrysopsinae. J Vector Ecol 37:216–220

Mutinelli F, Rademacher (2002) European legislation governing the use of drugs in bee colonies to control varroaosis. Reg Affairs 13:401–406

Mysterud A et al (2016) Phenology of deer ked (*Lipoptena cervi*) host-seeking flight activity and its relationship with prevailing autumn weather. Parasite Vectors 9:95

Nagy N et al (2007) Investigations on the life cycle and morphology of Tunga penetrans in Brazil. Parasitol Res 101(Suppl 2):S233–242

Nematollahi A et al (2013) External parasite infection of common carp (*Cyprinus carpio*) and big head (*Hypophthalmichthys nobilis*) in fish farms of Mashhad, northeast of Iran. J Parasite Dis 37:131–133

Newman CM et al (2016) Decreased flight activity in *Culex pipiens* (Diptera: Culicidae) naturally infected with *Culex* flavivirus. J Med Entomol 53:233–236

Niebuhr CN et al (2013) Off-host collection methods of the *Otobius megnini* (Acari: Argasidae). J Med Entomol 50:994–998

Niebuhr CN et al (2014) Efficacy of chemical repellents against *Otobius megnini* (Acari: Argasidae) and three species of ixodid ticks. Exp Appl Acarol 64:99–107

Niederegger S et al (2016) Muscle attachment site (MAS) patterns for species determination in five species of *Sarcophaga* (Diptera: Sarcophagidae). Parasitol Res 115:241–247

Okorie PN et al (2015) Insecticide susceptibility of *Anopheles coluzzii* and *Anopheles gambiae* mosquitoes in Ibadan, Southwest Nigeria. Med Vet Entomol 29:44–50

Osman SA et al (2006) Clinical and therapeutic studies on mange in horses. Vet Parasitol 141:191–195

Oteo JA et al (2014) '*Candidatus Rickettsia asemboensis*' and *Wolbachia* spp. in *Ctenocephalides felis* and *Pulex irritans* fleas removed from dogs in Ecuador. Parasite Vectors 7:455

Panakkool-Thamban A et al (2016) *Mothocya renardi* (Bleeker, 1857) (Crustacea: Isopoda: Cymothoidae) parasitising *Strongylura leiura* (Bleeker) (Belonidae) off the Malabar coast of India: redescription, occurrence and life-cycle. Syst Parasitol 93:583–599

Parcell BJ et al (2013) Conjuctivitis induced by a red bodied mite, *Neotrombicula autumnalis*. Parasite 20:25

Paterson S, Coumbe K (2009) An open study to evaluate topical treatment of equine chorioptic mange with shampooing and lime sulphur solution. Vet Dermatol 20:623–629

Pence DB, Casto S (1975) Two new species and new records of nasal mites of the genus *Sternostoma* (Acarina: Rhinonyssinae) from birds in Texas. J Parasitol 61:360–368

Peterson JK et al (2015) *Rhodnius prolixus* life history outcomes differ when infected with different *Trypanosoma cruzi* strains. Am J Trop Med Hyg 93:564–572

Phasuk J et al (2016) A comparison of attractants for sampling *Stomoxys calcitrans* (Diptera: Muscidae) on dairy farms in Saraburi Province, Thailand. J Econ Entomol 109:942–946

Piou V et al (2016) Impact of the phoretic phase on reproduction and damage caused by *Varroa destructor* (Anderson and Trueman) to its host, the European honey bee (*Apis mellifera* L.). PLoS One 11(4):e0153482

Pourseyed SH et al (2010) *Metarhizium anisopliae* (Ascomycota: Hypocreales): an effective alternative to chemical acaricides against different developmental stages of fowl tick *Argas persicus* (Acari: Argasidae). Vet Parasitol 172:305–310

Radfar MH et al (2012) Biodiversity and prevalence of parasites of domestic pigeons. Trop Anim Health Prod 44:225–229

Ramesh A et al (2016) The impact of climate on the abundance of *Musca sorbens*, the vector of trachoma. Parasite Vectors 9:48

Rameshkumar G et al (2016) Distribution of isopod parasites in Carangid fishes from Parangipettai, Southeast coast of India. J Parasite Dis 40:124–128

Ravera I et al (2015) Serum detection of IgG antibodies against *Demodex canis* in healthy dogs and dogs with juvenile demodicosis. Res Vet Sci 101:161–164

Rehbein S et al (2013) The efficacy of eprinomectin extended-release injection against *Hypoderma* spp. (Diptera: Oestridae) in cattle. Vet Parasitol 192:353–358

Rendle DI et al (2007) Comparative study of doramectin and fipronil in the treatment of equine chorioptic mange. Vet Rec 161:335–338

Ribeiro JM et al (2015) An insight into the sialome of the horse fly, *Tabanus bromius*. Insect Biochem Mol Biol 65:83–90

Rodríguez-Planes LI et al (2016) Selective insecticide applications directed against *Triatoma infestans* (Hemiptera: Reduviidae) affected a nontarget secondary vector of Chagas disease, *Triatoma garciabesi*. J Med Entomol 53:144–151

Rojas M et al (2012) Molecular study of three morphotypes of *Demodex* mites from dogs. Parasitol Res 111:2165–2172

Rosenkranz et al (2010) Biology and control of *Varroa destructor*. J Invertebr Pathol 103(Suppl 1):S96–S119

Rudolf I et al (2016) Molecular survey of arthropod-borne pathogens in sheep keds (*Melophagus ovinus*), Central Europe. Parasitol Res. 2016 Jun 20. [Epub ahead of print]

Rzotkiewicz S et al (2015) Novel evidence suggests that a '*Rickettsia felis*-like' organism is an endosymbiont of the desert flea, *Xenopsylla ramesis*. Mol Ecol 24:1364–1373
Salant H et al (2014) Ectoparasites in urban stray cats. Med Vet Entomol 28:314–318
Sands B et al (2016) Residual and ovicidal efficacy of essential oil-based formulations in vitro against the donkey chewing louse *Bovicola ocellatus*. Med Vet Entomol 30:78–84
Santos HA et al (2010) The influence of the fasting period on the number of nymphal instars and on the sex ratio of *Argas* (*Persicargas*) miniatus (Acari: Argasidae). Rev Bras Parasitol Vet 19:164–168
Sarre C et al (2015) Comparative immune responses against *Psoroptes ovis* in two cattle breeds. Vet Res 46:131
Schaefer M (ed) (2010) Brohmer – Fauna von Deutschland, 23rd edn. Quelle & Meyer, Wiebelsheim
Schilliger LH et al (2013) *Cheyletus eruditus* (*taurrus*): an effective candidate for the biological control of the snake mite (*Ophionyssus natricis*). J Zoo Wildl Med 44:654–659
Sekeyová Z et al (2012) Identification of *Rickettsia africae* and *Wolbachia* sp. in *Ceratophyllus garei* fleas from Passerine birds migrated from Africa. Vector Borne Zoonotic Dis 12:539–543
Sheele JM, Ridge GE (2016) Toxicity and potential utility of ivermectin and moxidectin as xenointoxicants against the common bed bug, *Cimex lectularius* L. Parasitol Res. doi:10.1007/s00436-016-5062-x
Shoracki M et al (2014a) Syringophylid mites associated with the rails. Syst Parasitol 88:227–232
Shoracki M et al (2014b) The new quill mite species of the family Syringophilidae parasitizing the house sparrow. Zootaxa 3765:194–200
Silbermayr K et al (2015) Phylogenetic relationships and new genetic tools for the detection and discrimination of the three feline *Demodex* mites. Parasitol Res 114:747–752
Six RH et al (2016) Efficacy of solaner, a novel oral isoxazoline against two common mite infestations in dogs: *Demodex* spp. and *Otodectes cynotis*. Vet Parasitol 222:62–66
Smith WA et al (2013) Phylogenetic analysis of symbionts in feather-feeding lice of the genus *Columbicola*: evidence for repeated symbiont replacements. BMC Evol Biol 13:109
Sokół R, Michalski MM (2015) Occurrence of *Hippobosca equina* in Polish primitive horses during the grazing season. Ann Parasitol 61:119–124
Sonenshine DE, Roe RM (eds) (2014) Biology of ticks, vol 2. Oxford University Press, Oxford
Song SD et al (2014) Variation in mitochondrial minichromosome composition between blood sucking lice of the genus *Haematopinus* that infest horses and pigs. Parasite Vectors 7:144
Steinke S et al (2016) Emergence of *Culicoides obsoletus* group species from farm-associated habitats in Germany. Med Vet Entomol 30:174–184
Stekolnikov AA, Klimov PB (2010) A revision of chiggers of the minuta species group (Acari, Trombiculidae: *Neotrombicula*). Syst Parasitol 77:55–69
Stekolnikov AA et al (2014) *Neotrombicula inopinata* : a possible causative of trombiculiasis in Europe. Parasites Vectors 7:90
Stoeckli MR et al (2013) The effect of *Psoroptes ovis* infection on ovine epidermal barrier function. Vet Res 44:11
Suh SJ et al (2012) Louse flies (Diptera: Hippoboscidae: Ornithomyinae) of the Republic of Korea: an updated checklist, including two new records of bird louse flies. J Med Entomol 49:231–236
Sukontason KL et al (2014) *Sarcophaga* (*Liosarcophaga*) *dux* (Diptera: Sarcophagidae): a flesh fly species of medical importance. Biol Res 47:14–17
Swe PM et al (2014) Scabies mites alter the skin microbiome and promote growth of opportunistic pathogens in a porcine model. PLoS Negl Trop Dis 8(5):e2897
Szpila K et al (2015) Flesh flies (Diptera: Sarcophagidae) colonising large carcasses in Central Europe. Parasitol Res 114:2341–2348
Tabari MA et al (2015) Carvacol as a potent natural acaricidal against *Dermanyssus gallinae*. Parasitol Res 114:2801–3806

Taborda NL et al (2016) A new species of *Ergasilus* (Copepoda: Ergasilidae) from *Geophagus altifrons* and *G. argyrostictus* (Perciformes: Cichlidae) in the Brazilian Amazon. Acta Parasitol 61:549–555

Tidemann SC et al (1992) Parasitism of wild Gouldian finches (*Erythrura gouldiae*) by the air-sac mite *Sternostoma tracheacolum*. J Wildl Dis 28:80–84

Ughasi J et al (2012) *Mansonia africana* and *Mansonia uniformis* are vectors in the transmission of *Wuchereria bancrofti* lymphatic filariasis in Ghana. Parasite Vectors 5:89

Underwood RM, Currie RW (2009) Indoor winter fumigation with formic acid for control of *Acarapis woodi* (Acari: Tarsonemidae) and *Nosema* disease, *Nosema* sp. J Econ Entomol 102:1729–1736

Ural K et al (2008) Eprinomectin treatment of psoroptic mange in hunter/jumper and dressage horses. Vet Parasitol 156:353–357

Uspensky I (2014) Tick pests and vectors (Acari: Ixodoidea) in European towns: Introduction, persistence and management. Ticks Tick Borne Dis 5:41–47

Valim MP, Cicchino AC (2015) A remarkable new genus and a new species of chewing louse (Phthiraptera, Ischnocera, Philopteridae) from Brazil. Zookeys 541:57–70

Veneziano V et al (2013) Field efficacy of eprinomectin against the sucking louse *Haematopinus asini* on naturally infested donkeys. Vet J 197:512–514

Vidal-Quist JC et al (2015) Allergen expression in the European house dust mite. Med Vet Entomol 29:137–146

Vieira VR et al (2015) Ecological aspects of phlebotomine sand flies (Diptera, Psychodidae, Phlebotominae) in areas of American cutaneous leishmaniasis, in the municipality of Paraty, Rio de Janeiro, Brazil I-index of abundance by location and type of capture. J Med Entomol 52:886–895

Visser M et al (2013) The treatment of bovine sarcoptic mange (*Sarcoptes scabiei* var. *bovis*) using eprinomectin extended-release injection. Vet Parasitol 192:359–364

Vobis M et al (2004) Molecular phylogeny of isolates of *Ctenocephalides felis* and related species based on analysis of ITS1, ITS2 and mitochondrial 16S rDNA sequences and random binding primers. Parasitol Res 94:219–226

Vobis M et al (2005) Experimental quantification of the feline leukaemia virus in the cat flea (*Ctenocephalides felis*) and its faeces. Parasitol Res 97:S102–S106

Volfova V et al (2008) Hyaluronidase of bloodsucking insects and its enhancing effect on leishmania infection in mice. PLoS Negl Trop Dis 2(9):e294

Wamwiri FN, Changasi RE (2016) Tsetse flies (*Glossina*) as vectors of human African trypanosomiasis: a review. Biomed Res Int 2016:6201350

Weitzel T et al (2015) *Culex pipiens* s.l. and *Culex torrentium* (Culicidae) in Wrocław area (Poland): occurrence and breeding site preferences of mosquito vectors. Parasitol Res 114:289–295

Wen H et al (2010) The effect of self-licking behaviour on pharmacokinetics of eprinomectin and clinical efficacy against *Psoroptes cuniculi* in topically administered rabbits. Parasitol Res 106:607–613

Werblow A et al (2015) Hyperparasitism of mosquitoes by water mite larvae. Parasitol Res 114:2757–2765

Werner D, Kampen H (2015) *Aedes albopictus* breeding in southern Germany, 2014. Parasitol Res 114:831–834

Werner D et al (2016) First record of *Aedes koreicus* (Diptera: Culicidae) in Germany. Parasitol Res 115:1331–1334

Westfall MC et al (2008) Patterns of trombiculid mite (*Hannemania dunni*) parasitism among plethodontid salamanders in the western Piedmont of North Carolina. J Parasitol 94:631–634

Yiguan W et al (2016) Influence of bloodmeal host on blood feeding, egg production, and offspring sex ratio of *Ctenocephalides felis felis* (Siphonaptera: Pulicidae). J Med Entomol 53:888–893

Yoon KS et al (2015) In vitro and in vivo evaluation of infestation deterrents against lice. J Med Entomol 52:970–978

Zarroug IM et al (2016) Seasonal variation in biting rates of *Simulium damnosum* sensu lato, vector of *Onchocerca volvulus*, in two Sudanese foci. PLoS One 11(3):e0150309

Zeegers T, Müller GC (2014) A review of the *Tabanus semiargenteus*-subgroup as part of the *Tabanus bovinus* species-group (Diptera: Tabanidae) with the description of two new species for science. Acta Trop 137:152–160

Zhao YE et al (2014a) Molecular identification on phylogenetic study of *Demodex caprae*. Parasitol Res 113:3601–3608

Zhao YE et al (2014b) Population identification of *Sarcoptes hominis* and *S. canis* in China using DNA sequences. Parasitol Res 114:1001–1010

Addendum A: Antiparasitic Drugs Used for Animals

Each product can only be used according to dosages and descriptions given on the leaflet within each package.

Table A.1 Selection of drugs against protozoan diseases of dogs and cats (these compounds are not approved in all countries but are often available by import)

Parasites	Active compound	Dosage (mg/kg body weight)	Application
Isospora species	Toltrazuril	**D**: 10.00	1× per day for 4–5 d; p.o.
Toxoplasma gondii (acute infection)	Clindamycin	**D**: 12.5 **C**: 12.5–25	Every 12 h for 2–4 weeks; o. Every 12 h for 2–4 weeks; o.
Neospora sp. (systemic infection)	Clindamycin + Sulfadiazine/ Trimethoprim	**D**: 12.5	2× per d for 4–8 weeks; o.
Giardia species	Fenbendazol	**D/C**: 50.0	1× per day for 3–5 days; o.
Babesia species	Imidocarb	**D**: 3–6	Possibly repeat after 12–24 h; s.c.
Leishmania species	Allopurinol	**D**: 20.0	1× per day for months up to years; o.
Hepatozoon species	Imidocarb (I) + Doxycycline (D)	**D**: 5.0 (I) + 5.0 (D)	(I) 2× in intervals of 2 weeks; s.c. plus (D) 2× per day on 7 days; o.

C cat, *D* dog, *d* day, *kg* kilogram, *mg* milligram, *o.* orally, *s.c.* subcutaneously

Table A.2 Selection of drugs against nematodes of dogs and cats (unfortunately not effective against a broad spectrum of parasites)

Active compounds	Trade names	Dosage (mg/kg body weight)	Application
Fenbendazole	Panacur®	**D**: 50.0 for 3 d **C**: 50.0 for 3 d	o.
Flubendazole	Flubenol®	**D**: 22.0 for 3 d **C**: 22.0 for 3 d	o.

(continued)

© Springer International Publishing Switzerland 2016
H. Mehlhorn, *Animal Parasites*, DOI 10.1007/978-3-319-46403-9

Table A.2 (continued)

Active compounds	Trade names	Dosage (mg/kg body weight)	Application
Mebendazole	Telmin® KH	**D/C**: 2×/d 50–100 mg/5 d	o.
Selamectin	Stronghold®	**D/C**: 6.0	Spot on
Epsiprantel	Cestex®	**D**: 5.5 **C**: 2.75	o.
Nitroscanate	Lopatol®, Excanat®	**D**: 50.0	o.
Pyrantelpamoate	Banminth®	**D**: 5.0 **C**: 20.0	o.
Pyrantel + Febantel	Welpan®	**D**: 5.0 (B) + 15.0	o.
Nitroscanate	Lopatol®	**C**: 50.0	o.

B base, *C* cat, *D* dog, *d* day, *kg* kilogram, *mg* milligram, *o.* orally

Table A.3 Combination drugs against cestodes and nematodes of dogs and cats (selection)

Active compound	Trade name	Dosage (mg/kg body weight)	Application
Fenbendazole (F) + Praziquantel (P)	Caniquantel® Plus	**D/C**: 50.0 (F) for 2–3 d + 5.0 (P)	o.
Fenbantel + Praziquantel pyrantelembonate	Drontal® Plus	**D**: 15.0 + 5.0 + 5.0	o.
Pyrantelembonate + Praziquantel	Drontal®	**C**: 20.0 + 5.0	o.
Pyrantelembonate + Oxantel + Praziquantel	Plerion®	**D**: 25–50 (P), 100–200 (O), 25–50 (PR)	Chewable tablet
Moxidectin + Imidacloprid (not against TW)	Advocate®	**D**: 2.5 + 10.0 **C**: 1.0 + 10.0	Spot on
Emodepside + Praziquantel	Profender®	**D**: 10.0 + 50.0 **C**: 3.0 + 12.0	D: o. C: Spot on
Milbemyxinoxim + Praziquantel	Milbemax®	**D/C**: 0.5 + 5.0	o.

C cat, *D* dog, *d* day, *kg* kilogram, *mg* milligram, *o.* orally, *TW* tapeworms

Table A.4 Antiparasitics acting against ectoparasites of dogs and cats (selection)

Active compound	Trade name	Dosage/host (mg/kg body weight)	Application	Target group
Fipronil Fipronil + Methopren	Frontline® Frontline Combo®	**D/C**: 6.7 **D/C**: K: 5 + 6	Spot-on	Ticks, fleas, lice, *Cheyletiella* mites
Imidacloprid + Permethrin	Advantix®	**D**: 10 + 50	Spot-on	Ticks, fleas, sand flies, mosquitoes, lice
Metaflumizon + Amitraz	ProMeris Duo®	**D**: 20 + 20	Spot-on	Ticks, fleas
Permethrin	Exspot®	**D**: 50	Spot-on	Ticks, fleas, sand flies

(continued)

Table A.4 (continued)

Active compound	Trade name	Dosage/host (mg/kg body weight)	Application	Target group
Pyriprol	Prac-tic®	**D**: 12.5	Spot-on	Ticks
Selamectin	Stronghold®	**D/C**: 6.0	Spot-on	Fleas, sarcoptic mites
Imidacloprid + Moxidectin	Advocate®	**D**: 10 + 2.5 **C**: 10 + 1.0	Spot-on	*Sarcoptes*, *Demodex*, *Otodectes* mites, lice, fleas
Metaflumizon	ProMeris®	**C**: 40	Spot-on	Fleas
Imidacloprid	Advantage®	**D/C**: 10	Spot-on	Fleas
Permethrin + Pyriproxyfen	Duowin®	**D**: 94 + 1	Spray	Ticks
Lufenuron	Program®	**D**: 10 **C**: 10	o. s.c.	Fleas
Nitenpyram	Capstar®	**D/C**: 1	o.	Fleas
Deltamethrin	Scalibor®	**D**: neclace	Neclace	Fleas, sand flies, ticks
Flumethrin + Propoxur	Kiltix®	**D**: neclace	Neclace	Fleas, ticks
Imidacloprid + Flumethrin	Serestro®	**C**: neclace	Neclace	Fleas
Margosa extract	Vermin shampoo Wash Away®	**D/C**: Washing: 50 ml shampoo for small animals	Single washing	Ticks, mites, fleas, lice, mallophages
Margosa extract	MiteStop® Dog/cat	**D/C**: dilute extract 1:40 with tap water	Dip into the diluted extract or wet the fur	Ticks, mites, fleas, mallophages, especially autumn mites
Amitraz	Ectodex®	**D**: 0.05 %	Bath every 7 d	Skin mites
Fenthion	Tiguvon®	**C**: 100 mg/ml	Pour-on	Fleas

C cat, *d* dog, *kg* kilogram, *mg* milligram, *o.* orally, *s.c.* subcutaneously

Table A.5 Anthelmintics of cattle for oral or pour-on application (selection)

Active compound	Trade name	Dosage (mg/kg body weight)	Application	Waiting time/days (Germany)
(Pro)Benzimidazole				
Febantel	Rintal®	7.5	o.	DA = 2 d; E = 14 d
Fenbendazole	Panacur®	7.5	o.	DA = 6 d; E = 7 d
Albendazole	Valbazen®	7.5	o.	DA = 5 d; E = 28 d
Neotobimin	Hepadex®	7.5	o.	DA = 5; E = 10
Oxfendazole	Systamex®	4.5	o.	DA = 5; E = 10
Macrocyclic lactones				
Doramectin	Dectomax®	0.2 (0.5)	s.c. (pour-on)	NL: 50 (35) d
Eprinomectin	Eprinex®	0.5	Pour-on	DA = 0 d; E = 30 d
Ivermectin	Ivomec®	0.2 (0.5)	s.c. (pour-on)	NL; E = 28

(continued)

Table A.5 (continued)

Active compound	Trade name	Dosage (mg/kg body weight)	Application	Waiting time/days (Germany)
Moxidectin	Cydectin®	0.2 (0.5)	s.c. (pour-on)	DA = 0 d; E = 14 d (pour-on) resp. 65 d (s.c.)
Imidazothiazole				
Levamisole	Citarin®	7.5; 0.5; 10	p.o.; i.m.; pour-on	NL; E = 8 d; 8 d; 22 d

d day, *DA* dairy animals, *E* edible tissue, *i.m.* intramuscularly, *kg* kilogram, *mg* milligram, *NL* not for lactating animals, *o.* orally, *s.c.* subcutaneously

Table A.6 Anthelmintics for cattle as bolus against gastrointestinal strongylids

Active compound	Trade name	Dosage	Waiting time	Duration of effect
Oxfendazole (7.5 g)	Systamex Intervall-Bolus forte®	Every 23 d 1250 mg each	E = 100 d, NL	150 d
Fenbendazole (12 g)	Panacur-SR®-Bolus	Daily 0.2–0.4 mg/kg body weight	E = 100 d, NL	Up to 140 d
Levamisole (18.8 g)	Chronomintic®-Bolus	(a) 2 g at the beginning (b) then 220 mg/d	E = 140 d, NL	Up to 90 d

d days, *E* waiting time for edible tissues, *g* gram, *mg* milligram, *NL* not for lactating animals

Table A.7 Nematocidal anthelmintics for sheep and goats (selection)

Active compound	Trade name	Dosage (mg/kg body weight)	Application	Waiting time/days (Germany)
(Pro)Benzimidazoles				
Mebendazole	Ovitelmin®	15–20	o.	DA = 0 d; E = 2 d
Fenbendazole	Panacur®	5.0	o.	DA = 5 d; E = 21 d
Albendazole	Valbazen®, Albazol®	3.8	o.	DA = 5 d; E = 10–14 d
Febantel	Rintal®	5.0	o.	DA = 2 d; E = 14 d
Netobimin	Hepadex®	7.5	o.	DA = 5 d; E = 5 d
Oxfendazole	Systamex®	5.0	o.	NL; E = 14 d
Macrocyclic lactones				
Doramectin	Dectomax®	0.2	i.m.	NL; E = 70 d
Ivermectin	Ivomec®, Virbamec®	0.2	s.c.	NL; E = 42 d
Eprinomectin	Eprinex®	1.0	Pour-on	DA = 0; E = 15 d
Moxidectin	Cydectin®	0.2	o.	DA = 5 d; E = 14 d
Imidazothiazole				
Levamisole	Citarin® Chronomintic®	8.0	s.c.	NL; E = 14 d

d day, *DA* dairy animal, *E* edible tissue, *i.m.* intramuscularly, *kg* kilogram, *mg* milligram, *NL* not for lactating animals, *o.* orally, *s.c.* subcutaneously

Table A.8 Selection of remedies against ectoparasites of ruminants (check for approval for animal species and efficacy)

Active compound	Trade name	Dosage (mg/kg body weight)	Application	Waiting time/ days
Amitraz	Taktic®	0.025 % Spray	Spray	4 d
Cyfluthrin	Bayofly®	0.2	Pour-on	0 d
Cyhalothrin	Cyhalothrin®	0.2–0.4	Pour-on	0 d
Cypermethrin	Flectron-Earclip®	Earclip	Attachment	0 d
Deltamethrin	Butox®	0.75	Pour-on	18 d; DA = 0 d
Flumethrin	Bayticol®	2.0	Pour-on	5 d; DA = 8 d
Doramectin	Dectomex®	0.2–0.5	Pour-on	35 d; NL
Ivermectin	Ivomec®	0.2–0.5	Pour-on	28 d; NL
Eprinomectin	Eprinex®	0.5	Pour-on	15 d; DA = 0 d
Moxidectin	Cydectin®	0.5	Pour-on	14 d; DA = 0 d
Phoxim	Sebacil®	Spray	Spray	35 d; NL

d day, *DA* dairy animal, *kg* kilogram, *mg* milligram, *NL* not for lactating animals

Table A.9 Selection of remedies against ectoparasites of horses

Active compound	Trade name	Dosage (mg/kg body weight)	Application	Parasite
Ivermectin	Ivomec P®, Eraquell®	0.2	Orally	*Gasterophilus* larvae
Ivermectin + Praziquantel	Equimax®, Eqvalan®	0.2 + 1.5	Orally	*Gasterophilus* larvae
Moxidectin	Equest®	0.4	Orally	*Gasterophilus* larvae
Moxidectin + Praziquantel	Equest Pramox®	0.4 + 2.5	Orally	*Gasterophilus* larvae
Permethrin	Wellcare® Emulsion	4.2	Application on the fur every 14 d	Flies, gadflies
Neem Seed Extract + Saltidin	TaonX®	Spray until fur is moistened	Spray	Gadflies, flies, midges, black flies, mosquitoes, ticks, mites
Neem Seed Extract + Margosa	MiteStop® Horse	Dilution of 100 ml extract in 2 l of H_2O and application on the fur with a brush or sponge	Brush or sponge	Ticks, mites, mallophages, gadflies, flies, midges, black flies, mosquitoes

d day, *kg* kilogram, *mg* milligram

Table A.10 Anthelmintics for horses against nematodes

Active compound	Trade name	Dosage (mg/kg body weight)	Application	Waiting time/ days (Germany)
(Pro)Benzimidazoles				
Febantel Febantel + Metrifonate	Rintal® Rintal® Plus	6.0	Orally	E = 20 d; NL
Fenbendazole	Panacur®, Equivermex®	10.0	Orally	E = 7 d; NL
Mebendazole Mebendazole + Metrifonat	Telmin®, Telmin® Plus	10.0	Orally	E = 7 d; NL
Oxibendazole	Equitac®	10.0	Orally	E = 8 d
Macrocyclic lactones				
Ivermectin	Ivomec®, Eqvalan®, Equell®	0.2	Orally	E = 21 d; NL
Moxidectin	Equest®	0.4	Orally	E = 32 d; NL
Pyrimidine				
Pyrantelembonate	Banminth®, Strongid-P®	6.6	Orally	E = 1 d; NL

d day, *E* edible tissues, *kg* kilogram, *mg* milligram, *NL* not for lactating animals, *o.* orally

Table A.11 Anthelmintics for pigs

Active compound	Trade name	Dosage (mg/kg body weight)	Application	Waiting time/days (Germany)
(Pro)Benzimidazoles				
Fenbendazole	Panacur®	5.0	p.o.	E = 5 d
Febantel	Rintal®	5.0	p.o.	E = 6 d
Flubendazole	Flubenol®	5.0	p.o.	E = 14 d
Macrocyclic lactones				
Doramectin	Dectomax®	0.3	i.m.	E = 56 d
Ivermectin	Ivomec®, Virbamec®	0.3	s.c.	E = 14 d
Imidazothiazole				
Levamisole	Belamisol®, Citarin®	7.5	p.o.	E = 8–15 d

d day, *E* edible tissues, *i.m.* intramuscularly, *kg* kilogram, *mg* milligram, *o.* orally, *s.c.* subcutaneously

Table A.12 Selection of remedies against ectoparasites of pigs (*Sarcoptes*, *Haematobia*)

Active compound	Trade name	Dosage (mg/kg body weight)	Application	Waiting time/ days
Phoxim	Sebacil®	4 ml 7.5 %/10 kg	Pour on	16 d
Amitraz	Tactic®	0.05 % as spray	Spray	1 d

(continued)

Table A.12 (continued)

Active compound	Trade name	Dosage (mg/kg body weight)	Application	Waiting time/ days
Ivermectin	Ivomec®, Virbamec®	0.3	s.c.	13 d
Doramectin	Dectomax®	0.3	i.m.	56 d
Levamisole	Belamisol®, Citarin®	7.5	o.	15 d

d day, *i.m.* intramuscularly, *kg* kilogram, *mg* milligram, *o.* orally, *s.c.* subcutaneously

Table A.13 Selection of remedies against parasites of poultry

Active compound	Trade names/ target group of parasites	Dosage (mg/kg body weight)	Application	Hosts/ waiting time
Toltrazuril	Baycox® 2.5 % Coccidia	With drinking water	With drinking water	Chickens/ 10 d
Monensin-Na	Monensin®, Elancoban® Coccidia	100	In food	Chickens/ 3 d
Lasalocid-Na	Lasalocid®, Avatec® Coccidia	75–125	In food	Chickens/ 5 d
Salinomycin-Na	Salinomycin®, Sacox® Coccidia	60–70	In food	Chickens/ 5 d
Narasin	Narasin®, Monteban® Coccidia	60–70	In food	Chickens/ 1 d
Diclazuril	Appertex®, Clinacox® Coccidia	1	In food	Pigeons/5 d
Levamisole	Concurat®-L 10 % Nematodes	20–30	In food	Chickens, turkeys/ 14 d; NL
Flubendazole	Flubendazol®, Flubenol® Cestodes and nematode	20–60 ppm for 7 d	In food	Poultry/0 d
Neem Seed Extract (Margosa)	MiteStop® Chicken mites, mallophages	Dilute 1:40 with H_2O	Spraying: stables and animals	Poultry/0 d
Phoxim	ByeMite® Mites	Dilute 1:250 with H_2O	Spraying of cages after food and eggs have been removed	Poultry/1 d

d day, *kg* kilogram, *mg* milligram, *NL* not for laying hens, *ppm* parts per million

Table A.14 Selection of remedies against parasites of rabbits

Active compound	Trade names/target group of parasites	Dosage (mg/kg body weight)	Application	Hosts/ waiting time
Toltrazuril	Baycox®/*Eimeria* species	25 ppm in drinking water at 2 d; repeating	Drinking water	n/a
Salinomycin-Na	Sacox®/*Eimeria* species	25 ppm in food	Food	5 d
Robenidin	Cycostat®/Coccidia	60 ppm in foof	Food	5 d
Praziquantel	Droncit®/cestodes	10.0	o.	5 d
Ivermectin	Ivomec®/nematodes	0.2–0.4	s.c.	n/a
Fenbendazole	Panacur®/nematodes	5–10.0	o.	n/a
Selamectin	Stronghold®/ nematodes	6.0	Spot-on	n/a
Imidacloprid	Advantage®/fleas	10.0	Spot-on	n/a

d Tag, *kg* kilogram, *mg* milligram, *n/a* not applicable, *o.* orally, *ppm* parts per million

Addendum B: Diagnostic Stages

© Springer International Publishing Switzerland 2016
H. Mehlhorn, *Animal Parasites*, DOI 10.1007/978-3-319-46403-9

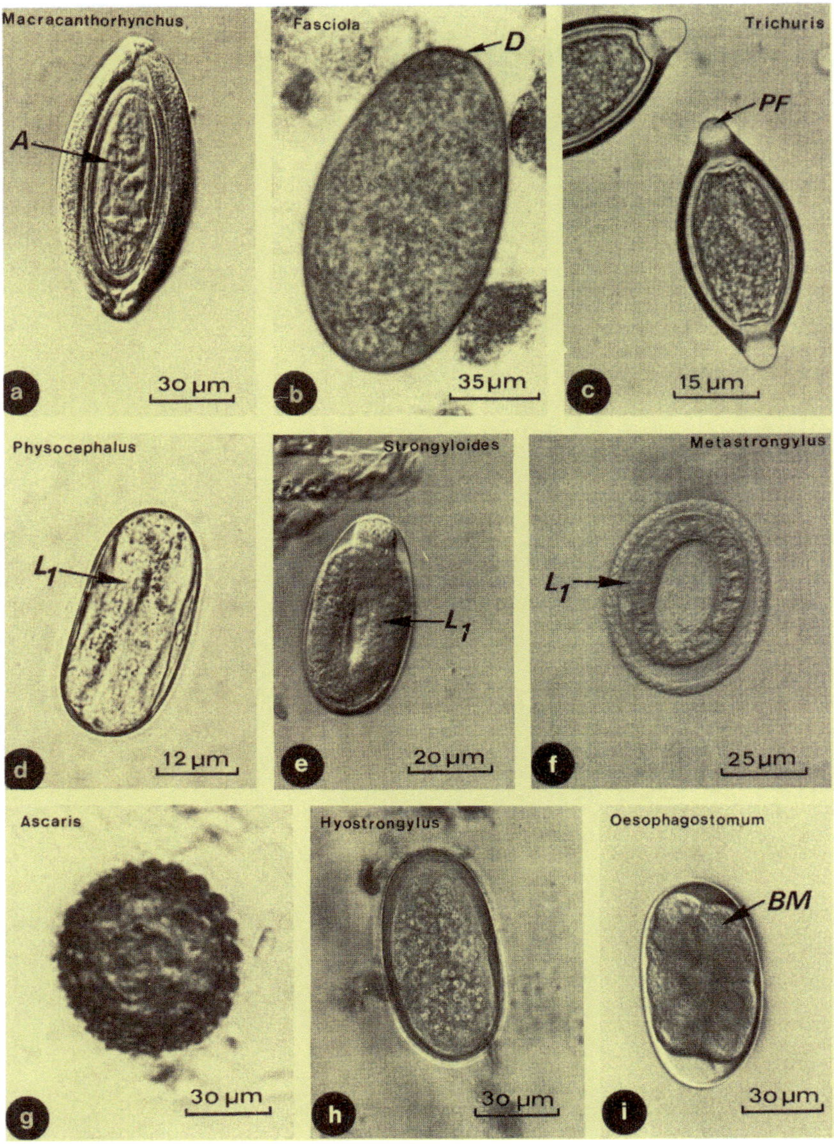

Fig. B.1 Worm eggs of pigs. (**a**) *Macracanthorhynchus hirudinaceus*: acanthocephalan eggs contain a hooked larva (A = acanthor). (**b**) *Fasciola hepatica*: egg of the liver fluke (D = lid). (**c**) *Trichuris suis*; with light bipolar plugs (PF), content is brownish. (**d**) *Physocephalus sexalatus* (with larva = L_1). (**e**) *Strongyloides* sp. (always containing a larva = L_1). (**f**) *Metastrongylus* sp.; the eggs of lungworms have a thick, dark grey and wrinkly shell and contain a larva (L_1). (**g**) *Ascaris suum*; the egg appears golden brown with a very wrinkly surface; the content is unembryonated. (**h**, **i**) *Hyostrongylus rubidus* and *Oesophagostomum* sp.; these eggs are not distinguishable among each other and also not from those of the hookworms of the genus *Globocephalus*. Species determination is possible by larva cultivation. The L_3 of *Hyostrongylus* show a long tail (from anus to the terminal end: about 70 μm) and those of *Oesophagostomum* a short tail (about 50 μm); those values do not refer to the always existing sheath but only on the body of the worm (BM = blastomeres)

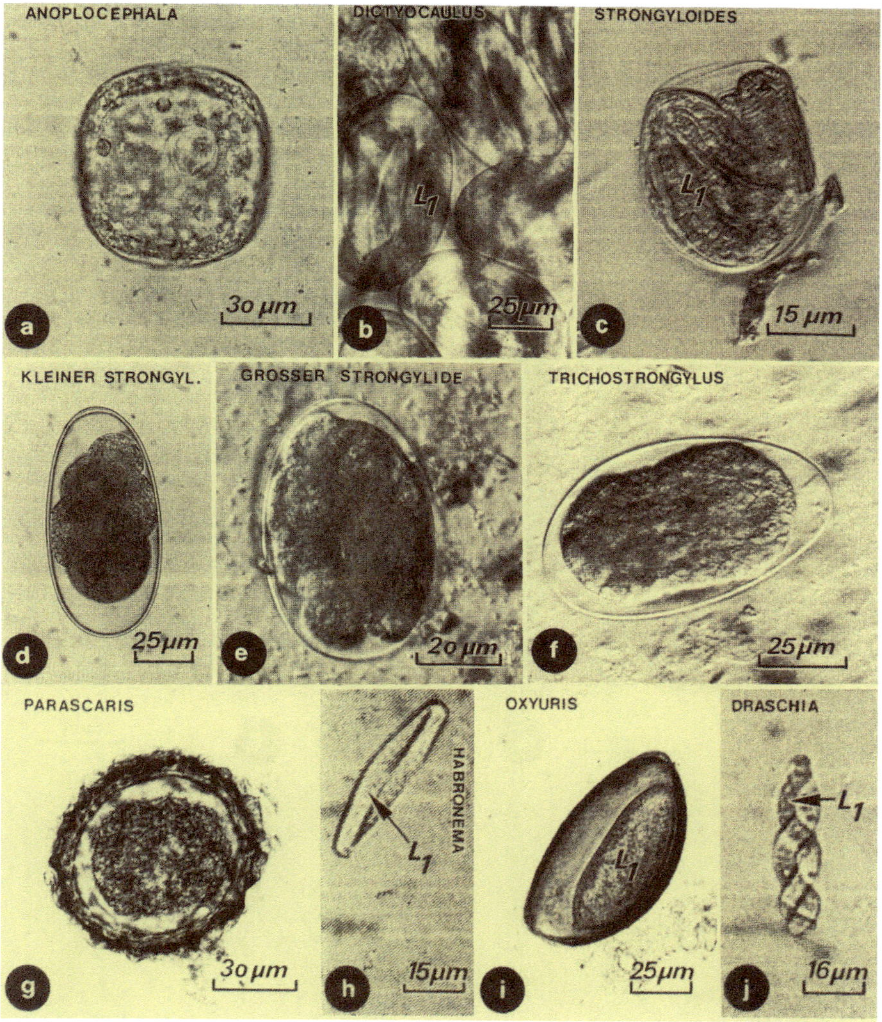

Fig. B.2 Worm eggs of equids. (**a**) *Anoplocephala perfoliata* (cestode egg with oncosphaera). (**b–j**) Nematode eggs. (**b**) *Dictyocaulus arnfieldi*; larva 1 (L_1; 300–480 µm) is hatching from the egg very quickly. (**c**) *Strongyloides westeri*; the (normal) ovoid egg (with equal poles) contains a short, thick larva 1 (L_1). (**d**) Type of eggs of small strongylids; the eggs measure more than 100 µm in length, have parallel side walls and contain only a few big blastomeres; the width of the eggs is smaller than half of the length. (**f**) *Trichostrongylus* sp.; the egg appears as an irregular ellipse (=poles are unequal). (**g**) *Parascaris equorum*; the spherical, brown-yellowish egg is covered by a wrinkly and albuminous (shrinking) layer. (**h**) *Habronema* sp.; larva 1 (L_1) inside the egg mostly appears folded once. (**i**) *Oxyuris equi*; the asymmetrical egg mostly contains a larva 1 (or at least a quite developed embryo). (**j**) *Draschia* sp.: larva 1 is rolled up spirilla-like

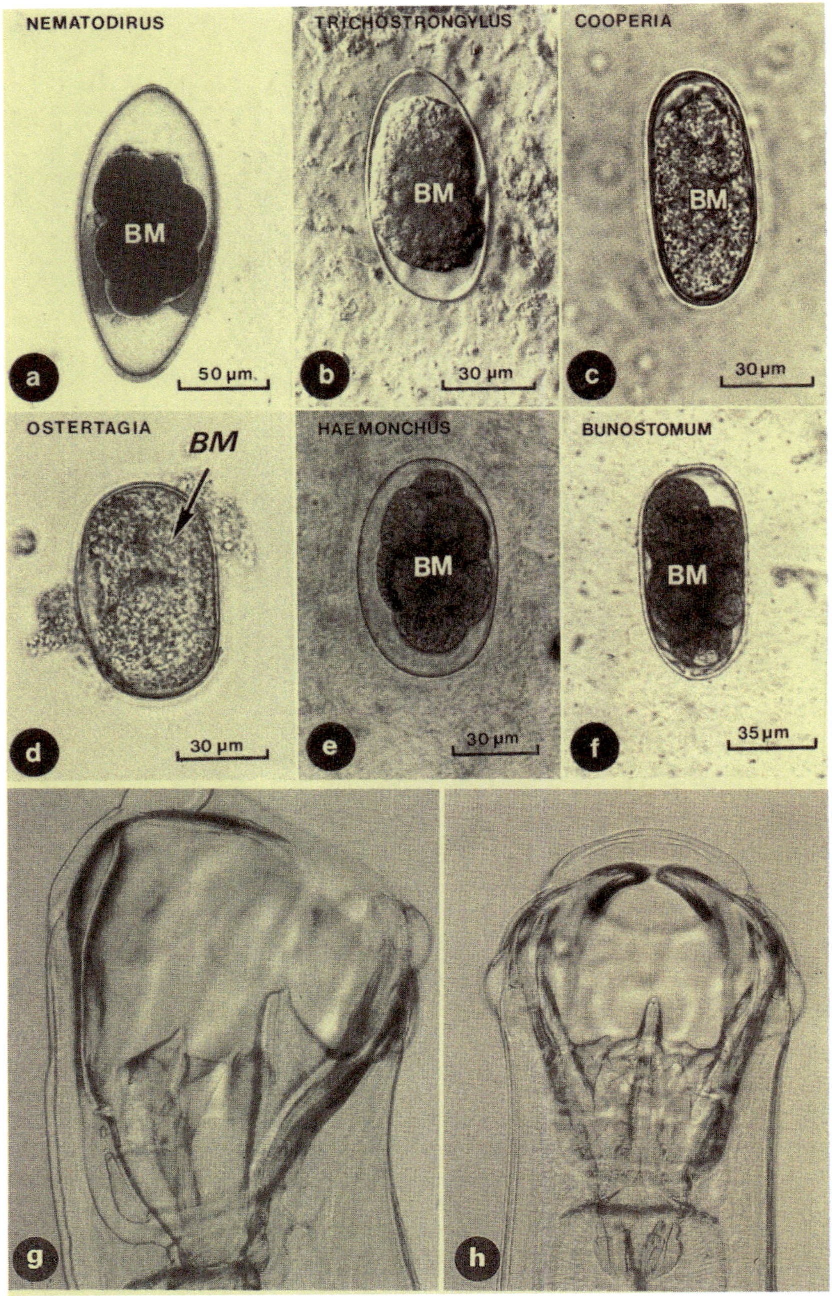

Fig. B.3 Stages of nematodes of ruminants.(**a–f**) Light micrographs of eggs of different nematode genera, which (except for *Nematodirus*) (**a**) are hardly distinguishable (the eggs origin from defined females). (**g**, **h**) Light micrographs of the anterior end of the hookworm *Bunostomum phlebotomum*; in dorsal and lateral view, the teeth appear at the base of the oral cavity; in (**g**) two teeth are shown in lateral view; in (**h**) one tooth is shown from above. Unstained, BM = blastomeres (in different numbers)

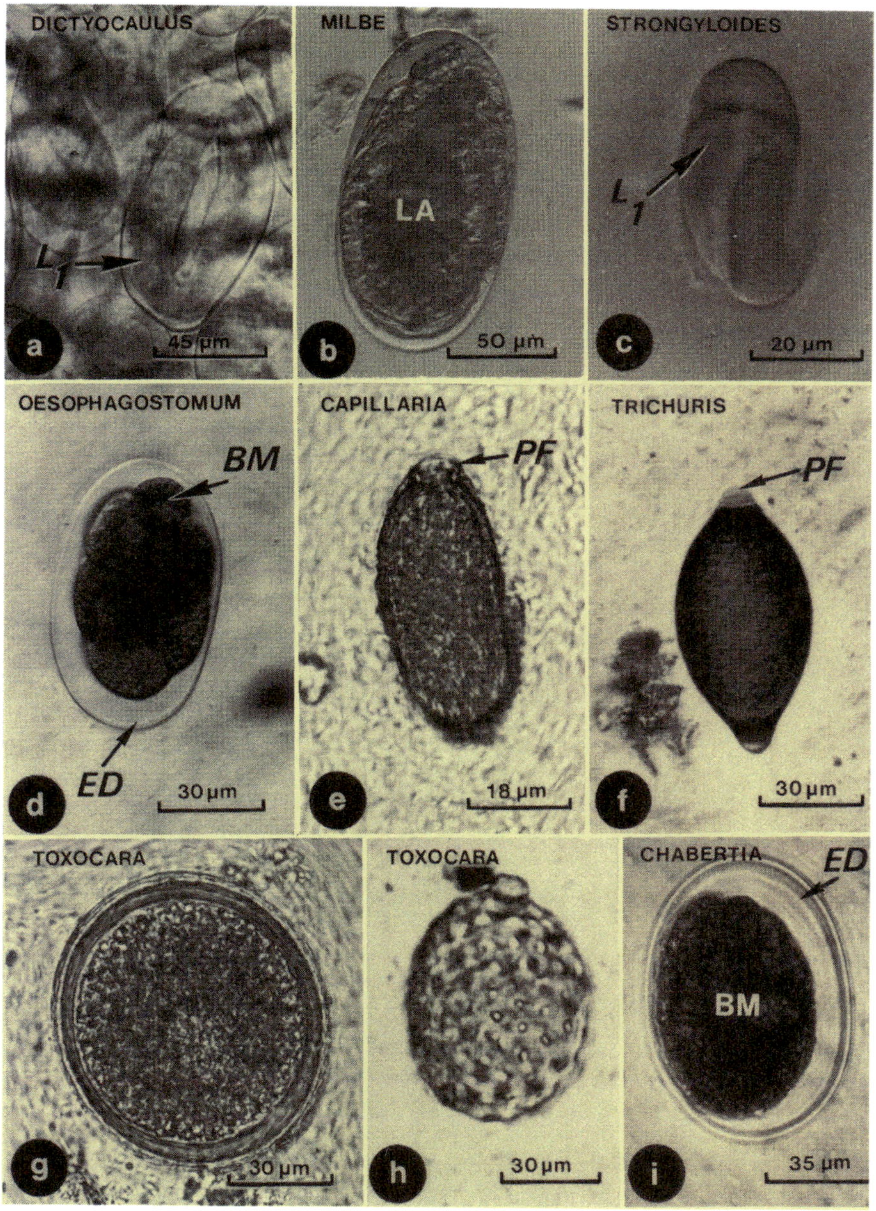

Fig. B.4 Light micrographs of nematode and mite eggs.(**a**) *Dictyocaulus viviparus*; eggs are rare; normally larvae are found (L_1 = larva 1). (**b**) Mite egg; it ended up with the food in the intestine of the cattle (LA = larval stage). (**c**) *Strongyloides papillosus*; the egg shell is thin and uncoloured; the larva hatches on the day of depositing the feces. (**d**) *Oesophagostomum* sp.; thin egg shell with a smooth exterior; the interior of the shell is covered by a yolk membrane (ED); eggs contain 16–32 blastomeres (BM) in fresh feces. (**e**) *Capillaria* sp.; thick, wrinkly shell, polar plugs (PF) hardly protruding. (**f**) *Trichuris ovis*; eggs often dark brown, lemon shape with protruding translucent polar plugs (PF). (**g**, **h**) *Toxocara vitulorum*; ascarid egg in central view (**g**) and from above (**h**); the shell shows fine dents (sculptured). (**i**) *Chabertia ovina*; egg with equally rounded poles, smooth surface; 16–32 blastomeres (BM) inside

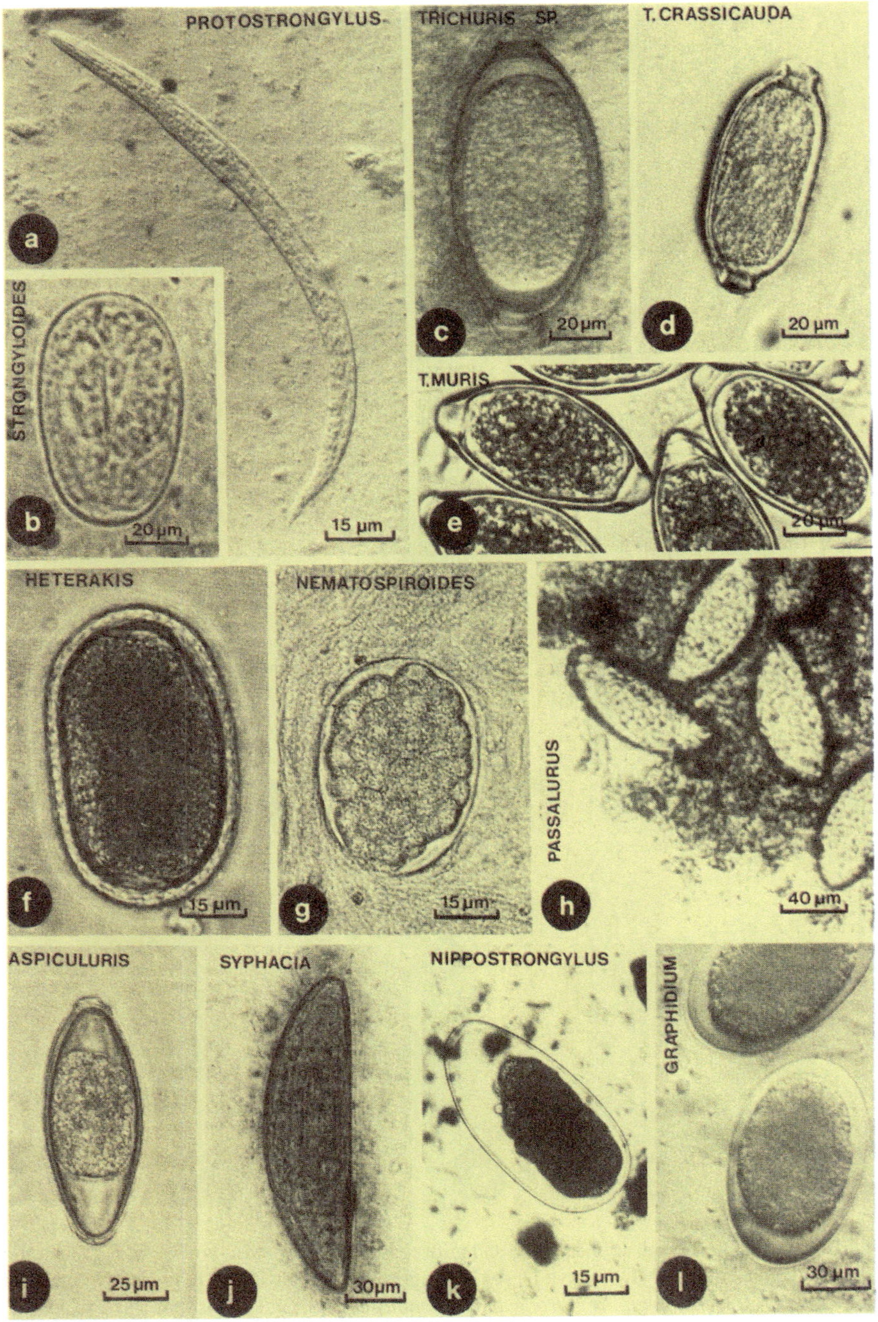

Fig. B.5 Stages of nematodes of hares, rabbits and laboratory rodents. (**a**) *Protostrongylus* sp.; larva of lung worms. (**b**) *Strongyloides* sp.; larva-containing egg. (**c**, **e**) *Trichuris* species in feces with clearly protruding polar plugs. (**d**) *Trichosomoides crassicauda*; this egg is found in the urine of rats and contains mostly a larva; the polar plugs are flattened; the shell is thick and brown. (**f**) *Heterakis spumosa*. (**g**) *Heligmosomoides* (syn. *Nematospiroides*) *dubius*. (**h**) *Passalurus ambiguus*. (**i**) *Aspiculuris tetraptera*. (**j**) *Syphacia obvelata*. (**k**) *Nippostrongylus* sp. (**l**) *Graphidium* sp.

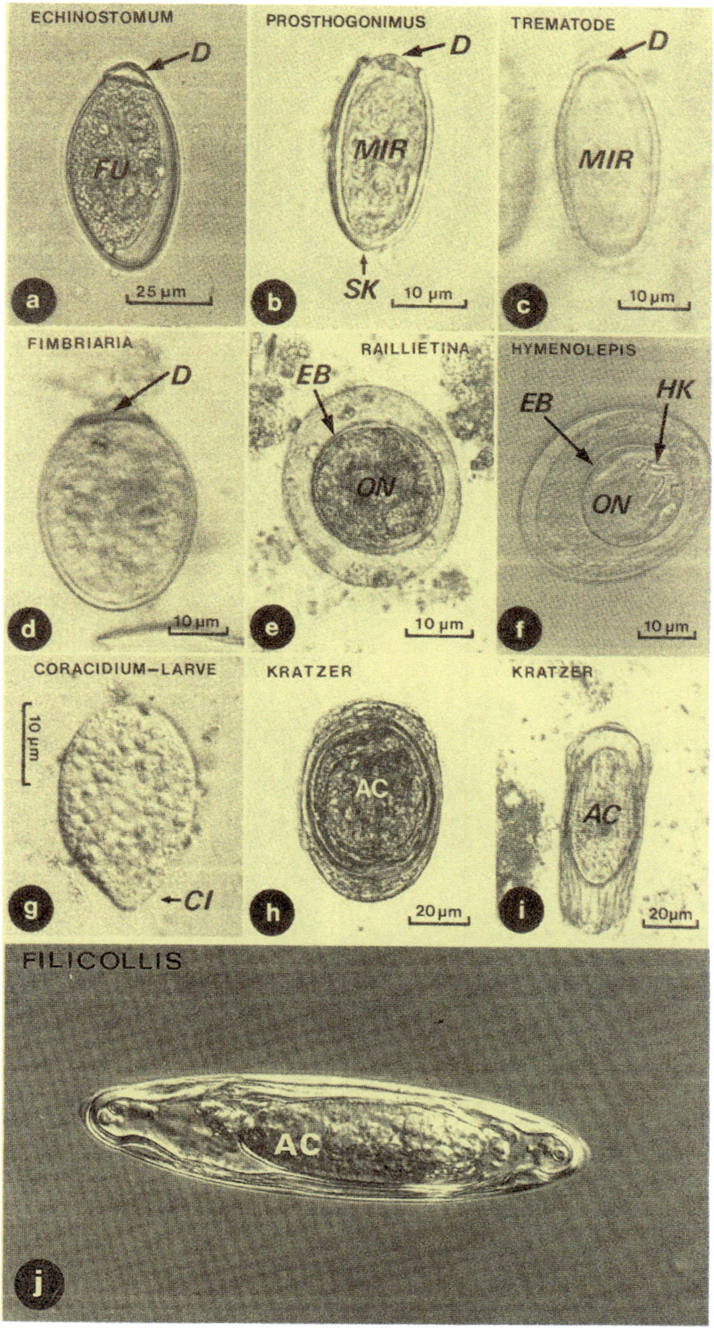

Fig. B.6 Worm eggs of birds. Light micrographs of worm eggs (trematodes, cestodes, acanthocephalans) from bird feces; (**g**) shows a coracidium larva of a pseudophyllidean tapeworm (*Fibriaria*). (**h–i**) show eggs of acanthocephalans; only *Fillicollis anatis* (**j**) origins from experimental transmission. AC = acanthor; CI = cilia; D = operculum, lid; EB = embryophore; FU = embryonic cells; HK = hook; MIR = miracidium; ON = oncosphaera; SK = shell buttons

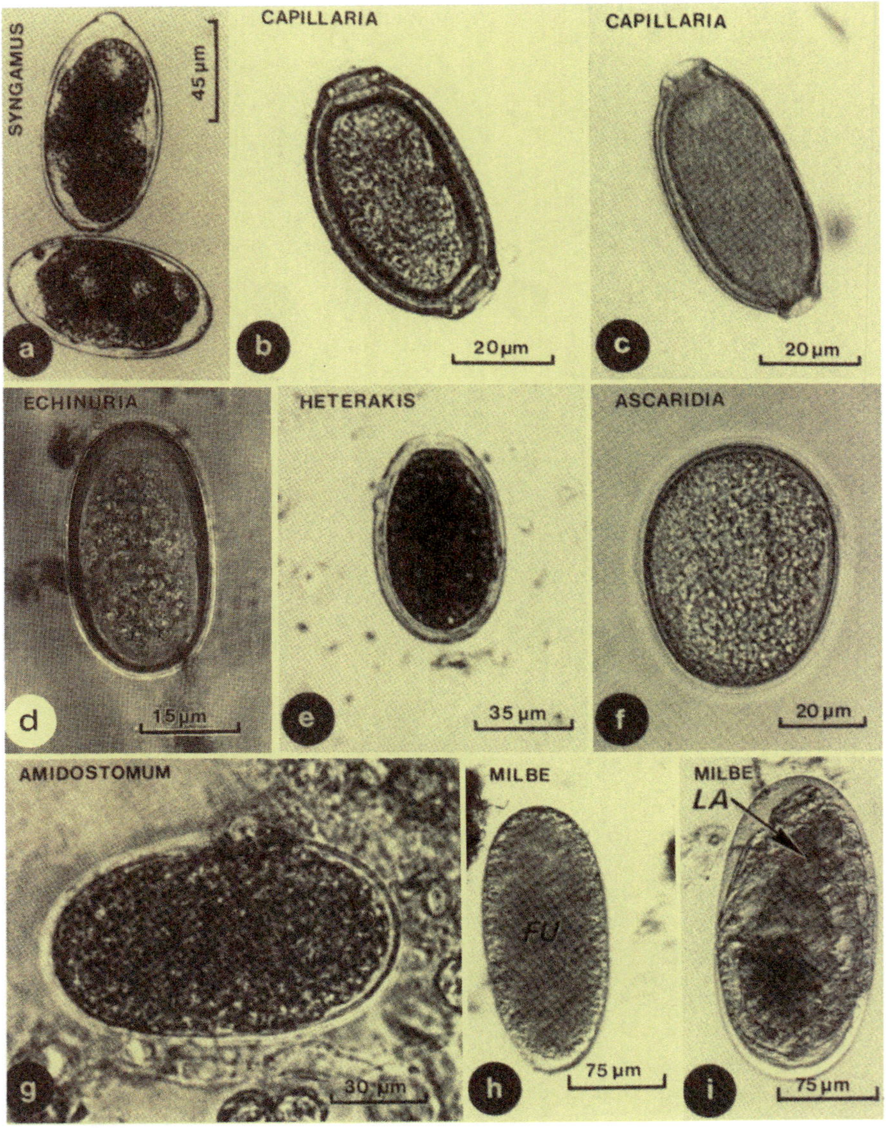

Fig. B.7 Nematode eggs of birds. Light micrographs of eggs of different nematode genera (**a**–**g**) and food mites, which may occur in feces (**h**, **i**). FU = cleavage stages; LA = larva

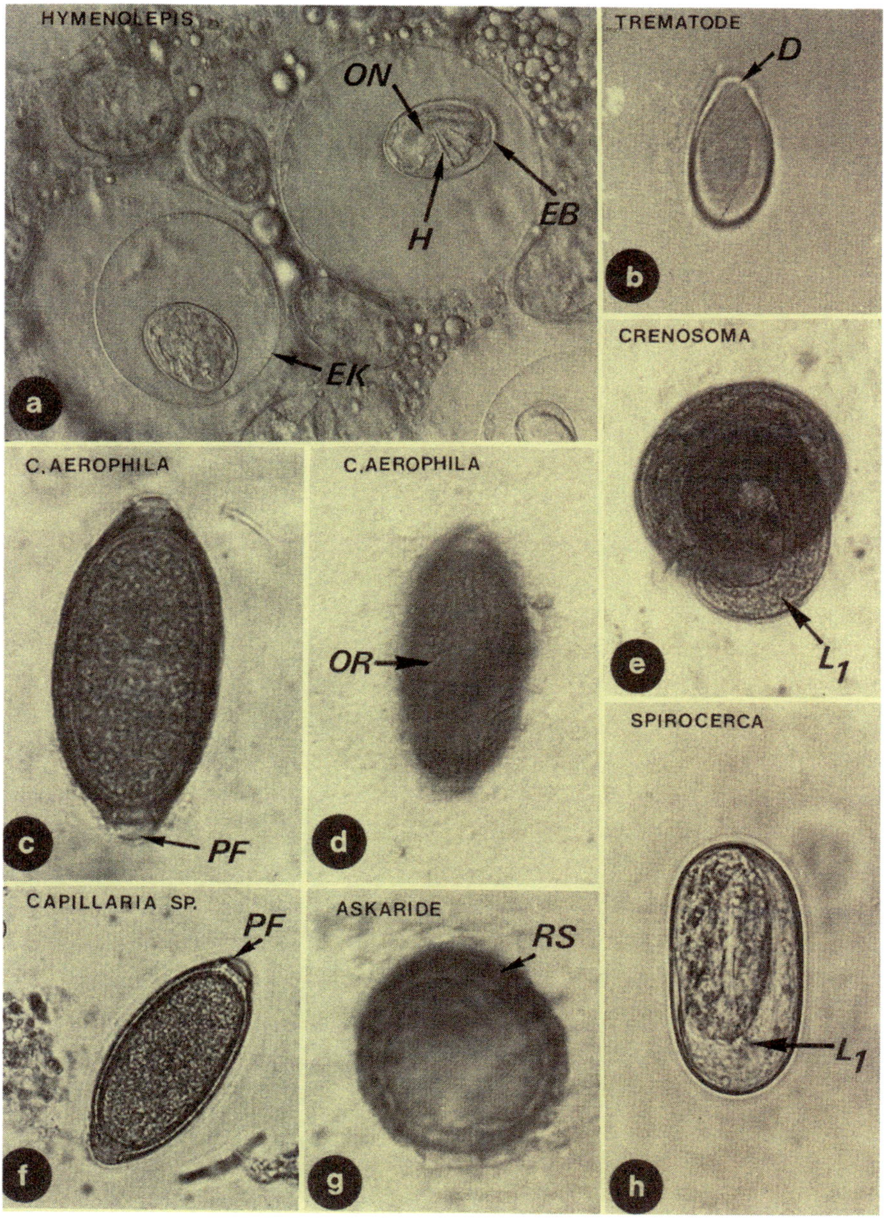

Fig. B.8 Worm eggs of hedgehogs. (**a**) Tapeworm eggs (*Hymenolepis erinacei*). (**b**) Trematode egg (*Brachylaemus erinacei*). The excretion of eggs stopped after treatment with praziquantel, even though no adult stage was found. (**c**, **d**) *Capillaria aerophila* (lung hairworm); the rough surface of the shell (OR) is characteristic. (**e**) *Crenosoma striatum* (lung worm); egg with hatching larva (L_1). Often only larvae of this worm are found. (**f**) *Capillaria* sp.; egg of an intestinal hairworm; characteristic is the smooth surface. (**g**) Ascarid egg from hedgehog feces, not determined. (**h**) *Spirocerca* sp.; egg of a spiruid containing already a larva (L_1). D = operculum, lid; EB = embryophore (=inner egg membrane); EK = egg capsule; H = hooks (3 pairs); L_1 = larva 1; ON = oncosphaera; OR = superficial wrinkle; PF = polar plug; RS = rough shell

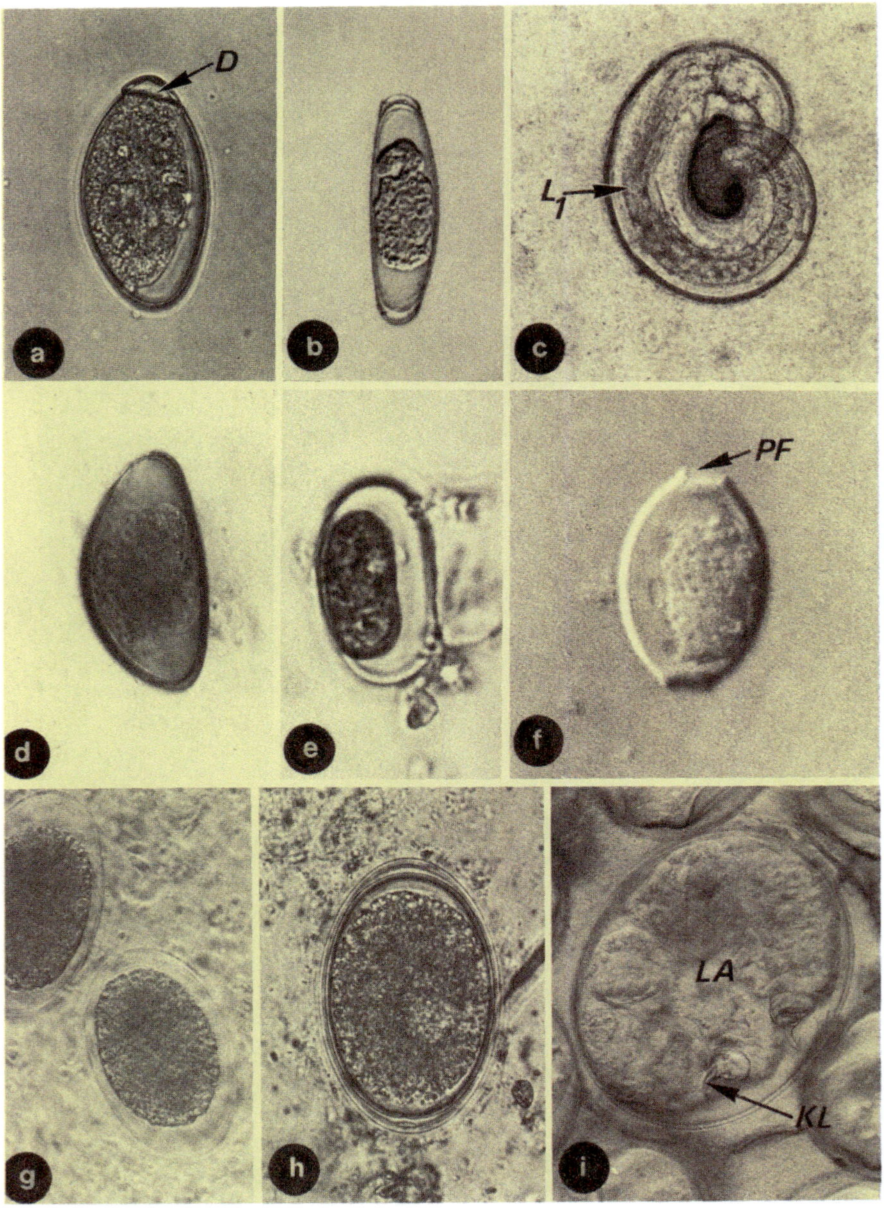

Fig. B.9 Worm eggs of reptiles and amphibians. (**a**) Trematode egg. (**b**, **d**) Oxyurid egg. (**c**, **e**) Spiruoidea egg. (**f**) Trichuroidea egg. (**g**, **h**) Strongyloid egg. (**i**) Pentastomid egg. D = operculum, lid; KL = claws of the extremities; L1 = larva 1 of nematodes; LA = larva of pentastomids; PF = polar plug

Addendum C: Questions to Test Obtained Knowledge

Only one answer is correct!

1. What is a final host?
   (a) The final host exclusively carries the asexual stages of a parasite.
   (b) The final host carries nothing but the sexual stages of a parasite.
   (c) The final host is also called "paratenic host".
   (d) In the final host the transmission of parasites stops, because it is a failing host.
   (e) After penetration into the final host the parasite is bound to die.
2. What is a habitat?
   (a) The body appearance of a parasite (e.g. front tip, apex upside down).
   (b) The lifetime of a parasite in a host.
   (c) The penetration process into the host by the parasite.
   (d) The attachment to a host's surface by a parasite.
   (e) The environment of a plant or animal species.
3. What is a monoxenous life cycle and where does it occur?
   (a) Each host is afflicted by one parasite only (tapeworms).
   (b) The whole life cycle takes place in one cell only (Microsporidia).
   (c) The parasite species develops all its stages in one host only (Coccidia).
   (d) An interrupted process (tapeworm larvae).
   (e) The species of a parasite group derive from one existing family group only (Monogenea).
4. Where do the stages of the genus *Eimeria* occur?
   (a) Exclusively in carnivores.
   (b) In herbivores as well as in some omnivores.
   (c) Extracellularly in the intestinal lumen of their hosts.
   (d) Exclusively in snakes.
   (e) Inside the muscle fibres of warm-blooded animals.
5. *Eimeria tenella* parasitizes in cells...
   (a) ... of the intestine of cattle,
   (b) ... of the appendix of sheep,
   (c) ... of the caecum of chickens,
   (d) ... of the colon in horses,

© Springer International Publishing Switzerland 2016

H. Mehlhorn, *Animal Parasites*, DOI 10.1007/978-3-319-46403-9

   (e) ... of the stomach of earthworms which later have to be eaten by birds (e.g.blackbirds).
6. What are globidia?
   (a) Another name for Sarcosporidia.
   (b) Giant schizonts in the intestine cells of ruminants.
   (c) Another name for globular oocysts.
   (d) Wandering stages of hookworms.
   (e) Appendage to male nematodes.
7. Which combination is correct considering the hosts?
   (a) *Eimeria mitis – E. acervulina – E. maxima.*
   (b) *Eimeria scabra – E. stiedai – E. zürnii.*
   (c) *Eimeria ninakohlyakimovae – E. leuckarti – E. parva.*
   (d) *Eimeria bovis – E. alabamensis – E. praecox.*
   (e) *Eimeria brunetti – E. intestinalis – E. columbarum.*
8. Which symptom will never occur in case of coccidiosis?
   (a) Haemorrhage.
   (b) Catarrhal enteritis.
   (c) Nectrotizing enteritis.
   (d) Muscular dystrophy.
   (e) Loss of weight.
9. What is the meaning of morbidity?
   (a) The number of infections per time unit in a defined number of animals or humans.
   (b) The number of infections in an outbreak of disease.
   (c) The number of actual deaths in an outbreak of disease.
   (d) The number of reinfections in an outbreak of disease.
   (e) The infection potential of a pathogen.
10. How can lethality be explained?
   (a) The relationship between the actual deaths and the number of infected animals or humans.
   (b) The number of infections due to one developmental stage of a pathogen.
   (c) Another word for mortality.
   (d) A camouflaged phase of an infection.
   (e) The time before the outbreak of symptoms.
11. How can a sporulated oocyst of *Eimeria* be characterized? Its interior contains...
   (a) 2 sporocysts with 4 sporozoites each,
   (b) 2 sporocysts with 2 sporozoites each,
   (c) 4 sporocysts with 2 sporozoites each,
   (d) 4 sporocysts with 4 sporozoites each or
   (e) 2 sporocysts with 8 sporozoites each.
12. Sporulated *Toxoplama gondii* oocysts can be characterized as follows:
   (a) 4 sporocysts with 2 sporozoites each—occurring in the feces of cats.
   (b) 2 sporocysts with 4 sporozoites each—occurring in the feces of chickens.
   (c) 2 sporocysts with 4 sporozoites each—occurring in the feces of felidae.
   (d) 2 sporocysts with 2 sporozoites each—occurring in the feces of cats.
   (e) 2 sporocysts with 2 sporozoites each—occurring in the feces of dogs.

13. The final host of the pathogen of neosporosis is...
    (a) the dog,
    (b) the cat,
    (c) the horse,
    (d) the cow, or are
    (e) humans.
14. Which combination of pathogen and parasitism is *wrong*?
    (a) *Sarcocystis*—occurring intracellularly in muscle fibres.
    (b) *Trichinella*—occurring intracellularly in muscle fibres.
    (c) *Theileria*—occurring intracellularly in salivary glands of ticks.
    (d) *Besnoitia*—occurring in the interstitial tissue of the skin of cattle.
    (e) *Cystoisospora*—occurring intracellularly in intestinal cells of cattle.
15. What is a schizont?
    (a) Another name for meront.
    (b) A transmittable stage of *Eimeria* species.
    (c) A skin disturbing intruder.
    (d) A penetrating parasite.
    (e) A sexual stage in the microgamogony.
16. What are Sarcosporidia?
    (a) Species of Microsporidia with infestation in muscle cells.
    (b) Species of Myxozoa in the muscle of fish.
    (c) Species which afflict carnivores as final hosts and herbivores as intermediate hosts.
    (d) Spores of flagellates.
    (e) Spores of amoebae.
17. Which combination of vector, agent and host is *correct*?
    (a) Flea—*Babesia bovis*—cattle.
    (b) Raptor bug—*Theileria cruzi*—sheep.
    (c) Soft tick *(Argasidae)*—*Borrelia burgdorferi*—cattle.
    (d) Wood tick (*Ixodes ricinus*)—*Theileria equi*—horse.
    (e) Sand fly—*Trypanosoma equiperdum*—horse.
18. Which combination of disease, pathogen and vector is *correct*?
    (a) Babesiosis—*Babesia canis*—*Rhipicephalus*.
    (b) Isosporosis—*Isospora felis*—*Amblyomma*.
    (c) Hepatozoonosis—*Hepatozoon canis*—*Phlebotomus*.
    (d) Toxoplasmosis—*Toxoplasma gondii*—*Rattus maleficus*.
    (e) Elephantiasis tropica—*Brughia malayi*—*Chrysops*.
19. What is a micropyle?
    (a) An aperture in the sporocyst.
    (b) An aperture in the sarcocyst.
    (c) A thinner portion of the oocyst wall (e.g. *Eimeria*).
    (d) A thinner portion in the cyst wall of *Giardia*.
    (e) A thinner portion in the cyst wall of *Amoeba*.
20. Where does the primary cyst wall occur?
    (a) In each cell being infected with Sporozoa.

   (b) In the oocyst of malaria agents.
   (c) In the sporocyst of *Eimeria* oocysts.
   (d) As border of the tissue cyst in case of *Toxoplasma gondii.*
   (e) As boundary of the endoerythrocytic stages of plasmodia.
21. Which statement is *correct*?
   (a) *Cryptosporidium* species occur in animals only.
   (b) *Cryptosporidium* species have oocysts containing 2 sporocysts with 2 sporozoites each.
   (c) Piroplasms occur exclusively in animals.
   (d) *Leucocytozoon* species are the agents of piroplasmosis.
   (e) The cyst of *Giardia* contains 4 nuclei.
22. Which statement is *wrong*?
   (a) *Theileria* species multiply only in lymphocytes.
   (b) *Babesia* species multiply in dogs exclusively in the erythrocytes.
   (c) East coast fever is a disease caused by *Theileria* living in ticks.
   (d) Red water fever is caused by an infection with *Babesia.*
   (e) *Balantidium coli* produces vegetative stages with 2 types of nuclei.
23. Where can *Buxtonella sulcata* be found?
   (a) In the colon of pigs.
   (b) In the colon of cattle.
   (c) In the caecum of cattle.
   (d) In the caecum of chickens.
   (e) In the colon of odd-toed ungulates (*Perissodactyla).*
24. Where does *Entamoeba invadens* occur?
   (a) In ungulates.
   (b) In equine species.
   (c) In rodents.
   (d) In chickens.
   (e) In reptiles.
25. How do microsporidia develop?
   (a) Intracellularly.
   (b) Extracellularly in the body cavity.
   (c) In tissue cysts.
   (d) In an ameboma.
   (e) In threads of slime.
26. Endo- and exospores can be found in…
   (a) Oocysts of *Eimeria* species.
   (b) Sporocysts of *Eimeria* species.
   (c) Spores of Microsporidia.
   (d) Spores of caryospores.
   (e) Sporangia of *Plasmodium* species.
27. What is the outer boundary of flukes/trematodes and tapeworms?
   (a) A one-layer epithelium.
   (b) A squamous epithelium.
   (c) A cellular cuticle.

(d) A non-cellular, tanned layer of proteins.
(e) A tegument without inner cell membranes and nuclei.

28. Which stages occur in the life cycle of *Fasciola hepatica*?
    (a) Free miracidium—sporocyst—redia—cercaria—metacercaria.
    (b) Free miracidium—sporocyst—cercaria.
    (c) Free miracidium—sporocyst—redia—cercaria.
    (d) Not free miracidium—sporocyst—cercaria.
    (e) Free miracidium—sporocyst—cercaria—metacercaria.
29. Which adult parasites roam in the bile ducts of their hosts?
    (a) *Fasciolopsis buski*
    (b) *Paramphistomum cervi*
    (c) *Schistosoma bovis*
    (d) *Opisthorchis felineus*
    (e) *Heterophyes heterophyes*
30. Where do so-called Monogenea occur?
    (a) In the intestine of ruminants.
    (b) In the intestine of equids.
    (c) In the area of the lungs in birds.
    (d) In the interior of lungs of reptiles.
    (e) On the gills of fish.
31. How does the surface of tapeworms look like?
    (a) The monolayer epithelium helps to enlarge its own surface with microvilli.
    (b) The multilayer epithelium shows so-called microtriches.
    (c) The syncytial tegument prosesses so-called bosses.
    (d) The multilayer tegument shows lots of thorns.
    (e) The tegument is covered by a monolayer membrane and produces many piercy microtriches.
32. How can a dog become infected by *Dipylidium caninum*?
    (a) By oral uptake of of a cysticercus in raw meat.
    (b) By oral uptake of a cysticercus in raw fish.
    (c) By oral uptake of plerocercoids in small crabs out of the water of brooks or rivers.
    (d) By ingestion of the liver of sheep.
    (e) By swallowing infected fleas.
33. Which length may be reached by equine tapeworms?
    (a) Up to 4.5 cm.
    (b) Up to 80 cm.
    (c) Up to 1.50 m.
    (d) Up to 3 m.
    (e) Up to 16 m.
34. What is characteristic for adult equine tapeworms?
    (a) The proglottids are square shaped.
    (b) The body appears wedge shaped.

(c) They possess rectangular proglottids in the size of 0.8 cm × 0.2 cm.
(d) They show 4 suckers and 1 crown of hooks.
(e) They show proglottids looking like a string of pearls.

35. What is a metacestode?
(a) A primitive tapeworm species.
(b) The section of the tapeworm situated posterior to the scolex.
(c) The terminal proglottids dropped by the tapeworm.
(d) The tapeworm larva in the intermediate host.
(e) A tapeworm after an unsuccessful/unsufficient medical treatment.

36. What is an oncosphaera?
(a) An infectious larva of *Taenia* tapeworms.
(b) The 10-hook larva of monogeneous gill worms.
(c) The larva of a hookworm.
(d) The infectious larva of pentastomids.
(e) The infectious larva of trematodes.

37. Which are the final hosts of *Taenia pisiformis*?
(a) Hominids.
(b) Canids.
(c) Ungulates.
(d) Equids.
(e) Hares, rabbits.

38. Which length has *Echinococcus multilocularis*, the fox tapeworm?
(a) Up to 1 m.
(b) Up to 1.50 m.
(c) Up to 3 mm.
(d) Up to 1.5 cm.
(e) Up to 0.8 cm.

39. Which one of the following statements on *Schistosoma* species is *correct*?
(a) The female *Schistosoma* stage envelops the smaller male with its widened side parts.
(b) The schistosomula needs to reach the blood of snails.
(c) The eggs of *Schistosoma* have to be ingested by snails.
(d) The rediae of *Schistosoma* species produce cercariae in snails.
(e) The schistosomes mate in the blood vessels of their host. They stay in pairs for the rest of their life and lay eggs.

40. How do the eggs of *Schistosoma haematobium* reach the lumen of the bladder?
(a) Via urethra.
(b) They penetrate the bladder wall with the help of their egg sting.
(c) Via intestine.
(d) The eggs are deponed in the bladder.
(e) They penetrate through the bladder wall with the help of tissue-induced inflammations.

41. Which of the following statements is *wrong*? *Echinococcus granulosus* ....
(a) does not occur in Europe any more.
(b) reaches the final host in its egg stage via so-called smear infections.

   (c) can develop in sheep only to cyst larvae but not to sexual mature worms.
   (d) occurs as cyst larva in numerous herbivores.
   (e) Adults constrict motile proglottids at their terminal pole, although they possess only 3–4 proglottids.

42. Which of the following statements on the biology of the cattle tapeworm is *wrong?*
   (a) The cyst larva penetrates the intestine wall of cattle.
   (b) Mature proglottids with their eggs can leave the intestine via anus actively.
   (c) The more terminal proglottids are fertilized by the younger ones.
   (d) The cattle tapeworm does not possess a crown of hooks at its scolex.
   (e) The cyst larva of the cattle tapeworm is smaller than 10 mm.
43. How do the tapeworms take up their food?
   (a) Via the proboscis.
   (b) With its mouth with the help of its surrounding sucker.
   (c) Through the tegument.
   (d) Via phagocytosis of peculiar cells.
   (e) By its ventral mouth.
44. Which one of the following statements on trichines is *correct*?
   (a) Worms are transmitted from carnivore to carnivore by ingesting muscle trichines.
   (b) Worms are transmitted regularly from herbivore to carnivore by ingesting muscle trichines.
   (c) Eggs are excreted within the feces of herbivores.
   (d) Eggs are excreted within the feces of carnivores.
   (e) The transmission occurs orally by female parasites being hatched in the intestine.
45. Which one of the following statements on roundworms is *wrong?*
   (a) After eggs are laid in the mesenteria of the intestine or the bladder, they reach the lumen of the intestine or the bladder due to inflammatory processes.
   (b) Eggs need a time of staying in the open air.
   (c) Larvae can already slough their skin in the egg shell.
   (d) Larvae undergo a passage through the portal vein.
   (e) Muscle cells of the adults form protrusions to the nerves.
46. Which one of the following statements is *correct?*
   (a) *Ascaris* larvae pentrate the skin to enter their host.
   (b) An infection with *Oxyuris* species can be induced either by taking up eggs orally or by invasion of larvae into the anus.
   (c) All female filariae try to leave their host via skin in order to lay their eggs into the water.
   (d) Animals are infected by *Ancylostoma* by ingestion of eggs.
   (e) Infections with nematodes happen only by eating food contaminated with eggs.

47. The infection with hookworms occurs…
    (a) through penetration of cercariae into the skin while bathing.
    (b) during the bite of blackflies.
    (c) by ingestion of fertile eggs with the food.
    (d) by ingestion of raw meat.
    (e) by active penetration of filariform larvae.
48. The infection of cattle with *Trichinella spiralis* occurs…
    (a) not at all.
    (b) by uptake of worm eggs within contaminated food.
    (c) via mosquito bites.
    (d) by ingestion of metacercariae from grass blades.
    (e) by viviparous larvae, which penetrate into the skin.
49. Acanthocephalans are…
    (a) worms without intestine.
    (b) ectoparasites of amphibians.
    (c) a special group of trematodes, the so-called monogeneans.
    (d) cestodes, belonging to the group of Caryophyllidea.
    (e) none of these.
50. Elephantiasis tropica is induced by …
    (a) *Wuchereria bancrofti.*
    (b) *Loa loa.*
    (c) *Dracunculus medinensis.*
    (d) *Leishmania.*
    (e) *Furunculosa vitiosa.*
51. The agents of the plague are transmitted by…
    (a) bites of ticks.
    (b) feces of lice.
    (c) stings of the rat flea.
    (d) stings of ants.
    (e) feces of sand fleas.
52. Which of the following statements is *correct*?
    (a) Lice can only survive with symbionts, which therefore have to be transferred into the eggs by the female louse.
    (b) The trichobothria of the pygidial plate of fleas are used for the perception of attractants being excreted by the host.
    (c) In case of the mosquitoes the first maxilla forms a saliva channel and the second maxilla is the tube for bloodsucking.
    (d) The females of the mosquitoes have no wings.
    (e) Body lice transmit the agent of spotted fever exclusively while sucking blood.
53. Which agents of disease are *not* transferred during blood sucking?
    (a) The pathogen of the Chagas disease by bed bugs.
    (b) Microfilariae by tabanids or mosquitoes.
    (c) Agents of encephalitis by ticks.
    (d) Rickettsiae by ticks.

(e) Agents of plague by fleas.

54. Which combination of vectors and diseases is *correct*?
    (a) Flea: plague, scabies, spotted fever.
    (b) Tick: texas fever, tularemia, trichomoniasis.
    (c) Bed bug: typhus, plague, trichomoniasis.
    (d) Mosquitoes: filariosis, yellow fever, malaria.
    (e) Pubic louse: typhus, syphilis, maroditis.
55. The distribution of body lice increases. Which determining feature combination is significant for these parasites?
    (a) A maximum of 3 pairs of legs and a Haller's organ.
    (b) A maximum of 3 pairs of legs with claws and absence of wings (apterism).
    (c) "Piercing-sucking mouthparts and halteres.
    (d) Pupae appearing as tons and adults have no wings.
    (e) Mycetomes with symbionts and parthenogenesis.
56. What is a myiasis?
    (a) A virus infection transmitted by mosquitoes.
    (b) A bacterial infection transmitted by biting house flies (e.g. *Stomoxys*).
    (c) Rickettsiosis.
    (d) Disease induced by wandering larvae of flies.
    (e) Disease caused by wandering worm larvae.
57. Which determining features are significant for adult ticks?
    (a) 3 pairs of legs and the absence of wings (apterism).
    (b) Wings and piercing mouthparts existing.
    (c) Piercing mouthparts and 4 pairs of legs.
    (d) 4 pairs of legs and 2 pairs of maxillae.
    (e) Tracheoles and halteres.
58. Which one of the following statements is *wrong*?
    (a) Male mosquitoes transmit malaria when bloodsucking.
    (b) Females of the head louse attach their eggs to the head hair of humans.
    (c) Female and male ticks of *Ixodes ricinus* can transmit pathogens of encephalitis.
    (d) Male fleas suck blood, too.
    (e) Female body lice transmit symbiontic bacteriae into their eggs.
59. Scabies mites are actually progressing. Which of the following statements is *correct*?
    (a) Scabies mites live on the skin and feed on it.
    (b) They have piercing mouthparts and suck blood.
    (c) They dig tunnels in the skin.
    (d) They are surrounded by a skin wall only when sucking blood and lead to tissue inflammation.
    (e) They feed on house dust.
60. Which one of the following statements is *correct?* The flour mite transmits...
    (a) the agents of toxoplasmosis.
    (b) the agents of typhus.
    (c) the agents of scabies.

(d) the agents of amoebic dysentery.
(e) no pathogens.

61. Which one of these statements is *wrong*?
    (a) Hydatids are the larvae of the pig tapeworm *Taenia solium.*
    (b) Sporocysts are multiplication stages of trematodes in snails.
    (c) The metacercariae of some trematodes can be observed in the muscle of fish.
    (d) The larva 3 of *Necator americanus* lives outside of the body.
    (e) *Echinococcus multilocularis* worms may occur in large numbers in the intestine of fox, dog and cat.
62. Which combination of parasitic stage and afflicted organ is *wrong*?
    (a) Liver: malaria schizonts and eggs of schistosomes.
    (b) Liver: cysts of *Entamoeba* and larvae of *Ascaris.*
    (c) Liver: *Clonorchis sinensis* and oxyures.
    (d) Eye: adults of *Loa loa* and larvae of *Onchocerca volvulus.*
    (e) Skin: *Sarcoptes scabiei* and larvae of *Onchocerca volvulus.*
63. Which statement is *wrong*?
    (a) Cestodes feed through their sucker.
    (b) The intestine of the trematodes is bifurcated and terminally closed.
    (c) Schistosomes live in the vein system of their hosts.
    (d) *N*ematodes can be transmitted by mosquitoes.
    (e) *Ancylostoma* larvae penetrate actively into the human skin.
64. The so-called *Cysticercus cellulosae* is. . .
    (a) . . . the larva of the dog tapeworm *Echinococcus granulosus.* It can be situated in the muscles of sheep.
    (b) . . . the larva of the dog tapeworm *Taenia pisiformis.* It can occur in the liver of sheep.
    (c) . . . the larva of the tapeworm *Taenia solium.* It can be found in multiple organs of humans.
    (d) . . . the larva of schistosomes. It can be situated in the human liver.
    (e) . . . the larva of the Chinese liver fluke. It can be found exclusively in the muscle of fish.
65. The infection with roundworms of the genus *Ascaris* takes place by . . .
    (a) ingestion of freshly laid eggs with contaminated salad a.s.o..
    (b) ingestion of eggs having been stored outdoors for a long while.
    (c) the ingestion of larvae in unsufficiently cooked meat.
    (d) percutaneous penetration of rhabditiform larvae.
    (e) percutaneous penetration of filariform larvae.
66. How can an infection with *Schistosoma haematobium* be diagnosed?
    (a) By the proof of eggs with a lateral spike in human feces.
    (b) By the proof of eggs with a terminal spike in human urine.
    (c) By the proof of eggs with a terminal spike in human sputum.
    (d) By the proof of eggs with a terminal spike in the feces.
    (e) By the proof of eggs with a lateral spike in the urine.
67. Which statement is *correct*?

(a) Flies never suck blood while all mosquitoes do it.
(b) Only female fleas suck blood, but they transmit the pathogen of the plague via feces.
(c) Males and females of the body lice suck blood.
(d) Both males and females of *Anopheles* mosquitoes transmit the pathogen of malaria.
(e) Males and females of the body lice transmit spirilles while sucking.

68. Which combination is *correct*?
(a) Leishmaniasis—phlebotomids—Ile de France
(b) Maroditis perniciosa—louse—South East Bavaria
(c) Cysticercosis—mosquitoes—Venezuela
(d) Leishmaniasis—sand fly—Baleares, Spain, Italy
(e) Filariosis—tick—Germany

69. Which one is known as the vector of the agent of Lyme borreliosis?
(a) All argasid ticks.
(b) The pigeon tick.
(c) The castor bean tick.
(d) The bed bug.
(e) The brown dog tick.

70. Which statement is *correct*?
(a) The egg shell of Ascarids is thick. The egg contains a larva already when deponed.
(b) The eggs of the so-called whipworms contain a larva already when deponed.
(c) The eggs of *Trichuris* species possess bulbus-like protrusions at their poles.
(d) The eggs of filariae are undeveloped when deponed.
(e) The egg shell of hookworms is thick and shows a wrinkled surface.

71. Which statement is *wrong*?
(a) In the case of species of the genus *Strongyloides* both females and males can be found in the intestine.
(b) The stages of *Strongyloides* species found in the intestine of pigs are always females.
(c) *Strongyloides* species are normally not longer than a few millimeter.
(d) *Strongyloides* species have a free-living generation.
(e) Larvae of *Strongyloides* can survive in tissue for long.

72. What are hypobiotic stages?
(a) Sleeping stages with arrested development.
(b) Undersized stages.
(c) Fully developed stages.
(d) Stages which penetrate into lower parts of the body.
(e) Stages which cannot multiply any more.

73. What is the meaning of prepatency?
(a) The time of life of the pathogen after getting mature.
(b) The time of life of the host before showing symptoms.

(c) The time of life of the host before new infectious pathogens are developed.
(d) The time just before death.
(e) The time before the penetration into a host.

74. The larva 4 of the so-called small *Strongyloides* can be observed in the...
(a) ileum.
(b) caecum.
(c) dorsal colon.
(d) duodenum.
(e) abomasum.

75. What are spicula?
(a) Thorns in a trematode's segment.
(b) Copulation organs of the male fleas.
(c) Paired copulatory organs in nematodes.
(d) A produdible accessory copulation system in case of tapeworms.
(e) Inner stabilization elements in the body of arthropods.

76. Who does *not* live in permanent copulation?
(a) Hookworms.
(b) Schistosomes.
(c) *Diplozoon paradoxum.*
(d) *Syngamus trachea.*
(e) *Strongylus vulgaris.*

77. How can specimens of the genus *Ollulanus* be transmitted?
(a) By ingestion of vomited intestinal fluids.
(b) Via penetration of free-living larvae.
(c) By ingestion of larvae containing eggs.
(d) By a mosquito bite.
(e) By the sting of ticks.

78. Adult stages of *Dictyocaulus* live...
(a) in the lung system.
(b) in the rectum.
(c) in the ceacum.
(d) in the stomach.
(e) in the trachea.

79. The intermediate hosts of protostrongylids are...
(a) ants.
(b) larvae of fleas.
(c) mallophages.
(d) terrestrial slugs.
(e) water snails.

80. The *Filaroides* larvae of canids live...
(a) in skin nodules.
(b) nodes in the lung.
(c) nodes in the wall of the abomasum.
(d) nodes in the trachea.
(e) nodes in the liver.

81. The species of *Anisakis* use as final hosts. . .
    (a) marine mammals.
    (b) birds.
    (c) fish of prey.
    (d) crustaceans.
    (e) canids.
82. Species of the genus *Habronema* of the horse are transmitted by. . .
    (a) flies of the genus *Musca*.
    (b) flies of the genus *Stomoxys*.
    (c) midges.
    (d) black flies.
    (e) mosquitoes.
83. The adults of *Dirofilaria immitis* can be found in. . .
    (a) the skin.
    (b) the liver.
    (c) the kidneys.
    (d) in the arterial system of the lungs.
    (e) in the rectum.
84. *Dirofilaria immitis* is transmitted. . .
    (a) by oral uptake of larvae containing eggs.
    (b) by penetration of free-living larvae.
    (c) by the bite of *Culex* mosquitoes.
    (d) by the bite of sand flies.
    (e) by the sting of midges.
85. *Onchocerca* species of cattle are transmitted. . .
    (a) via oral uptake of larvae containing eggs.
    (b) by oral uptake of larvae on grass.
    (c) by the sting of midges or blackflies.
    (d) by the bite of house mosquitoes.
    (e) by licking of calves at their mothers.
86. What is a scutum?
    (a) Back shield in the case of ticks.
    (b) Ventral plate in the case of ticks.
    (c) Covering plate in case of lice eggs.
    (d) Head plate in case of mites.
    (e) Overwintering stage of mites.
87. Where is the Haller's organ located?
    (a) On the pedipalps of argasids.
    (b) On the basic palpus of insects.
    (c) On the front tarsus of Ixodidae.
    (d) On the palps of female mosquitoes.
    (e) On the pedipalps of midges.
88. *Ixodes hexagonus* is . . .
    (a) the Brown dog tick.
    (b) fox tick.

(c) the mouse tick.
(d) Castor bean tick (wood tick).
(e) hedgehog tick.

89. The species of the genus *Ornithodoros* are ....
(a) Argasid ticks.
(b) Ixodid ticks.
(c) Vectors of borreliosis (Lyme disease).
(d) Vectors of the East Coast Fever.
(e) Vectors of the Texas Fever.

90. *Demodex* mites live ...
(a) in channels of the epidermis.
(b) in sebaceous glands of the skin.
(c) in quills of birds.
(d) in the subcutis.
(e) right on the skin.

91. *Dermanyssus gallinae* mites feed...
(a) by sucking blood.
(b) skin cells and flakes.
(c) by sucking lymph.
(d) as predators.
(e) on detritus.

92. *Varroa* mites can be found ...
(a) in bird nests.
(b) in foxholes.
(c) on the skin of hedgehogs.
(d) in hair follicles.
(e) on bee brood.

93. *Neotrombicula autumnalis* ....
(a) sucks blood.
(b) sucks lymph.
(c) drills channels in the epidermis.
(d) feeds on hairs in the fur.
(e) afflicts the lungs of birds.

94. *Notoedres* species...
(a) transmits the pathogens of borreliosis.
(b) transmits the pathogens of hepatozoonosis.
(c) leads to scabies-like lesions in the skin (similar to scabies) in cats.
(d) leads to summer wounds in horses.
(e) destroys food.

95. Mallophages belong to the groups of
(a) bugs.
(b) lice.
(c) fleas.
(d) mites.
(e) tabanids.

96. The bluetongue virus of ruminants is transmitted by...
    (a) fleas.
    (b) bugs.
    (c) midges.
    (d) Simuliidae.
    (e) Muscidae.
97. Moth flies belong to the family of...
    (a) Muscidae.
    (b) Culicidae.
    (c) Simuliidae.
    (d) Psychodidae.
    (e) Ceratopogonidae.
98. Blackflies develop in...
    (a) loose sand.
    (b) tranquil waters.
    (c) fast running waters.
    (d) detritus in stables.
    (e) in nests of animals.
99. Which of the following statements is *correct*?
    (a) In the case of horse flies both genders suck blood.
    (b) In the case of tabanids only females suck blood.
    (c) Horse flies possess long antennae with many segments.
    (d) Horse flies develop in clear fast flowing waters.
    (e) Horse flies feed on fruits.
100. Larvae of myiasis flies can easily be identified ...
    (a) by their squat legs.
    (b) by the shape of their legs.
    (c) by the shape of the slits on their terminal plates (stigmata).
    (d) by their head shape.
    (e) by the number of their body segments.

Solutions

1b; 2e; 3c; 4b; 5c; 6b; 7a; 8d; 9a; 10a; 11c; 12c; 13a; 14d; 15a; 16c; 17d; 18a; 19c; 20d; 21e; 22a; 23c; 24e; 25a; 26c; 27e; 28a; 29d; 30e; 31e; 32e; 33a; 34b; 35d; 36a; 37b; 38c; 39e; 40e; 41a; 42a; 43c; 44a; 45a; 46b; 47e; 48a; 49a; 50a; 51c; 52a; 53a; 54d; 55b; 56d; 57c; 58a; 59c; 60e; 61a; 62c; 63a; 64c; 65b; 66b; 67c; 68d; 69c; 70c; 71a; 72a; 73c; 74b; 75c; 76e; 77a; 78a; 79d; 80d; 81a; 82a; 83d; 84c; 85c; 86a; 87c; 88e; 89a; 90b; 91a; 92e; 93b; 94c; 95b; 96c; 97d; 98c; 99b; 100c

Addendum D: Origin of Figures

The following colleagues contributed pictures to this and the preceding volumes:

Dr. Bonin (Frankfurt): Figs. 5.110, 6.30, 6.33, 6.93, 6.119a and 6.120.
Dr. Düwel (Dänischenhagen): Figs. 5.43, 5.45, 5.47, 5.53, 5.56, 5.57, 5.61, 5.65, 5.100, 5.109, 5.112, 5.126 and 5.128.
Dr. Förster (Düsseldorf): Figs. 6.103d, e.
Prof. Dr. Heydorn (Berlin): Figs. 4.22, 4.24, 4.27, 4.32, 4.43, 4.65, 4.67, 4.73, 4.87, 4.88, 4.94, 4.95, 4.99, 4.116 and 4.117.
Prof. Dr. Horchner (Berlin): Fig. 5.22c.
Prof. Dr. Klimpel (Frankfurt): Fig. 6.74b.
Prof. Dr. Raether (Dreieich): Figs. 4.10, 4.39 and 4.46.
Prof. Dr. Schein (†) (Berlin): Figs. 4.98, 6.23, 6.24 and 6.27.
Prof. Dr. Stoye (†) (Hannover): Fig. 5.101.
Prof. Dr. Taraschewski (Karlsruhe): Fig. 5.148.
All other figures originate from the author.

© Springer International Publishing Switzerland 2016

H. Mehlhorn, *Animal Parasites*, DOI 10.1007/978-3-319-46403-9

Author Index

© Springer International Publishing Switzerland 2016
H. Mehlhorn, *Animal Parasites*, DOI 10.1007/978-3-319-46403-9

Subject Index

© Springer International Publishing Switzerland 2016
H. Mehlhorn, *Animal Parasites*, DOI 10.1007/978-3-319-46403-9

GPSR Compliance

The European Union's (EU) General Product Safety Regulation (GPSR) is a set of rules that requires consumer products to be safe and our obligations to ensure this.

If you have any concerns about our products, you can contact us on ProductSafety@springernature.com

In case Publisher is established outside the EU, the EU authorized representative is:

Springer Nature Customer Service Center GmbH
Europaplatz 3
69115 Heidelberg, Germany

Batch number: 10583317

Printed by Printforce, the Netherlands